Engineering Materials

This series provides topical information on innovative, structural and functional materials and composites with applications in optical, electrical, mechanical, civil, aeronautical, medical, bio- and nano-engineering. The individual volumes are complete, comprehensive monographs covering the structure, properties, manufacturing process and applications of these materials. This multidisciplinary series is devoted to professionals, students and all those interested in the latest developments in the Materials Science field, that look for a carefully selected collection of high quality review articles on their respective field of expertise.

Indexed at Compendex (2021) and Scopus (2022)

Rishabha Malviya · Sonali Sundram
Editors

Engineered Biomaterials

Synthesis and Applications

Springer

Editors
Rishabha Malviya
Department of Pharmacy
Galgotias University
Greater Noida, Uttar Pradesh, India

Sonali Sundram
Department of Pharmacy
Galgotias University
Greater Noida, Uttar Pradesh, India

ISSN 1612-1317 ISSN 1868-1212 (electronic)
Engineering Materials
ISBN 978-981-99-6697-4 ISBN 978-981-99-6698-1 (eBook)
https://doi.org/10.1007/978-981-99-6698-1

This Springer imprint is published by the registered company Springer Nature Singapore Pte Ltd.
The registered company address is: 152 Beach Road, #21-01/04 Gateway East, Singapore 189721, Singapore

Paper in this product is recyclable.

Dear Healthcare Professionals,

We are dedicating this book to you. Our Love for profession shall live forever.

This book is also dedicated to friendship of editors. We work together and edited a fruitful book.

Foreword

Biomaterials are materials with novel properties that can enable them to come in contact with the living tissue. They can be synthesised by a variety of processes and engineered according to a specific use and application in order to mimic biological phenomena or else be employed for some other process based on its use. There are immense benefits of Biomaterials particularly in improving the quality of life benefits apart from others. The progress in this field is dependent on interdisciplinary research with scientists from diverse backgrounds like chemists, physicists, biologists, pharmaceutical scientists, medical doctors, engineers, material producers and manufacturers etc. For medicinal purposes it can either be a diagnostic purpose or therapeutic purpose. The main rationale behind writing this book is to cover all the diverse aspects and the scientific knowledge associated with engineering of biomaterials as well as its utilisation in the healthcare field. This book attempts to explain in detail the principles and foremost methods for the synthesis of biomaterials and its application interconnected with prevention, mitigation, diagnosis and treatment purposes. The readers of this book will get an idea about how biomaterials are growing wider in terms of application as well as the increasing future opportunities associated with them. This book can also serve as a useful source of information for studying engineering biomaterials. Various methods for the synthesis and development of biomaterials are provided in the contributed chapters. These methods include those related to genetic engineering for the development of novel biomaterials, environmentally friendly approaches for the development of biomaterials based on bone and tissue engineering, gold nanoparticles derived from microorganisms as a synthetic method for developing biomaterials, and biomimetic approaches for producing biomaterials.

Each chapter of this book highlights methods of synthesizing biomaterials and how to apply them in healthcare. Not only the opportunities and advantages have been covered but the possible challenges have been addressed so that the readers and researchers get an idea about the obstacles and their means to overcome them.

It is my pleasure to present this forward for this book entitled *Engineered Biomaterials: Synthetic Approach and Applications* Edited by Dr. Rishabha Malviya.

Different chapters in this book have been contributed by the experts in their respective fields.

I highly acknowledge all the authors as well as the editors for their endeavour and efforts to compile this book and wish them all the very best.

Prof. Roop Krishen Khar
Director
B. S. Anangpuria Educational Institutes
Faridabad, India

Director
Formulators Koncept Pvt. Ltd.
New Delhi, India

Director
Unicure India Ltd.
New Delhi, India

Former Dean and HOD
Jamia Hamdard University
New Delhi, India

Preface

Biomaterial can be defined as a material which is engineered for interacting with the biological system for the healthcare purpose in diagnosis, mitigation and treatment motives. Engineering biomaterials means the study of biomaterials. Biomaterials have a property of biocompatibility which makes them suitable as well as safe for use. Methods for the synthesis of biomaterials and all the aspects associated with it are covered in this book. In the initial chapter information is given about the biomaterials which are derived naturally, its advances as well as opportunities. Then various techniques are given for synthesis and development of biomaterials which includes techniques associated with genetic engineering for the development of novel biomaterials, green methods for the development of bone and tissue engineering based biomaterials, gold nanoparticles from a microorganism—a synthetic approach for development of biomaterials and the biomimetic approaches for manufacturing of biomaterials. A detail description about genetically induced biomaterial advances in medical science is given in one of the chapters. Nowadays biomaterials are applied widely in healthcare applications which are growing wider along with time. Along with the synthetic approaches, the applications of biomaterials are also covered in this book. The usage of nano-sized biomaterials in drug delivery systems, recent trends associated with it as well as the future opportunities are covered here. Usage of biomaterials in implant devices, stimuli responsive materials in controlled release of drug, advanced tissue engineering with novel engineered biomaterials, collagen based nanomaterials for delivering drug, photo responsive material for 4D printing in tissue engineering, surface modified biomaterials in developing medical devices, nano-porous metal foams as versatile nanoplatforms for drug delivery applications and its properties, recent progress as well as challenges related to it, development of cardiovascular biomaterials from collagenous tissues, stimuli responsive material in controlled release of drug, gold nanoparticles along with their clinical applications and morphological study of silver nanoparticles supported on ti1-xcexo2, their synthesis and antimicrobial applications are covered in the chapters of this book. The students as well as researchers reading this book will get a detailed idea about what are biomaterials, how they are synthesized and how they can be applied in healthcare for medical purposes. The readers will find the information useful and precise.

Careful attention has been given by the authors to avoid making the information complex so that readers do not get confused while reading the book. Authors have tried to cover recent points linked to biomaterials so that latest information can be conveyed through this book. We thank the authors for putting their efforts in writing this book.

Greater Noida, India

Rishabha Malviya
Sonali Sundram

Contents

About the Editors

Dr. Rishabha Malviya completed B. Pharmacy from Uttar Pradesh Technical University and M. Pharmacy (Pharmaceutics) from Gautam Buddha Technical University, Lucknow Uttar Pradesh. His Ph.D. (Pharmacy) work was in the area of Novel formulation development techniques. He has 12 years of research experience and is presently working as Associate Professor in the Department of Pharmacy, School of Medical and Allied Sciences, Galgotias University for the past eight years. His area of interest includes formulation optimization, nanoformulation, targeted drug delivery, localized drug delivery and characterization of natural polymers as pharmaceutical excipients. He has authored more than 200 research/review papers for national/international journals of repute. He has 58 patents (19 grants, 38 published, 1 filed) and publications in reputed National and International journals with total of more than 200 cumulative impact factor. He has also received an Outstanding Reviewer award from Elsevier. He has authored/edited/editing 48 books (Wiley, CRC Press/Taylor and Francis, Springer, River Publisher, IOP publishing and OMICS publication) and authored 31 book chapters. His name has been included in word's top 2% scientist list for the year 2020 and 2021 by Elsevier BV and Stanford University. He is Reviewer/Editor/Editorial board member of more than 50 national and international journals of repute. He is the invited author for *Atlas of Science* and pharma magazine dealing with industry (B2B) "Ingredient south Asia Magazines".

Prof. Sonali Sundram completed B.Pharm. and M.Pharm. (Pharmacology) from AKTU, Lucknow. She has worked as a research scientist in an ICMR project at the King George's Medical University, Lucknow after which she joined BBDNIIT and currently working in Galgotias University, Greater Noida. Her Ph.D. (Pharmacy) work was in the area of Neurodegeneration and Nanoformulation. Her area of interest is neurodegeneration, clinical research, artificial intelligence. She has authored/edited/editing more than 15 books (Wiley, CRC Press/Taylor and Francis, IOP publishing, Apple Academic Press/Taylor and Francis, Springer Nature and River Publisher) She has attended as well organized more than 15 national and international seminar/conferences/workshops. She has more than eight patents, national and international, to her credit. She has published six SCI indexed manuscripts (cumulative impact factor: 20.71) with reputed international publishers. She has delivered oral presentations in international conferences organized in different European countries.

Approach for the Synthesis of Biomaterials

Naturally Derived Biomaterials: Advances and Opportunities

Ainil Hawa Jasni, Azlin Suhaida Azmi, Noor Illi Mohamad Puad, Fathilah Ali, and Yusilawati Ahmad Nor

Abstract Biomaterials are materials that have been formed from or created by biological organisms such as plants, animals, bacteria, fungus, and other forms of life are referred to as biologically derived materials. Biomaterials are normally designed to interface with biological systems, for the treatment, augmentation, or replacement of biological functions. Across billions of years, life has been composed of and existed within these biomaterial molecules, monomers, and polymers. For instance, biomaterials of polysaccharides are sugars or starch polymers. Cellulose is the most ubiquitous and abundant polysaccharide. Polysaccharides are found in the tissues of both trees and humans. Meanwhile, natural biomaterials are substances that are derived from natural sources such as plants, animals, or minerals, and are used in medical and healthcare applications. Examples of natural biomaterials include collagen, chitosan, silk, cellulose, hyaluronic acid, and bone minerals such as hydroxyapatite. These materials are attractive in the field of regenerative medicine and tissue engineering due to their biocompatibility and biodegradability. Additionally, some natural biomaterials can mimic the physical and chemical properties of the body's natural tissues, making them ideal for use in implants and scaffolds. Recent advances in the production of natural biomaterials include the development of more efficient and scalable manufacturing processes, which has made them more widely available and accessible for use in medical applications. In addition, advances in the understanding of the biological interactions between these materials and the body have allowed for the development of new and improved medical devices and therapies. The use of natural biomaterials also provides unique opportunities for customization and personalization in medical treatment. For example, natural biomaterials such as collagen and hyaluronic acid can be engineered to meet specific patient needs, such as tissue repair and regeneration, wound healing, and drug delivery. Overall, natural biomaterials have shown great promise in many fields. This chapter's goal is to give readers a quick introduction to naturally derived biomaterials and their advances and opportunities. For example, recent developments in the production of natural biomaterials have made them more widely available and accessible for use

A. H. Jasni (✉) · A. S. Azmi · N. I. M. Puad · F. Ali · Y. A. Nor
Department of Chemical Engineering and Sustainability Kulliyyah of Engineering, International Islamic University Malaysia, Jalan Gombak, 53100 Kuala Lumpur, Malaysia
e-mail: a.hawa.jasni@gmail.com

R. Malviya and S. Sundram (eds.), *Engineered Biomaterials*, Engineering Materials,
https://doi.org/10.1007/978-981-99-6698-1_1

in medical applications, and advances in the understanding of the biological interactions between these materials and the body have allowed for the development of new and improved medical devices and therapies. In the coming years, the adoption of new advanced experimental methodologies, such as bioengineering approaches, will alter the practice of medicine in the applications using natural derived biomaterials. Tissue engineering, a multidisciplinary field of research involving the principles of materials science, engineering, biological sciences, and medical research, is a clear illustration of this.

Keywords Biocompatibility · Biomaterials · Natural biomaterials · Regenerative medicine · Tissue engineering

1 Introduction

Natural biomaterials have developed into an increasingly popular area of study in recent years. These materials are derived from natural sources, such as plants or animals, and can be used for a variety of applications in the medical field. The fact that naturally derived biomaterials are biocompatible, or do not have negative effects when inserted into the body, is one of their main benefits. This makes them perfect for application in tissue engineering, medical devices, and drug delivery systems. One promising application of naturally derived biomaterials is in regenerative medicine. Utilizing biological components, regenerative medicine aims to replace or repair damaged tissues and organs.

Naturally derived biomaterials offer many advantages over synthetic materials in this context, including better compatibility with living cells and tissues. For example, researchers have developed scaffolds made from collagen, a protein found in skin and connective tissues, which can support the growth of new tissues. Other naturally derived biomaterials being investigated for regenerative medicine include chitosan, silk fibroin, and hyaluronic acid.

Another exciting opportunity for naturally derived biomaterials is in the development of sustainable materials. Many conventional materials used in the industry are derived from non-renewable resources, which can have negative environmental impacts. By contrast, naturally derived biomaterials can often be produced sustainably and may even be biodegradable [1]. Overall, advances in naturally derived biomaterials hold great promise for improving human health and reducing our impact on the environment. Naturally derived biomaterials are offering exciting opportunities when it comes to regenerative medicine. The ability to regenerate tissues and organs using biological materials has been a long-standing goal for medical researchers, and with the development of naturally derived biomaterials, this dream is becoming a reality. One of the primary advantages of these materials over synthetic ones is their compatibility with living cells and tissues. Additionally, advances in 3D printing technology have made it possible to create complex structures that mimic the architecture of natural tissues.

Naturally derived biomaterials also offer promising solutions for sustainable materials. Many conventional materials used in industry rely on non-renewable resources and often result in negative environmental impacts. In contrast, biomaterials can be produced sustainably and may even be biodegradable. By leveraging these materials, we can reduce our impact on the environment while still producing high-quality products. Advances in naturally derived biomaterials represent an incredible opportunity for improving human livelihoods while also protecting our environment. As research continues to explore the full range of applications for these materials, we can look forward to even more exciting discoveries and innovations in the years to come. It's clear that the possibilities are truly remarkable, and we should all be excited about what the future holds for this cutting-edge field.

2 Definition of Naturally Derived Biomaterials

Naturally derived biomaterials are materials that are sourced from biological sources such as plants, animals, and microorganisms [1]. These materials are usually biocompatible and biodegradable, meaning they can be safely used in living organisms without causing harm and can eventually be broken down by biological processes. Biomaterials are materials that interact with biological systems for various purposes, for instance the delivery of drugs, medical devices, and tissue engineering [2]. Naturally derived biomaterials, obtained from natural sources such as animals, plants, and microorganisms [3] have undergone substantial research and are utilised in biomedical applications for their biocompatibility, biodegradability, and bioactivity.

3 The Importance of Naturally Derived Biomaterials in Various Applications

Here are some of the reasons why naturally derived biomaterials are important in various applications (Table 1).

4 Scope of the Chapter

This chapter covers a variety of closely connected topics to the development, properties, and applications of biomaterials that are derived from natural sources. The topics which are covered in this chapter include:

Table 1 The importance of naturally derived biomaterials in applications

Added values in naturally derived biomaterials	Definition and examples
1. Biocompatibility	Naturally derived biomaterials are often biocompatible, meaning they are not harmful to living tissues and can be used in medical applications without causing adverse reactions. For example, collagen and hyaluronic acid are naturally derived biomaterials that are often utilised in tissue engineering and regenerative medicine [4, 5]
2. Sustainability	Many naturally derived biomaterials are renewable and sustainable, making them environmentally friendly alternatives to synthetic materials. For example, cellulose, a naturally occurring polymer found in plants, can be used to make biodegradable plastics [6] and packaging materials
3. Diversity	Naturally derived biomaterials come in a wide variety of forms, allowing for customization to specific applications. For example, chitosan, a polysaccharide derived from crustacean shells, can be modified to have different properties, and used in wound dressings or drug delivery systems [7]
4. Biodegradability	Naturally derived biomaterials can be designed to be biodegradable, which is particularly important in applications where long-term implantation is not desired. For example, silk fibroin, a protein derived from silkworm cocoons, can be used as a biodegradable scaffold in tissue engineering [8]
5. Antibacterial properties	Some naturally derived biomaterials possess inherent antibacterial properties, making them useful in medical applications where preventing infection is critical [9]. For example, silver nanoparticles can be incorporated into chitosan to create a material which can be utilised in wound dressings due to its antimicrobial qualities. Other than that, natural antimicrobial peptides (AMPs) can be used in the treatment of bacterial infections [9]

1. Overview of naturally derived biomaterials: This include an introduction to the types of biomaterials that are commonly derived from natural sources. This chapter offers a summary of the physical and chemical properties of these biomaterials and their biological interactions.
2. Advances in the synthesis and processing of naturally derived biomaterials: This section focuses on recent developments in the synthesis and processing of natural biomaterials, including advances in biomaterials engineering and biotechnology. This includes discussions of new methods for isolating and purifying natural biomaterials, as well as approaches for modifying their chemical and physical properties.

3. Applications of naturally derived biomaterials: This section explores the diverse range of applications of natural biomaterials, including their use in wound healing, tissue engineering, drug delivery, and medical devices. It also highlights the advantages of using natural biomaterials over synthetic alternatives, such as their biodegradability, biocompatibility, and sustainability.
4. Challenges and opportunities in the field: This section examines some of the current challenges and opportunities in the field of naturally derived biomaterials. This includes discussions of regulatory issues related to the use of natural biomaterials in medical devices and pharmaceuticals, as well as challenges related to scaling up production and optimizing biomaterial properties for specific applications.

5 Classification of Naturally Derived Biomaterials

Naturally derived biomaterials can be categorised as several groups depending on their chemical structure and origin. Here are some examples:

1. Proteins: Proteins are a diverse group of biomolecules that are composed of amino acids [5]. They can be created with a variety of natural sources like animal tissues, plants, and microorganisms. Among the naturally derived protein biomaterials are collagen, gelatine, silk fibroin, and elastin.
2. Carbohydrates: Carbohydrates are a class of biomolecules that are made up of carbon, hydrogen, and oxygen [6]. They can be obtained from numerous natural sources, such as plants, algae, and microorganisms. Examples of naturally derived carbohydrate biomaterials include chitosan, hyaluronic acid, and cellulose.
3. Lipids: Lipids are a class of biomolecules that are composed of fatty acids and glycerol. They can come from a range of natural sources, such as animal tissues and plant oils. Examples of naturally derived lipid biomaterials include phospholipids, triglycerides, and waxes [7].
4. Nucleic acids: Nucleic acids are a class of biomolecules that are composed of nucleotides [8]. There are numerous natural sources from which they can be obtained, such as animal and plant tissues, as well as microorganisms. Examples of naturally derived nucleic acid biomaterials include DNA, RNA, and oligonucleotides [9].
5. Extracellular matrix (ECM) components: The extracellular matrix consists of an intricate network of proteins and carbs that provides structural support and regulates cell behaviour in tissues [10]. ECM components can be extracted from natural sources, including animal tissues or plant materials. Examples of naturally derived ECM biomaterials include collagen, hyaluronic acid, and decellularized tissues [11].
6. Bio-based polyester: Common polyester is not naturally derived; it is possible to create bio-based polyesters by using sustainable resources, including plant-based oils and sugars. Bio-based polyesters have a variety of prospective uses in the field of biomaterials. For example, they can be used to create biodegradable sutures,

implants, and drug delivery systems. One advantage of bio-based polyesters is that they can be designed to break down over time, reducing the risk of long-term harm to the environment.

These biomaterials can be categorised further depending on their physical and chemical characteristics. properties, such as their solubility, biodegradability, and mechanical strength. This classification can help researchers select the appropriate biomaterials for specific applications in tissue engineering, drug delivery, and other biomedical fields.

6 Polysaccharides

Polysaccharides are lengthy chains of monosaccharide (simple sugar) molecules that make up complex carbohydrates [12]. They are found in many natural sources such as plants, animals, and microorganisms [13]. Polysaccharides play a variety of roles in living organisms, such as energy storage, structural support, and cellular communication [14]. Here are some examples of polysaccharides and their chemical, physical properties, and functions.

6.1 Starch

Starch is a polysaccharide that acts as the main component for storing energy in plants. Amylose and amylopectin, two different forms of glucose polymers, make up the substance [15]. Amylopectin is a branching chain of glucose molecules, whereas amylose is a linear chain [16]. Starch is a naturally occurring polymer composed of glucose units that can be found in a variety of plant-based products, including corn, wheat, rice, potatoes, and cassava [17]. It is frequently utilised as a thickening, stabilizer, and gelling agent in the food sector because of its unique chemical and physical properties [18].

6.1.1 Chemical Properties of Starch

Starch is insoluble in cold water but can be partially solubilized by heat or acid hydrolysis, breaking down the starch into smaller glucose units. Starch has a high molecular weight, which makes it viscous and difficult to dissolve [19] especially in the electrospinning and membrane casting applications. Starch can be effectively dissolved using ionic liquids, 4-methylmorpholine 4-oxide (NMMO), and dimethyl sulfoxide (DMSO), with DMSO being the most cost-effective and soluble [20]. Raw starch must be chemically changed to make it more hydrophobic, stronger film-forming abilities, and higher tensile resilience to broaden the applications of starch.

6.1.2 Physical Properties of Starch

Starch has a white, odourless, and tasteless appearance [21]. Starch has a large capacity to store water and can hold up to 20 times its weight in liquid [22]. Starch can form a gel when heated in water, allowing it to thicken food products and increase their viscosity [23]. Starch is a renewable and biodegradable resource, making it an attractive material for use in various applications beyond food. Natural derived biomaterials derived from starch can include:

1. Bioplastics: As an alternative, starch-based bioplastics can be used to replace traditional petroleum-based plastics and can be used in packaging materials [24], disposable utensils, and other single-use products
2. Adhesives: Starch-based adhesives are used in paper and cardboard production and can replace synthetic adhesives [25].
3. Textiles: Starch-based materials can be used in textiles as a sizing agent, which helps prevent shrinkage during washing and ironing [26].
4. Medicine: Starch can be used as a binder in tablet formulations and as a disintegrant to help tablets dissolve in the body [27].
5. Agriculture: Materials made of starch can be utilised as a biodegradable substitute to synthetic mulch films used in agriculture [28].

6.2 Glycogen

Glycogen is a polysaccharide that is the principal energy storage molecule in animals [29]. It's similar to starch. but is more highly branched and can be broken down more rapidly to provide energy for cellular processes. Because of its unique features, glycogen has sparked increased interest as a biomaterial in recent years. It is biocompatible, biodegradable, and readily available in large quantities, making it an attractive alternative to synthetic materials suitable for a variety of uses. One potential use of glycogen is in drug delivery. Because glycogen is a natural product of the body and in comparison, to synthetic materials, it is less likely to elicit an immunological response or induce toxicity. Additionally, glycogen can be easily modified to allow for controlled release of drugs. Consequently, it is an excellent option for targeted medication delivery systems [30].

Glycogen also has the potential of a wound-healing material. It can be used to create a scaffold for tissue regeneration [31], as well as to promote the growth of new blood vessels. Because glycogen is highly branched, it can form a porous structure that allows for the infiltration of cells and nutrients, aiding in the regeneration process. For instance, Alkaline phosphatase (ALP) activity and gene expression investigations revealed that collagen/glycogen/hydroxyapatite (HAP) components influenced the differentiation of bone mesenchymal stem cells into osteoblasts or chondrocytes [31].

Overall, the use of glycogen as a biomaterial is still in its early stages, and more research is needed to fully understand its potential. However, glycogen's unique qualities make it an intriguing contender for a variety of applications including medication delivery, tissue engineering, and wound healing.

6.3 Peptidoglycan

Peptidoglycan is a polysaccharide found in bacterial cell walls [32]. For bacterial cells, it offers structural stability and defence. It is made up of two sugars, N-acetylglucosamine (NAG) and N-acetylmuramic acid (NAM), which are repeated and connected by short peptides to form a mesh-like structure [33]. There has recently been an increase in interest in using peptidoglycan as a biomaterial because of its unique attributes. It is biocompatible, biodegradable, and can be easily isolated from bacterial cell walls, making it an attractive alternative to a variety of applications for synthetic materials.

One potential use of peptidoglycan is in tissue engineering. It can be used as a scaffold for cell development, promoting the regeneration of the damaged tissue [34]. Additionally, peptidoglycan can be modified to provide growth factors or additional bioactive compounds that can improve tissue regeneration [35]. Peptidoglycan also has the potential as an antibacterial material [34]. It can be used to create coatings for medical devices or implants, which can prevent bacterial colonization and biofilm formation. Additionally, peptidoglycan can be modified to incorporate antimicrobial peptides, which can further enhance its antibacterial properties.

6.4 Cellulose

Cellulose is a polysaccharide that makes up the structural component of plant cell walls. It is composed of cross-linked, lengthy chains of glucose molecules. to form a strong and rigid structure [36]. Cellulose is widely used as a biomaterial due to its renewable nature, biocompatibility, and biodegradability [37]. Recent advancements in the use of cellulose as a biomaterial have focused on developing new processing techniques, modifying its chemical and physical properties, and exploring its potential applications in various fields.

One recent advancement is the development of nanocellulose, which refers to cellulose fibres that have been broken down into smaller nanoscale dimensions. Nanocellulose has several special merits including high stiffness, strength, and surface area, making it suitable for use in a variety of applications [38] including packaging materials, medical devices, and drug delivery systems.

Another recent advancement is the development of cellulose-based hydrogels, which are water-swollen networks of crosslinked cellulose fibres. These hydrogels have shown promise in a range of applications, including tissue engineering, drug delivery, and wound healing. The use of bacterial cellulose multifunction bandage with potent antibacterial capabilities and wound healing for angiogenesis stimulators had been reported by [39].

Cellulose has also been explored for use in 3D printing, where it can be used as a feedstock material. By combining cellulose with other materials, such as chitin or lignin, 3D printed structures with enhanced mechanical and physical properties have been developed. For instance, the greatest capability was shown by methylcellulose and wood glue, with the former having the benefit of being natural and producing a bio-based composition. The result was a homogeneous paste with moderate adhesion and viscosity, like the clay frequently used in 3D printing, that could be used with a caulking gun and a syringe to create multi-layered extrusions that were stable and smooth. Although it lacked the strength and rigidity of wood, it performed better than other bio-based new materials currently being developed thanks to its mechanical qualities, which put it on par with rigid polymer foams frequently used as insulation boards [40].

6.5 Chitin and Chitosan

Insects, crustaceans, and other arthropods' exoskeletons are made of the carbohydrate chitin It is also present in some fungi cell walls [41]. Chitin provides strength and protection to these organisms. Chitin can be processed into nanofibers, nanoparticles, and other nanostructures, which have special characteristics including a large surface area, mechanical strength, and biocompatibility [42]. These nanomaterials have potential applications in wound healing, tissue engineering, and drug delivery.

Another recent development in the use of chitin is when it comes to food packing [43]. In the not-too-distant future, it has been reviewed by authors in [44] that a significant amount of chitin and chitosan can be derived from mushrooms [44]. There are bright prospects for the industrial use of fungus-sourced chitosan, especially within the food and pharmaceutical industries, as a result of the comparisons made between the physicochemical, functional, and biological traits of that substance and that of marine chitin and chitosan [44].

Chitin has also been explored for its potential use in agriculture as a biopesticide. Chitosan can act as a natural insecticide, disrupting the growth and development of pests while being safe for humans and environment [45]. Additionally, chitin can be used as a soil conditioner, enhancing soil fertility and plant growth [46]. It was found that the exuviae of a particular insect species (mainly chitin) had enhanced bacteria that grew after the topsoil underwent treatment with 10 g/kg in the rhizosphere of *B. oleracea*. However, at a ratio of 1 g/kg, there were no appreciable changes in bacterial abundance. The results are consistent with earlier research on chitin alterations, which has shown to increase bacterial population abundance [47].

Chitin, a material present within the exoskeletons of crustaceans including crabs, prawns and lobsters, is the source of chitosan, a naturally occurring polysaccharide [48]. Fruits and vegetables have been coated with chitosan to prolong their shelf life. The chitosan coating can slow down the rate of water loss and gas exchange, thereby extending the freshness and quality of the produce [49]. Chitosan coatings have been utilised to increase the implant surfaces' antibacterial characteristics and promote tissue integration [50]. Recent findings showed that the combination of chitosan with enzymes and antimicrobial peptides enhances its antibacterial activity and broadens its application in a variety of physiological circumstances. The combination of chitosan and polymers also produced coatings with improved antibacterial and anti-adhesive qualities to reduce the incidence of implant-associated infections [51].

6.6 Alginate

Brown seaweed is the source of the natural polymer known as alginate, such as kelp [52]. This environmentally friendly and renewable biomaterial exhibits several intriguing qualities, including good mild gelation when adding divalent cations (such as Ca^{2+}), biocompatibility, low toxicity, and cost effectiveness. High-molecular-weight alginate's poor solubility and high viscosity, the aqueous solution's polyelectrolyte nature, the lack of suitable organic solvents, the substantial amount of intra- and intermolecular hydrogen bonds, continue to be problems in alginate alone applications [53]. Thus, it must be coupled with other polymers to enhance its values [54] that had developed natural polymer high-strength hydrogels using gelatin and hydrazide alginate (HAlg), which were then crosslinked to mimic the structure of collagen and glycosaminoglycans (GAGs) within the extracellular matrix, respectively. The Gelatin-HAlg-DN hydrogels exhibit excellent biodegradability and swelling consistency in physiological conditions, as well as the capacity to facilitate cell binding and proliferation. In a rat model with critical size bone defects, psoralen-loaded gelatin-HAlg-DN hydrogels effectively induced bone regeneration offering intriguing potential as tissue engineering scaffolds [54].

6.7 Hyaluronic Acid

A polymer present in the connective tissues called hyaluronic acid (HA) aids in maintaining hydration and lubrication. It is also used in some cosmetic products for its moisturizing properties [55]. Bioink additions include HA and multi-walled carbon nanotubes (CNTs) in 3D bioprinting for fabrication of cartilage had been reported [56]. HA acts as a molecule that is both structural and signalling. It possesses extraordinary water-retention abilities that create a gel-like environment within the tissue and give the entire structure flexibility for the structure [57].

6.8 Others

6.8.1 Inulin

Inulin a polysaccharide generated from plants, is a bioinspired, adaptable, and useful biomaterial [58] In comparison to other biodegradable polysaccharides, inulin's distinctive and adaptable structure, stabilising and protecting properties, and capacity to target specific organs make it an ideal drug delivery carrier. Each fructose unit has three hydroxyl groups that act as an anchor for chemistry changes. Thus, this enhances cellular bioavailability and absorption, and accomplishing medicines' and biomolecules' targeted, prolonged, and regulated release [58]. Inulin used via solvent-casting technique was used to create sustainable water-based films for wound healing [59] employed inulin (INL), a carbohydrate-based matrix, and polyvinyl alcohol (PVA), a petroleum-free matrix. Due to its sustainability and antimicrobial qualities, pumpkin powder made from vegetable waste was employed in the study [59].

6.8.2 Xylan

Xylan is a fluoropolymer-based commercial coating that is mostly found in nonstick cookware, footware and automotive coatings due to its water repellent characteristic [60]. Click-chemistry techniques were used to create reactive xylan compounds that can be crosslinked chemically and functionalized further. Quantitatively, functional xylan carbamates (XCs) were created from xylan phenylcarbonates (XPCs) along with magnitude of substitution (DS) with propargyl amine (PA) or 6-azidohexan-1-amine (AA). Alkynyl-functionalized XCs were crosslinked with bisazide linkers using copper-catalyzed 1,3-dipolar cycloaddition within an organic medium, and it was demonstrated that the gelation process produced a reactive moiety plus effective solubility in water for additional appropriate applications [61].

6.8.3 Proteins

Proteins are substantial, intricate molecules consisting of long chains of amino acids [62]. They are found in every cell of every living organism and are involved in a variety of functions, such as providing structural support, catalysing chemical reactions, and transporting molecules within cells and throughout the body. A protein's linear sequence of amino acids makes up its primary structural component. These consistent folding structures, alpha helices and beta sheets, make up a protein's secondary structure. The collection of curves and folds in a single linear chain of amino acids, sometimes referred to as a polypeptide, is the tertiary structure of a protein. Last but not least, macromolecules having multiple polypeptide chain structures or subunits are known as proteins [63]. Figure 1 is the illustration each type of protein structure.

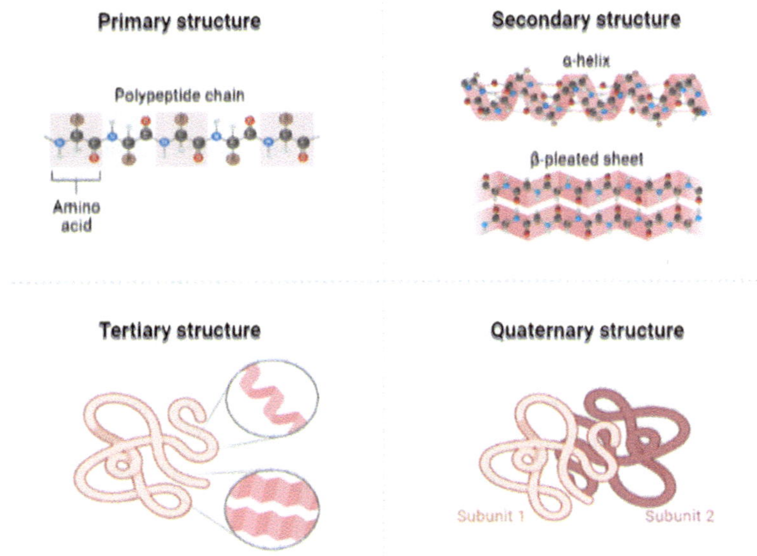

Fig. 1 Illustration of type of protein structure (Taken from open source, https://doi.org/10.1016/j.bea.2021.100021)

Here are some examples of protein-based naturally derived biomaterials and their functions:

- **Structural proteins**: Structural proteins, such as collagen, keratin and elastin, provide support and shape to tissues and organs in the body [64].
- **Transport proteins**: Transport proteins, such as haemoglobin and albumin, bind to and transport molecules, such as oxygen and nutrients, throughout the body. For example, one of the most prevalent proteins in blood plasma, serum albumin is crucial to all biological activities and has been used in several biomedical procedures. Human albumin, bovine albumin, and ovalbumin, which are used to manufacture biomaterials, are appropriate for use in bone regeneration because they have the right microstructure, hydrophilicity, and biocompatibility [65].
- **Contractile proteins**: Actin and myosin are examples of contractile proteins that cause muscles to contract and move.

In summary, proteins are necessary molecules that have several functions in living things. Their unique properties, such as their ability to provide structural support, and transport molecules, make them important in many material science applications, such as food, medicine, and biotechnology.

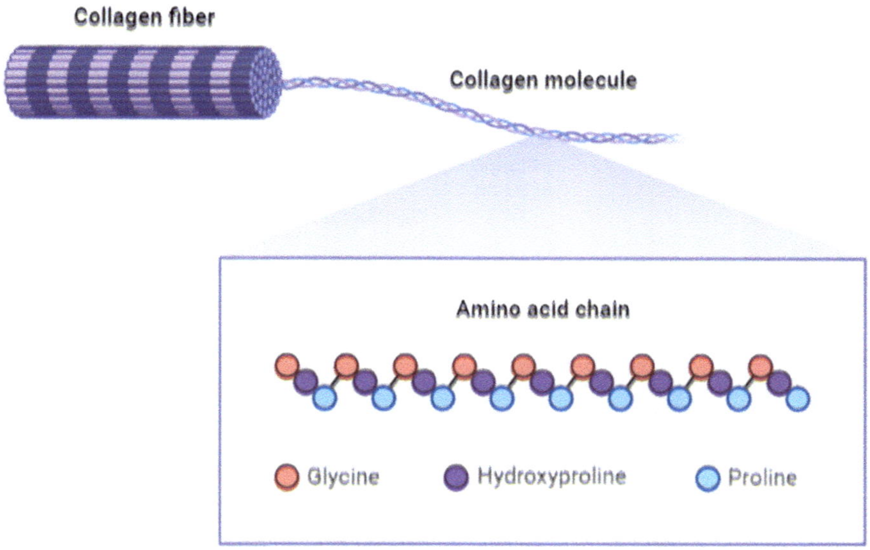

Fig. 2 Collagen fibre structure, from fibrils and molecules to chains, together with amino acid sequence of Hydroxyproline, Proline, and Glycine. (Created with biorender)

6.8.4 Collagen

The protein collagen, which is the most prevalent in the human body, gives numerous tissues their structural support [66]. It has been frequently employed beyond the field of tissue engineering owing to its remarkable biocompatibility, low immunogenicity, and ability to support cell adhesion and proliferation. Collagen is made up of three chains. The chains are linked to form a triple helix. Because glycine is the smallest amino acid, it permits the chain to form a tight shape and endure stress [67]. Figure 2 is the structure of the collagen fibre.

6.8.5 Silk Fibroin

Silk is a naturally occurring protein fibre produced by silkworms. It has been used in tissue engineering, implant coating and others due to its biocompatibility, biodegradability, and ability to form scaffolds with tuneable mechanical properties [68]. A recent study reported the possibility of encouraging the development of bone-like tissue from human dental pulp stromal cells (hDPSCs) both in in vitro as well as in vivo within porous *Bombyx Mori* silk structures by employing the specific HDAC2 and 3 inhibitor MI192. Fabricated and evaluated were scaffolds made of 2 and 5 wt.% silk. In contrast to the 2 wt% silk scaffolds, the scaffolds with a weight of 5% exhibit larger internal lamellae, slower rates of expansion and deterioration, and higher torsional moduli. The alkaline phosphatase activity of hDPSCs on a 5 wt%

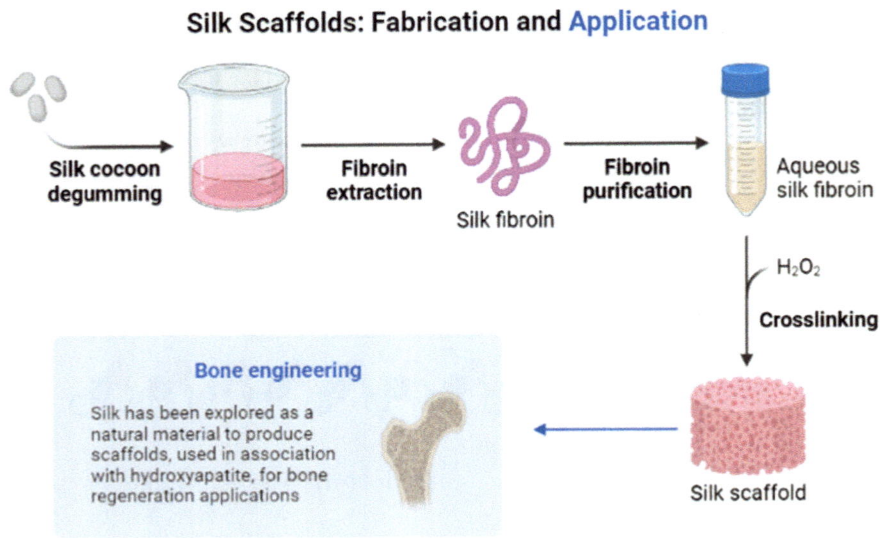

Fig. 3 The technique used for fabrication silk scaffolds for biomedical applications (adapted from open source, https://doi.org/10.3390/polym14153110)

silk scaffold was dramatically increased by pre-treatment with MI192 (ALP). Histological examination proved that, in contrast to the control group, the use of MI192-pretreated hDPSCs-silk scaffolds improved bone extracellular matrix (ALP, Col1a, and OCN) deposition and mineralization. After six weeks of subcutaneous implantation in mice, the MI192-pretreated hDPSCs-silk scaffold constructs increased vascularization and extracellular matrix mineralization compared to the untreated control. After six weeks of subcutaneous implantation in nude mice, the MI192-pretreated hDPSCs-silk scaffold constructs increased vascularization and extracellular matrix mineralization compared to the untreated control [69]. Figure 3 is the workflow of silk fabrication and application.

6.8.6 Gelatin

Gelatin is a substantially hydrolyzed version of collagen with characteristics similar to collagen. It can be easily modified to form hydrogels with tunable mechanical properties, making it a suitable material for tissue engineering.

6.8.7 Elastin

Elastin materials can be derived from organic, recombinant, or synthetic sources. Elastin can be recovered from elastin-rich tissues of animals by hydrolyzing a portion of the peptide linkages in the insoluble elastin [70].

6.9 *Others*

6.9.1 Fibrin

Fibrin is a naturally occurring protein that aids in the formation of clots and wound healing. It is utilized in tissue engineering to form hydrogels that can support cell growth and migration [71].

6.9.2 Lipids

Lipids are a wide class of naturally occurring macromolecules present in plants and animals [72]. They are distinguished by their hydrophobic characteristics and are used in a number of biomaterials research applications. One major use of lipids is in the creation of liposomes. Liposomes are spherical vesicles consisting of a bilayer of phospholipids that can be used to encapsulate drugs and other therapeutic molecules for targeted drug delivery [73]. The inside of the liposome is formed by the hydrophilic portions of the phospholipids, which contain the drug, while the hydrophilic head groups form the outer shell, allowing the liposome to interact with biological systems [74].

Lipids are also used in the creation of biodegradable polymers. Lipid-based polymers can be designed to degrade in vivo, making them suitable for use in the fields of tissue engineering and drug delivery [75]. They are additionally employed to produce micro- and nanoparticle for targeted medication delivery [76]. Another use of lipids is the formation of lipid membranes. Lipid membranes are used in biomimetic systems to mimic the properties of natural cell membranes [77]. They can be used for a variety of purposes, including water treatment, biological sensors, and drug detection [78]. Lipids are additionally useful to create hydrogels. Hydrogels comprise three-dimensional, water-swollen polymer networks that can be useful for drug delivery, wound care, and tissue engineering. Lipids can be incorporated into hydrogels to provide mechanical support and improve cell adhesion and proliferation [79].

Overall, lipids are a versatile biomaterial with many potential applications in biomaterial science. Their hydrophobic properties and ability to self-assemble make them appealing candidates for drug delivery, tissue engineering, and biomimetic system.

6.9.3 Fatty Acids

Fatty acids are naturally occurring biomolecules that are commonly found in plant and animal tissues. They are composed of long hydrocarbon chains with a carboxyl group at one end, making them amphipathic like phospholipids [72]. Fatty acids have many uses in biomaterial science due to their biocompatibility, biodegradability, and versatility.

One major use of fatty acids is in the creation of bio-based polymers. Fatty acids can be used as monomers to create polymers such as polyesters, polyamides, and polyurethanes. These polymers are used for drug delivery, tissue engineering, and exterior coatings thar are only a few of the potential applications [80]. Hydrogels are also manufactured using fatty acids. Hydrogels are cross-linked networks of liquid-swollen polymers which are able to be utilized for the administration of drugs, wound healing, and other applications. tissue engineering. Fatty acids can be incorporated into hydrogels to provide mechanical support and improve cell adhesion and proliferation. A recent study showed that hydrogenated phosphatidylcholine and oleic acid were used to create a highly hydrated hydrogel (95% water) for potential biomedical uses [81].

Another application of fatty acids is in the production of surfactants. Surfactants are molecules that can reduce the surface tension of liquids and increase their ability to mix with other substances. Fatty acids can be used to create natural surfactants that are biodegradable and non-toxic, making them suitable for use in personal care products, detergents, and other industrial applications. It has been known that oleochemical-based surfactants outperform petroleum-based ones in terms of biocompatibility and biodegradability [82]. Sorbitan esters are commercially known as "Span" in the contemporary industrial sector, specifically Span 80, a biodegradable surfactant made from oleic acid and sorbitol-based sugar alcohol. These esters are commonly used in the development and formulation of water-in-oil emulsions due to their relative hydrophobicity. Sorbitan esters are widely derivatized by combining with ethylene oxide to form sorbitan ester ethoxylates, which enhances their hydrophilicity. Polyethoxylated sorbitan monoesters, popularly known as "Tween" in the industry, are surfactants that are ideal for creating oil-in-water emulsions [83, 84].

Overall, fatty acids are a versatile biomaterial with many potential applications in biomaterials science. Their biocompatibility, biodegradability, and versatility make them an attractive option for the development of sustainable and biologically compatible materials.

6.9.4 Phospholipids

Phospholipids are naturally occurring biomolecules that are found in cell membranes and other biological structures. They are amphipathic, meaning that they have both hydrophilic (water-loving) and hydrophobic (water-repelling) regions [85]. This unique property makes phospholipids useful in a variety of applications in biomaterial science.

One major use of phospholipids is in drug delivery systems. Phospholipid-based liposomes can be used to encapsulate drugs and other therapeutic molecules, allowing for targeted delivery to specific tissues or cells. The liposome's outer shell is formed by the hydrophilic head regions of the phospholipids, while the hydrophobic tails form the interior, where the drug is contained. The liposome can then be designed to be released by the reaction to certain stimuli, such as changes in pH or temperature, and the medication is released. The optimum phospholipid for managing the release of the anaesthetic bupivacaine (BUP) medication from liposomal depots and controlling the regulated aggregation of the liposomes is phosphatidylglycerol (PG) [86].

Phospholipids are also used in the creation of biomimetic membranes, which are synthetic structures designed to mimic the properties of natural cell membranes. These membranes have a wide range of uses, including water filtration, biosensors, and drug screening. The amphipathic nature of phospholipids allows them to form bilayers, which can be used as the foundation for these biomimetic membranes. The best exemplar of this is a study done by [87] where the cell membrane phospholipids are functions as dentin's mineralization sites and potential raw materials for bottom-up strategies aiming for quicker and more intricate production of dentin-like structures.

Phospholipids are also used in tissue engineering applications. They can be added to scaffolds to offer structural support while also promoting cell adhesion and proliferation. Because of their improved blood compatibility and cytocompatibility, electrospun poly (lactic-co-glycolic acid) (PLGA) membranes were a suitable case in point [88]. Overall, phospholipids' distinct features make them flexible biomaterials with numerous uses in pharmaceutical delivery, tissue engineering, and biomimetic membrane design.

6.9.5 Waxes

Wax is a simple lipid that is generated by esterifying a long-chain alcohol with a fatty acid. Alcohol can have 12–32 carbon atoms. Waxes are found in nature as coatings on plants and stems. The wax protects the plant from excessive water loss [89]. In practise, cellulose combined with a trace amount of plant-derived wax (nonacosane-10-ol and nonacosane-5, 10-diol) has a higher mechanical strength and modulus of elasticity [90] than synthetic plastic. The diffusion coefficients of oxygen, nitrogen, and water molecules through this material were found to be at least 50% lower than those found in polyethylene. It is an environmentally friendly, long-lasting packaging material with outstanding mechanical, thermal, and barrier properties [90].

7 Synthesis and Modification of Naturally Derived Biomaterials

The synthesis of naturally derived biomaterials involves the extraction, purification, and modification of the natural source to obtain a suitable material for the intended application. The following approaches can be used for the synthesis of naturally derived biomaterials:

1. Extraction: The first step in the synthesis of naturally derived biomaterials is the extraction of the material from its natural source [91]. The extraction method varies depending on the source material and the intended application.

For example, collagen can be extracted from animal tissues using acid or enzymatic digestion [92], while cellulose can be extracted from plants using chemical or mechanical methods [93].
2. Purification: After extraction, the material is usually purified to remove impurities and obtain a pure form of the material. Purification methods include filtration, centrifugation, and chromatography [94].
3. Modification: Modification of naturally derived biomaterials is frequently required to modify their qualities and adapt them to certain uses. Modification can be achieved through chemical or physical methods. For example, cross-linking of collagen can improve its mechanical strength and stability, while surface modification of cellulose can improve its biocompatibility [95].
4. Characterization: Characterization of naturally derived biomaterials is essential to determine their properties and suitability for the intended application. Characterization methods include spectroscopy, microscopy, and mechanical testing.
5. Quantification: Quantification is the act of assigning a numerical value to a measurement of anything, or counting the quanta of whatever is being measured. Quantification creates a defined method of measurement that permits statistical operations and mathematical computations to be performed [96].
6. Pilot production: A pilot study, pilot project, pilot test, pilot production or pilot experiment is a small-scale preliminary study undertaken prior to the performance of a full-scale research project to evaluate feasibility, duration, cost, adverse occurrences, and to enhance the study design [97]. Figure 4 is the summary of biomaterials synthesis workflow.

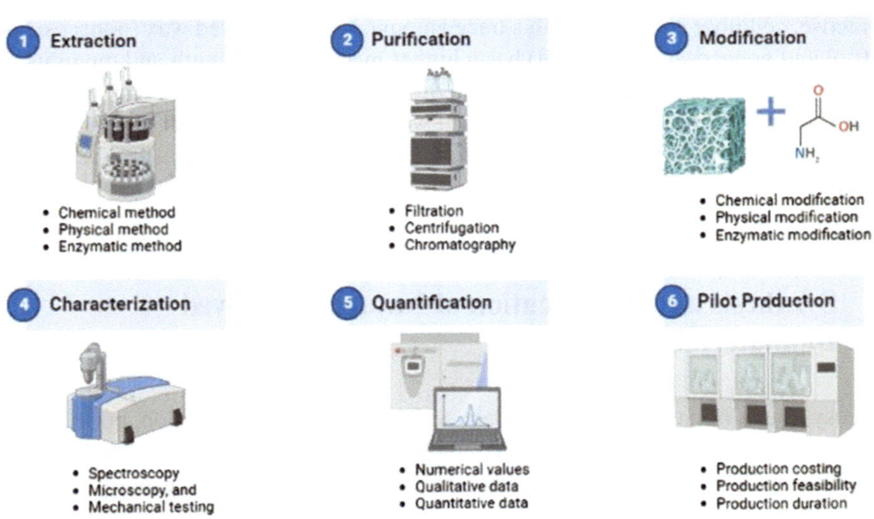

Fig. 4 Workflow and techniques in biomaterials synthesis

8 Chemical Modification

Chemical modification of naturally derived biomaterials is a crucial step in developing biomaterials with improved properties and functionality [98]. The chemical modification can be performed on different biomaterials such as proteins, polysaccharides, and lipids. Here are some examples of chemical modifications of naturally derived biomaterials:

1. Crosslinking: Crosslinking is the development of covalent connections between polymer chains in a biomaterial, resulting in improved mechanical properties and stability. Crosslinking can be achieved using various crosslinkers, such as glutaraldehyde, genipin, and formaldehyde [99]. Other than that hydrogel formation involves the crosslinking of a hydrophilic polymer to form a three-dimensional network, resulting in improved mechanical properties and biocompatibility [100]. Hydrogels can be formed from various biomaterials such as alginate, chitosan, and gelatin.
2. Acylation: Acylation involves the introduction of acyl groups into the biomaterial, resulting in improved solubility, stability, and biocompatibility. For example, acylation of chitosan can improve its solubility in organic solvents, making it easier to process for various applications [101].
3. Grafting: Grafting involves attaching a functional group or a polymer onto the surface of the biomaterial, resulting in improved properties such as biocompatibility and hydrophilicity. Grafting can be achieved using various methods such as surface-initiated polymerization, click chemistry, and plasma treatment [102].
4. Glycosylation: Glycosylation involves the addition of sugar moieties onto the biomaterial, resulting in improved bioactivity and biocompatibility. For example, glycosylation of proteins such as collagen and elastin can improve their interactions with cells and tissues. A case of this was the use of glycosylation in biomaterial engineering by adding saccharides to the catalogue of construction blocks to embellish it with adhesion properties. The use of a *Campylobacter jejuni* glycosylation circuit to de novo glycosylate the *Bacillus subtilis* amyloid-like biofilm protein TasA was reported as a unique biomaterial engineering technique for increasing the adhesiveness of TasA fibrils [103].

In conclusion, chemical modification of naturally derived biomaterials can improve their properties and functionality, making them suitable for various biomedical applications. The method of modification used is determined on the biomaterial and the desired application as well.

9 Physical Modification

Physical modification of biomaterials refers to the changes made to the physical properties of a material without altering its chemical composition [104]. This can include changes to the surface topography, roughness, stiffness, and shape of the material. Some common physical modifications of biomaterials include:

1. Surface modification: This involves changing the topography and roughness of the surface of a material to improve its interaction with cells and tissues. Surface modification techniques can include plasma treatment, micro- and nano-patterning, and coatings [105].
2. Mechanical modification: This involves altering the stiffness and elasticity of a material to mimic the mechanical properties of natural tissues. This can be achieved through various methods of mechanical modification such as mechanical stretching, compression, and shear forces [106].
3. Shape modification: This involves changing the shape and size of a material to better fit a specific application or tissue. Techniques can include moulding, cutting, and 3D printing [105].
4. Porosity modification: This entails altering a material's pore size and shape to promote cell infiltration and tissue ingrowth. This can be accomplished using a variety of procedures such as salt leaching, solvent casting, and gas foaming [107].

Physical modification of biomaterials can greatly enhance their biocompatibility, bioactivity, and mechanical properties, making them better ideal for a wide range of biomedical applications, including tissue engineering, medication delivery, and medical devices.

10 Enzymatic Modification

Enzymatic modification of biomaterials refers to the surface modification with enzymes or bulk properties of biomaterials without changing their chemical composition [108]. Enzymes are biological catalysts that can selectively modify the surface functional groups of biomaterials to add new functions, improve biocompatibility, and enhance the ability of the material to interact with cells and tissues. Some common enzymatic modifications of biomaterials include:

1. Surface functionalization: This entails using enzymes to change the outermost layer of a biomaterial in order to introduce new functional groups that can facilitate cell adhesion and proliferation. For example, enzymes like horseradish peroxidase [109] and tyrosinase can be used to generate reactive oxygen species [110] that can covalently bind proteins and peptides to the surface of the biomaterial.

2. Degradation: This involves the use of enzymes to selectively degrade the biomaterials to control the rate of release of bioactive molecules, drugs, or cells from the biomaterial. For example, enzymes like metalloproteinase [111] collagenase, hyaluronidase, and chondroitinase can be used to selectively degrade collagen, hyaluronic acid, and chondroitin sulphate, respectively [112].
3. Cross-linking: This involves the use of enzymes to cross-link biomaterials to enhance their mechanical and structural properties. For example, enzymes like transglutaminase can be used to cross-link proteins, such as gelatin and collagen, to form stable hydrogels [113].
4. Surface cleaning: This involves the use of enzymes to remove contaminants and impurities from the surface of the biomaterial to improve its biocompatibility. For example, enzymes like lipase and protease can be used to remove lipids and proteins from the surface of the biomaterial [114].

Enzymatic modification of biomaterials is an intriguing approach for improving biocompatibility and performance for a variety of purposes, namely drug delivery, tissue engineering, and medical devices.

11 Applications of Naturally Derived Biomaterials

11.1 Medical and Biomedical Applications

Because of their capacity for biocompatibility, biodegradability, and propensity to replicate the extracellular matrix (ECM) of natural tissues, naturally produced biomaterials have been extensively researched for use in tissue engineering. Because of their biocompatibility, biodegradability, and ability to package and transport medications to specific tissues, naturally produced biomaterials have showed considerable promise in drug delivery. Some naturally generated biomaterials that have been employed in biomedical applications are listed in Table 2.

Naturally derived biomaterials offer many advantages over synthetic materials in tissue engineering, drug delivery and implant coating, including their capacity to replicate the ECM of living tissues and promote cell adhesion and proliferation. However, there are also some limitations, such as batch-to-batch variability, difficulty in controlling their mechanical properties, and potential immunogenicity. Overall, naturally derived biomaterials hold great promise for the development of safer products for these constructs that can more closely resemble natural tissues.

12 Industrial Applications

Naturally derived biomaterials have several industrial applications, including:

Table 2 Recent discoveries of naturally produced biomaterials used in biomedical applications

No.	Type of biomaterials	Findings	Study
Tissue engineering			
1	Fibrin: – Is a naturally occurring protein that helps the body's blood clot and repair wounds – It has been used in tissue engineering to support cell growth and migration	• A wound dressing consisting of sodium carboxymethylcellulose (Hcel® NaT), fibrin, and in vitro-seeded dermal fibroblasts • Fibrin promotes healthy cell adhesion and spreading, and aids in cell migration and subsequent proliferation in a wound	[115]
2	HA+Chitosan	• Degradable hydrogel from N, O-Carboxymethyl Chitosan (NOCC), Aldehyde-Hyaluronic Acid, and the addition of *Allium sativum* (garlic oil) • A breakthrough for preventing intraperitoneal adhesion following surgery	[116]
3	Chitosan+collagen type 1, hyaluronic acid, Poly(L-lacticacid)/gelatin/β-tricalcium phosphate, gamma-poly [glutamic acid] polyelectrolyte/titanium alloy, modified Poly(L-Lactide-co-Epsilon-Caprolactone), calcium phosphate, β-glycerophosphate hydrogel/calcium phosphate cement (CPC), and CPC-Chitosan-RGD	• Improved cartilage and bone cell lesions repair when mesenchymal stem cells (MSCs) are combined with chitosan-based scaffolds • The scaffold's synergistic effects on the proliferation and differentiation of MSCs into bone and cartilage tissue	[117]
Drug delivery			
1	Albumin	• Flexible hollow human serum albumin (HHSA) loaded with the chemotherapy medication doxorubicin (DOX) and the photosensitizer chlorin e6 (Ce6) for synergistic cancer treatment • Stronger therapeutic effects without radiation development of 4T1 breast tumours • Surprisingly, remarkable biocompatibility has been demonstrated both in vitro and in vivo	[118]

(continued)

Table 2 (continued)

No.	Type of biomaterials	Findings	Study
2	Collagen	• Collagen nanostructures using the scales of fish were produced for the very first time using desolvation techniques • According to the histological and macroscopical study, the synthetic fish scale's collagen nanomaterials aided in the healing process with little toxicity compared to the saline group	[119]
3	Alginate-chitosan+HA	• Alginate-chitosan microbeads and thermo-responsive hyaluronic acid-poly(N-isopropylacrylamide) (HA-pNIPAM) hydrogels have been created as bioresorbable local bacteriophage (*Staphylococcus aureus* phage ISP and *Pseudomonas aeruginosa* phage LUZ19) delivery systems to permit both delayed and quick phage release • Both phages were gradually released hydrogels HA-pNIPAM during the course of 21 days. The short-tailed LUZ19 was continuously released from the microbeads for 21 days, whereas the long-tailed ISP released in bursts at first and then at a declining rate over time	[120]
4	Fibrin	• Research on the drug release and antibacterial properties of platelet-rich fibrin (PRF), a natural vehicle for the administration of antibiotics • Following oral surgery, using antibiotic-loaded PRF may reduce the risk of post-operative infection, replace or enhance systemic antibiotic therapy, and maintain the healing properties of PRF	[121]

(continued)

Table 2 (continued)

No.	Type of biomaterials	Findings	Study
5	Cellulose	• Using an in-situ co-precipitation approach, Magnetic iron oxide (MIO)-incorporated waste tissue sheets (WTP) and sugarcane bagasse (SCB) were incorporated to make WTP/MIO and SCB/MIO nanocomposite particles (NCPs) • By including MIO-NPs into the cellulose matrix, the swelling capacity, drug loading capacity, and drug release time were all improved	[122]
Implant coatings			
1	Laminin	• A poly (D, L-lactide) (PDLLA)-Laminin 332 (LN332) hybrid film that displayed timed release of LN332for up to 28 days was successfully created using a layer-by-layer manufacturing process • It accelerated gingival mesenchymal stem cells' epithelial growth and promoted their adhesion, dissemination, and proliferation	[123]
2	Hydroxyapatite	• 10 implants featuring 10 implants having a dual acid-etching (DAA) surface and a nanostructured hydroxyapatite coating (HAnano). was given to ten lambs (aged 2–4) • Low-density bone in sheep after 28 days with HAnano enhances bone development as compared to DAA surface	[124]
3	HA	• Tannic acid (TA) and hyaluronic acid (HYA) use is necessary to lower the rate of corrosion of implants made of magnesium alloys • The magnesium alloy's corrosion rate was lowered by the TA-HYA coating from 7.379 mm/year to 0.204 mm/year in in vitro corrosion testing employing Tafel polarisation	[125]

(continued)

Table 2 (continued)

No.	Type of biomaterials	Findings	Study
4	Silk	• FibroFix Cartilage P™ (FibroFix™)—commercial brand name • Substantial bone repair and strong implant tissue integration, which, in our opinion, will speed up the healing process (promote tissue regeneration)	[126]
5	Collagen	• An innovative injectable medicinal product on primary cultures of human gingival fibroblasts (hGF), dental SKIN BioRegulation (Guna S.p.A., Milan, Italy) was put to the test • It is made of type I collagen of porcine origin • Significant improvement in 48-h viability of hGF-grown cells ($p = 0.05$) and 24-h wound healing ($p = 0.001$)	[127]

1. Food industry: Natural polymers with gelling, thickening, and stabilising qualities, like cellulose, chitosan, and alginate, are utilised as food additives.
2. Agriculture: Natural biopolymers such as starch and cellulose are used as plant growth promoters and soil conditioners.
3. Textile industry: Natural polymers such as silk, cotton, and wool are used to make textiles.
4. Packaging industry Biodegradable packaging materials are created using natural polymers like cellulose, chitosan, and starch.
5. Cosmetic industry: Natural polymers with moisturising and anti-aging qualities, like collagen, elastin, and hyaluronic acid, are employed in cosmetic products.
6. Medical industry: Medical uses for naturally generated biomaterials include tissue engineering, medication delivery, and wound healing. Examples include collagen, chitosan, and hyaluronic acid.
7. Construction industry: Building materials are produced using natural biopolymers like cellulose and lignin for insulation, cement composite, and adhesives.

The use of naturally derived biomaterials in these industries offers several advantages over synthetic materials, including their biodegradability, biocompatibility, and renewability. Their use in industry is also in keeping with the rising demand for environmentally friendly and sustainable materials. Table 3 summarizes the recent discoveries within industrial applications. Composite made up of Oyster shell is shown in Fig. 5.

Table 3 Recent industrial applications of naturally derived biomaterials

No.	Sample type	Findings	References
Food industry			
1.	Polylactic acid (PLA)	• Bilayer films were developed that included a PLA layer with a layer made of washed cottonseed meal (CSM) • The elongation at break has decreased while the Young's modulus and tensile strength have both increased	[128]
Agriculture industry			
2.	Chitosan	• Hydrogels made of chitosan that have embedded mineral fertilisers were developed • It was discovered that the experimental group's seedling survival rate was 40% greater than the control group	[129]
3.	Chitosan	• Vegan chitosan from *Cunninghamella echinulate* was extracted • It has been utilised as an alternative to chemical treatments to manage plant diseases	[130]
Textile industry			
4.	Bacterial cellulose	• Adapted microbial weaving process via *Komagataeibacter rhaeticus* bacteria • The incredibly tiny nanocellulose fibres are eight times stiffer and stronger than steel	[131]

(continued)

Table 3 (continued)

No.	Sample type	Findings	References
5.	Cellulose	• Bio sequins fabricated from bioplastic derived from tree woods namely 'Bio Iridescent' • Traditional sequins are produced of polyvinyl chloride (PVC), a polyester film, often known as Mylar, which poses serious environmental and health dangers by creating hazardous, bioaccumulate compounds, including hormone disruptors and carcinogens like phthalates	[132]
Packaging industry			
6.	Glycol	• monoethylene glycols (meg) and monopropylene glycols (mpg) from woods • a genuine greener option for making recyclable pet, polyesters, and industrial liquids	[133]
Cosmetic industry			
7.	Glycol	• A fully environmentally friendly option for the manufacture of carrier liquids. the secret to providing a generation of resins, detergents, de-icing agents, and cosmetics that is sustainable and renewable	[133]
8.	PLA+Polyhydroxyalkanoates (PHA)	• The cosmetic case is made of polylactic acid (PLA) and amorphous PHA technology for a private brand, 'Wakemake' • An alternative substitute of acrylonitrile butadiene styrene (ABS) casing	[134]

(continued)

Table 3 (continued)

No.	Sample type	Findings	References
Medical industry			
9.	HA	• HA-coated gold nanoparticles (AuNP-HA) synthesised in one pot. Usage as a drug (Sulfasalazine) (SSZ), delivery vehicle • AuNP-HA exhibits a promising material as a modern medicine delivery method with characteristics of controlled release	[135]
10.	Collagen+Glycogen+Hydroxyapatite	• Composite hydrogels functions as bone repair scaffold • When cultivated on the composite hydrogels, bone mesenchymal stem cells displayed favourable cell adhesion, vitality, and proliferation	[31]
Construction industry			
11.	Cellulose	• Food leftovers are combined with water and seasoning, then heated up and pressed into a mould. It is called 'Fabula' • Has a bending strength four times more than that of concrete and is primarily made of vacuum-dried, ground-up fruit peels and coffee grounds	[136]
12.	Alginate+Oyster shells	• 'Ostra'-A biomaterial resembling ceramic and cement that is created from used oyster shells and seaweed extract • Sourced from local seafood restaurants in Italy	[136]

Fig. 5 Composite made of oyster shell [137]

13 Challenges and Future Opportunities

Naturally derived biomaterials hold great promise for a variety of commercial and biological uses. However, there are also several concerns that must be completely addressed to realize their potential.

13.1 Challenges in the Use of Naturally Derived Biomaterials

Some of the challenges and future opportunities associated with naturally derived biomaterials are:

- Standardization and quality control: Naturally derived biomaterials can have batch-to-batch variability, which can affect their properties and performance. Standardization and quality control measures need to be developed to ensure consistent quality of naturally derived biomaterials [138].
- Cost-effectiveness: Naturally derived biomaterials can be expensive to produce and process, which can limit their use in industrial applications [139]. Improvements in production efficiency and processing methods can help to reduce costs and increase the scalability of naturally derived biomaterials.
- Mechanical and chemical properties: The mechanical and chemical properties of naturally derived biomaterials can vary widely, which can make it difficult to control their properties for specific applications. Advances in processing methods and functionalization techniques can help to overcome these limitations [140].

- Biodegradability and stability: The biodegradability of naturally derived biomaterials can be both an advantage and a challenge, as it can affect their stability and longevity in applications where long-term stability is required. Research on stabilizing naturally derived biomaterials while maintaining their biodegradability can help to overcome this challenge.
- Biocompatibility: While naturally derived biomaterials are generally biocompatible, some may cause an immune response or other adverse reactions in certain individuals. This is to better comprehend the biocompatibility of naturally derived biomaterials and to create new strategies to overcome any potential limitations [141].

13.2 Future Opportunities in Research and Development

Advances in biotechnology and material science are likely to drive new opportunities for the development of naturally derived biomaterials. For example, genetic engineering and synthetic biology approaches can be used to tailor the natural biomaterials' characteristics for particular uses, while utilising novel processing techniques that can help to overcome challenges related to scalability and processing efficiency.

Genetic engineering allows for the modification of an organism's DNA to produce specific traits or characteristics that can be beneficial for certain applications [142]. For example, scientists can engineer bacteria to produce large quantities of proteins or other biomolecules that can be used in medical or industrial settings. Similarly, genetic engineering can be used to modify the properties of natural biomaterials such as silk or collagen, allowing them to be tailored for specific applications. Synthetic biology, on the other hand, involves creating new biological systems and designing existing ones or organisms with desired properties or functions [143]. This approach can be used to create entirely new biomaterials that do not exist in nature, such as biofuels or biodegradable plastics.

Both genetic engineering and synthetic biology can be used to improve the properties of natural biomaterials for specific applications. For example, scientists can use genetic engineering to modify the properties of silk to make it more elastic or stronger. Synthetic biology can also be used to create new biomaterials with unique properties, such as self-healing materials or materials that respond to changes in their environment.

In addition to genetic engineering and synthetic biology, novel processing techniques can also be used to enhance the functionality and performance of natural biomaterials. For example, scientists can use electrospinning or 3D printing to construct scaffolds for delivery of drugs or tissue engineering. These methods enable fine-grained control over the composition and characteristics of the biomaterials, enabling them to be tailored for specific applications.

The capacity of natural biomaterials to combine with nanotechnology is perhaps the most fascinating part of them. By combining the unique properties of natural materials with the precision of nanotechnology, researchers can create smart biomaterials that having the capacity to revolutionize a vast area of industries. The possibilities are truly endless.

13.3 What Are Among the Crucial Areas for Biomaterials Research in the Future?

1. Immunomodulation is the process of adjusting the immune reaction to a specific level. Immunomodulating biomaterials could aid in the fight against common chronic illnesses like type 1 diabetes, an autoimmune condition in which the body's defences kill the insulin-producing cells within the pancreas. Injectable synthetic biomaterial that treated type 1 diabetes in non-obese diabetic mice was recently developed by researchers. This is a significant step in creating a sustainable platform to help control the disease's effects [144].
2. Injectable biomaterials are indeed being employed more frequently to deliver therapeutic agents such drugs, genetic material, and proteins. They provide focused distribution while preventing immune system absorption, opening the possibility of treating a number of illnesses. Injectable biomaterials that are now being researched and made from both artificial and naturally present materials might someday be utilised to treat heart attacks, cancer, and bone abnormalities [144].
3. Supramolecular biomaterials, which are collections of molecules that go beyond the capabilities of individual molecules, have the capacity to feel and react, making them perfect materials for treating disease or injury. Supramolecular biomaterials that imitate natural biological signalling or that can be switched either on or off regarding physiological cues are being investigated by researchers [144].
4. Quorum sensing in synthetic biology which are needed in areas like bioterrorism, combat, food and water safety, therapeutic properties, and fuel reliability are pertinent to work on exploiting the indigenous microbial communication mechanism involving the signalling molecules Autoinducer-2 (AI-2) [145]. It has been suggested that the signal molecule AI-2, which is created by the enzyme S-Ribosyl homocysteinase (LuxS), is present in many bacterial species, and enables interspecies communication [145]. To create a reliable and sensitive biosensing device, the researchers use synthetic biology to design a bacterial sensor and rewire the communication system of bacteria. It would be possible to create more advanced and/or enhanced systems for a larger range of applications with the aid of additional theoretical studies in genome circuits [146].
5. Photonic biomaterials: new fields including neuromorphic photonics, biodegradable photonics, and Artificial Intelligence (AI) design are currently the subject of progressive investigation. Combined, these new inventions—from an all-optical internet to quantum communications—will help further transform the world today [147].

6. Long-lived biomaterial-based fuel cells and batteries for wearable and implantable electronics or high-power microbial fuel cells. Without any commercial success, biofuel cells and biobatteries have only ever been considered laboratory marvels. The primary difficulty is because historically, research has been conducted by scientists and research organizations, and this decentralized method encourages an imbalanced and deceptive image of biofuel cells that suggests itself more to science fiction than an instant answer [148].
7. Food biopolymers for cultured food production: Due to the rise of societal and commercial interest, cultured meat has gradually become well-known. The cultured meat method comprises the laboratory cultivation of designed muscle tissue as opposed to the traditional meat manufacturing method, which relies on animals. Unfortunately, bioengineers have limited experience in engineering muscle tissue for its post-mortem features, which largely determine how consumers define meat quality. Until now, they have only built tissues to fulfil biological objectives. Furthermore, the technical viability and industrial scalability of current tissue engineering technologies for the manufacture of cultured meat face basic hurdles. Thus, new explorations are needed in this area [149].
8. Grave to cradle concept of recycling waste as beneficial biomaterials for multiple application. Reprocessing liquid and solid waste is an emerging research area for the development of novel biomaterials with potential industrial applications as wastes are abundant and cheap. This concept is aligned with the bio circular economy model and is sought globally.

14 Conclusion

In summary, while there are challenges associated with naturally derived biomaterials, ongoing research and development efforts hold significant promise for the future of these materials in an extensive array of manufacturing and biomedical industries. Overall, this chapter delivers a summary of the synthesis of natural biomaterials and their possible applications, emphasizing the need for continued research in this field. Regardless of the approach used, sustainability is always an important consideration when synthesizing biomaterials. By utilizing renewable resources and minimizing environmental impact, researchers can create more sustainable biomaterials that benefit both the users and the planet. This is particularly significant given the growing concern about environment variation besides the requirement for more eco-friendly solutions in all areas of life. As more research is conducted, we will undoubtedly discover even more applications for these incredible materials, making a positive impact on human health and our planet's future.

References

1. Coppola, G., Gaudio, M.T., Lopresto, C.G. et al.: Earth Syst. Environ. **5**, 231–251 (2021)
2. Sciences, Penn State College of Agricultural. What is biomaterial? Department of Agricultural Economics, Sociology and Education (2023). https://aese.psu.edu/teachag/curriculum/modules/biomaterials/what-is-a-biomaterial#:~:text=A%20material%20derived%20from%2C%20or,injury%20or%20growing%20biological%20cells
3. Leong, K., Quek, C.H., Basu, B., Chan, B., Goodman, S.B., Le Visage, C., Liang, X.-J.: Science Direct. Biomaterials (2023). https://www.sciencedirect.com/journal/biomaterials
4. Birajdar, M.S., Joo, H., Koh, W.G. et al.: Biomater. Res. **8**, 25 (2021)
5. Fowler, S.: Introduction to the Chemistry of Life. Charles Molnar and Jane Gair. Concepts of Biology (2021)
6. Cole, L., Kramer, P.R.: Macronutrients. Human Physiology, Biochemistry and Basic Medicine, pp. 157–164. Academic Press (2016)
7. Biology. Lipid. Biology Online (2022). https://www.biologyonline.com/dictionary/lipid
8. Ngwuluka, N.C.: Responsive polysaccharides and polysaccharides-based nanoparticles for drug delivery. [book auth.] Nedal Y. Abu-Thabit Abdel Salam Hamdy Makhlouf. Stimuli Responsive Polymeric Nanocarriers for Drug Delivery Applications, pp. 531–554. Woodhead Publishing (2018)
9. Wang, F., Li, P., Chu, H.C., Lo, P.K.: Biosensors (Basel) **12**(2), 93 (2022)
10. Muncie, J.M., Weaver, V.M.: Curr. Top. Dev. Biol. **130**, 1–37 (2018)
11. Badylak, L.T., SaldinMadeline, C., CramerSachin, S., VelankarLisa, J., WhiteStephen, F.: Elsevier BV, pp. 1–46 (2017)
12. Chempages. Biomolecules: Carbohydrates. Polysaccharides (2023). https://www2.chem.wisc.edu/deptfiles/genchem/netorial/modules/biomolecules/modules/carbs/carb6.htm
13. Mohammed, A.S.A., Naveed, M., Jost, N.: J. Polym. Environ. **29**(8), 2359–2371 (2021)
14. Helmenstine, A.M.: Polysaccharide Definition and Functions. Thought Co. (2020). https://www.thoughtco.com/polysaccharide-definition-and-functions-4780155
15. Zhang, Y.H.P.: 9—Starch: A high-density chemical hydrogen storage compound for PEM fuel cells. [book auth.] Antonio M. Chaparro Paloma Ferreira-Aparicio. Portable Hydrogen Energy Systems, pp. 161–173. Academic Press (2018)
16. Kadokawa, J.: Enzymatic synthesis of functional amylosic materials and amylose analog polysaccharides. [book auth.] Katja Loos Nico Bruns. Methods in Enzymology, pp. 189–213. Academic Press (2019)
17. Falua, K.J., Pokharel, A., Babaei-Ghazvini, A., Ai, Y., Acharya, B.: Polymers **14**(11), 2215 (2022)
18. Omoregie Egharevba, H.: IntechOpen (2020)
19. Tian, J., Deng, H., Huang, M., Liu, R., Yi, Y., Dong, X.: Electrospun Nanofibers for Food and Food Packaging Technology. [book auth.] Xianfeng Wang, Jianyong Yu Bin Ding. Micro and Nano Technologies: Electrospinning: Nanofabrication and Applications, pp. 455–516. William Andrew Publishing (2019)
20. Ren, F., Wang, J., Yu, J., Zhong, C., Xie, F., Wang, S.: ACS Omega **6**(41) (2021)
21. Cengage. Encyclopedia.com. Organic Chemistry: Starch (2018). https://www.encyclopedia.com/science-and-technology/chemistry/organic-chemistry/starch
22. Annie, M.: Moisture and Mosses. Mountain Moss (2023). https://www.mountainmoss.com/pages/moisture-and-mosses
23. Bakerpedia. What is Starch? Starch (2023). https://bakerpedia.com/ingredients/starch/
24. do Val Siqueira, L., La Fuente Arias, C.I., Maniglia, B.C., Tadini, C.C.: Current Opinion in Food Science 122–130 (2021)
25. Gadhave, R., Gadhave, C.: Open J. Polym. Chem. **12**, 55–79 (2022)
26. Leverette, M.M.: How to Use Laundry Starch and Sizing? The Spruce, New York (2021)
27. Apeji, Y.E., Kaigama, R.T., Ibrahim, S.H., Anyebe, S.N., Abdussalam, A.O., Oyi, A.R.: Turk. J. Pharm. Sci. **19**(5), 513–520 (2022)

28. Gamage, A., Liyanapathiranage, A., Manamperi, A., Gunathilake, C., Mani, S., Merah, O., Madhujith, T.: Sustainability **14**(10), 6085 (2022)
29. FlexBooks. Polysaccharides. Chemistry For High School (2023). https://flexbooks.ck12.org/cbook/ck-12-chemistry-flexbook-2.0/section/26.3/primary/lesson/polysaccharides-chem/
30. Gagliardi, A., Giuliano, E., Venkateswararao, E., Fresta, M., Bulotta, S., Awasthi, V., Cosco, D.: Front. Pharmacol. **12**, 601626 (2021)
31. Zhang, X., Zhou, J., Ying, H., Zhou, Y., Lai, J., Chen, J.: ACS Sustain. Chem. Eng. **8**(4), 2106–2114 (2020)
32. Salton, M.R.J., Kim, K.S.: Structure. In: Baron, S. (Ed.) Medical Microbiology, 4th edn. Galveston (TX): [book auth.] Baron S. Medical Microbiology 4th Edition. University of Texas Medical Branch at Galveston (1996)
33. Kim, S.J., Chang, J., Singh, M.: Biochim. Biophys. Acta **1848**(1 Pt B), 350–62 (2015)
34. Garde, S., Chodisetti, P.K., Reddy, M.: EcoSal Plus **9**(2) (2021)
35. Davis, K.M., Weiser, J.N.: Infect. Immun. **79**(2), 562–70 (2011)
36. Godoy, M.G., Amorim, G.M., Barreto, M.S., Freire, D.M.G.: Agricultural Residues as Animal Feed: Protein Enrichment and Detoxification Using Solid-State Fermentation. [book auth.] Christian Larroche, Carlos Ricardo Soccol Ashok Pandey. Current Developments in Biotechnology and Bioengineering, pp. 235–256. Elsevier (2018)
37. Klemm, D., Heublein, B., Fink, H.-P., Bohn, A.: Angewandte Chemie (International ed. in English) **44**, 3358–93 (2005)
38. Norizan, M.N., Shazleen, S.S., Alias, A.H., Sabaruddin, F.A., Asyraf, M.R.M., Zainudin, E.S., Abdullah, N., Samsudin, M.S., Kamarudin, S.H., Norrrahim, M.N.F.: Nanomaterials (Basel) **12**(19), 3483 (2022)
39. Deng, L., Huang, Y., Chen, S., Han, Z., Han, Z., Jin, M., Qu, X., Wang, B., Wang, H., Gu, S.: Carbohydr. Polym. **308**, 120647 (2023)
40. Bierach, C., Coelho, A.A., Turrin, M. et al.: Archit. Struct. Constr. (2023)
41. Elieh-Ali-Komi, D., Hamblin, M.R.: Int. J. Adv. Res. (Indore) 411–427 (2016)
42. Arockianathan, M.: 4 - Chitin-based nanomaterials. [book auth.] Ashok Kumar, Tuan Anh Nguyen, Swati Sharma, Yassine Slimani Shamsher Kanwar. In: Micro and Nano Technologies: Biopolymeric Nanomaterials, pp. 61–99. Elsevier (2021)
43. Stefanowska, K., Woźniak, M., Dobrucka, R., Ratajczak, I.: Materials, vol. 16, p. 1579 (2023)
44. Alimi, B.A., Pathania, S., Wilson, J., Duffy, B., Frias, J.M.C.: Int. J. Biol. Macromol. **237**, 124195 (2023)
45. Saberi Riseh, R., Hassanisaadi, M., Vatankhah, M., Babaki, S., Ait Barka, E.: Int. J. Biol. Macromol. **220** (2022)
46. Application of Chitosan in Agriculture. Linkedin. (2022). https://www.linkedin.com/pulse/application-chitosan-agriculture-cathy-liu
47. Wantulla, M., Joop J.A. van Loon, Dicke, M.: Appl. Soil Ecol. **188** (2023). ISSN 0929-1393
48. Iber, B., Kasan, N., Torsabo, D., Omuwa, J.: J. Renew. Mater. **10**, 42–49 (2021)
49. Parvin, N. et al.: **9**, 64 (2023)
50. López-Valverde, N., Aragoneses, J., López-Valverde, A., Rodríguez, C., Macedo de Sousa, B., Aragoneses, J.M.: Front. Bioeng. Biotechnol. **10**, 907589 (907589)
51. Teixeira-Santos, R., Lima, M., Gomes, L.C., Mergulhão, F.J.: iScience **24**(12), 103480 (2022)
52. Pereira, L., Cotas, J.: Introductory Chapter: Alginates—Alginates—Recent Uses of This Natural Polymer. [A General Overview [Internet].] IntechOpen (2020). https://doi.org/10.5772/intechopen.88381
53. Farshidfar, N., Iravani, S., Varma, R.S.: Alginate-Based Biomaterials in Tissue Engineering and Regenerative Medicine **21**(3), 189 (2023). Marine Drugs. https://doi.org/10.3390/md21030189
54. Wu, T., Liu, L., Gao, Z., Cui, C., Fan, C., Liu, Y., Mingyuan Di, A., Yang, Q., Xu, Z., Liua, W.: Biomaterials Science (2023)
55. Best Summer Skincare Products of 2023. Outlook India (2023). https://www.outlookindia.com/business-spotlight/best-summer-skincare-products-of-2023-news-271961
56. Szymański, T., Semba, J.A., Mieloch, A.A. et al.: Sci. Rep. **13**, 646 (2023)

57. Kwon, H., Brown, W.E., Lee, C.A. et al.: Nat. Rev. Rheumatol. **15**, 550–570 (2019)
58. Hussain, M.A., Haseeb, M.T., Muhammad, G., Tahir, M.N.: Inulin Type Fructan: A Versatile Functional Material for Food and Healthcare. [book auth.] M., Sheardown, H., Al-Ahmed, A. Jafar Mazumder. Functional Biopolymers. Polymers and Polymeric Composite (2019)
59. Parın, F.N. et al.: Polymers **15**, 1002 (2023)
60. Paz-Gómez, G., del Caño-Ochoa, J.C., Rodríguez-Alabanda, O., Romero, P.E., Cabrerizo-Vílchez, M., Guerrero-Vaca, G., Rodríguez-Valverde, M.A.: Coatings **9**, 5–29 (2019)
61. Martin Gericke, L.H.S., Heinze, T.: Carbohydrate Polymers **300**, 120251 (2023). ISSN 0144-8617
62. What are proteins and what do they do? Medline Plus (2023). https://medlineplus.gov/genetics/understanding/howgeneswork/protein/#:~:text=Proteins%20are%20made%20up%20of,combined%20to%20make%20a%20protein
63. Education, Nature. Protein Structure. Scitable (2023). https://www.nature.com/scitable/topicpage/protein-structure-14122136/#:~:text=The%20overall%20structure%20of%20the,and%20beta%20sheets%20(red).&text=The%20primary%20structure%20of%20a,protein's%20unique%20three%2Ddimensional%20shape
64. Elastin. Cleveland Clinic (2023). https://my.clevelandclinic.org/health/body/22482-elastin
65. Mengmeng Jin, I., Zhu, S., Hou, Y.: ACS Biomaterials Science and Engineering Article ASAP (2023)
66. What is collagen, and why do people use it? Medical News Today. [Online] Healthline Media (2022). https://www.medicalnewstoday.com/articles/262881
67. Wu, M., Cronin, K., Crane, J.S.: Biochemistry, Collagen Synthesis. Treasure Island (FL). StatPearls Publishing (2023)
68. Manoukian, O.S., Sardashti, N., Stedman, T., Gailiunas, K., Ojha, A., Penalosa, A., Mancuso, C., Hobert, M., Kumbar, S.G.: Biomaterials for Tissue Engineering and Regenerative Medicine. [book auth.] Roger Narayan. Encyclopedia of Biomedical Engineering. Elsevier (2019)
69. Man, K., Joukhdar, H., Manz, X.D. et al.: Cell Tissue Res. **388**, 565–581 (2022)
70. Loureiro dos Santos, L.A.: Natural Polymeric Biomaterials: Processing and Properties. Reference Module in Materials Science and Materials Engineering. Elsevier (2017)
71. Murphy, K.C., Whitehead, J., Zhou, D., Ho, S.S., Leach, J.K.: Acta Biomater. 64, 176–186 (2017)
72. Ahmed, S., Shah, P., Ahmed, O.: Biochemistry, Lipids. StatPearls Publishing, Treasure Island (FL) (2023)
73. Nsairat, H., Khater, D., Sayed, U., Odeh, F., Al Bawab, A., Alshaer, W.: **8**(5), e09394 (2022). Heliyon
74. Nakhaei, P., Margiana, R., Bokov, D.O., Abdelbasset, W.K., Jadidi Kouhbanani, M.A., Varma, R.S., Marofi, F., Jarahian, M., Beheshtkhoo, N.: Front. Bioeng. Biotechnol. (2021)
75. Farasati Far, B., Naimi-Jamal, M.R., Sedaghat, M., Hoseini, A., Mohammadi, N., Bodaghi, M.: J. Funct. Biomater. **14**(2), 115 (2023)
76. Musielak, E., Feliczak-Guzik, A., Nowak, I.: Materials (Basel) **15**(2), 682 (2022)
77. Luchini, A., Vitiello, G.: Biomimetics (Basel) **6**(1), 3 (2021)
78. Nikoleli, G.P., Nikolelis, D.P., Siontorou, C.G., Nikolelis, M.T., Karapetis, S.: Membranes (Basel) **8**(4), 108 (2018)
79. Syed Azhar, S.N.A., Ashari, S.E., Zainuddin, N., Hassan, M.: UPM, Serdang, Selangor. Molecules 27(1), 289 (2022)
80. Maisonneuve, L., Lebarbé, T., Cramail, E.G.H.: Structure–properties relationship of fatty acid-based thermoplastics as synthetic polymer mimics (2013)
81. Talló, K., Bosch, M., Pons, R., Cocera, M., López, O.: J. Mater. Chem. B **8**(1), 61–167 (2020)
82. Nagtode, V.S., Cardoza, C., Yasin, H.K.A., Mali, S.N., Tambe, S.M., Roy, P., Singh, K., Goel, A., Amin, P.D., Thorat, B.R., Cruz, J.N., Pratap, A.P.: ACS Omega **8**(13), 11674–99 (2023)
83. Hill, K., Rhode, O.: Lipid/Fett **101**(1), 25–33 (1999)
84. Allen, D.K., Tao, B.Y.: Carbohydrate-alkyl ester derivatives as biosurfactants. J. Surfact. Deterg **2**(3), 383–390 (1999)

85. Alberts, B., Johnson, A., Lewis, J., et al.: Molecular Biology of the Cell, 4th edn. Garland Science, New York (2002)
86. Aleandri, S., Rahnfeld, L., Chatzikleanthous, D., Bergadano, A., Bühr, C., Detotto, C., Fuochi, S., Weber-Wilk, K., Schürch, S., van Hoogevest, P., Luciani, P.: Eur. J. Pharmaceut. Biopharmaceut. **181**, 300–309 (2022). ISSN 0939-6411
87. Anada, R., Hara, E.S., Nagaoka, N., Okada, M., Kamioka, H., Matsumoto, T.: J. Mater. Chem. B (2022)
88. Mao, Y., Guidoin, R., li, Y., Brochu, G., Zhang, Z., Wang, L.: Mater. Des. **205**, 109737 (2021)
89. Wax. Chemistry. [Online] LibreText (2022). https://chem.libretexts.org/Bookshelves/Biological_Chemistry/Supplemental_Modules_(Biological_Chemistry)/Lipids/Non-glyceride_Lipids/Wax#:~:text=A%20wax%20is%20a%20simple,coatings%20on%20leaves%20and%20stems
90. Madhuranthakam, C.M.R., Fernandes, S.Q., Piozzi, A., Francolini, I.: Int. J. Mol. Sci. **23**(16), 9501 (2022)
91. Zhang, Q.W., Lin, L.G., Ye, W.C.: Chin. Med. **13**, 20 (2018)
92. Matinong, A.M.E., Chisti, Y., Pickering, K.L., Haverkamp, R.G.: Biology (Basel) **11**(6), 905 (2022)
93. Ranganagowda, R.P.G., Kamath, S.S., Bennehalli, B.: Mat. Sci. Res. India **16**(1) (2019)
94. Sepmag. Purification Techniques. Purification Techniques. [Online] (2023). https://www.sepmag.eu/blog/purification-techniques
95. Jiang, Y.H., Lou, Y.Y., Li, T.H., Liu, B.Z., Chen, K., Zhang, D., Li, T.: Am. J. Transl. Res. **14**(2), 1146–1159 (2022)
96. Quantifiers and Quantification. Stanford Encyclopedia of Philosophy. [Online] Stanford Education (2022). https://plato.stanford.edu/entries/quantification/#:~:text=Quantifier%20expressions%20are%20marks%20of,most%20common%20examples%20of%20quantification
97. Lewis, M., Bromley, K., Sutton, C.J., McCray, G., Myers, H.L., Lancaster, G.A.: Pilot Feasibil. Stud. **7**, 1 (2021)
98. Troy, E., Tilbury, M.A., Power, A.M., Wall, J.G.: Polymers (Basel) **13**(19), 3321 (2021)
99. Carmona, P., et al.: Gels **7**, 186 (2021)
100. Taaca, K.L.M., Prieto, E.I., Vasquez, M.R.: Polymers 14 **13**, 2560 (2022)
101. Aranaz, I., Alcántara, A.R., Civera, M.C., Arias, C., Elorza, B., Heras Caballero, A., Acosta, N.: Polymers (Basel) **13**(19), 3256 (2021)
102. Purohit, P., Bhatt, A., Mittal, R.K., Abdellattif, M.H., Farghaly, T.A.: Front. Bioeng. Biotechnol. **10**, 1044927 (2023)
103. Sahin Kehribar, E., Isilak, M.E., Bozkurt, E.U., Adamcik, J., Mezzenga, R., Seker, U.O.S.: Biomater. Sci. **9**(10), 3650–3661 (2021)
104. Bose, S., Robertson, S.F., Bandyopadhyay, A.: Acta Biomater. **66**, 6–22 (2018)
105. Jasni, A.H.: Fabrication of nanostructures by physical techniques. [book auth.] Preetha Balakrishnan Sabu Thomas. In: Micro and Nano Technologies Nanoscale Processing. Elsevier (2021)
106. Gopi, S., Balakrishnan, P.: Liposomal nanostructures: Properties and applications. [book auth.] Preetha Balakrishnan Sabu Thomas. In: Micro and Nano Technologies Nanoscale Processing. Elsevier (2021)
107. Ilangovan, R., Subha, V., Earnest Ravindran, R.S., Kirubanandan, S., Renganathan, S.: Nanomaterials: Synthesis, physicochemical characterization, and biopharmaceutical applications. [book auth.] Preetha Balakrishnan Sabu Thomas. In: Micro and Nano Technologies. Elsevier (2021)
108. Gerwig, G.J., Poele, E.M., Dijkhuizen, L., Kamerling, J.P.: Stevia Glycosides: Chemical and Enzymatic Modifications of Their Carbohydrate Moieties to Improve the Sweet-Tasting Quality,. [book auth.] David C. Baker. Advances in Carbohydrate Chemistry and Biochemistry. Academic Press (2016)
109. Giri, B. : Simultaneous Determination of Protein and Glucose in Urine Sample Using a Paper-Based Bioanalytical Device. Laboratory Methods in Microfluidics. Elsevier (2017)

110. Perluigi, M., Marco, F., Foppoli, C., Coccia, R., Blarzino, C., Marcante, M., Cini, C.: Biochemical and biophysical research communications **3** (2003)
111. Brouns, J.E.P., Dankers, P.Y.W.: Biomacromolecules **22**(1), 4–23 (2021)
112. Cell Dissociation Methods for Disaggregation of Tissue: Mechanical vs Enzymatic vs Chemical. Akadeum Life Science. [Online] Mteric Marketing, March 2021. https://www.akadeum.com/blog/cell-dissociation/#:~:text=Enzymatic%20dissociation%20uses%20specific%20proteins,combination%20leads%20to%20optimal%20results
113. Duarte, L., Matte, C.R., Bizarro, C.V. et al.: World J. Microbiol. Biotechnol. **36**, 11 (2020)
114. Chandra, P., Enespa, S.R., Arora, P.K.: Microb. Cell Fact. **19**(1), 169 (2020)
115. Bacakova, M., Pajorova, J., Sopuch, T., Bacakova, L.: Materials (Basel) **11**(11), 2314 (2018)
116. Widiyanti, P., Priskawati, Y.C.A.: Int. J. Biomater. (2023)
117. Qi, P., Ning, Z., Zhang, X.: IET Nanobiotechnol. 1–8 (2022)
118. Kang, W., Shi, Y., Yang, Z., Yin, X., Zhao, Y., Weng, L., Teng, Z.: RSC Adv. **13**, 5609–5618 (2023)
119. Shalaby, M., Ghareeb, A.Z., Khedr, S.M., Mostafa, H.M., Saeed, H., Hamouda, D.
120. Rotman, S.G., Post, V., Foster, A.L., Lavigne, R., Wagemans, J., Trampuz, A., Gonzalez Moreno, M., Metsemakers, W.-J., Grijpma, D.W., Richards, R.G., Eglin, D., Moriarty, T.F.: J. Drug Deliv. Sci. Technol. **79**, 103991 (2023). ISSN 1773-2247
121. Bennardo, F., Gallelli, L., Palleria, C., Colosimo, M., Fortunato, L., De Sarro, G., Giudice, A.: BMC Oral Health **23**(1), 134 (2023)
122. Naznin, A., Dhar, P.K., Dutta, S.K., Chakrabarty, S., Karmakar, U.K., Kundu, P., Hossain, M.S., Barai, H.R., Haque, M.R.: Pharmaceutics **15**(3), 732 (2023)
123. Liu, L., Wang, J., Li, Y., Liu, B., Zhang, W., An, W., Wang, Q., Xu, B., Zhao, L., Ma, C.: Regenerat. Biomater. **9**, rbac054 (2022)
124. Almeida, D., Sartoretto, S.C., Calasans-Maia, J.D.A., Ghiraldini, B., Bezerra, F.J.B., Granjeiro, J.M., et al.: PLoS ONE **18**(2), e0282067 (2023)
125. Salsabila, A., et al.: Metals **13**, 494 (2023)
126. Centre, Bristol Biomedical Research. First in human study to assess knee cartilage repair implant launches at Southmead Hospital. NIHR. [Online] July 13, 2022. https://www.bristolbrc.nihr.ac.uk/news/first-in-human-study-to-assess-knee-cartilage-repair-implant-launches-at-southmead-hospital/.
127. Romasco, T., et al.: Biomedicines **11**, 786 (2023)
128. Biswas, A., et al.: Polymers **15**, 1425 (2023)
129. Zhatkanbayev, Y., Zhatkanbayeva, Z., Iskakova, Z., Kolpek, A., Serikov, A., Moldagulova, N., Danlybayeva, G., Sarsenova, A.: Int. J. Biomater. Hindawi (2023)
130. Castim, D.: Vegan Biomaterials Could Replace Chemicals In Crop Management. Vegan Biomaterials Could Replace Chemicals in Crop Management (2022)
131. Carlson, C.: Modern Synthesis uses bacteria to create biomaterial fabric. Dezeen. [Online] (2023). https://www.dezeen.com/2023/04/03/modern-synthesis-bacteria-biomaterial-fabric/
132. Finney, A.: Stella McCartney releases jumpsuit made with iridescent BioSequins. Dezeen. [Online] (2023). https://www.dezeen.com/2019/08/02/bio-iridescent-sequin-elissa-brunato-sustainable-fashion/
133. Future proof your product and elevate your brand - with woodbased renewable Glycols. UPM Biochemicals. [Online] (2023). https://www.upmbiochemicals.com/glycols/
134. Mueller, J.: CJ Biomaterials Develops Cosmetic Case Featuring Bio-sourced Materials for CJ Olive Young. Global Cosmetic Industry (2022). https://www.gcimagazine.com/packaging/containers/news/22618509/cj-bio-cj-biomaterials-inc-develops-cosmetic-case-featuring-bio sourced-materials
135. Kim, H.-M., Park, J.H., Choi, Y.J., Ohb, J.-M., Park, J.: RSC Adv. 8 (2023)
136. Bagshaw, E.: Top 23 biomaterial designers to watch in 2023. Mater. Source (2022)
137. Markos: Oyster shell composite. Material exploration for moulding. Instagram, Italy (2020)
138. Troy, E., et al.: Polymers **13** (2021)
139. O'Brien, F.J.: Mater. Today **14**(3), 88–95I (2011). SSN 1369-7021

140. Yusoff, N.H.M., Chong, C.H., Wan, Y.K., Cheah, K.H., Wong, V.-L.: J. Water Process Eng. **51**, 103410 (2023). ISSN 2214-7144
141. Mariani, E., Lisignoli, G., Borzì, R.M., Pulsatelli, L.: Int. J. Mol. Sci. **20**(3), 636 (2019)
142. Genetic Engineering. National Human Genome Research Institute. [Online] USA (2023). https://www.genome.gov/genetics-glossary/Genetic-Engineering
143. Joyce, S., Mazza, A.-M., Kendall, S.: Rapporteurs. Synthetic Biology: Science and Technology for the New Millennium. Positioning Synthetic Biology to Meet the Challenges of the 21st Century: Summary Report of a Six Academies Symposium Series. National Academic Press (2013)
144. Majumder, K.: Biomaterials. Times of India (2021)
145. National Academies of Sciences, Engineering, and Medicine. Evaluation of the Army Research Laboratory: Interim Report. The National Academies Press, Washington (2014)
146. Pereira, C.S., Thompson, J.A., Xavier, K.B.: FEMS Microbiol. Rev. **37**(2), 156–81 (2013)
147. What Does The Future Have In Store For Photonics?. Stensborg. [Online] Stensborg (2022). https://www.stensborg.com/post/what-does-the-future-have-in-store-for-photonics#:~:text=The%20future%20of%20photonics&text=Right%20now%2C%20research%20is%20ongoing,optical%20internet%20to%20quantum%20communications
148. Choi, S.: Batteries **9**(2), 119 (2023)
149. Ng, S., Kurisawa, M.: Acta Biomaterialia 108–129 (2021). ISSN 1742-7061

A. H. Jasni a former graduate of Monash University (MU) and National Defence University of Malaysia (NDUM). She just completed her Ph.D. in Engineering at Faculty of Engineering, International Islamic University Malaysia. She obtained her Bachelor of Science in Biotechnology from MU and Master of Science in Biology from NDUM. She had been working under the area of nanotechnology, material science, biotechnology in military applications including microbiology, biosensor, proteomic, and electrospinning since 2013.

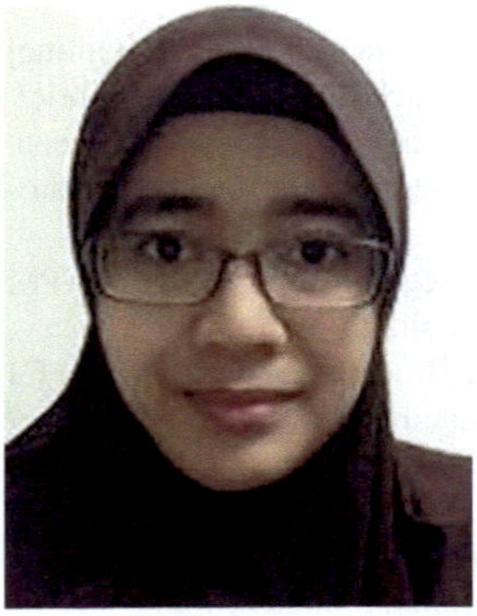

A. S. Azmi a former graduate of the Widener University, Pennsylvania, USA started her career as a Chemist at First Malaysia Coating Sdn. Bhd. on April 1999. On September 1999, she received an offer from UniversitiTeknologi PETRONAS (UTP) and working as trainee lecturer. She continued her study in MSc in Process Integration at University of Manchester Institute of Science and Technology (UMIST), Manchester, UK in 2001 for one year. Early 2002 she was back to UTP and being appointed and served as Lecturer until October 2006. On November 2006, she moved to International Islamic University Malaysia (IIUM) as lecturer before she pursued her study at University of Malaya (UM), Malaysia on December 2007. She obtained her Ph.D. in Bioprocess in Chemical Engineering on September 2012. Currently she is attached as an Associate Professor to the Department of Chemical Engineering and Sustainability, International Islamic University Malaysia, Kuala Lumpur.

N. I. M. Puad graduated with B. Eng (Biochemical-Biotechnology) (Honors) from International Islamic University Malaysia (IIUM) in 2007. She was then appointed as an Assistant Lecturer at the Calatrava Department of Biotechnology Engineering, IIUM in the same year. Later in 2011, she obtained her Ph.D. in Chemical Engineering and Analytical Science from The University of Manchester, UK. Her research interest is mainly on Plant Cell Culture Technology, Flux Balance Analysis, Kinetic Modelling and Simulation, Plant Secondary Metabolite and Natural Products, Bioprocess and Renewable Energy. Presently, she is an Assistant Professor at the Department of Chemical Engineering and Sustainability, Faculty of Engineering, International Islamic University Malaysia (IIUM).

F. Ali is an Assistant Professor at the Department of Biotechnology Engineering, Faculty of Engineering, International Islamic University Malaysia (IIUM), Gombak 50728, Kuala Lumpur. She obtained her Ph.D. in Chemical and Biomolecular Engineering from Korea Advanced Institute of Technology (KAIST) in 2013. Her research interest is in the area of Polymer Blend and Composites, Polymer Nanocomposites, Synthesis of polymers, block copolymers and polyurethane, Synthesis of Nanoparticles using natural sources, Polymers in Sensors, Polymers for packaging materials, Functional Polymers for lithography, Polymerization of biopolymer from the monomers produced from fermentation process, and Separation and Purification Techniques. She has nearly two years' experience in teaching on Process Plant and Design, Biopharmaceutical Engineering, Seminar for undergraduates and Fluid Mechanics. Dr. Fathilah also serves as a reviewer for polymer related journals. She published 11 papers in National and International conferences, 6 papers in SCOPUS and ISI journals and 2 book chapters. She was secretary for Nano Research Group and currently the Deputy Dean of Student Affairs (January 2015–present).

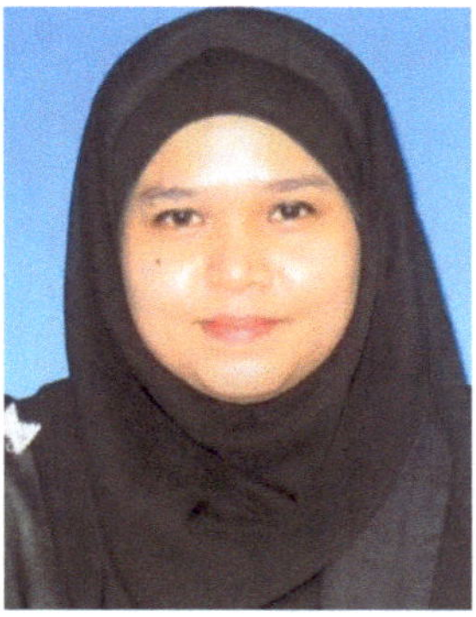

Y. A. Nor obtained her qualifications of Chemical and Biological Science (Nanotechnology) for her Doctor of Philosophy from University of Queensland and she acquired her Biotechnology Engineering Masters Degree from the International Islamic University Malaysia (IIUM). She is now working as an Assistant Professor at the Department of Chemical Engineering and Sustainability, Faculty of Engineering IIUM.

Different Techniques of Genetic Engineering Used for the Development of Novel Biomaterials

Aashveen Chhina, Vridhi Sachdeva, and Shubham Thakur

Abstract Biomaterials are garnering huge success in the modern era of unforeseen disease conditions. They find their utility in specifically interacting with biological systems for therapeutic or diagnostic purposes. They help patients recuperate from disease or injury by restoring tissue function thereby, sometimes also leading to regeneration. Some of the biomaterials are living (microorganisms based) and can be prepared through microencapsulation, 3D printing, coating, and spinning by which they can be inculcated into matrices. On the other hand, microbes can even produce their own matrix which can be genetically engineered to have control over the responses of the body. Genetic engineering and biotechnological approaches are an ideal amalgamation of systems for preparing biomaterials that handle accuracy and complexity quite conveniently. The biomaterials generated by the use of genetically engineered techniques have proven to be a boon for the regenerative medicines market, gene delivery systems, tissue regeneration platforms, and controlled drug delivery systems. In the current scenario, researchers have focused their attention on molecular evolution for the preparation of advanced designs of genetically engineered biomaterials and the development of inert biomaterials that can be properly integrated into devices. The advancements and research carried out in genetic engineering continue to assist researchers to gain insights into a specific change in the structure of employed molecules for the preparation of biomaterials and the way it will affect the development of biomaterials and their function in the longer run. The manufacturing sector of biomaterials is a domain that has significantly evolved and is still being researched.

Keywords Genetic engineering · Biomaterials · Recombinant DNA technology · Genetically engineered biomaterials · CRISPR- Cas9 · mRNA

A. Chhina · V. Sachdeva · S. Thakur (✉)
Department of Pharmaceutical Sciences, Guru Nanak Dev University, Amritsar, Punjab 143005, India
e-mail: shubhamdpharma.rsh@gndu.ac.in

R. Malviya and S. Sundram (eds.), *Engineered Biomaterials*, Engineering Materials,
https://doi.org/10.1007/978-981-99-6698-1_2

Abbreviations

DNA	Deoxyribonucleic Acid
GF	Growth Factors
ECM	Extracellular Matrix
GR and CNT	Graphene and Carbon Nanotubes
ELP	Elastin-like Proteins
SLP	Silk-like Proteins
SELP	Silk-elastin-like Proteins
HDP	Host Defense Peptides
CPP	Cell-penetrating Peptides
AMP	Antimicrobial Peptides
mRNA	Messenger RNA
SDS	Sodium Dodecyl Sulfate
CDI	Carbodiimide
HMDC	Hexamethylene Diamine Carbamate
EtO	Ethylene Oxide
PAA	Peracetic Acid
GI	Gamma Irradiation
dECM	Decellularized Extracellular Matrix
PBS	Phosphate Buffer Saline
VEGF	Vascular Endothelial Growth Factor
bFGF	Basic Fibroblast Growth Factor
BMP	Bone Morphogenetic Proteins
TENG	Triboelectric Nanogenerator
PENG	Piezoelectric Nanogenerator
PyENG	Pyroelectric Nanogenerator
Bio	TENG, Biodegradable Nanogenerator
PTFE	Polytetraflouroethylene
AFR	Aniline Formaldehyde Resin
S-TENG	Starch Triboelectric Nanogenerator
CD-TENG	Chitosan-diatom-based biomaterial TENG
TDD	Transdermal Drug Delivery
HAP	Hydroxyapatite
CRISPR/CAS 9	CRISPR, Associated Protein 9
HFG	Hepatocyte Growth Factor
sgRNA	Single Guide RNA
PLGA	Polylactic Glycolic Acid
GCE	Genetic Code Expansion
HDACi	HDAC Inhibitors
CAR T-cell	Chimeric Antigen Receptor T-cell

1 Introduction

The main goal of medical research aims to convert multifaceted engineering science along with the biological science into procedures for the substitution, rejuvenate, cell repair, tissue repair, or organs in order to restore compromised function. Biomaterials are any substances, whether organic or inorganic, living or dead, typically composed of several parts that engage with the body system. In medical uses, such intelligent materials are frequently used to supplement or take the place of a natural function. To state it in a more simplified manner, a biomaterial is a substance that has been developed to take on a shape that, either on its own or as part of a complex system, is utilized to regulate interactions with living systems and has the potential aid in the growth of any curative or clinical method. Smart biomaterials are used in medicinal uses to brace, improve, and restore a biological functionality or tissue that's damaged. They can be natural or synthetic. Contrary to what the term "biomaterial" might imply, a biomaterial need not be biological or composed of materials linked to life. The actual substance may be made of metal, plastic, or different types of composites, but it may also be bioinspired and derived from natural sources. They are intended to interact with the body or live in a biological environment. Biomaterial sciences deal with investigations regarding biomaterials. They have expanded steadily and significantly during the course of their existence as a result of various enterprises making considerable financial expenditures in the development of new stuff. Aspects of tissue engineering, biology, chemistry, and materials science are all included in the discipline of biomaterials science.

Due to the incorporation of genetic engineering, cell-based biomaterials have attracted attention from all over the globe for their potential use in a variety of medical disciplines. These engineered biomaterials can change shape or mechanical characteristics, release bioactive ions, and actively discharge chemicals as a result of physiological triggers. All of these material-induced activities may result in native cells and influence the immune system These designed biomaterials can be utilized in the deposition of extracellular matrix (ECM) to aid in the repair of tissues, and enhance cell adhesion and proliferation by incorporating biophysical and biochemical signals [1]. Evaluation of the short- and long-lasting cellular and when creating the next generation of bio-responsive materials, understanding how materials interact at the molecular level is essential as material capabilities increase [2].

Medications, genetic materials, and proteins are all delivered therapeutically through the use of injectable biomaterials. They provide the opportunity to address a range of conditions by delivering medications precisely and preventing immune system uptake. Injectable biomaterials with both synthetic and organically derived components are being investigated for use in managing heart attacks, cancer, and bone defects. Widespread chronic illnesses like type 1 diabetes, an autoimmune disease in which the body's defenses attack pancreatic cells that produce insulin, may be combated with the aid of immunomodulating biomaterials. An injectable synthetic biomaterial recently created by researchers reversed type 1 diabetes in non-obese diabetic mice. This is a significant move toward creating a biodegradable platform to help manage the disease's effects. Applications for genetically engineered materials include the delivery of drug, nanoparticle coatings, carriers that are macromolecular, and hydrogels. Engineering of the tissue has a significant influence on novel cutting-edge approaches to gene engineering. Studies supported various approaches to obtaining, creating, and using these materials while highlighting the evolved functions and properties.

2 Biomaterials: Structural and Functional Roles

Biomaterials are more frequently integrated into devices than they are used as simple materials in medical uses. Supramolecular biomaterials, which are collections of molecules that go beyond the capabilities of individual molecules, have the capacity to detect and react, making them perfect materials for treating disease or injury. Supramolecular biomaterials that can respond to physiological signals by activating or deactivating or that mimic biological signaling are being investigated by researchers.

The method of creating biomaterials has changed over time. Many of the biomaterials that are used in medicine today were not intended to be used in medicine; instead, they were discovered by clinicians to help treat patients. Thus, cellulose acetate, a common plastic, was used to create the first dialysis conduit. Initial vascular grafts employed polymers like Dacron, which were taken from fabrics. Initially, polyurethanes of industrial quality were used as the basis for the materials for artificial hearts. These resources made it possible to handle critical medical issues. However, they also brought about challenges. Platelets and the complement system may be activated by dialysis tubing; vascular grafts made of dacron that are at least 6 mm thick can be used; otherwise, biological processes at the blood-material and tissue-material interfaces could result in occlusion; additionally, blood-material interactions could cause clot development in an artificial heart, which would raise the risk of stroke and other complications [3]. Recent studies are based on drug and cell transporters made of a variety of biomaterials. In tissue engineering, it may be necessary to combine medications, cells, and an appropriate carrier with a tailored degradation profile, specialized macroscopic features, and specific biological cues to promote healthy tissue growth [4]. Biological activity can be added to biomaterials by incorporating oligopeptides that promote binding.

Traditional biomaterials like polytetrafluoroethylene, silicone rubber, or polyethylene are implicitly recognized by cells *in vivo*. Non-specific protein adsorption from bodily secretions occurs on the surface of surfaces, and adsorbed homologous adhesive proteins encourage adherence of the cell by binding to cell surface receptors. The explicit power of adhesion of cells on intelligent materials is possible by preventing nonspecific protein adhesion on the material surface and utilizing adhesion-promoting peptides that are unique to particular cell types [5]. Short primary segments of these peptides are derived from the domains that are receptor bound with the adhesion proteins. The tri-peptide sequence RGD is the adhesion peptide that is

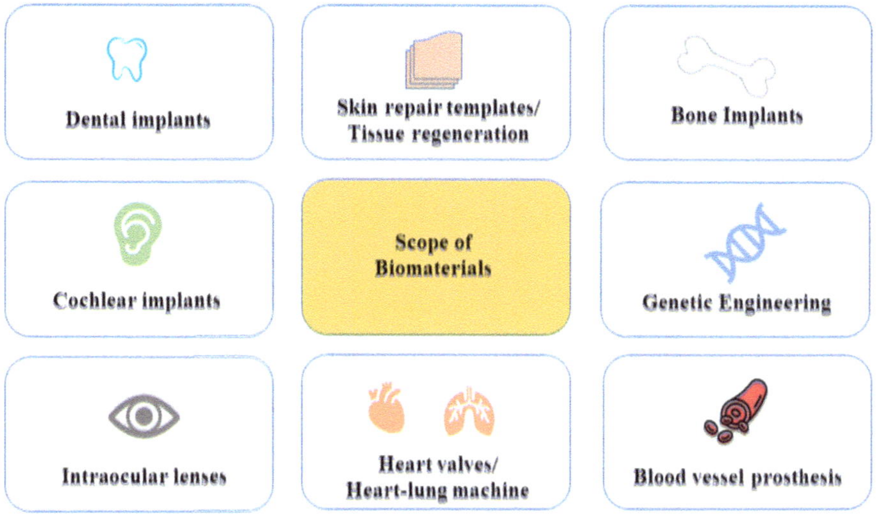

Fig. 1 The scope of biomaterials in various fields

most frequently investigated [6]. Figure 1 illustrates the scope of biomaterials in various fields. Over the last ten years, significant advancements have been made in our knowledge of how cells interact with biomaterials [7, 8]. There is growing proof that biomaterials do more than just act as a scaffold that is temporary since they facilitate the coordination of tissue morphogenesis and cell adhesion [9, 10]. Biocompatibility is a frequently employed notion involving determining the viability of biomaterials in therapeutic implementation for tissue repair and recovery. Biocompatibility, however, reveals some limitations in the ability to describe and define some crucial characteristics of superior biomaterials should have as immunology and biology science advance.

A specific biomaterial must enable long-term use in any application without being rejected by the host organism or losing its intended purpose. The design and selection of biomaterials must take a variety of properties into consideration to achieve this. Although these characteristics can be broken down into mechanical, physical, molecular, and biological groups, in reality, these divisions are frequently entwined. The ideal properties of biomaterials are always evaluated in the context of their particular use because the requirements for various biomaterials are frequently different. To prevent disruption, biomaterials must take into consideration the structure and operation of the tissues and systems in the vicinity. According to an implantation region and possibly even the patient's medical history, the properties of an ideal biomaterial may vary. Bifunctionality is required in biomaterials [11]. This implies that they must have the necessary qualities to perform their particular role, as implants or other types of devices. It might restore a diseased or injured tissue's lost functionality, completely replace that function, or be used to make a diagnosis. A sustainable

Table 1 Applications of biomaterials

S.no	Nature of biomaterial	Organ system	Application
1	Ceramics (Bioactive ceramics, non-oxide bioinert ceramics, glass ceramics)	Bones, heart, joints, teeth	Joint replacement, heart valve, dental implant
2	Metals and alloys (Zinc, magnesium, copper, titanium, platinum, high-entropy alloys)	Heart, bones, teeth, ears	Stents, orthopedic screws, dental implant, cochlear replacement
3	Polymers (Poly (α-hydroxyesters), polylactic acid, polyglycolic acid and their co-polymers (PLGA, polycaprolactone, silicones, polyurethane)	Skin, eyes, bones, heart, kidney	Surgical sutures, skin repair templates, ocular implants, corneal bandage, cartilage, contact lenses, bone cement, heart–lung machine, blood vessel prosthesis, artificial kidney
4	Genetically engineered biomaterials (Collagen, silk, gelatin, chitin, cellulose, glucose)	Bones, skin	Elastin-like-polypeptides based Hydroxyapatite matrices, tissue regeneration

biomaterial is a relationship due to which we have a great opportunity to build innovative sustainable development strategies in the upcoming years owing to the synergy between biomaterials and renewable natural resources. Significant efforts and steps are being made now to create and design materials using sustainable resources that will eventually replace conventional materials. We are currently at a critical juncture where we require sustainable energy, which is only made feasible by advancements in green technologies [12]. The applications of various types of biomaterials have been listed in Table 1.

3 Need for Genetic Engineering of Biomaterials

Genetically engineered biomaterials have been a topic of research ever since recombinant DNA technology has come into play. Such biomaterials can be of animal or plant origin with properties of strength, firmness, and ability to adapt to the environment. Where genetically engineered techniques aid in developing new, relevant, and functional biomaterials, there they also help in developing a better version of already existing biomaterials along with enhancing chemical and physical qualities leading to an increase in the applicability of biomaterials which have a significant influence on biomedical applications such vaccine production, engineering of tissue and regeneration, drug administration, and zoological interpretations, etc. In the creation of hydrogels, development of films, formulation of matrices, development of pharmaceuticals, as well as several physiological and anatomical implants, the use of genetically altered proteins can provide good quality materials with excellent multi-functional properties, broadening the range of applications and need for developing better genetically engineered biomaterials [13]. Moreover, electrical functions can also be inculcated into the biomaterials through the use of technology which helps in excellent antibacterial activities which could be in vitro or in vivo.

Multi-stage moderation of smart biomaterials, which includes gene alteration (e.g., modifying gene sequence, molecular weight, and chain size), chemical conjugation (e.g., adding light-sensitive molecules, oil-loving and/or water-loving side chains), and mesoscopic blending (e.g., doping with metallic nanoparticles, GR, and CNT), allows for the integration of biochemical and electrical capabilities into the material system based on biopolymers [14]. To circumvent the limitations of viral vectors for NK cell engineering, attempts have been made to substitute nanoparticles for viral vectors as they did not show any efficacy due to a transduction efficiency of less than 20%. Due to the distinct physicochemical features of genetic modification, nanoparticles can add imaging modality to the vector while avoiding many of the possible safety problems associated with viral vectors [15].

Living-engineered biomaterials that have microorganisms involved have the capacity to respond in complicated ways to environmental cues, and they can be genetically modified to enable user control of behavior and the incorporation of a variety of inputs. The altered microbes can either produce their matrix, as in biofilms, or they can be integrated into matrices utilizing numerous innovations, including coating, 3D printing, spinning, and microencapsulation [16]. In addition, the use of biomaterials in clinical environments may be hampered by a few drawbacks that have been observed. For instance, a significant hurdle impeding effective tissue repair and regeneration is biomaterial-mediated inflammation; as a result, biocompatibility is being continuously researched and enhanced which can be fulfilled through the process of genetic engineering [17]. Various sorts of immunomodulatory genes were introduced for alteration to stop biomaterials from being immunologically rejected. Transmembrane proteins, cytokines, and genes genes-producing all have immunomodulatory effects and other proteins have been proven in several investigations and these genes acted in a variety of ways to exert influence that is suppressive or agitative on the immune system elements [18].

Live cells can now be incorporated into structures and allowed to flourish owing to developments in 3D bioprinting. Also, chimeric biomaterials can be 3D printed more accurately and intricately. The conventional biomaterials used as non-conventional biomaterials can be produced with more precision and finer qualities and hence can also help in saving cost [19]. Consequently, there are many uses for intelligent biomaterials, ranging from health (such as engineering of the tissues, and delivery of the drugs, and biosensors) to more newly researched environmental uses like ecosystem restoration (e.g., coral reefs and environmental remediation) [20]. The development of physicochemical features of protein-based drug delivery devices that are accurately calibrated is made possible by genetic engineering approaches, offering a higher degree of customization than is possible with synthetic polymers. Elastin-like proteins (ELP), silk-like proteins (SLP), and silk-elastin-like proteins (SELP) due to their inbuilt physical and chemical attributes and the simplicity of engineering made possible by recombinant DNA technology, a particular set of substitutes for creating drug delivery systems [21].

4 Techniques of Genetic Engineering

4.1 Scaffolding

As a result of the variety of biomaterial amenities, including the adaptable nature of protein-engineered materials can be independently tailored, offering a desirable option to harvested natural scaffolds or synthetic polymeric scaffolds. Different modular peptide domains are created by encoding specific molecular sequences in a DNA plasmid, transfecting the plasmid into a target organism, expressing the plasmid, and purifying the resulting biopolymer definition make up protein-engineered biomaterials [22]. By combining a variety of peptide sequences allowing in a single, modular biomaterial, the scaffolds can be created to imitate many properties of the natural extracellular matrix, including cell adhesion, cell signaling, elasticity, and biodegradability. The usefulness and scope of protein-engineered biomaterials have recently been increased by the inclusion of functional elements that are uncommon in the extracellular matrix.

Each functional peptide module's location and density can be precisely adjusted thanks to the genetically encoded DNA sequence being used as a template to create these protein-based materials, allowing the independent tailoring of numerous material properties. For each engineered biomaterial, a precise DNA template needs to be made for the sake of completely realizing the potential of this design approach at molecular level. Therefore, meticulous control of DNA template design is essential for enabling the adaptability of biomaterials made of modified proteins. For the prevention of DNA recombination when generating extreme repetition of amino acid patterns, a variety of codons must be used [23]. The smart biomaterial is engineered through protein in which a DNA template is created and it is then turned into a recombinant plasmid. The chosen host cell is transfected with the plasmid, which causes the chosen host cell to translate the genetic information involved with the desired protein. A protein-engineered scaffold is created once the protein has been processed after being refined in step five. The resulting scaffold is examined using methods involving in vivo and in vitro techniques, which reveal data on how to enhance the scaffold's characteristics and lead to remodeling the biomaterial and altering the encoding template's DNA. Figure 2 elucidates the steps involved in the scaffolding technique.

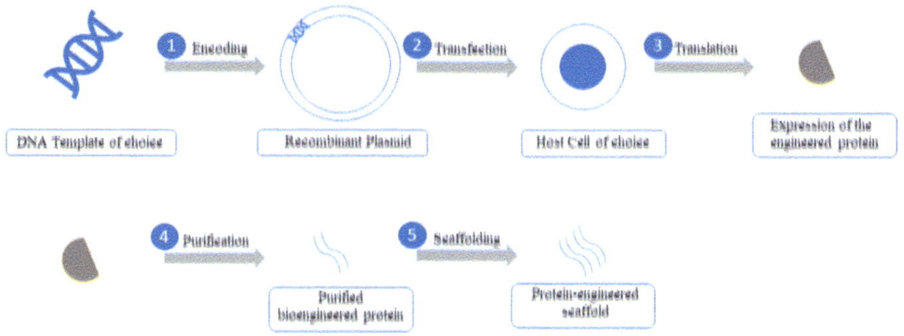

Fig. 2 The steps involved in the scaffolding technique

4.2 Employment of Sequences

Both organic and synthetic nanoengineered biomaterials that are polymeric are employed in biomedical utilization like targeted and smart drug delivery with prolonged and controlled release. Natural polymeric nanocomposites are frequently employed in medicine because they are biodegradable, nearly biocompatible, and not poisonous. Applications for synthetic biopolymer nanocomposites are numerous. By combining with biomolecules, they can be altered to become more biocompatible and harmless to the body [24].

Sequencing in the genetic engineering of biomaterials is one of the significant techniques as the properties of biomaterials depend on the sequences that are employed in the biomaterials which is an emerging topic for research recently. Examples of synthetic methods for sequence, control include iterative methods where each monomer unit is added sequentially, regulated polymerization techniques using both chain and step-growth mechanisms, and bioinspired and supported methodologies like molecular imprinting and templated syntheses. In the case of genetic engineering of biomaterials. Biomaterials made of particular peptidomimetics and statistical copolymers have both been made using the comparatively simple process of sequence selection known as "bioabstraction," or replicating minimum functional DNA or protein sequences. Synthetic analogs have been produced using this technique which are biologically important compounds, including host defense peptides (HDPs), cell-penetrating peptides (CPPs), and antimicrobial peptides (AMPs) [25].

4.3 Recombinant DNA Technology

Protein-based biomaterials are created using recombinant DNA technology, or genetic engineering. Because of precise control and the modification of the DNA structure that encodes the protein structure, it is possible to alter the basic structure, layout, and length of the protein biomaterial's chain. This quickly developing technology has given rise to effective tools for creating unique smart biomaterials with preset 3D structures [26]. However, several issues need to be resolved to synthesize protein chains using genetic engineering techniques: Repetitive DNA sequences may be altered and deleted in various ways, sequences of messenger RNA (mRNA) necessary for a certain codon for a particular amino acid residue. The host cell may break down the target protein before sufficient levels are synthesized, reducing the yield. Biomaterials that have been genetically modified have many benefits. By carefully regulating the length, stereochemistry, and 3D structure of the polymer chain, it is possible to create materials with specific, distinctive characteristics.

4.4 Decellularization

Recently, novel methods based on genetic changes have been developed, including decellularized biomaterials. Decellularization is the elimination of cells and constituents involving the nuclear ones, of the tissue, using a variety of chemicals while conserving the ultrastructure of the collagen, elastin, microfibrils, proteoglycans, glycosaminoglycans (GAGs), and other growth factor components found in the extracellular matrix [27]. Decellularization seeks to reduce the immunogenicity, calcification-inducing potential, and cytotoxicity of native extracellular matrix (ECM) by removing the majority of cells from it with the aid of physical, chemical, and biological methods [28]. Extracellular matrix (ECM)-based biomaterials now have new prospects in the transfusion of organs and tissue regrowth in preclinical and clinical settings because of advances in decellularization techniques. The resultant dECM scaffold should ideally maintain its functional content and three-dimensional (3D) structure for tissue restoration purposes. Before dECM scaffold transplantation, it was important to recellularize the dECM with a particular cell type to increase cell survival and reduce immunological rejection [29]. Steps in the decellularization are given in Fig. 3.

Decullarization occurs either through physical methods such as freeze-thaw, hydrostatic pressure, ultrahydrostatic pressure, and oscillation, chemical methods like hypertonic/hypotonic solutions, detergents namely sodium dodecyl sulfate (SDS), alcohols or triton X-100 and enzymatic methods such as RNases, DNases, and trypsin which aids in separating a tissue's extracellular matrix (ECM) from the cells that reside there, leaving behind an ECM scaffold that can be employed in artificial organ and tissue regeneration [29]. Crosslinking after decellularization occurs through a number of crosslinking agents which can be physical, chemical, or natural. Thermal dehydrogenation and photo-oxidation crosslinking are the two basic types of physical crosslinking. Thermal dehydrogenation was frequently utilized in the beginning and now, this approach is frequently employed as a backup strategy because of the difficult crosslinking conditions. The crosslinking conditions are difficult to manage using the photo-oxidation approach. This makes it more frequently employed in the decellularization of tumor tissue, which is typically done for disease modeling

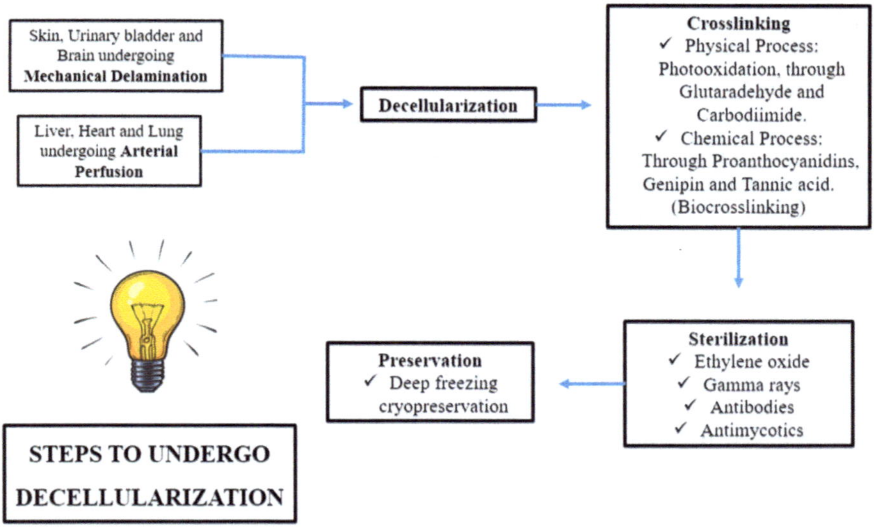

Fig. 3 Steps involved in the decellularization process

purposes. The dECM of the tumor was crosslinked by Lü et al. using photo-oxidation mediated by methylene blue [14]. This was the first study to show whether photo-oxidative crosslinking affected the tumor dECM scaffold's structural, physical, and biological characteristics. Epoxy compounds, carbodiimide (CDI), glutaraldehyde, and hexamethylene diamine carbamate (HMDC) are examples of crosslinkers which are employed in chemical crosslinking but have huge side effects due to which better crosslinking agents need to be used [29].

Sterilization being the third step is the most crucial step when decellularization is involved as the applicability of different biomaterials is put off due to limited sterilization methods for a period that is more than expected. The structure of the ECM will inevitably change as a result of current sterilization techniques, such as exposing the material to radiations such as ethylene oxide, gamma, or electron beam, and this could negatively impact the biomaterial's mechanical properties [29]. With respect to the list provided above, a more suitable option is provided by the development of supercritical carbon dioxide. [30]. The impact of three widely used sterilizing techniques, including ethylene oxide (EtO), peracetic acid (PAA), and gamma irradiation (GI), on material qualities, was examined by Matuska et al. In this procedure, they studied the early cellular contacts and discovered that, unlike GI and EtO, PAA showed advantages in cell adhesion but did not significantly alter the structure of the cell. According to this, the choice of sterilizing technique is important for modifying the properties of biomaterials in a way that promotes cellular adherence and is highly relevant to the use of biomaterials in vivo [31]. There is currently no sterilization method that can completely stop the incidence of undesirable consequences [32].

Consequently, selecting the best sterilizing process is a crucial step for using decellularized scaffolds in the future for therapeutic applications, and more research is still required.

Future uses of the decellularized matrix-based materials will depend on the final step of preservation in the process of decellularization as well. The manufactured decellularized scaffolds were now typically stored in PBS alongside antibiotics and antimycotics at a temperature of 4°C for a brief length of time, on standby, or at 20 °C for long-term preservation [29, 33]. The degradation of active substances and structural damage, however, seems to be unavoidable at the moment and require urgently more work. The ideal characteristics of a tissue for repair in wound repair include strong bioactivity, adequate mechanical characteristics, excellent degradation, and outstanding biocompatibility are all desirable characteristics in a scaffold for tissue repair. Decellularized ECM (dECM) is a three-dimensional natural scaffold produced from autologous, allogeneic, and xenogenic tissues that retains its intrinsic tissue structure despite having had the cellular components removed. The dECM is promising in translational medicine [34] because of exceptional physiological activity, high biocompatibility, lack of immunogenicity, and use as a diverse source of basic materials [35].

New advancements in the biomaterial's domain in the realms of regenerative medicine and engineering of tissue, dECM-based smart biomaterials have had great success. Also, scientists were inspired to create a novel standard for the application of dECM biomaterials, such as constructing a safe and biocompatible method for delivering therapeutic agents like drugs and cells. For instance, dECM can be ground into nano- or microparticle size and used as a therapeutic delivery equipment to repair anastomoses [29]. Clinically, both Axogen's Avance nerve transplant and AlloDerm, an acellular skin matrix from Biohorizon is a regenerative tissue matrix intended to hasten skin regeneration. are used to treat wounded nerves and are two examples of emerging clinical applications of dECM. DECM can also be utilized to make bioink-related hydrogels. Of the inks now on the market for 3D bioprinting, bioinks based on dECM exhibit the highest level of biomimetic capability. The building block for 3D printing a functioning human organ is made of dECM bioinks in conjunction with the recent development of bioprinters. It is fascinating that multiple studies have been published employing ECM hydrogels synthesized from almost every organ where dECM bioinks are employed for bioprinting. ECM hydrogels could be employed in minimally invasive deliveries as an easily administered pre-gel viscous solution into a patient utilizing a catheter or syringe [36].

4.5 Physical Adsorption

Combining protein-based biomaterials with bioactive molecules can improve their properties and increase either in live conditions or the laboratory. Surfaces generated from proteins can be changed by chemical, physical, or encapsulating procedures. Using these methods, proprotein-based materials can be functionalized with various antimicrobials and growth factors are two examples of biologically active chemicals that can improve cellular and tissue reactions [27]. When introducing bioactive compounds, such as growth hormones or extracellular matrix proteins, through dip coating, to the surface of scaffolds physical adsorption, a straightforward immobilization technique, is commonly used. Topography of surface, functional groups, pH, wettability, and electrical charge of the substance are just a few examples of the physical and chemical characteristics that affect adsorption efficiency. Since many nanomaterials are hydrophobic, techniques are required to increase their wettability and turn them into hydrophilic materials. Chemical interactions between carbon and non-carbon atoms are broken down using physical techniques like ion bombardment, UV light, and plasma modification. As a result, oxygen reacts with unsaturated bonds and radicals to promote hydrophilicity and reactivity toward substances that are biological. Since they contain a large number of reactive chemical groups (hydroxyl, carboxyl, and amide), natural polymers have the benefit of being less hydrophobic and able to interact with physiological molecules. Materials composed of proteins, such as collagen and silk, have been customized by the adsorption of biologically active compounds, such as vascular endothelial growth factor (VEGF) [37], basic fibroblast growth factor (bFGF) [38], and bone morphogenetic proteins (BMPs) [39], and pharmaceuticals like antibiotics, and heparin [40].

Van der Waals, hydrogen, and hydrophobic interactions serve a purpose in adsorption while they are electrostatic in nature, external factors can affect how stable the adsorbed molecules' bond is relatively weak or moderate. In this manner, uncontrolled release of the immobilized species can be brought on by changes in the pH, strength of ions, and the number of adsorbed species in the surrounding media. For instance, large doses of bone morphogenetic proteins (BMPs) in order to trigger the correct osteogenic response, substances that diffuse away from the fracture location are required. The BMP-2 release profile from collagen sponges exhibits an early spike within the first ten minutes, during which 30% of the BMP-2 is lost by the messenger carrier, and is thereafter sluggish over the course of the next 3–5 days. Clinical conditions like soft tissue hematomas, ectopic bone growth, and receding bones may result from this original burst release [41].

4.6 Triboelectric Nanogenerators (TENG)

TENG is one of three types of nanogenerators in which the concepts of triboelectrification and electrostatic induction work together to transform mechanical energy, including that from very low frequencies to electrical power. In the "triboelectric effect," the charges are transferred during mutual contact between materials having varying triboelectric polarities (distinct capacities for electron binding). TENG makes use of this phenomenon by using an external force to separate two materials which further creates a potential difference hence, creating an electric field [42]. Another type is piezoelectric nanogenerators (PENGs), which function by interacting between the mechanical and electrical states linearly while responding to mechanical stress and are made of crystalline materials lacking inversion symmetry [43]. Pyroelectric nanogenerators (PyENGs), which transform variations in heat energy into electrical energy, come in third [44]. Recently, it was shown that TENGs and PyENGs can work together by harnessing the human body's mechanical and thermal energy to generate electrical signals [45]. Nonetheless, triboelectric nanogenerators are frequently chosen in comparison to their counterparts piezoelectric nanogenerators or pyroelectric nanogenerators due to their conditions of operation for regions of low frequency/amplitude, ecosystem quality, more extensive resources, and material selection [46].

Polymers, such as polyvinylidene fluoride, polytetrafluoroethylene, polydimethylsiloxane, and others, make up the majority of triboelectric nanogenerators because they all include the element fluorine, which has a great affinity for electrons. Although these polymers can produce significant triboelectricity, they can be challenging to handle and decompose, and their accumulation and disposal could be hazardous to both human health and the environment. The ecologically friendly and biodegradable TENGs (bio-TENGs) will create smart portable tools to address these issues, setting a new standard for the establishment of a smart and green society. TENGs in which the triboelectric layer components are partially or entirely made of natural substances or their constituents are classified as bio-TENGs. Firstly, biomaterials having high biocompatibility and biodegradability are obtained from natural species and they can be decomposed by microorganisms after being disposed of in nature. Additionally, because TENGs have an innate ability to generate green energy from their environment and biologic processes, these intelligent bio-TENGs can be used to power autonomous driving sensors or smart electronics. Finally, there are numerous sources of biological materials, ranging from marine products like shellfish and seaweed to agricultural by-products like straw, rice husks, bark, and bagasse [47]. The majority of biodegradable polymers have low toxicity and strong biocompatibility, which makes BD-TENGs effective in biomedical applications. The implantations for medical devices should be removed promptly when their purpose has been served in the human body. Biodegradable devices are typically employed to treat disorders of the heart and there is no requirement of additional surgery, in contrast to non-degradable devices that can only be removed through procedures, which causes subsequent injury to the patients [48]. They are also employed in tissue regeneration.

For the current triboelectric series to be expanded, new triboelectric materials must be developed. The design of positive and negative materials for placement in the triboelectric series is under the limelight. The synthesis of novel materials containing significantly more accurate halogenic elements is the main focus of techniques for creating novel tribo-negative materials. As an illustration, Lee et al. stated that fluorinated polymeric sulfur outperformed commercial PTFE, which is ranked the worst in the series of triboelectric materials. A disadvantage associated with tribo-negative materials is their insulation. The creation of materials that are tribo-positive is inspired by the emergence of the highest positive charges in polymers comprising nitrogen and oxygen with pyridine amide, amine, or hydroxyl groups. Based on this hypothesis, Zhao et al. prepared an aniline formaldehyde resin (AFR) with sufficient mechanical strength and controlled microstructures for TENG device applications [43].

By capturing high entropy energy, TENGs offer a fresh approach to the problem of supply of energy. Nevertheless, flexible, humidity-resistant, and affordable TENG must meet strict standards for wearable electronic devices. Here, a straightforward and environmentally benign process was used to create the multipurpose wheat starch TENG (S-TENG). The S-TENG opens the door for the mass production of multifunctional biomaterials-based TENG as well as the real-world use of wearable electronics and self-powered sensors [49].

There is a reported way for easily improving the quality of a chitosan-diatom-based biomaterial TENG (CD-TENG) by controlling the weight ratio of the diatom frustule, which can increase the density of charge in the chitosan-diatom composite film [50]. This is useful to apply in wearable devices that might be easily attached to human skin and will not be associated with any adverse implications. Nanocarriers are stimuli-responsive biopolymers that are commonly used in invasive and transdermal drug delivery (TDD) systems that are stimulated and regulated by TENGs power supply. Other common drug-delivery agents include nanoparticles, microneedles, and liposomes. TENG is therefore utilized to deliver drugs with organelles and biomaterials for offering a target-oriented controlled drug release [51]. Paper-based cellulose TENGs, nano cellulose TENGs, and micro/nano composite TENGs are other advancements in biomaterials based on the triboelectric nanogenerator principle. The advantages of cellulose as a naturally occurring polymer are cost effectiveness, processability, good mechanical flexibility, biocompatibility, and biodegradability. Additionally, it may display a special combination of chemical, structural, dielectric, and optical characteristics. These benefits may transform cellulose-based functional materials into appealing substrates or TENG elements [52].

The creation of a durable, stable encapsulating covering is necessary for the commercialization of biomaterial TENG devices as clinical implants since it will shield the device from the hostile in vivo environment. This is because overcoming the immune system's response to a foreign body is challenging, and fibroblast growth may eventually cause the implanted TENGs to respond to motion stimuli less effectively [53]. To construct well-designed TENG sensors, researchers will eventually need to use fabric, shape-memory polymers made of latex or rubber, hydrogel, and other cutting-edge functional materials. In addition, significant progress in the area of flexible electrodes is required to increase the wearability of TENGs both inside and outside the living systems. Hard and soft materials, including carbonaceous nanoparticles, polymeric substances, 2D materials, inorganic and ceramic materials, and polymers, are used to create today's naturally flexible electrodes [54]. The TENG-based healthcare system benefits greatly from TENG-based biomaterials, which have a bright future in terms of consistency, sustainability, comfort, and adaptability.

4.7 Elastin like Polypeptides (ELPs)-Based Hydroxyapatite Composites

Organic matrix macromolecules are essential for improving the mechanical characteristics of biomineralized compounds like bone and teeth in nature. Because there is still a lot to learn about how natural matrix components work, creating artificial matrix counterparts is both exciting and difficult. Biomimetic matrices can be created utilizing genetically produced elastin-like polypeptides (ELPs) instead of natural components for the design of mechanically robust ELP-hydroxyapatite (HAP) composites. ELPs with clearly defined backbone charge distributions can be made by intermittently including negative, positive, or neutral side chains or HAP-binding octa glutamic acid patterns at one or both protein termini. ELPs have sequence-specific properties that affect how they interact with ions, bind HAP, and spread HAP nanoparticles. With ELPs that bind with HAP, calcium phosphate cement can be further improved to provide materials with enhanced mechanical strength, injectability, and anti-washout properties [55].

Elastin-like polypeptides possess some adaptable properties which enable them to form different nanostructures, including nanoparticles, nanofibers, and nanocomposites [56, 57]. When utilized in drug delivery, these nanotechnologies increase the effectiveness of the therapy because the nanostructure can target the molecule, increasing its stability and increasing contact surface. This could boost these structures' activity, which is essential for maintaining their thermodynamic properties [58, 59].

4.8 CRISPR-Cas 9

Streptococcus thermophilus, a bacterium frequently employed in the dairy industry, was utilized to identify the CRISPR/Cas mechanism system for the initial time in 2007. By functioning as a highly effective "antiviral system" to prevent viral infections, this system serves as an adaptive immunological mechanism that combats this gram-negative bacterium. Later, it was found that prokaryotes, which comprise 84% of archaea and 45% of bacteria, are the organisms where CRISPR is currently most common. The term "CRISPR/Cas system" refers to the locus, which is perpetually active in the presence of CRISPR-associated (Cas) genes which generate the endonuclease enzyme. CRISPR's operons of Cas genes, spacers, and repetitions are its essential components [60]. Table 2 describes the biomaterial vectors along with their Cas9 cargoes with their applications and advantages.

Since viral predation poses a constant threat to bacteria, they have developed a range of defense mechanisms, including CRISPR/Cas systems. CRISPR-Cas 9 operates by modifying DNA. The nuclease protein (Cas9) of the Streptococcus pyogenes Type II-A system addresses a particular sequence of DNA under the direction of an

Table 2 Biomaterial vectors along with their Cas9 cargoes: Applications and Advantages [63]

Cas9 Cargo	Delivery System	Application	Advantages
DNA	Palmstearin derived fatty acyl chains	Analyzing nanoparticle molecular architecture effect on Cas9 plasmid delivery and characterization of delivery to HEK-293 cells	The PS-Lips cationic lipid nanocarrier system's liposomes successfully carried genome-editing tools containing CRISPR/Cas9 encoded pDNA
DNA	Turbofect	Identification of Nf-1 pain-related therapeutic targets	Effective pain management
mRNA	Extracellular vesicles derived RBCs	Targeting in MOLM13 cells	In bothhuman cells and xenograft animal models, RNA medication delivery with RBCEVs exhibits highly strong microRNA inhibition and CRISPR-Cas9 genome editing with no detectableharm
Protein	Nanoclews involving yarn-like DNA	Disruption of EGFP in U2OS-EGFP cells in vitro and in vivo	Enhanced delivery of CRISPR-Cas9
DNA	Liposome-Exosome hybrids	Lipofectamine 2000 and HEK293FT-derived exosome performing targeting in mesenchymal stem cells	Exosome-liposome hybrid nanoparticles are promising for in vivo gene editing since they can transport the CRISPR-Cas9 system to MSCs
mRNA	Zwitter-ionic lipids amino	Lung cancer cell delivery and in vivo targeting	Improvement in safety and effective usage of gene editing
Protein-tagged NLS and glutamic acid	Arginine coated cationic gold nanoparticles (ArgNPs)	Works by targeting Hela cells and disrupting signals for cancer survival	The method assures the development of "weaponized" macrophages for cancer immunotherapy due to the enhanced attack and destruction of cancer cells
mRNAwith modified sgRNA	Cholesterol,DSPC lipid, PEG-2000 DMG and LP01 lipid	In TTR amydoilosis for gene targeting	Helps in reducing Transthyretin (TTR) serum protein concurrently with in vivo genome-editing levels that are clinically significant

(continued)

Table 2 (continued)

Cas9 Cargo	Delivery System	Application	Advantages
Highly negative GFP fused protein	Lipofectamine 2000, RNAiMAX	Delivery which is in vivo to mice cochlea	When compared to DNA transfection, delivery of unaltered complexes resulted in genomic modifications of up to 80%
DNA	Cationic polypeptide made with surface PEGylation	Disrupting GFP and targeting in HeLa tumor-bearing mice	The system provides a flexible gene-editing platform for biological research and therapeutic applications, enabling multiplex gene knock-out, gene knock-in, and gene activation in vitro and in vivo

associated RNA sequence, is the system that is used for the majority of CRISPR applications in bacteria. This method causes an incision in the designated area, allowing for DNA editing. The DNA that recognizes the protospacer is bound by the Cas9-sgRNA complex, which leads to a slow unraveling of the DNA. The CRISPR-Cas9 technology has the advantage of enabling researchers to more precisely control how the CRISPR/Cas systems are activated by targeting transcriptional activators to specific genomic regions. The latest advances in the interaction of *Porphyromonas gingivalis* (*P. gingivalis*), *Enterococcus faecalis* (*E. faecalis*), and *Streptococcus mutans* (*S. mutans*) with bacterial functions and CRISPR/Cas systems are relevant to dentistry [61].

Delivering the CRISPR-Cas complex to particular cells or tissues is challenging and laborious because its components must transcend tissue and cell membrane barriers to enter the nucleus, where they can affect the nuclear genome. The two broadly used physical techniques for introducing the CRISPR-Cas complex into cells are electroporation and single-cell microinjection. These techniques are crucial for editing genes in developing embryos and creating transgenic animals [62].

Due to their adaptability, biocompatibility, and rising transfection effectiveness, biomaterials are increasingly being used as non-viral vectors. Capacity restrictions are not an issue, and nanoparticles can be further tailored to enhance nuclear transport and tissue selectivity. The capacity of biomaterials to be tuned offers considerable potential for enhancing CRISPR-Cas9 delivery for in vivo gene editing. Since there are practically no restrictions on how biomaterials can be customized, this gives physical and viral delivery a particular advantage. This is due to the fact that several organ systems along with their types of cells will eventually need gene editing methods that are specially tailored to their microenvironments [63]. By incorporating various biomaterials into them, Cas9 can be supplied in the form of DNA, plasmids, mRNA, and proteins.

4.9 3D Printing

The methods of genetic engineering are utilized to extract and mix particular DNA from diverse materials to create chimera DNA. When the chimeric DNA is sent back with the help of plasmids to bacterial cells, it can then be duplicated. The survival of cells carrying chimeric DNA can be backed by bioprinting techniques like inkjet-, extrusion-, and laser-based printing. They can design chimeric biomaterials for 3D printing in a controlled and cost-effective way. Recombinant DNA technology allows engineers to employ chimeric biomaterials in a number of biological research and development applications by taking their desirable traits from nonhuman origin. The creation of fake CAR T cells for use in adoptive cell therapy for the treatment of cancer is the most well-known application of chimeric biomaterials. Particularly, Fc-chimeric proteins may be utilized for the identification of desired stem cells and as test compounds in advanced 3D printers. Chimeric organoids that mimic the tumor microenvironment can be designed by bioprinting cancer cells and healthy mammary epithelial cells. Furthermore, chimeric biomaterials can be included in drug delivery systems to improve the therapeutic potency of previously accessible proteins and drugs. Recently, attempts have been made to combine chimera biomaterials with biopolymers [19]. Despite being incredibly effective in CAR T-cell therapy as a customized approach to cancer treatment, chimeric biomaterials can potentially have unfavorable side effects and flaws in patients. Additionally, because of its high cost and drawn-out approach, CAR T-cell treatment is not commonly available. The ethical ramifications of research using human organoids and/or the introduction and integration of chimera material into a host mammal are also complex. Sound ethical principles must therefore be upheld as chimeric biomaterials are increasingly used in modern science. It is necessary to harvest DNA from numerous genomes in order to create chimeric biomaterials, which calls for the employment of additional 3D printing settings that allow high cell survival and density. Across all printing techniques, there is a definite trade-off between cell survival, resolution, and price.

Chimeric biomaterials are designed using recombinant DNA technology, which reconstructs genes from many species into desired effects by programming them at specified gene sequences. Chimeric polymers have a much greater degree of control over the amino acid sequences than their chemically synthesized equivalents, which is much greater than what is possible with solution or solid-phase peptide synthesis. Incorporating stimuli-responsive motifs into a chimeric polymer's backbone allows for accurate phase changes in response to changing temperature, pH, and ionic strength. Additionally, this addition might enhance interactions between an implant and already-existing cells. Chimeric polymer-based biodegradable structures can also maintain their temporal and spatial precision so that breakdown only happens in the presence of specific enzymes. This can be achieved through the addition of amino acids which are prone to breakdown [64]. Engineers could advance the creation of treatments that address a variety of unresolved medical issues that have plagued society since the dawn of humankind by leveraging desirable properties from other sources.

4.9.1 Epigenetic Modifications/Omics-Based Approaches

Recent advances in biology technology, notably in the domain of omics approaches, have enabled functional high throughput reading of the cell's genome, epigenome, transcriptome, proteome, and metabolome. The capacity of omics-based techniques in forecasting the biological response to designed biomaterials is the primary checkpoint in developing the next generation of biomaterials, and it has the potential to revolutionize biomedical research and replace the traditional "trial and error" method. These omics-based tools can also be applied to the study of the in vivo behavior of biomaterials, a field that is based on semi-quantitative imaging-based techniques instead of precise computational techniques [65].

In order to make hydrogels whose mechanical characteristics depend on calcium concentration, calmodulin, a protein that changes conformation in order to bind calcium ions, was used and is one of the biomaterials that is prepared through a protein [2]. Successfully combining hydrogel encapsulation and mRNA transfection opens up new avenues for the prospective clinical delivery of mRNA-engineered cell products. Primarily, improved cell attachment is necessary for significant secretion of the IVT-mRNA encoded protein, a recently discovered synergism between mRNA transfection of cells and their microenvironment. The increased secretion of indigenous HGF and mRNA-overexpressed VEGF in collagen hydrogels indicates that material-mediated effects can further modulate the cells' secretome in concert with IVT-mRNA overexpression [2]. Depending on the method of nuclease delivery, gene editing needs different regulatory provisions than other gene therapy medications. The location where nuclease genes are inserted into the genome, how the genes are delivered, how they are administered, and what kinds of gene editing nucleases and sgRNA are used can all affect the level of risk and the categories of risk factors are one of the challenges when a gene needs to be modified [66].

The most extensively researched epigenetic modifications are DNA methylation and histone protein alterations, that are directly involved in regulation of the gene expression levels and defining the characteristics of the cell. Methylation of DNA alters gene transcription by preventing transcription factors from attaching to the DNA. The same is true for post-translational histone modifications, which are crucial for determining chromatin structure and regulating DNA expression. Enzymatic proteins that are involved in these epigenetic changes can also reverse them. These changes are practical because they can be undone, which opens the door to using them for *in situ* tissue regeneration. In fact, chromatin structure and epigenetic processes in cells have been regulated by nanomaterials [67]. In a recent study, epigenetic regulators were employed to stimulate osteogenic differentiation in human MSCs113. These regulators included gemcitabine, decitabine, I-CBP112, chidamide, and SIRT1/2 inhibitor IV. Notably, stem cells obtained from elderly adult donors showed a fivefold increase in osteogenic efficacy when treated with gemcitabine and decitabine. This research emphasizes the function of the nucleosome remodeling and epigenetics in cellular differentiation as well as their potential application to in situ tissue regeneration [68]. The surface organization and material energy of biomaterials influence the epigenetic patterns of cells, which can subsequently influence the gene expression of cells that come into contact with the material. Substances in tissue engineering that may affect epigenetic pathways are titanium, silica, PLGA, bioglass, and ceramics [69]. Another organic material that has undergone epigenetic functionalization is silk. James et al. examined an anti-sense-miRNA-214 silk device utilizing surface coating, which led to a continuous release of miRNA inhibitors up to 7 days in vitro and increased the expression of osteogenic genes in the human mesenchymal stem cells implanted on these devices [70]. These findings would suggest that this unique technique could be helpful for localized bone tissue creation and for promoting osteogenesis at the implant surface. Collagen sponges and macroporous biphasic calcium phosphate scaffolds combined with HDAC inhibitors (HDACi) caused the creation of woven bone and newly produced bone at the scaffold's interface, which is another epigenetic mechanism for collagen functionalization that has been explored [71]. Although Cre recombinases are widely utilized, optogenetics has been found to be beneficial for a variety of phylogenetic histone deacetylases, methyltransferases, acetyltransferase inhibitors, and proteins that recruit these enzymes. Included in this group of proteins are KYP, TgSET8, NUE, PHF19, Sin3a, Sirt3, NcoR, HDAC8, RPD3, and Sir2a [72].

Genetic code expansion (GCE), a related approach, is comparable to it. Next-generation protein engineering, in particular GCE, enables heterologous expression of proteins with NCAA/UNAA inclusion in proteins in both prokaryotic and eukaryotic organisms [73]. GCE promotes biomedical and tissue engineering research by developing innovative protein-based biopolymers with broad structural and functional flexibility. Silk, elastin, collagen, and fibrin are used to create a variety of biomaterials that are created using the GCE process. The ability to enhance the production of biomaterial proteins by GCE is expected to usher in a new era of tissue engineering and biotechnology. According to recent advancements in tissue engineering, customized tissue constructs with anisotropic architecture, heterogeneous cells, and composite nano/biomaterials are being created for regenerating functional tissue and organotypic tissue models [74]. Because they enable precise control over the loading of bioactive moieties, microstructure, and handling, biomaterials with spatiotemporal variations are essential for 3D manufacturing. GCE improves the possibility of producing biomaterials made of growth factors and congener proteins that act as brokers for interactions between cells and the matrix. GCE presents a unique opportunity for the creation of bioinspired proteins and customized biomaterials that show potential for use in precision and personalized medicine. GCE creates protein biomaterials with all-inclusive qualities to ensure enhanced interactions between the components of the tissue engineering trio [74].

5 Pre-Clinical and Clinical Relevance of Genetically Engineered Biomaterials

Biomaterials can serve a variety of purposes in cartilage repair and regeneration, including mechanical support systems (such as visco supplementation, defect plugs), delivery vehicles for the delivery of therapeutic molecules as well as scaffolds for cell-driven repair of tissues. Hydrogels, 3D polymer networks respond viscoelastically and are the optimal materials for cartilage repair since they are high in water content, like healthy cartilage [75]. Drug delivery, the development of cell-rich bioengineered devices, and tissue and immunological engineering are some possible applications for smart materials [76]. The mechanical properties of hydrogels can be reversibly, wavelength-specifically, dose- and space-controlled tuned by incorporating a protein under optical control into polymeric polymeric materials. These light-controlled poly (ethylene glycol) (PEG) hydrogels were first made successful, according to a PhoCl protein, which raised the potential of their practical implementation [77].

Biomaterials that are appropriate for human use and of neurosurgical grade may provide a feasible replacement. In particular, the neurosurgical grade using the biomaterial Duragen PlusTM matrix that has been approved by regulators and is mostly utilized in duraplasty treatments. Type I bovine collagen is used to make it. Here, it permits fibroblast percolation to restore and heal the dura mater after an intraoperative breach, and it is said to be resorbed after six to eight weeks. According to reports, the substance is conformable, clean, and biocompatible as well as lacking immunogenicity, cytotoxicity, and pyrogenicity. It has been demonstrated that Duragen PlusTM, which has been transplanted into the nervous system, stimulates neural cell growth, which might involve that of rat cortical neurons, without producing any adverse side effects [78].

dECM-based biomaterials enhance the advancement of more precise delivery of a certain drug at the intended region and facilitate chemical component distribution as carriers. Decellularized ECM biomaterials have also been investigated for growth factor delivery due to the intrinsic qualities of ECM and its natural interactions with various growth factors. For instance, the decellularized matrix containing hydrogel was designed to load bFGF to treat spinal cord injury since ECM is essential in directing growth factor signaling in vivo for bone regeneration [79]. The delivery of biological cues using dECM-based material is advantageous because it increases in vivo utilization and effectiveness. As a number of ECM goods (AlloDerm, LifeCell Corp.), as well as injectable ECM materials (ACell's CytalTM Wound Matrix, MicroMatrix®, and CorMatrix®), have already passed the commercialization stage, injectable ECM hydrogels may soon follow this trend. It is crucial to remember that musculoskeletal dECM biomaterials have had a tremendous impact on clinic tissue repair all over the world. Applications for dECM biomaterials in preclinical studies have included urethral reconstruction, bone tissue healing, and muscle regeneration [79, 80].

Biomaterials can be used in a variety of ways to induce immunological tolerance because they offer reliable control over immune cells' interactions with specific antigens. They can therefore make excellent use of the immune system's numerous internal regulating mechanisms. One way to promote tolerance is through inducing energy or deletion of self-reactive lymphocytes. Additionally, targeting helper T cell activation toward the regulatory pathway can help prevent the activation of other T cells' undesired effector functions.

Biomaterials allow for engineered solutions to complex biological issues, as was the case with vaccine development, and they hold hope for the future of our ability to effectively treat autoimmune diseases and aberrant immune activity [81]. Macroscale scaffolds based on biomaterials can be utilized to disseminate immunomodulatory elements like proteins (like cytokines and antibodies), oligonucleotides (such silencing RNA and plasmid DNA), and scavengers that eliminate or reroute harmful substances (like ROS) locally. Future biomaterials-based cell carriers and niches may be created to monitor certain immune responses for diagnostic, investigation and observation, and clinical study purposes together with their use for therapeutic goals. For instance, the delivery of a pancreatic-islet lysate to gather, separate, and analyze the repertoire of T cells that are diabetogenic has been investigated using a PLGA-based microporous scaffold. Similar to this, a locally applied adjuvant- and antigen-loaded alginate microneedle patches can attract T cells indicating active systemic responses. [82].

By combining a suitable biomaterial (which is an essential part of the extracellular matrix involving the airway system) using airway organoids harboring disease-relevant cell types, it is possible to construct a structure like the tissues, with favorable functional and morphological attributes. Lung-on-a-chip systems also need to be enhanced and optimized in order to replicate and imitate the critical pulmonary tissue microenvironment conditions. This could aid in the discovery of innovative medications and therapeutics for challenging lung diseases like the newer SARS-CoV-2. A promising method for determining potential side effects of recently created SARS-CoV-2 medications and vaccinations is to combine lung-on-a-chip technology with heart, liver, and kidney organs-on-a-chip systems [83]. An acellular dermal matrix created by the usage of skin from the human deceased when mixed fat grafts in skin graft and breast reconstruction surgeries, to enhance the localization and survival of transplanted fat cells procedures to improve the localization and survival of transplanted fat cells. In the commercially available product Epigraft, collagen has been used to enable the organization of composite cell types within tissue grafts, such as skin grafts. Lately, cardiomyocytes derived from stem cells and collagen sheet-laden epicardium cells have been used in combination for myocardial tissue healing following infarction [84].

Hence it is believed that biomaterials will eventually be used for more sophisticated purposes, such as signal transducers that trigger cellular roles. Early viability studies in animals suggest that remote control of cell activity is possible using external actuators including ultrasonic, magnetic fields, and electronic inputs. These external actuators generate forces that the body can withstand [1]. These technologies are in their nascent stages, so advancements in the production of external signals will be required for propelling their development, and therefore, genetically modified biomaterials will reach heights in the near future.

6 The Future of Genetically Engineered Biomaterials

Over the past fifteen years, there has been a lot of research into genetically modified biomaterials, and CRISPR has already begun to play a significant part in the advancement of this field. With the advancement of manufacturing techniques that produce genetically engineered biomaterials, new ways have been paved for harnessing this potential of the manufacturing sector and formulating genetically engineered biomaterials from it. Scalable production has provided hope for the synthesis of genetically engineered nanofibrous materials. With features like biocompatibility, biodegradability, prospective energy collection, thermal, optical, and carcinogenic detection potentialities, among others, these methodologies may arise in the not too-distant future. Intelligent and genetically generated biomaterials and medicine delivery systems have created tremendous strides in the last few years. It is being tested to create new materials with unique properties. Researchers presently concentrate mostly on designs found in nature, such as spider silk motifs, in order to develop materials with outstanding mechanical properties. The design of novel materials, not present in nature, that are based on an understanding of the relationship between the structure of protein-based materials and biorecognition, self-assembly, and characteristics has the most potential, though [1]. Bio adaptability would be the best factor for selecting the optimal biomaterials that prove to be useful in certain tissue repair. Biomaterials with exceptional bio adaptability must possess some essential qualities, such as adaptable components that mimic cell response and tissue reconstruction, matched mechanical properties with natural tissues, suitable surfaces with advantageous tissue reaction, and biomimetic multi-level structures.

Tissue engineering and regenerative medicine, two techniques that are now in development, will someday transform medical implants. Tissue engineering will make it possible to create functional, living alternatives for many tissues and organs. Tissue engineering will benefit from developments in biodegradable polymers, fast prototyping, drug administration, cell culture, stem cell procedures, angiogenesis, and biomimetic approaches for building extracellular, matrix-like biomimetics. The potential applications of clever systems and micro- and nanofabrication are almost endless. Future in vitro systems will be built on a sophisticated knowledge of how biological systems engage with synthetic surfaces. These novel technologies will also be carefully and morally addressed because they have the potential to alter expectations for humans. Another area of study that will improve the functionality of implants is lowering the risk of infection. This can be assisted by antibacterial surface coatings, non-fouling surfaces, and antibiotic-controlled release mechanisms. The development of devices that function as both a device and a drug will also be made possible by advancements in controlled drug release in general.

7 Conclusion

This chapter offers a comprehensive overview of the biomaterials industry. It is meant to give the reader a vantage position from which they can start to situate all the subthemes within the context of the bigger picture. To reiterate a crucial point, the study of biomaterials may be the one that is most interdisciplinary. As a result, in order to be proficient in their area, biomaterials scientists need to be knowledgeable in a wide range of science, technology, engineering, and medical specialties. Being involved in a mentally challenging project that advances our knowledge of fundamental sciences and also helps to alleviate human suffering is the reward for mastering this volume of material. The journey of biomaterials can surely be considered a success from the significance they hold for contemporary medical treatments, the market's economic growth, and steady development of the industry over the past 50 years. The body will require fewer synthetic elements as a result of regenerative medicine. We anticipate the need for genetically engineered biomaterials to last well into this century and that many uses will continue to call for synthetic materials. Modern biomaterial development necessitates the design of materials that are bioinert and at the same time, also engage with and even react to their surrounding living environment. Biological cues that may communicate cell adhesion, migration, proliferation, differentiation, or even apoptosis must be provided by bioactive materials. The materials can communicate with the nearby biological environment in this way to convey their messages, for instance, by sending a signal of growth to encourage tissue repair. Bioactive substances with built-in responses to stimuli may respond to nearby physiological activity, such as by releasing a drug or breaking down to respond to a biological stimulus. Future genetically engineered intelligent biomaterials will depend heavily on biomaterials' capacity to detect changes in their surroundings or biological demand.

Conclusively, this is the stage where our knowledge of cellular biological sciences and biochemistry allows us to incorporate particular biologic functions into genetically engineered biomaterials. As new information is discovered, it can be immediately incorporated into a biomaterial to govern specific amino acid, lipid, or carbohydrate sequences that regulate cell differentiation, immune responses, or other biological phenomena. Thus, a key activity in the future will be the integration of materials science and cell biology knowledge to create a new class of materials that can truly facilitate desired medical outcomes.

References

1. Bashor, C.J., Hilton, I.B., Bandukwala, H., Smith, D.M., Veiseh, O.: Engineering the next generation of cell-based therapeutics. Nat. Rev. Drug Discov. **21**(9), 655–675 (2022)
2. Kersey, A.L., Nguyen, T.U., Nayak, B., Singh, I., Gaharwar, A.K.: Omics-based approaches to guide the design of biomaterials. Mater. Today. 2023.

3. Peppas, N.A., Langer, R.: New challenges in biomaterials. Science **263**(5154), 1715–1720 (1994)
4. Ratner, B.D., Bryant, S.J.: Biomaterials: where we have been and where we are going. Annu. Rev. Biomed. Eng. **6**(1), 41–75 (2004)
5. Purbrick, M.D.: Book review: Surface modification of polymeric biomaterials. Edited by B. D. Ratner and D. G. Castner. Plenum Publishing Corporation, New York, 1997. Polym Int. 46(1): 78–78 (1998)
6. Pierschbacher, M.D., Ruoslahti, E.: Cell attachment activity of fibronectin can be duplicated by small synthetic fragments of the molecule. Nat. **309**(5963), 30–33 (1984)
7. Sikavitsas, V.I., Temenoff, J.S., Mikos, A.G.: Biomaterials and bone mechanotransduction. Biomaterials **22**(19), 2581–2593 (2001)
8. Lloyd, A.W., Faragher, R.G., Denyer, S.P.: Ocular biomaterials and implants. Biomaterials **22**(8), 769–785 (2001)
9. Ziats, N.P., Miller, K.M., Anderson, J.M.: In vitro and in vivo interactions of cells with biomaterials. Biomaterials **9**(1), 5–13 (1988)
10. Hench, L.L.: Biomaterials: a forecast for the future. Biomaterials **19**(16), 1419–1423 (1998)
11. Patel, N.R., Gohil, P.P.: A review on biomaterials: scope, applications & human anatomy significance. Int. J. Emerg. Technol. Adv. Eng. 2(4):91–101 (2012)
12. Biswal, T., BadJena, S.K., Pradhan, D.: Sustainable biomaterials and their applications: a short review. Mater. Today: Proc. **1**(30), 274–282 (2020)
13. Chutia, S.J., Yashwanth, B.S., Kumar, M., Nath, R.J., Bora, G., Sarma, D.K.: Adv. Genet. Eng. (2020)
14. Zhang, Y., Zhou, Z., Sun, L., Liu, Z., Xia, X., Tao, T.H.: "Genetically engineered" biofunctional triboelectric nanogenerators using recombinant spider silk. Adv. Mater. **30**(50), 1805722 (2018)
15. Kim, K.S., Han, J.H., Park, J.H., Kim, H.K., Choi, S.H., Kim, G.R., Song, H., An, H.J., Han, D.K., Park, W., Park, K.S.: Multifunctional nanoparticles for genetic engineering and bioimaging of natural killer (NK) cell therapeutics. Biomaterials **1**(221), 119418 (2019)
16. Aleixandre, R.N., Shrikrishnan, S., Dalby, M.J., Salmeron-Sanchez, M.: Engineered living biomaterials. Nat. Rev. Mater. **6**(12), 1175–1190 (2021)
17. Zhu, D., Jiang, Z., Li, N., Wang, X., Ren, L., Ye, Y., Pan, Y., Yang, G.: Insights into the use of genetically modified decellularized biomaterials for tissue engineering and regenerative medicine. Adv. Drug Delivery Rev. **28**, 114413 (2022)
18. Jiang, Z., Fu, M., Zhu, D., Wang, X., Li, N., Ren, L., He, J., Yang, G.: Genetically modified immunomodulatory cell-based biomaterials in tissue regeneration and engineering. Cytokine Growth Factor Rev. 2022
19. Dobres, S., Mula, G., Sauer, J., Zhu, D.: Applications of 3D printed chimeric DNA biomaterials. ER **3**(1), 13–23 (2022)
20. de León, E.H., Valle-Pérez, A.U., Khan, Z.N., Hauser, C.A.: Intelligent and smart biomaterials for sustainable 3D printing applications. Curr. Opin. Biomed. Eng. **4**, 100450 (2023)
21. Chambre, L., Martín-Moldes, Z., Parker, R.N., Kaplan, D.L.: Bioengineered elastin-and silk-biomaterials for drug and gene delivery. Adv. Drug Delivery Rev. **1**(160), 186–198 (2020)
22. Sengupta, D., Heilshorn, S.C.: Protein-engineered biomaterials: highly tunable tissue engineering scaffolds. Tissue Eng. B Rev. **16**(3), 285–293 (2010)
23. Bzymek, M., Lovett, S.T.: Instability of repetitive DNA sequences: the role of replication in multiple mechanisms. Proc. Natl. Acad. Sci. U.S.A. **98**(15), 8319–8325 (2001)
24. Maurya, A.K., Mishra, A., Mishra, N.: Nanoengineered polymeric biomaterials for drug delivery system. In: Nanoengineered Biomaterials for Advanced Drug Delivery. Elsevier; pp. 109–143 (2020)
25. Austin, M.J., Rosales, A.M.: Tunable biomaterials from synthetic, sequence-controlled polymers. Biomater. Sci. **7**(2), 490–505 (2019)
26. McGrath, K.P., Fournier, M.J., Mason, T.L., Tirrell, D.A.: Genetically directed syntheses of new polymeric materials. Expression of artificial genes encoding proteins with repeating-(AlaGly) 3ProGluGly-elements. J. Am. Chem. Soc. **114**(2), 727–33 (1992)

27. Choudhury, D., Yee, M., Sheng, Z.L., Amirul, A., Naing, M.W.: Decellularization systems and devices: State-of-the-art. Acta Biomater. **1**(115), 51–59 (2020)
28. Kawecki, M., Łabuś, W., Klama-Baryla, A., Kitala, D., Kraut, M., Glik, J., Misiuga, M., Nowak, M., Bielecki, T., Kasperczyk, A.: A review of decellurization methods caused by an urgent need for quality control of cell-free extracellular matrix'scaffolds and their role in regenerative medicine. J. Biomed. Mater. Res. Part B Appl. Biomater. **106**(2), 909–23 (2018)
29. Yao, Q., Zheng, Y.W., Lan, Q.H., Kou, L., Xu, H.L., Zhao, Y.Z.: Recent development and biomedical applications of decellularized extracellular matrix biomaterials. Mater. Sci. Eng. C **1**(104), 109942 (2019)
30. Lü, W.D., Sun, R.F., Hu, Y.R., Lu, J.R., Gu, L., Liu, Z.G., Lei, G.Y., Qiang, Z., Cai, L.: Photooxidatively crosslinked acellular tumor extracellular matrices as potential tumor engineering scaffolds. Acta Biomater. **15**(71), 460–473 (2018)
31. Matuska, A.M., McFetridge, P.S.: The effect of terminal sterilization on structural and biophysical properties of a decellularized collagen-based scaffold; implications for stem cell adhesion. J. Biomed. Mater. Res. Part B Appl. Biomater. **103**(2):397–406 (2015)
32. Keane, T.J., Swinehart, I.T., Badylak, S.F.: Methods of tissue decellularization used for preparation of biologic scaffolds and in vivo relevance. Methods **1**(84), 25–34 (2015)
33. Zou, J.L., Liu, S., Sun, J.H., Yang, W.H., Xu, Y.W., Rao, Z.L., Jiang, B., Zhu, Q.T., Liu, X.L., Wu, J.L., Chang, C.: Peripheral nerve-derived matrix hydrogel promotes remyelination and inhibits synapse formation. Adv. Funct. Mater. **28**(13), 1705739 (2018)
34. Yu, R., Zhang, H., Guo, B.: Conductive biomaterials as bioactive wound dressing for wound healing and skin tissue engineering. Nano-Micro Lett. **14**, 1–46 (2022)
35. Zhang, X., Chen, X., Hong, H., Hu, R., Liu, J., Liu, C.: Decellularized extracellular matrix scaffolds: recent trends and emerging strategies in tissue engineering. Bioact. Mater. **1**(10), 15–31 (2022)
36. Gomes, S., Leonor, I.B., Mano, J.F., Reis, R.L., Kaplan, D.L.: Natural and genetically engineered proteins for tissue engineering. Prog. Polym. Sci. **37**(1), 1–17 (2012)
37. Kleinheinz, J., Jung, S., Wermker, K., Fischer, C., Joos, U.: Release kinetics of VEGF165 from a collagen matrix and structural matrix changes in a circulation model. Head Face Med. **6**(1), 1–7 (2010)
38. Wongpanit, P., Ueda, H., Tabata, Y., Rujiravanit, R.: In vitro and in vivo release of basic fibroblast growth factor using a silk fibroin scaffold as delivery carrier. J. Biomater. Sci. Polym. Ed. **21**(11), 1403–1419 (2010)
39. Friess, W., Uludag, H., Foskett, S., Biron, R., Sargeant, C.: Characterization of absorbable collagen sponges as recombinant human bone morphogenetic protein-2 carriers. Int. J. Pharm. **185**(1), 51–60 (1999)
40. Wang, X., Zhang, X., Castellot, J., Herman, I., Iafrati, M., Kaplan, D.L.: Controlled release from multilayer silk biomaterial coatings to modulate vascular cell responses. Biomaterials **29**(7), 894–903 (2008)
41. Shields, L.B., Raque, G.H., Glassman, S.D., Campbell, M., Vitaz, T., Harpring, J., Shields, C.B.: Adverse effects associated with high-dose recombinant human bone morphogenetic protein-2 use in anterior cervical spine fusion. Spine J. **31**(5), 542–547 (2006)
42. Lyu, Y., Wang, Y.: Output optimization of biodegradable triboelectric nanogenerators. Nano Energy **16**, 107811 (2022)
43. Briscoe, J., Dunn, S.: Piezoelectric nanogenerators–a review of nanostructured piezoelectric energy harvesters. Nano Energy **1**(14), 15–29 (2015)
44. Korkmaz, S., Kariper, İA.: Pyroelectric nanogenerators (PyNGs) in converting thermal energy into electrical energy: Fundamentals and current status. Nano Energy **1**(84), 105888 (2021)
45. Jiang, D., Su, Y., Wang, K., Wang, Y., Xu, M., Dong, M., Chen, G.: A triboelectric and pyroelectric hybrid energy harvester for recovering energy from low-grade waste fluids. Nano Energy **1**(70), 104459 (2020)
46. Ahmed, A., Hassan, I., Helal, A.S., Sencadas, V., Radhi, A., Jeong, C.K., El-Kady, M.F.: Triboelectric nanogenerator versus piezoelectric generator at low frequency (< 4 Hz): a quantitative comparison. Iscience. **23**(7), 101286 (2020)

47. Li, X., Jiang, C., Ying, Y., Ping, J.: Biotriboelectric nanogenerators: materials, structures, and applications. Adv. Energy Mater. **10**(44), 2002001 (2020)
48. Zhou, Y., Deng, W., Xu, J., Chen, J.: Engineering materials at the nanoscale for triboelectric nanogenerators. Cell Rep. Phys. Sci. 1(8) (2020)
49. Zheng, N., Xue, J., Jie, Y., Cao, X., Wang, Z.L.: Wearable and humidity-resistant biomaterials-based triboelectric nanogenerator for high entropy energy harvesting and self-powered sensing. Nano Res. **15**(7), 6213–6219 (2022)
50. Kim, J.N., Lee, J., Go, T.W., Rajabi-Abhari, A., Mahato, M., Park, J.Y., Lee, H., Oh, I.K.: Skin-attachable and biofriendly chitosan-diatom triboelectric nanogenerator. Nano Energy **1**(75), 104904 (2020)
51. Adhikary, P., Mahmud, M.P., Solaiman, T., Wang, Z.L.: Recent advances on biomechanical motion- driven triboelectric nanogenerators for drug delivery. Nano Today **1**(45), 101513 (2022)
52. Niu, Z., Cheng, W., Cao, M., Wang, D., Wang, Q., Han, J., Long, Y., Han, G.: Recent advances in cellulose-based flexible triboelectric nanogenerators. Nano Energy **1**(87), 106175 (2021)
53. Correa-Gallegos, D., Jiang, D., Rinkevich, Y.: Fibroblasts as confederates of the immune system. Immunol. Rev. **302**(1), 147–162 (2021)
54. Nidhi, S., Joseph, N., Vighnesh, N.P., Sabarinath, P.J., John, J., John, H., Padmanabhan, N.T.: Recent updates on triboelectric nanogenerator based advanced biomedical technologies: A short review. RINENG 100782 (2022)
55. Wang, E., Lee, S.H., Lee, S.W.: Elastin-like polypeptide based hydroxyapatite bionanocomposites. Biomacromol **12**(3), 672–680 (2011)
56. Peddi, S., Roberts, S.K., MacKay, J.A.: Nanotoxicology of an elastin-like polypeptide rapamycin formulation for breast cancer. Biomacromol **21**(3), 1091–1102 (2020)
57. Lin, Y., Jin, W., Qiu, Y., Zhang, G.: Programmable stimuli-responsive polypeptides for biomimetic synthesis of silica nanocomposites and enzyme self-immobilization. Int. J. Biol. Macromol. **1**(134), 1156–1169 (2019)
58. Butcher, N.J., Mortimer, G.M., Minchin, R.F.: Unravelling the stealth effect. Nat. Nanotechnol. **11**(4), 310–311 (2016)
59. Schöttler, S., Becker, G., Winzen, S., Steinbach, T., Mohr, K., Landfester, K., Mailänder, V., Wurm, F.R.: Protein adsorption is required for stealth effect of poly (ethylene glycol)-and poly (phosphoester)-coated nanocarriers. Nat. Nanotechnol. **11**(4), 372–377 (2016)
60. Nidhi, S., Anand, U., Oleksak, P., Tripathi, P., Lal, J.A., Thomas, G., Kuca, K., Tripathi, V.: Novel CRISPR–Cas systems: an updated review of the current achievements, applications, and future research perspectives. Int. J. Mol. Sci. **22**(7), 3327 (2021)
61. Chavez-Granados, P.A., Manisekaran, R., Acosta-Torres, L.S., Garcia-Contreras, R.: CRISPR/Cas gene-editing technology and its advances in dentistry. Biochim. **1**(194), 96–107 (2022)
62. Bhattacharjee, R., Jana, A., Nandi, A., Sinha, A., Bhattacharjee, A., Mitra, S., Kar, S., Dey, A., Singh, S.K., Varma, S.R., Panda, P.K.: Synergy of Nanocarriers with CRISPR-Cas9 in an Emerging Technology Platform for Biomedical Appliances: Current Insights and Perspectives. Mater. Des. **24**, 111415 (2022)
63. Eoh, J., Gu, L.: Biomaterials as vectors for the delivery of CRISPR–Cas9. Biomater. Sci. **7**(4), 1240–1261 (2019)
64. Ghandehari, H.: Recombinant biomaterials for pharmaceutical and biomedical applications. Pharm. Res. **25**, 672–673 (2008)
65. Shadish, J.A., DeForest, C.A.: Site-selective protein modification: from functionalized proteins to functional biomaterials. Matter. **2**(1), 50–77 (2020)
66. Shim, G., Kim, D., Park, G.T., Jin, H., Suh, S.K., Oh, Y.K.: Therapeutic gene editing: delivery and regulatory perspectives. Acta Pharmacol. Sin. **38**(6), 738–753 (2017)
67. Gaharwar, A.K., Singh, I., Khademhosseini, A.: Engineered biomaterials for in situ tissue regeneration. Nat. Rev. Mater. **5**(9), 686–705 (2020)
68. Dhaliwal, A., Pelka, S., Gray, D.S., Moghe, P.V.: Engineering lineage potency and plasticity of stem cells using epigenetic molecules. Sci. Rep. **8**(1), 16289 (2018)
69. Larsson, L., Pilipchuk, S.P., Giannobile, W.V., Castilho, R.M.: When epigenetics meets bioengineering—A material characteristics and surface topography perspective. J. Biomed. Mater. Res. Part B Appl. Biomater. **106**(5), 2065–71 (2018)

70. James, E.N., Van Doren, E., Li, C., Kaplan, D.L.: Silk biomaterials-mediated miRNA functionalized orthopedic devices. Tissue Eng. Part A **25**(1–2), 12–23 (2019)
71. Asa'ad, F., Pelanyte, G., Philip, J., Dahlin, C., Larsson, L.: The role of epigenetic functionalization of implants and biomaterials in osseointegration and bone regeneration—A review. Molecules **25**(24), 5879 (2020)
72. Hu, W., Li, Q., Li, B., Ma, K., Zhang, C., Fu, X.: Optogenetics sheds new light on tissue engineering and regenerative medicine. Biomaterials **1**(227), 119546 (2020)
73. Williams, T.L., Iskandar, D.J., Nödling, A.R., Tan, Y., Luk, L.Y., Tsai, Y.H.: Transferability of N-terminal mutations of pyrrolysyl-tRNA synthetase in one species to that in another species on unnatural amino acid incorporation efficiency. Amino Acids **53**, 89–96 (2021)
74. Sisila, V., Indhu, M., Radhakrishnan, J., Ayyadurai, N.: Building biomaterials through genetic code expansion. Trends Biotechnol. (2022)
75. Haq-Siddiqi, N.A., Britton, D., Montclare, J.K.: Protein-engineered biomaterials for cartilage therapeutics and repair. Adv. Drug Deliv. Rev. **9**, 114647 (2022)
76. Kowalski, P.S., Bhattacharya, C., Afewerki, S., Langer, R.: Smart biomaterials: recent advances and future directions. ACS Biomater. Sci. Eng. **4**(11), 3809–3817 (2018)
77. Lyu, S., Fang, J., Duan, T., Fu, L., Liu, J., Li, H.: Optically controlled reversible protein hydrogels based on photoswitchable fluorescent protein Dronpa. ChemComm. **53**(100), 13375–13378 (2017)
78. Finch, L., Harris, S., Solomou, G., Sen, J., Tzerakis, N., Emes, R.D., Lane, C.S., Hart, S.R., Adams, C.F., Chari, D.M.: Safe nanoengineering and incorporation of transplant populations in a neurosurgical grade biomaterial, DuraGen Plus^{TM}, for protected cell therapy applications. JCR **10**(321), 553–563 (2020)
79. Diomede, F., D'Aurora, M., Gugliandolo, A., Merciaro, I., Orsini, T., Gatta, V., Piattelli, A., Trubiani, O., Mazzon, E.: Biofunctionalized scaffold in bone tissue repair. Int. J. Mol. Sci. **19**(4), 1022 (2018)
80. Ribeiro-Filho, L.A., Sievert, K.D.: Acellular matrix in urethral reconstruction. Adv. Drug Deliv. Rev. **1**(82), 38–46 (2015)
81. Thomas, S.N., Archer, P.A., Manspeaker, M.P.: Biomaterials for Immunoengineering. In: Biomaterials Science. Academic Press, pp. 1199–1215 (2020)
82. Dellacherie, M.O., Seo, B.R., Mooney, D.J.: Macroscale biomaterials strategies for local immunomodulation. Nat. Rev. Mater. **4**(6), 379–397 (2019)
83. Seyfoori, A., Amereh, M., Dabiri, S.M., Askari, E., Walsh, T., Akbari, M.: The role of biomaterials and three dimensional (3D) in vitro tissue models in fighting against COVID-19. Biomater. Sci. **9**(4), 1217–1226 (2021)
84. Kopeček, J.: Smart and genetically engineered biomaterials and drug delivery systems. Eur. J. Pharm. Sci. **20**(1), 1–6 (2003)

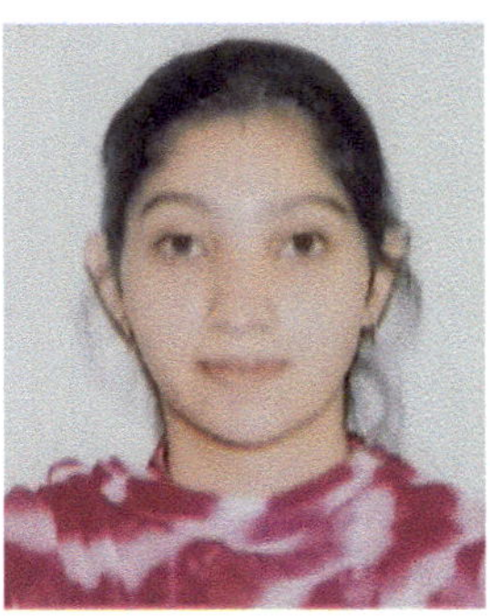

Aashveen has a master's degree in pharmaceutical sciences and a keen interest in developing novel drug delivery systems. She has worked on projects involving medical writing skills involving the knowledge of polymer sciences and how they work in various delivery systems along with their roles. Some of these articles have been published and some are in the process of publishing.

Vridhi is a Formulation Scientist with experience in the Research & Development sector of the pharmaceutical industry. Having a Master's degree in Pharmaceutical Sciences, she works on novel drug delivery systems to enhance active pharmaceutical ingredients' therapeutic potential and reduce the associated side effects. She is actively involved in the publication of scientific and health-related articles.

Shubham Thakur completed his undergraduate and postgraduate studies in pharmacy at Guru Nanak Dev University in Amritsar in 2016 and 2018 respectively. Presently, he serves as an Indian Council of Medical Research-Senior Research Fellow and is pursuing a Ph.D. in the Department of Pharmaceutical Sciences at Guru Nanak Dev University, Amritsar. His research focuses on the advancement of novel drug delivery systems, aiming to enhance the kinetics, efficacy, and safety of existing drugs. Specifically, his current project involves developing a safe oral formulation for pregnant women. Till now he has published 25 articles and 5 book chapters in Scopus indexed journals and books with h-index of 5.

Green Methods for the Development of Bone and Tissue Engineering-Based Biomaterials

Avipsa Hazra, Gowrav Baradwaj, A. S. Dhanu, Gobianand Kuppannan, Malarvizhi Arthanari, and B. M. Kanthesh

Abstract Bioengineering of bones and tissues includes the culmination of principles of bioactive compounds, scaffoldings, and developing materials for the treatment of body cells and organs that are required to be replaced due to damage of some kind. However, synthetic materials used are known to cause extensive pollution and thus to develop methods with no or lesser pollution multiple green manufacturing technologies are being designed. Due to their outstanding biodegradability and biocompatibility in biological media, green materials are proven to be a useful factor in medicine. Though this technology is still in its cradle stage, there have been multiple kinds of materials like chitosan, collagen, alginate, and corals that are being implemented for manufacturing composite for hard tissue implants such as artificial bone, bone cement and for the construction of multiple-generation biomaterials. This chapter will cover the various materials and methodologies that are already in practice or are being developed to regenerate bones and tissues, keeping in sight the safety of the environment and to give us a pollution free world.

Keywords Bioengineering · Biomaterials · Scaffolds · Bone tissue engineering

Abbreviations

TE	Tissue Engineering
BE	Bone Engineering

A. Hazra · G. Baradwaj · A. S. Dhanu · B. M. Kanthesh (✉)
Division of Molecular Biology, School of Life Sciences, JSS Academy of Higher Education and Research, Mysuru, Karnataka 570015, India
e-mail: kantheshmb@jssuni.edu.in

G. Kuppannan
Department of Microbiology, Noorul Islam College of Dental Sciences, Aralummoodu, Thiruvananthapuram, Kerala 695123, India

M. Arthanari
Vivekanandha College of Arts and Sciences for Women (Autonomous), Elayampalayam, Tiruchengode, Tamil Nadu 637205, India

R. Malviya and S. Sundram (eds.), *Engineered Biomaterials*, Engineering Materials,
https://doi.org/10.1007/978-981-99-6698-1_3

PGA	Polyglycolic Acid
PEG	Polyethylene Glycol
PCL	Polycaprolactone
3D	Three-dimensional
BTE	Bone Tissue Engineering
ECM	Extracellular Matrix
ALP	Alkaline Phosphatase
GlcNAc	N-acetyl Glucosamine
GAGs	Glycosaminoglycans
PMMA	Polymethyl methacrylate
PLLA	Poly-L-Lactic acid
MSC	Mesenchymal Stem Cell
DNA	Deoxyribonucleic Acid
HAP	Hydroxyapatite
HFIP	Hexafluoro-2-propanol
GHG	Greenhouse Gases
HEMA	2-Hydroxyethyl methacrylate
PEO	Polyethylene oxide
PVA	Polyvinyl alcohol

1 Introduction

With an increase in the rates of diseases affecting various organs of the human body, and the rate expected to double by the end of 2030, the need for the ability to replace any such organ has become imperative [1]. It has led to the expansion of the field of medicine known as regenerative medicine, which involves tissue engineering (TE), regeneration of bones, and self-healing through activation of the host's immune system [2]. It is a multidisciplinary discipline that integrates physics concepts with principles derived from life science to provide clinical treatment strategies. The concept of regenerative medicine emerged in the early 1990s, and since then, efforts have been made to strengthen research in the field.

Although bone bioengineering is a type of tissue engineering, it is considered distinct from tissue engineering of other organs in today's context, with bone engineering being one of the most developed arenas. The creation of bio-active artificial cells and tissues for the substitution and regeneration of faulty tissues or organs is the focus of tissue engineering [3]. This includes isolation of suitable cells, development of compatible scaffolds, and implantation of the molecules into the tissues to alter functions or initiate repair of damaged organs and tissues [2]. Bone engineering (BE) involves the synergistic effect of engineered biomaterials with the host's recovery factors to regenerate bone tissues. It appears to have no side effects so far and allows for healing of bone fractures and other flaws using the host's cells [4]. However, with the progress and improvement in the field, there has been a distinct rate of pollution because of it and has led to the environmental depletion.

The raw materials being used, the chemicals being depleted, and the destruction of faulty end products have led to soil erosion as well as major water pollution. It has also been known to affect the health of individuals being involved in the process which has now led to the thought of developing a process that is responsible toward human health and environment- green practice. Green practices involve the implementation of resource- and environmentally conscious methods throughout the entire lifecycle, from design to construction, operation, maintenance, renovation, and destruction [5]. The traditional considerations of economy, usefulness, durability, and comfort are expanded and are now being complemented with green practices.

The primary concept of green practices was presented by Anastas and Warner who put forward the foundation based on 12 principles, which revolves around reducing waste, hazardous products and derivatives, balancing atom economy and energy efficiency, developing designs that make use of catalysis and based on the usage of benign compounds and renewable sources, and always having a detailed plan about degradation of the product, pollution control, and accident prevention. Thus, in the light of 2023, today scientists from all fields are working on improving the production of biomaterials such that it keeps the environment stress free [6].

2 Usual Methods and Their Issues

To develop a system which was economically sustainable, environmental wellness had often got a backseat and so has been the case with the initial bioengineering methods. The prime concern for the methodologies was the disposal of waste—as most of it is unusable, they have often been disposed of into landfills. Alongside the foul odor and land wastage, the waste has even been recorded to yield pathogens [7]. Even though theoretically the use of nanoparticles to maintain cell viability seemed like a sublime idea, over the years the use of it in medicine and under human biology has posed multiple adverse effects due to the variation in manufacturing, particle size, chemical composition, and materials involved in the production of the nanoparticles [8–10]. Poly-glycolic acid (PGA), lactic acid, polyethylene glycol (PEG), and polycaprolactone (PCL) were used as the sources to produce synthetic polymers for the fabrication of scaffolds for tissue engineering [11–14]. However, the biological and mechanical functioning of the said chemicals proved to be weakened when used singly in tissue engineering and demanded reinforced materials for the increase of the functioning properties [10].

Usually, scaffolds in tissue engineering have been created and produced using methods such as freeze-drying, salt leaching, gas forming, and phase separation; however, the precision of this method to develop complex structures and fit the internal architecture is insufficient. Owing to these issues, production of scaffolds with the use of 3D printing came into play as it was able to mold materials into the desired shape [15–17]. Nevertheless, the use of 3D printers is limited because not every substance can be modified in them, reducing the range of sources for producing scaffolds in tissue engineering [18]. Among each of the other tissue engineering processes, bone regeneration is the most critical one as it is founded on numerous principles of molecular and cellular biology, biological development, and structural biology, all of which are monitored by biomechanics and bioengineering. This is because bone tissues are responsible for dynamics of the body and must be adjusted according to its different locations. The most difficult aspect of bone tissue development is retaining vascularization of blood vessels by making the scaffolding porous and integrating it with the host system. With current synthetic mechanisms, only the periphery of a scaffolding is suitable to be used as an implantation thus limiting its use. Owing to these reasons, the synthetic mechanisms for production of bone tissues are not widely in use [19].

Usually, the current strategies are based on either the cellular or source material mechanism, thus for the usage to be widespread all the factors must be collaborated. Bone engineering is also an expensive process due to it being patient-dependent and poses health issues due to erroneous cell isolations, seeding, culturing, and implantation [20]. Methods need to be established such that all the factors can be culminated, and they impose no adverse effects on human health [4]. Moreover, the growth in the field of tissue regeneration has been quite stagnant over the past two decades. The clinical translations of the basic research for bone, cartilage, and ligament regeneration have met a roadblock ever since the involvement of stem cells had been implemented. Though the synthetic polymers at one time seemed like the most promising solution to tissue engineering, in the current scenario there seems to be an unresolved failure at the translation stage for application of the methods [4, 21]. Also, for the current approaches to be effective a proper cell for the source must be determined for the model to be designed. However, the separation of human cells and culturing it in vitro such that its sensitivity is maintained, and its differentiation capacity is not reduced is a tedious task. Also, for some cells to work they need to be converted into their embryonic forms. Furthermore, the different methods are influenced by the tissue origin, the patient's age, and the individual's health condition. Also, the environment around the cell must be replicated exactly to establish the correct behavior of the polymer of the scaffold. Thus is the need of designing materials from natural sources which can be used as scaffolds by exploiting their chemical, physical, and biological properties [22–29]. These natural fibers can also be combined with synthetic materials depending on the need of the tissue at the site [30].

Another issue that crops up while developing regenerated tissues is the innervation of those tissues. For the structures to attain complete functionality after implantation, the structures need to get consistent with other growing tissues in the region [31]. These requirements of innervations have been seen to be very critical for regions like bladders and small intestines [32]. This association not only develops an interlink with the nervous system but also rejuvenates the tissue regenerated [33, 34]. For the bioengineering of peripheral nerves, an important criterion is the inculcation of neurotrophic factors which would ultimately lead to synapse formation, even muscle tissues require electric impulse for stimulation. These are critical steps and are difficult to implement as well [35, 36]. Some grafted organs require neovascularization to occur, such that the branches developed after morphogenesis can help in acceptance of the implantation, but it is more than often seen to be not there with synthetic polymers. However, in other cases the synthetic materials sometimes exceed the functionality rate of native tissues, leading to some complications [37, 38]. Overall many trials with synthetic fibers have been postponed owing to the financial factors as some of the synthetic fibers which are best suited for human use have a high cost [39].

3 Development of Green Methods and Its Types

The idea of developing environmentally friendly techniques for bone and tissue engineering arose because of the dangers that synthetic techniques posed to the human body and the environment [40]. This resulted in the creation of engineered biomaterials and scaffolds composed of natural polymers that offer little to no risk to the environment or human health. Green approaches involve using healthy, non-toxic materials and procedures to lessen the hazardous and damaging effects of scientific study. The reduction of pollution's harmful impacts on the environment and the human body is a key objective of green technologies and practices [41]. The goal of bone tissue engineering (BTE) is to create new, functional bone tissues using knowledge of the structure of bones, bone mechanics, and tissue building. To put it another way, if you want to efficiently regenerate as well as repair bone, you must first understand the biology of bone and how it develops. The ideas for biologically active chemicals, scaffolds, engineering, and material production are also combined to build, repair, or treat injured tissues or organs [42–44].

Because of their inherent bioactivity, including cell adhesion and proliferation, as well as the fact that they are harmless to native, healthy tissue, natural biomaterials are critical for tissue engineering. Biomaterials made from natural substances include those made from glycosaminoglycans, cellulose, chitin, collagen, chitosan, elastin, alginate, fibrinogen, dextran, gelatin, laminin, and silk. Among these materials, collagen, fibrinogen, glycosaminoglycans, elastin, and laminin are mimics of original extracellular matrix (ECM) components. Extracellular matrix is crucial for tissue regeneration because it supports cells physically and controls cellular processes including growth, proliferation, differentiation, and homeostasis, enabling the body to repair tiny flaws in cells, tissues, and organs [45, 46].

The human body is unable to regenerate deficiencies because of a shortage of extracellular matrix (ECM) to make up large defects. Natural biomaterials were created in this manner to be employed in applications involving tissue engineering. These natural biomaterials are employed in hydrogels, that additionally serve as an excellent tool for tissue engineering. Hydrogels are capable of being made from both synthetic and naturally occurring biomaterials. Artificial polymers include poly

HEMA, PEO, and PVA. Naturally occurring biomaterials used in hydrogel preparation include collagen, chitosan, gelatin, fibrin, alginate, agarose, and many more [47].

3.1 Collagen

In mammals, the protein collagen is a particularly abundant structural protein in the extracellular space. It accounts for approximately 30% of total protein weight. These are fibrous proteins which are major building block components in most of the animals (invertebrates, vertebrates) and sponges [44]. The chemical makeup, microscopic patterns, and mechanical strength of collagen fibrils all work together to preserve the structural integrity of each connecting tissue. Collagen, a component of the ECM, is found naturally in human tissues such as bone, skin, ligaments, and other types of connective tissue. The first type of collagen fibrils is relatively homogeneous in size, ranging from 50 to 500 nm [48–50]. Collagen is among the most used biomaterial for nanofibrous scaffolds, and several recent research articles have reported on the use of collagen and collagen-based composites in bone tissue engineering. According to some studies, the collagen/chitosan/hydroxyapatite composite is extremely effective in promoting osteogenic differentiation of mesenchymal stem cells [51].

Collagen is the triple helix structure made up of 2 $\alpha 1$ chains and one $\alpha 2$ chains, each chain weighing approximately about 100 kDa. Among the 29 types of collagens, type I is the most used for bone tissue engineering. The connective tissues of the vertebrae almost all contain collagen type I, while tendon and bone contain the most. The ability of this collagen to produce insoluble fibers also gives tissues a high tensile strength and stability [52]. Numerous discoveries proved that type I collagen can be isolated from different animal species and tissues. The most common source of its extraction is from cattle, which could have concerns including spreading bovine spongiform encephalopathy. Human recombinant collagen, on the other hand, has been used in tissue engineering processes [53]. It is possible to change the mechanical properties and collagen rate of degradation by adjusting various characteristics, including the amount of cross-linking.

Collagen is currently reported to be used in a variety of scaffolds for bone tissue engineering. Freeze drying is one of the most common and widely used processes for producing collagen scaffolds, in which acetic acid-based collagen suspensions are made and collagen slurries are allowed to develop prior to the freeze-drying procedure [54]. Another method included mineralization of collagen fibrils. This method included precipitation of collagen and their mineralization using amorphous calcium. These fibrils are then subjected to freeze drying to produce scaffolds.

Type I collagen is known to be an important component of bone mineralization since it is known to be secreted by osteoblasts during ossification. As a result, when collagen scaffolding is used in vivo, improved physiological integration with the surrounding tissue is possible. Several commercial products used in bone tissue

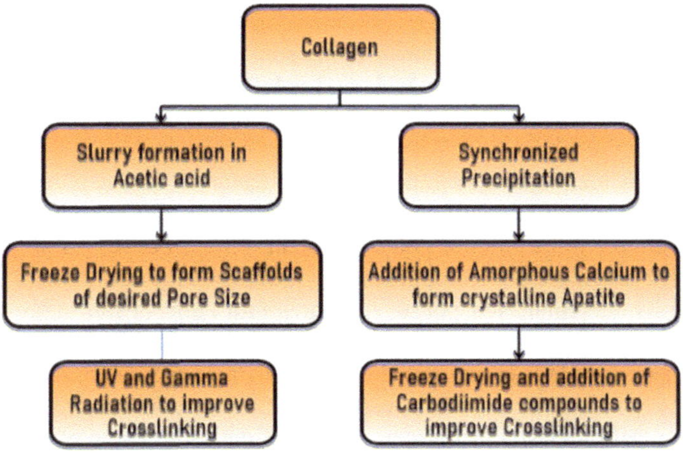

Fig. 1 Development of Scaffolds from Collagen

engineering are based on type I collagen. Collagen/cellulose scaffolds were shown to support cellular attachment and the phenotypic preservation of cultivated human osteoblasts in vitro. High levels of alkaline phosphatase (ALP), an osteogenic enzyme, were also observed [55, 56]. Development of scaffold from collagen is shown in flowchart in Fig. 1.

3.2 Chitosan

After cellulose, chitosan is the most abundant biopolymer. It is frequently found in invertebrates as shells or cuticles in crustaceans or insects. It is also found in some yeast and algae cell walls [56]. It is a deacetylated derivative from chitin. It is a polysaccharide that is made up of N-glucosamine and N-acetyl glucosamine (GlcNAc). It has a deacetylation degree of 30–95% and a molecular weight range of 300–1000 kDa [57, 58]. Chitosan is undissolved in aqueous solutions above the pH of 7. Nonetheless, free amino groups become protonated in weak acids, and the molecule becomes completely soluble below a pH of 5. Chitosan is a promising biomaterial in tissue engineering due to its antimicrobial properties and induced minimal foreign body response. Another important reason for using chitosan in tissue engineering is its structural similarity to GAGs, as it contains glucosamine, a key component of ECM. One of GAG's characteristics is that it has several specialized interactions with growth factors, receptors, and adhesion proteins. Chitosan's similar structure may also have comparable bioactivities. Most importantly, chitosan is cationic in nature, which aids in the retention or accumulation of molecules due to interactions with anionic GAGs and proteoglycans [59]. Chitosan is also appealing as a material for bone scaffolds because it promotes osteoblast cell adhesion and proliferation as well as the development of mineralized bone matrix. Because chitosan is a polymer composed of amino acids and polysaccharides, it contains breakable glycosidic bonds. In vivochitosan degradation is typically carried out by the enzyme lysozyme [60].

Chitosan has been combined with a variety of delivery components, including alginate, hydroxyapatite, hyaluronic acid, calcium phosphate, PMMA, poly-l-lactic acid (PLLA), and growth factors, for potential use in orthopedics. Chitosan, in general, offers a wide range of options for cell-based tissue creation. Chitosan can be made into 3D highly interconnected, porous structures with a high degree of porosity using a variety of technologies, allowing cells to proliferate and move nutrients. Porous scaffolds, for example, can be created by lyophilizing a frozen chitosan solution in acetic acid [61]. Chitosan and acetic acid may phase split and form ice crystals when frozen. The scaffold develops holes as a result of the sublimation of acetic acid ice crystals during the freeze-drying process. Another method is to use porogens (salts, sugar) in conjunction with chitosan in a melt-based technique [62].

Rapid prototyping technology can also be used to create chitosan scaffolds. This is a computer-aided method [63]. The use of chitosan-based scaffolds that mimic some bone characteristics can improve osteoblast adhesion and proliferation. For example, chitosan scaffolds combined with bioactive ceramics can impart osteoconductive and even osteo-inductive properties to the final structure, directing bone development engineering [64]. Chitosan, in combination with hydroxyapatite and tricalcium phosphates, has shown promising results in mesenchymal stem cell (MSC) adhesion, proliferation, and differentiation, as well as intriguing results in bone tissue engineering [65]. Chitosan has been used to improve the biological and mechanical properties of marine-derived polymers in order to provide an alternative to collagen derived from animal sources. When combined with polyesters, chitosan increased cell viability and adhesion [62]. Development of scaffolds from chitosan is in flowchart shown in Fig. 2.

3.3 Gelatin

Gelatin is a naturally occurring biopolymer created when animal collagen from skin, bones, and tendons is partially hydrolyzed in an acid or alkaline solution, and the structure is very similar to collagen. It is comprised of various proteins and amino acids. Regardless of how collagen is hydrolyzed to create gelatin, both polymers include up to 20 distinct amino acids in varying amounts in their basic structure. This basic structure provides a three-amino-acid detection sequence for integrin-mediated adhesion of cells. The properties of gelatin are determined by the source of collagen, the type of collagen, and the method of collagen conversion to gelatin. Various methods are implemented for the creation of gelatin scaffolds. One such process includes electrospinning. The production of polyelectrolyte colloidal sol makes it very challenging to electro-spun gelatin using water as the solvent.

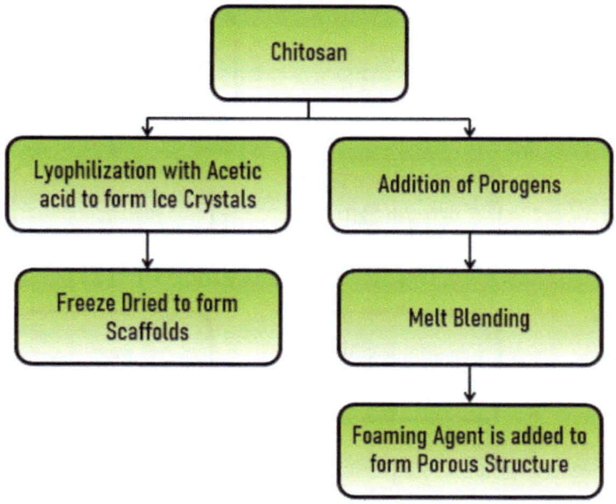

Fig. 2 Development of Scaffolds from Chitosan

The silky gelatin nanofibers were created using a variety of polar organic solvents, including trifluoroethanol and hexafluoro isopropanol [66]. Biomaterials made of gelatin are widely used in tissue engineering. Gelatin scaffolds demonstrated high DNA, GAG, and collagen contents and enabled the development of chondrogenic cells. Gelatin used with chitosan was used in skin tissue engineering [67]. Much interest has been shown in electro-spun nanofibrous scaffolds as prospective platforms for bone repair and regeneration. Degradable polymers, whether they are synthetic or natural, were used as a starting point for research into the advantageous properties of nanofibrous matrices. Many synthetic polymers' hydrophobic properties restrict cell affinities and early cell adhesion. Another method to increase hydrophilicity and, consequently, cell compatibility in polymers is to combine synthetic and natural polymers. This is done using a composite of polycaprolactone-gelatin scaffolds. Gelatin-apatite composite was also used since it mimicked the properties of bone ECM [68].

3.4 Silk Fibroin

Silk fibroin has been extensively researched for tissue engineering applications due to its superior biological and mechanical strength, as well as biocompatibility and biodegradability.

Silk is made up of fibrous protein (SF) and sericin. Silk is produced by a variety of arthropods, including spiders and silkworms with the most used silk produced by the silkworm *Bombyx mori*. SF has been shown to be very biocompatible. After observing its attachment to fibroblast cells and growth on SF matrices, its role as a biomaterial was evaluated. Silk fibroin can be easily separated from sericin by using a steaming alkaline or surfacting solution (degumming). The resulting fibers have a

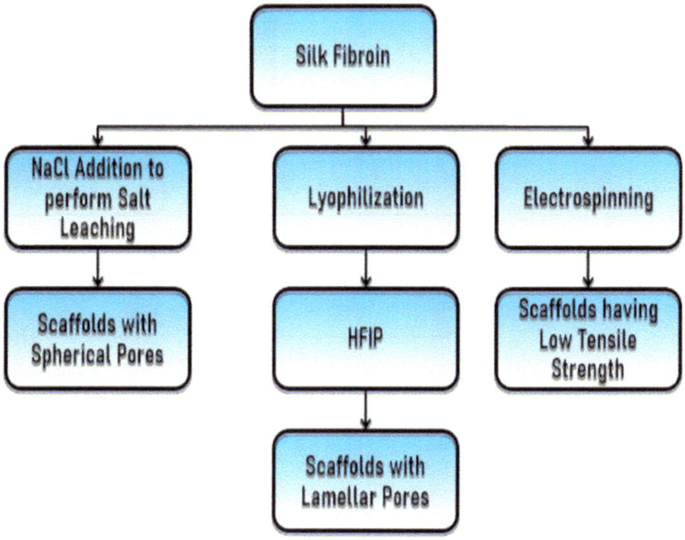

Fig. 3 Development of Scaffolds from Silk Fibroin

diameter of 10–25 mm and are composed of a hydrophobic heavy chain (350 kDa) and a hydrophilic light chain (25 kDa).

Depending on the type of scaffold required, silk fibroin can be created using a variety of techniques. One such technique is salt leaching using NaCl particles for scaffolds with spherical pores and for scaffolds with lamellar pores a process involving freezing, and lyophilizing was used which resulted in the formation of beta sheets. Another technique included electrospinning which produced fibers with low tensile strength [69]. Silk fibroin is a very useful biomaterial in tissue engineering. By making the fibers stiffer, it was seen to increase osteogenic induction. Silk used along with calcium phosphate in engineering of bone showed increased stability, and bioactivity with no toxicity. An osteo-inductive composite was created by combining silk and HAp, the important mineral seen in bone, to replicate the composition of bone. HAp improved the scaffolds' compressive strength as well as the ability of cells to proliferate and produce bone. Other biofunctional substances for restoration or therapy of tissues of bone are titanium dioxide (TiO_2) nanoparticle and natural silk fibroin-based nanocomposite scaffolds [70, 71]. The development of scaffolds from silk fibroin in flowchart is shown in Fig. 3.

3.5 *Starch*

Starch is a polysaccharide that plants and algae produce as an energy-storing compound in a variety of granules. Amylose and amylopectin are the two subunits of starch. Native starch includes waxy starch, which contains little amylose, regular starch, which contains 15–30% amylose, and higher content of amylose, which contains more than 50% amylose. Crystallinity can range from 15% in high-amylose starch to 50% in waxy starch [72]. Starch's brittleness prevents it from being used in most situations. As a result, starch must be modified or combined with other polymers to improve its properties.

Starch scaffolds can be produced by traditional casting in solvent procedures. In this procedure, the starch is dissolved in chloroform and cast into a mold. Cellulose and starch were combined by leaching with salt and freeze-drying methods in a different approach. Wet spinning method is also used to create scaffolds of desired pore size which include mixing of starch with other polymers. Chitosan-starch microparticles can be obtained from water in oil emulsification techniques. Starch can also be used along with HAp and chitosan which is processed by co-precipitation [73]. Several studies on bone tissue engineering have used starch-based scaffolds. When inserted into a critical-sized defect, scaffolds made of polycaprolactone, starch, and stem cells boosted the repair of bone. In comparison to scaffolds loaded with no differentiated cells, scaffolds loaded with osteoblastic cells produced more new bone. Modified starch scaffolds and polycaprolactone with silanol group showed improved osteogenic differentiation and bone healing. 3D scaffolds promoted adhesion and recruitment of cells and promoted tissue regeneration [74]. The development of scaffolds from starch in Flowchart is shown in Fig. 4.

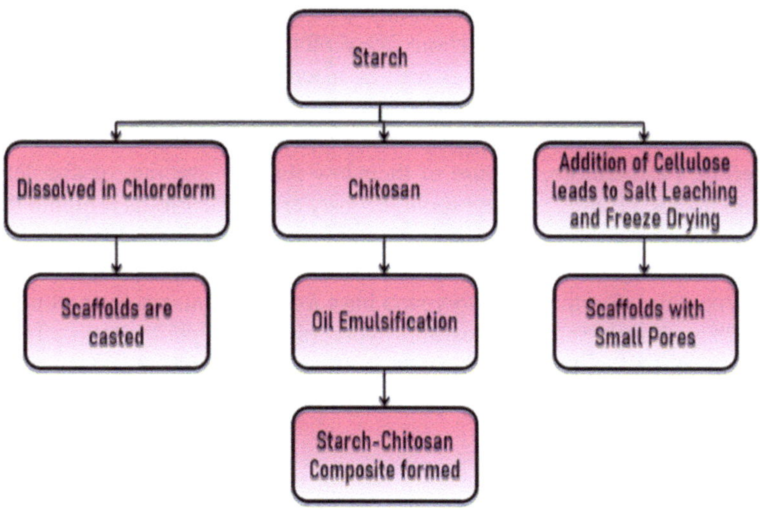

Fig. 4 Development of Scaffolds from Starch

4 Advantages and Disadvantages of Green Methods

The goal of numerous studies is to create techniques that are cost-effective, low-energy, or sustainable to produce various tissues while decreasing the consumption of harmful ingredients. One of the safest ways to create tissues and organs is to apply natural substances taken from plants or animal bones [75]. Synthetic materials exhibit physicochemical properties that may be regulated during the production process. They are also robust, affordable, dependable, frequently easily electro spinnable, and strong. Their lack of cell-recognition sites, however, results in a low affinity for cell attachment.

Thus, natural polymers are favored because they resemble the macromolecular components of the human body [76]. The main benefits of bio-based and biodegradable polymers include reducing GHG emissions, utilizing local resources, reusing by products, and preserving fossil fuels. Without any prior processing, biopolymers print beautifully and have good scent and fat barriers. Polymers made of starch offer excellent surface finishing and antistatic characteristics. Naturally occurring biopolymers (collagen, chitosan, starch, silk) have a better role in cell adhesion, proliferation, and less cytotoxicity, hence are better materials for scaffold preparation for tissue engineering. Polymers made from natural materials outperform semi-synthetic or synthetic polymers in replicating the ECM and interacting with tissues, owing to their high resemblance to the tissue environment [77].

Despite having the above said advantages, natural polymers do have some disadvantages, such as often high production costs (collagen, hyaluronic acid). Its complex structural and chemical makeup, as well as macromolecular complexity, architecture and shape, unpredictable hydration rate, resource limitations, and potential for microbial deterioration are further disadvantages that may limit their use in tissue engineering. Furthermore, because of their low stability, the rate of disintegration and catabolization is high of some spontaneously created scaffolds is higher than the pace of regeneration of the host tissue. Certain natural polymers, like cellulose and chitosan, have limited mechanical characteristics and some have poor processability e.g., polypeptides. Synthetic biopolymers have a few advantages over natural polymers, such as a tunable and engineerable hydrophilic/hydrophobic ratio, a faster rate of breakdown, and better mechanical properties. Another disadvantage of natural biopolymers is degradation, so it can't be used for a long period of time [78].

5 Challenges and Future Scope for Green Methods

Though there has been overwhelming progress in the field of green bioengineering, some constraints have always been present. The acquisition of the natural sources and its behavior has always been a point of criticality. Current focus is being shifted to use perennial plants as source material so that the restriction due to seasonal changes can be avoided [79]. This would be helpful as the production would not be limited by time frames. Also, plant sources are being tried to be extracted from plants which have broader availability, that it is found in multiple regions to make the production uniform [80, 81]. Also, to make the environment a safer place and reduce economical stress, waste utilization has cropped up as a concept over the years, where agricultural wastes are being worked upon to lead to the conformation of different scaffolding frameworks [82].

Also, currently most production processes require high energy, which differs depending on the source material that would be extracted. One of the current aims is to develop methodologies or extract forms that can be molded at less than 100 °C [83, 84]. Despite a lot of challenges in the field, the major challenges that need to be combated include, developing the natural polymer in a way that it imitates all structural, functional, and morphological properties of the said tissue, and can also show protein adsorption as well as inter- and intra-cellular interactions. Also, they must be suitable for drug delivery as well as having a window for chemical and biomechanical modifications [85, 86]. Thus, though the research in the field has made remarkable discoveries with respect to the use of natural polymers, they are now trying to make it more feasible and extraction and development to be easier as well as affordable [87].

6 Conclusion

Current research in this field has dealt with the extraction, development, and modification of natural polymers to produce particulars with concerns to bone and engineering of tissue. The use of polysaccharides and hydrogels has given a new dimension to discovery in this field and has contributed toward reducing environmental damage and quite a bit of the economic load. These new compositions have been seen to have better congregation with human tissues and thus can work toward the replacement of faulty ones. The progression of them into organ transplantation remains to be envisaged into. The multiple methodologies that have been developed for regeneration are still left to be optimally standardized and its reach to be increased. The novel strategies and newly constructed biomaterials have played a pivotal role in turning engineered templates for tissue replacement, into fully functional independent systems. They possess all characteristics including biocompatibility, conductivity of impulses, cellular interactions, and integration with host's native cells. Further research aims at investing the natural polymers into other complicated surgeries and formulations of degradable, injectable, and mechanically stronger components for restoration of functionality of various organs.

References

1. Baroli, B.: From natural bone grafts to tissue engineering therapeutics: brainstorming on pharmaceutical formulative requirements and challenges. J. Pharm. Sci. **98**(4), 1317–1375 (2009)
2. Carleton, P.F., Schachter, S., Parrish, J.A., Collins, J.M., Crocker, J.B., Dixon, R.F., Edgman-Levitan, S., Lewandrowski, K.B., Stahl, J.E., Klapperich, C., Cabodi, M.: National institute of biomedical imaging and bioengineering point-of-care technology research network: advancing precision medicine. IEEE J. Trans. Eng. Health Med. **16**(4), 1–4 (2016)
3. Panebianco, C.J., Dutta, P., Frost, J.R., Huang, A., Kim, O.S., Iatridis, J.C., Vernengo, A.J., x7uWeiser, J.R.: Teaching tissue repair through an inquiry-based learning bioadhesives module. Biomed. Eng. Educ. **4**, 1–4 (2022)
4. Amini, A.R., Laurencin, C.T., Nukavarapu, S.P.: Bone tissue engineering: recent advances and challenges. Crit. Rev. Biomed. Eng. **40**(5) (2012)
5. Anastas, P.T., Warner, J.C.: Principles of green chemistry. Green Chem. Theory Pract. **29** (1998)
6. DeVierno, K.A., House-Knight, T., Whitford, J., Ponnusamy, E., Miller, P., Jesse, N., Rodenborn, R., Sayag, S., Gebel, M., Aped, I., Sharfstein, I.: A method for assessing greener alternatives between chemical products following the 12 principles of green chemistry. ACS Sustain. Chem. Eng. **5**(4), 2927–2935 (2017)
7. De Angelis, G., Medeghini, L., Conte, A.M., Mignardi, S.: Recycling of eggshell waste into low-cost adsorbent for Ni removal from wastewater. J. Clean. Prod. **15**(164), 1497–1506 (2017)
8. Oberbek, P., Bolek, T., Chlanda, A., Hirano, S., Kusnieruk, S., Rogowska-Tylman, J., Nechyporenko, G., Zinchenko, V., Swieszkowski, W., Puzyn, T.: Characterization and influence of hydroxyapatite nanopowders on living cells. Beilstein J. Nanotechnol. **9**(1), 3079–3094 (2018)
9. Dulińska-Molak, I., Mao, H., Kawazoe, N., Chen, G.: Effect of single-wall carbon nanotubes on mechanical property of chondrocytes. J. Nanosci. Nanotechnol. **14**(3), 2459–2465 (2014)
10. Hembrick-Holloman, V., Samuel, T., Mohammed, Z., Jeelani, S., Rangari, V.K.: Ecofriendly production of bioactive tissue engineering scaffolds derived from egg-and seashells. J. Market. Res. **9**(6), 13729–13739 (2020)
11. Huang, R., Zhu, X., Tu, H., Wan, A.: The crystallization behavior of porous poly (lactic acid) prepared by modified solvent casting/particulate leaching technique for potential use of tissue engineering scaffold. Mater. Lett. **1**(136), 126–129 (2014)
12. Patricio, T., Domingos, M., Gloria, A., Bártolo, P.: Characterisation of PCL and PCL/PLA scaffolds for tissue engineering. Proc. Cirp. **1**(5), 110–114 (2013)
13. Johari, N., Fathi, M.H., Golozar, M.A.: Fabrication, characterization and evaluation of the mechanical properties of poly (ε-caprolactone)/nano-fluoridated hydroxyapatite scaffold for bone tissue engineering. Compos. B Eng. **43**(3), 1671–1675 (2012)
14. Zhu, X., Zhong, T., Huang, R., Wan, A.: Preparation of hydrophilic poly (lactic acid) tissue engineering scaffold via (PLA)-(PLA-b-PEG)-(PEG) solution casting and thermal-induced surface structural transformation. J. Biomater. Sci. Polym. Ed. **26**(17), 1286–1296 (2015)
15. Loh, Q.L., Choong, C.: Three-dimensional scaffolds for tissue engineering applications: role of porosity and pore size. Tissue Eng. Part B Rev. **19**(6), 485–502 (2013). https://doi.org/10.1089/ten.TEB.2012.0437
16. Shuai, C., Liu, G., Yang, Y., Qi, F., Peng, S., Yang, W., He, C., Wang, G., Qian, G.: A strawberry-like Ag-decorated barium titanate enhances piezoelectric and antibacterial activities of polymer scaffold. Nano Energy **1**(74), 104825 (2020)
17. Yang, Y., Lu, C., Peng, S., Shen, L., Wang, D., Qi, F., Shuai, C.: Laser additive manufacturing of Mg-based composite with improved degradation behaviour. Virtual Phys. Prototyp. **15**(3), 278–293 (2020)
18. Yang, S., Leong, K.F., Du, Z., Chua, C.K.: The design of scaffolds for use in tissue engineering. Part II. Rapid prototyping techniques. Tissue Eng. **8**(1), 1–1 (2002)
19. Hollister, S.J., Murphy, W.L.: Scaffold translation: barriers between concept and clinic. Tissue Eng. Part B Rev. **17**(6), 459–474 (2011)

20. Alman, B.A., Kelley, S.P., Nam, D.: Heal thyself: using endogenous regeneration to repair bone. Tissue Eng. Part B Rev. **17**(6), 431–436 (2011)
21. Quarto, R., Giannoni, P.: Bone tissue engineering: past–present–future. Mesenchymal Stem Cells Meth. Protoc. 21–33 (2016)
22. Baino, F., Vitale-Brovarone, C.: Three-dimensional glass-derived scaffolds for bone tissue engineering: current trends and forecasts for the future. J. Biomed. Mater. Res. A **97**(4), 514–535 (2011). https://doi.org/10.1002/jbm.a.33072
23. Baino, F., Novajra, G., Vitale-Brovarone, C.: Bioceramics and scaffolds: a winning combination for tissue engineering. Front. Bioeng. Biotechnol. **17**(3), 202 (2015)
24. Boffito, M., Sartori, S., Ciardelli, G.: Polymeric scaffolds for cardiac tissue engineering: requirements and fabrication technologies. Polym. Int. **63**(1), 2–11 (2014)
25. Boffito, M., Bernardi, E., Sartori, S., Ciardelli, G., Sassi, M.P.: A mechanical characterization of polymer scaffolds and films at the macroscale and nanoscale. J. Biomed. Mater. Res., Part A **103**(1), 162–169 (2015)
26. Chiono, V., Sartori, S., Calzone, S., Boffito, M., Tonda-Turo, C., Mattu, C., Gentile, P., Ciardelli, G.: Synthetic biodegradable medical polyurethanes. In: Science and Principles of Biodegradable and Bioresorbable Medical Polymers, pp. 189–216. Woodhead Publishing (2017)
27. Sartori, S., Boffito, M., Serafini, P., Caporale, A., Silvestri, A., Bernardi, E., Sassi, M.P., Boccafoschi, F., Ciardelli, G.: Synthesis and structure–property relationship of polyester-urethanes and their evaluation for the regeneration of contractile tissues. React. Funct. Polym. **73**(10), 1366–1376 (2013)
28. Silvestri, A., Sartori, S., Boffito, M., Mattu, C., Di Rienzo, A.M., Boccafoschi, F., Ciardelli, G.: Biomimetic myocardial patches fabricated with poly (ε-caprolactone) and polyethylene glycol-based polyurethanes. J. Biomed. Mater. Res. B Appl. Biomater. **102**(5), 1002–1013 (2014)
29. Sionkowska, A.: Current research on the blends of natural and synthetic polymers as new biomaterials. Prog. Polym. Sci. **36**(9), 1254–1276 (2011)
30. Caddeo, S., Boffito, M., Sartori, S.: Tissue engineering approaches in the design of healthy and pathological in vitro tissue models. Front. Bioeng. Biotechnol. **26**(5), 40 (2017)
31. Oberpenning, F., Meng, J., Yoo, J.J., Atala, A.: De novo reconstitution of a functional mammalian urinary bladder by tissue engineering. Nat. Biotechnol. **17**(2), 149–155 (1999)
32. Nakase, Y., Nakamura, T., Kin, S., Nakashima, S., Yoshikawa, T., Kuriu, Y., Miyagawa, K., Sakakura, C., Otsuji, E., Ikada, Y., Yamagishi, H.: Endocrine cell and nerve regeneration in autologous in situ tissue-engineered small intestine. J. Surg. Res. **137**(1), 61–68 (2007)
33. Auger, F.A., Berthod, F., Moulin, V., Pouliot, R., Germain, L.: Tissue-engineered skin substitutes: from in vitro constructs to in vivo applications. Biotechnol. Appl. Biochem. **39**(3), 263–275 (2004)
34. Suuronen, E.J., McLaughlin, C.R., Stys, P.K., Nakamura, M., Munger, R., Griffith, M.: Functional innervation in tissue engineered models for in vitro study and testing purposes. Toxicol. Sci. **82**(2), 525–533 (2004)
35. Burdick, J.A., Ward, M., Liang, E., Young, M.J., Langer, R.: Stimulation of neurite outgrowth by neurotrophins delivered from degradable hydrogels. Biomaterials **27**(3), 452–459 (2006)
36. Dennis, R.G., Dow, D.E.: Excitability of skeletal muscle during development, denervation, and tissue culture. Tissue Eng. **13**(10), 2395–2404 (2007)
37. Risau, W.: Mechanisms of angiogenesis. Nature **386**(6626), 671–674 (1997)
38. Lutolf, M.P., Gilbert, P.M., Blau, H.M.: Designing materials to direct stem-cell fate. Nature **462**(7272), 433–441 (2009)
39. Olson, J.K., Boldyrev, A.I.: Ab initio characterization of the flexural B3H8-anion found in the reversible dehydrogenation. Comput. Theor. Chem. **967**(1), 1–4 (2011)
40. Chahal, S., Kumar, A., Hussian, F.S.: Development of biomimetic electrospun polymeric biomaterials for bone tissue engineering. a review. J. Biomater. Sci. Polym. Ed. **30**(14), 1308–1355 (2019)

41. Yu, X., Tang, X., Gohil, S.V., Laurencin, C.T.: Biomaterials for bone regenerative engineering. Adv. Healthcare Mater. **4**(9), 1268–1285 (2015)
42. Li, J.J., Kaplan, D.L., Zreiqat, H.: Scaffold-based regeneration of skeletal tissues to meet clinical challenges. J Mater Chem B. **2**, 7272–7306 (2014)
43. Dimitriou, R., Jones, E., McGonagle, D., Giannoudis, P.V.: Bone regeneration: current concepts and future directions. BMC Med. **9**(1), 1–0 (2011)
44. Iravani, A., Akbari, M.H., Zohoori, M.: Advantages and disadvantages of green technology; goals, challenges and strengths. Int. J. Sci. Eng. Appl. **6**(9), 272–284 (2017)
45. Mukherjee, S., Reddy Venugopal, J., Ravichandran, R., Ramakrishna, S., Raghunath, M.: Evaluation of the biocompatibility of PLACL/collagen nanostructured matrices with cardiomyocytes as a model for the regeneration of infarcted myocardium. Adv. Func. Mater. **21**(12), 2291–2300 (2011)
46. Theocharis, A.D., Skandalis, S.S., Gialeli, C., Karamanos, N.K.: Extracellular matrix structure. Adv. Drug Deliv. Rev. **1**(97), 4–27 (2016)
47. Ullah, S., Chen, X.: Fabrication, applications and challenges of natural biomaterials in tissue engineering. Appl. Mater. Today **1**(20), 100656 (2020)
48. Brodsky, B., Werkmeister, J.A., Ramshaw, J.A.: Collagens and gelatins. Biopolymers online. Wiley-VCH Verlag GmbH & Co. KgaA, Germany (2005)
49. Matthews, J.A., Wnek, G.E., Simpson, D.G., et al.: Electrospinning of collagen nanofibers. Biomacromol **3**, 232–238 (2002)
50. Parry, D.A., Craig, A.S.: Collagen fibrils during development and maturation and their contribution to the mechanical attributes of connective tissue. In: Collagen: Volume II Biochemistry and Biomechanics, pp. 1–23. CRC Press (2018)
51. Xie, J., Lou, X., Wang, X., Yang, L., Zhang, Y.: Electrospun nanofibers of hydroxyapatite/collagen/chitosan promote osteogenic differentiation of the induced pluripotent stem cell-derived mesenchymal stem cells. J. Control. Release Off. J. Control. Release Soc. **213**, e53 (2015)
52. Maia, F.R., Correlo, V.M., Oliveira, J.M., Reis, R.L.: Natural origin materials for bone tissue engineering: properties, processing, and performance. In: Principles of Regenerative Medicine, pp. 535–558. Academic Press (2019)
53. Liu, D., Nikoo, M., Boran, G., Zhou, P., Regenstein, J.M.: Collagen and gelatin. Ann. Rev. Food Sci. Technol. **6**, 527e57 (2015)
54. Croisier, F., Jérôme, C.: Chitosan-based biomaterials for tissue engineering. Eur. Polymer J. **49**(4), 780–792 (2013)
55. Aravamudhan, A., Ramos, D.M., Nip, J., Harmon, M.D., James, R., Deng, M., et al.: Cellulose and collagen derived micro-nano structured scaffolds for bone tissue engineering. J. Biomed. Nanotechnol. **9**(4), 719e31 (2013)
56. Miguel, F.B., de Almeida Barbosa Ju´nior, A., de Paula, F.L., Barreto, I.C., Goissis, G., Rosa, F.P.: Regeneration of critical bone defects with anionic collagen matrix as scaffolds. J. Mater. Sci. Mater. Med. **24**(11), 2567e75 (2013)
57. Venkatesan, J., Vinodhini, P.A., Sudha, P.N., Kim, S.K.: Chitin and chitosan composites for bone tissue regeneration. Adv. Food Nutr. Res. **73**, 59e81 (2014)
58. Rodríguez-Vázquez, M., Vega-Ruiz, B., Ramos-Zúñiga, R., Saldaña-Koppel, D.A., Quiñones-Olvera, L.F.: Chitosan and its potential use as a scaffold for tissue engineering in regenerative medicine. Biomed. Res. Int. **4**, 2015 (2015)
59. Aranaz, I., Gutiérrez, M.C., Ferrer, M.L., Del Monte, F.: Preparation of chitosan nanocomposites with a macroporous structure by unidirectional freezing and subsequent freeze-drying. Mar. Drugs **12**(11), 5619–5642 (2014)
60. Aranaz, I., Mengíbar, M., Harris, R., Paños, I., Miralles, B., Acosta, N., Galed, G., Heras, Á.: Functional characterization of chitin and chitosan. Curr. Chem. Biol. **3**(2), 203–230 (2009)
61. Jana, S., Florczyk, S.J., Leung, M., Zhang, M.: High-strength pristine porous chitosan scaffolds for tissue engineering. J. Mater. Chem. **22**(13), 6291e9 (2012)
62. Costa-Pinto, A.R., Vargel, I., Tuzlakoglu, K., Correlo, V.M., Sol, P.C., Faria, S., et al.: Influence of scaffold composition over in vitro osteogenic differentiation of hBMSCs and in vivo inflammatory response. J. Biomater. Appl. **28**(9), 1430e42 (2014)

63. Li, L., Li, B., Zhao, M., Ding, S., Zhou, C.: Single-step mineralization of woodpile chitosan scaffolds with improved cell compatibility. J. Biomed. Mater. Res. B Appl. Biomater. **98B**(2), 230e7 (2011)
64. Mota, J., Yu, N., Caridade, S.G., Luz, G.M., Gomes, M.E., Reis, R.L., et al.: Chitosan/bioactive glass nanoparticle composite membranes for periodontal regeneration. Acta Biomater. **8**(11), 4173e80 (2012)
65. Shavandi, A., Bekhit Ael, D., Ali, M.A., Sun, Z., Gould, M.: Development and characterization of hydroxyapatite/beta-TCP/chitosan composites for tissue engineering applications. Mater. Sci. Eng. C Mater. Biol. Appl. **56**, 481e93 (2015)
66. Hoque, M.E., Nuge, T., Yeow, T.K., et al.: Gelatin based scaffolds for tissue engineering—a review. Polym. Res. J. **9**, 15 (2015)
67. Awad, H.A., Wickham, M.Q., Leddy, H.A., Gimble, J.M., Guilak, F.: Chondrogenic differentiation of adipose-derived adult stem cells in agarose, alginate, and gelatin scaffolds. Biomaterials **25**(16), 3211–3222 (2004)
68. Aldana, A.A., Abraham, G.A.: Current advances in electrospun gelatin-based scaffolds for tissue engineering applications. Int. J. Pharm. **523**(2), 441–453 (2017)
69. Correia, C., Bhumiratana, S., Yan, L.-P., Oliveira, A.L., Gimble, J.M., Rockwood, D., et al.: Development of silk-based scaffolds for tissue engineering of bone from human adipose-derived stem cells. Acta Biomater. **8**(7), 2483e92 (2012)
70. Huang, X., Bai, S., Lu, Q., Liu, X., Liu, S., Zhu, H.: Osteoinductive-nanoscaled silk/HA composite scaffolds for bone tissue engineering application. J. Biomed. Mater. Res. Part B Appl. Biomater. **103**(7), 1402e14 (2015)
71. Yang, J., Yang, X., Wang, L., Zhang, W., Yu, W., Wang, N., Peng, B., Zheng, W., Yang, G., Jiang, X.: Biomimetic nanofibers can construct effective tissue-engineered intervertebral discs for therapeutic implantation. Nanoscale **9**(35), 13095–13103 (2017)
72. Alca´zar-Alay, S.C., Meireles, M.A.A.: Physicochemical properties, modifications and applications of starches from different botanical sources. Food Sci. Technol. (Campinas) **35**, 215e36 (2015)
73. Balmayor, E.R., Baran, T.E., Unger, M., Marques, A.P., Azevedo, H.S., Reis, R.L.: Presence of starch enhances in vitro biodegradation and biocompatibility of a gentamicin delivery formulation. J. Biomed. Mater. Res. B Appl. Biomater. **103**(8), 1610e20 (2015)
74. Requicha, J.F., Moura, T., Leonor, I.B., Martins, T., Munoz, F., Reis, R.L., et al.: Evaluation of a starch-based double layer scaffold for bone regeneration in a rat model. J. Orthop. Res. **32**(7), 904e9 (2014)
75. Jahangirian, H., Lemraski, E.G., Rafiee-Moghaddam, R., Webster, T.J.: A review of using green chemistry methods for biomaterials in tissue engineering. Int. J. Nanomed. **13**, 5953 (2018)
76. De Vrieze, S., Van Camp, T., Nelvig, A., Hagström, B., Westbroek, P., De Clerck, K.: The effect of temperature and humidity on electrospinning. J. Mater. Sci. **44**, 1357–1362 (2009)
77. Malafaya, P.B., Silva, G.A., Reis, R.L.: Natural–origin polymers as carriers and scaffolds for biomolecules and cell delivery in tissue engineering applications. Adv. Drug Deliv. Rev. **59**(4–5), 207–233 (2007)
78. Abbasian, M., Massoumi, B., Mohammad-Rezaei, R., Samadian, H., Jaymand, M.: Scaffolding polymeric biomaterials: Are naturally occurring biological macromolecules more appropriate for tissue engineering? Int. J. Biol. Macromol. **1**(134), 673–694 (2019)
79. Sharmila, G., Fathima, M.F., Haries, S., Geetha, S., Kumar, N.M., Muthukumaran, C.: Green synthesis, characterization and antibacterial efficacy of palladium nanoparticles synthesized using Filicium decipiens leaf extract. J. Mol. Struct. **15**(1138), 35–40 (2017)
80. Edison, T.N., Lee, Y.R., Sethuraman, M.G.: Green synthesis of silver nanoparticles using Terminalia cuneata and its catalytic action in reduction of direct yellow-12 dye. Spectrochim. Acta Part A Mol. Biomol. Spectrosc. **5**(161), 122–129 (2016)
81. Rostami-Vartooni, A., Nasrollahzadeh, M., Alizadeh, M.: Green synthesis of seashell supported silver nanoparticles using Bunium persicum seeds extract: application of the particles for catalytic reduction of organic dyes. J. Colloid Interface Sci. **15**(470), 268–275 (2016)

82. Wei, Y., Fang, Z., Zheng, L., Tsang, E.P.: Biosynthesized iron nanoparticles in aqueous extracts of Eichhornia crassipes and its mechanism in the hexavalent chromium removal. Appl. Surf. Sci. **31**(399), 322–329 (2017)
83. Naseem, T., Farrukh, M.A.: Antibacterial activity of green synthesis of iron nanoparticles using Lawsonia inermis and Gardenia jasminoides leaves extract. J. Chem. **1**, 2015 (2015)
84. Nasrollahzadeh, M., Sajadi, S.M., Maham, M.: Green synthesis of palladium nanoparticles using Hippophae rhamnoides Linn leaf extract and their catalytic activity for the Suzuki-Miyaura coupling in water. J. Mol. Catal. A: Chem. **1**(396), 297–303 (2015)
85. Kargozar, S., Ramakrishna, S., Mozafari, M.: Chemistry of biomaterials: prospects. Curr. Opinion Biomed. Eng. **1**(10), 181–190 (2019)
86. Mohammadinejad, R., Maleki, H., Larraneta, E., Fajardo, A.R., Nik, A.B., Shavandi, A., Sheikhi, A., Ghorbanpour, M., Farokhi, M., Govindh, P., Cabane, E.: Status and future scope of plant-based green hydrogels in biomedical engineering. Appl. Mater. Today **1**(16), 213–246 (2019)
87. Ebrahiminezhad, A., Zare-Hoseinabadi, A., Berenjian, A., Ghasemi, Y.: Green synthesis and characterization of zero-valent iron nanoparticles using stinging nettle (Urtica dioica) leaf extract. Green Process. Synth. **6**(5), 469–475 (2017)

A. Hazra is a postgraduate student under Dr. Kanthesh BM an Associate Professor and Coordinator Division of Molecular Biology, in JSS Academy of Higher Education & Research. She is currently pursuing her master's in Molecular Biology. Before joining JSS, she was in St. Xavier's College, Mumbai for her undergraduate where she got a double major in Microbiology and Biochemistry. She has also worked as a research trainee at ACTREC, Tata Memorial Centre in Bose Lab, an Integrated Biophysics and Structural Biology Lab. After completing her postgraduate studies, Avipsa wants to pursue Ph.D. and continue with research in the field of Genetics and Cancer. She is a research-oriented individual who is always looking to expand her reach of science.

G. Baradwaj is a postgraduate student under Dr. Kanthesh BM, an Associate Professor and Coordinator of Division of Molecular biology, in JSS Academy of Higher Education & Research. He is currently pursuing his master's in Molecular biology. He has completed his undergraduate in NITTE University Centre for Science Education and Research, Mangaluru in Biomedical Science Hons. After completion of postgraduate studies, Gowrav wants to pursue his Ph.D. and continue with research in the field of Stem cell research and cancer.

A. S. Dhanu is a Research Scholar under Dr. Kanthesh M Basalingappa, an Associate Professor and course coordinator at Division of Molecular Biology, School of Life Sciences, JSS Academy of Higher Education & Research-Mysore. Her current research interest is oriented towards Traditional medicine in Triple negative Breast cancer. In the year 2020, she graduated from JSS College of Arts, Commerce, and Science—Mysore with a UG degree in Biochemistry, Microbiology, and Biotechnology. In the year 2022, she was awarded a postgraduate degree, an M.Sc. in Biotechnology from Shivagangotri, Davangere University—Davangere. The author is a gold medalist, holds the first rank from Davangere University, Shivagangotri, Karnataka, in the academic year 2022. The author is currently pursuing her Ph.D. at Division of Molecular Biology, School of Life Sciences, JSS Academy of Higher Education & Research—Mysore. She is a young Molecular Biology researcher with a passion for invention and innovation.

G. Kuppannan is a Reader of Microbiology at the Noorul Islam College of Dental Sciences in Aralumoodu, Thiruvanthapuram, India, which is affiliated with the Kerala University of Health Sciences in Thrissur. He is focusing his research interests in the fields of microbiology, bacteriology, virology, stem cell research, and innovative medication creation using natural resources. He is now working on an environmental and soil microbiology as main research project that will lead to an epidemiological investigation of organic and inorganic fertilisers used in diverse crop cultivation soil samples. With the current research's integrated inquiry seeking to create the bioconsortium utilising a substantial strains for all crop sorts. Dr. Gobianand Kuppannan earned his doctorate from the University of Madras in 2007, and then worked as a Senior Researcher in the Department of Medical Microbiology at the University of Malaya in Kuala Lumpur, Malaysia (2007–2009), before being awarded as a Researcher by the National Institute of Animal Sciences, Central Government of South Korea (2009–2012). He joined the Department of Microbiology, K. S. Rangasamy College of Arts and Science, Tiruchengode, as an Assistant Professor with a research background.He awarded as Senior Researcher by the Malaysian Agricultural Research and Development Institute in connection Universiti Malaya (2007). In 2009, He was awarded as Post-Doctoral research fellow by the Central Government of South Korea. He has been devoted to academics and research for the last 20 years, and he has produced 23 research articles and two review papers during his career. He is a Life Member of the Indian Association of Applied Microbiology (IAAM), the Indian Association of Biomedical Scientists (IABMS), and the Association of Microbiologists of India (AMI). He contributes as an editor, technical editor, associate editor, and reviewer for a number of prestigious journals for its publications.

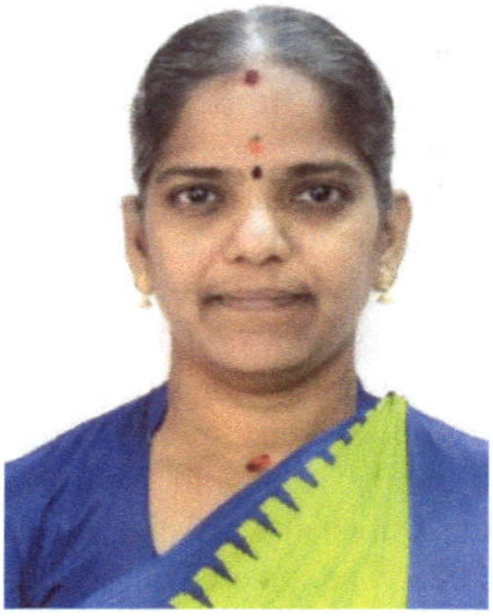

M. Arthanari is serving as a Head and Assistant Professor, Department of Microbiology, Vivekanandha College of Arts and Sciences for Women (Autonomous), Elayampalayam, Tiruchengode, Namakkal (DT), Tamilnadu, India, affiliated under Peiyar University, Salem. Her field of research is mainly bacteriology and virology. Currently her research focuses on gut microbiome associated with diabetes and its complications, mood swinging and behavioral changes. She is mainly working on biodegradation of crude oil spills on marine environment, also she studied phytochemical compounds extracted from macroalgae and worked on their applications. She is aiming her research interest that focusing on environmental degradation of oil spills and concentrating to elucidate, identify and elucidate the bioactive compounds from macroalgal. Dr. Malarvizhi Arthanari rearmed her doctorate from Bharathiyar University in the year of 2014. She is serving as active teaching faculty in Vivekanandha College of Arts and Sciences for women (Autonomous), Elayampalayam, Tiruchengode, Namakkal (DT), Tamilnadu for more than 16 years of experience. She is actively involved in academic responsibilities, and she is serving as Board of Studies Member in reputed Universities/Higher Education Institutes. She has published 22 research/review manuscripts in reputed journals. She is actively connecting and occupied with the Scientific associations/bodies with her allocated responsibilities with very sincere and punctual. She has the role in her present institution as Mentor and Coordinator for various activities in which she is regularly monitoring, serving, and recording to the respective responsibilities to bring it to the maximum height. She is supervising two scholars for their Doctoral Degree, registered with Periyar University.

B. M. Kanthesh is an Associate Professor of Molecular Biology at the School of Life Sciences, JSS Academy of Higher Education and Research, Mysuru, India. His goal of research is to determine the role of RNA Binding proteins in tumor progression and metastasis. Post-transcriptional regulation of gene expression by RNA binding protein is a crucial mechanism in regulating the timing and the amount of expression of genes. Growing evidence indicate that the alteration of the expression and function of RNA binding proteins could potentially play a role in inflammation and cancer. Dr. Kanthesh BM did is Ph.D. from University of Madras (2005), He also did postdoctoral research at the University Malaya, Kuala Lumpur, Malaysia (2007–2009); West Virginia University, Morgantown, USA (2090–2011) and University of Oklahoma Health Sciences, Oklahoma, USA (2011-2014). He received Malaysia Prestigious Bio-Malaysia Gold medal Award (2008). For his research area is Arbovirus infections, in that they done patented work on "Early detection of BK virus using molecular methods". He also Received Dr. Wilson Aruni "Best Research Mentor and Teacher Gold medal Award" from the Indian Association of Applied Microbiology (IAAM) (2018). He has been engaged in teaching and research in Microbiology and Molecular Biology for the past 22 years. He has published over 95 original research papers, 15 book chapters, and 15 review articles. He is also Professional and Scientific Memberships in, American Association for Cancer Research (AACR), Life Member of Indian Association of Applied Microbiology (IAAM), Life Member of Indian Association of Biomedical Scientists (IABMS), Indian Association of Medical Microbiologist (IAMM). He Received Fellowship Award from Indian Association of Applied Microbiology (FIAAM). At present he is a having collaboration with Royal Research Foundation, a research institute in India.

Genetically Induced Biomaterial Advances in Medical Sciences

Eva Kaushik and Rohit Kaushik

Abstract Protein-based monomers are among the most alluring options for achieving improved results with cutting-edge biomaterials since recent advances in bioartificial and microbiology technologies enable the very complicated, accurate design, and fabrication of protein-based biomaterials. These sequences are easily enhanced with bioactive motifs that improve their functionalities, material-host interactions, and basic biological requirements since their sequences are built from architectural protein-based modules. These polypeptides have gained increasing interest for usage in biomedical activities such as cell cultures, bioengineering, separation, and purification design is vital, as well as controlled drug administration because of their versatility, self-assembly behavior, cued, and biocompatibility. The biopolymers discussed in this piece are biomimetic materials and genetic algorithms created from elastin-derived proteins. These genetically programmed polymers' design, manufacturing, and characterization as well as their potential uses in regenerative medicine, bioinformatics, targeted medication delivery, and stem cell therapy will also be covered.

Keywords Extracellular matrix · Tissue engineering · Regenerative medicine · Stem cell · Biomaterial · Protein engineering

1 Introduction

Medical science, biology, chemistry, materials science, and engineering are all interrelated fields in the study of biomaterials. It has been said that "biopolymer is utilized to create systems that substitute an element or a function that the body performs in a secure, dependable, cost-effective, and physiologically reasonable way." An

E. Kaushik (✉)
Information Technology and Engineering Department, Dr. Akhilesh Das Gupta Institute of Technology & Management, New Delhi, India
e-mail: Kaushikeva0026@gmail.com

R. Kaushik
CS and Mathematics Department, Data Analytics Program, University of Illinois, Illinois, USA

R. Malviya and S. Sundram (eds.), *Engineered Biomaterials*, Engineering Materials,
https://doi.org/10.1007/978-981-99-6698-1_4

unworkable resource utilized in a pharmaceutical drug that is meant to engage with biochemical processes is a biological material. Biomaterials are widely employed in medicine, and some are even used in medical equipment. Dental implants, artificial hip joints, intraocular lenses, replacement heart valves, vascular grafts, kidney dialysis, and heart–lung machines are only a few examples of medical technology. All three types of materials—metals, ceramics, and polymers—are utilized as biomaterials [1]. Metallic nanomaterials must be utilized in load-bearing applications and have enough fatigue strength to withstand regular activities such as walking and chewing to be employed in implant placement and prosthetic hip joints. Ceramic biomaterials are often used for applications like joint surfaces in bones and teeth and bone adhesive interface in implants due to their wear resistance and hardness. Polymeric materials, which are frequently used for their suppleness and stability, have also been used to generate low-friction articulated surfaces. Engineers must have a thorough understanding of biomaterials since they must choose their materials and create their designs in accordance with the needs of the biomaterials. The choice of material and design is crucial because either one gone wrong could result in a person's demise [2]. Metals are used in load-bearing applications, but because most metals react with water and the human body contains a lot of water, when metals interact with human tissue, they form oxide layers. Ceramic coating is used on the surface of metals to prevent this because ceramic does not interact with body cells. Medical device design utilizing biomaterials is extremely constrained and requires great accuracy and precision because the design must fit with the following body parts and engage with human body parts to carry out the desired operation. As a result, without knowledge of biomaterials, engineers who create or build biomaterial devices are utterly helpless. In the field of biomedicine, novel nanotechnology with cutting-edge and complex capabilities has found specific applications and can be particularly significant in fields like controlled drug delivery tissue repair and bioengineering in general. Classes of biomaterials are shown in Fig. 1.

For specialized use as a carrier for the dispersal of bioactive chemicals in controlled medication delivery, new biomaterials are required. Only a small portion of treatments administered really reach the sites for which they were intended, making improving the effectiveness of chemotherapy medications a key area of research. Standard chemotherapeutic drugs' negative impacts on cells beyond the target cells limit their efficacy [3, 4]. Because they are created to lessen the noxious effect and enhance the solubility, consistency, pharmacokinetics, and pharmacodynamics of conventional pharmaceuticals, these novel types of drugs can therefore offer significant benefits about the toxic metabolites and degenerative diseases brought on by the distribution of the free drug. By responding to specific cues, efficient pharmaceutical systems are developed to control the dosage of therapies and their associated side effects. This reduces the dispersion of medications into target tissues. Innovative biomaterials with the ability to organize supramolecular at higher orders and self-assemble in a manner that is bio-responsive will therefore be required. Physical aid to the healing process, proactive engagement with biochemical processes, and preservation or enhancement of tissue function are all expected from these tools. The innovation of these systems frequently coincides with an improved understanding

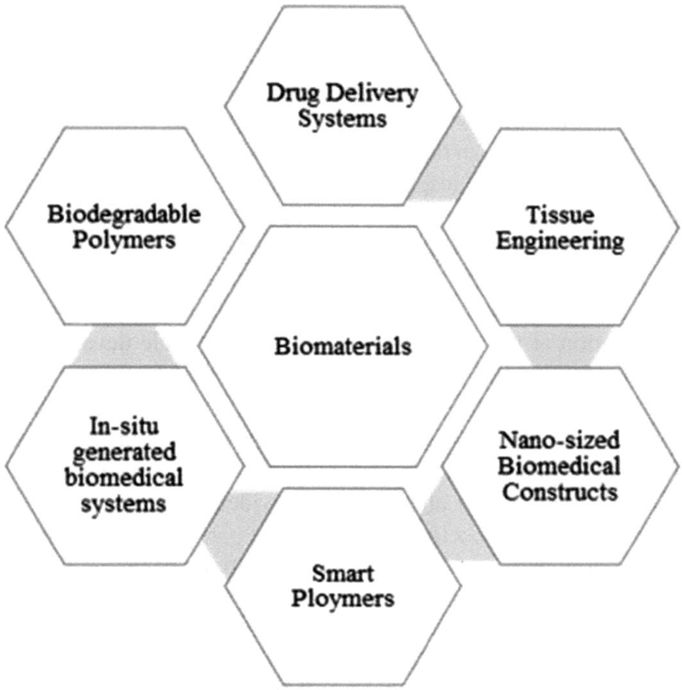

Fig. 1 Biomaterials classification

of the mechanisms dictating the body's systems and advances in genomic science, allowing the use of refined methods that take inspiration from the natural world to mimic the mechanical and biological attributes of the tumor site while ensuring security. Depending on the purpose, biomedical technologies for muscle restoration and medication development could be produced or natural, degradable, or not. Protein-based galactose, which was inspired by naturally occurring proteins discovered in the ECM components, stands out as a leading candidate for use as essential components of cutting-edge medical technology and improved replacements in tissue regeneration because their characteristics are determined by the physicochemical characteristics and loop of constituent monomers. Genetic manipulation still offers significant control over the component organic chemicals used to make delivery, transportation, and skin substitutes, despite the significant advances made in the field of polyelectrolyte materials in terms of advancements in the effectiveness of polymeric materials research methods and lower polydispersity's [5]. According to sequence-structure interactions, the length and concentration of genetically created drug carriers affect a variety of macro, micro, and nanoscale characteristics. Protein polymers made of elastin-like polypeptides are used as well-developed genetic engineering drug carriers as well as personalized supports for tissue restoration. Polymers known as elastin-like peptides (ELPs) have brief pentapeptide repetitions that mimic those in real elastin. They are influenced biologically. The most frequent recurrence of any of these amino acid residues is VPGXG, where X can be any other important ingredient than L-proline. The term "elastin-like recombinases" (ELRs), a recent invention, is already in use to describe genetically modified ELPs. With absolute control over the sequence of amino acids and the possibility to include a range of functions and bioactive sequences, ELRs can be made utilizing DNA technology. Because of their biological and thermal properties, such as biomaterials, degradability, and behavior that is responsive to temperatures and environment, ELRs, which are microbiologically biomaterials, present an appealing alternative for the development of innovative biomedical devices [6, 7]. ELRs also can self-assemble sterically over a temperature slowly referred to as the low ambient temperatures (Tt), which causes a conformational reorganization at the cellular scale and the reversible development of steric effects. The inverse thermal transformation (ITT) of ELRs is the name given to this characteristic. Numerous elements, such as variations in temperature, pH, light, ionic imbalances, etc., can worsen the polymer backbone. Furthermore, because it is controlled by the molecular size and make-up of the recombinant, the activity that responds to stimuli can be altered. They are ideal materials for the majority of materials technology and bioscience applications, particularly for controlled tissue engineering and drug delivery because of their metabolic behavior, flexible mechanical characteristics, heat sensitivity, and capacity for self-assembly. In this study, we explore the application of ELRs as cutting-edge drug delivery systems to safely deliver therapeutic drugs to targeted areas. ELRs' diverse topologies, sequencing, and assembly enable the creation of specialized pharmacological devices such as glucose molecules, nanoparticles, and three-dimensional assemblages. It will also be discussed how monomeric ELRs can be used as aqueous delivery methods for certain peptides that act as knh or cell-penetrating sections. Such systems are suitable for systemic distribution since they are designed to self-assemble in response to environmental cues. They are very beneficial in tumor tissue therapy because of their thermoresponsive activity. ELRs can self-assemble similarly to how they can produce nanoparticles (NPs). Thus, the usage of bases and hydrogels as a 3D framework for drug-delivery devices as well as the applications of NPs as drug nanomaterials will be discussed [8]. Utilization of NPs as drug solid lipid nanoparticles when delivering therapeutic approaches, proteins, or signaling molecules to the skin's surface, or those arrangements that can internalize the medication, bioactive constituents, or even DNA to maximize absorption and persistence effects. Problems with inadequate drug absorption loading can be solved by building macroscopic scaffolds using chemical and physical ELR chain bridging, which also enables controllable localization of the reactive drug release region and rate.

1.1 Regenerative Medicine and Tissue Engineering

Tissue engineered and tissue regeneration are very beneficial methods for the preservation, repair, or enhancement of functional body materials or complete systems after they have exhausted their viability. The development of support systems that are "allowed to exist" when applied directly to injured tissues but are unable to recreate on their own has progressed in recent years in the field of nanomaterials for tissue regeneration. Systems that can interact with the organism and reproduce biological environments are being created using contemporary technology. To build functional tissues, these diverse fields of nanomaterials research are progressively combining the use of cutting-edge scaffolds, the right cells, and bioactive chemicals [9, 10]. For the construction of a functional scaffold, a few components have been identified as being crucial, including the following: Factors include the needed three-dimensional graphene, its porosity, its unique material performance, its water sorption and permeable, epithelial connection via chemical signals, its hydrophilicity and permeation, and its length of residence before host tissue remodeling.

Elastin-like scaffolds have been created with the goal to mimic some of the characteristics of the ECM, including the architectural surface features, appropriate firmness and porosity, improvement of ligand binding and viability, sensitivity to cell signaling, processability, and cued to variations in their environment, among others. The skeletal system, the vascular system, as well as the rejuvenation of soft tissues are among the areas of the body where elastin has been designed to obtain specialized support. These methods are covered in this study. The following discussion examines a few recent cases.

1.2 Cardiovascular System Regeneration

One of the main challenges in tissue healing is managing vascularization. In this context, artificial tissue-engineered scaffolds must allow for appropriate oxygen and nutrient transport as well as the expulsion of catabolic waste. Therefore, it is essential that the scaffold develop an effective network of capillaries and blood vessels. ELR Hydrogels offer a viscous, bioactive substance that is both biocompatible and bioactive despite the wide variety of polymers that are now available. Human adipose tissue-derived SVF cells are transplanted in a variety of ELR hyaluronan, from non-functionalized composite (NF-hydrogels) to a dendrimer gelatin incorporating two cell adherence integrin-mediated motifs, to promote in vitro vascularization (REDV is specific for endothelial cells). The SEM photos of a non-functional system revealed that there was little cell attachment on the flat, elongated parallel granules of the RGD/REDV agarose. The anti-systems had poor cell adherence, while the RGD/REDV polymers had flat, elongated symmetry cells that were well-organized [11]. Moreover, endothelial cells gathered in the RGD/REDV hydrogel but not in the non-functionalized hydrogel, according to CD-31 labeling. Notably, cells contained in NF-hydrogels produced 79% more anti-angiogenic factors than proangiogenic factors, including FGF or IL-8, which were each up 21%. One of these molecules that prevent angiogenesis is vasohibin, along with IGFBP-3, prolactin, depositors, and others. One of the more powerful proangiogenic agents, VEGF, was studied for its release pattern. It of released, and it was found that while it was maintained within the polymeric and was not liberated in the case of NF hyaluronic acid, it was liberated into the precipitation of RGD/REDV hydrogels. RGD/REDV hydrogels sustained blood vessel penetration and were penetrable by host vascularization, according to an in vivo experiment performed on naked rats. NF-hydrogels also showed a decrease in host cell penetration in conjunction with not displaying any additional vascularization. As a result of the VGVAPG sequence's ability to attract monocytes and its vulnerability to proteolysis, RGD/RED hydrogels break down, allowing fibroblasts and endothelial cells to populate the matrix. Even though vascular repair is challenging, some research groups are trying to speed up the procedure by stimulating small-caliber vascular regeneration. ELR-based scaffolds were implanted in the rat's abdomen. To investigate the healing of blood vascular tissue, an ELR-RGD was selected. After the temperature approaches 12 °C, the VPGIG-based hydrophobic sequence of the ELR begins to assemble into nanomaterials up to 10 m in diameter [12, 13].

In vitro, studies showed that HUVEC adhesion to the scaffold was boosted by the RGD motif, not just the ELR by itself. In the in vivo testing, rats that got the ELR-RGD scaffold showed evidence of arterial tissue regeneration, but rats that obtained an ELR scaffold did not. One of the most frequent negative effects of many illnesses is intracranial hemorrhage (ICH), which results in brain edema and harms brain tissue. It is true that patients with ICH have high rates of morbidity and mortality. Cardiac tissue engineering (CTE) and their transplantation is shown in Table 1.

To address the severe symptoms of ICH, it is crucial to create both novel therapeutic approaches and new materials. In this instance, an ELR with RGD has been recommended to manage intracerebral hemorrhage while it is still acute. The effectiveness of the medication was evaluated in vivo tests by measuring the thrombus diameter and the inflammatory response. 6 h after the intracranial hemorrhage, the edema mass was lower in the group that received RGD-ELR as opposed to those getting RGD alone [14]. Moreover, it protected the damaged blood vessels to stop the subsequent loss of blood components, especially IgG. Von Willebrand factor (vWF),

Table 1 Cardiac tissue engineering (CTE) and their transplantation

Cells	CTE strategy	Scaffold	Transplantation
Fetal rat CM	Classical	Gelatine mesh	Avascular graft
Hesc-cm + human umbilical vein EC + embryonic fibroblasts	Classical	Poly-1-lactic acid + Matrigel	Vascularized graft
Neonatal rat CM	EHT	Collagen type 1 from rat tails + Matrigel	Avascular graft
Neonatal rat CM + human umbilical vein EC	Myocardial micro and macro tissue	Cell generated	Vascularized graft
Neonatal rat CM + rat aortic EC	Decellularized heart	Decellularized rat heart scaffold	Organ
Fetal rat CM	Classical	3D porous alginate	Avascular graft
Neonatal rat CM	In vivo flow through pedicle chamber	Fibrin Gel	Flap

a vital part of the coagulation system required for platelet adherence to collagen, was detected using immunohistochemistry testing. Additionally, the immunological response brought on by post-ICH activated microglia was dramatically diminished by the RGD-containing ELR therapy, as well as vWF overexpression.

1.3 Skeletal System

These substances constitute a significant portion of the bone and ligaments that make up the bones and joints, of the body's skeleton. Upon injury or in response to ongoing modifying stimuli, the mature bone does have the innate capacity to repair. Despite this ability to regenerate, some conditions or diseases that affect how efficiently the body can heal itself, such as inflammation brought on by the severity of the injury or oncogenic or pathologic bone problems, may impede the process. Tissue-engineered supports are necessary for these circumstances because of their osteogenic, osteoinduction, and osteogenic components, which provide structural functioning and promote bone healing. Due to their amphiphilic characteristics, which enable them to attach to thermo-responsive hydrogels in physiological circumstances, two distinct ELRs have been produced. The domains from self-aware fibrin typically contain the Arg-Gly-Asp (RGD) themes in addition to the osteoblastic and osteogenic bone technologist and technology protein-2 (BMP-2) motif. Elastase-sensitive action scenes are included in the segments of these ELRs to investigate the rate of degradation of the ELR molecules. While the RGD sequencing helped host cells adhere to the substrate and increase cell survival, the BMP-2 component encouraged osteoblast development. The deterioration of the artificial scaffold's organic composition occurred concurrently with bone regeneration, which was used in this work to show how host bone remodeling affects how rapidly the scaffold dissolves. In vivo experiments on female Zealand white bunnies revealed hematologic marrow and fibroblast layer infiltration and multiplication, suggesting a large potential for bone graft recovery and outstanding biocompatibility. The V40C2 ELR block, a pentapeptide made of valine and disulfide, was previously combined with human recombinant BMP-2 to increase the osteoblastic and osteoinductive capabilities of the ELR scaffold. At 37 °C, this ELR block is identified [15, 16]. The main goal was to maintain the bioactivity of BMP-2 in the fusion protein. It was demonstrated that despite the protein's diminished biocompatibility, however, neither ELR gene product nor the commercially available pure BMP-2 lost any of their cytocompatibility. BMP-V40C2 showed stronger calcium deposition than the control during a 21-day treatment regimen, as well as increased mRNA concentrations of the transcription factors Osterix (OSX), which is only expressed in osteoblasts. The production of artificial bone tissue utilizing hydrogel hydroxyapatite (Si-HA) scaffolding and ELRs is an additional intriguing method. Due to its osteoblasts, osteogenic, and accessibility, which encourage bone development and osseointegration, HA—a vital element of vertebrate bone and teeth—is widely utilized in orthopedic, dental, and cranial operations. The 3D degradable scaffolds employed in this investigation were coated with two bioactive ELRs that were produced and analyzed. It has been demonstrated that these ELRs mediate biochemical mechanisms at the interface of minerals and have a strong affinity for calcium phosphorus, much like HA. The production of the osteoblast markers OC and ALP, however, showed that the Si-HA-SNA15-containing scaffold was particularly effective at directing the BMSCs toward the osteoblastic lineage when osteoblast differentiation was assessed [17]. All studied scaffolds allowed cells to colonize and proliferate, but the functionalized ELR generated a biochemical pathway that markedly enhanced the scaffolds' ability to stimulate osteoconduction. Unlike bone, some tissues, like enamel, cannot regenerate. An ELR, a tiny protein that stimulates enamel remineralization, contains statherin. In contrast to the untreated substrates, the resultant STNA15-ELR formed mineral platelets after 8 days of growth in a mineralization solution at 37 °C. ELRs may therefore be crucial for preserving the mineral's stability. On day 8, SEM pictures verified the presence of a sodium fluoride and fluorapatite combination [18]. Fluorapatite was also visible in TEM images. This exploratory study might lead to the creation of a useful enamel biomineralization substance. One of the most challenging academic problems has long been developing the finest biomechanically feasible skeletal scaffolding for tissue engineering. To produce a hydrogel with sufficient wear resistance over time, cysteines and the domain peptide were included in this case. After 21 days in culture, the human mesenchymal stem cells (hMSCs) on the biomaterials appeared to still be alive, proving that contraction rather than erosion was the cause of the loss of volume.

These hMSCs also had higher levels of osteogenesis markers such as osteopontin and osteocalcin than hMSCs that were placed on glass slides. Chondrocytes were put to the gel after 28 days of growth and displayed 86% more vitality, respectable architecture, and elevated activity of alkaline phosphatase than the control [19, 20]. As we mentioned at the beginning, one of the most intriguing applications of elastin-like particles is the creation of scaffolding that mimics biological structures.

The final design of a 3D spheroid using an ELR chemically coupled to polyethyleneimine (PEI) was compared with that of a 2D monolayer culture, even though 2D cultures cannot replicate the physiological 3D environment. To create 3D microspheres as a 3D model of culture, these scientists plated hybrid fat-derived stem cells (hASCs) on normal styrene coated with the ions with a positive charge ELR-PEI. Viability tests were performed after both had been grown in differentiation media, and they revealed the same number of cells as for the 2D layer. Scaffolds or scaffolding for glycosidic bonds are also produced as composites to regulate bone regeneration because of the advantages mentioned above. The mechanical properties of the scaffolds can be improved by the stiffness and flexibility of ELRs [21]. Four distinct composites with various ELR/collagen ratios were examined in this experiment. The composite with the highest ELR/collagen content (18 mg/mL) had the highest levels of ALP activity, OCN transcription, and stiffer matrix even though all these composites contained hASCs that were proliferating and differentiating. Composites constructed entirely of chondro had substantially less matrix and took longer to mature than those with lower ELR/collagen ratios. Currently, one of the most prevalent causes of discomfort and incapacity in the elderly is the degeneration of joint cartilage.

Due to collagen's perivascular nature, it is challenging to identify articular structures that are harmed because of trauma or aging joints because collagen has a limited capacity to rebuild [22, 23].

Like this, hydrogels are intriguing scaffolds because of their mechanical properties and ease of manipulation, especially those that can self-gelate under physiological conditions and may precisely integrate a device in situ. Available internal and external ELR-HA hydrogel foundations have been created to compare scaffolds with varying component ratios. The stiffness of the supports remained consistent for all the ratios that were taken into consideration. When HA concentration in chondrocytes increased, there was a dose-dependent rise in the transcription of cartilage marker genes like aggrecan, SOX9, and type 2 collagen. The importance of these chondrogenic indicators must be stressed because chondrocytes are the only cells present in the knees that have the charge of controlling the extracellular matrix. The hydrogel's chondrocytes also showed increasing levels of type 1 collagen, a marker of an undesirable fibrous connective tissue, and metallopeptidase-13, a marker of matrix disintegration and remodeling, when the concentration of HA was reduced [24]. Moreover, chondrogenic multiplication was boosted in biomaterials with lower HA concentrations, whereas sGAG (sulfated glycosaminoglycan) deposition was promoted in hydrogels with higher HA contents. This system (5% HAhydrogel) was shown to be the best scaffold since it had the highest peak of cartilage indicators, the largest sGAG deposit, the least amount of cellular proliferation, and the fewest of type 1 collagen and fibrous tissue phenotypic development.

2 Advanced Medicine Delivery System from ELRs

The various ELR architectures are examined in this chapter as the building blocks for numerous drug delivery systems. In these subchapters, devices spanning between monomeric ELRs to amphoteric ELRs that can create nanoparticles have been discussed. The chapter is concluded with macroscale ELRs for the delivery of drugs in the forms of depots or hydrogels.

2.1 Drug Nanocarriers Made of Monomeric ELRs

Drug nanocarriers made of monomeric ELRs have been suggested as prospective drug carriers because they improve the size and durability of medical molecules, which are typically tiny and have short half-lives because they promote quick glomerular filtration. Simple elastin-based polypeptides called monomeric elastin-like recombinases can be modified and used as drugs [25]. As a result, they can be used as solvent delivery systems when connected to biopharmaceuticals or as pharmacokinetics boosters when incorporated into fusion proteins that make medications. A monomeric ELR carrier provides unique benefits over the free drug by improving the pharmacokinetics and drug load pharmacokinetic profile. Incorporating target peptides, such as knh or cell-penetrating motifs, as well as reactive areas useful for the strong covalent conjugation of medications is also made possible by genetically designing elastin-like recombinases. ELRs can be engineered to self-assemble in response to external stimulation, and they can be employed to deliver therapeutic medications to injured tissue both systemically and locally. Beneficial clinical drug delivery methods include monomeric ELRs coupled to drug-delivery molecules with known therapeutic potential [26, 27]. A drug is joined to a macromolecule during the thermoplastic coupling process; the most popular options are proteins and synthetic substances. The insertion of sensitive residues that can be used to further compound other medications, like the hydroxyl group in cysteines or the amine groups in lysines, is made possible by the recombinant origin of ELRs.

If the loading is a protein, it is possible to create the ELR-load complex through genetic engineering. The most common disease in the world is cancer, however, the majority of therapeutic drugs employed for treating it have molecular masses of less than 500 Da, are hydrophilic, have significant toxicities, have poor pharmacokinetic, and have low therapeutic indices. The therapeutic index is the ratio between the lowest dose necessary to achieve the desired result in 50% of a society and the dose necessary to kill 50% of a community (effective dose). Interaction among triad is shown in Fig. 2.

Additionally, several chemotherapy medications with anti-cancer effects, including Paclitaxel, Doxorubicin, Camptothecin, or Melphalan, have a significant drawback known as toxicity in normal healthy cells. Future uses of the thermal capacities of ELRs for therapeutic hypothermia have a decent probability of becoming successful. ELRs are anti-cancer medication delivery devices [28]. Increasing the local, regional, or bodily temperature above the physiological level (37 °C) to roughly 42 °C is known as hyperthermia. Under a specific transition temperature (Tt), elastin-like recombinamers are disordered and water-soluble, but as the temperature rises over this Tt, they amass and collect. This property makes them intrigue as drug delivery vehicles. Because ELRs are liquid at 37 °C, the body temperature, can move freely. These compounds become intractable when they circulate in tumor

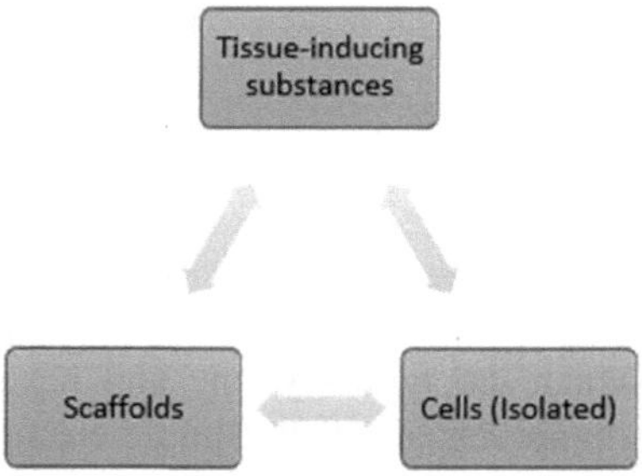

Fig. 2 Interaction among the triad resulting in the variation in functional, survival, and aesthetic qualities

tissue and concentrate in the tumor, even though hypothermia is characterized as the substance being warmer than the transition temperature. This suggests that using the phase change of these polypeptides could facilitate the localized drug delivery to tissues (Tt). The plasma membrane is one of the most challenging problems a system for drug delivery must solve. The plasma membrane, a permeable barrier found on living cells, is essential to their existence and ability to operate. Consequently, the main challenge to assuring intracellular drug administration is getting a bioactive molecule across the plasma membrane. Small compounds can pass right through the lipid bilayer of this membrane, while protein-based therapies must first penetrate the membrane via mobile transport. To increase the effectiveness of a medicinal agent's entry, cell-penetrating peptides (CPPs), among the most well-liked and effective vectors for improving cellular absorption, could be used. Surface functionalization of the ELR with aCPP can facilitate non-specific cell viability of the load via pathways, simplifying the cell survival of ELRs. CPPs are 5–30 peptide polypeptides that can pass through cell membranes. As a result, a variety of CPPs have been included in the bioactive of ELR drugs to enhance the targeted distribution of anticancer drugs in the tumor [29, 30]. CPPs are also used to reduce the rapid breakdown and insufficient infiltration of tumor cells by these drugs. Frankel and Green, who discovered that the HIV activator of transcription (TAT) peptide can traverse cell membranes and was effectively absorbed in vitro, reported the first CPP in 1988.

2.2 *Nanoparticles Based on ELR as Pharmaceutical Delivery*

The utilization of nanoparticles as having to cut drug delivery systems is covered in this chapter along with a variety of ELR-based architectures that can produce them. We discuss a variety of devices, including triblock copolymers, cell-specific ligands, hybrid diblock ELRs, and ELRs-blocks.

2.3 *Amphiphilic ELRs Nanoparticles*

Amphiphilic ELR-based diblocks can self-assemble to produce structures like nanoparticles, suggesting that there are several ways to induce or reinforce their self-assembly depending on the polymer design. The hydrophobic portion of the particle has also received several bioactive domains that are accessible at the subatomic particle shell. ELRs can release drugs under regulated conditions in reaction to pathological changes in temperature because of their thermosensitivity. Anticancer therapy benefits from heat generation at the tumor site because ELR-based methods are efficient at creating well-organized nanoparticles at the infection site. The pH differential between normal tissue and the cancerous environment is another factor that aids in the therapy of cancer. Callahan and colleagues created an ELR block copolymer (ELRBC), specifically [VG7A8]-80/[VH4]-100, to create nanoparticles with a pH-responsive component that can also be stabilized with Zn 2 + ions [31, 32]. When compared to ELRBCs made of [VG7A8]-64/[V]-120, a non-pH-sensitive ELR block copolymer, in vivo penetration, and distribution in tumors. The solid tumor tissues usually have extracellular pH ranges of 6.2–6.9 as opposed to the normal tissue's circulatory pH range of 7.2–7.4, therefore the ionic intensity ELR-based polyamide microspheres break down at this lower pH. The 60 pentapeptide repeats in the diblock, which include the hydroxyl group as a guest residue, form a hydrophobic block, whereas the 60 core protein residues, which contain glutamine and glycine in a 1:1 ratio, form a hydrophilic block. Then, Silaffin R5 and the hydrophilic region of the ELR were genetically connected. When phosphorus ions are present in phosphate-buffered saline, the ELR-R5 is crosslinked, resulting in the formation of silica nanoparticles. Afterward, silicic acid is added to the solution of the made-up templates to make the silica polycondensation. Many researchers have used the strain-promoted azide-alkyne cycloaddition (SPAAC) method to enhance the formation procedure by connecting the chain at the particle's outer shell. To ensure peripheral crosslinking, a clearly defined ELR was created by joining the more hydrophobic block to a lysine-rich area. The arginine amine groups were converted into azide units using the diazo-transfer method, and the NPs were subsequently stabilized through exergonic crosslinking using the chemical bis-cyclooctyne. Additional approaches involve employing membrane transporters made of polymer nanoparticles and genetically altered drug receptors after the prescription has been coupled to the receptor [33]. The basis of this innovative encapsulation method, developed by MacKay and colleagues, is the high selectivity between a given molecular medication and its protein sequence targets attached to the corona of protein polymer nanoparticles. The drug is administered gradually once it has been securely attached to the carrier. Rapamycin (Rapa), an immunosuppressive macrolide antibiotic, is applied over stents to prevent organ transplant rejection. Rapa's anti-proliferative qualities, constrained dispersion, low oral absorption, and quick systemic clearance made it the ideal candidate for these studies. 23 days after the initial dose, the rodents lost over 15% of their entire body mass, demonstrating the high toxicity of free Rapa. The drug for the prostate xenograft mouse model showed superior antitumor action, showing notably lower cytotoxicity in contrast to free Rapa, without exhibiting any symptoms of behavioral issues or body weight loss. Eventually, diverse proteins and small molecules will be encapsulated, targeted, and released using this technique.

2.4 *Modern Techniques for the Synthesis of ELR Nanoparticles*

More and more sophisticated devices have been developed that use ELR-based technologies to recognize complex NPs that can transport a variety of small molecules to improve regulated medication delivery systems. One of the best ways to incorporate bits into macro material is by self-assembly, which can transform nanoscale into devices with unique capabilities. Coworkers created the H6-ELR-CP triblock polymer in this situation, which consists of a polypeptide block with thermal responsiveness, a hex histidine block with the potential to bind ionic species, and the capsid peptides of the cowpea erythema mottle virus, for whom the self-assembly may be pH-dependent [34, 35]. New nanocarriers built on recombinamers that mimic elastin require inventive ways. Active targeting abilities can also be added thanks to the customization of these ELRs with fibroblast ligands. To facilitate active cell-targeting, ELR with a polymeric acid tail (ELR-D) and then added epidermal growth factor EGF (ELR-D-E). The resultant nanoparticles have been shown to have a specific biological interaction, and in vitro results have been encouraging using A549 cells, a normal lung cancer epithelium line that pushes the limits of the EGF receptors. It looked at how to distribute suicide genes using a plasmid that contains the Rna for the toxin PAP-S.

With excellent selectivity for MCF-7 cancer cell lines as opposed to a MUC-1-negative tumor line, this technique effectively transported the plasmid into tumor cells while protecting normal human cells. This aptamer-ELR vector's ability to transfect cells, as demonstrated by micropinocytosis uptake, opens the door for the use of ELRs in suicidal genome therapy to cure cancer and further opens the door for their use in the biomedical sector.

2.5 *In-Situ Therapeutic Depots*

One form of treatment where a prolonged local treatment of the tumor is required for success is anticancer therapy. Therapeutic drugs must navigate several biological barriers on their way to the tumor to be successful, and systemic medicines usually have high rates of toxicity as a side effect. Thank goodness, localized anticancer therapy exposure and reduced drug storage inside the tumor can be achieved through intertumoral injection, minimizing negative side effects. An enticing alternative to local delivery is the creation of a biocompatible substance that may be injected as a medication delivery deposit at room temperature. The thermosensitive ELR sequence in this instance carried a biopolymer into a tumor when soluble, and because of air temp coacervation, produced a depot in situ [36]. Intertumoral radiation therapy has a very good response rate in solid tumors, especially aggressive prostate cancers (brachytherapy). By introducing radioactive "seeds" into or close to the tumor, this technique gives the tumor a significant dosage of radiation while limiting unneeded exposure to the neighboring healthy tissues. Despite the method's overall benefits, existing seed implantations are permanent and can only be used in the treatment of low- or intermediate-grade malignancies. They are ineffective for treating high-grade tumors. The development of polymeric devices that function in conjunction with irradiation has undergone several attempts. It is preferable to avoid the problems with conventional brachytherapy. Avoiding the issues with traditional brachytherapy is preferred-based systems that have recently been taken into consideration for topically applied delivery, particularly in the context of creating new capabilities considering the prevalence of eye-related disorders. Lacritin is closely connected to a group of contemporary elements that can enhance corneal mitogen action and encourage tear protein secretion (Lacrt) [37, 38]. To create a monomer that can create a base on inter cache at temperatures close to and protect Lacrt-mediated cell-signaling pathways, Lacrt was fused to an ELR made up of 96 repeats of the pentapeptide VPGVG. In a mouse model based on inter defect, the therapeutic efficacy of the extended infusion of IL-1a and/or tumor necrosis factor was assessed. IL-1 inhibition significantly reduces PTA-related indicators like synovial inflammation, bone repair, and chondro degradation, according to in vivo research. Yet, lowering TNF-, whether alone or in conjunction with IL1Ra, impairs bone structure and hastens articular cartilage degeneration.

This novel method dramatically minimizes clearance from the joint space while boosting the impacts of IL-1a and decreasing injection dose and frequency by combining xELR with IL-1a for continual intra-articular administration.

2.6 Hydrogels

These highly hydrated hydrophilic, three-dimensional, insoluble viscoelastic polymers can swell when in contact with liquid but do not dissolve it. Physical substances that are activated by a range of inputs, such as pH, temperature, or even sonar waves, can create these networks. Cross-linking during chemical processes can also produce them. These nanogels are structurally stable due to crosslinking or physical interaction between polymer strands. Composites can get over some delivery restrictions because they can be placed close to the area of issue and distribute drugs locally for a long time. The window for reducing systemic effects just at target site is expanded by this restricted exposure. Enzymatically cross-linked ELRs are one of the most cutting-edge components of hyaluronan delivery systems for drugs because they could release bioactive compounds in response to specific environmental stimuli. As was already mentioned, elastolytic activity has been connected to pathological disorders like atherosclerosis, cystic fibrosis, and chronic wounds. Crosslinking was made possible by the addition of microbial transglutaminases. After being subjected to *P. aeruginosa* or human granulocytes, the final hydrogels emitted the recombinant green fluorescent protein (eGFP) that was added during the gelation process over a longer period (PMNs). Similar results were obtained once an electron transport system was added [39]. ELRs are a potential biopolymer for making genuine pass depots or biodegradable under specific circumstances because of their adaptability. On the other hand, noncovalent cross-linking frequently leads to insufficient structural and mechanical stability, rendering them unsuitable for a variety of applications. An injectable, repairable, dimeric crucifix silicone gel with properties halfway in the range of coacervates and more durable chemical hydrogels was created to increase the results in the applicability of these biomaterials. These results demonstrate the effectiveness of nanomaterials for the in vitro gradual release of a reference protein as well as the injectable biomaterial's transport properties. The composite's two components interacted secondarily, causing the ELR-collagen combo hydrogels to release doxycycline more rapidly than gelatin hydrogels. Four bacterial strains that are frequently found in clinical settings were used in bioassays to determine the cytocompatibility of the doxycycline discharged from the hydrogels.

Staphylococcus aureus multidrug testing was carried out using the function. Knowing that all hyaluronic acid produced doxycycline, which was effective against all but four of the bacterial strains tested, it was demonstrated that the hydrogels had no impact on the medicine's bioactivity. Combining ELR with collagen hydrogels can hasten the healing of wounds by reducing postoperative infection and extending the duration of an antibiotic's local site of action.

3 Background Related Work

[1] Frappier et al. [40]**;** Atomically small material alterations enable a variety of nanoelectronics devices, such as resonant tunneling diodes (RTDs), quantum well-integrating photodetectors (QWIPs), subatomic good lasers, and heterojunction junction effect transistors (HFETs). For the development and creation of such devices, a fundamental knowledge of electron propagation in such wavelengths is required. Based on a basic nonequilibrium electron transport theory, NEMO is a flexible tool for designing and studying quantum devices. EMO was linked with the parallel processing evolutionary algorithm application PGAPACK to modify the properties of the structural materials to match a certain set of experimental data. A numerical experiment is done to generate structural parameters like layer thicknesses and doping levels to evaluate an experimental current–voltage characteristic. We find that the basic and doping characteristics of the heNEMO model parameters closely reflect the genetic algorithm. Synthesis is achievable with such precise agreement between theory and technique are known.

[2] Schloss et al. [41]**;** In order to create new kinds of allergy vaccinations, DNA Fragment was used in allergens research. Details on the genetic sequencing and makeup were made available. One technique that is widely employed in the creation of antimicrobial peptides that meet certain T- or B-cell epitopes. A new technique for developing hypersensitive vaccinations that can provide a better and much less replication of the epitopes is genetic engineering allergen synthesis. To show how well these hypo-allergens reduce allergenicity, numerous stimulating skin and nasal testing have been employed. The vaccines' capacity to generate anti-allergen IgG antibodies and sustain T fibroblast activity have both frequently been shown, despite the diminished immunoglobulin E (IgE)-binding reactivity. The main hypoallergenic have been polypeptide segments and tetrameric birch allergen structures.

[3] Olorunniji et al. [42]**;** Atomic size differences in the materials enable devices like resonance tunneling diodes (RTDs), quantum well-infrared photodetectors (QWIPs), classical good lasers, and heterojunction field effect transistors (HFETs). Such heterostructure devices call for an in-depth knowledge of electron transport in quantum states for their design and optimization. This problem is addressed by the overall design and analysis tool for nanodevices known as the Nanoelectronics Modeling Tool (NEMO). NEMO was used with PGAPACK, a library of parallel processing neural network methods, to improve structural and material properties. The quasi-band effects in the longitudinal and transverse dimensions as well as the Wkb charge personality are included in the electron transport calculations displayed here.

[4] Tang [43]**;** A popular biocompatible substance for stem-cell cartilage healing is elastin-like polypeptides (ELPs), a class of artificial polypeptides with special characteristics. The pentapeptide Val-Pro-Gly-Xaa-Gly makes up ELPs, and Xaa can be any other amino acid besides Pro that passes through an inverted temperature phase change. They can dissolve in water when it is below their transition temperature (T t). Yet, when the temperature of water surpasses their T t, they congeal. This study evaluates the rheological characteristics of an uncross-linked ELP below and above its T t as well as the effectiveness of ELP in triggering chondrogenesis in vitro. Recombinant DNA techniques were used to create an ELP with a T t of 35 °C.

4 Application

4.1 Peptide-Protein Interactions

Under conditions of kinetic and thermodynamic equilibrium, the intrinsic self-recognition and processes of biomolecules enable the formation of organized structures just at the nanoscale or even at the macroscopic scale. Peptides and peptidomimetics are significant because they may allow for the rational dissection of biological pathways to produce new medications, materials, catalysts, and other products. Van der Waals forces, hydrogen bonds, and electrostatic attraction are a few instances of interactions that regulate self-recognition or self-assembly processes. They also produce secondary structures like -helix, -sheet, and polyproline II helix, which are necessary for many biological functions. Here, we give some recent and noteworthy examples of how design has been successfully used to create functional structural themes. These investigations are crucial for comprehending the fundamental relationships governing biological operations and the emergence of numerous diseases. The types of secondary configurations that peptides take on during self-assembly play a key role in the properties of biomaterials, such as the contacts, encapsulating, non-covalent contacts, or covalent interactions that are eventually useful for drug delivery applications [40].

The definition of the chemical self-assembly process is the bidirectional and spontaneous grouping of some substances (small or macromolecules, peptides, or proteins) into a well-defined and stable configuration while in thermodynamic equilibrium. Both healthy and pathological processes result in the formation of several beneficial sub-cellular structures. A few examples include amyloids linked to disorders including Alzheimer's, Parkinson's, diabetes, amyotrophic lateral sclerosis, hemoglobin, and membrane transport channels. In the field of protein interaction (PPI), where the connection may be connected to a conformational change that initiates a biological pathway, the supramolecular association created by low intramolecular pressures is an essential mechanism both in chemical and biological recognition. For instance, the growth of fiber-like structures is driven by hydrogen bonds or interactions. If the dissociation energy is compared, hydrogen bonds only require 10–65 kJ/mol for homolytic breakdown while covalent bonds take 100–400 kJ/mol. Multiple interaction sites are frequently combined throughout the assembly process. Many organic compounds have been found to be capable of forming supramolecular polymers under specific conditions [41, 42]. The fibrillation process is likely the most remarkable biological example of peptide self-assembling, such as amelogenesis in Alzheimer's disease. Peptides represent a different class of identity and self-recognition molecules. Moreover, recently, peptidomimetics. Since the early 1990s, researchers have been studying self-assembling peptides. Self-assembling peptides can be used to create novel biological materials, coatings for surfaces and semiconductors, as well as a new group of antibiotics to prevent and treat antibiotic resistance. A self-assembling peptide (SAP) system uses organic or synthetic scaffolding that can display a variety of cell-interactive constituents in spatially defined networks. They do this by using biomolecules self-assembly or naturally occurring organized aggregates. Hydrophobic interactions, hydrogen bonds, van der Waals forces, and electrostatic contacts all contribute to the stability and loss of energy of protein self-assembled structures. The length of the peptides, the percentage of hydrophobicity, the I amino acid sequence, and the duration of the self-assembling process are all factors that influence the self-assembling processes. To be used in a variety of biomedical fields, these innovative SAP-based materials usually make hydrogels that may mimic the extracellular matrix and have synthesized scaffolds that may contain different cell-interactive components. Comparison of peptide array is shown in Table 2.

4.2 Silk Protein Sericin in Biomaterial

A natural polymeric protein called silk sericin is derived from the Bombyx mori silkworm. Sericin can be retrieved and used in other ways during the many processes used to make raw silk and textiles. Additionally, sericin recovery lessens the impact of silk production on the environment. Sericin protein has advantages due to its characteristics. The protein is UV-resistant, antimicrobial, and resistant to oxidation [43]. It also readily collects and releases moisture. To create products with better qualities, sericin proteins can be polymerized, cross-linked, and combined with other biomolecules, notably synthetic polymers. Additionally, the protein can be used to enhance or coat both natural and synthetic fibers, fabrics, and products. Bioplastics, biomedical

Table 2 Peptide array comparison

	Product	Technology	Support	Scanning
Pepscan	PepChip microarray	Standard Fmoc SPPS	Glass slides	Standard screening with antibody-based detection
LC Sciences	PepArray micro-array Chip	Photolithography using microfluidics	Chips	Microfluidic system
PEPperPRINT	PEPperCHIP microarray	SPPS amino acids scanned with the electrical field produced via laser printer	Chips/glass slides	Antibody-based detection and in-house services for staining
JPT	PepSpot Microarray	Standard Fmoc SPPS using the SPOT synthesis technique	Cellulose sheets	Electrotransfer blotting to PVDF membrane
Intavis	Celluspots microarray	SC2 Technique	Glass slides	Antibody-based detection directly on an array

components, articles-forming polymers, functional membranes, fibers, and textiles can all be made from sericin-modified components and sericin composites.

4.3 *Modulation of Bone Cells' Responsiveness to Plasma Treatment with Starch-Based Biomaterials*

Investigated were the effects of o2 radio wave glow discharge (rfGD) on the interfaces of various starch-based biomaterials (SBB) as well as the effects of protein adsorption on controlling bone-cell activity. Both simple and complicated protein systems used fibronectin, vitronectin, and bovine serum albumin. Surfaces coated with RfGD demonstrated greater hydrophilicity and interface energy as compared to untreated SBB. In cornstarch-based biodegradable polymeric blends, polycaprolactone (SPCL; 30/70 weight percent), poly (ethylene alcohol (SEVA-C; 50/50 weight percent), and methylcellulose (SCA; 50/50 weight percent) were investigated. The maximum degree of change was seen in SCA and SCA reinforced with 10% fluorapatite (HA) after rfGD treatment [44]. On SCA, it was discovered that cell adhesion and multiplication were increased compared to untreated surfaces, and the plasma modification had no effect on SCA + 10%HA. BSA, FN, and VN single solutions enhanced cell adherence to SCA surfaces, and a similar impact was observed for ternary systems. Moreover, when compared to the untreated surfaces, SEVA-C treated with blood shows better adhesion and proliferation. MG63 cell multiplication was clearly aided by plasma modification, even though adherence to treated and uncontrolled SPCL was roughly equal. It was demonstrated that the variance in MG63 cells' morphology on SEVA-C surfaces was primarily controlled by the peptide system, whereas it was primarily regulated by the treatments on SPCL surfaces.

4.4 Tissue Engineering

In vivo tissue-engineered uses elastin-like biomaterials, however novel ELRs are often developed and enhanced to work better. For instance, the interaction and preservation of mouse pluripotent cells were achieved using four distinct constructs. Most of these inventions were composed of the ELR sequence (APGVGV) 12, the cell-adhesion sequence (RGD), as well as the CBP peptide, a lawfully binding peptide (iPSCs). The best method for encouraging mouse iPSCs to adhere to and develop on untreated culture plates was determined to be the ELR construct (ERE-CBP), which contains both the lower perceived and the CBP sequence. Additionally, it was shown that CBP peptide enhanced the RGD gene's capacity to permit cellular adhesion [43]. Additionally, compared to ERE alone, ERE-CBP showed much-increased expression of the embryonic marker Oct3/4 and alkaline phosphatase activities. There were no obvious differences between the construction without RGD and ERE-CBP. Some application of tissue engineering is shown in Fig. 3.

In opposition to CBP itself, which could not bind cells but successfully sustained pluripotency, RGD showed its ability to increase cell attachment but not pluripotency maintenance. The ERE-CBP study demonstrated how important it is for RGD and CBP to coexist to preserve pluripotency. This unique recombinant ECM protein may be best used in future in vivo tissue-engineered studies. One of tissue engineering's primary goals is likely to be the establishment of an extracellular environment that is similar to the natural one. The growth and cell adhesion of fetal mouse fibroblasts were examined in one study using GPG, or fiber of an ELR that had a double-hydrophobic triblock structure (NIH-3T3). The scientists developed three different hydrophilic structures with the RGD motif known to enhance cell adhesion properties: GPG, GPG-KAAK, and GPG-KAAK-GRGDS. They are all possibly suitable for coating polystyrene surfaces after being cleaned with D-PBS, and over 90% of these nanostructures remained intact. A noteworthy technique in the field of tissue engineering is the development of cytocompatibility adhesives [45, 45]. Unexpectedly, mELY16 outperformed glass coverslips in terms of adsorption capacity. Both constructions (ELY and mELY16) outperformed BSA and commercial sealant in terms of good adhesion (>2 MPa under dry conditions and 0.24 MPa under wet conditions, respectively) This new biomaterial could therefore have a significant impact on regenerative medicine and tissue engineering.

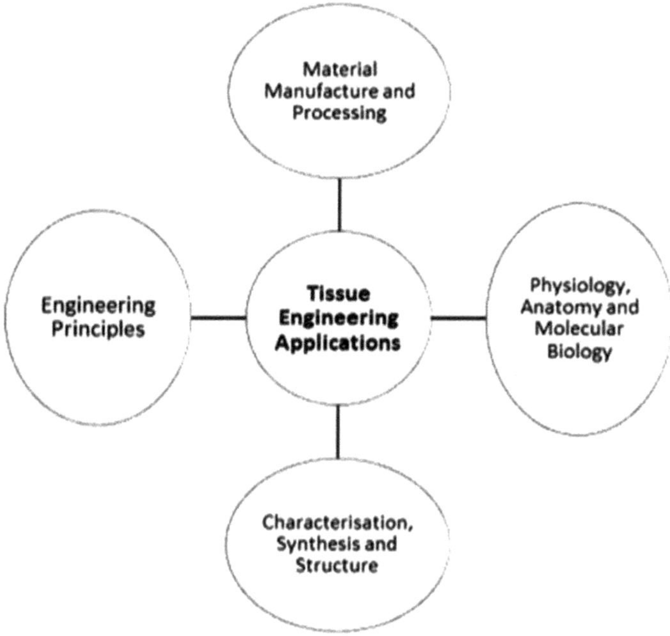

Fig. 3 Applications of tissue engineering

5 Case Study

5.1 Elastin-Based Biomaterials that Have Undergone Genetic Modification

One of the important features that distinguish the genetic algorithm from other algorithms like evolutionary algorithms is that its main search operators are crossover and mutation. The two main genetic operators utilized in genetic algorithms are crossover and mutation. The objective of the current study was to investigate the impacts of genetic algorithm operators such as mutation and crossover (P c & P), population size (n), and the number of iterations (I) used to forecast the biological material extruder's lowest hardness (N). The second-order polynomial regression equation developed for the extrudate characteristic hardness in terms of independent factors such as temperature, screw speed, the fish proportion of the feed, and feed moisture content was employed as the objective function in the GA analysis [47]. A simple genetic algorithm (SGA) based on crossover and mutation operators was used in the current study. A rank-based fitness selection algorithm was built into a C language program for an SGA. The maximum number of iterations and inhabitants was set to 100. It was discovered that a medium community of 50, iterations of the algorithm, and the lowest hardness values were feasible. According to the Pareto charts, P c

Table 3 Biomaterial state for myocardium tissue engineering

Biomaterial	Physical State
Natural	
Hyaluronan	Solid
Alginate mesh	Solid
Collagen	Liquid/gel
Synthetic	
PGA and copolymer with PLA	Solid
PPS	Solid

was found to have a greater impact when the population was below 50, while P m was crucial when the population was below ten. The threshold values for a crossover chance of less than 50% and a mutation probability of less than 5%. Physical state of biomaterial for myocardium tissue engineering is shown in Table 3.

5.2 Engineered Biomaterials to Enhance Stem Cell-Based Cardiac Tissue Engineering and Therapy

One of the biggest causes of death in the globe is cardiovascular disease. After a myocardial infarction, cardiac tissue containing deceased or damaged cardiac cells downstream of the occluded channel does not recover because adult cardiac cells have a restricted capacity for proliferating. The heart becomes feeble when non-functional fibrotic scar tissue takes the place of the original cardiac tissue [48]. Researchers are looking into the possibility of stem cells to regenerate damaged cardiac tissue because host cardiac cells have a restricted capacity for proliferation. This project has made tremendous progress. Currently, there is no agreement on the ideal stem cell type, matrix material, or microenvironmental stimuli for cells. This article provides a summary of the various biofunctional compounds and bioactive matrices that, when combined with stem cells, have demonstrated promise for the renewal and reinforcement of cardiac tissue. Engineered biomaterials are also used in cardiac tissue engineering, which uses stem cells and biomaterial scaffolds to generate tissues in vitro that can be used for drug testing or eventual implantation. This study highlights the advantages of repairing injured myocardium with biomaterials and stem cells and gives a quick overview of the characteristics of these polymers that make them such useful instruments in the field.

5.3 Recombinant Cellulose Crosslinking Protein

Cellulose-binding domains have been discovered in a variety of cellulases and proteins that lack hydrolytic function. The ability to alter the surfaces and mechanical properties of paper by a cellulose-binding domain was examined. Two cellulose-binding domains from *Clostridium cellulovorans* were combined to create a polypeptide that crosslinks cellulose (CCP). The mechanical characteristics of the synthetic polyfunctional cellulose-binding proteins were studied when it was submerged in Whatman cellulose filter paper after being expressed in *E. coli*. The extracted protein improved the tensile strength, fragility, Young's modulus, and energy to break off the treated paper. Furthermore, it was demonstrated that filter paper improved in water repellency following treatment with cellulose crosslinking protein. For the creation of a new class of paper-modifying materials, it is intriguing that cellulose-binding motifs can bind to cellulose under a variety of environmental conditions without the necessity for chemical interactions [49, 50].

6 Challenges

- **Bio Ceramic Challenge**

The long-term durability of tissue-engineered constructions after implantation depends on obtaining and sustaining a blood supply. Third-generation bioactive, resorbable composites have been employed in studies to improve the vascularization of a soft tissue construct made from regenerated tissue. To find the appropriate concentration of bioactive particles, the production of vascular endothelial factors was boosted while cell proliferation was observed using a rat fibroblast model. Adult rats were given subcutaneous implantation of samples of the biodegradable, active composite mesh. Embryonic stem cells and other cells were introduced into the composite meshes.

- **Stem Cell Engineering**

Stem cell therapy, a revolutionary twenty-first-century approach to healthcare, is built on the foundation of stem cells, which have the capacity for self-renewal and the potential for multiple lineages. The ability of cells to differentiate into a range of body tissues is widely known. The challenge is in getting stem cells to differentiate into the required lineage, obtaining higher purity populations of the desired cell phenotype, confirming that the cell line has no cancer-causing potential, and then implanting the selected cell bloodline in a way that ensures the organisms will multiply and replace or enhance the function of pathology or aging tissues. These cells can be collected, grown, and, if necessary, integrated into a tissue construction before being administered utilizing minimally invasive methods to the location that requires healing [51]. The tissue at the location will regenerate because of the activated stem cells. There are issues with accessibility, infrasound (in the marrow, there is one cell type for every 100,000 cells), and diminished developmental capacity with aging for some stem cell types. The control of stem cell development is a potential use of mechanistically customized bioactive materials that may also serve as the rationale for stem cell-based therapy.

- **Bio Ceramics: Control of Infection**

Due to bacterial adherence to biomaterials, which causes biomaterial-centered infection and insufficient tissue integration, several medical devices have a short lifespan. Globally, the problem of the spread of bacteria resistant to common antibiotics is getting worse. The increased incidence of chronic foot and limb lesions, which frequently require amputation, is a substantial barrier. The bacteria and inflammation that lead to cellular dedifferentiation must be controlled locally. AgBG concentrations of 0.05–0.20 mg per microliter of culture medium suppress the development of these bacteria. AgBG has a quick antibacterial effect in addition to being bacteriostatic. At AgBG concentrations of 10 mg ml1, a full bactericidal action was evoked within the first hours of incubation. 45S5 Both Bioglass and BG had no impact on the viability or proliferation of bacteria. Only the leaching of Ag+ ions from the glass matrix is responsible for AgBG's antibacterial activity [52]. Analytical measurements exclude any involvement of pH changes, ionic strength changes, or the dissolution of other ionic species from the biomaterials in AgBG-mediated bacterial death. The patterns of Ag+ dissolution from AgBG in the addition and exclusion of bacteria lend credence to the notion that the microbes are storing silver. According to XRD patterns, FTIR spectra, and ICP data, the Ag-doped gel glass exhibits the same bioactive behavior as bioactive gel glasses that have Approval from the FDA for use in bone repair.

- **Predicted in vitro testing of the biomaterials' and nanoparticles' toxicity and biocompatibility**

For both ethical and budgetary reasons, society is worried about the existing reliance on in vivo testing on animals to assess the safety of novel biomaterials, TE structures, and nanoparticles. The development of reliable and cost-effective predictive in vitro experiments based on human cells is a major challenge for the twenty-first century. There are a variety of issues with in vitro research that need to be resolved to ensure relevance to prospective in vivo applications [53]. A mature cell phenotypic in culture must, first and foremost, be representative of the same form of human cells in vivo. Cell culture research currently frequently uses eternal cell lines, which are unable to express the intricate protein arrays necessary for mature phenotypes. Third, while testing, the mature cell phenotype in the cultured cells must be preserved. The drug being tested might not be toxic enough to solely kill the cells; instead, it could cause de-differentiation and impair the tissues' ability to repair. These demand monitoring of cell morphologies, preferably in situ. Fourth, the in vitro testing should contain details regarding the molecular biological alterations that the chemical causes in the cells. Fifth, the in vitro tests ought to be able to do statistical analysis to discern between minute alterations in the cell population. Sixth, it's important to take pricing and usability into account.

7 Future Scope

Again, a biological basis to produce a generation of biomaterials provides the scientific foundation for the protein engineering of scaffolding for TE and for in situ organ regeneration and repair, preferably via minimally invasive surgical procedures. Every one of these cutting-edge approaches has significant financial advantages and could assist in resolving the problems related to caring for an aging population. A new class of marker nanoparticles that are uniquely adapted for specific patients and illness scenarios should be conceivable [54]. With the increased accessibility of predictive analytics approaches, innovative plans for universal healthcare can be developed. Until recently, this idea would have seemed ludicrous. It was difficult to envisage a material that living tissues wouldn't trash just 40 years ago. Millions of people have benefited from this therapeutic reality, and it most likely will inspire new ideas in the years to come.

8 Conclusion

It has also been studied how to distribute therapeutic drugs safely and effectively to certain places, as well as a variety of ELR-based architectures and their prospective application as superior drug delivery systems. Whether utilized with transgenic biopharma in a liquid delivery mechanism or as a pharmacokinetics enhancer, either via chemical linkage or by creating fusion proteins of the drug, monomeric elastin-like proteins that are implicated have been explored. These polypeptides can increase the half-life of medications when utilized as therapeutic systems. Genetic engineering also makes it possible for better cellular reception of nanostructures with therapeutic functions in damaged tissues by inserting targeting peptides like knh or fibroblast domains. Additionally, the right amino acid selection enables the existence of reactive groups that are beneficial for the covalent connections formed during the conjugation of molecules with proven therapeutic efficacy, with drugs and peptides being among the most popular choices. ELRs can be designed to identify in response to external stimuli, making them suited for systemic and local delivery of therapeutic drugs to the injured tissue. ELRs that are available at body temperature after thermoresponsive transition have been demonstrated to accumulate in cancerous tissue because of local overheating. In fact, by utilizing the porosity and perfusion of the vasculature, therapeutic hyperthermia has improved the local distribution of ELR prodrugs into tumor tissues. According to the diverse device structures, the biological applications of NPs—in particular, ELR-based nanoparticles having signaling pathways that can interact with organisms on their surfaces—have been studied. When used to heal brain injury, and persistent skin damage, or to promote bone regeneration, several of these have shown positive results. Lacritin nanomaterials are just one type of NP that have been employed as therapeutic agents and are coupled to proteins that really are visible on their surface. They support the integrity of the epithelium and corneal tissue healing. Last but just not least, studies on 3D ELR have been conducted. These include depots, hydrogels, and macromolecular carriers. Because they provide highly localized treatment, these are particularly promising for anticancer therapy. To create injectable depots for venous distribution, ELRs can collaborate. This enhances therapy options, localized anticancer medicine exposure, and drug stores in the tumor. As macromolecular carriers, fusion-based ELR depots can carry curcumin or glucagon-like peptide-1, which represents a substantial development in the treatment of t2dm and pro therapy, etc. [55, 56]. The benefit of utilizing biomaterials that have been chemically and structurally copolymerized is their ability to be positioned close to the area of interest and disperse embedded drugs over a long period of time locally, minimizing any negative effects. Because they enable the regulated release of several bioactive compounds like hormones, antibodies, signaling pathways, and other therapeutic biomolecules with possible biological uses, hydrogels containing a domain sensitive to proteolytic disintegration have drawn particular interest. In addition, the use of ELR-based 3D, notably hydrogels, has been investigated for tissue-engineered with an emphasis on skeletal regeneration. Numerous strategies have been examined, including the utilization of physiological conditions that replicate the affected body part, such as the musculoskeletal or circulatory systems, or the use of special scaffolds created from amphiphilic ELRs to restore soft tissues. Because they can introduce various bioactive patterns into the specifically provided sequencing or because they possess the appropriate cells for the regeneration of functional tissues, these creative and biomimetic scaffolds are very significant for tissue engineering. Studies have demonstrated that ELRs are excellent candidates for tissue engineering because of the synergy that occurs when they are combined with bicomponent scaffolds like ELR-collagen, ELR-silicon replacement perovskite, or ELR-hyaluronic acid hydrogel scaffolds. For bone reconstructive surgery, new methods based on precisely defined combinations of osteogenic materials, osteoinductive chemicals, and/or osteoblastic cells are highly desired. The objective is to effectively reproduce the original tissue without going through the bone harvesting process [57, 58]. To incorporate all the aforementioned qualities, a carefully controlled hybrid material must yet be developed. Further developments require cooperation between experts from a variety of disciplines, including stem cell biology genetics, diagnostic devices, material sciences, medicine release, synthetic biology, and surgery.

9 Discussion

Materials that have potential for use in biomedical and therapeutic applications are referred to as biomaterials [59]. They could be any or all the following, depending on the application:

- **Bioactive**: They actively interact with bodily systems. This is mostly considered in applications for tissue repair.

- **Bioinert**: This phrase implies that the substance in question does not affect or upset the body system.
- **Bioresolvable/Biodegradable**: To prevent environmental pollution, the material used for clinical or medicinal usage must be easily decomposed after use.
- **Biocompatible:** The material must be appropriate for its use, which is the most crucial requirement of all [60, 60]. The material that is best for a certain application will have mechanical characteristics that are like those of the real body component that is being used or replaced.
- **Mechanical Properties**: The mechanical qualities of a biomaterial reveal the molecule's biocompatibility with the body system by describing the material's toughness, strength, and ductility.

References

1. Taberna, T.B., Zandstra, P.W.: Hematopoietic stem/progenitor cells and engineering: Producing T cell competent blood progenitor cells from pluripotent stem cells in scalable dynamic suspension culture. Cytotherapy, **25**(6). https://doi.org/10.1016/s1465-3249(23)00385-7
2. Acknowledgment to the reviewers of Applied Microbiology in 2022.: Appl. Microbiol., **3**(1), 76–78 (2023). https://doi.org/10.3390/applmicrobiol3010006
3. Ahearne, M.: Mechanical testing of Hydrogels. Mech. Hydrogels, 73–90 (2022). https://doi.org/10.1016/b978-0-08-102862-9.00003-8
4. Zhang, Q.: Multi-scale instrumented indentation of Hydrogels. Mech. Hydrogels, 91–118 (2022). https://doi.org/10.1016/b978-0-08-102862-9.00001-4
5. Gao, M., Xie, X., Huang, T., Zhang, N., Wang, Y.: Glutaraldehyde-assisted crosslinking in regenerated cellulose films toward high dielectric and mechanical properties. Cellulose **29**(15), 8177–8194 (2022). https://doi.org/10.1007/s10570-022-04785-2
6. Singh, R., Malhotra, A., Bansal, R.: Synthetic cytotoxic drugs as cancer chemotherapeutic agents. Med. Chem. Chemother. Agents, 499–537 (2023). https://doi.org/10.1016/b978-0-323-90575-6.00010-7
7. Medicinal Chemistry of Chemotherapeutic Agents. (2023). https://doi.org/10.1016/c2020-0-03892-2
8. Importance of human anatomy and physiology in biomaterial science.: Biomater. Sci., 1–44 (2022). https://doi.org/10.1515/9783110655377-001
9. Peutzfeldt, A., Dahl, J.: Editorial announcement. Biomater. Investig. Dent., **9**(1), 110–110 (2022). https://doi.org/10.1080/26415275.2022.2152939
10. Paulussen, F.M., Grossmann, T.N.: Peptide-based covalent inhibitors of protein–protein interactions. J. Pept. Sci., **29**(1), (2022). https://doi.org/10.1002/psc.3457
11. Numata, K.: Biopolymer material and composite. Biopolym. Sci. Proteins Pept., 205–246 (2021). https://doi.org/10.1016/b978-0-12-820555-6.00009-4
12. Processing sericin.: Sustainable Use of Byproducts from Silk Processing, pp. 39–99. https://doi.org/10.1002/9783527828760.ch2
13. Peutzfeldt, A., Dahl, J.E.: Editorial. Biomater. Investig. Dent., **8**(1), 180–180 (2021). https://doi.org/10.1080/26415275.2021.2010366
14. Nanoparticle-functionalized poly-methyl methacrylate bone cement for sustained chemotherapeutic drug delivery.: OrthoMedia. (2021). https://doi.org/10.1302/3114-210284
15. González de Torre, I., González-Pérez, M., Alonso, M., Rodríguez-Cabello, J.C.: Elastin-like recombinamers (elrs) for biomedical applications. Soft Matter Biomed. Appl., 205–235 (2021). https://doi.org/10.1039/9781839161124-00205

16. Miguel, G.A., Álvarez-López, C.: Extraction and antioxidant activity of sericin, a protein from silk. Braz. J. Food Technol., **23**. https://doi.org/10.1590/1981-6723.05819
17. Pharmacodynamics.: Definitions. (2020c). https://doi.org/10.32388/yfuyn4
18. Wei, G.: Characterization techniques of protein and peptide nanofibers: Self-assembly kinetics. Artif. Protein Pept. Nanofibers, 99–118 (2020). https://doi.org/10.1016/b978-0-08-102850-6.00005-x
19. Krishnan, U.M.: Protein and peptide nanostructures for drug and Gene Delivery. Artif. Protein Pept. Nanofibers, 279–325 (2020). https://doi.org/10.1016/b978-0-08-102850-6.00013-9
20. Brodin, P.: Cell microbiology interview. Cell. Microbiol., **23**(3), (2020). https://doi.org/10.1111/cmi.13288
21. Chemotherapeutic perfusion.: Definitions. (2020). https://doi.org/10.32388/x77s0i
22. Chemotherapeutic topical agent.: Definitions. (2020b). https://doi.org/10.32388/hz6rdt
23. Peutzfeldt, A., Dahl, J.E.: Editorial. Biomater. Investig. Dent., **7**(1), 158–158 (2020). https://doi.org/10.1080/26415275.2020.1831296
24. Petersen, H.S.: Non-specific protein adsorption on polyetheretherketones -challenges for application as biomaterial. Mod. Approaches Mater. Sci., 1(4). https://doi.org/10.32474/mams.2019.01.000119
25. Lary, J.M.: Hyperthermia and Teratogenicity. Hyperth. Cancer Treat., 107–126 (2019). https://doi.org/10.1201/9780429266539-6
26. Horbett, T.A.: Protein adsorption to Hydrogels. Hydrogels Med. Pharm., 127–171 (2019). https://doi.org/10.1201/9780429285097-6
27. Levanič, J., Gericke, M., Heinze, T., Poljanšek, I., Oven, P.: Stable nanocellulose gels prepared by crosslinking of surface charged cellulose nanofibrils with di- and triiodoalkanes. Cellulose, **27**(4), 2053–2068 (2019). https://doi.org/10.1007/s10570-019-02947-3
28. Qiang, Y., Patel, A., Manas-Zloczower, I.: Enhancing microfibrillated cellulose reinforcing efficiency in epoxy composites by graphene oxide crosslinking. Cellulose **27**(4), 2211–2224 (2019). https://doi.org/10.1007/s10570-019-02916-w
29. Diehl, B.G., Watts, H.D., Kubicki, J.D., Regner, M.R., Ralph, J., Brown, N.R.: Correction to: Towards lignin-protein crosslinking: Amino acid adducts of a lignin model Quinone Methide. Cellulose **26**(11), 7025–7025 (2019). https://doi.org/10.1007/s10570-019-02493-y
30. Biophysical investigations of molecular mechanisms of chemotherapeutic agents action. 1. chemotherapeutic and antiviral agents (review).: Biophys. Bull., (42), (2019). https://doi.org/10.26565/2075-3810-2019-42-02
31. Nutt, D.J., Nestor, L.J.: Pharmacodynamics of addictive substances. Oxf. Med. Online (2018). https://doi.org/10.1093/med/9780198797746.003.0006
32. Tarr, G.E.: Manual methods for protein/peptide sequence analysis. Protein/Pept. Seq. Anal.: Curr. Methodol., 35–48. https://doi.org/10.1201/9781351076081-2
33. Krutzsch, H.C.: Protein/peptide sequence analysis by mass spectrometry. Protein/Pept. Seq. Anal.: Curr. Methodol., 161–180 (2018). https://doi.org/10.1201/9781351076081-9
34. Hayashi, R.: Enzymatic methods of protein/peptide sequencing from Carboxyterminal End. Protein/Pept. Seq. Anal.: Curr. Methodol., 145–160 (2018). https://doi.org/10.1201/9781351076081-8
35. Mazin, A.V., Mazina, O.M.: RAD51 and dmc1 recombinases. Mol. Life Sci., 1009–1016 (2018). https://doi.org/10.1007/978-1-4614-1531-2-67
36. Manian, A.P., Aldred, A.K., Lenninger, M., Bechtold, T.: Alkali pretreatments and crosslinking of Lyocell Fabrics. Cellulose, **24**(9), 3991–4002 (2017). https://doi.org/10.1007/s10570-017-1384-9
37. Aranda-Espinoza, H., Adlerz, K.: The cell as an inspiration in biomaterial design. Biomater. Mech., 103–116 (2017). https://doi.org/10.1201/9781315152585-5
38. K, S.:. Tamoxifen: Pharmacokinetics and pharmacodynamics. Open Access J. Pharm. Res., **1**(8), (2017a). https://doi.org/10.23880/oajpr-16000143
39. Shahbaz, K.: Tamoxifen: Pharmacokinetics and pharmacodynamics. Open Access J. Pharm. Res., **1**(8), (2017). https://doi.org/10.23880/oajpr-16000143

40. Frappier, V., Duran, M., Keating, A.E.: PixelDB: Protein-peptide complexes annotated with structural conservation of the peptide binding mode. Protein Sci. **27**(1), 276–285 (2017). https://doi.org/10.1002/pro.3320
41. Schloss, P.D.: Preprinting Microbiology. (2017). https://doi.org/10.1101/110858
42. Olorunniji, F.J., Rosser, S.J., Marshall Stark, W.: Purification and in vitro characterization of zinc finger recombinases. Methods Mol. Biol., 229–245 (2017). https://doi.org/10.1007/978-1-4939-7169-5-15
43. Tang, J.C., Rudolph, S., Cepko, C.L.: Viral delivery of GFP-dependent recombinases to the Mouse Brain. Methods Mol. Biol., 109–126 (2017). https://doi.org/10.1007/978-1-4939-7169-5-8
44. Voziyanova, E., Anderson, R.P., Voziyanov, Y.: (2017). Dual recombinase-mediated cassette exchange by tyrosine site-specific recombinases. Methods Mol. Biol., 53–67 (2017). https://doi.org/10.1007/978-1-4939-7169-5-4
45. Liu, W., Tuck, L.R., Wright, J.M., Cai, Y.: Using purified tyrosine site-specific recombinases in vitro to rapidly construct and diversify metabolic pathways. Methods Mol. Biol., 285–302 (2017). https://doi.org/10.1007/978-1-4939-7169-5_18
46. Biopolymers, **106**(1), (2016). https://doi.org/10.1002/bip.v106.1
47. Vaclaw, M.C.: Linking protein-catechol-chitin interactions to the physical properties of beetle elytral cuticle, a multicomponent biomaterial. In: 2016 International Congress of Entomology. (2016). https://doi.org/10.1603/ice.2016.95574
48. Abdalla, M.: Important factors influencing protein crystallization. Glob. J. Biotechnol. Biomater. Sci., 025–028 (2016). https://doi.org/10.17352/gjbbs.000008
49. Li, W., Su, X., Zhong, Q., Liu, Z., Cai, Y., Yao, J.: Influence of reaction conditions on the self-assembly of the natural silk sericin protein. Microsc. Res. Tech., 80 (3), 298–304 (2016). https://doi.org/10.1002/jemt.22666
50. David, K.: Advances and challenges with fibrous protein biomaterial designs. Front. Bioeng. Biotechnol., **4**. https://doi.org/10.3389/conf.fbioe.2016.01.02613
51. Chemotherapeutic.: Encycl. Nanotechnol., 524–524 (2016). https://doi.org/10.1007/978-94-017-9780-1_100176
52. Market scenario of biomaterial-based devices.: Smart Biomater. Devices, 207–217 (2016). https://doi.org/10.1201/9781315371559-10
53. Badylak, S.F.: In Host response to biomaterials: The impact of host response on biomaterial selection. Elsevier/AP, Amsterdam (2015)
54. Xiang, Z., Anthony, R., Lan, W., Runge, T.: Glutaraldehyde crosslinking of Arabinoxylan produced from corn ethanol residuals. Cellulose **23**(1), 307–321 (2015). https://doi.org/10.1007/s10570-015-0828-3
55. Donev, R.: Advances in protein chemistry and Structural Biology. Academic Press is an imprint of Elsevier, Waltham, MA (2015)
56. Fusion protein (fused protein, chimeric protein, fused peptide, fusion peptide, hybrid protein).: Dict. Genomics, Transcr. Proteomics, 1–1 (2015). https://doi.org/10.1002/9783527678679.dg04683
57. Makvilay, S., Sanongraj, W., Khamwichit, W.: Turbidity removl using silk sericin and silk sericin powder as coagulant aid. Advanced Materials Research **931–932**, 276–280 (2014). https://doi.org/10.4028/www.scientific.net/amr.931-932.276
58. Dunn, B.: Preface. Protein Pept. Lett. **20**(1), 1–1 (2013). https://doi.org/10.2174/092986613804096883
59. Tatemastu, K.: Utilization of transgenic silkworms for recombinant protein production. J. Biotechnol. & Biomater., *s9*(01), (2012). https://doi.org/10.4172/2155-952x.s9-004
60. Protein adsorption at the biomaterial–tissue interface.: An Introd. Biomater., 139–156 (2011). https://doi.org/10.1201/b11561-11
61. Pereira, P.: Peptide and protein application in tissue repair and regeneration. Pept. Protein Deliv., 291–311 (2011). https://doi.org/10.1016/b978-0-12-384935-9.10012-4

Eva Kaushik is an IT professional with extensive experience in research. As a researcher, she is well-versed in varied domains, including fintech, haptics, astrophysics, financial economics, bioinformatics, and genomics. She has published research manuscripts, book chapters, and articles in renowned journals supporting her interest. She has been a founding pioneer at Dexigner, a co-founder at FiCord community, and even augmented the FinTech niche via various initiatives. Her belief in hard work and perfection has led her to prominent positions. She has served as chairperson of IEEE ADGITM, an educational activist in IEEE USA, a core committee member of the IEEE QT3 Series, an editorial coordinator at IEEE WIE DS, and a public speaker at YPLO. She was chosen by Microsoft for the start-up program "Binance Build for Bharat" and received the IEEE WIE Affinity Group Award. She is serving as an advisor at FrontForumFocus (a non-profit organization). She has volunteered with MHRD, WWF, IEEE MOVE India, and the World Youth Alliance. Along with this, she is spreading cognition about technology at different platforms for the advancement of society. Precisely, being a woman of serious potential, she believes in achieving accuracy and perfection.

Rohit Kaushik is a graduate student at the University of Illinois, majoring in data analytics. His research interests lie in the fields of machine learning, mathematical computation, statistical analysis, artificial intelligence, and data mining. His previous research has focused on the integration of technology to diminish societal issues, leading to betterment. Rohit has enormous potential and a burning desire to accomplish his pursuits, along with a creative mind filled with new ideas.

Adding new heights to his research, he has been selected by the Illinois Department of Public Health (IDPH) to research the motor vehicle data linkage project in collaboration with the Illinois Department of Transportation (IDOT). He has been awarded the title of Director of Technology at "DREAMtorous," a business networking conclave and brand of Global Conflux of Stalwarts (GCS). His dynamic personality leads him to new opportunities, like getting invited as a guest speaker at the IEEE Student Branch (2021) to deliver a session on data science, machine learning, and deep learning. He has amazing analytical and management skills, that directed him to serve as a community manager at non-profit communities to support and enhance their development using precise statistical data insights.

Biomimetic Approaches for Biomaterials Development

Sudipta Choudhury, K. R. Arjun, M. N. Ramesh Bharadwaj, M. Maghimaa, and Kanthesh M. Basalingappa

Abstract In recent times, the concept of biomimetics has gained widespread acceptance across different industries. This interdisciplinary field combines principles from biology, engineering, and chemistry to develop materials, devices, or artificial systems that imitate biological processes. In the medical field, Biomaterials play an important part in the recovery process by restoring function and speeding up healing. These components, which might be natural or synthetic, are utilized to preserve, enhance, or replace damaged tissues or biological processes. Additionally, biomimetic technology has highlighted the importance of organelle attachment and detachment in an organism's ability to adapt to its environment. Tissue engineering's major goal is to make successful tissue grafts that can replace or repair damaged or deteriorated tissues and organs. Currently, there are ongoing pre-clinical studies and clinical applications of engineered tissues such as bones, cartilage, skin, skeletal muscles, blood vessels, and bladder. In this chapter, the interconnection between regenerative biology and engineering is examined, with emphasis on the utilization of biomimetics in tissue engineering and the creation of functional tissue transplants for regenerative medicine.

Abbreviation

BME	Biomedical engineering
BID	Biologically inspired design
ECM	extracellular matrices
3D	3-dimensional
M	meniscus

S. Choudhury · K. R. Arjun · M. N. R. Bharadwaj · K. M. Basalingappa (✉)
Division of Molecular Biology, School of Life Sciences, JSS Academy of Higher Education and Research, Mysuru, Karnataka, India
e-mail: kantheshmb@jssuni.edu.in

M. Maghimaa
Department of Microbiology, Muthayammal College of Arts and Science, Rasipuram, Namakkal, Tamil Nadu, India

R. Malviya and S. Sundram (eds.), *Engineered Biomaterials*, Engineering Materials,
https://doi.org/10.1007/978-981-99-6698-1_5

T	tendon
L	ligament
TEN	tissue-engineered nerve
CRISPR	clustered regularly interspaced short palindromic repeats
Cas9	clustered regularly interspaced short palindromic repeats-associated 9
ZIF-8	zeolitic imidazolate framework
ELPs	Elastin-like polypeptides
PGA	poly-glycolic acid
PLA	poly-lactic acid
OATS	osteochondral autograft transfer system
NYSERDA	New York State Energy Research and Development Authority
BIONIS	Biomimetics Network for Industrial Sustainability
PLMC	Poly D, L-lactide-co-trimethylene carbonate

1 Introduction

Human design has been "inspired by nature" for over three thousand years [1]. The phrase "learning from nature" can be referred to in a variety of ways, and as a systematic methodology, it is still in its infancy as a subject of study, particularly in engineering design. Biomimetics covers a wide range of study areas, has an impact on a number of application areas, and is thought to significantly improve the quality of life on the scientific, social, and economic levels [2]. The majority of important findings, however, have remained in their respective fields because of the large and dispersed nature of research topics. According to studies, individual parties rather than institutions have tended to practice biomimetics [3].

Nature has long served as an inspiration for the creation of useful materials and systems [4]. Biomimetics involves drawing inspiration, adapting, or deriving design ideas from nature. It is a process of developing innovative technologies by studying biological systems. Through biomimetics, concepts from the field of biology are applied to engineering, enabling the development of technologies that benefit from millions of years of development evolution in natural biological systems as a result of natural selection [2]. It denotes the investigation and imitation of natural processes, systems, and approaches [5].

Biomimetics is a relatively recent topic of study that entails applying biological scientific concepts and functions to engineering, design, chemistry, electronics, and other disciplines [1]. Biologists, physicists, chemists, material scientists, engineers, and other professionals combine to create biomimetics. Exploring the functions, structures, and principles of several natural things, as well as engineering and engineering materials and techniques with commercial uses, are all part of this multidisciplinary discipline [6]. In 1974, the term "biomimetics" first appeared in print in Webster's dictionary [7].

In biological components, two forms of polymers, namely proteins and polysaccharides (mostly composed of six-carbon sugars or hexoses), are commonly found. Interestingly, When the densities of biological materials are evaluated, their mechanical characteristics are equivalent to those of man-made materials [7]. Biomimetics can provide valuable insights for the design of composite materials, drawing inspiration from various natural examples such as the honeycomb beehive structure, The fiber arrangement incorporates wood, spider silks, nacre, bones, and hedgehog quills. Biomimetics explores solutions to human issues by studying nature's models, systems, processes, and components [8]. When using biomimetic membranes, the immobilization of biorecognition molecules may not significantly affect the affinity constant between the molecule and target analyte. This is because biomimetic membranes imitate a natural environment that can contain numerous biorecognition molecules [9].

Biomimeticists have traditionally concentrated on duplicating or copying biosystems with synthetic components and traditional processes, deriving inspiration from biological architecture and functions. However, recent advances in molecular and nanoscale engineering in the physical sciences and molecular biology have enabled biomimetics to be applied at the molecular level. Molecular biomimetics is a hybrid approach that combines synthetic nanoscale structures with natural molecular tools [4].

2 Terminology

2.1 *Bioinspiration*

"Based on observations of biological systems, an innovative method was developed" [3]. Bioinspiration is utilized as a means to develop intricate and advanced engineering models across different length scales, with the aim of harnessing the vast information available to address the urgent challenges confronting humanity [10].

2.2 *Biomimicry*

The principles of sustainable development can be approached through the application of interdisciplinary design techniques and philosophical frameworks that draw inspiration from nature as a model for addressing social, environmental, and economic concerns [3]. The convergence of biology and engineering, which encompasses biomimicry, is a prevailing trend that is fuelling innovation in the twenty-first century. This convergence presents opportunities for the transfer of knowledge across different domains, the emergence of novel areas of expertise, and the transportation of physical materials [11]. The field of engineered biomimicry is rapidly evolving and involves various disciplines, including biology, materials science, and manufacturing, making it a challenging area of study [12].

2.3 *Biomimetics*

The collaboration among diverse fields such as biology, technology, and other innovative areas has the goal of addressing practical issues by analyzing biological systems, creating models of them, and transferring and implementing these models to discover solutions [3]. "Biomimetics" refers to the imitation of life or nature. It comes from the Greek word bio mimesis [6].

2.4 *Bionics*

The term "bionics" was coined by Jack Steele of the US Air Force Medical Division in 1960 by combining the words "biology" and "technics". Bionics refers to a technical domain that aims to replicate, enhance, or replace biological functions with electronic and/or mechanical equivalents [13]. In addition, there is a growing interest in researching the improved regenerative abilities of bionic tissue-engineered nerve (TEN) grafts, with a specific focus on the bionics of their structure and components [14].

2.5 *Biomedicine*

Biomedicine is a medical field that applies biological and biochemical concepts to medical research or practice. It is distinctive within a culture because it incorporates specialized knowledge and unique practices based on that knowledge. Biomedicine also involves a hierarchical division of labor and specific principles or guidelines for social and professional interactions. The human body is the primary focus of biomedicine [15].

2.6 *Biomedical Engineering (BME)*

The application of engineering principles to the medical field is the focus of biomedical engineering, which is an engineering discipline that has gained increasing acceptance within the field [16]. Another fast-developing field of biomedical engineering is tissue engineering [17].

2.7 *Biologically Inspired Design (BID)*

BID is a prominent trend in contemporary design that involves the incorporation of biological functions and mechanisms to tackle human-related problems. Though not mandatory, BID is often associated with engineering because it deals with design issues akin to those encountered by engineers in various domains, such as mechanical and aerospace engineering, electrical and computer engineering, chemical engineering, and biomedical engineering [18].

2.8 *Biomechatronics*

"Biomechatronics refers to the utilization of medical and biological expertise to enhance mechatronic procedures and products" [19]. Unlike biomimetics, biomechatronics has a strong basis in engineering. Although it incorporates biomimetics, which is a significant part of Bio4Eng (biology for engineering) for human–machine interaction, it is not restricted to it and is considered a related field [20]. Interlink between Biology and Biomimetic approaches is shown in Fig. 1.

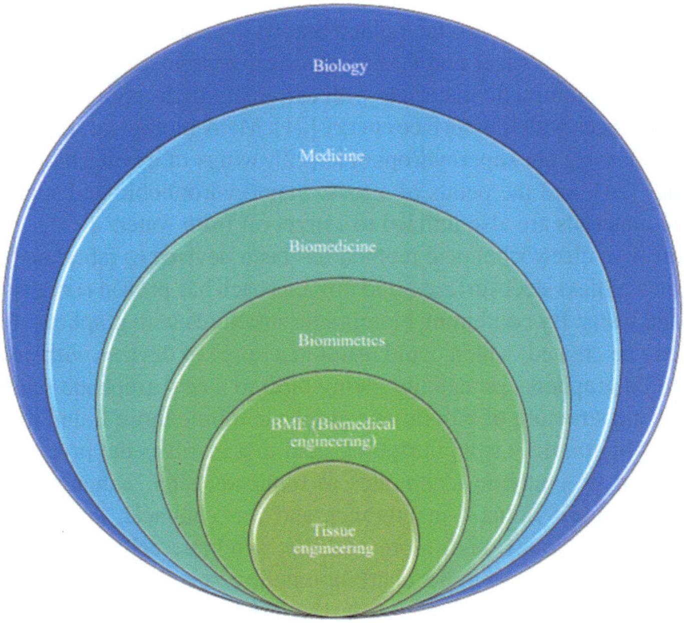

Fig. 1 Interlink between biology and biomimetic approaches (Made via powerpoint)

3 Background to Biomimetics

Biomimetics holds the potential for application in various biological research domains, including ecology, zoology, botany, molecular biology, and other related subfields. Research areas with high promise include structure and systems design, learning and memory, self-assembly and self-repair, perception and sensory systems, movement and locomotion, control structures and self-regulation, and new physiologically active materials. At the nanoscale level, biomimetic research is expected to have positive implications for component miniaturization, energy efficiency, and self-configuration. The development of biomimetic robots that can operate in various settings, including water, land, and air, is a significant area of attention. Additionally, modeling brain systems, known as neuromimetics, is a crucial field of study, particularly for comprehending the foundation of intelligent behavior. The integration of neuromimetic controllers in hardware, called "neuromorphic," and in the control frameworks of robots, referred to as neurorobotics, are emerging research fields in the domain of biomimetics [2].

During the 1950s, Otto Schmitt introduced the term "biomimetics" to differentiate between biological and engineering/physics approaches to science, which he referred to as "biomimetics" and "biophysics," respectively. Schmitt is also recognized as the founder of biomedical engineering, which encompasses biomaterials and is closely linked to biomimetics. The term "bionics," coined by Jack Steele of the US Air Force, pertains to the process of reproducing and modifying natural concepts [2].

According to the Wenzel and Cassie theories, superhydrophobic surfaces may be generated by biomimetic materials with extremely hydrophobic qualities, which need rough surfaces with low surface energy [21]. Many biological surfaces in nature, such as lotus leaves, are superhydrophobic [22], wings of a butterfly [23], feet of a water strider [24], and the petals of a rose. Superhydrophobic and superoleophilic biomimetic materials are also utilized to extract oil from water.

An innovative electrochemical method has been devised to fabricate biomimetic graphene on stainless steel surfaces, and this approach has proven successful in separating oil and water. By combining biomimetic structures with graphene, biomimetic graphene can be created, which could result in a range of devices with diverse properties based on graphene, potentially giving birth to a new graphene scientific field. The surface and structure of biomimetic graphene mimic those found in nature [8].

Molecular biomimicry is an intriguing field that involves the precise and robust implementation of synthetic techniques to replicate, alter, enhance, and analyze biological systems. Synthetic membranes and ion channels, artificial photosynthesis, synthetic enzymes, and, more recently, synthetic cellular life systems are some of the current areas of molecular biomimetic research. Additionally, subfields in materials research have expanded over the last few decades to include bioengineered and hybrid-polymer materials, adhesives, biomineralization, and imitations of extracellular matrix in terms of function, surface, and morphology [25].

4 Biological Materials

The body is a chemical laboratory that converts natural chemicals into energy, construction materials, trash, and a variety of multifunctional structures [26]. Humans have long recognized natural materials as sources of food, clothing, comfort, and other essentials [27]. Despite their tiny size, animals and insects may create huge quantities of natural products such as fur, leather, honey, wax, milk, and silk to suit human requirements [5]. Due to their closely regulated structure and better-oriented connections, many biological materials are less dense than synthetic materials with equivalent qualities [7].

For centuries, natural materials have been utilized due to their appealing characteristics like strength, durability, and beauty. In order to replicate these benefits, attempts have been made to produce artificial imitations of these materials in any quantity required. Natural materials possess unique features such as self-replication, self-healing, reconfigurability, multifunctionality, and chemical equilibrium. Unlike man-made materials that are typically processed through heating and pressurization, natural materials are created under ambient conditions. Materials derived from biological sources result in minimal waste and pollution, and the final product is usually biodegradable and recyclable by nature. Mastering the handling of these materials will expand our range of material options and enable us to produce more environmentally friendly and recyclable materials [5].

Biomaterials play a crucial role in wound healing by acting as carriers for protein and gene delivery, as well as providing scaffolds for cell attachment, growth, and differentiation. They have been utilized to transport growth factors, small organic compounds, and genes for the treatment of various diseases and wound healing [28]. Spider with their incredible webs is shown in Fig. 2.

5 The Two Approaches of Biomimetics

Recent advancements in biomedical research have been greatly aided by biomimetic methods [29]. Generally speaking, there are two ways to practice biomimetics: Problem-driven or solution-oriented. Both problem-driven and solution-based methodologies have distinct starting points and design features (Figs. 3 and 4) [3].

5.1 The Solution-Based Approach

The solution-based approach involves using the knowledge of biological systems to design and develop technology that replicates the functions of these systems. To do so, a thorough understanding of the biological system is necessary in order to identify

Fig. 2 Spiders are impressive "manufacturing engineers" of the natural world, able to create materials with remarkable efficiency, as evidenced by their incredible webs [5] The silk material used in the spider's web is five times stronger than steel, when compared by weight. (Original picture captured by author)

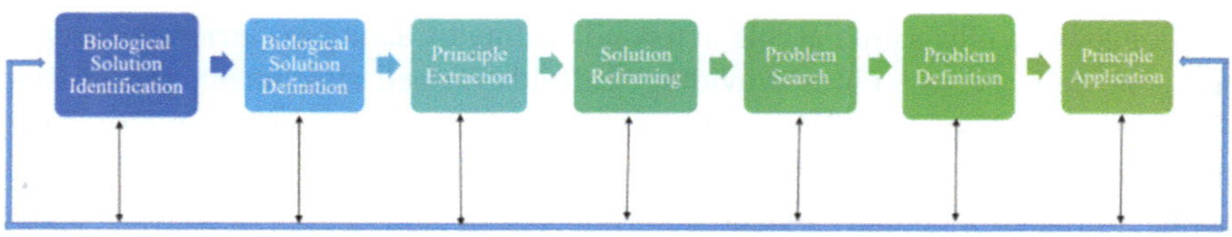

Fig. 3 The biologically inspired design employs two analogical processes: **a** An analogy based on a solution

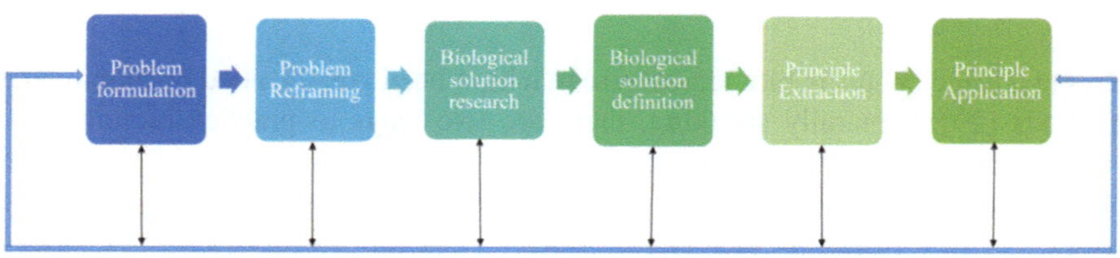

Fig. 4 The biologically inspired design employs two analogical processes: **b** analogy based on an issue

potential design solutions. The principles learned through fundamental research can then be applied to technology. The transfer of scientific knowledge to industry is closely tied to this approach [3] (Fig. 3).

The subsequent steps in the design process that is driven by solutions and inspired by biology [30]:

Step 1: Recognition of Biological Solutions

Designers start with a certain biological solution in mind [30]

Step 2: Definition of Biological Solution

Step 3: Extraction of the Principle

Step 4: Reframing the Problem

Creators are compelled to reinterpret their work in order to assess how consumers will appreciate the biological purpose attained [30].

Step 5: Look for a Problem

In contrast to search in the biological world, which entails searching through some finite space of biological answers, it may encompass inventing whole new challenges [30].

Step 6: Difficulties Identification

Step 7: Implementation of the Principle

5.2 The Problem-Driven Approach

The problem-driven approach in biomimetic development centers around identifying real-world issues and finding solutions to them, with the first step being the identification of a problem. By recognizing biological systems that accomplish a certain function or mechanism, it is feasible to integrate new or improved functions into technology by transferring and abstracting these principles. Despite the fact that the problem-solving method and problem-driven approach are closely related, they have inherent differences, and a detailed analysis is necessary to gain a deeper understanding of each process [3].

In biologically inspired design that is problem-driven, the process follows a pattern of non-linearity and dynamics, where the results of later stages can impact the outcomes of earlier stages. This pattern enables iterative loops of feedback and improvement [30].

Step 1: Formulation and definition of a problem

After being pushed to find or construct an issue to remedy, designers were asked to define their problem as a function [30].

Step 2: Problem Reframing

Designers reframed their problems in biological terms in order to find biological analogies for solutions [30].

Step 3: Search for Biological Solutions

If the problem is clearly mentioned as "staying cool," alter the limitations to "thermoregulation" to broaden the search range [30].

Multifunctionality: Look for animals or systems that can solve many problems at once [30].

Champion Adapters: Locate a system or organism that can withstand the most severe form of the issue under investigation [30].

Step 4: Definition of Biological Solution

Initially, designers uncovered biological system structures and surface mechanisms related to the reframed function [30].

Step 5: Extraction of the Principle

The identification of significant principles from a solution required a description that minimized specific structural and environmental limitations once a proper understanding of it was achieved [30].

Step 6: Application of the Principle

Once a thorough comprehension of a solution had been achieved, it was necessary to provide a description that eliminated as many particular structural and environmental constraints as feasible to identify significant principles. This procedure requires a translation from one domain area, such as biology, to another, such as mechanical engineering [30] (Fig. 4).

6 Biomimetic Materials Applications

6.1 Bioinspired Materials for Tissue Engineering

Tissue engineering enables the creation of biological substitutes for diseased and damaged tissues using a number of techniques [31]. To attain the desired outcome, numerous approaches are being investigated in musculoskeletal tissue engineering. Among these, implanting 3D synthetic polymeric scaffolds at the site of tissue damage is a widely used method. These scaffolds provide a framework that facilitates cell attachment, proliferation, and extracellular matrix (ECM) synthesis [32], to accelerate tissue regeneration [33]. Many researchers have created scaffolds using standard approaches such as phase separation salt leaching [34, 35].

The discovery of biomimetic scaffolds that mimic the architecture of the extracellular matrix (ECM) has recently emerged as a promising strategy for tissue regeneration. To fabricate these 3D biomimetic scaffolds, advanced techniques such as electrospinning and rapid prototyping have been employed. These scaffolds possess a hierarchical 3D organization and are composed of collagen as the organic component and carbonated hydroxyapatite nanoparticles as the inorganic component. The human skeletal system is predominantly composed of bone, which serves as a biological composite [8]. However, bone tissue engineering makes treating harvested and injured tissues much easier [36].

6.2 Bioinspired Components Used in Genome Technologies

The CRISPR (clustered regularly interspaced short palindromic repeats)-associated 9 (Cas9) gene-editing technology is widely employed in plants, model organisms, bacteria, and human cells (8). Cas9 allows for the disruption of a specific gene at a specific site and the activation of the endogenous repair process by inserting a specific piece of DNA, which can change the underlying cause of genetic illnesses [37]. CRISPR/Cas9 delivery systems are based on physical or viral techniques with limited applicability [38]. Ongoing research is being conducted to develop novel delivery methods for CRISPR/Cas9, which entails investigating synthetic delivery vehicles and metal-organic frameworks. Some examples of these include DNA nanoclews, gold nanowires, gold nanoparticles, and the zeolitic imidazolate framework [ZIF-8] [8].

Biomimetic nanoparticles, which blend synthetic cores with natural cell membranes, are commonly used in cancer therapy and immunization. Specifically, CC-ZIFs (zeolitic imidazolate frameworks enclosing CRISPR/Cas9) are nanoparticles coated with cancer cell membranes, enhancing the precision of cell-targeted gene editing by utilizing the homotypic binding events of tumor cells. Consequently, the creation of CC-ZIF metal-organic frameworks represents a major advancement in genome editing technologies, particularly in the domain of cell-specific cancer treatment [8].

6.3 Bioinspired Components for Ultrasound Imaging

Compared to other imaging techniques like single-photon emission tomography, positron emission tomography, and magnetic resonance imaging, ultrasound imaging is generally known for producing images of lower quality. This is owing in part to the input of echogenicities from adjacent tissues, which might result in poor picture contrast [39]. When circulation proteins are adsorbed onto a material, it can alter its physicochemical properties, including surface charge, surface composition, and size. This process can affect the material's cellular uptake, circulation duration, toxicity, and tendency to agglomerate.

Changes in biological environments can negatively impact the ability of particles to reach their targets effectively, resulting in suboptimal outcomes. A new technique called biomimetic coating or cloaking has been developed to address this problem. This technique involves using cell membranes obtained from lymphocytes, macrophages, thrombocytes, and other sources to overcome the challenges related to protein adsorption [8]. The biological composition of the cell membrane surface is crucial in controlling the adsorption of biological components on particles, which affects the circulation time, immune-mediated degradation, and biocompatibility of the particles [40].

6.4 Bioinspired Materials Used in Actuators

Stretchable and flexible pressure sensors rely on biomimetic materials with micro- and nano-hierarchical structures to work. These sensors have high sensitivity, stability, and a wide pressure range [8]. Stimuli-responsive actuators, which can transform diverse forms of energy such as thermal, optical, and electrical energy into mechanical energy, are another application of biomimetic materials. These actuators are highly versatile and can perform intricate mechanical movements in a precise sequence, making them useful for bionic devices, artificial muscles, and intelligent robots [41]. In recent times, there has been a growing focus on multi-stimulus actuators instead of single-stimulus actuators [41]. Bilayer actuators are an attractive candidate for biomimetic applications because of their outstanding mechanical characteristics and significant photo-induced stress [8].

7 Tissue Engineering

One of the most successful approaches for replacing or repairing damaged skin and other biological components is tissue engineering [42]. In the artificial tissue regeneration process, both in vivo and in vitro procedures are utilized [43]. It offers a significant treatment alternative by combining the patient's own cells with a scaffold for bone healing [42]. The purpose of biomimetic tissue engineering approaches is to use biomaterials to actively drive tissue repair and regeneration [44]. To guarantee effective regeneration of the target tissue, the biomaterials used in tissue-engineered constructs (TECs) and drug delivery must fulfill particular biological and mechanical parameters [45].

Biomimetic materials that mimic the characteristics of the natural extracellular matrix (ECM) have been created with the help of recent advancements in biomaterials and tissue engineering [46].

Scaffolding plays a crucial role in facilitating cell adhesion, growth, and proliferation during both in vivo and in vitro artificial tissue regeneration procedures [42]. The intricate nature of tissue engineering interfaces involves a multi-scale organization of various tissue types, which necessitates collaboration among professionals from diverse fields such as biomedical engineering, materials science, and orthopedic surgery [47].

Cell behavior such as migration, spreading, differentiation, and tissue formation is regulated by ECM proteins, which interact with cell-surface receptors, specifically integrins. The interaction between receptors and binding sites on ECM proteins is tightly controlled and affected by factors such as the binding motif's structural properties, ligand density, spatial orientation, and the overall integrity of the ECM structure. This precise regulation enables targeted communication between cells within a specific tissue [46]. The term "alive machines" is used to describe biomimetic systems due to their ability to replicate biological processes and provide a framework for investigating biological theories and addressing scientific and technical challenges [2].

Tissue engineering has enabled the exploration of the crucial biological roles played by musculoskeletal progenitors, stem cells, and endogenous cells in both natural and artificial microenvironments. While the skeleton offers structural support, it also acts as an endocrine organ. Ligaments and tendons serve as connectors between bones, permitting muscles to apply force on the skeleton, enabling smooth movement of joints that rely on articular cartilage and intervertebral discs. Although reproducing the growth and maturation of the musculoskeletal system in a laboratory environment is challenging, tissue engineering advancements have expanded our comprehension of these intricate processes, including mechanisms for repair and regeneration [48].

Damaged tissue caused by many factors such as severe trauma, disease, or forceful tumor excision (leading to protracted denervation) needs replacement or restoration with healthy tissue [49]. Tissue transplantation is the primary treatment for these issues, although it has substantial downsides. Tissue engineering is based on emulating organogenesis, which has lately been shown to be successful, and it serves as an alternate approach [32]. In the regeneration of the musculoskeletal system, including cartilage, bone, skeletal muscle, tendon, and ligament, synthetic biomaterials are instrumental, serving as tailored scaffolds that can be derived from both natural and synthetic sources [49]. Some biomaterials for tissue engineering is shown in Fig. 5.

One biomimetic approach to developing tissue-engineered products involves the use of natural skeletons; however, these must first be treated to remove immunogenic proteins and foreign materials to ensure safety before being seeded with human cell populations. Researchers are investigating ways to leverage the superior architecture of both soft (collagenous) and hard (ceramic) materials by incorporating physical or chemical spatial and surface cues to enhance tissue regeneration [12].

Polymers are frequently employed in industry due to their flexibility, simplicity of manufacture, and lightweight properties [50]. In bioinspired material science, scientists are investigating the synthesis and potential uses of biopolymers, particularly naturally occurring degradable polymers like chitosan, hyaluronan, collagen, and gelatin. When compared to synthetic polymers, biopolymers offer advantages such as gradual degradation, hydrophobicity, neutral charge distribution, and limited mechanical properties, which render them suitable for the manufacture of scaffolds [8].

The creation of composites with bone regeneration characteristics remains a serious problem in tissue engineering. Although several polymers have been

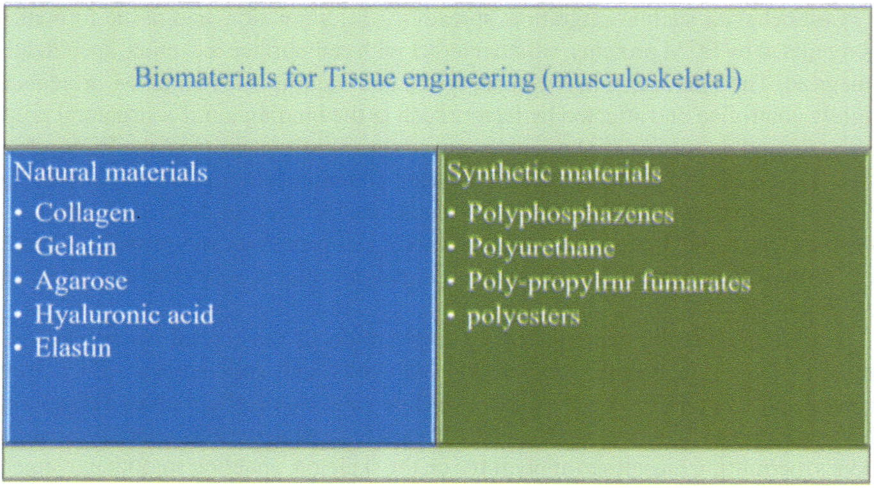

Fig. 5 Biomaterials, both natural and synthetic, for tissue engineering (Made via PowerPoint)

employed in bone tissue engineering, there is a growing need for modern biomaterials with improved mechanical and functional properties to overcome emerging challenges. Aliphatic polycarbonates are one such promising biomaterial, possessing desirable characteristics like biodegradability and biocompatibility, which make them increasingly valuable in medical applications [8].

8 Types of Biomaterials Used in Tissue Engineering

Based on their biophysical properties, biomaterials employed in musculoskeletal systems may be categorized into three major categories: Biomaterials that are flexible/ elastic, hard, and soft [49].

8.1 Flexible/ Elastic Biomaterials

The mechanical qualities of the musculoskeletal system's meniscus (M), tendon (T), and ligament (L) tissues are elastic and flexible (Fig. 6). Due to their relatively weaker circulatory system, these tissues require a lower amount of oxygen and nutrients for their regeneration and repair, in contrast to other tissues [47]. Due to the poor ability for spontaneous healing in these tissues, surgical treatments such as autografts and allografts are frequently required in situations of damage [51]. Traditional treatments have drawbacks, such as graft rejection and poor health consequences have prompted the investigation of biomaterial engineering for M/T/L tissues as a viable alternative.

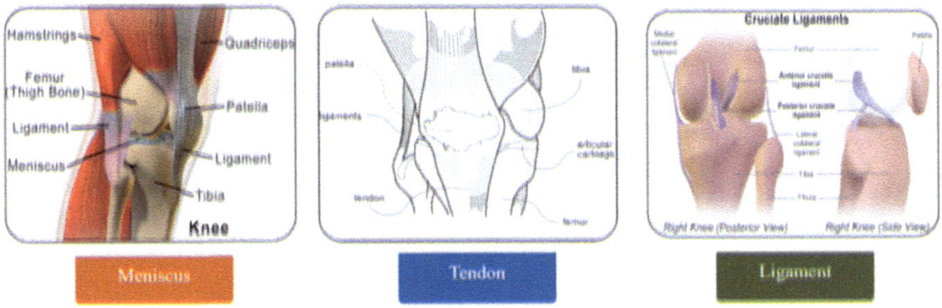

Fig. 6 Meniscus (M), tendon (T), and ligament (L) are all elastic tissues in the musculoskeletal system (L). (Images from Google used under Creative Commons Licenses)

The stabilization of joints and the restriction of bone movement are critical functions of ligaments, while tendons transmit forces generated by contracting muscles to bones. These tissue types' interfaces or junctions, which anchor soft tissue to bone, are critical components of this process [51].

Biomimetic biomaterials such as collagen and elastin are frequently used in the engineering of elastic tissues [49]. Elastin is a highly desirable biomaterial because of its exceptional elasticity, durability, and responsiveness to stimuli such as LCST. Elastin-like polypeptides (ELPs) are extensively generated through chemical synthesis and recombinant DNA expression, and serve as valuable model systems for research and biomaterials in numerous biomedical domains, including protein purification, drug delivery, and tissue engineering. By combining or crosslinking these polymers, biomimetic materials with varying levels of crosslinking or combination can be produced [25].

8.2 *Hard Biomaterials*

Researchers in the field of materials made by natural organisms are particularly interested in studying biological tissues that exhibit exceptional stiffness and strength, which are commonly referred to as "hard" tissues (Fig. 7). These tissues serve various purposes, such as providing structural support (bones), acting as tools for cutting, tearing, and crushing (teeth), or offering protective covering (seashells). Although stiffness is a more precise term to describe their resistance to deformation, "hardness" remains the primary mechanical attribute associated with these tissues [52]. Calcified tissues, such as bone, dental enamel, dentin, and cementum, are considered hard tissues due to their high concentration of calcium phosphate minerals. As the world's population ages, there is a greater need for hard tissue repair and regeneration [53]. Smart biomaterials and smart tissue engineering constructions, for example, have a lot of exciting promise as autogenous bone transplant alternatives [53].

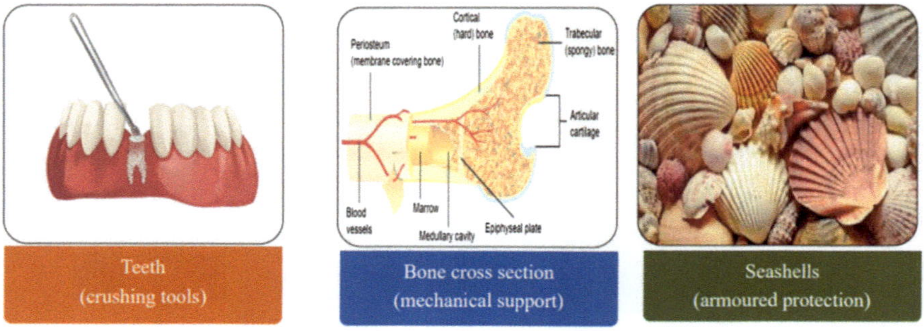

Fig. 7 Teeth, bone, and seashells are examples of hard tissues. (Pictures taken from Google under Creative Commons Licenses)

In addition to strength, toughness is a vital attribute that influences the tensile properties of hard materials. There have been about 60 biogenic minerals found, which are minerals that originate from biological processes. The most common biogenic minerals include hydroxyapatite (found in bones and teeth), calcium carbonate (found in seashells), and silica (found in sub-millimeter marine organisms such as radiolarians and diatoms). These minerals come in different shapes, amounts, and sizes [52].

Hard materials are essential in the design or restoration of various components of the musculoskeletal system, such as bone tissue. Orthopedic treatments, which are becoming increasingly prevalent, employ a variety of materials, each with its own set of benefits and drawbacks. Ceramics and bioglass were the initial hard biomaterials used in hard tissues [54, 55]. Alternatives include biomaterials made of calcium sulfate and calcium phosphate, which are biocompatible and absorbable. Several calcium and phosphate combinations have been investigated for orthopedic purposes, including bone cement. As these materials decompose, they produce safe ions such as calcium, phosphate, and sulfate that are already present in the body. Among the different forms of calcium phosphate, hydroxyapatite has gained more attention. Consequently, researchers have developed composite scaffolds by blending different forms of hydroxyapatite with organic or inorganic biodegradable polymers for use in hard tissues such as bone and osteochondral [49] (Fig. 7).

8.3 *Soft Biomaterials*

Several issues, such as tissue resorption and implant rupture or contracture, limit the use of autologous tissue transplants and synthetic implants in current soft tissue reconstruction techniques [56]. In order to construct soft tissue structures like muscle and cartilage in the musculoskeletal system, a combination of natural and synthetic biomaterials is utilized (Fig. 8). Collagen, gelatin, hyaluronic acid, chitosan, and acellular matrix are some of the most often employed natural compounds for this

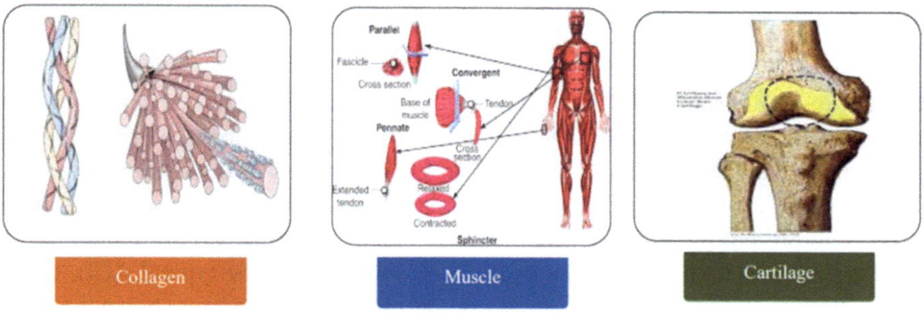

Fig. 8 Collagen is a natural substance found in soft tissues such as muscle and cartilage. (Picture taken from Google under creative commons licenses)

purpose [56]. Hydrogel structures and sponges made from a diverse range of materials, including alginate, agarose, collagen, hyaluronan, fibrin gels, poly (glycolic acid) (PGA), and poly (lactic acid) (PLA), are commonly utilized in cartilage tissue engineering [57].

The self-healing capacity of adult articular cartilage is limited, and even minor injuries or damage can lead to the degeneration of joints and result in severe pain and functional limitations. Despite advances in cartilage tissue engineering during the previous two decades, properly treating articular cartilage defects remains a substantial therapeutic problem. Despite considerable efforts to develop innovative biological solutions, there are currently limited therapeutic options available [57] (Fig. 8).

9 Tissue Grafting

Replacement of sick or damaged tissue removed from one portion of the body with healthy skin, bone, or other tissue.

There are three types of tissue grafts: autograft, allograft, and xenograft. Autograft involves transferring tissue from one area of the body to another, like using the semitendinosus tendon from the same patient to fix the anterior cruciate tendon. In contrast, allograft involves utilizing tissue from a donor to reconstruct and repair tissue in another person. Depending on the intended use, the donated tissue can be fresh, processed, or frozen. Strict screening, processing, and testing are conducted to ensure safety and minimize the risk of disease transmission, which may involve washing and irradiation [58].

10 Graft Strength and Its Properties

In order to ensure successful restoration of the ACL, it is important to choose a graft tissue that has sufficient strength to match the natural ACL. Biomechanical research has demonstrasted that different types of ACL grafts have varying loading strengths. The native ACL has a stiffness of approximately 242 N/mm and can withstand up to 2160 N before failing. Studies have indicated that a 10-mm BPTB graft has a higher load to failure, with a mean value of 2977 N, compared to the typical ACL. BPTB and hamstrings have similar graft stiffness, and clinical studies have shown that ACL reconstructions using either graft have similar or better stability outcomes. Although quadrupled hamstrings may have higher stiffness and strength in the laboratory, there is no evidence that this translates to better stability outcomes or increased survival rates. However, several recent registry studies have reported higher revision rates with hamstrings compared to BPTB [16, 28, 59].

It is important to consider that the size of hamstring grafts can vary considerably from person to person, and utilizing a smaller-diameter graft may increase the risk of ACL failure. Studies have shown that using an HT graft smaller than 8 mm can lead to a higher failure rate of 61%. The failure rate of ACL surgery after only 2 years is approximately 7%. However, the extensor mechanism has sturdy supporting retinacular structures that can strengthen the remaining tendon and enable acceptance of a 10-mm central graft. In our opinion, BPTB is the preferred choice due to its consistent graft size, which is not influenced by patient or tendon size, and excellent clinical outcomes [7].

11 Regenerative Medicine and its Effects on Tissue Grafting

Tissue regeneration is a novel technique to tissue repair or replacement, made possible by the advent of second and third-generation bioactive materials such as bioglass. This strategy seeks to regenerate tissues as opposed to merely replacing them, made feasible by the exceptional properties of these innovative materials [55]. The kind of tissue response at the implant interface is intimately related to the process of tissue attachment [54].

11.1 Regenerative Medicine

Stem cell therapy is concerned with the restoration of normal function by replacing or regenerating human cells, tissues, or organs. With the aid of pioneering methods, this field holds promise in effectively curing or treating ailments that were previously challenging to manage with conventional medications and procedures. As a result, regenerative medicine has the potential to transform human medicine [60].

Approximately six decades ago, Dr. Joseph Murray, John Merrill, and J. Hartwell Harrison spearheaded a team that executed the first triumphant organ transplant in Boston, effectively inaugurating this field. This momentous feat heralded a new age in organ transplantation and presented a remedy for patients with end-stage organ disease, an unprecedented achievement in human history [59].

In the late 1950s and 1960s, the first successful cell therapies, including bone marrow transplantation, were discovered [61].

Regenerative medicine encompasses several biomedical methodologies directed at mending injured organs or tissues, evoking both positive expectations and ethical controversies. The ethical predicaments surrounding this field have resulted in a schism between traditional Christian societies and post-Christian secular societies. In addressing bioethical disagreements in regenerative medicine, it is essential to recognize the distinct moral and metaphysical perspectives that underpin these debates. Each side presents individual interpretations of human nature. While mortality persists as an inevitability, regenerative medicine holds promise in prolonging human lifespans [9].

Research in regenerative medicine includes the study of self-healing, where the body employs its own mechanisms, possibly with the help of foreign biological material, to restore tissues and organs. As the focus of the field moves toward developing cures for complex and chronic illnesses, The phrases "tissue engineering" and "regenerative medicine" have grown more interchangeable [62].

While the regenerative medicine industry is developing, it has been noted that there are not only therapeutic applications but also non-therapeutic ones, including the use of tissues as biosensors to detect biological or chemical hazards, as well as tissue chips that can be used to assess the toxicity of new drugs [1].

11.2 Working of Regenerative Medicine

Cells are the fundamental unit of the human body, and they work together to form tissues. These tissues contain cells that produce an extracellular matrix or scaffold, which offers support to the cells and serves as a platform for signaling molecules to communicate with the cells. These signals trigger cellular responses that regulate their functions. By studying how cells interact with their environment, organize themselves into tissues, and respond to signals, scientists have been able to develop techniques to repair damaged tissues and generate new ones [63].

The first step in tissue engineering often involves the creation of a scaffold using a variety of materials, including proteins or polymers. Following the scaffold's construction, cells can be introduced either with or without growth factors, depending on the desired outcome. If the environmental conditions are favorable, the cells can self-organize into tissue. Alternatively, all three components—cells, scaffold, and growth factors—can be mixed simultaneously, resulting in self-assembly of the tissue [64].

Another approach for tissue engineering is to repurpose an existing scaffold. For instance, the collagen scaffold remaining after the cells are extracted from a donated organ can be repurposed for generating new tissue. Various organs, including the heart, liver, lung, and kidney, have been studied. Bioengineered successfully using this method. This technique has great potential to utilize human tissue scaffolding that would otherwise be discarded after surgical procedures and integrate it with a patient's own tissue.

The application of this technique holds immense promise for the development of personalized organs that can avoid immune rejection. By utilizing scaffolding derived from human tissue that is typically discarded after surgeries and combining it with the patient's cells, customized organs can be engineered [2].

11.3 Effects and Its Applications

Although tissue engineering has shown promise in generating implants for certain body parts such as bladders, tiny arteries, skin grafts, and even trachea, it is still not a widely used treatment in patient care due to its experimental nature and high costs. Although laboratory-engineered tissues such as heart, lung, and liver tissues have been developed, they are still far from being entirely reproducible for implantation into humans. However, these tissues can be beneficial for drug development research, allowing for the use of active human tissue to test drug candidates, saving money, reducing the usage of research animals, and potentially speeding up the development of personalized therapy [2].

12 Biomimetics–Promises and Obstacles with a Future Look

Biomimetics is gaining recognition as a technology with significant potential in Europe, Japan, and the USA, leading to increasing interest and investment in this field. To explore cutting-edge technology based on biological systems, major corporations, including Ford, General Electric, Herman Miller, HP, IBM, and Nike, have established research facilities and are collaborating with scientists.

The advantages of biomimicry, which include economic, environmental, and societal benefits, are becoming increasingly evident, and its scope is expanding to foster harmonious coexistence between humans and the environment. Developed nations are committing considerable resources to research, establishing a foundation for future progress and advancements in biomimicry [65].

Janine M. Benyus, the founder of the Biomimicry Institute and Biomimicry Guild, established Asknature.org in the United States to promote the progress of biomimetic technologies, serving as a platform for their development. Additionally, the New

York State Energy Research and Development Authority (NYSERDA) has made biomimicry a requirement to tackle several energy-related challenges [39].

Biomimetics has received significant attention in Germany, where 28 research institutions collaborate in BIOKON and the Federal Ministry of Education and Research oversees 35 initiatives. In the United Kingdom, BIONIS serves as a connector between academic institutions and businesses, while in Japan, the Century Center of Excellence, under the supervision of the Ministry of Education, Culture, Sports, Science, and Technology, focuses on biomimetic manufacturing and innovative use of biological resources in agriculture [33].

12.1 Biomedical Engineering Using Biomimicry

Biomimetics, which takes inspiration from living organisms, has led to the development of many useful products that have enhanced human life. The combination of biomedical engineering, medicine, science, and biomimetics is anticipated to have a significant impact on managing diseases, disabilities, and injuries in the future, particularly in fields such as tissue engineering and regenerative medicine. Biomimetic principles and functions, like the adhesive and regenerative properties of spider webs, leukocyte migration and adhesion during inflammation, and the regenerative capabilities of lizards and buckhorns, can be employed in biomedical engineering [59].

The emerging field of next-generation biomimetics is using a combination of biology and technology to tackle various issues. Nanotechnology is becoming increasingly important in understanding the structures of materials and accelerating the formation of secondary structures of proteins. To promote bone regeneration, researchers have designed multifunctional fibrous scaffolds that mimic the architecture of native tissue. One recent study examined the potential of using poly D, L-lactide-co-trimethylene carbonate (PLMC) nanofibers as scaffold materials for tissue repair and regeneration, and found that their biomimetic properties improved efficiency. By incorporating biomimetics into biomedical engineering, technology is being advanced in numerous ways [4].

13 Conclusion

Due to recent advancements in biology and engineering, biomimetics has emerged as an increasingly important field with the potential for further growth. Researchers are exploring the imitation of biological functions and developing alternative approaches. Tissue bionics, a new area within biomimetics, focuses on bioinspired design and synthesis to create human tissues, utilizing nature's intelligent design for new purposes. Collaboration between biomimetics, biomedical engineering, and biomechatronics could be mutually beneficial for the fields at the intersection of life sciences and engineering.

Biomimetic materials possess favorable properties such as structural performance, biocompatibility, and biodegradability, As a result, they are extremely valuable in a wide range of applications, including tissue engineering and ultrasonic imaging. These materials are useful for creating non-intrusive multimodal images that can be used to produce stimuli-responsive actuators. In addition, they are used to protect metals from corrosion and impurities, and in genome-editing and oil–water separation technologies. Despite ongoing efforts to expand biological synthesis techniques and develop new biomimetic materials with enhanced biological properties, significant challenges remain. Nevertheless, researchers continue to work toward overcoming these obstacles and creating innovative biomimetic materials.

There are numerous opportunities for advancing biomaterials, tissue engineering, and biomechanics in orthopedic interfacial tissue engineering. To achieve progress, we must enhance our knowledge of the function and production of entheses in vivo and our ability to design constructs that can replace these complex tissues. Future research should concentrate on comprehending the biological mechanisms involved in interface development, regeneration, and homeostasis, as well as the relationship between enthesis structure and function. More study is needed to understand how tissue boundaries, such as the four portions of the enthesis, form, retain, and mend themselves after damage.

Creating functional and integrated soft tissue repair largely depends on the successful restoration of musculoskeletal interfaces. A viable method for tissue engineering of these interfaces involves using multi-unit grafts, which utilize existing tissue engineering techniques to generate complex scaffolds for composite tissue regeneration. The latest advances in high-resolution 3D printing enable the construction of multi-layered and multi-cellular tissue composites with precise spatial organization of cells, minerals, and bioactive substances that closely mimic natural conditions.

Grafting is a captivating process that is currently being studied in depth. Its popularity is growing due to its capacity to boost vigor, disease resistance, and stress resilience without the need for genetic alterations or extensive plant breeding projects.

To better understand the developmental and regenerative processes, we can create precise experimental models that mimic the tissue environment in vitro. By using non-invasive longitudinal imaging techniques, we can study cellular responses to microenvironmental factors with accurate spatial and temporal precision as these constructs grow. However, the complex nature of cell responses to regulatory signals requires the development of suitable molecular markers and functional readouts based on expertise in developmental biology. The improvement of bioreactor imaging technologies can give useful information on the timing and combinations of biophysical inputs required to govern stem-progenitor cell fate choices and impact the development of tissues with various cellular phenotypes.

Biomimetics involves making important advances in science and technology by recognizing the amazing substances and structured surfaces that nature has created and refined. It is projected that biomimetic materials will be widely employed for the treatment of general and specialized medical diseases in roughly a decade. The influence of nature is predicted to continue driving technological innovation in many parts of our life. While some of the ideas may appear to be science fiction given our current technology constraints, these notions may get closer to reality than we realize as we develop a better knowledge of nature and enhance our skills.

References

1. Vincent, J.F., Bogatyreva, O.A., Bogatyrev, N.R., Bowyer, A., Pahl, A.K.: Biomimetics: its practice and theory. J. R. Soc. Interface. **3**(9), 471–482 (2006)
2. Lepora, N.F., Verschure, P., Prescott, T.J.: The state of the art in biomimetics. Bioinspir. Biomim. **8**(1), 013001 (2013)
3. Fayemi, P.E., Wanieck, K., Zollfrank, C., Maranzana, N., Aoussat, A.: Biomimetics: Process, tools and practice. Bioinspir. Biomim. **12**(1), 011002 (2017)
4. Sarikaya, M., Tamerler, C., Jen, A.K., Schulten, K., Baneyx, F.: Molecular biomimetics: nanotechnology through biology. Nat. Mater. **2**(9), 577–585 (2003)
5. Bar-Cohen Y.: Biomimetics: biologically inspired technologies. CRC press (2005)
6. Bhushan, B.: Biomimetics: lessons from nature–an overview. Philos. Trans. R. Soc. A: Math., Phys. Eng. Sci. **367**(1893), 1445–1486 (2009)
7. Vincent, J.F.: Biomimetics—a review. Proc. Inst. Mech. Eng. [H] **223**(8), 919–939 (2009)
8. Suresh Kumar, N., Padma Suvarna, R., Chandra Babu Naidu, K., Banerjee, P., Ratnamala, A., Manjunatha, H.: A review on biological and biomimetic materials and their applications. Appl. Phys. A. **126**(6), 445 (2020)
9. Ferrari, M., Martin, D.K.: Nanobiotechnology of biomimetic membranes. Springer Science (2007)
10. Naik, R.R., Singamaneni, S.: Introduction: bioinspired and biomimetic materials. Chem. Rev. **117**(20), 12581–12583 (2017)
11. Cohen, Y.H., Reich, Y.: Biomimetic design method for innovation and sustainability. Springer, Berlin, Germany (2016)
12. Martín-Palma, R.J., Lakhtakia, A.: Progress on bioinspired, biomimetic, and bioreplication routes to harvest solar energy. Appl. Phys. Rev. **4**(2), 021103 (2017)
13. Iouguina, A.: Biologically informed disciplines: a comparative analysis of terminology within the fields of bionics, biomimetics, and biomimicry (Doctoral dissertation, Carleton University)
14. Qi, T., Zhang, X., Gu, X., Cui, S.: Experimental study on repairing peripheral nerve defects with novel bionic tissue engineering. Adv. Healthcare Mater. **5**, 2203199 (2023)
15. Gaines, A.D., Davis-Floyd, R.: Biomedicine. Encycl. Med. Anthropol. **1**, 95–109 (2004)
16. Cha, C., Shin, S.R., Annabi, N., Dokmeci, M.R., Khademhosseini, A.: Carbon-based nanomaterials: multifunctional materials for biomedical engineering. ACS Nano **7**(4), 2891–2897 (2013)
17. Melchels, F.P., Feijen, J., Grijpma, D.W.: A review on stereolithography and its applications in biomedical engineering. Biomaterials **31**(24), 6121–6130 (2010)
18. Vattam, S.S., Helms, M.E., Goel, A.K.: Compound analogical design: interaction between problem decomposition and analogical transfer in biologically inspired design. Des. Comput. Cogn. **23**(8), 377–396 (2008)
19. Witte, H., Lutherdt, S., Schilling, C.: Biomechatronics: how much biology does the engineer need? In: Proceedings of the 2004 IEEE International conference on control applications, vol. 2, pp. 944–948. IEEE (2004)
20. Witte, H., Fremerey, M., Weyrich, S., Mämpel, J., Fischheiter, L., Voges, D., Zimmermann, K., Schilling, C.: Biomechatronics is not just biomimetics. In: 9th International workshop on robot motion and control, pp. 74–79. IEEE (2013)
21. Cassie, A.B., Baxter, S.: Wettability of porous surfaces. Trans. Faraday Soc. **40**, 546–551 (1944)
22. Zhang, J., Sheng, X., Jiang, L.: The dewetting properties of lotus leaves. Langmuir **25**(3), 1371–1376 (2009)

23. Bixler, G.D., Bhushan, B.: Bioinspired rice leaf and butterfly wing surface structures combining shark skin and lotus effects. Soft Matter **8**(44), 11271–11284 (2012)
24. Panchanathan, D., Rajappan, A., Varanasi, K.K., McKinley, G.H.: Plastron regeneration on submerged superhydrophobic surfaces using in situ gas generation by chemical reaction. ACS Appl. Mater. Interfaces **10**(39), 33684–33692 (2018)
25. Kushner, A.M., Guan, Z.: Modular design in natural and biomimetic soft materials. Angew. Chem. Int. Ed. **50**(39), 9026–9057 (2011)
26. Mann, S., (ed.).: Biomimetic materials chemistry. John Wiley & Sons, (1996)
27. Carlson, J., Ghaey, S., Moran, S., Tran, C.A., Kaplan, D.L.: Biological materials in engineering mechanisms. In: Biomimetics, pp. 383–398, CRC Press, (2005)
28. Andreadis, S.T., Geer, D.J.: Skin structure and physiology. Trends Biotechnol. **7**(24), 331–337 (2006)
29. Zhang, G.: Biomimicry in biomedical research
30. Helms, M., Vattam, S., Goel, A.K., Yen, J., Weissburg, M.: Problem-driven and solution-based design: twin processes of biologically inspired design. In: ACADIA08 Conference, (2008)
31. Salgado, A.J., Coutinho, O.P., Reis, R.L.: Bone tissue engineering: state of the art and future trends. Macromol. Biosci. **4**(8), 743–765 (2004)
32. Agrawal, C.M., Ray, R.B.: Biodegradable polymeric scaffolds for musculoskeletal tissue engineering. J. Biomed. Mater. Res.: Off. J. Soc. Biomater., Jpn. Soc. Biomater., Aust. Soc. Biomater. Korean Soc. Biomater. **55**(2), 141–150 (2001)
33. Loh, Q.L., Choong, C.: Three-dimensional scaffolds for tissue engineering applications: role of porosity and pore size
34. Diba, M., Kharaziha, M., Fathi, M.H., Gholipourmalekabadi, M., Samadikuchaksaraei, A.J.: Preparation and characterization of polycaprolactone/forsterite nanocomposite porous scaffolds designed for bone tissue regeneration. Compos. Sci. Technol. **72**(6), 716–723 (2012)
35. Golafshan, N., Gharibi, H., Kharaziha, M., Fathi, M.: A facile one-step strategy for development of a double network fibrous scaffold for nerve tissue engineering. Biofabrication **9**(2), 025008 (2017)
36. Hardy, J.G., Ghezzi, C.E., Saballos, R.J., Kaplan, D.L., Schmidt, C.E.: Supracolloidal assemblies as sacrificial templates for porous silk-based biomaterials. Int. J. Mol. Sci. **16**(9), 20511–20522 (2015)
37. Rouet, R., Thuma, B.A., Roy, M.D., Lintner, N.G., Rubitski, D.M., Finley, J.E., Wisniewska, H.M., Mendonsa, R., Hirsh, A., de Onate, L., Compte, B.J.: Receptor-mediated delivery of CRISPR-Cas9 endonuclease for cell-type-specific gene editing. J. Am. Chem. Soc. **140**(21), 6596–6603 (2018)
38. Wells, D.J.: Gene therapy progress and prospects: electroporation and other physical methods. Gene Ther. **11**(18), 1363–1369 (2004)
39. Stride, E., Saffari, N.: Microbubble ultrasound contrast agents: a review. Proc. Inst. Mech. Eng. [H] **217**(6), 429–447 (2003)
40. Su, G., Zhou, X., Zhou, H., Li, Y., Zhang, X., Liu, Y., Cao, D., Yan, B.: Size-dependent facilitation of cancer cell targeting by proteins adsorbed on nanoparticles. ACS Appl. Mater. Interfaces **8**(44), 30037–30047 (2016)
41. Hou, Y., Chang, H., Song, K., Lu, C., Zhang, P., Wang, Y., Wang, Q., Rao, W., Liu, J.: Coloration of liquid-metal soft robots: From silver-white to iridescent. ACS Appl. Mater. Interfaces **10**(48), 41627–41636 (2018)
42. Ng, J., Spiller, K., Bernhard, J., Vunjak-Novakovic, G.: Biomimetic approaches for bone tissue engineering. Tissue Eng. Part B Rev. **23**(5), 480–493 (2017)
43. Reddy, R., Reddy, N.: Biomimetic approaches for tissue engineering. J. Biomater. Sci. Polym. Ed. **29**(14), 1667–1685 (2018)
44. Thibeault, S.L., Klemuk, S.A., Smith, M.E., Leugers, C., Prestwich, G.: In vivo comparison of biomimetic approaches for tissue regeneration of the scarred vocal fold. Tissue Eng. Part A **15**(7), 1481–1487 (2009)
45. Bas, O., Catelas, I., De-Juan-Pardo, E.M., Hutmacher, D.W.: The quest for mechanically and biologically functional soft biomaterials via soft network composites. Adv. Drug Deliv. Rev. **1**(132), 214–234 (2018)

46. Rahmany, M.B., Van Dyke, M.: Biomimetic approaches to modulate cellular adhesion in biomaterials: A review. Acta Biomater. **9**(3), 5431–5437 (2013)
47. Boys, A.J., McCorry, M.C., Rodeo, S., Bonassar, L.J., Estroff, L.A.: Next generation tissue engineering of orthopedic soft tissue-to-bone interfaces. MRS Communs. **7**(3), 289–308 (2017)
48. Gingery, A., Killian, M.L.: * Special focus issue on strategic directions in musculoskeletal tissue engineering. Tissue Eng. Part A **23**(17–18), 873 (2017)
49. Del Bakhshayesh, A.R., Asadi, N., Alihemmati, A., Tayefi Nasrabadi, H., Montaseri, A., Davaran, S., Saghati, S., Akbarzadeh, A., Abedelahi, A.: An overview of advanced biocompatible and biomimetic materials for creation of replacement structures in the musculoskeletal systems: focusing on cartilage tissue engineering. J. Biol. Eng. **13**, 1–21 (2019)
50. Camargo, P.H., Satyanarayana, K.G., Wypych, F.: Nanocomposites: synthesis, structure, properties and new application opportunities. Mater. Res. **12**, 1–39 (2009)
51. Patel, S., Caldwell, J.M., Doty, S.B., Levine, W.N., Rodeo, S., Soslowsky, L.J., Thomopoulos, S., Lu, H.H.: Integrating soft and hard tissues via interface tissue engineering. J. Orthop. Research®. **36**(4), 1069–1077 (2018)
52. Barthelat, F.: Biomimetics for next generation materials. Philos. Trans. R. Soc. A: Math., Phys. Eng. Sci., **365**(1861), 2907–2919 (2007)
53. Zhang, K., Wang, S., Zhou, C., Cheng, L., Gao, X., Xie, X., Sun, J., Wang, H., Weir, M.D., Reynolds, M.A., Zhang, N.: Advanced smart biomaterials and constructs for hard tissue engineering and regeneration. Bone research. **6**(1), 31 (2018)
54. Hench, L.L.: Bioceramics: from concept to clinic. J. Am. Ceram. Soc. **72**, 93–98 (1993)
55. Hench, L.L.: Opening paper 2015-some comments on bioglass: four eras of discovery and development. Biomed. Glass. **1**(1), (2015)
56. Bressan, E., Favero, V., Gardin, C., Ferroni, L., Iacobellis, L., Favero, L., Vindigni, V., Berengo, M., Sivolella, S., Zavan, B.: Biopolymers for hard and soft engineered tissues: application in odontoiatric and plastic surgery field. Polymers **3**(1), 509–526 (2011)
57. Johnstone, B., Alini, M., Cucchiarini, M., Dodge, G.R., Eglin, D., Guilak, F., Madry, H., Mata, A., Mauck, R.L., Semino, C.E., Stoddart, M.J.: Tissue engineering for articular cartilage repair—the state of the art. Eur. Cell. Mater. **25**(248), e67 (2013)
58. Jones, L.C., Topoleski, L.T., Tsao, A.K.: Biomaterials in orthopaedic implants. In: Mechanical testing of orthopaedic implants, pp. 17–32. Woodhead Publishing, (2017)
59. Parker, A.R., Townley, H.E.: Biomimetics of photonic nanostructures. Nat. Nanotechnol. **2**(6), 347–353 (2007)
60. Hua, Q., Sun, J., Liu, H., Bao, R., Yu, R., Zhai, J., Pan, C., Wang, Z.L.: Skin-inspired highly stretchable and conformable matrix networks for multifunctional sensing. Nat. Commun. **9**(1), 244 (2018)
61. Mao, A.S., Mooney, D.J.: Regenerative medicine: Current therapies and future directions. Proc. Natl. Acad. Sci. **112**(47), 14452–14459 (2015)
62. Pangarkar, S., Pham, Q.G., Eapen, B.C. (eds.): Pain care essentials and innovations E-Book. Elsevier Health Sciences, (2020)
63. Stearns-Reider, K.M., Hicks, M.R., Hammond, K.G., Reynolds, J.C., Maity, A., Kurmangaliyev, Y.Z., Chin, J., Stieg, A.Z., Geisse, N.A., Hohlbauch, S., Kaemmer, S.: Myoscaffolds reveal laminin scarring is detrimental for stem cell function while sarcospan induces compensatory fibrosis. NPJ Regenerative Medicine. **8**(1), 16 (2023)
64. Zydney, J.M.: Scaffolding. Encycl. Sci. Learn., 2913–2916 (2012)
65. Hwang, J., Jeong, Y., Park, J.M., Lee, K.H., Hong, J.W., Choi, J.: Biomimetics: forecasting the future of science, engineering, and medicine. Int. J. Nanomed., 5701–5713 (2015)
66. Wu, M., Liu, X., Bai, H., Lai, L., Chen, Q., Huang, G., Liu, B., Tang, G.: Surface-layer protein-enhanced immunotherapy based on cell membrane-coated nanoparticles for the effective inhibition of tumor growth and metastasis. ACS Appl. Mater. Interfaces **11**(10), 9850–9859 (2019)
67. Sridhar, K., Bouhallab, S., Croguennec, T., Renard, D., Lechevalier, V.: Application of high-pressure and ultrasound technologies for legume proteins as wall material in microencapsulation: new insights and advances. Trends Food Sci. & Technol., (2022)

Sudipta Choudhury is an M.Sc. student at the School of Life Sciences, JSS Academy of Higher Education and Research, Mysuru where she is pursuing her postgraduate degree in Molecular Biology. In 2022, she received her UG degree from Jyoti Nivas College Autonomous in Bangalore with a triple major in Biotechnology, Biochemistry, and Genetics.

The author was able to complete one project during her undergraduate studies that dealt with the pathogenesis of obesity research by inducing obesity in danio rerio with the administration of a high protein diet and were able to give an oral presentation at the 18th International Conference organized by the Department of Animal Science & Indian Association of Applied Microbiologists (IAAM) on "Converging Microbiological Innovation for Application in Animal, Plant, Environment and Health care (ICCMIA–2023)" on 27th and 28th January, 2023.

During her under graduation, the author has been done an internship at EDUFABRICA in association with PULSE, AIIMS DELHI on "Food Microbiology" in 2021. She had taken some other short-term courses and been certified in Medical Laboratory Technology, Clinical Research, Nutrition and Dietetics.

K. R. Arjun is an M.Sc. student at the School of Life Sciences, JSS Academy of Higher Education and Research, Mysuru where he is pursuing his postgraduate degree in Molecular Biology. In 2022, he received his UG degree from Government Science College Bangalore Autonomous in Bangalore with a triple major in Biochemistry, Zoology and Genetics.

Arjun believes in acquiring practical knowledge, and has thus worked on a project that dealt with the 'Cryopreservation and Insemination of HF cow' during his field visit to Danis farm, Hessarghatta. He has also been an active member of Mini Symposium on 'Genetics: Basics and applied' organized by Indian Academy sciences, Bengaluru on 30th of May 2022. Mr. Arjun KR also represented Bangalore Urban district for Power Point Presentation on Topic : 'e- Waste' at District Level Gnana Vignana Mela organized by Vidhya Bharathi, Karnataka at Jaigopal Garodia Rashtrotthana Vidya Kendra, Bangalore on 11th august 2018 during his Pre-University days. He has also been University Student coordinator of National Service Scheme and successfully completed 4 university level camps and 2 state level camps. He was also the Bangalore Malleshwaram Division Convener at ABVP Karnataka during 2021.

Arjun endeavors to pursue his PhD in Genome Sequencing and has great interest the field of Bioinformatics. He wants to continue in the field of research work to make the country a better place for the mankind.

Mr. M. N. Ramesh Bharadwaj Ph.D. aspirant perceiving under the guidance of Dr. Kanthesh M Basalingappa an Associate Professor and Course Coordinator of Division of Molecular Biology, JSS Academy of Higher Education and Research-Mysuru. His area of Research Interest is currently aimed towards inhibition of Molecular signaling pathways involved pancreatic cancer, diabetis and development of drugs with the help of Phyto-chemicals, and genetics. He graduated his UG with a triple major of Microbiology, Biotechnology and chemistry in the year 2019 from Ramaiah College of arts science and commerce, Bengaluru University, jnana Bharathi campus, Bengaluru. He has been awarded with the Post graduate degree, an M.Sc. in Molecular biology from School of Life Sciences, JSS Academy of Higher Education and Research–Mysuru in the year 2021.The author is currently pursuing his Ph.D. under the Division of Molecular Biology in JSS Academy of Higher Education and Research- Mysuru.

During his undergraduate the author did his project related to Yeast producing oils. During his post-graduation the author did his dissertation project related to diabetes with the plant extract.

Dr. M. Maghimaa is an Assistant Professor of Microbiology, Muthayammal College of Arts & Science, Rasipuram, Namakkal, Tamilnadu, India. Her robust experience in the domains of teaching and research spans for 18 years. She began her career in 2004 and still it goes on relentless. She was conferred with a Ph.D. by Periyar University in 2014. Her research interests cover microbiology, biotechnology, and nanotechnology.

She is the author of more than 41 papers in international peer-reviewed Science Indexed (SCI) journals (SCI total impact factor is 50.535) and published two books and a book chapter.

She has to her credit an Australian patent, German patent and published two Indian patent. She root caused to generate major, minor, and student project grants like DBT, NewDelhi (8 Lakhs), Science & Technology Project TNSCST (DST-Grant) (4.2 Lakhs). (SPS projects (6 Nos) TNSCST, Tamil Nadu (0.5 Lakhs), and currently leading a collaborative research (CRS) project of (UGC-DAE-CSR, Indore) (4.54 Lakhs). She has seat-anchored to gear up grants from DBT Star-Scheme (104 Lakhs), and DST–FIST, (40 Lakhs).

She won laurels like National level first prize with a cash award for her presentation on Innovation Science Teaching during pandemic at National Science Teachers Congress (NSTC), 6th India International Science Festival (IISF-2020), organized by Ministry of Science & Technology, Government of India. She has been honoured with "Best Researcher Award from Periyar University among the 116 affiliated colleges of, young scientist Award, Best young researcher Award, Best Young Woman Scientist Award in Microbiology-2020" and more than 27 awards.

Dr. Kanthesh M. Basalingappa is an Associate Professor of Molecular Biology at the School of Life Sciences, JSS Academy of Higher Education & Research, Mysuru, India. His goal of research is to determine the role of RNA Binding proteins in tumor progression and metastasis. Post-transcriptional regulation of gene expression by RNA binding protein is a crucial mechanism in regulating the timing and the amount of expression of genes. Growing evidence indicate that the alteration of the expression and function of RNA binding proteins could potentially play a role in inflammation and cancer.

Dr. Kanthesh BM did is PhD from University of Madras (2005), He also did postdoctoral research at the University Malaya, Kuala Lumpur, Malaysia (2007–2009); West Virginia University, Morgantown, USA (2090–2011) and University of Oklahoma Health Sciences, Oklahoma, USA (2011–2014).

He received Malaysia Prestigious Bio-Malaysia Gold medal Award (2008). For his research area is Arbovirus infections, in that they done patented work on "Early detection of BK virus using molecular methods". He also Received Dr. Wilson Aruni "Best Research Mentor and Teacher Gold medal Award" from the Indian Association of Applied Microbiology (IAAM) (2018).

He has been engaged in teaching and research in Microbiology & Molecular Biology for the past 22 years. He has published over 95 original research papers, 15 book chapters, and 15 review articles. He is also Professional and Scientific Memberships in, American Association for Cancer Research (AACR), Life Member of Indian Association of Applied Microbiology (IAAM), Life Member of Indian Association of Biomedical Scientists (IABMS), Indian Association of Medical Microbiologist (IAMM). He Received Fellowship Award from Indian Association of Applied Microbiology (FIAAM). At present he is a having collaboration with Royal Research Foundation, a research institute in India.

Plant-Based Biomaterials in Tissue Engineering and Drug Delivery Systems

Azadeh Izadyari Aghmiuni, Arezoo Ghadi, and Elmira Azmoun

Abstract Nowadays, tissue engineering, regenerative medicine, and targeted drug delivery systems are considered three important topics in health sciences. However, the existence of a micro-environment for therapeutic cloning which can control cellular functions plays an important role in the control of cellular interactions, the differentiation into the target cells, or the treatment of diseases. In this field, biomaterials are known as one of the logical choices to design such micro-environments, as substrates or carriers of cells/drugs, and simulate damaged tissue functions. Hence, this chapter aims to review the types of biomaterials (protein-based, polysaccharide-based, plant-derived, and extracellular matrix-derived biomaterials), their therapeutic applications (advances and opportunities), and the study of methods for optimizing properties of some biomaterials (morphologically and mechanically), for individual applications.

Keywords Biomedical applications · Biomedicine · Tissue engineering biomaterials · Drug delivery systems · Extracellular matrix-derived biomaterials

A. Izadyari Aghmiuni (✉)
Department of Nanobiotechnology, Pasteur Institute of Iran, 1316943551 Tehran, Iran
e-mail: azadeh.izadyari@gmail.com

Department of Tissue Engineering and Applied Cell Science, School of Advanced Technologies in Medicine, Shahid Beheshti University of Medical Sciences, 19839-63113 Tehran, Iran

A. Ghadi
Department of Chemical Engineering, Ayatollah Amoli Branch, Islamic Azad University, 678 Amol, Iran

E. Azmoun
Department of Chemical Engineering, Islamic Azad University, Gaemshahr Branch, 163 Mazandaran, Iran

R. Malviya and S. Sundram (eds.), *Engineered Biomaterials*, Engineering Materials,
https://doi.org/10.1007/978-981-99-6698-1_6

1 Introduction

Nowadays, the successes of materials-based therapeutic methods depend so much on the origin and characteristics of materials used in this field. In this regard, there are 2 groups of biomedical materials, synthetic and natural; so that, nonbiodegradable/biodegradable polymers, ceramics, metals, etc., are considered synthetic biomaterials, likewise proteins, polysaccharides, herbal active ingredients, and decellularized tissue-derived biomaterials are known as natural biopolymers that each of them can be effective on the specific therapeutic method [1, 2]. Although, significant advances are observed in the field of synthetic biomaterials which can overcome their disadvantages, however, naturally derived biomaterials have still attracted more interest of scientists.

Generally, naturally derived materials (natural biomaterials) can play a crucial role in these approaches owing to their wide properties like biocompatibility, biodegradability, mimicking environmental conditions of the body, remodeling tissue, etc.) compared to synthetic materials/biomaterials [1–4].

Based on the reports, these biomaterials can be applied in biomedicine as carriers or therapeutic agents for treating diseases or regenerating tissues [4–6]. Indeed, they possess the ability to adequately support cellular adhesion, improve cell-to-cell/substrate interactions, and promote cell proliferation, differentiation, and migration [7–10]. Hence, these biomaterials not only can induce an extracellular matrix (ECM) similar to the native ECM of tissues but also lead to an increase in tissue repair/disease treatment process [11, 12]. Moreover, some of the other biomaterials can be suitable options to apply in drug delivery systems or medical devices [9, 13, 14].

According to the importance of the subject, it seems that the effective cognition of natural biomaterials (in terms of structure, features, etc.) can create broad research and therapeutic fields.

Hence, first, this chapter aims to review the types of biomaterials and then their therapeutic applications. In the following, we have discussed the methods for optimizing the properties of some biomaterials (morphologically and mechanically) for biomedical applications.

2 Naturally Derived Biomaterials

In the biomedical field, natural biomaterials based on their origin include the following 5 categories (Fig. 1).

- Polysaccharides: chitin, chitosan, cellulose, agarose, etc.
- Proteins: silk (fibroin and sericin), collagen, etc.
- Glycosaminoglycans
- Cell/organ-derived matrices such as decellularized skin, liver, heart valves, etc.
- Herbal active ingredients

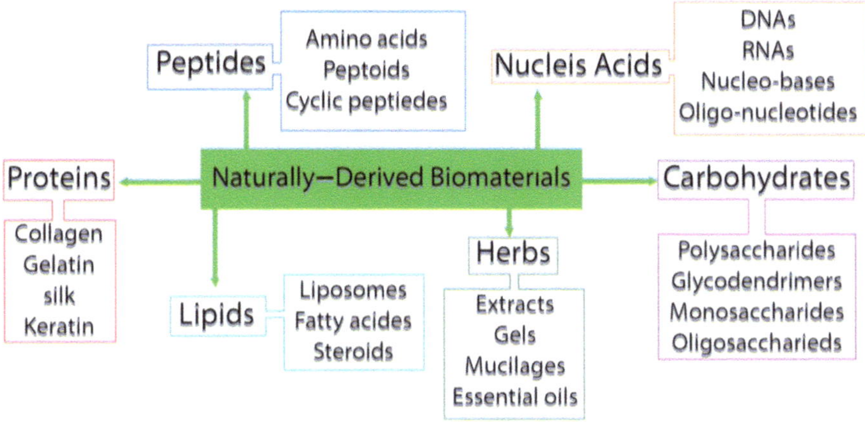

Fig. 1 Some of the natural biomaterials

All these groups play an impact on the decrease of the use of xenografts, allografts, and autografts, and can overcome the limitations such as the reduction of donor sites, and the creation of immune reactions. In this regard, decellularization is one of the methods that can be effectively used as homografts and xenografts [15]. In this method, the tissues or the organs lose all their cell components and leave intact ECMs [12]. In these conditions, the ECM can maintain its role in mechanical supports, signaling, and improvement of cell adherence [16]. Such matrices can be applied in regenerating tissues. Moreover, the design of natural polymer-based substrates, substitutes, or scaffolds can be also useful in achieving such a matrix that can imitate the function and structure of native ECMS [8].

These biopolymers also can act as carriers of drugs, genes, growth factors, macromolecules, etc. for biomedical applications [14, 17]. Notably, medicinal herbs themselves can be used as therapeutic agents or factors for improving cellular behavior.

However, despite the nature of these materials, natural biomaterials possess unique properties which led to an increase in their use of biomedical, tissue engineering technologies, and drug delivery systems. The characterizations and classification of these biomaterials have been listed in Table 1.

3 Applications of Naturally Derived Biomaterials

3.1 Drug Delivery Systems (Advances and Opportunities)

Nowadays, side effects of chemical/synthetic drugs have necessitated the use of methods and materials which can lead to a decrease in pharmaceutical problems (such as over-therapeutic concentrations) and effective and targeted delivery of drugs

Table 1 Classification of naturally derived biomaterials

Naturally-derived biomaterials	Categories	Structure	Sources	Refs.
Polysaccharides	Alginate	This anionic biopolymer with molecular weights of 10–1000 kDa is one of the hydrophilic polysaccharides that can act similarly to the ECM. The primary structure of alginate gels consists of the varying proportions of 1,4-linked β-Dmannuronic acid (M) and 1,4-linked α-L-guluronic acid (G) as M-blocks[a], G-blocks[b], and MG-blocks[c]	Marine brown algae/brown seaweed	[18–20]
	Dextran	This homopolymeric is an extracellular bacterial polysaccharide and includes α-(1 → 3)/(1 → 6) glycosidic linkages with the linear chain and branches. Notably, the properties of this bioactive vary based on the microbial strain. Dextran possesses wide applications in various industries such as cosmetics, food, biomedical, etc. Moreover, dextran can be digested in the colon due to microbial enzyme dextranases, hence a suitable carrier for delivering drugs to the colon region and its related diseases	lactic acid bacteria	[21–26]
	Chitin	This biopolymer is a hard and hydrophobic polysaccharide that consists of (1 → 4)-β-N-acetyl-D-glucosamine units. In this field glyco-chitin and carboxymethyl chitin are its derivatives	exoskeleton and/or internal structures of shellfish, arthropods, and insects (invertebrates)	[27, 28]

(continued)

Table 1 (continued)

Naturally-derived biomaterials	Categories	Structure	Sources	Refs.
	Chitosan	This natural polymer is insoluble in neutral (like water) or basic solvents, however, it can be solved in dilute acids. Notably, the deacetylation degrees of chitin play a crucial role in the modification of the solubility of this biopolymer and solution properties. Some of the chitosan derivatives include N-phthaloylchitosan (phthalic anhydride-treated chitosan), dendronized chitosan, methylthiocarbamoyl, and phenylthiocarbamoyl chitosan, etc.	Deacetylated derivative of chitin	[27, 29, 30]
	Cellulose	This linear polysaccharide consists of D-glucose residues with β-(1 → 4)-glycosidic bonds. Moreover, this organic compound is considered a common polymer in many plants (~33%), wood (~50%), and cotton (~90%) that can provide excellent thermomechanical properties	Cell walls of plants, algae, and bacteria (Acetobacter xylinum, Cluconacetobacter xylinus, and Gluconacetobacter xylinus)	[27, 31–36]
	Agarose	This biomaterial is a main constituent of agar that possesses beneficial features similar to alginate and can act as a differentiation medium for stem cells, as well as an excellent candidate to apply in neural and cardiac tissue engineering	Red algae and seaweed	[37–41]

(continued)

Table 1 (continued)

Naturally-derived biomaterials	Categories	Structure	Sources	Refs.
	Starch	This bioactive is the herbal polysaccharide derived from potato, corn, wheat, rice, etc. This polysaccharide consists of homopolymers of amylose and amylopectin and mechanically possesses poor strength. Hence, its mechanical properties are generally modified by blending with other polymers	Potato, corn, wheat, rice, etc.	[42–44]
Proteins	Collagen	This biomaterial possesses a triple helix [collagen I, II, and III (80–90% of structure)]. At a cellular level, this protein is secreted by fibroblast and epithelial cell and can play a vital role in regulating cellular functions. Collagen IV also supports tissue due to mechanical and biological functions (like tensile strength, tissue development, maintenance, regeneration, etc.) and leads to the improvement of the structural integrity of ECMs	Amniotic membrane (as an attractive source), fibrous tissue (like skin, tendon, ligament), bone, cartilage, corneas, dentin (in teeth), etc.	[19, 45]

(continued)

Table 1 (continued)

Naturally-derived biomaterials	Categories	Structure	Sources	Refs.
	Gelatin	This biopolymer possesses great solubility in aqueous systems and is considered as the major component of some tissues. This biomaterial possesses amino acids of glycine-XY triplets; X and Y are proline and hydroxyproline, respectively that their sequence leads to the formation of triple helical structures and consequently gels	Hydrolysis of collagen	[27, 46, 47]
	Silk (fibroin and sericin)	This biopolymer consists of two fibrous proteins of fibroin and sericin which create an ECM-like network due to the β-sheet structure	Silkworms (silkworm Bombyx mori), scorpions, spiders, etc.	[8]
Glycosaminoglycans	Heparan sulfate	This biomaterial is one of the other linear polysaccharides. Given the extraordinary structural diversity of heparan sulfate chains that can be due to the positions of the sulfate group, the types of proteins, and growth factors can be bonded to its chains	Animal tissues and epimerization of glucuronic acid to iduronic acid	[48–52]
	Chondroitin sulfate	This biomaterial is comprised of repeated di-saccharide units, as well as sulfated at positions 4 or 6. This biopolymer can play a key role in treating cartilage	Extracted and purified from bovine, shark, porcine, chicken, and skate cartilage	[53–55]

(continued)

Table 1 (continued)

Naturally-derived biomaterials	Categories	Structure	Sources	Refs.
	Keratan sulfate	This material includes repeated di-saccharide units of N-acetylglucosamine and galactose and plays an important role in decreasing cartilage tumors	Tissues	[55–57]
	Dermatan sulfate	This material consists of L-idouronic acid in the glycan-chain itself	Tissues	[55, 58, 59]
	Hyaluronic acid	This biomolecule includes repeated disaccharide units; the suitable candidate for biomedicine applications like dermatological fillers, the injection to ameliorate osteoarthritis, preparation of bioscaffolds for tissue repair, etc.	The connective tissues and some body fluids	[19, 60–62]
Cell/organ-derived matrices	Cell-Derived Matrices	These natural materials are acellular complexes of natural fibrillar proteins and macromolecules of the matrix which improve the composition of the native ECM microenvironment. These ECM-derived materials can also provide biomechanical support that leads to the promotion of cellular functions. Generally, these biomaterials are produced by the decellularization strategies such as mechanical, enzymatic, chemical, physical methods, or their combination	Acellular complexes	[63–65]

(continued)

Table 1 (continued)

Naturally-derived biomaterials	Categories	Structure	Sources	Refs.
	Decellularized cardiac valves matrices	The source of these matrices can come from heart valves obtained from humans or porcine, although, do not represent the completely non-immunogenic matrices. However, they can act as structures that possess suitable microarchitectural, biochemical, and biomechanical properties	Heart valves	[66, 67]
	Acellular Dermis	These matrices are another sample of the natural ECMs based on a multi-step process (the removal of the epidermis, disinfection, sterilization)	Human skin tissue	[68, 69]
	Analogs in vitro	These analogs are decellularized ECMs and are known as natural solid organs that not only can act as scaffolds for seeding cells but also lead to the removal of tissue transplantation. Indeed, the process of decellularized ECM and then recellularization of the whole construct can provide the same functions and structures of native tissue	ECM tissue (Human and animal)	[70–77]

(continued)

Table 1 (continued)

Naturally-derived biomaterials	Categories	Structure	Sources	Refs.
	Small intestinal and bladder submucosa	These natural biomaterials are one of the types of naturally occurring ECMs and possess unique combinations such as growth factors, glycosaminoglycans, cytokines, and proteins (collagen, vitronectin, and fibronectin). They can also provide biological signals and improve structural strength	Porcine small intestine or urinary bladder	[78, 79]
Herbal active ingredients	Flavonoid	These bioactive ingredients are one of the phytochemical compounds present in many herbs and fruits which possess potential medicinal applications. Anthoxanthins, narigenin, flavanonols, ugonin-K, hesperetin, flavanones, flavans, iso-flavonoids, flavanone glycoside, etc., are the samples of subgroups of this bioactive that are found in mulberry, orange, lemon, grape, and other citrus fruits, as well as the plants such as *Helminthostachys zeylanica* L., *Herba epimedii*, *Psoralea corylifolia* L., Gu sui-bu herb, *Poncirus trifoliata*	Plants, fruit and vegetables	[80–92]

(continued)

Table 1 (continued)

Naturally-derived biomaterials	Categories	Structure	Sources	Refs.
	Phytoestrogens	These herbal compounds are very similar to estrogens derived from plants. In this field, some phenolic compounds such as stilbene (cabbage, spinach, nuts, broccoli), lignin (wheat flour, flaxseed, peanuts, tea, coffee, and fruits), coumestan, genistein (lupin, fava beans, and soya-beans), and isoflavone are known as phytoestrogens. However, most studies are on isoflavones such as daidzein, biochanin A, glycitein, formononetin, and genistein which are derived from peanuts, soybeans, red clover, and other legumes and play a key role in the body's health	Nuts, seeds, fruits, and vegetables	[93–96]
	Iridoid glycosides	These ingredients are glycosides derived from plants and possess a form of cyclo-pentopyran with an iridodial molecular structure. This bioactive has anti-proliferative effects in cancer models of leukemic and lung cancer	Plants	[97–100]
	Phenolic glycoside	This bioactive includes a sugar unit bound to a phenol aglycone that possesses pharmaceutical properties and can be isolated from the methanol extract of some plants such as the bark of *Magnolia officinalis* and *Gastrodia elata*	Plants	[101, 102]

(continued)

Table 1 (continued)

Naturally-derived biomaterials	Categories	Structure	Sources	Refs.
	Steroidal glycoside	This bioactive is a steroidal skeleton that glycosidically binds to sugar moieties. This ingredient possesses wide pharmaceutical properties such as anti-cancer, cholesterol-lowering, immunoregulatory, anti-diabetic, etc.	Plants such as Radix Ophiopogon japonicas, and Starfish Choriaster granulates	[103–106]
	Monoterpene	These bioactive are found in essential oils and are responsible for the smell and flavor/scents of plants or fruits. In this field, Citral and Thymol are constituents of the smell and flavor of lemon and oranges, respectively. Likewise, limonene, 1,8-cineole, a-pinene, camphor, geraniol, fenchone, and geranyl are constituents of the scent of flowers. Generally, monoterpenes have many therapeutic properties like anti-microbial, anti-cancer, etc.	Plants, fruits, and vegetables	[107–109]

(continued)

Table 1 (continued)

Naturally-derived biomaterials	Categories	Structure	Sources	Refs.
	Triterpene	These bioactive are derived from active isoprene and their oxidation forms various structures. Many triterpenes play a crucial role in decreasing inflammatory diseases and immune responses and can act as inhibitors of hypersensitivity and agents of anti-diabetic and anti-viral	Medicinal plants	[110–115]
	Triterpenoids	These bio-actives (as the largest phytochemical group) are a metabolite of isopentenyl pyrophosphate oligomers that are found in apples, figs, olive, rosemary, lavender, thyme, oregano, etc. Cycloartanes, isomalabaricanes, lupanes, limonoids, oleananes, and sqalenes are triterpenoidal groups	Seaweeds, fruits, and medicinal plants	[116–118]

[a] Homogeneous M-M segments
[b] Homogeneous G-G segments
[c] Alternating M-G segments

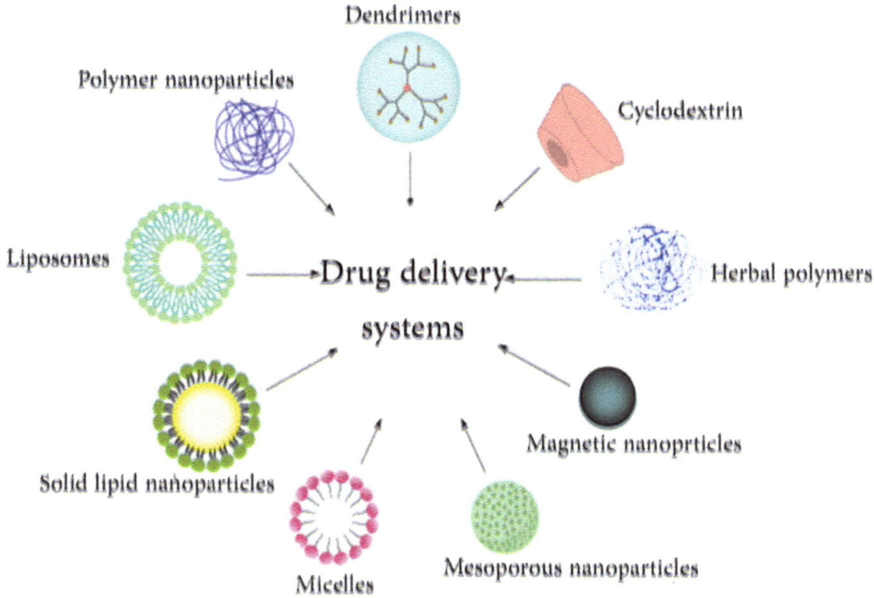

Fig. 2 The common carriers for biomedical applications

[119]. Pharmaceutical systems are considered the main approaches in this field that can be systemically and orally administered to deliver hydrophobic and hydrophilic therapeutic agents. These systems as effective therapeutic actions play a key role in controlled drug release, especially for low concentrations of drugs. A drug carrier in these systems provides a rate of controlled release and complete rehabilitation of the drug in more time [120–122]. These carriers can be injectable and/or implanted in the damaged or diseased tissue/cell to improve delivery (Fig. 2) [123].

One of the methods for achieving this aim is the surface modification of carriers or the design of hybrid/composite carriers to prevent the rapid destruction/degradation of the carrier, encapsulate the drug for a lengthened release, and sustain drug stability [124, 125]. In this regard, biomaterials and their products can be suitable options due to their biocompatible properties and nontoxic, lower processing costs, etc.

Many studies in this field indicate the interest of researchers in using these attractive materials. The study of Kilicarslan et al. is one of these cases that is about complex films prepared with alginate-chitosan to treat periodontal diseases [126]. Accordingly, it is known that the increase in alginate concentration and chitosan molecular weight can increase the adhesiveness of film content. Their reports indicated that a 3:1 concentration of alginate: chitosan along with the lower molecular weight of chitosan can provide more controlled release.

The release of ciprofloxacin antibiotic from alginate–gelatin microspheres is another study performed by Islan et al. [127]. They stated that the encapsulation of this antibiotic can decrease gastrointestinal irritations related to its oral administration. Based on their results, these microspheres indicated ~80% encapsulation efficiency and release of less than 10% in 15 min (under simulated gastric conditions). Moreover, the results of this study indicated that Ca^{2+}/glutaraldehyde-crosslinked alginate-gelatin microspheres can provide more suitable surfaces for decorating bioactive molecules.

Chen et al. also illustrated that chitosan/alginate-coated complexes can provide the controlled release of insulin and overcome challenges of insulin delivery [128]. Based on their results, the self-assembly of chitosan and alginate-coated nanoparticles through the double emulsion method provides nanoparticles in the 200–300 nm range and prolonged insulin release.

There are many pieces of research on the matter of targeted drug delivery for diabetes treatment, and the development of metformin nano-emulsions is one of these studies. Accordingly, Kumar et al. stated that alginate nanocapsules-loaded metformin with an encapsulation efficiency of 78% can improve gastrointestinal absorption and intestinal permeability and promote the efficacy of the drugs (3 times higher than that of pure metformin) [129].

Varadharaj et al. demonstrated that starch-based nanoparticles possess anti-diabetic and antioxidant properties. Such that 100 μg/mL concentrations indicated the highest antioxidant properties (~75%) [130]. They reported that synthesized nanoparticles possess potential properties against diabetes due to their high activity in alpha-amylase inhibition (58.56%).

Starch-modified nanoemulsions loaded with borage seed oil are another study for targeted delivery systems that can be applied in food industries and the production of dietary supplements [131]. Accordingly, Rehman et al. illustrated that such nano-emulsions along with peppermint oil can be used as a natural antioxidant (maximum antioxidant capacity: 99.42 μg Trolox/mL).

The development of polymer micelles loaded with curcumin is another study on treating diabetic ulcers [132]. In this study, Usman Akbar et al. showed that these polymer micelles as therapeutic carriers can overcome limitations in the use of curcumin like poor solubility and low bio-availability. Moreover, the curcumin-loaded micelles can accelerate the wound healing process compared to pure curcumin. Yang et al. have also reported that insulin-loaded polysaccharide hydrogels can be effective in treating type 2 diabetes mellitus [133]. Accordingly, the authors stated that oral administration of encapsulated insulin not only can significantly reduce the disease symptoms in diabetic mice but also improve insulin sensitivity and regulate lipid metabolism.

Some food gums such as Salcan can be also suitable candied for oral drug delivery. The study of Qi et al. is one sample in this field to deliver insulin orally [134]. Based on their reports, the acidic pHs led to an inhibition of insulin release, while neutral conditions significantly increased drug release. Furthermore, oral administration of these hydrogels illustrated a sustained reduction in the level of fasting blood glucose in diabetic rats (after 6 h).

Studies have demonstrated that herbal polysaccharides-based nano-hydrogels can also play an impact on dermal regeneration. In this regard, bio-composite nano-hydrogels based on bamboo prepared by Padwad et al. were known as a suitable therapeutic factor for healing wounds in streptococci that can regulate the expression level of inflammatory cytokines and can lead to an increase in collagen formation and epithelialization [135].

Moreover, the nano-micro-hydrogels based on animal and herbal proteins are known as high-throughput biomaterials that can be used in targeted drug delivery as the bioactive carrier. Such that, Abaee et al. stated that whey and soy proteins as the coating materials play an impact on the therapeutic agent encapsulation [136]. They also stated that the gelation methods and hydrogel network of protein hydrogels possess a significant effect on the liberation features and drug bioactivity.

Hou et al. also demonstrated that hydroxypropyl-β-cyclodextrin-based nano-emulsions containing cinnamon essential oil can act as slow-release systems with high antibacterial properties [137]. The authors stated that these nano-emulsions not only are used as an alternative to the synthetic preservative but also provide a long-term antibacterial effect (stable at 4–40 °C, for 60 days).

Moreover, some bio-molecules such as hyaluronic acid exhibit excellent function in delivering drugs with small molecular weight. In this field, Cai et al. reported that nano-conjugation of hyaluronic acid and doxorubicin can be effective in localized delivery systems of doxorubicin to the breast tissue for treating breast cancer [138]. They also found that the conjugation process can lead to a decrease in dose-limiting cardiac toxicity and inhibition of cancer progression. Finally, the authors stated that localized chemotherapy with the nano-carriers may be a promising strategy to treat some early-stage cancers.

Nowadays, some platforms have developed as suitable tumor-targeted drug delivery systems that apply in imaging magnetic resonance along with photothermal-chemotherapy. The reports of Zheng et al. indicated that hyaluronic acid-C16 co-polymers can well act for such aims [139]. They found that multifunctional micelles containing docetaxel can provide a uniform nanosize along with colloidal stability in aqueous solutions. Moreover, the creation of a magnetic field and its combination with photothermal therapies and drug-loaded multifunctional micelles can lead to an increase in the targeting efficacy and therapeutic synergist.

The adjustment and modification of silk fibroin surface features via covalent functionalization is another approach for using silk proteins. Accordingly, Zhang et al. illustrated that the use of covalently coupling silk fibroin with insulin can improve insulin loading efficiency and be led to drug release under certain environmental conditions [140]. They found that this method can provide a 2.1-fold and 1.7-fold increase in drug half-life in vitro than that of conjugated albumin-insulin and native insulin, respectively. It can be explained by intensive β-sheet structures in silk fibroin for stabilizing insulin.

Given the limitations of anti-cancer methods such as systemic administration or frequent administration intravenously, Gangrade et al. provided silk-based nano-hybrid injectable hydrogels for localized delivery of anti-cancer drugs [141]. They indicated that the incorporation of silk proteins and single-walled carbon nanotubes which functionalized with doxorubicin-loaded folic acid can increase the mechanical strength of hydrogel and be led to a slow sustained release of doxorubicin during 14 days. Moreover, such hydrogels possess easier injectables due to the viscoelastic property of silk and can be injected near or intratumoral tumors.

A multifunctional silk fibroin-manganese oxide nanoparticle-based platform was also synthesized by Yang et al. as a promising strategy for tri-modal therapy of cancer assisted by MR/fluorescence Imaging [3]. These photodynamic agent/doxorubicin-coloaded nanoparticles indicated the suitable reactive into the tumor microenvironments and provide more strong and stable photothermal effects. More studies in this field have been summarized in Table 2.

3.2 *Tissue Engineering (Advances and Opportunities)*

The reports have indicated that natural polymers and bioactive materials can be promising strategies in biomedical applications due to their role in the improvement of cell–cell biological responses and the imitation of functional and structural properties of the target tissues [163]. Indeed, recent efforts in the fabrication of naturally derived biomaterials-based substitutes [164–169] have provided significant opportunities to generate novel tissue substitutes [15, 170].

Medicinal herbs are known as one of these biomaterials which gaining significant attention among researchers due to their potential for increasing cell proliferation and controlled differentiation [171, 172]. The biopolymers similar to signaling biomolecules of tissue extracellular matrix (ECM) (i.e. glycoproteins, proteoglycans, and glycosaminoglycans) are other samples from these materials. In this field silk fibers [173] and hyaluronic acid [174–176] can imitate the behavior of glycosaminoglycans and proteoglycans in the tissue ECMs and be led to an increase in cell-to-cell or cell-to-scaffold signaling [175, 177, 178].

According to studies, naturally derived biomaterials can be effective in cellular functions like signaling, growth, differentiation, etc. Generally, the tremendous differentiation of stem cells acts as a double-edged sword, hence therapeutic strategies that led to the controlled differentiation of cells play a crucial role in this field. One of these strategies includes the use of biomaterials to design tissue substitutes [2, 11]. Such an approach provides suitable physicochemical and mechanical properties for cell differentiation and results in a sustained delivery and local of differentiated cells to the targeted tissues.

In this regard, the studies of Izadyari Aghmiuni et al. indicate that herbs-derived biomaterials possess high therapeutic potential [171]. Although, herbal ingredients-based different formulas such as gels, creams, ointments, etc., cannot be directly used in damaged tissues like bone, cartilage, nerve, cornea, tendon, etc. However, this research team showed that some mucilage such as quince seed possesses good restorative potential when used in the structure of tissue substitutes. Our results illustrated that quince seed mucilage-based hybrid scaffolds not only can provide a better porous network than chitosan-based scaffolds but also support dermal fibroblasts and improve the absorption capacity of water or wound secretions on the scaffold. They also found that the combination of quince seed mucilage and PEG can increase the transduction of mechanical force-induced signals and promotes the biological signals to induce stem cell differentiation into targeted cells such as dermal cells.

Table 2 Some of the carriers applied in various industries

Industry	Carrier/entrapped material	Function	Refs.
Pharmaceutical	Layer-by-layer coated anionic nano-liposome along with electrostatic interactions with chitosan/Insulin	Oral delivery of insulin; the increase of drug uptake and transport via epithelial cells; release into simulated intestinal and gastric fluids was 2% (in 2 weeks) and ~6% (in 1 h) respectively	[142]
	Hyaluronic acid-based nanoparticles/Insulin	The optimal drug loaded into the nano-carrier was 10.61 ± 1.16% (w/w); the release of insulin from the nano-carrier was 22.01 ± 3.84% and 63.23 ± 4.10% at pHs 1.2 and 7.4, respectively; relative bioavailability was measured 14.62%	[143]
	Modified banana starch nanoparticles/Curcumin	Acetylated nanoparticles provide a stronger hydrogen bond between the starch and curcumin, as well as increase the encapsulation efficiency of curcumin (>80%)	[144]
	Succinylated chitosan-alginate core–shell nanoparticles/Quercetin	Encapsulation of quercetin into core–shell nanocarrier was performed by ionic cross-linking. The encapsulation was ~95% along with the kinetics of the non-fickian trend. This carrier can be effective in oral quercetin delivery for treating diabetes	[145]
	Gelatin and carboxymethyl cellulose gel/Herbal extract and salicylic acid	Antioxidant and anti-inflammatory properties; suitable for wound dressing, moisture retention of the wound; wound exudate absorption, and biocompatibility	[146]
	Gelatin-polyacrylic acid nanoparticles/Oxytetracycline	This formula is a suitable method for delivering controlled dosage of antibiotics and decreasing ophthalmic infections that can increase therapeutic efficacy and the ocular residence time of antibiotics and improve the sustained effectiveness against keratitis	[147]
	Hydrogels based on Chitosan–gelatin/Curcumin-loaded nanoparticles	Reduction of oxidative stress in the trabecular network via curcumin nanoparticles with an optimal concentration of 20 μM; the decrease of latanoprost side effects; creation of the sustained-release in both therapeutic agents; a suitable method for treating glaucoma	[148]

(continued)

Table 2 (continued)

Industry	Carrier/entrapped material	Function	Refs.
	Eudragit RL-100/ gelatin-based nanoparticles/ Ketorolac tromethamine	Increase of controlled drug release and improvement of ketorolac tromethamine bioavailability for ocular applications	[149]
	Gelatin/Silver nanoparticles	The increase in stability and antibacterial activities against *Staphylococcus aureus*; inhibition of angiogenesis disturbance and corneal neovascularization; a dual functional nanotherapeutic (i.e. antimicrobial and antiangiogenic) to treat eye-related microbial infections	[150]
	Egg-phosphatidyl choline-derived Liposomes/ Latanoprost	A long-term sustained release of the drug via subconjunctival injection; lowers intraocular pressure for up to 90 days	[151]
	Carbopol 934 gel-based self-nano emulsifying/ 6-Gingerol	The clear, stable, and transparent nanoemulsion with spreadability of ~234 g cm/sec, an increase of permeation of skin and solubility; better topical application; management of wound healing due to the better anti-inflammatory properties	[152]
Food	Nanoemulsions/Essential oils	The increase in stability of essential oils; supports effective delivery systems	[153]
	Oil-in-water nanoemulsions/ Eugenol oil	Physically stable up to more than 1 month as an antifungal agent against *Fusarium oxysporum* f. sp. vasinfectum	[154]
	Nanoemulsion/Oregano oil	The stability of the oil; inactivating and controlling the growth of foodborne bacteria sustainably	[155]
	Soybean-based polysaccharide emulsions/ Thyme oil	Thyme oil was stable for 60 days (21 °C); thyme oil emulsion possessed a better antimicrobial property compared to pure oil of this herb; such emulsions can act as antimicrobial delivery systems for increasing the microbiological safety of products in the food industry	[156]

(continued)

Table 2 (continued)

Industry	Carrier/entrapped material	Function	Refs.
Cosmetic	Nanostructured lipid carriers based on Glyceryl palmito-stearate and fatty acid triglycerides/Idebenone	Liquid lipids such as fatty acid triglyceride provide a higher solubility for idebenone compared to solid lipids; this process can help the stability of drugs and be led to an increase in their applications in the cosmetic industry as the materials of antioxidant, anti-aging, anti-acne, etc.	[157]
	Nanostructured lipid carriers based on the liquid oil/ Coenzyme Q10	Improvement of the water solubility and photo-stability of Coenzyme Q10 in aqueous solution (>80%) than Coenzyme Q10 itself (<33.5%), so that photo-stability of this antioxidant was ~65% after 5 months; hence the nanostructured system can considerably increase the stability of materials for various applications	[158]
	Nanoparticles and nanoemulsion based on oleic acid/Lornoxicam	The size of nanoparticles and nano-emulsion were 141–295 nm with physically stable for 6 months (various temperatures) with a release mechanism of Fickian drug diffusion along with a high rate of drug penetration. Such systems can be effective in the transdermal delivery of drugs to treat inflammatory and painful diseases of the skin	[159]
	Nanostructured lipid carriers based on glycerin monostearate, glyceryl triacetate/ Alpha-lipoic acid	These nanostructured lipids possessed high photo-stability (93.9%) for 30 days that free alpha-lipoic acid (42.5%); such systems can act as carriers for loading unstable lipophilic active ingredients such as antioxidant materials and increase their effectiveness in cosmetic productions	[160]
	Lipid nanocarriers based on stearic acid, Virgin coconut oil/Phenylethyl Resorcinol	Such carriers can be led to a high photo-stability (~88%) in 90 days and an increase in the retention amount of Phenylethyl Resorcinol in the skin	[161]
	Nanostructured lipid carriers/ Podophyllotoxin	Such lipid nanocarriers can provide high dermal targeting efficiency and safety, as well as improve the penetration of hydrophobic drugs for topical application	[162]

Another study in this field is aloe vera-based substrates [179, 180]; the reports of Kallyanashis Paul et al. indicate that aloe vera-based hydrogels can alleviate maternal birth injuries [181]. Accordingly, aloe vera-alginate hydrogels can act as an immediate treatment by delivering stem cells and regenerating damaged tissue. Indeed, the authors stated local injection of such hydrogels (with or without stem cells) can be significantly effective in repairing birth injuries and act as a therapeutic strategy for preventing pelvic organ prolapse or healing birth injuries.

Suganya et al. demonstrated that fibers of poly(caprolactone) along with aloe vera powder (10 wt%) have better hydrophilic properties, higher tensile strength (6.28 MPa), and elastic modulus similar to dermal tissue. Our reports indicated that the electrospun meshes of aloe vera can increase cell proliferation, secretion of collagen, and expression of F-actin compared with polycaprolactone-collagen fibers [182].

Based on the studies of Tahmasebi et al., nano-fibrous polymer substrates blended with aloe vera gel can be also used in hard tissue engineering [183]. The results illustrated the better biocompatibility of the scaffold when blended with aloe vera. Moreover, they stated that aloe vera gel possesses osteoinductive potential and can be used for acceleration of bone regeneration as a bio-implant.

Based on the studies, the use of aloe vera along with other herbal compounds can accelerate the wound healing process. Hybrid nano-fibrous scaffolds based on Aaloe vera and curcumin are the sample in this field that was designed by Ezhilarasu et al. These scaffolds can provide a synergistic effect on the fibroblast behavior and improve the antimicrobial properties of the scaffold [184].

According to the studies, herbal active ingredients can play an impact in improving the biological characteristics of substrates, when these bioactive ingredients were used along with other naturally derived biomaterials. Such that, Zadegan et al. illustrated that silk fibroin-base nano-fibers containing *Urtica dioica* L. (nettle) possess better water uptake and cellular attachment, as well as higher cellular proliferation than that silk fibroin nano-fiber [185]. Based on the reports of Ghiyasi et al., scaffolds containing Nettle/ZnO-nanoparticle can also provide the synergy effects for the increase of antibacterial activities. They indicated that the addition of nettle and ZnO-nanoparticles to poly(caprolactone)-based substrates can increase the tensile strength up to 2.54 MPa, improve water uptake ability, and promote fibroblast L929 cell proliferation [186].

Hajiali et al. also indicated that sodium alginate/lavender essential oil-based dressings led to a high efficacy for the treatment of UVB-induced dermal burn [187]. Such dressings are also able to decrease and control the inflammatory responses, induced in the skin fibroblasts due to UVB exposure.

Some reports illustrate that animal fats can also improve the differentiation and proliferation of cells; so that, they can be a suitable candidate to fabricate wound dressings and/or bio-engineered substitutes containing stem cells for skin tissue repair [188]. There are many studies on this field, some examples have been listed in Table 3.

Table 3 Bioactive ingredients-based substitutes to improve cellular interactions and functions

Herb	Scaffolds, substrates, or gels	Function	Refs.
Curcumin	Electrospun poly(ε-caprolactone)-poly(ethylene glycol)-poly(ε-caprolactone) fibrous mat containing curcumin	An increase of wound closure and suitable candidates for wound dressings	[189]
	Hydrogels containing encapsulated curcumin	The increase of antioxidant and anti-inflammatory properties, acceleration in wound closure, and creation of collagen content	[190]
	Poly(ε-Caprolactone) nano-fibers containing curcumin	The increase in cell viability, anti-inflammatory properties, and antioxidant activities, the decrease of interleukin-6 release, as well as the increase in the rate of wound closure	[191]
	Chitosan–Gelatin Composite Sponge containing curcumin	High capacity for water absorption, the increase of antibacterial properties, and wound closure	[192]
	Fish scale collagen-based scaffold containing curcumin	Suitable antimicrobial properties, acceleration of wound healing, and suitable for dermal tissue engineering applications	[193]
	Collagen-alginate scaffold containing Curcumin-loaded chitosan nanoparticles	Acceleration in wound closure; the complete epithelialization along with the formation of thick granulation tissue; the decrease of inflammation; regeneration of diabetic wounds	[194]
The combination of herbal extracts	Electrospun polycaprolactone scaffolds containing herbal extracts	The increase in the proliferation of fibroblasts and epidermal differentiation; the suitable for dermal tissue engineering	[195]

(continued)

Table 3 (continued)

Herb	Scaffolds, substrates, or gels	Function	Refs.
Xylan	Polyvinyl alcohol-based nano-fibers containing xylan	The improvement of fibroblast proliferation and adhesion, mechanical properties, and the biodegradable rate of the scaffold; the suitable for skin tissue regeneration	[196]
Cissus quadrangularis	Polycaprolactone scaffold incorporated with Hydroxyapatite and Cissus quadrangularis	The increase of osteogenic activities and hard tissue regeneration, improvement of cell growth, and mineralization; the suitable for bone tissue regeneration	[197]
Spinacia oleracea	Alginate-carboxymethyl cellulose scaffold containing Spinacia oleracea extract	The improvement of mechanical stability and biocompatibility of scaffold, the control of degradation rate; the increase in the proliferation of human osteosarcoma cells, suitable for bone tissue regeneration	[198]
Gum tragacanth and curcumin	Electrospun nano-fibers containing Gum tragacanth-curcumin	The increase of the antibacterial properties, a decrease of the epithelial gap; acceleration in wound closure, as well as, the formation of granulation tissue, hair follicles, and sweat glands	[199]
Arnebia euchroma	Polycaprolactone-chitosan–polyethylene oxide scaffolds containing arnebia euchroma	High potential in burn wound healing, controlled degradation rate; improvement of swelling and the mechanical property; the enhancement of antibacterial activity, suitable for applications of dermal tissue engineering	[200]

3.3 Combination of Tissue Engineering and Drug Delivery

Tissue engineering is a potential strategy to address the demand for tissue substitutes resulting from trauma and diseases. However, the use of therapeutic agents in combination with this strategy plays a key role in the creation of new generations of engineered substrates that can regenerate the speed and quality of the formation of new tissues (Fig. 3). Some of the recent advances resulting from the combination of drug delivery systems and tissue engineering strategy have been described in this section. The studies of Karimi et al. are a sample in this field. They reported that hydrogel dressings containing silk fibroin/polydopamine nanoparticles as a combined method (drug delivery and tissue engineering) can play an impact on delivering antibiotics such as Ciprofloxacin [201]. The results of their study indicated such dressings can preserve the bioactivity of the released antibiotics (>80%) and be led to inhibition of the bacteria growth. These hydrogel dressings can also support fibroblast cells due to the existence of various nanoparticles.

Moreover, the electrospun scaffolds have recently received attention in topical drug delivery. Such an approach not only can significantly decrease the systemic absorption of the pharmaceutical agents but provides localized therapeutic effects at lower concentrations, as well.

Core–shell electrospun scaffolds are another sample that developed due to their therapeutic application. Indeed, according to this strategy, one component of the fiber or some biologically active molecules can be encapsulated inside another component [202] and the shell not only controls the release kinetics of the therapeutic agent but protects the core biomaterial as well [203, 204].

Zhang et al. assessed the protein encapsulation into PCL nanofibers by the electrospinning technique [205]. They reported that the core–shell nanofiber suppresses

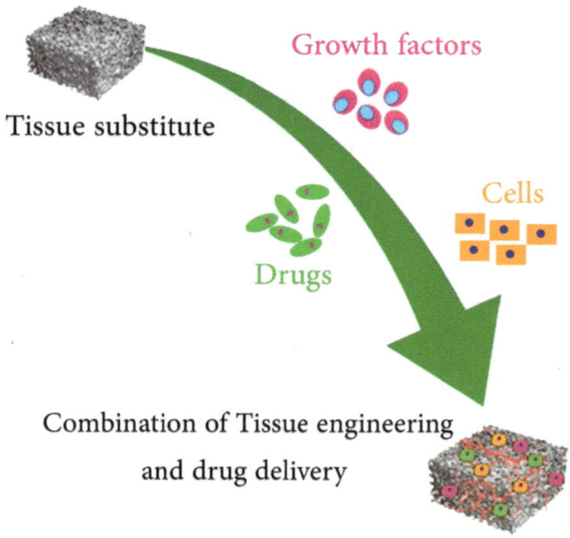

Fig. 3 The combination of delivery of therapeutic agents with tissue engineering for biomedical applications

the burst release of the drug and be led to sustained release. Notably, the release rate was highly dependent on the degradation rate of the outer fiber, the shell.

Emulsion electrospinning also is known as a new technique that can create core–shell fibers via stable polymeric emulsion. Yang et al. indicated that the use of the emulsion electrospinning method plays a crucial role in using core–shell-based ultrafine fibers as the carrier for therapeutic agents, such as proteins [206]. This technique leads to the creation of porous fibrous mats with bead-free, integrally core–shell structures.

The core–shell electrospun fibrous also can provide a composite carrier by forming a polymeric core and an inorganic shell. In this matter, Cui et al. demonstrated that the incubation of PDLLA electrospun fibers containing calcium nitrate in phosphate solution form calcium phosphate with a good dispersion on the fibers [207]. They stated that this core–shell nano-composite fibrous as a scaffold or filler possesses potential applications in tissue engineering and biomedical.

Grafting the amino groups on the mentioned fibers is another sample in this matter, performed by Cui et al. via an optimized aminolysis process [208]. The authors found that the incubation of PDLLA fibers and gelatin containing hydroxyapatite into the simulated body fluid can provide the core–shell composites with hydroxyapatite nucleation and growth. They also indicated that the core–shell composite of calcium phosphate is created through layer-by-layer self-assembly of polymer fibers and gelatin. It seems that such composites can be effective in bone regeneration and mineralization [209, 210].

Nowadays, there are many polymers (degradable and non-degradable) in this field that can act as carriers for local or systemic delivery of drugs by electrospinning process. Indeed, the advantage of this technique is in maintaining the bioactivities and molecular structures of the incorporated drugs or biomolecules that can be due to process conditions.

Hence, it seems that a combination of both tissue engineering and drug delivery techniques can be a promising approach to treat diseases or damaged tissues through local, systemic, or implant treatments.

4 Optimization of Biomaterials Properties

The physiological and biomechanical similarities of the biomaterials-based scaffolds/substrates to the native ECMs are the most important factors governing their ability to achieve tissue functions. Accordingly, native ECM of damaged tissue provokes a more natural healing response compared to synthetic scaffolds [211]. Moreover, biomaterials can promote cellular infiltration, proliferation, differentiation, and migration into targeted tissues and provide the mimicking substrates of the ECM structure and function [8, 12].

Given that many biomaterials can be used for the design of hydrogels, scaffolds, or substitutes of tissue, the preservation of the physical structure and their

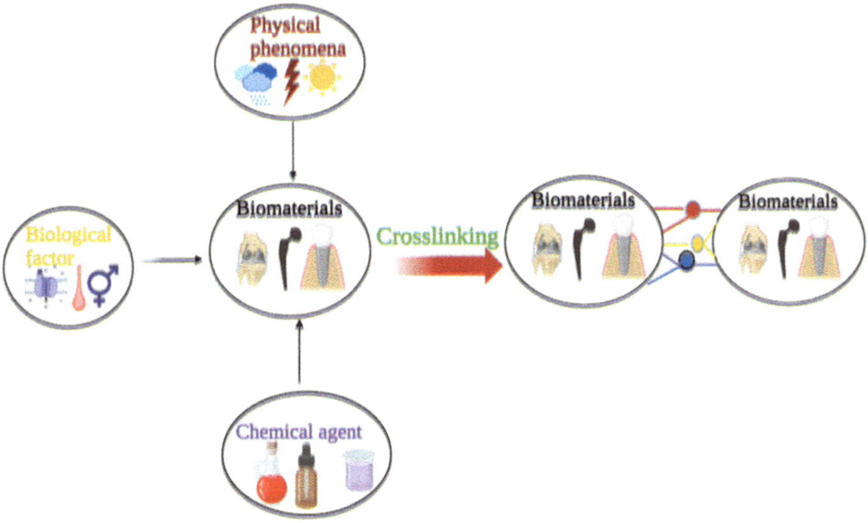

Fig. 4 The methods for modification of polymer matrices/structures

strength is paramount [211] because one of the problems of naturally derived biomaterials is rapid biodegradation rate and low stability in an aqueous and physiological environment.

In this field crosslinking method of the materials as the most common approach plays a key role in achieving the modified polymeric matrices/structures (Fig. 4). [212]. Glutaraldehyde is one of these crosslinkers that although led to cytotoxicity or undesirable changes in some of the biomaterial functions, however, can create permanent crosslinks between proteins, decrease the antigenicity activity of ECM biomaterials, and result in the non-biodegradable matrix [211, 213–216]. Hence, before seeding the cell, synthesized matrices with this crosslinker should be washed. Notably, such matrices are usually used for permanent implants in the body.

In this regard, the use of other chemical crosslinkers like sodium metaphosphates, epichlorohydrins, carbodiimides, polyepoxy, and carboxylic acids such as citric acid can provide suitable mechanical strength and stiffness along with the decrease of cytotoxic effect, as well as improvement of cell attachment and stability of matrix [211, 214, 217]. Although, in-vivo responses to such chemically altered matrice are much different compared to native ECM.

Physical crosslinking is another method for modifying matrice properties and comprises several steps of heating or compression. Although the advantage of this method is the absence of chemical residue in the matrix structure, however, physical crosslinks are not as stable as ones made with other methods [211].

Enzymatic crosslinking also is a natural method to create crosslinks between polymers. In this matter, transglutaminase (TGase) is known as one of the crosslinkers used by many organisms [211].

Table 4 The design of new biomaterials by crosslinking

Crosslinking biomaterials to form various structures	Description/Reason for using the crosslinking technique	Most widely used biopolymers and related crosslinkers	Effects	Refs.
Films and membranes	The easiest biomaterial structures for application of tissue engineering and controlled release/The improvement of mechanical property of substrate and decrease of degradation rate	Biomaterials: collagen Crosslinkers: EDC/NHS, acid chlorides	The increase in tensile strength (57%) and modulus (~17-fold), as well as a decrease in swelling of the structures via controlling crosslinking conditions	[218, 220, 221]
Porous and sponge-liked structure	These structures are known as composite/hybrid substrates and are used in the applications of hard tissue engineering, as well as the design of wound dressings and periodontal matrices/To improve thermal stability and increase the rate of biodegradation	Biomaterials: collagen, hyaluronic acid Crosslinkers: glutaraldehyde, EDC/NHS	The interconnected 3D structures, a decrease in swelling and enzymatic degradation without affecting the cell viability	[222–225]
Biopolymeric hydrogels	The structures with a high capacity to retain water and apply in biomedicine/To avoid the dissolution of hydrogels	Biomaterials: gelatin, collagen, chitosan Crosslinkers: glutaraldehyde, carboxylic acid, dextran aldehyde EDC/NHS	The decrease of enzymatic degradation, an increase of the viscoelastic properties and the breaking modulus of the gel, and promotion of adhesion and growth of cell	[214, 226–229]
Regular fibers via wet/melt spinning	The fibers with different diameters for biomedical applications and tissue engineering	Biomaterials: collagen, herbal proteins like and soy proteins, gelatin, silk fibroin Crosslinkers: EDC/NHS, carboxylic acids	The improvement of mechanical properties and cell adhesion	[230–232]

(continued)

Table 4 (continued)

Crosslinking biomaterials to form various structures	Description/Reason for using the crosslinking technique	Most widely used biopolymers and related crosslinkers	Effects	Refs.
Ultrafine fibers	These fibers possess unique properties for biomedical applications and are very similar to ECM networks/To mimic the structure and fabrications of ECMs, as well as increase tensile strength	Biomaterials: proteins and polysaccharides (like starch, pullan, dextran) Crosslinkers: EDC/NHS, glutaraldehyde, citric acid, glycerol, glycerol phosphate, tripolyphosphate, tannic acid, acetic acid, trisodium metaphosphite	Promotion of cellular attachment and proliferation, a decrease of degradation rate	[216, 218, 233–237]
Micro/nanoparticles	These particles derived from biomaterials are used for *in-vivo* pharmaceutical delivery and other therapeutic agents/The improvement of the performance of nanoparticles	Biomaterials: chitosan, carboxymethyl pullulan, hyaluronic acid, starch, alginate, collagen, gelatin Crosslinkers: glutaraldehyde, sulfate anions, sodium metaphosphate, tripolyphosphate, epichlorohydrin, carbodiimide	The increase in stability, a decrease of particle size and degradation rate, modification of physicochemical properties of particles like low viscosity, swelling, controlled release of drugs	[238–242]

Generally, crosslinking methods are used for loading drugs, nutraceuticals, genes, growth factors, etc., as well as designing electrospun structures, fibers, films, hydrogels, 2D/3D scaffolds or substrates, and micro/nano-particles from biomaterials for *in-vitro*/*in-vivo* applications [214, 218, 219]. Table 4 indicates the chemicals and techniques used for crosslinking biomaterials and medical applications.

Generally, the design of ideal matrices (scaffolds, substrates, substitutes, carriers, etc.) for applications of tissue engineering or drug delivery systems can play a key role in the regeneration of damaged tissues and the treatment of diseases. These matrices should be able to facilitate cell attachment, induce cell proliferation and differentiation, as well as form extracellular matrix (ECM) deposition, as well as lead to a controlled release of therapeutic agents in drug delivery systems [243]. In this matter, mechanical stability, modification of function and structure of nano/micro-carriers, as well as delivering and localizing the biomolecules (cells, growth factors, and bio-actives into damaged tissue) are the main properties of the designed matrices.

Some of the design criteria of such matrices for biomedical applications involve the following:

4.1 Biocompatibility.
4.2 Ability to adhere, differentiate, proliferate, and deliver cells.
4.3 Biodegradability.
4.4 Controlled release.
4.5 Mechanical properties depended on the biomaterial properties.
4.6 Porosity of matrix [the creation of porous structures (>90% along with a pore size of 300–500 μm) to penetrate and grow cells, and deliver nutrients for applications of tissue engineering, and nano/micro-size for drug delivery systems.
4.7 Fabrication capability (easy formation of biomaterials).

Table 5 indicates some of these parameters that can be effective in the design of matrices.

5 Conclusion and Future Perspective

Nowadays, biomedical sciences have excitingly led to excellent results in repairing and treating diseases. Such methods can overcome the limitations of some techniques such as grafts of autologous, allograft, or xenograft tissue, surgeries, and cell transplantation. In this regard, naturally derived biomaterials can play a crucial role in the process of regenerative medicine or the treatment of diseases; so that, if protocols or methods can be created via the use of these materials to proliferate and differentiate cells into targeted cells in damaged tissues or control the drug release, it will offer hope to cure many incurable diseases. Hence, it seems that a combination of modern and traditional medicine (the use of biomaterials along with herbal bioactive compounds) can lead to new developments in the field of therapeutic techniques. This

Table 5 Important parameters in the design of matrices

Bio-matrices	Pore size, porosity	Mechanical properties	Morphology	Refs.
Chitosan–gelatin/β-tricalcium phosphate	Pore size: 323 to 355 μm with a porosity of 92–98%	Yield compressive strength in the range of 0.3–0.88 MPa and modulus of 3.94–10.88 MPa	Macro-porous scaffolds	[244]
PLLA-polydopamine and quercetin	The mesh with square pores and interconnected pores (350 μm thread diameter, 460 μm pore sizes, and 56.9% porosity)	The dry compression strengths of 14.52 MPa and wet compression strengths of 10.81 MPa	3D-printed scaffolds with a rough surface	[245]
Poly (*L*-lactide) and chitosan containing quercetin	Fibers with 410 μm diameter, 60–65% theoretical porosity, 390 μm transverse distance between fibers, along with a longitudinal distance of 380 μm and irregular nano-fibers with a range of 80–600 nm diameter	Compressive strength (dry and wet): 13.09 MPa and 12.42 MPa respectively. With a modulus of 0.112 GPa	Micro/nano-fiber structures with pores of square-like	[246]
PLGA microspheres and calcium phosphate	Pore size: >100 μm along with 75% porosity	~65 MPa Modulus	The porous 3D scaffolds	[247]
Chitosan/γ-poly(glutamic acid) modified by elastin, poly-L-lysine, and albumin	The porosity of 90% with pore size ≥ 100 μm and biodegradation of 30–60%	Compressive strength and Young's modulus >4 MPa	3D scaffolds	[248]
Nano-hydroxyapatite and polyamide	Pore size: 50–500 μm and the porosity of 52–70%	Compressive strength and Modulus in the ranges of 13.20–33.90 MPa and 0.29–0.85 GPa, respectively	Macro-pores composite scaffolds	[249]

(continued)

Table 5 (continued)

Bio-matrices	Pore size, porosity	Mechanical properties	Morphology	Refs.
Gelatin-PCL along with nanohydroxyapatite	The average fiber diameter of ~615 nm with a pore size of ~4.7 μm	The increase in mechanical properties with adding PCL and nanohydroxyapatite to gelatin	Fibrous scaffolds	[250]
PLAGA-hydroxyapatite	Particle size in the range of 53 μm to 150 μm along with pore diameter of ~152–183 μm for scaffold	Compressive strength: 120 MPa; with an elastic modulus: 300 MPa	Microsphere matrices (cylindrical and tubular)	[251]
Chitosan/PEG_{4000}-Silk $fibroin_{(3\%)}$	Pore diameters in the range of 2.3–5.9 μm	Increase in mechanical properties (4.5 MPa elastic modulus along with 95% elongation at break %)	3D scaffolds	[8]
PCL and hydroxyapatite	Group1: Pore size: 750 μm Porosity: 70% Group 2: Pore size: 450 μm Porosity: 60%	Group 1: compressive modulus: 76 MPa Group 2: compressive modulus: 84 MPa	3D composite scaffold with internal pores and pore interconnectivity to regenerate bones	[252]
Chitosan-based scaffold containing silk fibroin, collagen, and hyaluronan	Pore diameter range of 8–52μm and porosity of 97.5%,	Water absorption%: 75.3%; tensile strength: 38.1MPa; modulus: 0.32 MPa	3D hybrid scaffold	[12]
PCL and octa-calcium phosphate (OCP)	Fiber diameter distribution: 0.1–1.7 μm Average fiber diameter: ~0.52 μm	Average ultimate tensile strength: ~4.34 MPa Maximum tensile strain: ~182.01% Young's modulus: ~5.59 MPa	Nano-fibrous scaffold	[253]
Collagen/chitosan modified by PEG/PCL composite	53.7 μm pore size, a porosity of ~92, and water absorption of fourfold	Average elastic modulus of ~18 MPa and Average tensile strength of 13.8 MPa	Hybrid scaffold	[15]

(continued)

Table 5 (continued)

Bio-matrices	Pore size, porosity	Mechanical properties	Morphology	Refs.
Nanohydroxyapatite with PCL and PCL–Polyethyleneglycol (PEG)–PCL	Average pore size: 20 μm; fiber diameter: 0.63 μm; porosity: 92%	Tensile strength of 12 MPa, ~269% elongation at breaking point, and Young's modulus of ~17.2 MPa along with 10.41 MPa storage modulus at 37 °C	Composite scaffolds	[254]
quince seed mucilage along with PCL/PEG copolymer	Pore diameter: 16–57 μm; porosity: 92.3%	22.8 MPa Elastic modulus, 13.8 MPa Tensile strength, and ~3 mm Elongation at the breakpoint	3D hybrid scaffold	[16]
PCL and PLGA along with biphasic tricalcium phosphate (BCP)	Pore size: 50–100 μm	–	Porous scaffolds	[255]
PCL and cuttlefish bone-derived hydroxyapatite	Pore dimension: 200–300 μm; porosity: ~85%	The compressive modulus of 0.46 PMa	Composite porous scaffold	[256]

not only can develop novel drugs or therapeutic methods with easier availability and least/no side effect but also decreases the therapeutic economic burden and health care problems.

References

1. Izadyari Aghmiuni, A., Heidari Keshel, S.: The Rol of the Extracellular Matrix (ECM) and ECM-Like Polymeric Substrates in Health and Disease. In: Advances in Medicine and Biology. pp. 145–175 (2021)
2. Aghmiuni, A.I., Ghadi, A., Azmoun. E., et al.: Electrospun Polymeric Substrates for Tissue Engineering: Viewpoints on Fabrication, Application, and Challenges. In: Electrospinning - Material Technology of the Future. InTech, pp. 1–26 (2022)
3. Yang, R., Hou, M., Gao, Y., et al.: Biomineralization-inspired Crystallization of Manganese Oxide on Silk Fibroin Nanoparticles for in vivo MR/fluorescence Imaging-assisted Tri-modal Therapy of Cancer. Theranostics **9**, 6314–6333 (2019)
4. Pilehvar-Soltanahmadi, Y., Dadashpour, M., Mohajeri, A., et al.: An Overview on Application of Natural Substances Incorporated with Electrospun Nanofibrous Scaffolds to Development of Innovative Wound Dressings. Mini Rev. Med. Chem. **18**, 414–427 (2018)
5. Cheng, Y., Deng, S., Chen, P., et al.: Polylactic acid (PLA) synthesis and modifications: a review. Front. Chem. China **4**, 259–264 (2009)
6. Shefa, A.A., Sultana, T., Park, M.K., et al.: Curcumin incorporation into an oxidized cellulose nanofiber-polyvinyl alcohol hydrogel system promotes wound healing. Mater. Des. **186**, 108313 (2020)

7. Huang, G., Li, F., Zhao, X., et al.: Functional and Biomimetic Materials for Engineering of the Three-Dimensional Cell Microenvironment. Chem. Rev. **117**, 12764–12850 (2017)
8. Aghmiuni, A.I., Keshel, S.H., Rezaei-tavirani, M., et al.: Effect of PEG Molecular Weight and Volume Ratio of Chitosan/PEG and Silk Fibroin on Physicomechanical Properties of Chitosan/PEG-SF Scaffold as a Bio-mimetic Substrate in Skin-tissue Engineering Applications. Fibers Polym **23**, 3358–3368 (2022)
9. Sengupta, P., Prasad, B.L.V.: Surface Modification of Polymeric Scaffolds for Tissue Engineering Applications. Regenerative Engineering and Translational Medicine **4**, 75–91 (2018)
10. Türkkan, S., Atila, D., Akdağ, A., et al.: Fabrication of functionalized citrus pectin/silk fibroin scaffolds for skin tissue engineering. J Biomed Mater Res Part B Appl Biomater **106**, 2625–2635 (2018)
11. Izadyari Aghmiuni A, Heidari Keshel S. The Role of the Extracellular Matrix (ECM) and ECM-Like Polymeric Substrates in Health and Disease. In: Berhardt L V. (ed) *Advances in Medicine and Biology*. NOVA Medicine and Health, 2022, pp. 145–175.
12. Izadyari Aghmiuni, A., Heidari Keshel, S., Sefat, F., et al.: Fabrication of 3D hybrid scaffold by combination technique of electrospinning-like and freeze-drying to create mechanotransduction signals and mimic extracellular matrix function of skin. Mater. Sci. Eng. C **120**, 111752 (2021)
13. Farhadihosseinabadi B, Farahani M, Tayebi T, et al. Amniotic membrane and its epithelial and mesenchymal stem cells as an appropriate source for skin tissue engineering and regenerative medicine. *Artif Cells, Nanomedicine, Biotechnol* 2018; 0: 1–10.
14. Ghiasi, M., Kalhor, N., Tabatabaei Qomi, R., et al.: The effects of synthetic and natural scaffolds on viability and proliferation of adipose-derived stem cells. Front Life Sci **9**, 32–43 (2016)
15. Aghmiuni, A.I., Baei, M.S., Keshel, S.H., et al.: Design of Novel 3D-Scaffold as a Potential Material to Induct Epidermal-Dermal Keratinocytes of Human-Adipose-Derived Stem Cells and Promote Fibroblast Cells Proliferation for Skin Regeneration. Fibers Polym **21**, 33–44 (2020)
16. Izadyari Aghmiuni A, Heidari Keshel S, Sefat F, et al. Quince seed mucilage-based scaffold as a smart biological substrate to mimic mechanobiological behavior of skin and promote fibroblasts proliferation and h-ASCs differentiation into keratinocytes. *Int J Biol Macromol*. Epub ahead of print October 2019. https://doi.org/10.1016/j.ijbiomac.2019.10.008.
17. Lin K, Zhang D, Macedo MH, et al. Advanced Collagen-Based Biomaterials for Regenerative Biomedicine. *Adv Funct Mater* 2019; 29: 1804943 (1–16).
18. Kumar Giri, T., Thakur, D., Alexander, A., et al.: Alginate based hydrogel as a potential biopolymeric carrier for drug delivery and cell delivery systems: present status and applications. Curr. Drug Deliv. **9**, 539–555 (2012)
19. Brovold M, Almeida JI, Pla-Palacín I, et al. *Naturally-Derived Biomaterials for Tissue Engineering Applications*. 2020. Epub ahead of print 2020. https://doi.org/10.1007/978-981-13-0947-2_23.
20. Manoukian OS, Ahmad A, Marin C, et al. Bioactive nanofiber dressings for wound healing. In: *Wound healing biomaterials*. Elsevier, 2016, pp. 451–481.
21. Debele, T.A., Mekuria, S.L., Tsai, H.-C.: Polysaccharide based nanogels in the drug delivery system: Application as the carrier of pharmaceutical agents. Mater. Sci. Eng. C **68**, 964–981 (2016)
22. Ye, G., Li, G., Wang, C., et al.: Extraction and characterization of dextran from Leuconostoc pseudomesenteroides YB-2 isolated from mango juice. Carbohydr. Polym. **207**, 218–223 (2019)
23. Kothari, D., Tingirikari, J.M.R., Goyal, A.: In vitro analysis of dextran from Leuconostoc mesenteroides NRRL B-1426 for functional food application. Bioact carbohydrates Diet fibre **6**, 55–61 (2015)
24. Baruah R, Maina NH, Katina K, et al. Functional food applications of dextran from Weissella cibaria RBA12 from pummelo (Citrus maxima). *Int J Food Microbiol*. Epub ahead of print 2016. https://doi.org/10.1016/j.ijfoodmicro.2016.11.012.

25. Chen, F., Huang, G., Huang, H.: Preparation and application of dextran and its derivatives as carriers. Int. J. Biol. Macromol. **145**, 827–834 (2020)
26. Sood, A., Gupta, A., Agrawal, G.: Recent advances in polysaccharides based biomaterials for drug delivery and tissue engineering applications. Carbohydr Polym Technol Appl **2**, 100067 (2021)
27. Ha TLB, Quan TM, Vu DN, et al. Naturally Derived Biomaterials: Preparation and Application. In: *Regenerative Medicine and Tissue Engineering*. 2013, pp. 247–274.
28. Park, B.K., Kim, M.-M.: Applications of chitin and its derivatives in biological medicine. Int. J. Mol. Sci. **11**, 5152–5164 (2010)
29. Li, Z., Ramay, H.R., Hauch, K.D., et al.: Chitosan–alginate hybrid scaffolds for bone tissue engineering. Biomaterials **26**, 3919–3928 (2005)
30. Venkatesan, J., Kim, S.-K.: Chitosan composites for bone tissue engineering—an overview. Mar. Drugs **8**, 2252–2266 (2010)
31. Eo, M.Y., Fan, H., Cho, Y.J., et al.: Cellulose membrane as a biomaterial: from hydrolysis to depolymerization with electron beam. Biomater Res **20**, 1–13 (2016)
32. Sawatjui, N., Limpaiboon, T., Schrobback, K., et al.: Biomimetic scaffolds and dynamic compression enhance the properties of chondrocyte-and MSC-based tissue-engineered cartilage. J. Tissue Eng. Regen. Med. **12**, 1220–1229 (2018)
33. Kim, D.K., Kim, J.I., Sim, B.R., et al.: Bioengineered porous composite curcumin/silk scaffolds for cartilage regeneration. Mater. Sci. Eng. C **78**, 571–578 (2017)
34. Fu, L., Zhang, J., Yang, G.: Present status and applications of bacterial cellulose-based materials for skin tissue repair. Carbohydr. Polym. **92**, 1432–1442 (2013)
35. Mano, J.F., Silva, G.A., Azevedo, H.S., et al.: Natural origin biodegradable systems in tissue engineering and regenerative medicine: present status and some moving trends. J. R. Soc. Interface. **4**, 999–1030 (2007)
36. Kowalska-Ludwicka, K., Cala, J., Grobelski, B., et al.: Special paper–new methods modified bacterial cellulose tubes for regeneration of damaged peripheral nerves. Arch. Med. Sci. **9**, 527–534 (2013)
37. Awad, H.A., Wickham, M.Q., Leddy, H.A., et al.: Chondrogenic differentiation of adipose-derived adult stem cells in agarose, alginate, and gelatin scaffolds. Biomaterials **25**, 3211–3222 (2004)
38. Gao, M., Lu, P., Bednark, B., et al.: Templated agarose scaffolds for the support of motor axon regeneration into sites of complete spinal cord transection. Biomaterials **34**, 1529–1536 (2013)
39. Lynam, D.A., Shahriari, D., Wolf, K.J., et al.: Brain derived neurotrophic factor release from layer-by-layer coated agarose nerve guidance scaffolds. Acta Biomater. **18**, 128–131 (2015)
40. Zarrintaj, P., Bakhshandeh, B., Rezaeian, I., et al.: A novel electroactive agarose-aniline pentamer platform as a potential candidate for neural tissue engineering. Sci. Rep. **7**, 17187 (2017)
41. Han, S., Lee, J.Y., Heo, E.Y., et al.: Implantation of a Matrigel-loaded agarose scaffold promotes functional regeneration of axons after spinal cord injury in rat. Biochem. Biophys. Res. Commun. **496**, 785–791 (2018)
42. Park, S.-B., Lih, E., Park, K.-S., et al.: Biopolymer-based functional composites for medical applications. Prog. Polym. Sci. **68**, 77–105 (2017)
43. Stanisz, M., Klapiszewski, Ł, Jesionowski, T.: Recent advances in the fabrication and application of biopolymer-based micro-and nanostructures: A comprehensive review. Chem. Eng. J. **397**, 125409 (2020)
44. Pires PC, Mascarenhas-melo F, Pedrosa K, et al. Polymer-based biomaterials for pharamaceutical and biomedical applications : A focus on topical drug administration. *Eur Polym J*; 187. Epub ahead of print 2023. https://doi.org/10.1016/j.eurpolymj.2023.111868.
45. Niknejad H, Peirovi H, Jorjani M, et al. PROPERTIES OF THE AMNIOTIC MEMBRANE FOR POTENTIAL USE IN TISSUE.
46. C Echave M, S Burgo L, L Pedraz J, et al. Gelatin as biomaterial for tissue engineering. *Curr Pharm Des* 2017; 23: 3567–3584.

47. Chen, J., Liu, W., Zhao, J., et al.: Gelatin microspheres containing calcitonin gene-related peptide or substance P repair bone defects in osteoporotic rabbits. Biotechnol. Lett. **39**, 465–472 (2017)
48. Gonzalez, D., Ragusa, J., Angeletti, P.C., et al.: Preparation and characterization of functionalized heparin-loaded poly-Ɛ-caprolactone fibrous mats to prevent infection with human papillomaviruses. PLoS ONE **13**, e0199925 (2018)
49. Park, J., Kim, J., Hwang, S.R., et al.: Chemical conjugate of low molecular weight heparin and suramin fragment inhibits tumor growth possibly by blocking VEGF165. Mol. Pharm. **12**, 3935–3942 (2015)
50. Wan, X., Li, P., Jin, X., et al.: Poly (ε-caprolactone)/keratin/heparin/VEGF biocomposite mats for vascular tissue engineering. J Biomed Mater Res Part A **108**, 292–300 (2020)
51. Wang, W., Liu, S., Huang, Y., et al.: Biodegradable dextran vesicles for effective haemoglobin encapsulation. J Mater Chem B **3**, 5753–5759 (2015)
52. Anak, D., Davis, S., Parish, C.R.: Heparan sulfate : a ubiquitous glycosaminoglycan with multiple roles in immunity. Front. Immunol. **4**, 1–7 (2013)
53. Lee, C.-T., Kung, P.-H., Lee, Y.-D.: Preparation of poly (vinyl alcohol)-chondroitin sulfate hydrogel as matrices in tissue engineering. Carbohydr. Polym. **61**, 348–354 (2005)
54. Henson, F.M.D., Getgood, A.M.J., Caborn, D.M., et al.: Effect of a solution of hyaluronic acid–chondroitin sulfate–N-acetyl glucosamine on the repair response of cartilage to single-impact load damage. Am. J. Vet. Res. **73**, 306–312 (2012)
55. Xu L, Tang L, Zhang L. *Proteoglycans as miscommunication biomarkers for cancer diagnosis*. 1st ed. Elsevier Inc. Epub ahead of print 2019. https://doi.org/10.1016/bs.pmbts.2018.12.003.
56. Weyers, A., Linhardt, R.J.: Neoproteoglycans in tissue engineering. FEBS J. **280**, 2511–2522 (2013)
57. Yang, J., Shen, M., Wen, H., et al.: Recent advance in delivery system and tissue engineering applications of chondroitin sulfate. Carbohydr. Polym. **230**, 115650 (2020)
58. Muzzarelli, R.A.A., Greco, F., Busilacchi, A., et al.: Chitosan, hyaluronan and chondroitin sulfate in tissue engineering for cartilage regeneration: a review. Carbohydr. Polym. **89**, 723–739 (2012)
59. Lafuente-Merchan, M., Ruiz-Alonso, S., Zabala, A., et al.: Chondroitin and dermatan sulfate bioinks for 3D bioprinting and cartilage regeneration. Macromol. Biosci. **22**, 2100435 (2022)
60. Kim, S.S., Kang, M.S., Lee, K.Y., et al.: Therapeutic effects of mesenchymal stem cells and hyaluronic acid injection on osteochondral defects in rabbits' knees. Knee Surg Relat Res **24**, 164–172 (2012)
61. Migliore, A., Procopio, S.: Effectiveness and utility of hyaluronic acid in osteoarthritis. Clin. Cases Miner. Bone Metab. **12**, 31 (2015)
62. Park, S.H., Seo, J.Y., Park, J.Y., et al.: An injectable, click-crosslinked, cytomodulin-modified hyaluronic acid hydrogel for cartilage tissue engineering. NPG Asia Mater **11**, 30 (2019)
63. Macadam, S.A., Lennox, P.A.: Acellular dermal matrices: use in reconstructive and aesthetic breast surgery. Can. J. Plast. Surg. **20**, 75–89 (2012)
64. Entcheva, E., Bien, H., Yin, L., et al.: Functional cardiac cell constructs on cellulose-based scaffolding. Biomaterials **25**, 5753–5762 (2004)
65. Robb, K.P., Shridhar, A., Flynn, L.E.: Decellularized matrices as cell-instructive scaffolds to guide tissue-specific regeneration. ACS Biomater. Sci. Eng. **4**, 3627–3643 (2017)
66. Granados, M., Morticelli, L., Andriopoulou, S., et al.: Development and characterization of a porcine mitral valve scaffold for tissue engineering. J. Cardiovasc. Transl. Res. **10**, 374–390 (2017)
67. Rana, D., Zreiqat, H., Benkirane-Jessel, N., et al.: Development of decellularized scaffolds for stem cell-driven tissue engineering. J. Tissue Eng. Regen. Med. **11**, 942–965 (2017)
68. Bozuk, M.I., Fearing, N.M., Leggett, P.L.: Use of decellularized human skin to repair esophageal anastomotic leak in humans. JSLS J Soc Laparoendosc Surg **10**, 83 (2006)
69. Lin LM, Lin CC, Chen CL, et al. Effects of an education program on intensive care unit nurses' attitudes and behavioral intentions to advocate deceased donor organ donation. In: *Transplantation proceedings*. Elsevier, 2014, pp. 1036–1040.

70. Moroni, F., Mirabella, T.: Decellularized matrices for cardiovascular tissue engineering. Am J Stem Cells **3**, 1 (2014)
71. Methe, K., Bäckdahl, H., Johansson, B.R., et al.: An alternative approach to decellularize whole porcine heart. Biores Open Access **3**, 327–338 (2014)
72. Taylor, D.A., Sampaio, L.C., Gobin, A.: Building new hearts: a review of trends in cardiac tissue engineering. Am. J. Transplant. **14**, 2448–2459 (2014)
73. Manji, R.A., Menkis, A.H., Ekser, B., et al.: Porcine bioprosthetic heart valves: The next generation. Am. Heart J. **164**, 177–185 (2012)
74. Baptista, P.M., Siddiqui, M.M., Lozier, G., et al.: The use of whole organ decellularization for the generation of a vascularized liver organoid. Hepatology **53**, 604–617 (2011)
75. Jaramillo, M., Yeh, H., Yarmush, M.L., et al.: Decellularized human liver extracellular matrix (hDLM)-mediated hepatic differentiation of human induced pluripotent stem cells (hIPSCs). J. Tissue Eng. Regen. Med. **12**, e1962–e1973 (2018)
76. Kakabadze, Z., Kakabadze, A., Chakhunashvili, D., et al.: Decellularized human placenta supports hepatic tissue and allows rescue in acute liver failure. Hepatology **67**, 1956–1969 (2018)
77. Baptista PM, Vyas D, Moran E, et al. Human liver bioengineering using a whole liver decellularized bioscaffold. *Organ Regen Methods Protoc* 2013; 289–298.
78. Chang, J., DeLillo, N., Jr., Khan, M., et al.: Review of small intestine submucosa extracellular matrix technology in multiple difficult-to-treat wound types. Wounds **25**, 113–120 (2013)
79. Chun, S.Y., Lim, G.J., Kwon, T.G., et al.: Identification and characterization of bioactive factors in bladder submucosa matrix. Biomaterials **28**, 4251–4256 (2007)
80. Xiao, H.H., Gao, Q.G., Zhang, Y., et al.: Vanillic acid exerts oestrogen-like activities in osteoblast-like UMR 106 cells through MAP kinase (MEK/ERK)-mediated ER signaling pathway. J. Steroid Biochem. Mol. Biol. **144**, 382–391 (2014)
81. Don, M.J., Lin, L.C., Chiou, W.F.: Neobavaisoflavone stimulates osteogenesis via p38-mediated up-regulation of transcription factors and osteoid genes expression in MC3T3-E1 cells. Phytomedicine **19**, 551–561 (2012)
82. Lee, C.H., Huang, Y.L., Liao, J.F., et al.: Ugonin K-stimulated osteogenesis involves estrogen receptor-dependent activation of non-classical Src signaling pathway and classical pathway. Eur. J. Pharmacol. **676**, 26–33 (2012)
83. Yoon HY, Yun S Il, Kim BY, et al. Poncirin promotes osteoblast differentiation but inhibits adipocyte differentiation in mesenchymal stem cells. *Eur J Pharmacol* 2011; 664: 54–59.
84. Pang, W.Y., Wang, X.L., Mok, S.K., et al.: Naringin improves bone properties in ovariectomized mice and exerts oestrogen-like activities in rat osteoblast-like (UMR-106) cells. Br. J. Pharmacol. **159**, 1693–1703 (2010)
85. Chen KY, Lin KC, Chen YS, et al. A novel porous gelatin composite containing naringin for bone repair. *Evidence-based Complement Altern Med*; 2013. Epub ahead of print 2013. https://doi.org/10.1155/2013/283941.
86. Tsuchiya, S., Sugimoto, K., Kamio, H., et al.: Kaempferol-immobilized titanium dioxide promotes formation of new bone: Effects of loading methods on bone marrow stromal cell differentiation in vivo and in vitro. Int. J. Nanomedicine **13**, 1665–1676 (2018)
87. Carvalho, M.J.N., Ferreira, K.B., Augusti, R., et al.: LEISHMANICIDAL ACTIVITY OF FLAVONOIDS NATURAL AND SYNTHETIC: A MINIREVIEW. OPEN Sci Res **II**(2), 266–282 (2022)
88. Zhao, L., Yuan, X., Wang, J., et al.: A review on flavones targeting serine/threonine protein kinases for potential anticancer drugs. Bioorg. Med. Chem. **27**, 677–685 (2019)
89. Zhao, K., Yuan, Y., Lin, B., et al.: LW-215, a newly synthesized flavonoid, exhibits potent anti-angiogenic activity in vitro and in vivo. Gene **642**, 533–541 (2018)
90. Camero, C.M., Germanò, M.P., Rapisarda, A., et al.: Anti-angiogenic activity of iridoids from Galium tunetanum. Rev Bras Farmacogn **28**, 374–377 (2018)
91. Mazidi, M., Katsiki, N., Banach, M.: A higher flavonoid intake is associated with less likelihood of nonalcoholic fatty liver disease: results from a multiethnic study. J. Nutr. Biochem. **65**, 66–71 (2019)

92. Khalifa, I., Zhu, W., Li, K., et al.: Polyphenols of mulberry fruits as multifaceted compounds: Compositions, metabolism, health benefits, and stability—A structural review. J Funct Foods **40**, 28–43 (2018)
93. Liao, M.H., Tai, Y.T., Cherng, Y.G., et al.: Genistein induces oestrogen receptor-α gene expression in osteoblasts through the activation of mitogen-activated protein kinases/NF-κB/ activator protein-1 and promotes cell mineralisation. Br. J. Nutr. **111**, 55–63 (2014)
94. Sirtori, C.R., Arnoldi, A., Johnson, S.K.: Phytoestrogens: end of a tale? Ann. Med. **37**, 423–438 (2005)
95. Sirotkin, A.V., Harrath, A.H.: Phytoestrogens and their effects. Eur. J. Pharmacol. **741**, 230–236 (2014)
96. Poluzzi, E., Piccinni, C., Raschi, E., et al.: Phytoestrogens in postmenopause: the state of the art from a chemical, pharmacological and regulatory perspective. Curr. Med. Chem. **21**, 417–436 (2014)
97. Sun, H., Li, L., Zhang, A., et al.: Protective effects of sweroside on human MG-63 cells and rat osteoblasts. Fitoterapia **84**, 174–179 (2013)
98. Kim, C., Choi, K.: Potential Roles of Iridoid Glycosides and Their Underlying Mechanisms against Diverse Cancer Growth and Metastasis : Do They Have an Inhibitory Effect on Cancer Progression ? Nutrients **13**, 2–14 (2021)
99. Kouda, R., Yakushiji, F.: Recent advances in Iridoid chemistry: biosynthesis and chemical synthesis. Chem. Asian J. **15**, 3771–3783 (2020)
100. Castejón ML, Montoya T, Alarcón-de-la-Lastra C, et al. Potential protective role exerted by secoiridoids from Olea europaea L. in cancer, cardiovascular, neurodegenerative, aging-related, and immunoinflammatory diseases. *Antioxidants* 2020; 9: 149.
101. Huang, Q., Shi, J., Gao, B., et al.: Gastrodin: An ancient Chinese herbal medicine as a source for anti-osteoporosis agents via reducing reactive oxygen species. Bone **73**, 132–144 (2015)
102. Yan R, Liu H, Zhang J, et al. Phenolic glycosides and other constituents from the bark of Magnolia officinalis. *J Asian Nat Prod Res* 2013; 37–41.
103. Huang, Q., Gao, B., Wang, L., et al.: Ophiopogonin D: A new herbal agent against osteoporosis. Bone **74**, 18–28 (2015)
104. Ivanchina, N.V., Malyarenko, T.V., Kicha, A.A., et al.: A New Steroidal Glycoside Granulatoside C from the Starfish Choriaster granulatus, Unexpectedly Combining Structural Features of Polar Steroids from Several Different Marine Invertebrate Phyla Natalia. Nat. Prod. Commun. **12**, 1585–1588 (2017)
105. Liu, M., Kong, J.: The enzymatic biosynthesis of acylated steroidal glycosides and their cytotoxic activity. Acta Pharm Sin B **8**, 981–994 (2018)
106. Ito, Y., Nakashima, Y., Matsuoka, S.: Rice bran extract containing acylated steryl glucoside fraction decreases elevated blood LDL cholesterol level in obese Japanese men. J Med Investig **62**, 80–84 (2015)
107. Suh, K.S., Choi, E.M., Lee, Y.S., et al.: Protective effect of albiflorin against oxidative-stress-mediated toxicity in osteoblast-like MC3T3-E1 cells. Fitoterapia **89**, 33–41 (2013)
108. Zieli, M.: Monoterpenes and Their Derivatives—Recent Development in Biological and Medical Applications. Int. J. Mol. Sci. **21**, 1–38 (2020)
109. Lei, D., Qiu, Z., Qiao, J., et al.: Biotechnology for Biofuels Plasticity engineering of plant monoterpene synthases and application for microbial production of monoterpenoids. Biotechnol. Biofuels **14**, 1–15 (2021)
110. Muthukumar, T., Aravinthan, A., Sharmila, J., et al.: Collagen/chitosan porous bone tissue engineering composite scaffold incorporated with Ginseng compound K. Carbohydr. Polym. **152**, 566–574 (2016)
111. Wang Z, Jiang R, Wang L, et al. Ginsenoside Rg1 Improves Differentiation by Inhibiting Senescence of Human Bone Marrow Mesenchymal Stem Cell via GSK-3 β and β-Catenin. *Stem Cells Int*; 2020. Epub ahead of print 2020. https://doi.org/10.1155/2020/2365814.
112. Chen, W., Balan, P.: Review of Ginseng Anti-Diabetic Studies. Molecules **24**, 1–16 (2019)
113. Borzym-kluczyk, J.N.M.: The role of triterpenes in the management of diabetes mellitus and its complications. Phytochem. Rev. **14**, 675–690 (2015)

114. Darshani P, Sen S, Amit S, et al. *Anti-viral triterpenes : a review*. Springer Netherlands. Epub ahead of print 2022. https://doi.org/10.1007/s11101-022-09808-1.
115. Ríos, J.: Effects of triterpenes on the immune system. J Ethnopharmacol J **128**, 1–14 (2010)
116. Lee, S.U., Park, S.J., Kwak, H.B., et al.: Anabolic activity of ursolic acid in bone: Stimulating osteoblast differentiation in vitro and inducing new bone formation in vivo. Pharmacol. Res. **58**, 290–296 (2008)
117. Ge, Y.W., Lu, J.W., Sun, Z.Y., et al.: Ursolic acid loaded-mesoporous bioglass/chitosan porous scaffolds as drug delivery system for bone regeneration. Nanomedicine Nanotechnology, Biol Med **18**, 336–346 (2019)
118. Manuscript A. Triterpenoids as potential agents for the chemoprevention and therapy of breast cancer. *NIH Public Access* 2011; 980–996.
119. Schneider, C., Langer, R., Loveday, D., et al.: Applications of ethylene vinyl acetate copolymers (EVA) in drug delivery systems. J. Control. Release **262**, 284–295 (2017)
120. Park, K.: Controlled drug delivery systems: past forward and future back. J. Control. Release **190**, 3–8 (2014)
121. Yun, Y.H., Lee, B.K., Park, K.: Controlled Drug Delivery: Historical perspective for the next generation. J. Control. Release **219**, 2–7 (2015)
122. Aghmiuni, A.I., Keshel, S.H., Rahmani, A., et al.: Retinal Tissue Engineering: Regenerative and Drug Delivery Approaches. Curr. Stem Cell Res. Ther. **18**, 608–640 (2023)
123. Bertrand, N., Wu, J., Xu, X., et al.: Cancer nanotechnology: the impact of passive and active targeting in the era of modern cancer biology. Adv. Drug Deliv. Rev. **66**, 2–25 (2014)
124. Adeosun SO, Ilomuanya MO, Gbenebor OP, et al. Biomaterials for Drug Delivery: Sources, Classification, Synthesis, Processing, and Applications. In: *Advanced Functional Materials have*. 2020, pp. 1–25.
125. Kamaly, N., Xiao, Z., Valencia, P.M., et al.: Targeted polymeric therapeutic nanoparticles: design, development and clinical translation. Chem. Soc. Rev. **41**, 2971–3010 (2012)
126. Kilicarslan, M., Ilhan, M., Inal, O., et al.: Preparation and evaluation of clindamycin phosphate loaded chitosan/alginate polyelectrolyte complex film as mucoadhesive drug delivery system for periodontal therapy. Eur. J. Pharm. Sci. **123**, 441–451 (2018)
127. Islan, G.A., Castro, G.R., Islan, G.A., et al.: Tailoring of alginate–gelatin microspheres properties for oral Ciprofloxacin-controlled release against Pseudomonas aeruginosa. Drug Deliv. **21**, 615–625 (2014)
128. Carriers, N., Chen, T., Li, S., et al.: Self-assembly pH-sensitive Chitosan/Alginate Coated Polyelectrolyte Complexes for Oral Delivery of Insulin. J Microencapsul Micro Nano Carriers **2048**, 1–30 (2019)
129. Kumar S, Bhanjana G, Kumar R, et al. Metformin-loaded alginate nanoparticles as an effective antidiabetic agent for controlled drug release. *J Pharm Pharmacol* 2016; 1–8.
130. Varadharaj, V., Ramaswamy, A., Sakthivel, R., et al.: Antidiabetic and Antioxidant Activity of Green Synthesized Starch Nanoparticles : An In Vitro Study. J. Clust. Sci. **3**, 1–10 (2019)
131. Rehman A, Mahdi S, Tong Q, et al. International Journal of Biological Macromolecules Role of peppermint oil in improving the oxidative stability and antioxidant capacity of borage seed oil-loaded nanoemulsions fabricated by modi fi ed starch. *Int J Biol Macromol* 2020; 153: 697–707.
132. Usman, M., Mahmood, K., Sajid, M., et al.: In-vivo anti-diabetic and wound healing potential of chitosan /alginate /maltodextrin /pluronic-based mixed polymeric micelles : Curcumin therapeutic potential. Int. J. Biol. Macromol. **120**, 2418–2430 (2018)
133. Gannon, S., Chu, A.F.: SUDDEN UNEXPECTED DEATH IN EPILEPSY AND LONG QT SYNDROME. J. Am. Coll. Cardiol. **69**, 2137 (2017)
134. Qi X, Yuan Y, Zhang J, et al. Oral administration of salecan-based hydrogels for controlled insulin delivery. *J Agric Food Chem*. Epub ahead of print 2018. https://doi.org/10.1021/acs.jafc.8b02879.
135. Singla R, Soni S, Patial V, et al. In vivo diabetic wound healing potential of nanobiocomposites containing bamboo cellulose nanocrystals impregnated with silver nanoparticles. *Int J Biol Macromol* 2017; 1–40.

136. Abaee, A., Mohammadian, M., Mahdi, S.: Whey and soy protein-based hydrogels and nano-hydrogels as bioactive delivery systems. Trends Food Sci. Technol. **70**, 69–81 (2017)
137. Hou, K., Xu, Y., Cen, K., et al.: Nanoemulsion of cinnamon essential oil Co-emulsified with hydroxypropyl- β -cyclodextrin and Tween-80: Antibacterial activity, stability and slow release performance. Food Biosci. **43**, 101232 (2021)
138. Cai, S., Thati, S., Bagby, T.R., et al.: Localized doxorubicin chemotherapy with a biopolymeric nanocarrier improves survival and reduces toxicity in xenografts of human breast cancer. J. Control. Release **146**, 212–218 (2010)
139. Zheng AS, Han J, Jin Z, et al. Dual tumor-targeted multifunctional magnetic hyaluronic acid micelles for enhanced MR imaging and combined photothermal-chemotherapy. *Colloids Surfaces B Biointerfaces* 2018; 1–29.
140. Zhang Y, Ma Y, Xia Y, et al. Synthesis of Silk Fibroin-Insulin Bioconjugates and Their Characterization and Activities In Vivo. *J Biomed Mater Res Part B Appl Biomater* 2006; 275–283.
141. Gangrade, A., Mandal, B.B.: Injectable Carbon Nanotube Impregnated Silk Based Multi-functional Hydrogel for Localized Targeted and On-Demand Anticancer Drug Delivery. ACS Biomater. Sci. Eng. **5**, 2365–2381 (2019)
142. Zhang Y, Xiong GM, Ali Y, et al. Layer-by-layer coated nanoliposomes for oral delivery of insulin. *R cociety Chem* 2020; 1–14.
143. Wang, S., Meng, S., Zhou, X., et al.: PH-Responsive and Mucoadhesive Nanoparticles for Enhanced Oral Insulin Delivery: The Effect of Hyaluronic Acid with Different Molecular Weights. Pharmaceutics **15**, 820 (2023)
144. Acevedo-Guevara, L., Nieto-Suaza, L., Sanchez, L.T., et al.: Development of native and modified banana starch nanoparticles as vehicles for curcumin. Int. J. Biol. Macromol. **111**, 498–504 (2018)
145. Mukhopadhyay, P., Maity, S., Mandal, S., et al.: Preparation, characterization and in vivo evaluation of pH sensitive, safe quercetin-succinylated chitosan-alginate core-shell-corona nanoparticle for diabetes treatment. Carbohydr. Polym. **182**, 42–51 (2018)
146. Lin YY, Lu SH, Gao R, et al. A Novel Biocompatible Herbal Extract-Loaded Hydrogel for Acne Treatment and Repair. *Oxid Med Cell Longev*; 2021. Epub ahead of print 2021. https://doi.org/10.1155/2021/5598291.
147. Abbas, M.N., Khan, S.A., Sadozai, S.K., et al.: Nanoparticles Loaded Thermoresponsive In Situ Gel for Ocular Antibiotic Delivery against Bacterial Keratitis. Polymers (Basel) **14**, 1–19 (2022)
148. Cheng, Y.H., Ko, Y.C., Chang, Y.F., et al.: Thermosensitive chitosan-gelatin-based hydrogel containing curcumin-loaded nanoparticles and latanoprost as a dual-drug delivery system for glaucoma treatment. Exp. Eye Res. **179**, 179–187 (2019)
149. Osman, R., Fetih, G., Habib, F.: KETOROLAC TROMETHAMINE LOADED NANOPARTICLES FOR OCULAR DELIVERY: FORMULATION, IN-VITRO AND EX-VIVO EVALUATION. Bull Pharm Sci Assiut **43**, 79–94 (2020)
150. Luo, L.J., Lin, T.Y., Yao, C.H., et al.: Dual-functional gelatin-capped silver nanoparticles for antibacterial and antiangiogenic treatment of bacterial keratitis. J. Colloid Interface Sci. **536**, 112–126 (2019)
151. Natarajan, J.V., Ang, M., Darwitan, A., et al.: Nanomedicine for glaucoma: Liposomes provide sustained release of latanoprost in the eye. Int. J. Nanomedicine **7**, 123–131 (2012)
152. Ahmad N, Khalid MS, Khan MF, et al. Beneficial effects of topical 6-gingerol loaded nanoemulsion gel for wound and inflammation management with their comparative dermatokinetic. *J Drug Deliv Sci Technol*; 80. Epub ahead of print 2023. https://doi.org/10.1016/j.jddst.2022.104094.
153. Donsì, F., Ferrari, G.: Essential oil nanoemulsions as antimicrobial agents in food. J. Biotechnol. **233**, 106–120 (2016)
154. Abd-Elsalam KA, Khokhlov AR. Eugenol oil nanoemulsion: antifungal activity against Fusarium oxysporum f. sp. vasinfectum and phytotoxicity on cottonseeds. *Appl Nanosci* 2015; 5: 255–265.

155. Bhargava, K., Conti, D.S., da Rocha, S.R.P., et al.: Application of an oregano oil nanoemulsion to the control of foodborne bacteria on fresh lettuce. Food Microbiol. **47**, 69–73 (2015)
156. Wu, J.E., Lin, J., Zhong, Q.: Physical and antimicrobial characteristics of thyme oil emulsified with soluble soybean polysaccharide. Food Hydrocoll **39**, 144–150 (2014)
157. Li, B., Ge, Z.Q.: Nanostructured lipid carriers improve skin permeation and chemical stability of idebenone. AAPS PharmSciTech **13**, 276–283 (2012)
158. Wang, J., Wang, H., Zhou, X., et al.: Physicochemical characterization, photo-stability and cytotoxicity of coenzyme Q10-loading nanostructured lipid carrier. J. Nanosci. Nanotechnol. **12**, 2136–2148 (2012)
159. Gönüllü, Ü., Üner, M., Yener, G., et al.: Formulation and characterization of solid lipid nanoparticles, nanostructured lipid carriers and nanoemulsion of lornoxicam for transdermal delivery. Acta Pharm. **65**, 1–13 (2015)
160. Wang, J., Tang, J., Zhou, X., et al.: Physicochemical characterization, identification and improved photo-stability of alpha-lipoic acid-loaded nanostructured lipid carrier. Drug Dev. Ind. Pharm. **40**, 201–210 (2014)
161. Fan, H., Liu, G., Huang, Y., et al.: Development of a nanostructured lipid carrier formulation for increasing photo-stability and water solubility of Phenylethyl Resorcinol. Appl. Surf. Sci. **288**, 193–200 (2014)
162. Zhao J, Piao X, Shi X, et al. Podophyllotoxin-loaded nanostructured lipid carriers for skin targeting: In vitro and in vivo studies. *Molecules*; 21. Epub ahead of print 2016. https://doi.org/10.3390/molecules21111549.
163. Yu, Q., Chang, J., Wu, C.: Silicate bioceramics: from soft tissue regeneration to tumor therapy. J Mater Chem B **7**, 5449–5460 (2019)
164. Chaudhari, A.A., Vig, K., Baganizi, D.R., et al.: Future Prospects for Scaffolding Methods and Biomaterials in Skin Tissue Engineering : A Review. Int. J. Mol. Sci. **17**, 1974 (2016)
165. Shafei, S., Foroughi, J., Stevens, L., et al.: Electroactive nanostructured scaffold produced by controlled deposition of PPy on electrospun PCL fibres. Res. Chem. Intermed. **43**, 1235–1251 (2017)
166. Ahmed, L.A.: Stem cells and cardiac repair: alternative and multifactorial approaches. J Regen Med Tissue Eng **2**, 10–7243 (2013)
167. Talebian, S., Mehrali, M., Taebnia, N., et al.: Self-healing hydrogels: the next paradigm shift in tissue engineering? Adv. Sci. **6**, 1801664 (2019)
168. Zhao, Z., Vizetto-Duarte, C., Moay, Z.K., et al.: Composite hydrogels in three-dimensional in vitro models. Front Bioeng Biotechnol **8**, 611 (2020)
169. Wang, Y., Adokoh, C.K., Narain, R.: Recent development and biomedical applications of self-healing hydrogels. Expert Opin. Drug Deliv. **15**, 77–91 (2018)
170. Ramalingam, V., Raja, S., Sundaramahalingam, S., et al.: Chemical fabrication of graphene oxide nanosheets attenuates biofilm formation of human clinical pathogens. Bioorg. Chem. **83**, 326–335 (2019)
171. Izadyari Aghmiuni, A., Heidari Keshel, S., Sefat, F., et al.: Quince seed mucilage-based scaffold as a smart biological substrate to mimic mechanobiological behavior of skin and promote fibroblasts proliferation and h-ASCs differentiation into keratinocytes. Int. J. Biol. Macromol. **142**, 668–679 (2020)
172. Dey, S.K., Banerjee, D., Chattapadhyay, S., et al.: Antimicrobial activities of some medicinal plants of West Bengal. Int J Pharma Bio Sci **1**, 1–10 (2010)
173. Omenetto, F.G., Kaplan, D.L.: New opportunities for an ancient material. Science **329**, 528–531 (2010)
174. Kogan, G., Šoltés, L., Stern, R., et al.: Hyaluronic acid: a natural biopolymer with a broad range of biomedical and industrial applications. Biotechnol. Lett. **29**, 17–25 (2006)
175. Volpi, N., Schiller, J., Stern, R., et al.: Role, Metabolism, Chemical Modifications and Applications of Hyaluronan. Curr. Med. Chem. **16**, 1718–1745 (2009)
176. Petrey, A.C., de la Motte, C.A.: Hyaluronan, a crucial regulator of inflammation. Front. Immunol. **5**, 1–14 (2014)

177. Lam, J., Truong, N.F., Segura, T.: Design of cell-matrix interactions in hyaluronic acid hydrogel scaffolds. Acta Biomater. **10**, 1571–1580 (2014)
178. Dicker, K.T., Gurski, L.A., Pradhan-Bhatt, S., et al.: Hyaluronan: A simple polysaccharide with diverse biological functions. Acta Biomater. **10**, 1558–1570 (2014)
179. Darzi, S., Paul, K., Leitan, S., et al.: Immunobiology and application of aloe vera-based scaffolds in tissue engineering. Int. J. Mol. Sci. **22**, 1–19 (2021)
180. Rahman, S., Carter, P., Bhattarai, N.: Aloe vera for tissue engineering applications. J Funct Biomater **8**, 6 (2017)
181. Paul, K., Darzi, S., Del Borgo, M.P., et al.: Vaginal delivery of tissue engineered endometrial mesenchymal stem/stromal cells in an aloe vera-alginate hydrogel alleviates maternal simulated birth injury. Appl. Mater. Today **22**, 100890 (2021)
182. Suganya, S., Venugopal, J., Agnes Mary, S., et al.: Aloe vera incorporated biomimetic nanofibrous scaffold: a regenerative approach for skin tissue engineering. Iran. Polym. J. **23**, 237–248 (2014)
183. Tahmasebi, A., Shapouri Moghadam, A., Enderami, S.E., et al.: Aloe Vera-Derived Gel-Blended PHBV Nanofibrous Scaffold for Bone Tissue Engineering. ASAIO J. **66**, 966–973 (2020)
184. Ezhilarasu, H., Ramalingam, R., Dhand, C., et al.: Biocompatible Aloe vera and Tetracycline Hydrochloride Loaded Hybrid Nanofibrous Scaffolds for Skin Tissue Engineering. Int. J. Mol. Sci. **20**, 5174 (2019)
185. Zadegan S, Nourmohammadi J, Vahidi B, et al. An investigation into osteogenic differentiation effects of silk fibroin-nettle (Urtica dioica L.) nanofibers. *Int J Biol Macromol* 2019; 133: 795–803.
186. Ghiyasi, Y., Salahi, E., Esfahani, H.: Synergy effect of Urtica dioica and ZnO NPs on microstructure, antibacterial activity and cytotoxicity of electrospun PCL scaffold for wound dressing application. Mater Today Commun **26**, 102163 (2021)
187. Hajiali, H., Summa, M., Russo, D., et al.: Alginate–lavender nanofibers with antibacterial and anti-inflammatory activity to effectively promote burn healing. J Mater Chem B **4**, 1686–1695 (2016)
188. Pilehvar-Soltanahmadi, Y., Nouri, M., Martino, M.M., et al.: Cytoprotection, proliferation and epidermal differentiation of adipose tissue-derived stem cells on emu oil based electrospun nanofibrous mat. Exp. Cell Res. **357**, 192–201 (2017)
189. Fu, S.Z., Meng, X.H., Fan, J., et al.: Acceleration of dermal wound healing by using electrospun curcumin-loaded poly(ε-caprolactone)-poly(ethylene glycol)-poly(ε-caprolactone) fibrous mats. J Biomed Mater Res - Part B Appl Biomater **102**, 533–542 (2014)
190. Gong, C.Y., Wu, Q.J., Wang, Y.J., et al.: A biodegradable hydrogel system containing curcumin encapsulated in micelles for cutaneous wound healing. Biomaterials **34**, 6377–6387 (2013)
191. Merrell, J.G., Mclaughlin, S.W., Tie, L., et al.: Curcumin Loaded Poly(ε-Caprolactone) Nanofibers: Diabetic Wound Dressing with Antioxidant and Anti-inflammatory Properties NIH Public Access Author Manuscript. Clin. Exp. Pharmacol. Physiol. **36**, 1149–1156 (2009)
192. Nguyen VC, Nguyen VB, Hsieh MF. Curcumin-loaded chitosan/gelatin composite sponge for wound healing application. *Int J Polym Sci*; 2013. Epub ahead of print 2013. https://doi.org/10.1155/2013/106570.
193. Mitra, T., Manna, P.J., Raja, S.T.K., et al.: Curcumin loaded nano graphene oxide reinforced fish scale collagen-a 3D scaffold biomaterial for wound healing applications. RSC Adv. **5**, 98653–98665 (2015)
194. Venkata V, Reddy S, Kuppusamy G, et al. Curcumin loaded chitosan nanoparticles impregnated into collagen- alginate scaffolds for diabetic wound healing. *Int J Biol Macromol*. Epub ahead of print 2016. https://doi.org/10.1016/j.ijbiomac.2016.05.038.
195. Jin, G., Prabhakaran, M.P., Kai, D., et al.: Tissue engineered plant extracts as nanofibrous wound dressing. Biomaterials **34**, 724–734 (2013)
196. Krishnan, R., Rajeswari, R., Venugopal, J., et al.: Polysaccharide nanofibrous scaffolds as a model for in vitro skin tissue regeneration. J. Mater. Sci. Mater. Med. **23**, 1511–1519 (2012)

197. Suganya S, Venugopal J, Ramakrishna S, et al. Herbally derived polymeric nanofibrous scaffolds for bone tissue regeneration. *J Appl Polym Sci* 2014; 131: n/a-n/a.
198. Sharmila, G., Muthukumaran, C., Kirthika, S., et al.: Fabrication and characterization of Spinacia oleracea extract incorporated alginate/carboxymethyl cellulose microporous scaffold for bone tissue engineering. Int. J. Biol. Macromol. **156**, 430–437 (2020)
199. Ranjbar-Mohammadi, M., Rabbani, S., Bahrami, S.H., et al.: Antibacterial performance and in vivo diabetic wound healing of curcumin loaded gum tragacanth/poly(ε-caprolactone) electrospun nanofibers. Mater. Sci. Eng. C **69**, 1183–1191 (2016)
200. Asghari F, Rabiei Faradonbeh D, Malekshahi ZV, et al. Hybrid PCL/chitosan-PEO nanofibrous scaffolds incorporated with A. euchroma extract for skin tissue engineering application. *Carbohydr Polym* 2022; 278: 118926.
201. Karimi, T., Mottaghitalab, F., Keshvari, H., et al.: Carboxymethyl chitosan /sodium carboxymethyl cellulose /agarose hydrogel dressings containing silk fibroin /polydopamine nanoparticles for antibiotic delivery. J Drug Deliv Sci Technol **80**, 1–14 (2023)
202. Yang, Y., Li, X., Cui, W., et al.: Structural stability and release profiles of proteins from core-shell poly (DL-lactide) ultrafine fibers prepared by emulsion electrospinning. J Biomed Mater Res Part A An Off J Soc Biomater Japanese Soc Biomater Aust Soc Biomater Korean Soc Biomater **86**, 374–385 (2008)
203. Yang, Y., Li, X., Qi, M., et al.: Release pattern and structural integrity of lysozyme encapsulated in core–sheath structured poly (dl-lactide) ultrafine fibers prepared by emulsion electrospinning. Eur. J. Pharm. Biopharm. **69**, 106–116 (2008)
204. Jiang, H., Hu, Y., Zhao, P., et al.: Modulation of protein release from biodegradable core–shell structured fibers prepared by coaxial electrospinning. J Biomed Mater Res Part B Appl Biomater An Off J Soc Biomater Japanese Soc Biomater Aust Soc Biomater Korean Soc Biomater **79**, 50–57 (2006)
205. Zhang, Y.Z., Wang, X., Feng, Y., et al.: Coaxial electrospinning of (fluorescein isothiocyanate-conjugated bovine serum albumin)-encapsulated poly (ε-caprolactone) nanofibers for sustained release. Biomacromol **7**, 1049–1057 (2006)
206. Cui, W., Zhou, Y., Chang, J.: Electrospun nanofibrous materials for tissue engineering and drug delivery. Sci. Technol. Adv. Mater. **11**, 14108 (2010)
207. Cui, W., Li, X., Zhou, S., et al.: In situ growth of hydroxyapatite within electrospun poly (DL-lactide) fibers. J Biomed Mater Res Part A An Off J Soc Biomater Japanese Soc Biomater Aust Soc Biomater Korean Soc Biomater **82**, 831–841 (2007)
208. Cui, W., Li, X., Chen, J., et al.: In situ growth kinetics of hydroxyapatite on electrospun poly (DL-lactide) fibers with gelatin grafted. Cryst. Growth Des. **8**, 4576–4582 (2008)
209. Li, X., Xie, J., Yuan, X., et al.: Coating electrospun poly (ε-caprolactone) fibers with gelatin and calcium phosphate and their use as biomimetic scaffolds for bone tissue engineering. Langmuir **24**, 14145–14150 (2008)
210. Li, X., Xie, J., Lipner, J., et al.: Nanofiber scaffolds with gradations in mineral content for mimicking the tendon-to-bone insertion site. Nano Lett. **9**, 2763–2768 (2009)
211. Coburn, J., Pandit, A.: Development of Naturally-Derived Biomaterials and Optimization of Their Biomechanical Properties. Top Tissue Eng **3**, 1–33 (2007)
212. Martinez, A.W., Caves, J.M., Ravi, S., et al.: Effects of crosslinking on the mechanical properties, drug release and cytocompatibility of protein polymers. Acta Biomater. **10**, 26–33 (2014)
213. Jiang, Q., Reddy, N., Zhang, S., et al.: Water-stable electrospun collagen fibers from a non-toxic solvent and crosslinking system. J Biomed Mater Res Part A **101**, 1237–1247 (2013)
214. Hennink, W.E., van Nostrum, C.F.: Novel crosslinking methods to design hydrogels. Adv. Drug Deliv. Rev. **64**, 223–236 (2012)
215. Lai, J.-Y.: Interrelationship between cross-linking structure, molecular stability, and cytocompatibility of amniotic membranes cross-linked with glutaraldehyde of varying concentrations. RSC Adv. **4**, 18871–18880 (2014)
216. Meng, L., Arnoult, O., Smith, M., et al.: Electrospinning of in situ crosslinked collagen nanofibers. J. Mater. Chem. **22**, 19412–19417 (2012)

217. Gyawali, D., Nair, P., Zhang, Y., et al.: Citric acid-derived in situ crosslinkable biodegradable polymers for cell delivery. Biomaterials **31**, 9092–9105 (2010)
218. Reddy, N., Reddy, R., Jiang, Q.: Crosslinking biopolymers for biomedical applications. Trends Biotechnol. **33**, 362–369 (2015)
219. Reddy, N., Yang, Y.: Potential of plant proteins for medical applications. Trends Biotechnol. **29**, 490–498 (2011)
220. Cao, H., Xu, S.-Y.: EDC/NHS-crosslinked type II collagen-chondroitin sulfate scaffold: characterization and in vitro evaluation. J. Mater. Sci. Mater. Med. **19**, 567–575 (2008)
221. Mitra, T., Sailakshmi, G., Gnanamani, A., et al.: Cross-linking with acid chlorides improves thermal and mechanical properties of collagen based biopolymer material. Thermochim. Acta **525**, 50–55 (2011)
222. Grover, C.N., Cameron, R.E., Best, S.M.: Investigating the morphological, mechanical and degradation properties of scaffolds comprising collagen, gelatin and elastin for use in soft tissue engineering. J. Mech. Behav. Biomed. Mater. **10**, 62–74 (2012)
223. Sionkowska, A.: Current research on the blends of natural and synthetic polymers as new biomaterials. Prog. Polym. Sci. **36**, 1254–1276 (2011)
224. Bi, L., Cao, Z., Hu, Y., et al.: Effects of different cross-linking conditions on the properties of genipin-cross-linked chitosan/collagen scaffolds for cartilage tissue engineering. J. Mater. Sci. Mater. Med. **22**, 51–62 (2011)
225. David, G., Cristea, M., Balhui, C., et al.: Effect of cross-linking methods on structure and properties of poly (ε-caprolactone) stabilized hydrogels containing biopolymers. Biomacromol **13**, 2263–2272 (2012)
226. Kuo, C.K., Ma, P.X.: Maintaining dimensions and mechanical properties of ionically crosslinked alginate hydrogel scaffolds in vitro. J Biomed Mater Res Part A An Off J Soc Biomater Japanese Soc Biomater Aust Soc Biomater Korean Soc Biomater **84**, 899–907 (2008)
227. Duan, X., Sheardown, H.: Crosslinking of collagen with dendrimers. J Biomed Mater Res Part A An Off J Soc Biomater Japanese Soc Biomater Aust Soc Biomater Korean Soc Biomater **75**, 510–518 (2005)
228. Saito, H., Murabayashi, S., Mitamura, Y., et al.: Characterization of alkali-treated collagen gels prepared by different crosslinkers. J. Mater. Sci. Mater. Med. **19**, 1297–1305 (2008)
229. Wang, L., Stegemann, J.P.: Glyoxal crosslinking of cell-seeded chitosan/collagen hydrogels for bone regeneration. Acta Biomater. **7**, 2410–2417 (2011)
230. Reddy, N., Li, Y., Yang, Y.: Alkali-catalyzed low temperature wet crosslinking of plant proteins using carboxylic acids. Biotechnol. Prog. **25**, 139–146 (2009)
231. Reddy, N., Li, Y., Yang, Y.: Wet cross-linking gliadin fibers with citric acid and a quantitative relationship between cross-linking conditions and mechanical properties. J. Agric. Food Chem. **57**, 90–98 (2009)
232. Li, Y., Reddy, N., Yang, Y.: A new crosslinked protein fiber from gliadin and the effect of crosslinking parameters on its mechanical properties and water stability. Polym. Int. **57**, 1174–1181 (2008)
233. Lu, W., Sun, J., Jiang, X.: Recent advances in electrospinning technology and biomedical applications of electrospun fibers. J Mater Chem B **2**, 2369–2380 (2014)
234. Mekhail, M., Wong, K.K.H., Padavan, D.T., et al.: Genipin-cross-linked electrospun collagen fibers. J. Biomater. Sci. Polym. Ed. **22**, 2241–2259 (2011)
235. Kiechel, M.A., Schauer, C.L.: Non-covalent crosslinkers for electrospun chitosan fibers. Carbohydr. Polym. **95**, 123–133 (2013)
236. Shi, L., Le Visage, C., Chew, S.Y.: Long-term stabilization of polysaccharide electrospun fibres by in situ cross-linking. J. Biomater. Sci. Polym. Ed. **22**, 1459–1472 (2011)
237. Jiang, Q., Xu, H., Cai, S., et al.: Ultrafine fibrous gelatin scaffolds with deep cell infiltration mimicking 3D ECMs for soft tissue repair. J. Mater. Sci. Mater. Med. **25**, 1789–1800 (2014)
238. Hauert, S., Bhatia, S.N.: Mechanisms of cooperation in cancer nanomedicine: towards systems nanotechnology. Trends Biotechnol. **32**, 448–455 (2014)

239. Reddy, N., Shi, Z., Xu, H., et al.: Development of wheat glutenin nanoparticles and their biodistribution in mice. J Biomed Mater Res Part A **103**, 1653–1658 (2015)
240. Avvakumova, S., Colombo, M., Tortora, P., et al.: Biotechnological approaches toward nanoparticle biofunctionalization. Trends Biotechnol. **32**, 11–20 (2014)
241. Al-Remawi, M.M.A.: Properties of chitosan nanoparticles formed using sulfate anions as crosslinking bridges. Am. J. Appl. Sci. **9**, 1091 (2012)
242. Pichayakorn, W., Boonme, P.: Evaluation of cross-linked chitosan microparticles containing metronidazole for periodontitis treatment. Mater. Sci. Eng. C **33**, 1197–1202 (2013)
243. Hu, J., Ma, P.X.: Nano-fibrous tissue engineering scaffolds capable of growth factor delivery. Pharm. Res. **28**, 1273–1281 (2011)
244. Yin, Y., Ye, F., Cui, J., et al.: Preparation and characterization of macroporous chitosan-gelatin/β -tricalcium phosphate composite scaffolds for bone tissue engineering. J Biomed Mater Res - Part A **67**, 844–855 (2003)
245. Chen, S., Zhu, L., Wen, W., et al.: Fabrication and Evaluation of 3D Printed Poly(l -lactide) Scaffold Functionalized with Quercetin-Polydopamine for Bone Tissue Engineering. ACS Biomater. Sci. Eng. **5**, 2506–2518 (2019)
246. Zhu, L, Chen S, Liu K, et al. 3D poly (L-lactide)/chitosan micro/nano fibrous scaffolds functionalized with quercetin-polydopamine for enhanced osteogenic and anti-inflammatory activities. Chem Eng J; 391. Epub ahead of print 2020. https://doi.org/10.1016/j.cej.2019.123524
247. Khan, Y.M., Katti, D.S., Laurencin, C.T.: Novel polymer-synthesized ceramic composite-based system for bone repair: An in vitro evaluation. J Biomed Mater Res - Part A **69**, 728–737 (2004)
248. Kuo, Y.C., Ku, H.F., Rajesh, R.: Chitosan/γ-poly(glutamic acid) scaffolds with surface-modified albumin, elastin and poly-L-lysine for cartilage tissue engineering. Mater. Sci. Eng. C **78**, 265–277 (2017)
249. Wang, H., Li, Y., Zuo, Y., et al.: Biocompatibility and osteogenesis of biomimetic nano-hydroxyapatite/polyamide composite scaffolds for bone tissue engineering. Biomaterials **28**, 3338–3348 (2007)
250. Gautam, S., Sharma, C., Purohit, S.D., et al.: Gelatin-polycaprolactone-nanohydroxyapatite electrospun nanocomposite scaffold for bone tissue engineering. Mater. Sci. Eng. C **119**, 111588 (2021)
251. Li, M., Mondrinos, M.J., Chen, X., et al.: Elastin Blends for Tissue Engineering Scaffolds. J Biomed Mater Res Part A **79**, 963–973 (2006)
252. Shor, L., Güçeri, S., Wen, X., et al.: Fabrication of three-dimensional polycaprolactone/hydroxyapatite tissue scaffolds and osteoblast-scaffold interactions in vitro. Biomaterials **28**, 5291–5297 (2007)
253. Heydari, Z., Mohebbi-Kalhori, D., Afarani, M.S.: Engineered electrospun polycaprolactone (PCL)/octacalcium phosphate (OCP) scaffold for bone tissue engineering. Mater. Sci. Eng. C **81**, 127–132 (2017)
254. Remya, K.R., Joseph, J., Mani, S., et al.: Nanohydroxyapatite incorporated electrospun polycaprolactone/polycaprolactone-polyethyleneglycol-polycaprolactone blend scaffold for bone tissue engineering applications. J. Biomed. Nanotechnol. **9**, 1483–1494 (2013)
255. Thi Hiep, N., Chan Khon, H., Dai Hai, N., et al.: Biocompatibility of PCL/PLGA-BCP porous scaffold for bone tissue engineering applications. J. Biomater. Sci. Polym. Ed. **28**, 864–878 (2017)
256. Kim, B.S., Yang, S.S., Lee, J.: A polycaprolactone/cuttlefish bone-derived hydroxyapatite composite porous scaffold for bone tissue engineering. J Biomed Mater Res - Part B Appl Biomater **102**, 943–951 (2014)

Azadeh Izadyari Aghmiuni has a Ph.D. in Chemical Engineering-Tissue Engineering. Now, she is working on the stem cells, biological macromolecules & natural/synthetic polymers for construction of engineered scaffolds and smart biological substrates to regenerate hard and soft tissues. In this field, she has one patent entitled "Hybrid nano-substrate containing collagen to regenerate skin wounds, IR101120". Moreover, she investigates on the application of the herbal active ingredient and targeted drug delivery (TDD) method in tissue engineering and regenerative medicine to treat diseases.

Arezoo Ghadi has a Ph.D. in Chemical Engineering. She is working on the design of polymeric substrates and the differentiation of the stem cells into soft tissues. Now, she is Assist. Professor of Islamic Azad University- Ayatollah Amoli Branch, Department of Chemical Engineering, Amol- Iran.

Arezoo Ghadi has a B.Sc. in Chemical Engineering. She is working on the polymers and biopolymers for biomedical applications.

Gold Nanoparticles from a Microorganism: A Synthetic Approach

Anil Thakur, Shubham Thakur, and Sonia Sharma

Abstract Nanotechnology has observed an accelerated advancement in metallic and nn-metallic nanoparticles due to their unique physiochemical properties and biomedical applications as compared to their bulk material. Metallic nanoparticles like gold, silver, and platinum have been extensively explored and among them, gold nanoparticles (GNPs) not only hold the most ancient evidence of application but are also known to be the most stable. GNPs have been reported with applications in areas viz., biosensing and imaging, drug and gene delivery, biomaterials, cancer therapeutics, antimicrobials, bioremediation, and dye degradation. GNPs are biologically stable and their easily controllable surface chemistry enables modification with desired functionalities in various shapes and structures. GNPs can be prepared on large scale by reducing the gold oxidation state from Au^{+1}/Au^{+3} to Au^0 through physical and chemical methods. Harmful effects associated with these methods, like metal toxicity and by-products attached to nanoparticles, limit its use. Biological synthesis of GNPs has several advantages over physiochemically derived GNPs, which have motivated researchers to adapt to the green synthesis of GNPs. The biological method of synthesizing GNPs is rapid, environment-friendly cost-effective, and recognized as biocompatible. Green synthesis of GNPs includes the use of biological extracts from plants and microbes. Biological extracts act as stabilizers, reducing agents, and capping agents to synthesize highly dispersible and stable bioactive GNPs. Microbes have been used extensively for various biomolecules, bio-products, and even in the synthesis of nanoparticles and thus are also regarded as biofactories.

Keywords Gold nanoparticles · Microorganism · Biofactories · Reductases · Biomaterial · Therapeutics · Bacteria · Fungi · Actinobacteria

A. Thakur
University Institute of Pharmaceutical Sciences, Panjab University, Chandigarh, India

S. Thakur
Department of Pharmacy, Guru Nanak Dev University, Amritsar, India

S. Sharma (✉)
Microbial Type Culture Collection Centre and Gene Bank, CSIR-Institute of Microbial Technology, Chandigarh, India
e-mail: soniabt918@gmail.com

R. Malviya and S. Sundram (eds.), *Engineered Biomaterials*, Engineering Materials,
https://doi.org/10.1007/978-981-99-6698-1_7

1 Introduction

Nanotechnology is a known branch of science and technology since the last century. A famous lecture "There's Plenty of Room at the Bottom" presented by Nobel laureate Richard P. Feynman, in 1959, first introduced concept of "nanotechnology" [1]. Long before this conceptualization, nano-sized particle were in use in ancient times. People from ancient Egypt used black hair dye which was believed to be plant based product, henna but recent research of Egyptian burial sites revealed the hair dye was not henna instead it was paste of lime, leadoxide and water. They used hair dye paste in a way that reacts with keratin sulphur to produce lead-sulfide nanoparticles, which give an even and steady colour to hair [2]. This field of research generally plays at atomic, molecular scale to create and synthesize novel nano-scale structures, products and devices. The structures of range 1–100 nm scale are referred to as nanomaterials and these nanomaterials have some unique properties as compared to their larger dimensions. Surface effect and quantum effect are two major factors that provide unique and significant properties to the nanomaterial over their bulk counterparts and lead to enhanced mechanical, magnetic, catalytical, optical, electronic properties [3, 4]. Nanomaterial due to its small size, has high number of particles per mass unit and therefore have large surface area as compared to bulk material. Moreover the nanomaterials not only have an increased fraction of atom exposed on the surface but also have fewer direct neighbours and these differences impact the physiochemical properties of nanomaterials to a level that the melting point for gold nanoparticles could be brought to a significantly lower level as compared to large scale bulk gold material [3, 5]. Nanomaterials also display pronounced quantum effect with the decrease in size of material and these quantum structures contain all the charge carriers (electrons) confined within a specific physical dimension [6]. With this quantum effect, some non-magnetic materials in its large dimension would develop magnetic properties when brought to nanoscale [5]. The quantum effect of nanomaterial brings significant changes in its electron affinity, which in return directly dictates catalytic properties of the material. Due to such significant properties of nanomaterials, nanotechnology is considered multidisciplinary research area that is being explored for diverse purpose; agriculture, health, environment, catalysis, sensing, electronics, photonics [7, 8].

2 Dimensions of Nanotechnology

There are three dimensions of nanotechnology (Fig. 1); wet, dry, and computational. Wet nanotechnology, involves live organisms or biological components like whole cell, secreted enzymes, secreted compounds, to develop nanoproducts. Dry nanotechnology unlike wet nanotechnology, is related to the production of nanostructures using inorganic compounds like silicon and carbon. In contrast to both, computational nanotechnologies do not deal with nanostructure production but are associated with a very important aspect of development of nanostructure and

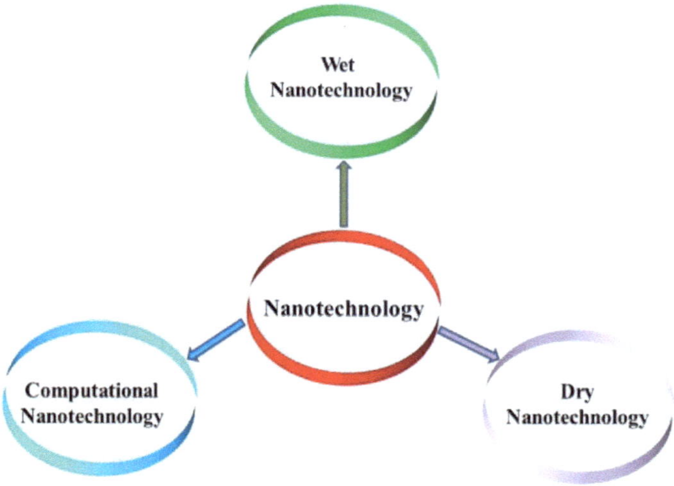

Fig. 1 Dimensions of nanotechnology

that is a simulation of nanometre-sized structures that could be helpful in studying physiochemical properties of desired design [9, 10].

3 Types of Nanomaterial

Dreaden group demonstrated the importance of size and shape of nanomaterial through difference observed among 20 nm particles made of gold, platinum, silver and palladium particles [11]. They were found to have characteristic wine colour for gold nanoparticle, red for platinum, yellowish grey for silver and black colour for palladium particles. It was observed that the nanomaterial shape is very much dependent on the size of the nanomaterial. At small size, nanomaterial can take up spherical shape but with an increase in size nanorod, nanotube nanosheets can be synthesized.

Based on dimension (size and shape), nanomaterials can be classified into four major types, these are zero-dimensional, one dimensional, two dimensional and three dimensional nanomaterials [3, 12] (Fig. 2). Zero-dimensional nanomaterials have all the dimensions along the three axis in the nanoscale range viz., nanoparticles, quantum dots and fullerenes. One-dimensional and two-dimensional are the nanomaterials with one or two of its dimensions above the nanoscale respectively. Nanorods, nanotubes, nanowires, nanofibers and nanohorns are few examples of one dimensional nanomaterial whereas nanosheets, nanofilms, and nanolayers are few representatives of two dimensional nanomaterials. The three dimensional nanomaterials are the ones with all three dimensions above nanoscale and this category includes an array of nanowires and nanotubes, powder and agglomeration of nanoparticles [3, 13].

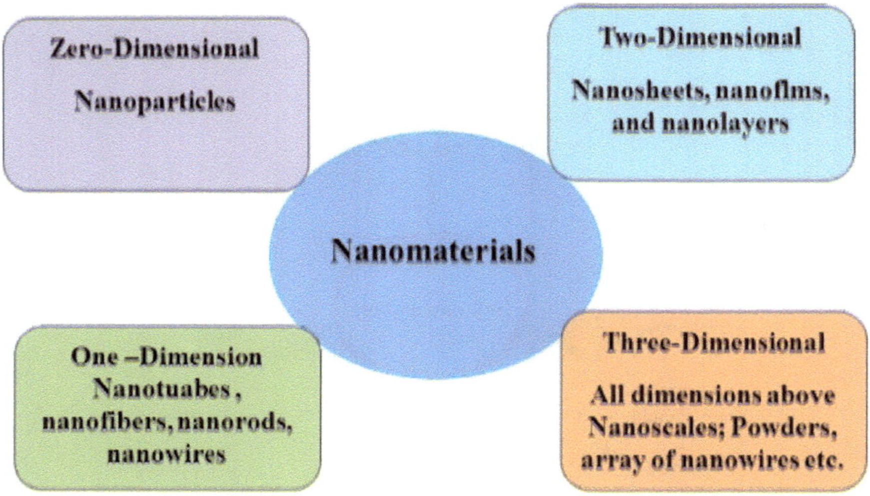

Fig. 2 Dimensions of nanomaterials

4 Nanoparticles (NPs)

Nanoparticles (NPs) are nano-objects that have all the three dimensions of the structure along the three axes, measures in nanoscale range, with no significant differences in the measure of the longest axis and the shortest axis. NPs can vary in their shapes, sizes, and structures and range from spherical, hollow core, spiral to irregular shape, cylindrical, tubular and conical [14]. NPs can measure from 1 to 100 nm in size and NPs lower than 1 nm in dimension, is termed as atom clusters. NPs can be crystalline or amorphous, uniform or multilayered and either loose or agglomerated [15]. NP is not just simple molecules but is formed of three or sometimes two layers. The three layered structure is composed of the outermost layer, i.e., surface layer, a middle shell layer and the innermost core. Surface layer is generally fabricated with a wide variety of molecules viz., metal ions, amino acid, peptides, polymers etc. The shell layer is a protective layer of core that is made of different material as compared to the core, and the core, which is the central most structure of nanoparticle is referred to as the NP itself which dictates the property of nanoparticle and thus the application [16].

5 Classification of NPs

NPs can be classified into three major classes, based on the components involved in synthesis of nanoparticles. These are organic, carbon-based, and inorganic (Fig. 3).

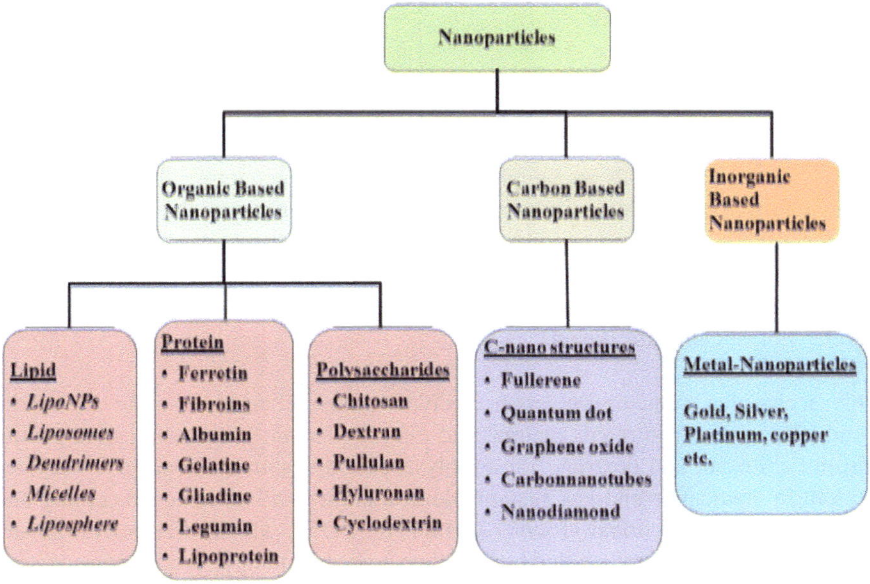

Fig. 3 Types of nanoparticles based on composition

5.1 *Organic Nanoparticles (ONPs)*

Organic nanoparticles are fabricated with organic compounds like, lipids, carbohydrates, proteins and peptides, and other organic compounds. They are non-toxic, biodegradable and biocompatible in nature and therefore are increasingly being used in medicine and food [17]. The main types of ONPs are lipid, protein and polysaccharide-based ONPs (Fig. 3).

5.1.1 Lipid Based NPs

Lipid-based systems mostly involve the use of phospholipids and are advantageous over other encapsulation methods as it uses natural ingredients which can entrap compounds of different solubility and can be targeted [17]. They are an attraction in pharmaceutical industry due to its favourable properties like biocompatibility and biodegradability. Liposomes, lipid nanoparticles, lipid nanotubes and lipid nano sphere, are some of the lipid-based nanostructures whereas the main lipid-based nanoencapsulation system includes nanoliposomes, nanoco-chelates, and archaeosomes [17–19].

5.1.2 Protein Based NPs

Protein-based organic nanoparticles involve use of protein for formulation and it possess unique property of forming gels and emulsions, which makes them a component of high choice with ability of encapsulation of bioactive compounds. Production of protein-based NPs are simple, reproducible and economically viable. It also offers multiple modifications, controlled release of the encapsulated bioactives. Few examples of protein-based nanoparticles are ferritin, lipoprotein, albumin, etc [20].

5.1.3 Polysaccharide Based

Polysaccharides are natural polymer made of monosaccharide linked by glycosidic bonds. It is present in plants, animal and microorganism few examples are pectin, insulin, chitosan, and dextran. These polysaccharides gained attention as they are present in abundance and has sustainable availability with easy of processing and presence of free functional groups (-OH and –COOH), which offer derivatization. Moreover it is a favoured choice for NPs because of its natural origin, non-toxicity, biocompatibility and biodegradability. In addition to this, polysaccharide NPs offer high drug loading efficiency, target delivery, fast release of drug, etc. Functional group derivatizations also allow tuning of its hydrophilicity and solubility and in return affects its application [20, 21]. The polysaccharide NPs forms a protective cover that helps in shielding the compound which gets cleaved into its unit saccharine by the microflora residing in colon. This gives an advantage of targeted delivery and slow release of drug entrapped. Chitosan, pectin is among the most used polysaccharide. Derivatives of polymers, poly-lactic acid as well as polyglycolic acid are brought into application as carrier for controlled drug release and proteins [21, 22].

5.2 Carbon-Based CNPs

CNPs are class of NPs that are made of carbon atoms and it includes carbon quantum dots, fullerenes, carbon black NPs, etc. [3]. Fullerenes are symmetrical closed-cage structure of carbon molecules and are formed with an arrangement of different number of carbon molecules, for instance C60 fullerenes is 60 carbon atoms arrangement that looks like soccer ball and few other examples include, C70 and C540 fullerenes. On the other hand, carbon black is arranged in grape-like structure with bunch of spherical particles fused together whereas carbon quantum dots are quasi-spherical carbon NPs [3, 23, 24]. Carbon-based NPs is the technology of putting together physio-chemical properties of sp2 -hybridized carbon bonds and nano-sized material. Carbon NPs have unique properties like electrical conductivity, electron affinity, high strength, thermal resistance, and adsorption which lead to its usage in a wide range of applications viz., bioimaging, delivery carrier for drugs and biomolecules, energy storage, photovoltaic devices and sensors for monitor microbial pathogens [25, 26].

5.3 Inorganic NPs

Nanoparticles, synthesized using metal (silver, gold, platinum) and metal oxide (oxide of iron and zinc) are known as inorganic nanoparticles. Inorganic nanoparticles posses localized surface plasmon resonance, unique thermal, magnetic, optical and electrical properties which makes them desirable material for development of nanodevices with wide variety of application in field of electronics, biomedicines physics and chemistry [3, 14, 16, 27]. In comparison amongst all metal NPs, nanoparticles fabricated with copper, mercury, silver, platinum, and gold, gold nanoparticles have gained researchers' attention all around the globe for its unique optical and physiochemical properties to investigate and probe gold NPs for different field applications [28].

6 Gold Nanoparticles (GNPs)

Gold nanoparticles (GNPs) are suspension of colloidal, nanometer dimension gold particles that are also referred to as colloidal gold. The first mentions of GNPs can be traced back to Roman times, where glass was stained for decorative purposes [29]. The ancient roman cup possessed an unusual colour changing property depending on the source of light falling on the cup. In presence of natural light, the cup looks green whereas when a lit candle placed within, the cup appears red (3, 31). In recent investigation of the cup, it was found that the cup was coated with Au and Ag nanoparticles in 50–100 nm range that made cup look differently coloured through the unique property of plasmon excitation of electrons in NPs [3, 21, 31]. Later the scientific report from Michael Faraday's work confirmed that colloidal gold or GNPs have some unique properties from the bulk gold [3]. GNPs have allured most attention amongst the other metal nanoparticles as it possesses unique physical, thermal, optical, chemical properties and are stable. GNPs can be synthesized in different shapes and size and their dimensions are decided by the condition of their synthesis and thus also dictate its use in as drug carriers for control release of drugs at different sites and other potential applications. It offers multiple modification and functionality that also determines its application [30]. GNPs functionalization with conjugated DNA, amino acid, protein and different biological molecules and their low toxicity and strong absorption spectrum is responsible for its use in biomedicine [32]. GNPs optical property is advantageous and widely used for coating to obtain high quality SEM images by enhancing the electronic stream [33].

7 Properties of GNPs

7.1 *Surface Plasmon Resonance*

Plasmon is oscillation of electron cloud under the exposure of light [34]. Plasmon resonance (PR) is an absorption phenomenon of material, which occurs with the oscillation of electrons produced by metal, on exposure to light [35]. The effect of incoming electromagnetic wave on the metal depends on the depth of the metal and generally decreases with increase in depth of the metal. This resonance on the metal surface is called surface plasmon resonance (SPR) [36]. SPR in small volume of nanoparticle leads to phenomenon of Localized Surface Plasmon Resonance (LSPR). LSPR is responsible for increased electromagnetic field and extinction coefficient and these two effects are further exploited for various biophysical and biochemical techniques like FRET, SEF, SERS, colorimetric methods, PTT and PDT [37].

7.2 *Surface-Enhanced Raman Spectroscopy*

Monochromatic incoming light on particle scatters with the same frequency in the same direction as the incident light. But when there is any adsorbed molecule on the surface of particles a deviation in the frequency of the scattered light is observed because of the additional vibration of the adsorbed molecule. This difference in the incident and scattered photon energies is called Raman scattering (RS) [38]. The change in energy is due to the energy of the vibrating adsorbed molecule. Raman image is known to provide a chemical fingerprint that is highly resolved ($\approx$0.1 nm) vibrational information but with the limitation of low signal sensitivity [39, 40]. Advancement in RS with surface-enhanced Raman Spectroscopy (SERS) is beneficial, which enhances the Raman signal up to 10^{15} from 10^6. This improved signal has led to its application in bio-sensing, diagnosis and imaging [41]. GNPs are being used as contrast agents of SERS, due to their optical properties for imaging tumors, distinguishing tumor cells and normal cells, tumor metabolism and tumor markers.

7.3 *Surface Enhanced Fluorescence*

GNPs due to their local surface Plasmon resonance, show two effects on the signal intensity of fluorescent molecules these are fluorescence quenching and amplification which in turn is dependent on the distance between the fluorescence molecules and GNPs. Fluorescence resonance energy transfer (FRET) is the main cause of the GNP's fluorescence quenching. If the distance is less than 5 nm, the excited fluorescent group transfers its energy to the GNPs resulting in fluorescence quenching [42].

The fluorescence quenching and enhancement can be used to study the interaction among the particles and therefore can be exploited in the field of diagnostics.

7.4 Photothermal Effect

Illumination of GNPs results in scattered motion of electrons and converts light energy into kinetic energy. Some of this kinetic energy is then converted into the vibration energy of the lattice. The vibration energy of the lattice then expresses itself as heat energy and is called the "photothermal effect". Since there is an obvious distinguish vascular characteristic in normal tissue and tumor tissue, the vascular variant of tumor tissue has lost vessel wall integrity and has many holes that cause blockage. This in turn affects the ability of tumor tissue to cool and the heat produced in the tumor accumulates and its temperature increase up to 46 °C. Since tumor tissue is less resistant to heat so the temperature above 42°C can prove lethal to tumor tissue. This fact is being used for bioimaging of tumor tissue as it can be distinguished from normal tissue. The photothermal effect of GNPs is applied in clinical diagnosis through photothermal imaging (PTI) and photoacoustic imaging (PAI) [43].

7.5 Photosensitization

GNPs upon excitation with photons, transfer energy to neighbour photosensitizer molecules like molecular oxygen, which leads to the generation of known cytotoxic agents like reactive oxygen species ($^1O_2$, O^{2-} and OH^-, etc.). In Photo Dynamic Therapy (PDT) of cancer, these species of oxygen play a very important role [44].

7.6 Colorimetric Responses

GNPs have a high molar absorption coefficient, which enhances their colorimeter sensitivity to the nanomole level. GNPs-based colorimetric assays are thus advantageous as it also takes into account the property of induced plasmon resonance wherein the change in GNPs- fabricated biomolecules (cytokines, proteins, nucleic acids, etc) aggregation corresponds to the change in endpoint colour from red to blue, purple or gray [45].

7.7 Increased X-ray Absorption Coefficient

Gold (Au, atomic number 79), with higher molecular weight shows an increase in the X-ray absorption coefficient and thus creates minimum damage to normal tissue with the minimal exposure time for X-ray and thus arose as a promising sensitizer for radiotherapy [46].

7.8 Functionalization

GNPs can be linked with S-and N-containing groups with a stable interaction which allows them to be fabricated with a wide variety of molecules and thereby bestow properties like biocompatibility, target therapy and delivery of drug and biomolecules (Fig. 4) [47].

8 Shape Size and Functionalization

The potential application of GNPs is depended upon their size, shape and functional groups attached (Fig. 4). There is a dramatic difference in colour observed for rod shaped GNPs than that for spheres. It is because the rod-shaped GNPs exhibit the plasmon oscillation along two axes of nanorod, one along the short axis and another along the long axis. In addition to this the resonance further depends strongly on the nanorod length-to-width ratio [48, 49], thus due to their unique optical properties, GNPs are subjected to research in bio-imaging, diagnosis and materials science

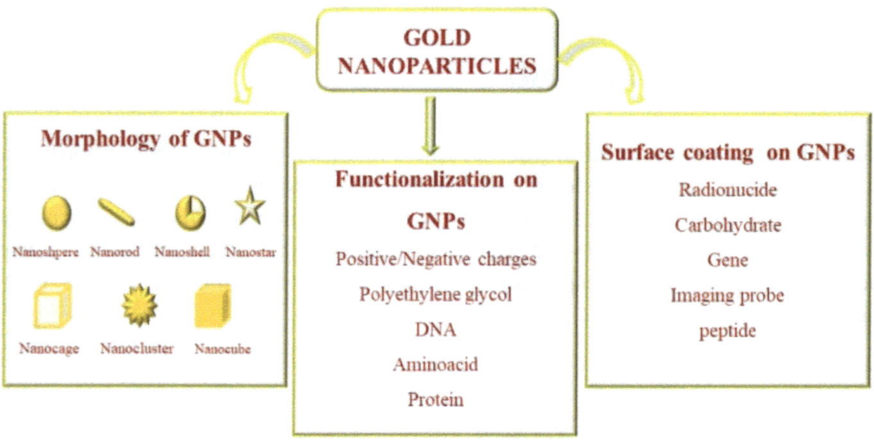

Fig. 4 GNPs morphology and functionalization

[50]. To develop GNPs for specific applications, specific methodologies with reproducibility in product formation have been employed and developed in the last couple of years [51, 52].

9 Synthesis of Nanoparticles

The two major approaches of synthesis of GNPs are top-down and bottom-up approach (Fig. 5). Top down approach is a subtractive process that involves slicing of starting bulk material into nanoscale structure. The different strategies of top-down approach include bulk laser ablation, UV-irradiation and IR-irradiation, ion sputtering and aerosol technology [53]. Large sized matter is subjected to these different strategies to form small particles but in contrast to this, bottom-up approach synthesizes GNPs starting from atomic level using gold of small scale. There are three distinct ways to synthesizes GNPs in bottom-up approach; physical method, chemical method and biological method.

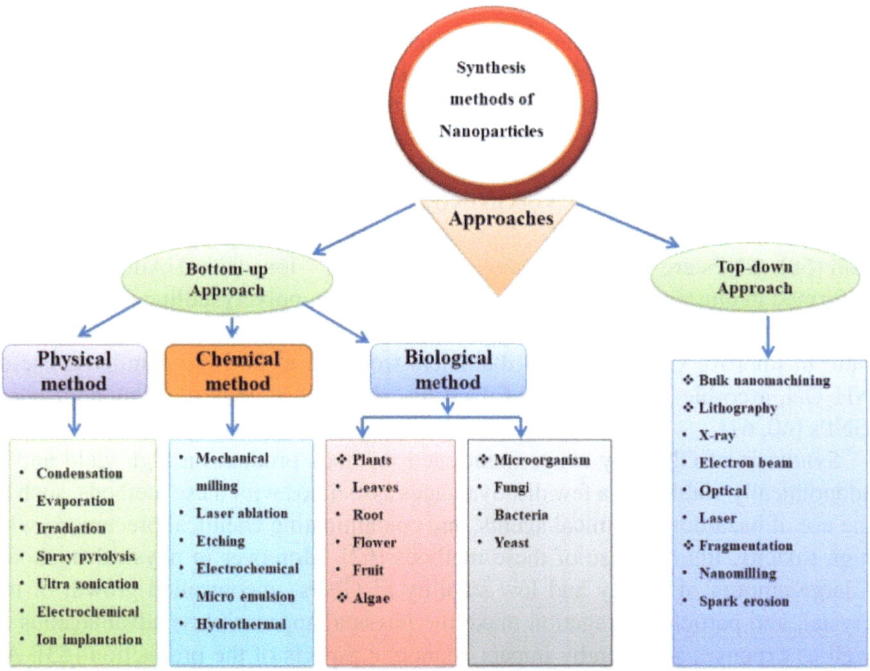

Fig. 5 Strategies for gold nanoparticles synthesis

9.1 Physical Method

The physical method employs strategies like laser ablation, gamma-irradiation, microwave irradiation, sputter deposition, ion implantation, lithography, UV and Ultrasound irradiation. The physical method generally requires maintaining controlled experiment factors with the required reducing and stabilizing agent and thus is free of contamination. Laser ablation is one of the most used methods and it alters the shape, surface and properties of GNPs [54, 55]. GNPs synthesized through Laser ablation are non-toxic and photothermally active but the yield was low and the method proved inconvenient for large scale purposes [56]. To improve scale up spherical, silica-coated GNPs were synthesized using pulsed laser ablation [57]. Another important method is ion implantation, lithography, irradiation etc. which involves high energy source thus an alternative method is in demand which lowers the demand of energy for synthesis of GNPs.

9.2 Chemical Methods of GNPs

This is the commonly used method of synthesis of GNPs and requires the use of chemical entities for gold ions reduction to nanoparticles and for stabilization of nanoparticles. The chemical agents mostly used are sodium citrate, sodium hydroxide, hydrogen tetrachloroaurate hydrate ($HAuCl_4$), SDS, sodium borohydride, cetyltrimethylammonium bromide, lithium aluminium hydride and ethyl glycol. The first report of the chemical synthesis of GNPs by Turkevich et al. [58] showed the synthesis of GNPs, on reduction of $HAuCl_4$ by sodium citrate in the single-phase aqueous solution [58]. GNPs are formed by the reduction of Au^{+3} ions from $HAuCl_4$ to Au^0. It is formed in clusters which then aggregate into large, polycrystalline form. In other reported methods, surfactant was used at varying concentrations of the gold-to-thiol ratio to improve GNPs range of diameters from 1.5 to 20 [59]. Similarly, use of NH_2OH, ascorbic acid (AA) and CTAB has shown the synthesis of monodisperse GNPs [60, 61].

Synthesis of GNPs by above-mentioned methods produces a high yield and is economically viable, but a few disadvantages associated with these methods, such as the use of hazardous chemical agents, and contaminating chemical precursors, with high toxicity, limit the use of these methods [62]. Moreover in physical methods, a large amount of energy and low stability of GNPs, uncontrolled growth of the crystal, and particle aggregation make the physical approach less advantageous as well as expenses and thereby impact economic aspects of the production [63]. An alternative to the physical and chemical method is the biological method which uses non-toxic agents and does not require the additional use of any external stabilizing agents and reducing agents. Moreover, it uses renewable materials with minimal energy expenditure and thus considered ecologically. To overcome these difficulties linked with physical and chemical approaches, enhanced interest in green synthesis is observed among researchers and they have investigated the biological production of GNPs through various biological means. Great diversity in plants and microorganisms provides not only huge options but also allows the production of specific GNPs using both whole organism and their products. The biomolecules not only synthesize GNPs but has a positive effect on the application of synthesized GNPs [63].

9.3 *Biological Method*

The biological method of GNPs is a simple process with no requirement of high temperature and pressure. In general, as the name suggests, the method involves the addition of biological extract (microorganism, plant, etc.) dropwise to the $HAuCl_4$ salt solution to initiate the GNPs synthesis. The change in colour is the indicator of the synthesis of GNPs [64]. The biological process of GNPs synthesis generally takes place in two steps: firstly, Gold Au^{3+} is reduced to Au^0 state and in the second step, GNPs agglomeration followed by stabilization [65, 66]. Biological extracts have different constituent molecules viz., enzymes, sugars, etc., which have the potential for gold reduction and stabilization and capping of GNPs [64].

9.3.1 Biosynthesis by Plants

There is a wide variety of plant species growing on earth amongst which there are a lot of plants that are well known in use for their medicinal value since ancient times. Therefore the potential of plants for GNPs has been explored for its economical as well as environment-friendly methodology. Different parts of plant, leaves, stem, root and fruit extract have been used to explore the production and application of GNPs [62]. Different constituents of plant extracts can synthesize GNPs like phenolic acids, flavonoids, terpenoids, polyphenols, etc., by reducing metal ions to corresponding nanoparticles [67]. Flavonoids and gold trivalent ion form an intermediate complex through flavonoid free radical which then results in the oxidization of flavonoid to Keto-forms and reducing trivalent gold to GNPs [68–71]. Salicin, a phenolic compound through its hydroxyl group and glucoside bonds participates in GNP synthesis and the stabilization, terpenoids can also synthesize nanoparticles by oxidizing itself (aldehyde group's oxidization to carboxylic acids) and reducing metal ions [71, 72]. The hydroxyl groups in the polyphenolic compound from *Mimosa tenuiflora* extract has been found to transfer an electron to trivalent gold ion, oxidizing themselves in the process to synthesize GNPs [73]. The role of tannins, alkaloids, polyols and plant extract sugar in reducing metal ions to nanoparticles is also observed [74–77]. In addition to the compounds, higher molecular weight proteins can also be considered important molecules for the synthesis of GNPs [78, 79]. Citrulline from watermelon is one such example [80].

9.3.2 Biosynthesis by Microorganism

But the use of microorganisms is advantageous over plants as they are easily cultivable and can be produced in large amounts in less time, in addition to this they do not require anything more than nutrients to grow at the temperature, pH and pressure in which they generally inhabit. The use of bacteria, fungi, yeast and algae has been reported in the synthesis of metal nanoparticles [64]. Synthesis of nanomaterials using microbial extracts or whole culture is actually the merging of the fields of microbiology, biotechnology, nanotechnology and giving rise to a new area of research and nano-biotechnology. It is a greener route of synthesis in which varied species of bacteria and fungi have been studied by researchers and have been reported [81, 82]. Microorganisms can synthesize nanoparticles using either extracellular or intracellular mechanisms. Extracellular synthesis basically involves adsorption of the metal ions at the microbial cell surface, followed by their direct reduction whereas in intracellular synthesis, metal ions are transported into the cell cytoplasmic region of the micro-organisms followed by their subsequent reduction of metal ions. The synthesized GNPs are then derived into the extract [64, 82].

9.3.3 Biosynthesis by Fungi

Fungi can withstand high metal ion concentrations as compared to bacteria and can secrete abundant of extracellular proteins, which can be instrumental in synthesize of GNPs. These secreted biomolecules have the ability to reduce soluble metal ions to their insoluble form and eventually to nanoparticles. The huge diversity of fungi and their potential to produce novel metal reductases for metal detoxification and bioreduction are potential enzymes for nanoparticle synthesis [83]. GNPs synthesis through fungi can take place in intracellular or extracellular space or by using fungal extract containing reducing agents. The extracellular mechanism of GNP formation in fungi relies on the entrapment of Au^{3+}ions by cell wall proteins, followed by reduction. According to the report on *Verticillium* sp., the extracellular interaction of $AuCl^{4+}$on the cell surface enzymes with groups having positive charges, (e.g. lysine) leads to reduction of $AuCl^{4+}$ to Au^0 [84, 85]. In either mechanism, NADPH-dependent oxidoreductases are the key enzymes in GNPs biosynthesis and NADH acts as a cofactor of the enzyme [86, 87]. In the intracellular GNP synthesis mechanism, gold (Au^{3+}) ions get diffused into the cell membrane where the ions are reduced by cystolic redox enzymes [86]. In the intracellular mechanism, reducing sugars, reductase enzymes such as, glyceraldehyde-3-phosphate dehydrogenase, ATPase and 3-glucan-binding proteins involved in the energy metabolism of fungal cells can catalyze the synthesis of nanoparticles [51, 84]. Au^{3+} diffuses into the cell membrane through transporter protein and is reduced by cytosolic prevalent redox enzymes. *Aspergillus fumigates* and *Aspergillus flavus* were reported to synthesize intracellular GNPs [83]. A glutathione-like compound, phytochelatin from *Candida albicans*, has been shown to directly catalyze the GNPs synthesis [88]. Gold ions are reduced to GNP by phytochelatin in the presence of glutathione [88].

Table 1 Gold nanoparticle synthesized by fungi

S. no	Fungi	Shape	Size	Application	References
1	*Pycnoporu ssanguineus*	Varied size; spherical,, triangular, pseudo-spherical, pentagonal, and hexagonal	~ 6	Catalytic activity of degrading Nitroaromatic compound 4-NA	[91]
2	*Pycnoporus sanguineus*	Spherical	6–25	Catalytic activity	[92]
3	*Magnusiomyces ingens*	Spherical, triangular, hexagonal	10–80	Catalytic activities for degrading nitrophenols	[93]
4	*Magnusiomyces ingensi* LH F1	Spherical, pseudo-spherical	28–20	Catalytic activities for nitrophenols	[94]
5	*Magnusiomyces ingensi*	Spherical,		Colorimetric detection of toxic heavy metal Hg with sensitivity and selectivity	[95]
6	*Trichoderma hamatum* SU136	Varied size: spherical, pentagonal and hexagonal	5–30	Antimicrobial activity	[96]
7	*Trichoderma sp.*		~ 19.7	Anti cancer activity against A549 and LN229	[97]
8	*Aspergillus flavus*	Spherical	12	Anti-cancer activity A549, HepG2 and MCF7	[98]
9	*Fusarium solani*	Spherical	40–45	Cytotoxicity against HeLa, MCF-7	[99]
10	*Alternaria*		28	Anticancer, antimicrobial and antioxidant activities	[100]

GNPs synthesis by phenol oxidases, Mn-peroxidases, laccases, and tyrosinases are remarkable examples of xylographic basidiomycetes, which uses both intracellularly and extracellularly mechanism [89]. Melanin was also reported in gold ion reduction to GNPs by *Yarrowia lipolytica* [90] (Table 1).

9.3.4 Synthesis by Bacteria

Bacteria can be considered a "biofactory" due to their potential for GNPs production. The mechanism for the biosynthesis of GNPs reported in bacteria is both extracellular and intracellular [82, 101]. The bacterial enzyme, NADH reductase and nitrate reductase were found to play a vital role in the gold ions reduction

[84, 102]. *Stenotrophomonas maltophilia*, *Pseudomonas putida*, *Rhodopseudomonas capsulate* and *Pseudomonas fluorescence* have been shown to synthesize GNPs through the enzyme NADPH-dependent reductase using electron transfer mechanism, which converts Au^{3+} to Au^0 [81]. In the intracellular mechanism, gold ions (Au^{3+}) can enter the cell via an ion transporter channel or proteins, into the cytoplasmic membrane and the cytoplasm, where it is reduced to GNPs enzymatic [105]. Metal ions bind to the negatively charged transporter protein, which then transports the metal ions inside the cell, passing through the cell wall surface inside the cytoplasm and subsequently catalyzing the synthesis of GNPs with the help of reductase enzymes in various dimensions [62, 103]. The transport of metal ions can also be facilitated through ion pumps, endocytosis, transporter proteins, lipid permeation and carrier-mediated transport [104]. GNPs from *Deinococcus radiodurans*, which are reported for their tolerance to radiation and oxidants, have been found to harbor antioxidants like carotenoid, pyrroloquinoline-quinone, and phosphoproteins which provide a protective shield against oxidative damage to nucleic acids and proteins to bacteria and in addition to this forms a potential microenvironment for the reduction of Au (III) and thereby GNP synthesis [105]. GNPs synthesis by *Lactobacillus kimchius*, showed the involvement of NADH oxidoreductase and sugars moieties embedded in the outer membrane of bacteria in Au^{3+} reduction, and also the proteins or amino acid present inside cells for stabilization of nanoparticles [63] (Table 2).

10 Applications

10.1 Visualization and Bioimaging

GNPs have high electron density which can be actively used for the identification and visualization of biological specimens through Transmission Electron microscopy (TEM), and high-resolution transmission electron microscope (HRTEM) with advanced systems of digital recording and the processing of images. The practical uses of GNPs in immune electron spectroscopy are in use for the identification of infectious agents and their surface antigens. Scanning probe microscopy, scanning electron microscopy, and fluorescence microscopy are also used for the same purpose. All types of GNPs ranging from conventional spherical GNPs to non-spherical particles, such as nanorods, nanoshells, nanocages, nanostars, and other types of particles are in use in diagnostics [113]. The biggest challenge to surgeons in tumor removal is to locate precise boundaries of tumor among healthy tissue. In this scenario, the removal of any less or more tissue can lead to leaving tumor tissue behind or removing healthy tissue that can be vital. Magnetic Resonance Imaging (MRIs) and Computed Tomography (CT) have a limitation of ability to detect tumors in millimeters or approximately 10 million cells. To overcome this limitation, photoimaging has emerged as a novel approach where functionalized GNPs are injected into the tumor which then binds to the cancer cells only and

Table 2 Gold nanoparticle synthesized by bacteria

S. no	Bacteria	Shape	Size (nm)	Application	References
1	*Pseudomonas putida and P. Fluorescences*	Spherical	10–50	Biofilm inhibition	[81]
2	*Stenotrophomonas maltophilia (AuRed02)*	Spherical	40		[102]
3	*Deinococcus radiodurans*	Varied shape; spherical, pseudo-spherical, truncated triangular and irregular	~ 44	Multifuctional; Antibacterial by cytoplasmic membrane damaging	[105]
4	*Marinobacter algicola*	Varied shape; spherical, triangular, pentagonal and hexagonal,	~ 74	Antibacterial activity	[106]
5	*Marinobactor pelagius*	Varied	2–6		[107]
6	*Baccilus subtilis*		7.6		[108]
7	Enterococcus sp.	Spherical	6–13	Anticancer activity against HepG2 and A549 cells	[109]
8	*Paracoccus haeundaensis* BC74171^T	Spherical	20.93	Antioxiant Anticancer aginstA549 and AGS cancer cells	[110]
9	*Enterococcus* sp.	Spherical	7	Anticancer activity against HepG2 cancer cells	[111]
10	*Vibrio alginolyticus*	Irregular	100–150	Anticancer activity HCA-7 cell line	[115]

shines, thereby distinguishing the tumor and healthy cells [114]. Qian et al., [115], reported the enhanced Raman signal and detection by conjugating GNPs to antibodies specific to epidermal growth factor receptors (EGFR), present on cancer cells and in xenograft tumor models and thereby help in efficient diagnosis [115]. Hossain and co-workers also used GNP-conjugated ITO substrate to enhance the Raman signals which can be used to differentiate characteristics sub-types of breast cancer cells whether from the same organ origin or different organs [116]. Lin et al., used GNP-based SERS for the diagnosis of colorectal tumor markers with enhanced sensitivity and specificity [117].

10.2 Antibacterial Activity

High resistance or escape pathogens are one of the serious problems in medicine as these microorganisms are ever evolving and thus very efficiently develop mechanisms to resistance to even the most advanced antibiotics. To overcome the issue newer technology is always employed to tackle such pathogens. Nanoparticles from various metals and especially gold are being explored. GNPs have been reported by many to show antimicrobial activities but antimicrobial activity depends on GNPs dimension and concentration [64, 118]. Any antimicrobial agent with a novel target and a broad range of activity both on gram positive and gram negative could be advantageous to tackle a broad range of pathogenic bacteria. The cell wall of bacteria is negatively charged which has a high affinity for positively charged GNPs. The difference in GNPs activity towards Gram negative and Gram positive is because of the cell wall structure difference. GNPs show better activity in Gram-negative bacteria as they have thinner cell walls which ease GNPs entry into the cell as compared to thick rigid layers of peptidoglycan in Gram-positive bacteria. Considerable difference in antibacterial effect has been observed between biosynthesized GNPs and chemically synthesized GNPs. This enhanced activity in biosynthesized antibacterial GNPs could be due to the activity of the capping agents, which showed an enhanced synergistic effect [28]. There are many reports showing the antimicrobial activity of GNPs against bacteria *(E. coli, Enterobacter sp., B. subtilis, etc.)* and bacterial biofilms. GNPs showed inhibition of virulence factors i.e. exopolysaccharide production and thereby weakening the biofilm formation of *Proteus* sp. [119]. GNPs conjugated with tetracycline demonstrated greater antibacterial activity as compared to tetracycline alone against both Gram-positive and Gram-negative bacteria [120].

10.3 Antifungal Activity

Like bacteria, pathogenic fungi also pose a serious threat to clinical research, due to the emerging antibiotic resistance. This new approach to combat these pathogenic fungi and their associated diseases is a need of the time. Nanotechnology is a promising area of research that is being explored for antifungal potential against pathogenic fungi like *Aspergillus* spp., *C. albicans*, *Trichoderma* spp., *Penicillium* spp., etc [121]. Nanoparticles of gold are being exploited as a potential as antifungal agent and have been observed as a potent agent. GNPs damage cell wall macromolecules and inhibit β-glucan synthase affecting the cell wall structure and its integrity and ultimately resulting in cell wall damage. Besides this antifungal activity of GNPs is also attributed to increased ROS in *C. albicans* [122].

10.4 Antiviral Activity

Viral diseases have gained a lot more attention these days as these are one of the greatest threats to humans and a great challenge to medical research. This is a very known fact that for many viral diseases for which there is no cure short drugs or vaccines developed. Therefore new approach like the use of metal nanoparticles has come up as a promising technology to fight these dangerous viruses [64]. GNPs are expected to show antiviral activity as it shows the potential to able to bind a viral particle and can block the viral receptors that lead to inhibition of the viral cycle. Moreover, GNPs have a charge density that on binding to viral particles' surface brings on a significant change in the membrane potential of the virus particle and thereby affects viral penetration into the cell [123]. GNPs have been reported as an effective agent against the measles virus (MeV) [124]. El-Sheikh et al. reported significant inhibition of Herpes Simplex (HSV-1) using GNPs in a dose-dependent manner [125].

10.5 Antioxidant Activity

Inflammation, aging and pathological conditions like atherosclerosis, cancer, and neurodegenerative diseases are mainly caused by oxidative stress occurred because of Reactive Oxygen Species (ROS) for example superoxide, hydroxyl, and singlet oxygen. Antioxidants help to relieve ROS-generated oxidative damage by directly reacting with free radicals or by inactivating radical-generating enzymes or increasing the expression of intracellular antioxidant enzymes [64, 126, 127]. Antioxidants can reduce cell damage, and DNA damage and thereby reduce the risk of various ROS-induced diseases. The increasing ROS leads to increased oxidative stress in the body which exceeds the antioxidant combat system and affects nucleic acid and enzymes in the body [64]. Intake of antioxidants in addition to existing intracellular antioxidant enzymes can help efficiently in combating increased ROS levels in the body [127]. GNPs synthesized using *Acanthopanacis* cortex extracts were reported to stimulate the expression of iNOS (Inducible nitric oxide synthase) and COX-2 (cyclooxygenase-2) protein in RAW 264.7 cells induced with LPS (lipopolysaccharides-induced) [128]. GNPs produced using the marine bacterium *Paracoccus haeundaensis* through biological method offers a promising antioxidant potential by scavenging free radicals [110]. According to the studies, various biomolecules grafted on the surface of GNPs, adds to their antioxidant property. GNPs produced using the marine bacterium *Vibrio alginolyticus* showed potent antioxidant activity [112].

10.6 Anticancer Activity

Cancer is a worldwide problem with an increasing number of cases in recent years. Most of the anticancer drugs used have a list of associated side effects, as the drugs affect normal cells as well. Therefore, there is continuous effect made by researchers to find a novel drug with minimal or no toxicity. The anticancer activity of GNPs has been reported in different human cancer cell lines such as MCF-7 (breast adenocarcinoma), A549 (human lung carcinoma), HCT-11 (colon carcinoma), Hep G2 (human liver cancer), ovarian adenocarcinoma (Caov-4), Hela N (Human cervix carcinoma) [129–134]. The anticancer activity of GNPs is significantly affected by their dimension and chemical composition [135–137]. Small sizes GNPs provide a larger surface area and have better anticancer activity than large-sized GNPs [110]. In addition to this, capping agents of GNPs also contribute to their anticancer activity by either modifying cell growth factors or by directly showing anticancer activity [110, 138]. Gold atom interaction with intracellular protein, and DNA through their functional group could attribute to the anticancer activity and influence the proliferation of cell line U87 (human primary glioblastoma cell line) [139, 140]. GNP-treated cells are found accumulating in the sub-G1 phase or G0/G1 phase of the cell cycle which shows the role of GNPs in cell cycle regulation by inducing apoptosis [141]. GNPs showed potential antiproliferative activity in MCF-7 and MDA-MB-231 and cells were found arrested in G0/G1 and S phases [140, 142]. Apoptosis is the most important mechanism of action of GNPs which are determined by morphological changes in cells such as cell shrinkage, extensive blebbing of the plasma membrane, nuclei fragmentation, etc [143]. GNPs activate programmed cell death through caspases-mediated apoptotic pathways [144–148]. GNPs biosynthesized using marine *Vibrio alginolyticus*, a marine bacterium showed anticancer activity against the human colon carcinoma cell line (HCA-7) through programme cell death (115).

10.7 Antidiabetic Activity

The GNPs have been found to offer antidiabetic activity in diabetic animals that were administered with oral GNPs injection and were found to regulate the metabolic process in addition to the restoration of cholesterol and triglycerides levels [149]. Another study in which GNPs treated rats was found to improve the body weight of rats by regulating insulin and glycemic levels [150]. Antidiabetic activity of GNPs synthesized from *Fritillaria cirrohosa* was observed in preclinical models [151]. GNPs from *Ziziphus jujuba* showed a decline in lipid peroxidation and oxidative stress which was fruitful in controlling diabetes [152].

10.8 Leishmanicidal Activity

GNPs from *Jasminum nervosum* leaf extract showed potent larvicidal activity against filarial and arbovirus vector *Culex quinquefasciatus* [142, 157]. Larvae and pupae of the malaria vector *A. stephensi* and the dengue vector *A. aegypti* were also reported to be managed by GNPs [154].

10.9 Drug Delivery

GNPs can be functionalized by binding molecules with various functional groups and this property provides GNPs with the ability to deliver various therapeutic agents. The bound molecules attribute to a wide variety of delivery applications of GNPs. GNPs can be linked through covalently and non-covalently linkages with other materials [101]. GNPs can be fabricated with compounds that have healing effects, molecules like PEG and BSA as a coating to provide a surface for attachment of specific cells [14] for instance PEGylated GNPs selectively minimize uptake of macrophages and monocytes and thereby prolong their availability of tumor tissue in concentration [155]. In addition to this, GNPs also offer delivery of biomolecules like DNA, RNA, peptides, and proteins [156]. Anticancer drugs, doxorubicin and 5-Fluorouracil were targeted by GNPs for delivery [157, 158].

10.9.1 Bone Tissue Engineering

Tissue Engineering is a field of study that aims to standardize tissue growth and differentiation, mimicking normal environment. Bone Tissue Engineering (BTE) aims at providing appropriate conditions for osteoblast differentiation and matrix mineralization of osteoblastic precursor cell lines [159–161]. Mesenchymal stem cells (MSCs) are precursor cells of bone tissue that further differentiate into multiple lineages which are osteoblasts, chondrocytes, and adipocytes [162]. GNPs have been shown to induce BTE by stimulating osteogenic differentiation. Yi et al. showed the interaction of GNPs with cytosolic protein and extracellular matrix, leading to the activation of the p38 MAPK signaling pathway [163]. Activation of this pathway causes up-regulation of osteogenic genes and simultaneous downregulation of adipogenic genes. Zhang et al. showed that GNPs activate the ERK/ MAPK pathway by increased ERK phosphorylation which is important for osteoblast differentiation and mineralization [164]. The studies showed the significant ability of GNPs and G / HA-NPs for differentiation of osteogenic cells and the ability to induce differentiation is very much dependent on the concentration, size and shape of the GNPs [165–167]. GNPs have been found to promoting not only cell proliferation and differentiation but also add mechanical strength.

11 Conclusion

Nanoparticles have attracted all spheres of research due to their unique physical and chemical properties. Nanoparticles can be subdivided on the basis of their composition into organic nanoparticles, carbon nanoparticles and inorganic nanoparticles. Amongst inorganic nanoparticles, gold is highly chosen metal for nanoparticle synthesis and characterization these days. Gold nanoparticles (GNPs) can be synthesized by physical methods, chemical methods and biological methods. These two methods are conventionally used methods but these methods have many disadvantages to the environment and health, like the efflux of hazardous chemicals which is not only cancer-causing but polluting the environment. Moreover, there is a high energy requirement in the physical method that renders it cost ineffective. An alternative method that could prove advantageous to the environment with minimal energy requirement and no risk to health is biological methods. In biological methods, plants, and microorganisms can be employed for the synthesis of GNPs. Plant extract from different parts can be used for synthesis whereas in microorganisms there are two different ways in which GNPs are synthesized; internal and external. Externally the gold ions (Au^+) are trapped by the membrane protein followed by membrane-reducing enzymes while in the internal method, the gold ions are firstly transported inside the cell through transported protein and then the cytosolic proteins reduced the gold ion into GNPs. There is a wide range of microorganisms employed to synthesize GNPs and researchers have explored a variety of applications of biogenic GNPs. GNPs have been reported to exhibit antimicrobial, anticancer, antioxidant, calorimetric detection of Hg, bioimaging and diagnosis, degradation of contaminants, etc.

References

1. Feynman, R.: There's plenty of room at the bottom. Engineering and science and science. Sci. Res. **3**(23), 22–36, 1960
2. Walter, P., Welcomme, E., Hallégot, P., Zaluzec, N.J., Deeb, C., Castaing, J., Veyssière, P., Bréniaux, R., Lévêque, J.L., Tsoucaris, G.: Early use of PbS nanotechnology for an ancient hair dyeing formula. Nano Lett. **6**(10), 2215–2219 (2006)
3. Joudeh, N., Linke, D.: Nanoparticle classification, physicochemical properties, characterization, and applications: a comprehensive review for biologists. J. Nanobiotechnology. **20**(1), 262 (2022)
4. Lines, M.G.: Nanomaterials for practical functional uses. J. Alloy. Compd. **449**(1–2), 242–245 (2008)
5. Roduner, E.: Size matters: why nanomaterials are different. Chem. Soc. Rev. **35**(7), 583–592 (2006)
6. Geoffrion, L.D., Guisbiers, G.: Quantum confinement: size on the grill! J. Phys. Chem. Solids **1**(140), 109320 (2020)
7. Bratovcic, A.: Diferent applications of nanomaterials and their impact on the environment. Int J Mater Sci Eng. **5**, 1–7 (2019)
8. Gajanan, K., Tijare, S.N.: Applications of nanomaterials. Materials Today: Proceedings. **5**(1), 1093–1096 (2018)

9. Khan, F., Shariq, M., Asif, M., Siddiqui, M.A., Malan, P., Ahmad, F.: Green nanotechnology: plant-mediated nanoparticle synthesis and application. Nanomaterials **12**(4), 673 (2022)
10. Chung, I.M., Park, I., Seung-Hyun, K., Thiruvengadam, M., Rajakumar, G.: Plant-mediated synthesis of silver nanoparticles: their characteristic properties and therapeutic applications. Nanoscale Res. Lett. **11**(1), 1–4 (2016)
11. Dreaden, E.C., Alkilany, A.M., Huang, X., Murphy, C.J., El-Sayed, M.A.: The golden age: gold nanoparticles for biomedicine. Chem. Soc. Rev. **41**(7), 2740–2779 (2012)
12. Bhatia, S., Bhatia, S.: Nanoparticles types, classification, characterization, fabrication methods and drug delivery applications. Nat. Polym. Drug Deliv. Syst.: Nanoparticles, Plants, Algae, 33–93 (2016)
13. Alex, S., Tiwari, A.: Functionalized gold nanoparticles: synthesis, properties and applications—a review. J. Nanosci. Nanotechnol. **15**(3), 1869–1894 (2015)
14. Ealia, S.A., Saravanakumar, M.P.: A review on the classification, characterisation, synthesis of nanoparticles and their application. In: IOP conference series: materials science and engineering, vol. 263, no. 3, p. 032019. IOP Publishing, (2017)
15. Machado, S., Pacheco, J.G., Nouws, H.P., Albergaria, J.T., Delerue-Matos, C.: Characterization of green zero-valent iron nanoparticles produced with tree leaf extracts. Sci. Total. Environ. **15**(533), 76–81 (2015)
16. Khan, I., Saeed, K., Khan, I.: Nanoparticles: Properties, applications and toxicities. Arab. J. Chem. **12**(7), 908–931 (2019)
17. Kumar, R., Lal, S.: Synthesis of organic nanoparticles and their applications in drug delivery and food nanotechnology: a review. J Nanomater Mol Nanotechnol 3: 4. of. 2014, 11:2.Verma et al., (2010)
18. Hughes, G.A.: Nanostructure-mediated drug delivery. Nanomedicine: Nanotechnol., Biol. Med. **1**(1), 22–30 (2005)
19. Mozafari, M.R.: Nanoliposomes: preparation and analysis. Liposomes: methods and protocols, vol. 1: Pharmaceutical Nanocarriers., 29–50 (2010)
20. Hong, S., Choi, D.W., Kim, H.N., Park, C.G., Lee, W., Park, H.H.: Protein-based nanoparticles as drug delivery systems. Pharmaceutics. **12**(7), 604 (2020)
21. Swierczewska, M., Han, H.S., Kim, K., Park, J.H., Lee, S.: Polysaccharide-based nanoparticles for theranostic nanomedicine. Adv. Drug Deliv. Reviews. **1**(99), 70–84 (2016)
22. Plucinski, A., Lyu, Z., Schmidt, B.V.: Polysaccharide nanoparticles: From fabrication to applications. J. Mater. Chem. B. **9**(35), 7030–7062 (2021)
23. Yuan, X., Zhang, X., Sun, L., Wei, Y., Wei, X.: Cellular toxicity and immunological effects of carbon-based nanomaterials. Part. Fibre Toxicol. **16**(1), 1–27 (2019)
24. Lu, K.Q., Quan, Q., Zhang, N., Xu, Y.J.: Multifarious roles of carbon quantum dots in heterogeneous photocatalysis. J. Energy Chem. **25**(6), 927–935 (2016)
25. Fang, F., Li, M., Zhang, J., Lee, C.S.: Different strategies for organic nanoparticle preparation in biomedicine. ACS Materials Letters. **2**(5), 531–549 (2020)
26. Mauter, M.S., Elimelech, M.: Environmental applications of carbon-based nanomaterials. Environ. Sci. & Technology. **42**(16), 5843–5859 (2008)
27. Mody, V.V., Siwale, R., Singh, A., Mody, H.R.: Introduction to metallic nanoparticles. J. Pharm. Bioallied Sci. **2**(4), 282 (2010)
28. Nagalingam, M., Kalpana, V.N., Panneerselvam, A.: Biosynthesis, characterization, and evaluation of bioactivities of leaf extract-mediated biocompatible gold nanoparticles from Alternanthera bettzickiana. Biotechnology Reports. **1**(19), e00268 (2018)
29. Giljohann, D.A., Seferos, D.S., Daniel, W.L., Massich, M.D., Patel, P.C., Mirkin, C.A.: Gold nanoparticles for biology and medicine. Angew. Chem. Int. Ed. **49**(19), 3280–3294 (2010)
30. Hammami, I., Alabdallah, N.M.: Gold nanoparticles: Synthesis properties and applications. J. King Saud Univ.-Sci. **33**(7), 101560 (2021)
31. Barber, D.J., Freestone, I.C.: An investigation of the origin of the colour of the Lycurgus Cup by analytical transmission electron microscopy. Archaeometry **32**(1), 33–45 (1990)
32. Das, M., Shim, K.H., An, S.S., Yi, D.K.: Review on gold nanoparticles and their applications. Toxicol. Environ. Health Sci. **3**, 193–205 (2011)

33. Zhang, J., Mou, L., Jiang, X.: Surface chemistry of gold nanoparticles for health-related applications. Chem. Sci. **11**(4), 923–936 (2020)
34. Maier, S.A., Atwater, H.A.: Plasmonics: Localization and guiding of electromagnetic energy in metal/dielectric structures. J. Appl. Phys. **98**(1), 10 (2005)
35. Jauffred, L., Samadi, A., Klingberg, H., Bendix, P.M., Oddershede, L.B.: Plasmonic heating of nanostructures. Chem. Rev. **119**(13), 8087–8130 (2019)
36. Schuller, J.A., Barnard, E.S., Cai, W., Jun, Y.C., White, J.S., Brongersma, M.L.: Plasmonics for extreme light concentration and manipulation. Nat. Mater. **9**(3), 193–204 (2010)
37. Bai, X., Wang, Y., Song, Z., Feng, Y., Chen, Y., Zhang, D., Feng, L.: The basic properties of gold nanoparticles and their applications in tumor diagnosis and treatment. Int. J. Mol. Sci. **21**(7), 2480 (2020)
38. Shuker, R., Gammon, R.W.: Raman-scattering selection-rule breaking and the density of states in amorphous materials. Phys. Rev. Lett. **25**(4), 222 (1970)
39. Zumbusch, A., Holtom, G.R., Xie, X.S.: Three-dimensional vibrational imaging by coherent anti-Stokes Raman scattering. Phys. Rev. Lett. **82**(20), 4142 (1999)
40. Nie, S., Emory, S.R.: Probing single molecules and single nanoparticles by surface-enhanced Raman scattering. Science **275**(5303), 1102–1106 (1997)
41. Lee, M., Lee, S., Lee, J.H., Lim, H.W., Seong, G.H., Lee, E.K., Chang, S.I., Oh, C.H., Choo, J.: Highly reproducible immunoassay of cancer markers on a gold-patterned microarray chip using surface-enhanced Raman scattering imaging. Biosens. Bioelectron. **26**(5), 2135–2141 (2011)
42. Yun, C.S., Javier, A., Jennings, T., Fisher, M., Hira, S., Peterson, S., Hopkins, B., Reich, N.O., Strouse, G.F.: Nanometal surface energy transfer in optical rulers, breaking the FRET barrier. J. Am. Chem. Soc. **127**(9), 3115–3119 (2005)
43. Huang, X., El-Sayed, I.H., Qian, W., El-Sayed, M.A.: Cancer cell imaging and photothermal therapy in the near-infrared region by using gold nanorods. J. Am. Chem. Soc. **128**(6), 2115–2120 (2006)
44. Lucky, S.S., Soo, K.C., Zhang, Y.: Nanoparticles in photodynamic therapy. Chem. Rev. **115**(4), 1990–2042 (2015)
45. Zhou, W., Gao, X., Liu, D., Chen, X.: Gold nanoparticles for in vitro diagnostics. Chem. Rev. **115**(19), 10575–10636 (2015)
46. Seltzer, S.M.: Tables of X-Ray mass attenuation coefficients and mass energy-absorption coefficients NISTIR 5632. Pramana **72**(2), 375–387 (1995)
47. Jans, H., Huo, Q.: Gold nanoparticle-enabled biological and chemical detection and analysis. Chem. Soc. Rev. **41**(7), 2849–2866 (2012)
48. Kuo, W.S., Chang, Y.T., Cho, K.C., Chiu, K.C., Lien, C.H., Yeh, C.S., Chen, S.J.: Gold nanomaterials conjugated with indocyanine green for dual-modality photodynamic and photothermal therapy. Biomaterials **33**(11), 3270–3278 (2012)
49. Wang, B., Wang, J.H., Liu, Q., Huang, H., Chen, M., Li, K., Li, C., Yu, X.F., Chu, P.K.: Rose-bengal-conjugated gold nanorods for in vivo photodynamic and photothermal oral cancer therapies. Biomaterials **35**(6), 1954–1966 (2014)
50. Terentyuk, G., Panfilova, E., Khanadeev, V., Chumakov, D., Genina, E., Bashkatov, A., Tuchin, V., Bucharskaya, A., Maslyakova, G., Khlebtsov, N., Khlebtsov, B.: Gold nanorods with a hematoporphyrin-loaded silica shell for dual-modality photodynamic and photothermal treatment of tumors in vivo. Nano Res. **7**, 325–337 (2014)
51. Wang, N., Zhao, Z., Lv, Y., Fan, H., Bai, H., Meng, H., Long, Y., Fu, T., Zhang, X., Tan, W.: Gold nanorod-photosensitizer conjugate with extracellular pH-driven tumor targeting ability for photothermal/photodynamic therapy. Nano Res. **7**, 1291–1301 (2014)
52. Bucharskaya, A., Maslyakova, G., Terentyuk, G., Yakunin, A., Avetisyan, Y., Bibikova, O., Tuchina, E., Khlebtsov, B., Khlebtsov, N., Tuchin, V.: Towards effective photothermal/photodynamic treatment using plasmonic gold nanoparticles. Int. J. Mol. Sci. **17**(8), 1295 (2016)
53. Khanna, P., Kaur, A., Goyal, D.: Algae-based metallic nanoparticles: Synthesis, characterization and applications. J. Microbiol. Methods **1**(163), 105656 (2019)

54. Correard, F., Maximova, K., Estève, M.A., Villard, C., Roy, M., Al-Kattan, A., Sentis, M., Gingras, M., Kabashin, A.V., Braguer, D.: Gold nanoparticles prepared by laser ablation in aqueous biocompatible solutions: assessment of safety and biological identity for nanomedicine applications. Int. J. Nanomedicine **9**, 5415 (2014)
55. González-Rubio, G., Guerrero-Martínez, A., Liz-Marzán, L.M.: Reshaping, fragmentation, and assembly of gold nanoparticles assisted by pulse lasers. Acc. Chem. Res. **49**(4), 678–686 (2016)
56. Vinod, M., Jayasree, R.S., Gopchandran, K.G.: Synthesis of pure and biocompatible gold nanoparticles using laser ablation method for SERS and photothermal applications. Curr. Appl. Phys. **17**(11), 1430–1438 (2017)
57. Riedel, R., Mahr, N., Yao, C., Wu, A., Yang, F., Hampp, N.: Synthesis of gold–silica core–shell nanoparticles by pulsed laser ablation in liquid and their physico-chemical properties towards photothermal cancer therapy. Nanoscale **12**(5), 3007–3018 (2020)
58. Turkevich, J., Stevenson, P.C., Hillier, J.: A study of the nucleation and growth processes in the synthesis of colloidal gold. Discuss. Faraday Soc. **11**, 55–75 (1951)
59. Leff, D.V., Ohara, P.C., Heath, J.R., Gelbart, W.M.: Thermodynamic control of gold nanocrystal size: experiment and theory. J. Phys. Chem. **99**(18), 7036–7041 (1995)
60. Khan, Z., Singh, T., Hussain, J.I., Hashmi, A.A.: Au (III)–CTAB reduction by ascorbic acid: Preparation and characterization of gold nanoparticles. Colloids Surf., B **1**(104), 11–17 (2013)
61. Ma, X., Wang, M.C., Feng, J., Zhao, X.: Aspect ratio control of Au nanorods via covariation of the total amount of HAuCl4 and ascorbic acid. J. Alloy. Compd. **15**(637), 36–43 (2015)
62. Ramalingam, V.: Multifunctionality of gold nanoparticles: Plausible and convincing properties. Adv. Colloid Interface Sci. **1**(271), 101989 (2019)
63. Iravani, S.: Green synthesis of metal nanoparticles using plants. Green Chem. **13**(10), 2638–2650 (2011)
64. Mikhailova, E.O.: Gold nanoparticles: biosynthesis and potential of biomedical application. J. Funct. Biomater. **12**(4), 70 (2021)
65. Gu, X., Xu, Z., Gu, L., Xu, H., Han, F., Chen, B., Pan, X.: Preparation and antibacterial properties of gold nanoparticles: A review. Environ. Chem. Lett. **19**, 167–187 (2021)
66. Sheny, D.S., Mathew, J., Philip, D.: Phytosynthesis of Au, Ag and Au–Ag bimetallic nanoparticles using aqueous extract and dried leaf of *Anacardium occidentale*. Spectrochim. Acta Part A Mol. Biomol. Spectrosc. **79**(1), 254–262 (2011)
67. Tamuly, C., Hazarika, M., Bordoloi, M.: Biosynthesis of Au nanoparticles by *Gymnocladus assamicus* and its catalytic activity. Mater. Lett. **1**(108), 276–279 (2013)
68. Rajan, A., Vilas, V., Philip, D.: Studies on catalytic, antioxidant, antibacterial and anticancer activities of biogenic gold nanoparticles. J. Mol. Liq. **1**(212), 331–339 (2015)
69. Kumar, P.V., Kala, S.M., Prakash, K.S.: Green synthesis of gold nanoparticles using *Croton Caudatus Geisel* leaf extract and their biological studies. Mater. Lett. **1**(236), 19–22 (2019)
70. El-Borady, O.M., Ayat, M.S., Shabrawy, M.A., Millet, P.: Green synthesis of gold nanoparticles using Parsley leaves extract and their applications as an alternative catalytic, antioxidant, anticancer, and antibacterial agents. Adv. Powder Technol. **31**(10), 4390–4400 (2020)
71. Islam, N.U., Jalil, K., Shahid, M., Rauf, A., Muhammad, N., Khan, A., Shah, M.R., Khan, M.A.: Green synthesis and biological activities of gold nanoparticles functionalized with Salix alba. Arab. J. Chem. **12**(8), 2914–2925 (2019)
72. Shankar, S.S., Ahmad, A., Pasricha, R., Sastry, M.: Bioreduction of chloroaurate ions by geranium leaves and its endophytic fungus yields gold nanoparticles of different shapes. J. Mater. Chem. **13**(7), 1822–1826 (2003)
73. Rodríguez-León, E., Rodríguez-Vázquez, B.E., Martínez-Higuera, A., Rodríguez-Beas, C., Larios-Rodríguez, E., Navarro, R.E., López-Esparza, R., Iñiguez-Palomares, R.A.: Synthesis of gold nanoparticles using *Mimosa tenuiflora* extract, assessments of cytotoxicity, cellular uptake, and catalysis. Nanoscale Res. Lett. **14**(1), 1–6 (2019)
74. Kumar, K.M., Mandal, B.K., Sinha, M., Krishnakumar, V.: Terminalia chebula mediated green and rapid synthesis of gold nanoparticles. Spectrochim. Acta Part A Mol. Biomol. Spectrosc. **1**(86), 490–494 (2012)

75. Balasubramanian, S., Kala, S.M., Pushparaj, T.L.: Biogenic synthesis of gold nanoparticles using *Jasminum auriculatum* leaf extract and their catalytic, antimicrobial and anticancer activities. Journal of Drug Delivery Science and Technology. **1**(57), 101620 (2020)
76. Muthuvel, A., Adavallan, K., Balamurugan, K., Krishnakumar, N.: Biosynthesis of gold nanoparticles using *Solanum nigrum* leaf extract and screening their free radical scavenging and antibacterial properties. Biomed. Prev. Nutr. **4**(2), 325–332 (2014)
77. Shankar, S.S., Rai, A., Ahmad, A., Sastry, M.: Rapid synthesis of Au, Ag, and bimetallic Au core–Ag shell nanoparticles using Neem (*Azadirachta indica*) leaf broth. J. Colloid Interface Sci. **275**(2), 496–502 (2004)
78. Ovais, M., Raza, A., Naz, S., Islam, N.U., Khalil, A.T., Ali, S., Khan, M.A., Shinwari, Z.K.: Current state and prospects of the phytosynthesized colloidal gold nanoparticles and their applications in cancer theranostics. Appl. Microbiol. Biotechnol. **101**, 3551–3565 (2017)
79. Mukherjee, S., Sushma, V., Patra, S., Barui, A.K., Bhadra, M.P., Sreedhar, B., Patra, C.R.: Green chemistry approach for the synthesis and stabilization of biocompatible gold nanoparticles and their potential applications in cancer therapy. Nanotechnology **23**(45), 455103 (2012)
80. Patra, J.K., Baek, K.H.: Novel green synthesis of gold nanoparticles using *Citrullus lanatus* rind and investigation of proteasome inhibitory activity, antibacterial, and antioxidant potential. Int. J. Nanomed. **10**, 7253 (2015)
81. Hosseini, M., Mashreghi, M., Eshghi, H.: Biosynthesis and antibacterial activity of gold nanoparticles coated with reductase enzymes. Micro & Nano Letters. **11**(9), 484–489 (2016)
82. Ghosh, S., Ahmad, R., Zeyaullah, M., Khare, S.K.: Microbial nano-factories: synthesis and biomedical applications. Front. Chem. **16**(9), 626834 (2021)
83. Zhang, X., He, X., Wang, K., Yang, X.: Different active biomolecules involved in biosynthesis of gold nanoparticles by three fungus species. J. Biomed. Nanotechnol. **7**(2), 245–254 (2011)
84. Menon, S., Rajeshkumar, S., Kumar, V.: A review on biogenic synthesis of gold nanoparticles, characterization, and its applications. Resour.-Effic. Technol. **3**(4), 516–527 (2017)
85. Durán, N., Marcato, P.D., Durán, M., Yadav, A., Gade, A., Rai, M.: Mechanistic aspects in the biogenic synthesis of extracellular metal nanoparticles by peptides, bacteria, fungi, and plants. Appl. Microbiol. Biotechnol. **90**, 1609–1624 (2011)
86. Das, S.K., Dickinson, C., Lafir, F., Brougham, D.F., Marsili, E.: Synthesis, characterization and catalytic activity of gold nanoparticles biosynthesized with *Rhizopus oryzae* protein extract. Green Chem. **14**(5), 1322–1334 (2012)
87. Scott, D., Toney, M., Muzikár, M.: Harnessing the mechanism of glutathione reductase for synthesis of active site bound metallic nanoparticles and electrical connection to electrodes. J. Am. Chem. Soc. **130**(3), 865–874 (2008)
88. Chauhan, A., Zubair, S., Tufail, S., Sherwani, A., Sajid, M., Raman, S.C., Azam, A., Owais, M.: Fungus-mediated biological synthesis of gold nanoparticles: potential in detection of liver cancer. Int. J. Nanomed. **12**, 2305–2319 (2011)
89. Vetchinkina, E., Loshchinina, E., Kupryashina, M., Burov, A., Pylaev, T., Nikitina, V.: Green synthesis of nanoparticles with extracellular and intracellular extracts of basidiomycetes. PeerJ **20**(6), e5237 (2018)
90. Apte, M., Girme, G., Bankar, A., RaviKumar, A., Zinjarde, S.: 3, 4-dihydroxy-L-phenylalanine-derived melanin from *Yarrowia lipolytica* mediates the synthesis of silver and gold nanostructures. J. Nanobiotechnology. **11**(1), 1–9 (2013)
91. Shi, C., Zhu, N., Cao, Y., Wu, P.: Biosynthesis of gold nanoparticles assisted by the intracellular protein extract of *Pycnoporus sanguineus* and its catalysis in degradation of 4-nitroaniline. Nanoscale Res. Lett. **10**, 1–8 (2015)
92. Liu, H., Zhu, N., Luo, D., Li, M., Wu, P., Mo, Y.: Spherical gold nanoparticles (AuNPs) formation by selected intracellular protein fraction of fungi. Beilstein. Arch. **31**(1), 34 (2019)
93. Zhang, X., Qu, Y., Shen, W., Wang, J., Li, H., Zhang, Z., Li, S., Zhou, J.: Biogenic synthesis of gold nanoparticles by yeast *Magnusiomyces ingens* LH-F1 for catalytic reduction of nitrophenols. Colloids Surf., A **20**(497), 280–285 (2016)

94. Qu, Y., You, S., Zhang, X., Pei, X., Shen, W., Li, Z., Li, S., Zhang, Z.: Biosynthesis of gold nanoparticles using cell-free extracts of *Magnusiomyces ingens* LH-F1 for nitrophenols reduction. Bioprocess Biosyst. Eng. **41**, 359–367 (2018)
95. Zhang, X., Qu, Y., Shen, W., You, S., Pei, X., Li, S., Wang, J., Zhou, J.: Colorimetric response of biogenetic gold nanoparticles to mercury (II) ions. Colloids Surf., A **5**(508), 360–365 (2016)
96. Abdel-Kareem, M.M., Zohri, A.A.: Extracellular mycosynthesis of gold nanoparticles using *Trichoderma hamatum*: optimization, characterization and antimicrobial activity. Lett. Appl. Microbiol. **67**(5), 465–475 (2018)
97. Saravanakumar, K., Sathiyaseelan, A., Mariadoss, A.V., Hu, X., Venkatachalam, K., Wang, M.H.: Nucleolin targeted delivery of aptamer tagged *Trichoderma* derived crude protein coated gold nanoparticles for improved cytotoxicity in cancer cells. Process Biochem. **1**(102), 325–332 (2021)
98. Abu-Tahon, M.A., Ghareib, M., Abdallah, W.E.: Environmentally benign rapid biosynthesis of extracellular gold nanoparticles using *Aspergillus flavus* and their cytotoxic and catalytic activities. Process Biochem. **1**(95), 1–1 (2020)
99. Clarance, P., Luvankar, B., Sales, J., Khusro, A., Agastian, P., Tack, J.C., Al Khulaifi, M.M., Al-Shwaiman, H.A., Elgorban, A.M., Syed, A., Kim, H.J.: Green synthesis and characterization of gold nanoparticles using endophytic fungi *Fusarium solani* and its in-vitro anticancer and biomedical applications. Saudi J. Biol. Sciences. **27**(2), 706–712 (2020)
100. Hemashekhar, B., Chandrappa, C.P., Govindappa, M., Chandrashekar, N.: Endophytic fungus *Alternaria* spp isolated from *Rauvolfia tetraphylla* root arbitrate synthesis of gold nanoparticles and evaluation of their antibacterial, antioxidant and antimitotic activities. Adv. Nat. Sci.: Nanosci. Nanotechnol. **10**(3), 035010 (2019)
101. Lee, K.X., Shameli, K., Yew, Y.P., Teow, S.Y., Jahangirian, H., Rafiee-Moghaddam, R., Webster, T.J.: Recent developments in the facile bio-synthesis of gold nanoparticles (AuNPs) and their biomedical applications. Int. J. Nanomed. **16**, 275–300 (2020)
102. Nangia, Y., Wangoo, N., Goyal, N., Shekhawat, G., Suri, C.R.: A novel bacterial isolate *Stenotrophomonas maltophilia* as living factory for synthesis of gold nanoparticles. Microb. Cell Fact. **8**, 1–7 (2009)
103. Khandel, P., Shahi, S.K.: Microbes mediated synthesis of metal nanoparticles: current status and future prospects. Int J Nanomater Biostruct. **6**(1), 1–24 (2016)
104. Issazadeh, K., Jahanpour, N., Pourghorbanali, F., Raeisi, G., Faekhondeh, J.: Heavy metals resistance by bacterial strains. Ann. Biol. Res. **4**(2), 60–63 (2013)
105. Li, J., Webster, T.J., Tian, B.: Functionalized nanomaterial assembling and biosynthesis using the extremophile *Deinococcus radiodurans* for multifunctional applications. Small **15**(20), 1900600 (2019)
106. Gupta, R., Padmanabhan, P.: Biogenic synthesis and characterization of gold nanoparticles by a novel marine bacteria *Marinobacter algicola:* progression from nanospheres to various geometrical shapes. J. Microbiol. Biotechnol. Food Sci. **6**(2021), 732–737 (2021)
107. Sharma, N., Pinnaka, A.K., Raje, M., Fnu, A., Bhattacharyya, M.S., Choudhury, A.R.: Exploitation of marine bacteria for production of gold nanoparticles. Microb. Cell Fact. **11**, 1–6 (2012)
108. Reddy, A.S., Chen, C.Y., Chen, C.C., Jean, J.S., Chen, H.R., Tseng, M.J., Fan, C.W., Wang, J.C.: Biological synthesis of gold and silver nanoparticles mediated by the bacteria *Bacillus subtilis*. J. Nanosci. Nanotechnol. **10**(10), 6567–6574 (2010)
109. Rajeshkumar, S.: Anticancer activity of eco-friendly gold nanoparticles against lung and liver cancer cells. J. Genet. Eng. Biotechnology. **14**(1), 195–202 (2016)
110. Patil, M.P., Kang, M.J., Niyonizigiye, I., Singh, A., Kim, J.O., Seo, Y.B., Kim, G.D.: Extracellular synthesis of gold nanoparticles using the marine bacterium *Paracoccus haeundaensis* BC74171T and evaluation of their antioxidant activity and antiproliferative effect on normal and cancer cell lines. Colloids Surf., B **1**(183), 110455 (2019)
111. Nandhini, J.T., Ezhilarasan, D., Rajeshkumar, S.: An ecofriendly synthesized gold nanoparticles induces cytotoxicity via apoptosis in HepG2 cells. Environ. Toxicol. **36**(1), 24–32 (2021)

112. Shunmugam, R., Balusamy, S.R., Kumar, V., Menon, S., Lakshmi, T., Perumalsamy, H.: Biosynthesis of gold nanoparticles using marine microbe (*Vibrio alginolyticus*) and its anticancer and antioxidant analysis. J. King Saud Univ.-Sci. **33**(1), 101260 (2021)
113. Khlebtsov, N., Dykman, L.: Biodistribution and toxicity of engineered gold nanoparticles: a review of in vitro and in vivo studies. Chem. Soc. Rev. **40**(3), 1647–1671 (2011)
114. Singh, P., Pandit, S., Mokkapati, V.R., Garg, A., Ravikumar, V., Mijakovic, I.: Gold nanoparticles in diagnostics and therapeutics for human cancer. Int. J. Mol. Sci. **19**(7), 1979 (2018)
115. Qian, X., Nie, S.: Surface-Enhanced raman nanoparticles for in-vivo tumor targeting and spectroscopic detection. In: AIP Conference Proceedings, vol. 1267, no. 1, pp. 81–81. American Institute of Physics. (2010)
116. Hossain, M., Cho, H.Y., Choi, J.W.: Gold nanosphere-deposited substrate for distinguishing of breast cancer subtypes using surface-enhanced raman spectroscopy. J. Nanosci. Nanotechnol. **16**(6), 6299–6303 (2016)
117. Lin, D., Feng, S., Pan, J., Chen, Y., Lin, J., Chen, G., Xie, S., Zeng, H., Chen, R.: Colorectal cancer detection by gold nanoparticle based surface-enhanced Raman spectroscopy of blood serum and statistical analysis. Opt. Express **19**(14), 13565–13577 (2011)
118. Ahmad, T., Wani, I.A., Manzoor, N., Ahmed, J., Asiri, A.M.: Biosynthesis, structural characterization and antimicrobial activity of gold and silver nanoparticles. Colloids Surf., B **1**(107), 227–234 (2013)
119. Samanta, A., Gangopadhyay, R., Ghosh, C.K., Ray, M.: Enhanced photoluminescence from gold nanoparticle decorated polyaniline nanowire bundles. RSC Adv. **7**(44), 27473–27479 (2017)
120. Naimi-Shamel, N., Pourali, P., Dolatabadi, S.: Green synthesis of gold nanoparticles using *Fusarium oxysporum* and antibacterial activity of its tetracycline conjugant. J. Mycol. Med. **29**(1), 7–13 (2019)
121. Yu, Q., Li, J., Zhang, Y., Wang, Y., Liu, L., Li, M.: Inhibition of gold nanoparticles (AuNPs) on pathogenic biofilm formation and invasion to host cells. Sci. Rep. **6**(1), 26667 (2016)
122. Wani, I.A., Ahmad, T.: Size and shape dependant antifungal activity of gold nanoparticles: a case study of *Candida*. Colloids Surf., B **1**(101), 162–170 (2013)
123. Lysenko, V., Lozovski, V., Lokshyn, M., Gomeniuk, Y.V., Dorovskih, A., Rusinchuk, N., Pankivska, Y., Povnitsa, O., Zagorodnya, S., Tertykh, V., Bolbukh, Y.: Nanoparticles as antiviral agents against adenoviruses. Adv. Nat. Sci.: Nanosci. Nanotechnology. **9**(2), 025021 (2018)
124. Meléndez-Villanueva, M.A., Morán-Santibañez, K., Martínez-Sanmiguel, J.J., Rangel-López, R., Garza-Navarro, M.A., Rodríguez-Padilla, C., Zarate-Triviño, D.G., Trejo-Ávila, L.M.: Virucidal activity of gold nanoparticles synthesized by green chemistry using garlic extract. Viruses **11**(12), 1111 (2019)
125. El-Sheekh, M.M., Shabaan, M.T., Hassan, L., Morsi, H.H.: Antiviral activity of algae biosynthesized silver and gold nanoparticles against Herps Simplex (HSV-1) virus in vitro using cell-line culture technique. Int. J. Environ. Health Res. **32**(3), 616–627 (2022)
126. Li, H., Ma, X., Dong, J., Qian, W.: Development of methodology based on the formation process of gold nanoshells for detecting hydrogen peroxide scavenging activity. Anal. Chem. **81**(21), 8916–8922 (2009)
127. Lü, J.M., Lin, P.H., Yao, Q., Chen, C.: Chemical and molecular mechanisms of antioxidants: experimental approaches and model systems. J. Cell Mol. Med. **14**(4), 840–860 (2010)
128. Ahn, S., Singh, P., Jang, M., Kim, Y.J., Castro-Aceituno, V., Simu, S.Y., Kim, Y.J., Yang, D.C.: Gold nanoflowers synthesized using *Acanthopanacis* cortex extract inhibit inflammatory mediators in LPS-induced RAW264. 7 macrophages via NF-κB and AP-1 pathways. Colloids Surf. B: Biointerfaces. **162**, 398–404 (2018)
129. Wang, L., Xu, J., Yan, Y., Liu, H., Karunakaran, T., Li, F.: Green synthesis of gold nanoparticles from *Scutellaria barbata* and its anticancer activity in pancreatic cancer cell (PANC-1). Artif. Cells, Nanomedicine, Biotechnol. **47**(1), 1617–1627 (2019)

130. Jafari, M., Rokhbakhsh-Zamin, F., Shakibaie, M., Moshafi, M.H., Ameri, A., Rahimi, H.R., Forootanfar, H.: Cytotoxic and antibacterial activities of biologically synthesized gold nanoparticles assisted by *Micrococcus yunnanensis* strain J2. Biocatal. Agric. Biotechnol. **1**(15), 245–253 (2018)
131. Ismail, E.H., Saqer, A.M., Assirey, E., Naqvi, A., Okasha, R.M.: Successful green synthesis of gold nanoparticles using a *Corchorus olitorius* extract and their antiproliferative effect in cancer cells. Int. J. Mol. Sci. **19**(9), 2612 (2018)
132. Mmola, M., Roes-Hill, M.L., Durrell, K., Bolton, J.J., Sibuyi, N., Meyer, M.E., Beukes, D.R., Antunes, E.: Enhanced antimicrobial and anticancer activity of silver and gold nanoparticles synthesised using *Sargassum incisifolium* aqueous extracts. Molecules **21**(12), 1633 (2016)
133. Kajani, A.A., Bordbar, A.K., Esfahani, S.H., Razmjou, A.: Gold nanoparticles as potent anticancer agent: green synthesis, characterization, and in vitro study. RSC Adv. **6**(68), 63973–63983 (2016)
134. Rajkuberan, C., Sudha, K., Sathishkumar, G., Sivaramakrishnan, S.: Antibacterial and cytotoxic potential of silver nanoparticles synthesized using latex of *Calotropis gigantea* L. Spectrochim. Acta Part A Mol. Biomol. Spectrosc. **5**(136), 924–930 (2015)
135. Dhas, T.S., Kumar, V.G., Karthick, V., Govindaraju, K., Narayana, T.S.: Biosynthesis of gold nanoparticles using *Sargassum swartzii* and its cytotoxicity effect on HeLa cells. Spectrochim. Acta Part A Mol. Biomol. Spectrosc. **10**(133), 102–106 (2014)
136. Boomi, P., Ganesan, R.M., Poorani, G., Prabu, H.G., Ravikumar, S., Jeyakanthan, J.: Biological synergy of greener gold nanoparticles by using *Coleus aromaticus* leaf extract. Mater. Sci. Eng., C **1**(99), 202–210 (2019)
137. Jeyaraj, M., Arun, R., Sathishkumar, G., MubarakAli, D., Rajesh, M., Sivanandhan, G., Kapildev, G., Manickavasagam, M., Thajuddin, N., Ganapathi, A.: An evidence on G2/M arrest, DNA damage and caspase mediated apoptotic effect of biosynthesized gold nanoparticles on human cervical carcinoma cells (HeLa). Mater. Res. Bull. **1**(52), 15–24 (2014)
138. Alkilany, A.M., Murphy, C.J.: Toxicity and cellular uptake of gold nanoparticles: what we have learned so far? J. Nanopart. Res. **12**, 2313–2333 (2010)
139. Li, L., Zhang, W., Desikan Seshadri, V.D., Cao, G.: Synthesis and characterization of gold nanoparticles from *Marsdenia tenacissima* and its anticancer activity of liver cancer HepG2 cells. Artificial cells, nanomedicine, and biotechnology. **47**(1), 3029–3036 (2019)
140. Jennifer, M., Maciej, W.: Nanoparticle technology as a double-edged sword: cytotoxic, genotoxic and epigenetic effects on living cells. J. Biomater. Nanobiotechnol. **4**, 53–63 (2013)
141. Ganeshkumar, M., Sathishkumar, M., Ponrasu, T., Dinesh, M.G., Suguna, L.: Spontaneous ultra fast synthesis of gold nanoparticles using *Punica granatum* for cancer targeted drug delivery. Colloids Surf., B **1**(106), 208–216 (2013)
142. Patra, S., Mukherjee, S., Barui, A.K., Ganguly, A., Sreedhar, B., Patra, C.R.: Green synthesis, characterization of gold and silver nanoparticles and their potential application for cancer therapeutics. Mater. Sci. Eng., C **1**(53), 298–309 (2015)
143. Vemuri, S.K., Banala, R.R., Mukherjee, S., Uppula, P., Subbaiah, G.P., AV, G.R., Malarvilli, T.: Novel biosynthesized gold nanoparticles as anti-cancer agents against breast cancer: Synthesis, biological evaluation, molecular modelling studies. Mater. Sci. Eng.: C. **99**, 417–429 (2019)
144. Kuno, T., Tsukamoto, T., Hara, A., Tanaka, T.: Cancer chemoprevention through the induction of apoptosis by natural compounds. J. Biophys. Checm. **3**(02), 156–173 (2012)
145. Qian, L., Su, W., Wang, Y., Dang, M., Zhang, W., Wang, C.: Synthesis and characterization of gold nanoparticles from aqueous leaf extract of *Alternanthera sessilis* and its anticancer activity on cervical cancer cells (HeLa). Artif. Cells, Nanomedicine, Biotechnol. **47**(1), 1173–1180 (2019)
146. Ramachandran, R., Krishnaraj, C., Sivakumar, A.S., Prasannakumar, P., Kumar, V.A., Shim, K.S., Song, C.G., Yun, S.I.: Anticancer activity of biologically synthesized silver and gold nanoparticles on mouse myoblast cancer cells and their toxicity against embryonic zebrafish. Mater. Sci. Eng., C **1**(73), 674–683 (2017)

147. Selim, M.E., Hendi, A.A.: Gold nanoparticles induce apoptosis in MCF-7 human breast cancer cells. Asian Pac. J. Cancer Prev. **13**(4), 1617–1620 (2012)
148. Baharara, J., Ramezani, T., Divsalar, A., Mousavi, M., Seyedarabi, A.: Induction of apoptosis by green synthesized gold nanoparticles through activation of caspase-3 and 9 in human cervical cancer cells. Avicenna J. Med. Biotechnol. **8**(2), 75 (2016)
149. Venkatachalam, M., Govindaraju, K., Sadiq, A.M., Tamilselvan, S., Kumar, V.G., Singaravelu, G.: Functionalization of gold nanoparticles as antidiabetic nanomaterial. Spectrochim. Acta Part A Mol. Biomol. Spectrosc. **1**(116), 331–338 (2013)
150. Daisy, P., Saipriya, K.: Biochemical analysis of Cassia fistula aqueous extract and phytochemically synthesized gold nanoparticles as hypoglycemic treatment for diabetes mellitus. Int. J. Nanomed. **7**, 1189–1202 (2012)
151. Guo, Y., Jiang, N., Zhang, L., Yin, M.: Green synthesis of gold nanoparticles from *Fritillaria cirrhosa* and its anti-diabetic activity on Streptozotocin induced rats. Arab. J. Chem. **13**(4), 5096–5106 (2020)
152. Javanshir, R., Honarmand, M., Hosseini, M., Hemmati, M.: Anti-dyslipidemic properties of green gold nanoparticle: Improvement in oxidative antioxidative balance and associated atherogenicity and insulin resistance. Clinical Phytoscience. **6**(1), 1 (2020)
153. Lallawmawma, H., Sathishkumar, G., Sarathbabu, S., Ghatak, S., Sivaramakrishnan, S., Gurusubramanian, G., Kumar, N.S.: Synthesis of silver and gold nanoparticles using Jasminum nervosum leaf extract and its larvicidal activity against filarial and arboviral vector Culex quinquefasciatus Say (Diptera: Culicidae). Environ. Sci. Pollut. Res. **22**, 17753–17768 (2015)
154. Murugan, K., Benelli, G., Panneerselvam, C., Subramaniam, J., Jeyalalitha, T., Dinesh, D., Nicoletti, M., Hwang, J.S., Suresh, U., Madhiyazhagan, P.: Cymbopogon citratus-synthesized gold nanoparticles boost the predation efficiency of copepod Mesocyclops aspericornis against malaria and dengue mosquitoes. Exp. Parasitol. **1**(153), 129–138 (2015)
155. Tiwari, P.M., Vig, K., Dennis, V.A., Singh, S.R.: Functionalized gold nanoparticles and their biomedical applications. Nanomaterials **1**(1), 31–63 (2011)
156. Kah, J.C., Wong, K.Y., Neoh, K.G., Song, J.H., Fu, J.W., Mhaisalkar, S., Olivo, M., Sheppard, C.J.: Critical parameters in the pegylation of gold nanoshells for biomedical applications: an in vitro macrophage study. J. Drug Target. **17**(3), 181–193 (2009)
157. Mukherjee, S., Sau, S., Madhuri, D., Bollu, V.S., Madhusudana, K., Sreedhar, B., Banerjee, R., Patra, C.R.: Green synthesis and characterization of monodispersed gold nanoparticles: toxicity study, delivery of doxorubicin and its bio-distribution in mouse model. J. Biomed. Nanotechnol. **12**(1), 165–181 (2016)
158. Thirumurugan, A., Blessy, V., Karthikeyan, M.: Comparative study on doxorubicin loaded metallic nanoparticles in drug delivery against MCF-7 cell line. In: Applications of nanomaterials, pp. 303–313. Woodhead Publishing, (2018)
159. Li, H., Pan, S., Xia, P., Chang, Y., Fu, C., Kong, W., Yu, Z., Wang, K., Yang, X., Qi, Z.: Advances in the application of gold nanoparticles in bone tissue engineering. J. Biol. Eng. **14**, 1–5 (2020)
160. He, Y., Li, Y., Chen, G., Wei, C., Zhang, X., Zeng, B., Yi, C., Wang, C., Yu, D.: Concentration-dependent cellular behavior and osteogenic differentiation effect induced in bone marrow mesenchymal stem cells treated with magnetic graphene oxide. J. Biomed. Mater. Res., Part A **108**(1), 50–60 (2020)
161. Ouchi, T., Nakagawa, T.: Mesenchymal stem cell-based tissue regeneration therapies for periodontitis. Regenerative Therapy. **1**(14), 72–78 (2020)
162. Dong, Y., Hong, M., Dai, R., Wu, H., Zhu, P.: Engineered bioactive nanoparticles incorporated biofunctionalized ECM/silk proteins based cardiac patches combined with MSCs for the repair of myocardial infarction: In vitro and in vivo evaluations. Sci. Total Environ. **10**(707), 135976 (2020)
163. Yi, C., Liu, D., Fong, C.-C., Zhang, J., Yang, M.: Gold nanoparticles promote Osteogenic differentiation of Mesenchymal stem cells through p38 MAPK pathway. ACS Nano **4**, 6439–6448 (2010)

164. Zhang, D., Liu, D., Zhang, J., Fong, C., Yang, M.: Gold nanoparticles stimulate differentiation and mineralization of primary osteoblasts through the ERK/ MAPK signaling pathway. Mater. Sci. Eng. **42**, 70–77 (2014)
165. Heo, D.N., et al.: Inhibition of osteoclast differentiation by gold nanoparticles functionalized with Cyclodextrin Curcumin complexes. ACS Nano **8**, 12049–12062 (2014)
166. Conners, C.M., Bhethanabotla, V.R., Gupta, V.K.: Concentration-dependent effects of alendronate and pamidronate functionalized gold nanoparticles on osteoclast and osteoblast viability. J. Biomed. Mater. Res. B Appl. Biomater. **105**, 21–29 (2017)
167. Li, J., Chen, Y., Yang, Y., Kawazoe, N., Chen, G.: Sub-10 nm gold nanoparticles promote adipogenesis and inhibit osteogenesis of mesenchymal stem cells. J Mater Chem B. **5**, 1353–1362 (2017)

Anil Thakur is pursuing his PhD from University Institute of Pharmaceutical Sciences, Panjab University, Chandigarh with the support of Industrial Fellowship. He has received master degree in Pharmaceutics from Lachoo Memorial College of Science and Technology, Jodhpur and bachelor degree from Govt. College of Pharmacy, Rohru, Shimla. He has recently published three papers in reputed UGC approved Journals. He is currently working on "Nano-Colloidal Drug Delivery system".

Shubham Thakur completed his undergraduate and postgraduate studies in pharmacy at Guru Nanak Dev University in Amritsar in 2016 and 2018 respectively. Presently, he serves as an Indian Council of Medical Research-Senior Research Fellow and is pursuing a Ph.D. in the Department of Pharmaceutical Sciences at Guru Nanak Dev University, Amritsar. His research focuses on the advancement of novel drug delivery systems, aiming to enhance the kinetics, efficacy, and safety of existing drugs. Specifically, his current project involves developing a safe oral formulation for pregnant women. Till now he has published 25 articles and 5 book chapters in Scopus indexed journals and books with h-index of 5.

Sonia Sharma did her Ph.D. from Guru Nana Dev University Amritsar in 2021. She studied chitinases and chitin extracting enzymes and explored these enzymes as an alternate to chemical methods. She is currently working as Senior Project Associate at MTTC, CSIR Institute of Microbial Technology, Chandigarh. She is experienced in microbiology and intends to explore the rich Indian microbial diversity for its abundant bio-molecules reserves, targeting therapeutics as well as industrial important enzymes. She has published seven research articles and one book chapter in SCOPUS indexed journals.

Applications of Biomaterials

Nanostructured Biomaterials in Drug Delivery

İbrahim Mizan Kahyaoğlu, Erdi Can Aytar, Alper Durmaz, and Selcan Karakuş

Abstract In recent years, there has been a significant advancement in the manufacturing of biomaterial-based nanoagents with distinctive characteristics such as biodegradability, biocompatibility, high efficacy, safety, and physicochemical properties. In particular, research on the performance and drug delivery mechanisms based on polymeric nanostructures is gaining interest in the field of health nanotechnologies. This chapter discusses all aspects of biomaterials-based nanoagent preparation methods, characterizations, and drug delivery mechanisms. Furthermore, there has been great attention to the preparation and performance of thermo-sensitive nanogels, biopolymeric nanofibers, green nanoparticles, and nanocomposites for health care systems. The unique advantages, present constraints, and potential future of green nanocarriers and target delivery of chemotherapeutic nanodrugs were highlighted, along with some novel approaches for nanoplatforms.

Keywords Drug delivery systems · Nanodrugs · Bio-based materials · Biomaterials-based nanoagents · Biocompatibility · Chemotherapeutic nanodrugs

Abbreviations

AA	Acrylic acid
AC	Acetyl curcumin
CAP	Cellulose acetate phthalate
CMC	Carboxymethylcellulose

İ. M. Kahyaoğlu
Faculty of Science, Department of Chemistry, Ondokuz Mayıs University, 55139 Samsun, Turkey

E. C. Aytar · A. Durmaz
Faculty of Science, Department of Biology, Ondokuz Mayıs University, 55139 Samsun, Turkey
e-mail: alper.durmaz@windowslive.com

S. Karakuş (✉)
Faculty of Engineering, Department of Chemistry, Istanbul University-Cerrahpaşa, 34320 Istanbul, Turkey
e-mail: selcan@iuc.edu.tr

R. Malviya and S. Sundram (eds.), *Engineered Biomaterials*, Engineering Materials,
https://doi.org/10.1007/978-981-99-6698-1_8

CNS	Central nervous system disorders
CVD	Chemical vapor deposition
DOX	Doxorubicin
EGCG	Epigallocatechin-3-gallate
EMA	European Medicines Agency
FDA	United States Food and Drug Administration
FTIR	Fourier-transform infrared spectroscopy
HA	Hyaluronic acid
HEC	Hydroxyethyl cellulose
HPC	Hydroxypropyl cellulose
HPMC	Hydroxypropyl methylcellulose
HSPs	Heat shock proteins
MG	Methylene green
MTT	(3-(4,5-Dimethylthiazol-2-yl)-2,5-diphenyltetrazolium bromide)
NIPAM	N-isopropylacrylamide
NLCs	Nanostructured lipid carrier molecules
NP	Nanoparticle
OVA	Antigen ovalbumin
PAA	Poly(acrylic acid)
PAMAM	Polyamidoamine
PBS	Polysorbate 80, poly(butylene succinate)
PCL	Polycaprolactone
PEG	Polyethylene glycol
PEI	Polyethyleneimine
PGA	Poly (glutamic acid)
PGS	Poly (glycerol sebacate)
PLA	Polylactic acid
PLGA	Poly (lactic acid-co-glycolic acid)
PMAA	Poly (methacrylic acid)
PNA	Poly (N-isopropylacrylamide-acrylic acid)
PNIPAM	Poly(N-isopropylacrylamide)
PNPs	Polymeric nanoparticles
PPDL	Poly (pentadecalactone)
PPF	Poly (propylene fumarate)
PPG	Poly (propylene glycol)
PPO	Poly (propylene oxide)
PSA	Poly (sebacic anhydride)
PVA	Polyvinyl alcohol
QDs	Quantum dots
ROS	Reactive oxygen
SMGO	Salep-modified graphene oxide
TEM	Transmission electron microscope
TPP	Tri-Phenyl-Phosphonium

UV-vis	Ultraviolet-visible spectroscopy
WHO	World Health Organization
XRD	X-ray diffraction

1 Bio-Based Nanostructures

A recently popular method for producing nanopharmaceuticals involves utilizing environmentally friendly technologies. Currently, the majority of pharmaceuticals are produced using traditional methods that are not environmentally friendly. However, the importance of utilizing green alternatives is growing significantly. The goal of the green pharmaceutical sector is to create products and procedures that completely eliminate or dramatically minimize the consumption and production of dangerous compounds, while also preventing any environmental or health-related consequences from the very beginning of the process. While there is a growing interest in exploring innovative environmental technologies and strategies in the pharmaceutical industry, the primary objective is to promote and optimize the utilization of new nanomaterials that are both environmentally friendly, sustainable, and economically advanced.

Nanomaterials that have at least one dimension in the range of 1–100 nm are considered to be in the nanoscale size range. Due to their small size, large surface area and excellent surface/volume ratio, these materials often demonstrate unique properties and performances, including significant physicochemical, optical, biological, morphological, structural, thermal, and mechanical characteristics that distinguish them from conventional materials. In addition to the particle size and material composition, the morphology and shape of nanostructures also play crucial roles in determining their characteristics and potential applications. Various materials, such as clays, metals, polymers, ceramics, and plant extracts, can be utilized to synthesize nanostructures. These nanomaterials may include components such as nanowires, nanotubes, nanoparticles (NPs), nanofibers, and other types of nanostructures that can be fabricated through either bottom-up or top-down processes during preparation (Fig. 1). Furthermore, the strategies used for developing such nanomaterials vary depending on the application and may include methods including sol–gel, hydrothermal, spray drying, and microfluidics [1]. Nanostructures are popular choices for a wide range of purposes with their special chemical and physical characteristics. Nanomaterials have been used to biosensor applications, imaging, and therapeutic agents in the discipline of biology. Nanomaterials' unique optical and electrical characteristics have been employed in electronics to improve the functioning of devices including electronics, monitors, and solar panels. Cells, battery storage, and photovoltaic arrays are just a few examples of the energy generating and storage solutions where materials may find use. Nanomaterials can also be applied as efficient adsorbent materials to purify the environment of pollutants. Their various uses highlight the adaptability of nanostructures and show how significant they are as an interesting subject for research and development. "Bio-based nanostructures" are nanoscale materials made from natural substances such polypeptides, polysaccharides, oils, and polynucleotides. These nanostructures have distinctive biochemical, structural, and physical characteristics that render them suited for a range of uses in sectors such as medicine, environmental sciences, and biotechnology. One of the applications of bio-based nanostructures is in products derived from renewable resources like wood/cotton, polymeric materials, and leaf extracts. Due to their extensive surface area, exceptional aspect ratio, and excellent mechanical properties, these bio-based nanostructures are well-suited for various applications, such as stabilizing fillers in nanocomposites, pharmaceutical drug carriers, and packaging materials. Alternative bio-nanomaterials encompass polypeptides, biopolymers, and extracts, among others, which possess unique biocompatible and biodegradable properties that render them suitable for a wide range of applications, such as green remediation applications, tissue engineering, cell therapy, bioengineering, biomedical sensors, agriculture, food packaging systems, renewable energy applications, bio-electrochemical platforms, point-of-care biosensing systems, diagnostics, treatments, and drug release systems. Lipid membranes, which are self-assembled films of biomembranes, are extensively used in pharmaceutical agents due to their ability to encapsulate and deliver therapeutics to specific targeted areas, such as living cells, organs or tissues. Bio-nanostructures are often preferred over synthetic ones due to their biocompatibility, renewability, sustainability, and recyclability, making this research area attractive for future applications in a wide range of disciplines. For usage in various fields of nanotechnology, a number of bio-based structures have been recommended, including polymeric nanoparticles (PNPs), nanostructured lipid carriers, protein-polysaccharide complexes, microemulsions, nanoemulsions, nanocapsules, and nano-hydrogels. In their 2016 work, Rashidinejad and colleagues proved that by encapsulating catechin and polyphenolic compounds in liposomes, the antioxidant activities of hard cheese with lower fat content can be effectively improved. In the in vitro gastrointestinal digestion, the antioxidants were released and recovered. The low-fat hard cheese exhibited improved antioxidant activity and total phenolics when epigallocatechin-3-gallate (EGCG) and liposome-encapsulated catechin were added at a dosage of 250 ppm, while maintaining its main structure and pH [2].

Given their distinctive characteristics and abilities, bio-based nanostructures are perfect for a variety of water filters and detoxification operations. These nanostructures have the capacity to function as adsorbent agents, collecting and reacting to contaminants in the environment and removing them from the mixture. Moreover, certain bio-based nanomaterials contain catalytic properties that enable them to decompose or degrade impurities through chemical changes. They can also be modified or synthesized and characterized to modify an impurity's behavior or improve its efficacy in applications that require treated wastewater. By utilizing bio-based nanostructures, it is possible to create more durable and cost-effective alternatives to current water and wastewater filtration strategies. Additionally, the development of bio-based nanostructures highlights their potential applications in addressing agricultural and environmental issues, such as the removal of pollutants and the enhancement of soil

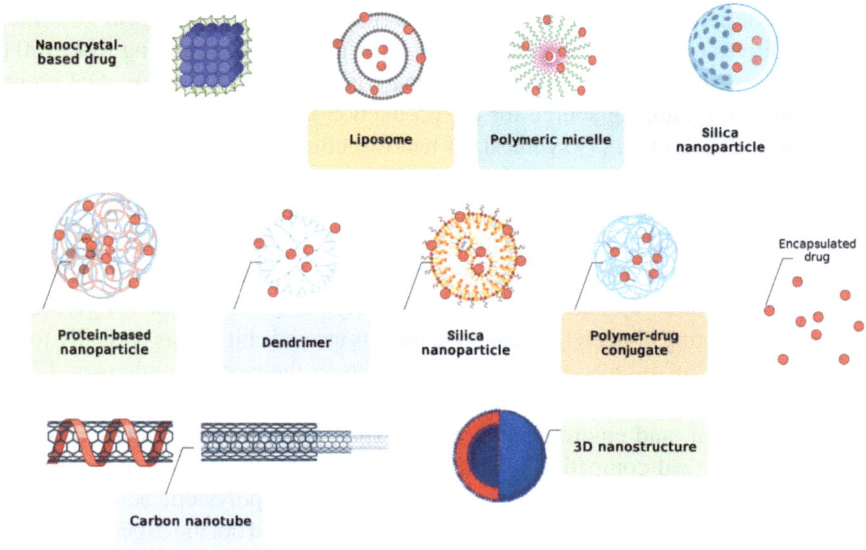

Fig. 1 Structures of nano drugs

efficiency in crop production fields. This is especially important in the case of heavy metal pollution, which can cause harm to both the ecosystem and crop production [3, 4].

Biopolymers are utilized in bio-nanostructures due to the fact that they are naturally occurring macromolecules that originate from renewable sources such as lignin, flowers, plants, leaves, cellulose, animals, shells, algeas, and microorganisms. The functional characteristics of the polysaccharide vary depending on its composition, structure, and type. For instance, carbohydrates are made up of simple glucose molecules and can have a wide range of shapes, such as linear, branching, or cyclical, but proteins are formed up of amino acids with a complicated three-dimensional architecture that affects their functionality. Moreover, biopolymers' characteristics can be altered chemically or by processing techniques to improve their compatibility for a number of different uses. Recently, there has been growing interest among researchers in these biostructures, mainly due to their abundant availability from natural resources. For example, lignin is a byproduct of paper production and is one of the polymeric materials found in plants. These materials, which make up approximately 70–80% of a plant's cell walls, are ubiquitous in the environment and are integral components of biomaterials. Lignin has been the subject of numerous studies investigating its potential benefits in bioscience. In a 2020 study, Lee et al. demonstrated the pharmaceutical applications of lignin by creating nanomaterials with a size of approximately 225 nm from rice-derived lignin. The study also identified for the first time lignin's inherent ability to provide UV protection, resulting in it being a viable option for usage as sunscreen [5]. In another study, Adil et al. demonstrated the antibacterial effect of the proposed nanostructure by mixing lignin with inorganic additives to produce silver nanoparticles with a diameter ranging from 100 to 300 nm [6]. Cellulose, the most abundant material in nature, has many -OH groups that make it a promising source for the production of beneficial nanomaterials. In their study, Zhang et al. [7] synthesized nanostructures using cellulose fibers from economically bleached wood cellulose pulp. The resulting spherical shapes were 50–150 nm in particle size and were evaluated for temperature changes and in vitro drug (5-fluorouracil, 5-FU) release effectivenes. The researchers found that the nanostructures exhibited a rapid 5-FU release pattern, with up to 60% of the drug released within the first hour [7]. An organic linear biopolymer consisting of 1,4-glycosidic linkages connecting N-acetylglucosamine units is termed chitin. It is typically found in the skeletons of insects, and fungus and also in their cell membranes. Chitin has a broad spectrum of potential benefits in a number of sectors, including healthcare, agricultural, and environmental engineering because of its special qualities, including biological compatibility and biodegradability. In their research, Maskur et al. [8] reinforced chitin derived from crab shells using polylactic acid to develop nanomaterials with a size distribution of 90 nm. They carried out the experiment using varying chitin ratios to determine durability and revealed that the durability reduced with increasing chitin concentration [8]. Because of chitin-chitin aggregating, which results in poor distribution of the nanostructures within the matrix phase, additional chitin nanofiber adding reduces the structural rigidity performance of the nanostructure. Chitin derived from shrimp shells is a polymer with antibacterial and antioxidant capabilities that are consistent with epithelial cells, suggesting potential therapeutic potential, according to Coltelli et al. [9]. A natural, renewable alkaline carbohydrate with high hydrating and adsorb characteristics is chitosan, a deacetylated derivative of chitin. Chitin is a naturally occurring substance that is plentiful but has less favorable best possible way than cellulose. Chitosan is also non-toxic, has really no known negative impacts, and is thought to be acceptable for use in a wide range of fields. Because to its advantageous characteristics, including bioavailability, nontoxicity, and chemical stability, chitosan has received considerable attention in a number of sectors. Chitosan's usage in both food and drugs has been authorized by the United States Food and Drug Administration (FDA), which has confirmed the nano/micromaterial's safety. Its certification enables the use of chitosan as a thickening agent, food stabilizer, component of nutritional supplements, and pharmaceutical delivery mechanism. Chitosan has the potential to be a major potential in a multitude of sectors, including agriculture, drugs, and nanotechnology, thanks to its broad application possibilities [10]. While chitosan offers numerous benefits, its ability to mitigate environmental contaminants is particularly noteworthy and has garnered significant research attention.

In 2022, Ansari et al. synthesized novel chitosan/Fe_2O_3/$NiFe_2O_4$ nanostructures and investigated their antibacterial effects. Their results showed that the adsorbent effectively eliminated 96.8% of the methylene green (MG) after 120 min of stirring the solution, which decreased to 77.22% after 240 min. These findings highlight the impressive ability of these nanoparticles to remove pollutants using methylene green adsorption [11]. Since the initially developed polymers were developed for drug release, chitosan has demonstrated exceptional biological and physicochemical characteristics for a variety of healthcare and manufacturing applications. Chitosan's cationic behavior, which originates from of the existence of amino groups, is its predominant component. Besides that, these amino groups provide chitosan unique aspects including transfection, enhanced permeation, and controlled administration of drugs [12]. In 2021, Gao and colleagues conducted a study on the use of nano delivery systems as immune active ingredients to improve vaccinations. Squalene, which is the functional unit of the authorized complement MF59, has the potential to trigger immunological reactions. The researchers developed a functionalized and cationic nanoparticulate delivery system by encapsulating the modeling antigen ovalbumin (OVA) in squalene-based nanostructured lipid carrier molecules (NLCs) and modifying the nanomaterials with the cationic polymer chitosan (OVA-csNLCs). Initially, the optimal csNLC composition was efficiently determined, and the nanostructures of 235.80 ± 5.99 nm were found. Furthermore, the mouse immunization experiment revealed that OVA-csNLCs have excellent compatibility and promote spleen lymphocyte proliferation [13]. The studies mentioned previously highlight the various disciplines in which bio-based nanomaterials could find applications. Technological improvements and solutions have enabled the development of bio-based nanostructures with a wide range of sizes, geometries, shapes, dimensions, physical and chemical characteristics. These low-cost synthesis methods and simple application methodologies offer these sustainable and environment nanomaterials an alternative current to conventional methods. The findings indicate the tremendous potential of bio-based nanoparticles across such a variety of industry sectors.

2 Synthesis of Nanostructured Biomaterials

The two most used methods for delivering pharmaceuticals specifically targeted to cancer patients are systemic chemotherapy administered in nanoagents and the designed release of treatment drugs to the patient's tissues and organs. Although chemotherapy drugs are an effective approach to cancer treatment, they can also be potentially dangerous to healthy cells in the body. Encapsulation is a method used to modify how these drugs are absorbed and employed by the human body. Surrounding them in nanocarriers can improve the bio-distribution, permeability, toxic effects, tolerability, and acceptance of the pharmaceuticals, increasing their therapeutic value and reducing any adverse reactions that may occur. By minimizing damage to healthy tissues, this strategy can make it easier to provide greater doses of chemotherapy treatments directly to cancerous cells. The structural affinities of nanostructures to organs and pathogens are crucial for effective interactions between nanomaterials and living organisms. When nanomaterials mimic the morphology and form of cellular organisms, they are more likely to be recognized and interacted with, leading to enhanced effectiveness and personalized solutions. The targeted administration and combination of pharmaceuticals within a person's cells and tissues, as well as the monitoring of illness concentrations in problematic fluids such as blood, saliva, and organs like the kidneys and liver, can all be accomplished through the application of nanomaterials. Pharmaceuticals efficiency can be improved while generating fewer risks to health when they are encapsulated in nano-based compounds. Furthermore, the application of nanotechnology can facilitate the development of pharmaceutical nanoformulations that function as local analgesics by evading the human body's immune system, including the phagocytic system, due to their unique morphology. This prevents potential side effects that could diminish positive health outcomes and results in a long-lasting effect. Various factors, such as poor colloidal stability, unfavorable hydrodynamic and pharmacodynamic properties, slow dissolution rate, and inadequate distribution, along with concerns about toxicity and stability, can contribute to an inadequate pharmacological response and minimal drug release at the receptor site. Despite the numerous neurobiological, pharmacological, and therapeutic challenges associated with intravenous administration, healthcare professionals and scientists are turning to nano-based carriers as a viable solution. Currently, various methods are utilized for manufacturing nanostructured biomaterials, including chemical vapor deposition (CVD), biological method, physical/mechanical method, self-assembly method, sol–gel method, ultrasound waves-based method, sono-solvo method, and hydrothermal method. These methods use different approaches to produce the desired nanomaterials, which can vary in terms of effectiveness, complexity, and cost. The synthesis of nanostructured biomaterials cannot be accomplished with a single technique. The optimum strategy is chosen by filtering the alternatives according to factors such as the final product, the starting raw material, the cost of the procedure, the application, and the quantity being manufactured. The procedures for nanoparticle synthesis are constantly evolving as new tactics and raw materials are discovered due to the advancement of innovation.

2.1 Hydrothermal Method

The hydrothermal methodology employs a great deal of pressure and temperature in a liquid media, usually water, to introduce a variety, in particular nanomaterials. In order to facilitate the manufacturing of the final products, the precursor components are put into a sealed vessel and heated at a specific pressure and temperature. Metallic nanostructures like metal oxides and sulfur compounds as well as natural nanostructures including polymeric material and carbon-based materials are also extensively fabricated to use this approach. The manufacturing of bio-nanomaterials from waste plant, animals, and microbiological sources also use the hydrothermal methodology. In a study published in 2019, Satiskumar and colleagues used the hydrothermal method for producing nanosized hydroxyapatite crystals from the scales of *Cirrhinus mrigala* fish. Having followed a step of protein removal, the dried fish scales were crushed and mixed for two hours at 100 °C and 500 rpm in an aqueous NaOH solution to synthesize nano-hydroxyapatite crystals. The nanostructures' biocompatibility was examined in vitro after they were characterized by Transmission Electron Microscopy (TEM), X-ray Diffraction (XRD), and Fourier

Transform Infrared Spectroscopy (FTIR) techniques. For nanomaterials in especially, whose characteristics are mainly controlled by their crystal structure, XRD is employed to analyze the crystalline structure of materials. High-resolution images of nanomaterials are given by TEM, allowing the analysis of their shape, scale, and structure. By determining the absorption and transmission of ultraviolet light, which can provide details about functional groups and chemical bonds, FTIR is used to examine the chemical composition of nanomaterials. These approaches used together can support researchers in understanding the properties of nanosystems and in developing them for various application. In their study, Satiskumar and colleagues demonstrated that the hydrothermal method yielded hydroxyapatite nanorods with a diameter range of 30–50 nm, which exhibited exceptional biocompatibility [14].

2.2 *Ultrasonic Method*

The ultrasonic method is a technique that uses high-frequency sound waves to create mechanical vibrations in a liquid solution. These vibrations can induce a range of physical and chemical effects, such as generating cavitation microstreaming, acoustic cavitation, and acoustic flowing which can lead to the formation of nanostructures of various shapes and sizes. The ultrasonic method uses a high-frequency ultrasonic instrument to generate sound waves that propagate through a liquid solution. The sound waves create alternating high/low pressure regions, causing liquid molecules to vibrate and collide with each other. This leads to the creation of cavitation bubbles, that can collapse violently, generating high temperatures and pressures. These extreme conditions can trigger a range of chemical reactions, including the formation of, nanowires, nanospheres, nanotubes, metal/metal oxide nanoparticles, polymeric nanoparticles, quantum dots, and nanocomposites. The ultrasonic sound methodology has been used for many kinds of applications, including cleaning, food processing, agriculture, medical imaging nanosystems, nanomaterial processing, nanoelectronic, nanocatalysis, nanodrug, chemical synthesis of nanostructures. It is an adaptable and effective way to produce nanomaterials. The acoustic method generates localized hot spots in the liquid, facilitating chemical reactions that would otherwise be difficult to perform under normal conditions due to the extreme temperatures and pressures within these hot spots. The hotspots produced by the method affect extremely high-pressure air bubbles, which rapidly reach temperatures of around 5000 K in nanoseconds. The size and shape of the final materials generated through the ultrasound process are significantly affected by the process's duration duration, ambient temperature, concentration, and other parameters [15]. In 2021, Dai et al. synthesized novel nanocarbon structures by modifying the conditions of ultrasonic irradiation on a KOH/ethanol mixture. By subjecting a KOH/ethanol medium with an amount of 210 mg mL^{-1} to ultrasonic irradiation for an hour at room temperature, followed by allowing it to rest at the same temperature for five days, they were able to generate carbon nanotubes, nano-sized graphene polyhedra, and single/core nanoonions. The ultrasonic synthesis method was used to synthesize different carbon structures under varying conditions. A multicore carbon nano-onion was produced by subjecting a KOH/ethanol medium with an amount of 105 mg mL^{-1} to ultrasonic irradiation for an hour, followed by allowing it to sit at room temperature for eight days. N-diamond nanocrystals were generated when solutions with concentrations of 210, 105, and 52.5 mg mL^{-1} were kept at room temperature for seven to fourteen days without ultrasonic irradiation. However, when the same process was applied after an hour of ultrasonic irradiation, graphene was synthesized. These results demonstrate how changing the conditions of sono-based synthesis can significantly impact the resulting products [16].

2.3 *Pyrolysis Method*

Pyrolysis is a process that involves heating precursor materials to high temperatures without the presence of air to create nanostructures. Typically, the precursor compounds undergo heat treatment in a vacuum within an oven or reactor, with temperatures ranging from a few hundred to over a thousand degrees Celsius. During combustion, the precursor compounds undergo chemical processes such as polymerization, cross-linking, and dehydrogenation, leading to the formation of solid materials that are carbon-rich. With further processing, these materials can be converted into various types of nanomaterials, such as carbon nanotubes, graphene, and metal oxide nanostructures. The primary benefit of the pyrolysis approach is that it enables exact control of the form, composition, and size of the resultant nanostructures by modifying a variety of processing factors such as temperature, pressure, and composition of the precursor materials. The process is also relatively simple to apply, adaptable, and has the potential of yielding huge quantities of nanostructures. The pyrolysis method involves applying heat energy to the material in order to produce and modify it. For the synthesis of nanomaterials, direct or microwave-assisted decomposition processes are commonly employed. This method is widely used in the production of quantum dots (QDs) due to its ability to produce QDs in large quantities. While there are more efficient methods for producing small-diameter QDs, their effectiveness decreases in large-scale applications [17]. In a study published in 2021, Agnol and colleagues used the pyrolysis method to synthesize quantum dots from microalgae called Spirulina. They applied the pyrolysis process to the microalgae at a temperature of 300 °C for a duration of 2 h. The average size of the synthesized QDs was 10 nm, and a quantum yield of 23% was calculated with regard to the synthesized QDs. The research presented here shows that pyrolysis may serve as a beneficial method to generate high-quality QDs from renewable resources, such as micro algae, with prospective applications in a variety of industries, including medical treatment, electronic devices, and energy [18].

2.4 Chemical Vapor Deposition (CVD) Method

CVD is a technique used to produce a variety of nanoscale components, including carbon structures. In CVD, several precursor gases are vaporized and injected into a reaction chamber, where they interact and settle onto a substrate to form a thin coating or nanostructure. To speed up the deposition process, the substrate may be heated or maintained at certain temperatures. Depending on the precursor gases and deposition conditions selected, the deposited material can have a wide range of forms and characteristics. For example, carbon nanotubes, diamond, quantum dot, and graphene are a few examples of the carbon-based materials that can be produced with CVD. The resulting structure, functionality and size of the carbon nanostructure can be influenced by the precursor gases and deposition conditions used. For instance, hydrocarbons and a metal-based catalyst, such as Ni, Fe, and Cu can be used to prepare carbon nanotubes, while high quality graphene can be produced using hydrocarbon gas and a substrate with a specific crystalline structure. Overall, CVD is a versatile method for producing various kinds of nano-sized materials, including carbon structures [19]. In a study published in 2019, Irfan et al. used a method known as Chemical Vapor Deposition to generate small particles that were composed of iron (Fe), cobalt (Co), and a mixture of both (Fe–Co). The method consisted of heating the gas to 1000 °C, which formed small nanoparticles covered in a covering of carbon. From 51 to 165 nm, the researchers were able to synthesize nanostructures and examined how the reaction temperature changed the size of the particles. It was observed that the size of the nanomaterials enhanced together with the reaction temperature. Overall, their investigation suggests that synthesizing nanostructures using CO_2 gas and CVD is successful and highlights the significance of reaction temperature control to achieve ideal size of the particles [20].

3 Drug Delivery Systems

Based on biologically physico-chemical, and mathematical theories, various methods have been used to deliver medicinal products to humans or animals in a way that is controlled. With this approach, the product is healthier, the patient has a more comfortable qualifications, and the therapy is a greater success. Maintaining drug levels in the circulatory system within the limits of optimal concentration and least toxic level is one of the fundamental objectives for controlling the release of medication since it is vital to gaining the intended beneficial effects. Nano-drug therapy is an advanced treatment strategy for many different diseases. It involves using drugs to treat a specific illness. The amount of the drug present at the site of the medical condition or damage is an essential factor in determining the effectiveness of the therapy. When drugs are administered, they enter the bloodstream and can circulate throughout the body. The goal of drug therapy is to maximize the release of the active ingredient at the site of the tumor or affected area where it is most needed while minimizing its release in other areas of the body. This is important because it can help to maximize the drug's therapeutic effects while minimizing potential side effects. Targeted drug delivery can be achieved through various strategies, including encapsulation or coating of drugs with nanomaterials, use of nanoparticles or liposomes, localized injection of the drug, or drug implantation. By controlling the release and distribution of drugs within the body, nanodrugs can be tailored to effectively manage a range of conditions. In recent years, the field of bio-nanostructured drug delivery systems has become increasingly fascinating. Compared to conventional biofilms, these systems have the potential to significantly improve the therapeutic effect of drugs by delivering them directly to the targeted region. Various nanomaterials, such as thermo-sensitive nanogels, biopolymeric nanofibers, nanocomposites, green-synthesized nanoparticles, nanotubes, and multilayered structures such as MXenes, have been extensively researched as drug delivery systems. Scientists aim to develop safer and more effective treatments for a wide range of diseases using these innovative approaches to drug delivery, ultimately improving patient health [21]. It is possible to deliver numerous dosages of drugs in a short amount of time without changing the amount of the pharmaceuticals in the bloodstream. Patients may fail to adhere to their drug plan as a result, though. To deal with this problem, it is necessary for researchers to create small vehicles (nanocarriers) that provide drugs continuously over an extended period, hence requiring minimal dosages. To achieve this, the drug should be released at a constant rate, which is called a zero-order drug release kinetic model. If a drug delivery system doesn't release the drug effectively, its potency may be reduced. Depending on the intended use of the system, drugs can be delivered in various ways. These delivery mechanisms can occur simultaneously or at different stages. To develop controlled-release systems, it's important to understand these kinetic mechanisms. Table 1 displays various kinetic models.

Table 1 Different kinetic models

Model	Formula	References
Zero-order	$\frac{M_t}{M_\infty} = kt^n$	[22]
Ritger-Peppas	$\frac{M_t}{M_\infty} = K_0 t$	[22]
First-order	$\ln(1 - \frac{M_t}{M_\infty}) = -K_1 t$	[23]
Higuchi	$\frac{M_t}{M_\infty} = K_H t^{0.5}$	[23]
Korsmeyer-Peppas	$\frac{M_t}{M_\infty} = K_K t^n$	[23]
Hixson-Crowell	$\left(1 - \frac{M_t}{M_\infty}\right)^{\frac{1}{3}} = -K_{HC} t$	[24]
Baker-Lonsdale	$\frac{3}{2}\left[1 - \left(1 - \frac{M_t}{M_\infty}\right)^{\frac{2}{3}}\right] - \frac{M_t}{M_\infty} = K_{BL} t$	[24]

where *t*: time, *Mt*: cumulative release of the drug in time t, and *M*∞: the theoretical maximum release of the drug. K: model release constant, and *n*: the diffusion exponent

Some of these mechanisms include erosion, swelling, partitioning, dissolution, and diffusion. Erosion and diffusion, which are commonly referred to as sustained release, are two of the most common mechanisms. The sustained release profile is particularly important for drugs that are rapidly metabolized and eliminated from the body, as it allows the drug to be administered at the determined optimum dose over an extended period. This controlled drug release mechanism helps to maintain a constant release rate and drug concentration in the bloodstream, which is crucial for effective cancer treatment. In contrast, immediate-release drugs cannot control the dose, leading to potential differences in drug concentration in both plasma and tissue. However, in some cases, drug-laden transporters may be released rapidly initially, which can cause potential problems. The targeting mechanism occurs as a pharmacological event across several biological regions, such as cancer cells, when targeted drugs are delivered with internal stimulus. Depending on how cancerous cells and the surrounding tissue react and respond during their interactions with suitable nanocarriers that many different types of pathways can be formed.

3.1 PH-Responsive Delivery Mechanisms

pH-responsive drug delivery mechanisms refer to the ability of certain nanocarriers to release pharmaceuticals in response to changes in pH levels. For example, when exposed to the acid environment of a cancer cell, a pH-sensitive nanocarrier can release its payload of drugs. This pH-responsive behavior is particularly useful in cancer treatment, as it enables targeted drug delivery to cancerous cells while minimizing damage to normal tissue. Different strategies can be used to achieve pH-responsiveness, such as incorporating pH-sensitive groups into the bio-based nanocarriers or designing the nanocarrier to respond to the pH gradient between the cancerous area and healthy tissue. The broad application of pH-responsive nanostructures in cancer treatment is a consequence of the distinctive acidic properties that tumors exhibit, which are not typically found in other drug delivery systems. pH-responsive nanostructures release drugs in response to changes in pH levels, making them particularly effective in targeting cancerous cells while minimizing damage to normal cells. These small particles react to the acidic environment by undergoing changes in their structure and releasing the drug payload. The ability to control the release of drugs in response to pH changes is what makes these nanostructures so useful in cancer therapy. By specifically targeting cancer cells and avoiding healthy tissue, pH-responsive nanostructures offer a promising strategy for drug delivery in cancer treatment. Pharmaceutical substances have been delivered to a target biological organ or cellular area more quickly by taking advantage of the changing pH levels shown in illnesses such as cancer or inflammatory. Use of endosomes or lysosomes is required in order to achieve this. Tumor cells and other disease-affected tissues, such as inflammatory areas, typically exhibit a low pH level (acidic) of about 6.5, which is nearly one pH unit less than the pH of normal blood (7.4). Furthermore, a decrease in pH levels between 5.5 and 5.0 is observed in intracellular vesicles, including lysosomes and endosomes. pH variations enable the development of delivery systems for drugs that maintain stability at biological pH levels, decreasing pharmaceutical loss during bloodstream circulation. The release of anticancer medicines at the site of action and the achievement of focused antitumor activity can occur as a result of these systems becoming much more disturbed in acid conditions, such as cancerous tissue. To target the acidic microenvironments in tumors, biopolymeric nanocarriers have been developed using two primary design approaches. The first method involves the use of pH-responsive biopolymers with chemical groups capable of donating or accepting ions. These polymers are typically deprotonated under normal physiological conditions but become protonated in acid environments, leading to morphological changes, modifications in hydrophobicity, and subsequently, the release of the pharmacological cargo. During the development of pH-responsive biobased nanocarriers, it is common to use biopolymers containing electroactive groups (such as weak acids or bases) to break down the biopolymeric structures. Microspheres are formed inside healthy cells or on the surface of cancerous cells in an acidic environment. pH-responsive biopolymeric nanocarriers are created using a variety of biopolymers. Some common examples of such biopolymers include poly(lactic acid-co-glycolic acid) (PLGA), starch, modified starch, poly(acrylic acid) (PAA), poly(methacrylic acid) (PMAA), polycaprolactone (PCL), sodium caseinate, poly (propylene fumarate) (PPF), xanthan gum, whey protein isolate, poly(glutamic acid) (PGA), poly(propylene oxide) (PPO), carboxymethylcellulose (CMC), gelatine, alginate, pectin, polylactic acid (PLA), poly (pentadecalactone) (PPDL), chitosan, cellulose acetate phthalate (CAP), hydroxypropyl methylcellulose (HPMC hydroxypropyl cellulose (HPC), hydroxyethyl cellulose (HEC), poly(glycerol sebacate) (PGS), polyamidoamine (PAMAM), polyethylene glycol (PEG), polyethyleneimine (PEI), poly(N-isopropylacrylamide) (PNIPAM), sulfated polysaccharides, poly(propylene glycol) (PPG), polyvinyl alcohol (PVA), polysorbate 80, poly(butylene succinate) (PBS), poly(N-isopropylacrylamide-acrylic acid) (PNA), poly(sebacic anhydride) (PSA), Hyaluronic acid (HA), polyaspartamide, phosphatidylcholine, dextran sulfate, carrageenan, cashew gum, guar gum, gellan gum, zein, pullulan, cyclodextrin, tea polyphenols, and green tea components. In Table 2, some examples of some bio-based delivery systems are given.

By protonating the carboxylate groups, the attractive forces between a drug (cationic) and an biopolymer (anionic) can be reduced, thereby enabling the release of the cargo at acidic pH values. The second design approach for pH-responsive nanostructures involves incorporating bifunctional acid-sensitive bonds into their composition. These bonds can be broken at acidic pH levels, causing the selective release of drug cargoes that are either enclosed within or attached to the nanostructures, through the disruption of interactions between the polymer and the drug or within amphiphilic block copolymers. In 2017, Bardajee et al. developed a novel thermo- and pH-sensitive nanogel that incorporated branched N-isopropylacrylamide (NIPAM), salep-modified graphene oxide (SMGO), and acrylic acid (AA). They investigated the consequences of loading Doxorubicin (DOX) onto the manufactured nanoformulations and discovered that in a low-temperature, pH-7.0 environment, drug (DOX) release was delayed. Yet, it was shown that a pH 7 environment and

Table 2 Examples of some bio-based delivery systems

Bio-based nanostructure	Delivery system
Cashew gum nanocarrier	Drug delivery system
Xanthan gum-based nanocarrier	Drug delivery system
Collagen/nano-hydroxyapatite gel	Drug delivery system
Alginate/chitosan-coated cellulose nanocrystal	Oral antigen delivery system/mucosal immune responses
Hyaluronic acid-magnetic nanocomposites	Gene delivery system
Gum polysaccharides based nano-composite	Drug delivery system
Propolis extract loaded nanoformulation	Drug delivery system
Gum arabic/nanomagnetite-*Dunaliella salina* extract	Oral drug delivery system
Chitosan-based nanodrug	Drug delivery system
PAA-CMC-halloysite nanotubes	Controlled drug delivery system
PEG- PAA	Drug delivery system
PVA/ PAA electrospun nanofiber	Drug delivery system
Guar gum/PAA-silver nanofilm	Drug delivery system
Zein-xanthan gum nanodrug	Drug delivery system
Graphene oxide-based nanodrug	Drug delivery system
$T_{i3}C_2Tx$ MXene-PEG-cystamine	Drug delivery system
Chitosan/hyaluronic acid-MXene/gold nanorods	Near infrared responsive drug delivery system
Hydroxyapatite-Ti_3C_2 MXene/gold nanorods	Triggered drug delivery system
ZnO nanoparticles-polyethyleneglycoldiacrylate cross-linked carboxymethyl tamarind kernel gum/polysodiumacrylate composite	Drug delivery system
Graphene oxide-gelatin-polyvinylpyrrolidone almond oil nanoemulsion	Drug delivery system

higher temperature dramatically improved drug release. Up to a dosage of 410 g/mL, the DOX-loaded nanomedicine had no negative effects on cell viability, according to the study. Furthermore, the nanomedicine had higher effectiveness when tested against HeLa cells than the equal amount of DOX. These results imply that the novel nanomedicine have a great potentiality for use as an effective drug delivery system for the fight against cancer [25]. In a work published in 2022, Gao et al. produced a novel kind of nanogel termed PTX/Bio-NG with the intention of selectively targeting breast cancer cell lines. Paclitaxel, a chemotherapeutic drug, was embedded in this hyaluronic acid-based nanogel. The spherical nanogel was developed to have dual targeting characteristics by incorporating CD44 and biotin receptors in order to improve the targeting performance. The scientists found that the amount of drug that was effectively captured within the spherical nanostructure (encapsulation efficiency), was extremely high for the PTX-loaded nanostructure. The PTX-loaded nanostructure exhibited an encapsulation efficiency of 90.17 ± 0.52%. Moreover, the capacity of the nanostructure for drug loading, was also found to be high at 15.28 ± 0.10%. These findings indicate that the PTX-loaded nanostructure has a high drug loading capacity, high entrapment efficiency, and dual targeting, making it a potential therapeutic delivery system for chemotherapy in breast cancer. The scientists discovered that in the presence of hyaluronidase and/or lipase, the PTX-loaded nanostructure they created rapidly released the drug. Based on research conducted on cell lines, they found that 4T1 cells absorbed the nanostructure through the CD44 receptor and bio-specific proteins. Additionally, they observed that the PTX-loaded nanostructure was more cytotoxic to 4T1 cells than PTX-loaded biotin-free nanostructure. During their in vivo studies using mice with 4T1 tumors, they observed a remarkable therapeutic efficacy of the PTX-loaded nanostructure. They claimed that the nanostructure exhibited better pharmacological properties than Taxol and the combined action of hyaluronic acid and biotin in the nanostructure produced an excellent tumor targeting effect. Based on these findings, the researchers concluded that PTX-loaded nanostructure could be an excellent candidate for breast cancer treatment [26]. To encourage healing process in skin problems, Zhang and colleagues designed a biocompatible nanomaterial with redox-sensitive functionalities. The radical emulsion polymerization strategy was employed by the researchers to synthesize redox-sensitive PNA-based nanostructure. Scientists found that the biocompatible membranes containing PNA enhanced cell adhesion as well as proliferation in compared to the natural PLLA membrane through in vitro biological assessments. They showed that the PNA-loaded nanomembranes' bonds are formed controlled the redox potential and adjusted the Reactive Oxygen Species (ROS) level in the damaged epidermis, speeding the process of healing in in vivo experiments using a mouse skin injury model. Consequently, the research showed that PNA-loaded nanomembranes could be an effective treatment for skin diseases by balancing ROS levels and promoting skin regeneration [27]. In 2017, Wang et al. synthesized novel hyaluronic acid-coated chitosan nanoparticles (NPs) to enhance the chemotherapeutic efficacy of the cancer drug 5-FU. The goal was to target therapeutic agents to cancerous cells using CD44. The researchers observed a stable release profile of 5-FU from the nanoformulation with 75% of the total 5-FU released in the first 48 h. They found that the interaction between HA and CD44 facilitated more effective drug delivery to cancerous cells when CD44 was overexpressed, resulting in greater drug accumulation in these cells compared to free pharmaceuticals and uncoated NPs. Based on these findings, the researchers concluded that HA-coated chitosan NPs could represent a promising approach for inducing apoptosis in tumor cells and treating cancer effectively. Finally, the study demonstrated the potential of HA-coated chitosan NPs for CD44-mediated drug delivery to cancer and improving the therapeutic efficacy of 5-FU and other anticancer agents [28]. In their 2022 study, Tak et al. developed a synthesis way to produce graphene quantum dots from of *Polygala tenuifolia* extracts for the inhibition of acetylcholine esterase. They then functionalized the graphene quantum dots with donepezil hydrochloride to create DO-loaded graphene quantum dots, which were compared to the conventional Alzheimer's drug. The scientists found that the novel DO-loaded graphene quantum dots demonstrated greater inhibitory activity on acetylcholine esterase compared to graphene quantum dots alone or the conventional medicine. Moreover, the study revealed a synergistic effect between the pharmaceutical delivery mechanism and the Polygala tenuifolia root extract in the Alzheimer's disease treatment process. The research concluded that the prepared DO-loaded graphene quantum dots represent a safe and effective drug delivery system for Alzheimer's treatment, with potential for future medical applications [29]. Reddy et al. developed a novel drug delivery system in 2019 by incorporating acetyl curcumin (AC) into PLGA liposome nanostructures. The researchers found that at its highest rate, 48% of the drug was released from the nanocarriers within approximately 4 days. The study also showed that the prepared nanostructure released more AC than the free drug due to functional group interactions between the lipid head groups and AC. Importantly, the researchers observed no cytotoxicity on the HDFa cell line, suggesting that the nanostructures are safe for normal tissues. Based on these results, the researchers suggested that the proposed nanostructure could serve as a promising drug delivery system for clinical applications [30]. In order to look into the possibility of employing exosomes loaded with lapatinib as a drug delivery vehicle, Deirmenci et al. conducted an investigation in 2022. Exosomes were extracted from normal breast epithelial cells, and it was found that their good biocompatibility, and low toxicity make them more appropriate as carriers than synthesized nanostructures. In HER2 (+) SKBR 3 cell line, and MCF10 cell line, the pharmacological effects of Lapatinib/exosomes, free Lapatinib, and uncharged exosomes were tested. In contrast to free Lapatinib, they observed that exosomes carrying Lapatinib were more effective at preventing division of cells and maintaining apoptosis in different types of cancer cells. The researchers hypothesized that using nanocarriers loaded with lapatinib would allow the use of drugs at smaller doses. Overall, the study suggests that exosomes loaded with Lapatinib have the potential to be used as a drug delivery tool in cancer treatment [31].

3.2 ROS-Responsive Delivery Mechanisms

Highly reactive chemicals known as ROS have the potential to harm cells if improperly controlled. Cells have created a number of pathways that can react to variations in ROS levels in order to protect themselves against damage that ROS causes. Superoxide dismutase, glutathione peroxidase, and catalase are a few antioxidant enzymes that cells contain that can reduce ROS by converting them into less dangerous molecules. Heat shock proteins (HSPs): HSPs have a role in preventing oxidative cellular damage and can be generated in response to ROS and other cellular stressors. HSP70, for instance, can inhibit particle agglomeration brought on by ROS. A few transcription factors, such NRF2, can activate the production of enzymatic antioxidants and other chemoprotective proteins in response to ROS. During the process of autophagy, cells can recycle undesirable or damaged parts of their structure, including ROS-affected parts. Together, these processes defend cell lines from ROS-caused harm and preserve cell balance. In 2023, Gandomi et al. developed a novel nano-sonosensitizer by coating metal/metal oxide hybrid-based nanocarriers with l-cysteine and graphene quantum dots. When exposed to internal and external factors such as sonic and pH modifications, the bio-nanoformulation was able to alter its surface characteristics in tumor cells and enhance its sonocatalytic properties. In contrast to the free form of curcumin, they observed that using the bio-nanoformulation as a drug delivery vehicle for curcumin in chemo-sonodynamic treatment resulted in a significant suppression of A549 cells. The l-cysteine/hybrid-based nanocarriers and bio-nanoformulation demonstrated high drug loading efficiencies of 61.3% and 72.2%, respectively [32]. Overall, the research indicates that the novel bio-nanoformulation is an effective and safe approach to treating cancer.

4 Target Delivery of Chemotherapeutic Nanodrugs

Cancer has complicated pathophysiological effects. Uncontrolled cell growth, the method by which cancer cells function, is currently incompletely defined. The copying of genomic DNA and the segregation of new cells, which takes place in multicellular organisms, are the two fundamental regions that are under the control of the cell cycle. Typically, cancerous cells move through cell cycle checkpoints, which is one of the precursors to the cancer process. Proliferative growth, apoptosis release, tumor growth, and angiography are the four main aspects that set cancerous cells apart from healthy cells (Fig. 2).

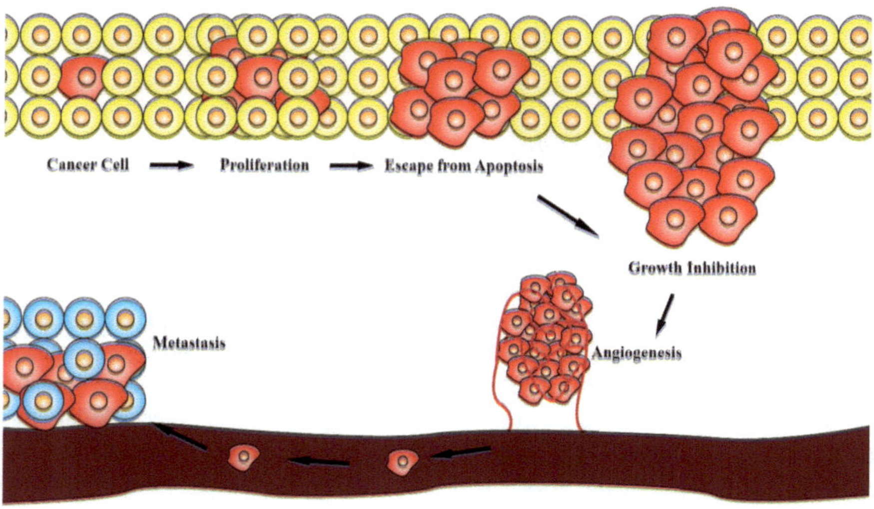

Fig. 2 Characteristics of cancer cells

It is essential to provide anticancer bio-based nanodrugs in a way that maximizes their bioavailability to cancerous cells and minimizes their side effects on normal tissues. Drug delivery techniques must be used for this to operate. These nanoparticles' shape, size, and morphology have a crucial role in how rapidly the therapeutics are delivered to the target cells. It is possible to boost therapeutic potential, lessen cytotoxicity to normal tissues, and enhance patient outcomes by applying suitable and selecting these properties. Because of this, researchers focus on developing nanodrug distribution strategies that can increase the bioavailability of anticancer drugs while reducing their adverse effects on immune tissues. Nanomaterials are currently being used in drug delivery systems for chemotherapy due to their favorable properties such as penetration and stability in target tissues (Fig. 3). Typically, bio-nanostructures ranging from 10 to 200 nm in size are selected for this purpose. As a result, nanotechnology has emerged as a promising area for cancer chemotherapy [33, 34]. The dimensions, shapes, and surface characteristics of nanoparticles play a role in drug delivery systems for biomedicine. Various techniques have been developed to target specific regions using nanomaterials. To enhance the effectiveness of bio-based nanodrug release, scientists are concentrating on antibodies or ligands that match the surfaces of abnormal increases on the endothelial cells of cancerous cells, which are targeted when aiming for tumor blood vessels. In contrast, normal cells show minimal increases in these surfaces. To increase of bio-based nanodrug release, an alternative method is to focus on tumor cell membranes that possess receptors that differ greatly from those present in normal cells. This is achieved by employing cell-binding ligands such as hyaluronic acid and folic acid, which can attach to these receptors. As a result, nanocarriers can bind to these areas and augment drug release. Further methods for subcellular organelle delivery involve targeting mitochondrial and nucleus in addition to cancerous cells. To induce death in cancer cells using the mitochondrial method, tri-Phenyl-Phosphonium (TPP) is absorbed into the cellular membrane during nanodrug trials, triggering Caspase9 and Caspase 3 pathways. In contrast, the goal of anticancer drugs in the core-targeted approach is to target the DNA and core-based enzymes. Yet, very small pores must be crossed in order for agents to go from the cell's cytoplasm to its nucleus. The advantage of nanoparticle drugs in this regard derives from their small dimensions. Furthermore, gene therapy strategies can also involve modifying the properties. By reducing the expression of gene mutations during treatments, regenerative medicine methods including plasmid DNA and miRNA can induce the elimination of cancerous cells [35]. New chemotherapeutic drug development is regarded as a top priority by pharmaceutical companies and researchers worldwide. Chemotherapy may be substantially more successful by novel therapeutic strategies that highlight bio-based chemotherapeutic nanodrugs. Hundreds more pharmaceuticals that bio-based chemotherapeutic nanodrugs are currently in preclinical or scientific research, and dozens have successfully entered various phases of clinical testing, which highlights the ongoing efforts of drug companies and researchers to develop novel chemotherapeutic drugs. Ivosidenib, Napabucasin, Venetoclax, and Vismodegib are the four targeting drugs whose commercial use has been authorized. By preventing the STAT3 protein from functioning, napabucasin is used to treat pancreatic and stomach cancer. Advanced basal cell carcinoma is treated with vismodegib, an inhibitor of the Hedgehog signaling system. Relapsed acute myelocytic leukemia is treated with the IDH1 inhibitor ivosidenib. Small lymphocytic lymphoma and chronic lymphocytic leukemia are both treated with venetoclax, a Bcl-2 inhibitor [36]. With the development of nanomedicine, studies on improving cancer chemotherapy are increasing day by day [37]. Table 3 lists some nanodrugs and cancer treatment.

By conjugating nano-drugs with ligands that target tumors and encourage their intracellular uptake, anti-cancer medications can be enhanced in their efficacy. Moreover, negatively charged nano-drugs often stay in the bloodstream for longer periods of time and are less likely to be metabolized by cancerous cells, which decreases intracellular concentration of the drug. For this reason, researchers are focused on overcoming the limitations associated with negatively charged surfaces to improve the efficacy of nano-drugs. To increase the amount of drug inside cancer cells, positively charged nano-drugs are preferred, as they have a higher chance of being taken up by these cells. There are several methods of producing positively charged nano-drugs, including conjugating them with ligands that target specific receptors. In 2022, Huang et al. developed a positively charged nanodrug to improve cancer intracellular uptake. They used carbodiimide, hydroxysuccinimide, and platinum. They synthesized metal-polymer-chlorin by reacting polyethyleneimine. In order to obtain the bio-nanostructure, with an average size of 160 nm and a spherical structure, the resulting polymer was dissolved in deionized water and filtered. Hyaluronic acid was used to coat the bio-nanostructure in order to give them a negatively charged surface potential and enhance the targeted delivery of bio-nanostructure to melanoma

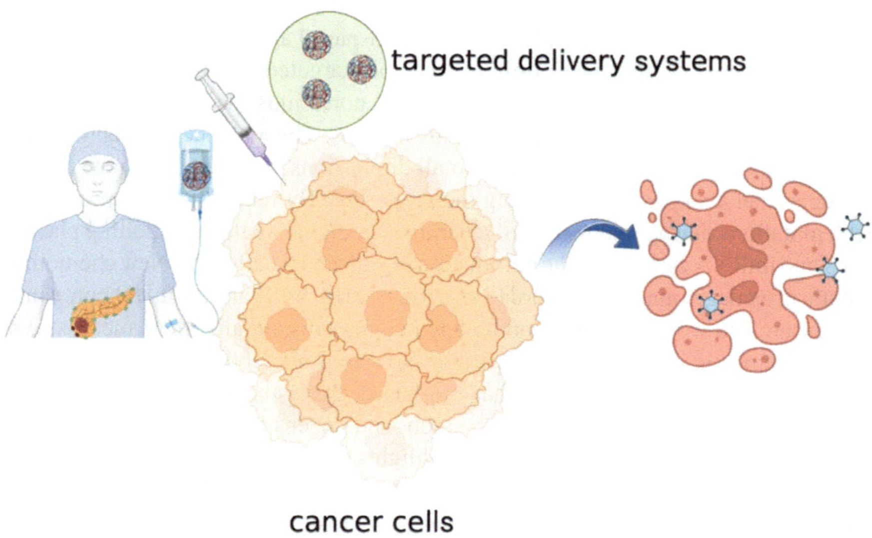

Fig. 3 Schematic presentation of targeted drug delivery systems

Table 3 The use of nanomedicine-based commercial formulations in cancer treatment

Name	Nanodrug-nanosystem type	Cancer type-approval date
Lipoplatin™	Cisplatin-Liposome	Nonsmall cell lung, head and neck, ovaria, gastric cancer-2005-Chinese
Lipusu®	Paclitaxel-Liposome	Nonsmall cell lung, breast, gastric cancer-2013-FDA
Myocet®	Doxorubicin-Liposome	Breast cancer-2000-EMA
Cornell dots	Silica NPs-cRGDY peptide	Brain cancer-clinical stage ongoing
AZD2811	Nanomicelles-Aurora kinase inhibitor	Solid tumors-clinical stage ongoing
NC-6004	Nanomicelles-cisplatin	Lung cancer, bladder, pancreatic cancer-clinical stage ongoing
Abraxane®	Paclitaxel-Albumin nanoparticle	Breast cancer-2005-FDA
AGuIX	Gadolinium based nanoparticle	Lung and pancreas cancer-clinical stage ongoing
Genexol-PM®	Paclitaxel-Micelle	Breast cancer-2007-FDA
Nano-QUT	Quercetin-PLGA-PEG nanoparticles	Oral cancer-clinical stage ongoing

cells that overexpress CD44. The hyaluronic acid was degraded by the positively charged bio-nanostructure which increased the absorption into tumor cells. In their in vivo study, they showed that bio-nanostructure has anti-cancer properties against melanoma cells in mice [38]. In another study, Mo et al. designed a nano drug system that combines drug delivery with cooperative chemo-photothermal anticancer potential. They used a hydrothermal method to produce molybdenum disulfide nanolayers and dispersed 50 mg of lipoic acid functionalized polymer (PEG) in 20 mL of molybdenum disulfide dispersion. They produced the molybdenum disulfide/PEG composite after 24 h of continuous stirring. To develop the nanocomposite, they added 3 integrin binding peptide to the composite. The anticancer drug (DOX) was used to manufacture the nano-drug in the final stage after the required centrifugation procedures. The resulting 200–300 nm-sized flower-like nanomaterials. Using the MTT assay, the produced nanocarrier MPRS's cytotoxicity against MCF-7 cells, VSMC cells, Hela cells, and H9c2 cells was assessed. The produced nano drug's release in the tumor microenvironment was boosted, increasing its ability to destroy tumor cells while decreasing its toxicity against healthy cells [39]. In another study, Skoll et al. investigated into the development of proteinaceous serum albumin nanocarriers with plant oil that do not contain hazardous cross-linkers or potentially harmful non-aqueous solutions. They used a sonochemical method with different plant oils to produce these nanomaterials, which produced particles with limited size distribution and stable suspensions that can be frozen to keep for a long period. Using bio-nanocarriers (molar ratio: 7/1), they were also succeeded in integrating wheat germ agglutinin as a target molecule into the shell. Researchers found that nanocarriers had up to a 55% higher cell binding potential than those without a targeter after conducting urothelial cell binding assays [40]. It shows that targeted protein nanocarriers could be an innovative pharmaceutical delivery approach. Recent studies have shown promising results in the field of nanomedicine for treating difficult conditions, particularly in the area of neurodegenerative diseases such as Parkinson's, Alzheimer's, and Huntington's. These problems are often challenging to treat effectively, and nanomedicine offers new approaches to combat them. Central nervous system disorders (CNS) are a group of diseases that also lack effective treatment options. Among these, gliomas and glioblastomas, two types of brain cancer, are the most common tumors in the central nervous system of adults. Systems for treating these disorders with nanomedicine should have no trouble crossing the blood–brain barrier. Peng et al. conducted a study in 2018 to assess the effectiveness of a combination therapy approach on glioma apoptosis and cancer progression inhibition. The therapy consisted of temozolomide and a folate-targeted nanocarrier that delivered anti-BCL-2 siRNA. The nanocarrier was designed to improve the permeability of the blood–brain barrier. In their in vitro cell study, the researchers observed a significant apoptotic response in cancer cells, which was attributed to the suppression of the BCL-2 gene and activation of the proapoptotic Bax gene [41]. In the last few years, the advancement of intelligent nano-drugs has led to a rise in research on reducing the insufficiency and toxic impacts of drugs used in tumor cells with a complex structure. As nano-based drugs advance to clinical trial stages, it is expected that more efficient mechanisms will be uncovered in the following years. The escalation in productive in vitro and in vivo studies demonstrates these outcomes.

5 Challenges and Prospects

Bio-based materials are now used in the development of delivery nanosystems because of the growing fascination with developing environmentally and biologically safe nano delivery systems. Drugs for both treating and diagnosing a wide range of illnesses and conditions can be carried by these bio-based systems. Bio-based nanomaterials are recommended since they often come from plants or animals and are less likely to have adverse effects on the body. Extensively biobased structures are frequently employed in tissue engineering. They are now being used more frequently in nanodelivery systems because of their biocompatibility and minimal cytotoxicity. It is suggested to investigate the effects of this nanodrug system on several kinds of cancer cells, including triple-negative and non-hormone sensitive breast cancer cell lines. Focusing the nanodrug conjugate on the location of the tumor and delivering it directly to the tumor tissue will help to reduce adverse effects. Bio-based nanomaterials can therefore be altered by an antibody to enhance its ability to target tumor cells. Based on the research that is currently available, it would seem that bio-nanocarriers are a good choice for encapsulation techniques because they can provide stable capture and can be customized for food packaging and drug applications. The development of bio-nanocarriers still faces some difficulties, nevertheless. Studies are being done to suggest low-cost, straightforward manufacturing methods that can be scaled up to satisfy market demand, notably in the pharmaceutical industries, in order to address these problems. To enhance the functionality of produced bio-nanocarriers, more research is required on the biomedical research of biopolymer derivatives in vivo as well as on enhanced combinatorial bio-nanocarriers.

References

1. Stanisz, M., Klapiszewski, Ł, Jesionowski, T.: Recent advances in the fabrication and application of biopolymer-based micro- and nanostructures: a comprehensive review. Chem. Eng. J. **397**, 125409 (2020). https://doi.org/10.1016/J.CEJ.2020.125409
2. Rashidinejad, A., Birch, E.J., Sun-Waterhouse, D., Everett, D.W.: Effect of liposomal encapsulation on the recovery and antioxidant properties of green tea catechins incorporated into a hard low-fat cheese following *in vitro* simulated gastrointestinal digestion. Food Bioprod. Process. **100**, 238–245 (2016). https://doi.org/10.1016/J.FBP.2016.07.005
3. Li, Y., Qiu, S., Sun, J., Ren, Y., Wang, S., Wang, X., Wang, W., Li, H., Fei, B., Gu, X., Zhang, S.: A new strategy to prepare fully bio-based poly(lactic acid) composite with high flame retardancy, UV resistance, and rapid degradation in soil. Chem. Eng. J. **428**, 131979 (2022). https://doi.org/10.1016/J.CEJ.2021.131979
4. Santhiago, M., Garcia, P.S., Strauss, M.: Bio-based nanostructured carbons toward sustainable technologies. Curr. Opin. Green Sustain. Chem. **12**, 22–26 (2018). https://doi.org/10.1016/J.COGSC.2018.04.009
5. Lee, S.C., Yoo, E., Lee, S.H., Won, K.: Preparation and application of light-colored lignin nanoparticles for broad-spectrum sunscreens. Polymers **12**, 699 (2020). https://doi.org/10.3390/POLYM12030699
6. Aadil, K.R., Mussatto, S.I., Jha, H.: Synthesis and characterization of silver nanoparticles loaded poly(vinyl alcohol)-lignin electrospun nanofibers and their antimicrobial activity. Int. J. Biol. Macromol. **120**, 763–767 (2018). https://doi.org/10.1016/J.IJBIOMAC.2018.08.109
7. Zhang, F., Wu, W., Zhang, X., Meng, X., Tong, G., Deng, Y.: Temperature-sensitive poly-NIPAm modified cellulose nanofibril cryogel microspheres for controlled drug release. Cellulose **23**, 415–425 (2016). https://doi.org/10.1007/S10570-015-0799-4/TABLES/1
8. Maskur, M.H., Wan Nawawi, W.M.F., Ali, F., Alkhatib, M.F.R.: Crustacean chitin nanomaterial as reinforcement for bio-based polymer 1–12 (2020). https://doi.org/10.1080/2374068X.2020.1793638
9. Coltelli, M.B., Morganti, P., Castelvetro, V., Lazzeri, A., Danti, S., Benjelloun-Mlayah, B., Gagliardini, A., Fusco, A., Donnarumma, G.: Chitin nanofibril-nanolignin complexes as carriers of functional molecules for skin contact applications. Nanomaterials **12**, 1295 (2022). https://doi.org/10.3390/NANO12081295
10. Wang, W., Meng, Q., Li, Q., Liu, J., Zhou, M., Jin, Z., Zhao, K.: Chitosan derivatives and their application in biomedicine. Int. J. Mol. Sci. **21**, 487 (2020). https://doi.org/10.3390/IJMS21020487
11. Ansari, M.J., Jasim, S.A., Bokov, D.O., Thangavelu, L., Yasin, G., Khalaji, A.D.: Preparation of new bio-based chitosan/Fe2O3/NiFe2O4 as an efficient removal of methyl green from aqueous solution. Int. J. Biol. Macromol. **198**, 128–134 (2022). https://doi.org/10.1016/J.IJBIOMAC.2021.12.082
12. Aranaz, I., Alcántara, A.R., Civera, M.C., Arias, C., Elorza, B., Caballero, A.H., Acosta, N., Velasco, H., Mecerreyes, D., Antonio, R., Gimeno, B., María Díez-Pascual, A., Moreno, V.C., Serra, A.: Chitosan: an overview of ıts properties and applications. Polymers **13**, 3256 (2021). https://doi.org/10.3390/POLYM13193256
13. Gao, X., Gong, J., Cai, Y., Wang, J., Wen, J., Peng, L., Ji, H., Jiang, S., Guo, D.: Chitosan modified squalene nanostructured lipid carriers as a promising adjuvant for freeze-dried ovalbumin vaccine. Int. J. Biol. Macromol. **188**, 855–862 (2021). https://doi.org/10.1016/J.IJBIOMAC.2021.08.074

14. Sathiskumar, S., Vanaraj, S., Sabarinathan, D., Bharath, S., Sivarasan, G., Arulmani, S., Preethi, K., Ponnusamy, V.K.: Green synthesis of biocompatible nanostructured hydroxyapatite from Cirrhinus mrigala fish scale—A biowaste to biomaterial. Ceram. Int. **45**, 7804–7810 (2019). https://doi.org/10.1016/j.ceramint.2019.01.086
15. Kim, K.H., Kim, K.B.: Ultrasound assisted synthesis of nano-sized lithium cobalt oxide. Ultrason. Sonochem. **15**, 1019–1025 (2008). https://doi.org/10.1016/J.ULTSONCH.2007.11.004
16. Dai, D., Li, Y., Fan, J.: Room-temperature synthesis of various allotropes of carbon nanostructures (graphene, graphene polyhedra, carbon nanotubes and nano-onions, n-diamond nanocrystals) with aid of ultrasonic shock using ethanol and potassium hydroxide. Carbon N Y. **179**, 133–141 (2021). https://doi.org/10.1016/J.CARBON.2021.04.038
17. Wang, C., Yang, M., Shi, H., Yao, Z., Liu, E., Hu, X., Guo, P., Xue, W., Fan, J.: Carbon quantum dots prepared by pyrolysis: Investigation of the luminescence mechanism and application as fluorescent probes. Dyes Pigm. **204**, 110431 (2022). https://doi.org/10.1016/j.dyepig.2022.110431
18. Agnol, L.D., Neves, R.M., Maraschin, M., Moura, S., Ornaghi, H.L., Dias, F.T.G., Bianchi, O.: Green synthesis of Spirulina-based carbon dots for stimulating agricultural plant growth. Sustain. Mater. Technol. **30** (2021). https://doi.org/10.1016/j.susmat.2021.e00347
19. Nurfazianawatie, M.Z., Omar, H., Rosman, N.F., Malek, N.S.A., Afaah, A.N., Buniyamin, I., Salifairus, M.J., Malek, M.F., Mahat, M.M., Rusop, M., Asli, N.A.: Forming multilayer graphene at different precursor temperature using cooking oil waste from AYAMAS via chemical vapour deposition. Mater Today Proc. **75**, 127–132 (2022). https://doi.org/10.1016/j.matpr.2022.10.169
20. Irfan, M.F., Arami-Niya, A.: Temperature effect on the synthesis of iron–cobalt nano-particles using catalytic chemical vapor deposition of CO_2 in thermo-gravimetric analyzer: analytical and thermodynamic studies. Nano-Struct. Nano-Objects **18**, 100261 (2019). https://doi.org/10.1016/j.nanoso.2019.100261
21. Wang, T., Cornel, E.J., Li, C., Du, J.: Drug delivery approaches for enhanced antibiofilm therapy. J. Control. Release **353**, 350–365 (2023). https://doi.org/10.1016/J.JCONREL.2022.12.002
22. Mariz, M., Murta, J., Gil, M.H., Ferreira, P.: An ocular insert with zero-order extended delivery: release kinetics and mathematical models. Eur. J. Pharm. Biopharm. **181**, 79–87 (2022). https://doi.org/10.1016/J.EJPB.2022.10.023
23. Yang, M., Abdalkarim, S.Y.H., Yu, H.Y., Asad, R.A.M., Ge, D., Zhou, Y.: Thermo-sensitive composite microspheres incorporating cellulose nanocrystals for regulated drug release kinetics. Carbohydr. Polym. **301**, 120350 (2023). https://doi.org/10.1016/J.CARBPOL.2022.120350
24. Zirak, N., Bolandparvaz Jahromi, A., Salahinejad, E.: Vancomycin release kinetics from Mg–Ca silicate porous microspheres developed for controlled drug delivery. Ceram Int. **46**, 508–512 (2020). https://doi.org/10.1016/J.CERAMINT.2019.08.290
25. Bardajee, G.R., Hooshyar, Z., Farsi, M., Mobini, A., Sang, G.: Synthesis of a novel thermo/pH sensitive nanogel based on salep modified graphene oxide for drug release. Mater. Sci. Eng. C **72**, 558–565 (2017). https://doi.org/10.1016/J.MSEC.2016.11.109
26. Gao, D., Asghar, S., Ye, J., Zhang, M., Hu, R., Wang, Y., Huang, L., Yuan, C., Chen, Z., Xiao, Y.: Dual-targeted enzyme-sensitive hyaluronic acid nanogels loading paclitaxel for the therapy of breast cancer. Carbohydr. Polym. **294**, 119785 (2022). https://doi.org/10.1016/J.CARBPOL.2022.119785
27. Zhang, S., Li, Y., Qiu, X., Jiao, A., Luo, W., Lin, X., Zhang, X., Zhang, Z., Hong, J., Cai, P., Zhang, Y., Wu, Y., Gao, J., Liu, C., Li, Y.: Incorporating redox-sensitive nanogels into bioabsorbable nanofibrous membrane to acquire ROS-balance capacity for skin regeneration. Bioact Mater. **6**, 3461–3472 (2021). https://doi.org/10.1016/J.BIOACTMAT.2021.03.009
28. Wang, T., Hou, J., Su, C., Zhao, L., Shi, Y.: Hyaluronic acid-coated chitosan nanoparticles induce ROS-mediated tumor cell apoptosis and enhance antitumor efficiency by targeted drug delivery via CD44. J. Nanobiotechnol. **15**, 7 (2017). https://doi.org/10.1186/s12951-016-0245-2

29. Tak, K., Sharma, P., Sharma, R., Dave, V., Jain, S., Sharma, S.: One-pot hydrothermal green synthesis of Polygala tenuifolia mediated graphene quantum dots for acetylcholine esterase inhibitory activity. J. Drug Deliv. Sci. Technol. **73**, 103486 (2022). https://doi.org/10.1016/J.JDDST.2022.103486
30. Reddy, A.S., Lakshmi, B.A., Kim, S., Kim, J.: Synthesis and characterization of acetyl curcumin-loaded core/shell liposome nanoparticles via an electrospray process for drug delivery, and theranostic applications. Eur. J. Pharm. Biopharm. **142**, 518–530 (2019). https://doi.org/10.1016/J.EJPB.2019.07.024
31. Değirmenci, N.S., Uslu, M., Kırbaş, O.K., Şahin, F., Önay Uçar, E.: Lapatinib loaded exosomes as a drug delivery system in breast cancer. J. Drug Deliv. Sci. Technol. **75**, 103584 (2022). https://doi.org/10.1016/J.JDDST.2022.103584
32. Gandomi, F., Rostami, M., Ahmadi, F., Mohammad Sorouri, A., Badiei, A., Fasihi-Ramandi, M., Reza Ganjali, M., Ehrlich, H., Rahimi-Nasrabadi, M.: ROS, pH, and magnetically responsive ZnFe2O4@l-Cysteine@NGQDs nanocarriers as charge-reversal drug delivery system for controlled and targeted cancer chemo-sonodynamic therapy. Inorg. Chem. Commun. **150**, 110544 (2023). https://doi.org/10.1016/J.INOCHE.2023.110544
33. Zhang, Y., Qian, J., Gu, C., Yang, Y.: Alternative splicing and cancer: a systematic review. Signal Transduct Target Ther. **6** (2021). https://doi.org/10.1038/S41392-021-00486-7
34. Qin, S.-Y., Zhang, A.-Q., Zhang, X.-Z., Qin, S.-Y., Q. A. Zhang, Zhang, Z. X.: Recent advances in targeted tumor chemotherapy based on smart nanomedicines. Small **14**, 1802417 (2018). https://doi.org/10.1002/SMLL.201802417
35. Wei, G., Wang, Y., Yang, G., Wang, Y., Ju, R.: Recent progress in nanomedicine for enhanced cancer chemotherapy. Theranostics **11**, 6370 (2021). https://doi.org/10.7150/THNO.57828
36. Yan, J., Long, X., Liang, Y., Li, F., Yu, H., Li, Y., Li, Z., Tian, Y., He, B., Sun, Y.: Nanodrug delivery systems and cancer stem cells: from delivery carriers to treatment. Colloids Surf. B Biointerf. **217**, 112701 (2022). https://doi.org/10.1016/J.COLSURFB.2022.112701
37. Khot, V.M., Salunkhe, A.B., Pricl, S., Bauer, J., Thorat, N.D., Townley, H.: Nanomedicine-driven molecular targeting, drug delivery, and therapeutic approaches to cancer chemoresistance. Drug Discov. Today **26**, 724–739 (2021). https://doi.org/10.1016/J.DRUDIS.2020.12.016
38. Huang, X., Mu, N., Ding, Y., Lam, H.W., Yue, L., Gao, C., Chen, T., Yuan, Z., Wang, R.: Targeted delivery and enhanced uptake of chemo-photodynamic nanomedicine for melanoma treatment. Acta Biomater. **147**, 356–365 (2022). https://doi.org/10.1016/J.ACTBIO.2022.05.015
39. Mo, C., Wang, Z., Yang, J., Ouyang, Y., Mo, Q., Li, S., He, P., Chen, L., Li, X.: Rational assembly of RGD/MoS2/Doxorubicin nanodrug for targeted drug delivery, GSH-stimulus release and chemo-photothermal synergistic antitumor activity. J. Photochem. Photobiol. B. **233**, 112487 (2022). https://doi.org/10.1016/J.JPHOTOBIOL.2022.112487
40. Skoll, K., Ritschka, M., Fuchs, S., Wirth, M., Gabor, F.: Characterization of sonochemically prepared human serum albumin nanocapsules using different plant oils as core component for targeted drug delivery. Ultrason. Sonochem. **76**, 105617 (2021). https://doi.org/10.1016/J.ULTSONCH.2021.105617
41. Peng, Y., Huang, J., Xiao, H., Wu, T., Shuai, X.: Codelivery of temozolomide and siRNA with polymeric nanocarrier for effective glioma treatment. Int. J. Nanomed. **13**, 3467 (2018). https://doi.org/10.2147/IJN.S164611

İbrahim Mizan Kahyaoğlu is currently an Research Assistant in the Department of Chemistry, at Ondokuz Mayis University, Turkiye. He received her Master of Science degree in physical chemistry from Ondokuz Mayis University in 2023, and he is in Ph.D. program in analytical chemistry at Ondokuz Mayis University now. He has research experience in drug carrier systems, nanoparticles, nanocomposites, copolymer blends, heavy metal detecting, analysis of organic and inorganic compounds. He has published several research articles and book chapters.

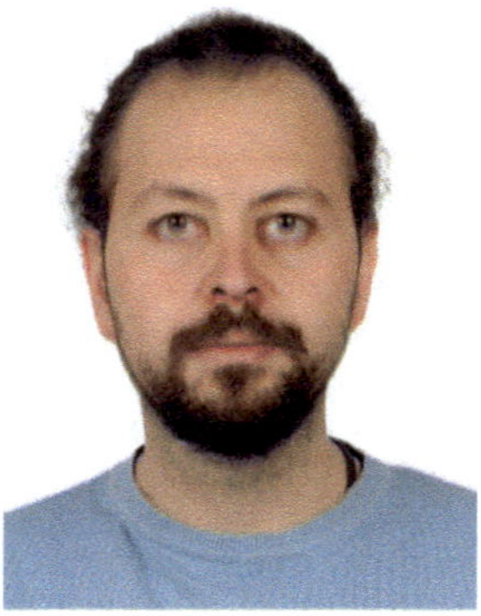

Erdi Can Aytar is currently an Research Assistant in the Department of Biology, at Ondokuz Mayis University, Turkiye. He received her Master of Science degree in Molecular Biology from Gazi University in 2016, and he is in Ph.D. program in Botany at Ondokuz Mayis University now. He has research experience in Antioxidant and antimicrobial activity, cancer, isolation of organic compounds, plant physiology and germination. He has worked on national and international projects. He has published several research articles.

Alper Durmaz received his Master of Science degree in botany from Ondokuz Mayis University in 2016, and his Ph.D in botany from Ondokuz Mayis University in 2021. He has research experience in plant systematic and plant ecology. He has worked on a variety of projects funded by Ondokuz Mayis University and has published several review and research articles.

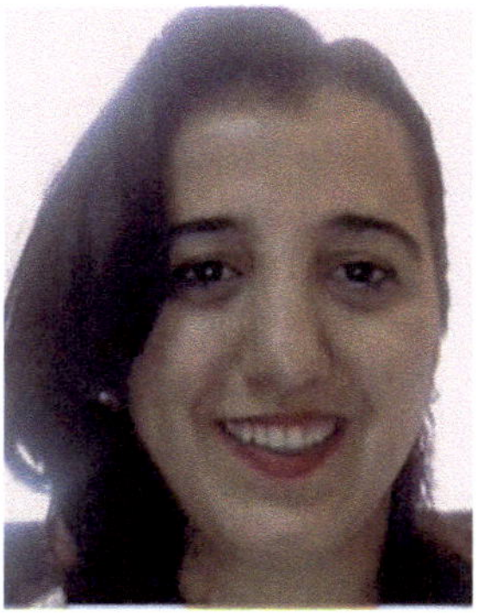

Selcan Karakuş is currently an Associate Professor in the Department of Chemistry, at Istanbul University-Cerrahpasa (IUC), Turkiye. She received her Master of Science degree in physical chemistry from Istanbul University (IU) in 2006, and her Ph.D. in physical chemistry from IU in 2011. She has been a visiting researcher at the University of Massachusetts, Department of Polymer Science and Engineering. She has research experience in drug carrier systems, nanoparticles, nanocomposites, nano-emulsion self-assembled polymeric nanostructures, and copolymer blends. She has worked on a variety of projects funded by IUC and has published several research articles and book chapters.

Nanostructured Biomaterials in Drug Delivery: Current Trends and Upcoming Possibilities

Pankaj Sharma

Abstract Materials in the nanoscale category are used for diagnostic purpose or to administer therapeutic material to specialized targeted places in a regulated manner in the relatively young but rapidly emerging field of nano-biomedicine. In the treatment of chronic human illnesses, nanotechnology has a number of advantages, including the targeted and site-specific administration of precise medications. The use of nano-biomedicine (chemotherapeutic substances, biologic substances, immunotherapeutic substances, etc.) in the therapy of many illnesses has recently resulted in an array of notable developments. By doing careful examination of the development and application of nano-biomaterials to enhance the effectiveness of both new and old drugs (such as natural therapeutics) and particular detection by means of illness marker molecules, this chapter gives a revised overview of current developments in the discipline of nano-biomedicines and nano centred systems for drug delivery. The benefits and drawbacks of employing nanomedicines for the therapeutic administration of pharmaceuticals derived from synthetic or natural sources are also discussed. Additionally, we provided details on the developments and aspirations in the field of nanomedicine.

Keywords Nano-biomaterials · Drug carriers · Biomaterials · Drug delivery · Nano-biomedicine

Abbreviations

5-FU	5-fluorouracil
CMC	Carboxymethylcellulose
XG	Xanthan gum
PEG	Polyethylene glycol
QDs	Quantum Dots

P. Sharma (✉)
Department of Pharmaceutics, ShriRam College of Pharmacy, Banmore, Morena, Madhya Pradesh, India
e-mail: pankajsharma223@gmail.com

R. Malviya and S. Sundram (eds.), *Engineered Biomaterials*, Engineering Materials,
https://doi.org/10.1007/978-981-99-6698-1_9

EPR Enhanced permeability and retention

1 Introduction

Humans have long employed natural remedies derived from plants to treat a variety of illnesses. The majority of modern medications are made from herbs using conventional wisdom and methods. On the market today, the most significant pharmaceutical compounds and their derivatives make up around 25% of the total [1, 2]. A starting point for the creation of novel pharmaceuticals is natural compounds with varied molecular ancestries. Designing lead compounds that are synthetically accessible and resemble their counterparts' chemistry is the latest fad in natural product-based pharmaceutical research [3]. Only a few of the astounding properties that biomaterials exhibit include remarkable chemical variation, biological and chemical potential including macromolecular accuracy, and decreased noxiousness. They do so because they see them as good prospects for developing new drugs [4]. Additionally, target-based drug deliveries, drug prediction at the cellular level, and other next-generation pharmaceutical breakthroughs have all been made possible thanks to computational techniques.

Despite the benefits, healthcare industries [5] are disinclined to pay-out additional funds on biomaterial-based medicaments development and medicaments delivery setup, attempt to uncover new medicines by searching over databases of chemical substances. However, natural moieties are now being used to treat a number of grave diseases, such as cancer, inflammatory, microbiological, and cardiovascular issues. The specific advantages that natural medicines offer like less toxicity and side responses, cheap cost, and potent therapeutic value are largely to blame for this. Nevertheless, it is more challenging to use natural compounds as medications because of concerns regarding their bioactivity and security. Numerous natural compounds are stalling in the clinical study stages as a result of these problems [6–8]. It is extremely difficult to use large-scale biomaterials for drug delivery because of their in vivo instability, poor body absorption, decreased solubility and bioavailability, difficulties with site-specific delivery, problems with tonic performance, and possibly even negative pharmacological effects. Therefore, a different approach to solving these urgent issues may be to employ cutting-edge drug delivery techniques to target treatments at specific body regions [9, 10]. Therefore, nanotechnology has a significant influence on the formulations of contemporary medications, as well as their successful delivery and regulated disintegration. In a diversity of medical domains, involving nano-medicaments and nano-scaled drug delivery networks, where such substances are of particular significance, the use of nanomaterials and nanophases has shown the ability of nanotechnology to establish a connection between the physical and biological sciences. Nano-biomaterials, which may be defined as substances with a diameter of 1–100 nm, have a tremendous influence on the future of nanomedicine, impacting everything from tissue engineering to microfluidics to drug delivery and biosensors

[11]. Therapeutic compounds are created at the nano-criterion to develop nanotechnology and nanomedicine. The advancement of nano-biotechnology, biosensors, medicaments delivery, and tissue engineering in the field of biomedicine has all been facilitated by nanoparticles. Because they are composed of substances that are formed at the atomic or subatomic level, nanoparticles frequently take the form of small nanospheres [11]. As a result, they can move through the body more freely than heavier things can. Nanoscale particles exhibit unique magnetic, biological, mechanical, chemical, and structural properties. Past developments have seen a rise in interest in nanomedicines as a result of the potential of nanostructures to work as delivery medium for pharmaceuticals by encasing them or adhering to them, allowing for more regulated distribution of the therapeutics to the intended tissues [10, 11]. In the developing nanomedicine discipline, clinical biology, preventative medicine, and treatment are all addressed using the expertise and techniques of nano-biomaterials. It suggests using nanodimensional substances to actuate components in living cells as well as nanosensors, nanorobots, and sensors for delivery, diagnostic, and receptive applications (Fig. 1). As instance, a nanoparticulate derived approach that included both cancer therapy and cancer detection imaging techniques has been established [12]. Micelles and liposomes, that has recently received FDA approval, were members of the initial phase of nanoparticle-based therapies. These vesicles may incorporate inorganic or magnetic nanoparticles, such as gold [13]. The use of inorganic nanoparticles has increased as a result of these properties, which are mostly focused on drug delivery, scanning, and therapeutic applications. Additionally, it is claimed that nanomaterials can help medicines that are only weakly water-soluble reach their intended areas and prevent gastrointestinal contamination of medications. Nanodrugs include a higher oral bioavailability because of their typical assimilative endocytotic uptake techniques.

Nanostructures allow the absorption of combination medications at the required dose since they rest in the blood circulation system for a lengthy time. As a result, they have less of a detrimental impact and produce fewer plasma oscillations. Due to their nanoscale, these structures may readily penetrate the tissue system, enhancing the effectiveness of pharmaceutical delivery and ensuring that the medication operates as intended. Compared to big materials between one and ten metres in size, nanostructures are significantly simpler for cells to absorb [14]. As a result, they collaborate immediately to heal the diseased cells more successfully and with little to no side effects.

Nanotechnology has been discovered to be helpful in gathering information at all phases of clinical procedures due to their application in several innovative tests to cure and diagnose illnesses. The major advantages of these nanomaterials are linked to unique surface characteristics because different proteins can attach to the surface. As example, gold nanoparticles are utilized for tumour tags and biomarkers in a variety of biomaterials monitoring procedures.

The choice of the nanoparticles is made and deployed on the physicochemical parameters of the medications when using nan-biomaterials for drug delivery. The utilization of bioactive natural chemicals in conjunction with nanotechnologies is highly appealing and has grown significantly in recent years. Whenever it refers to the

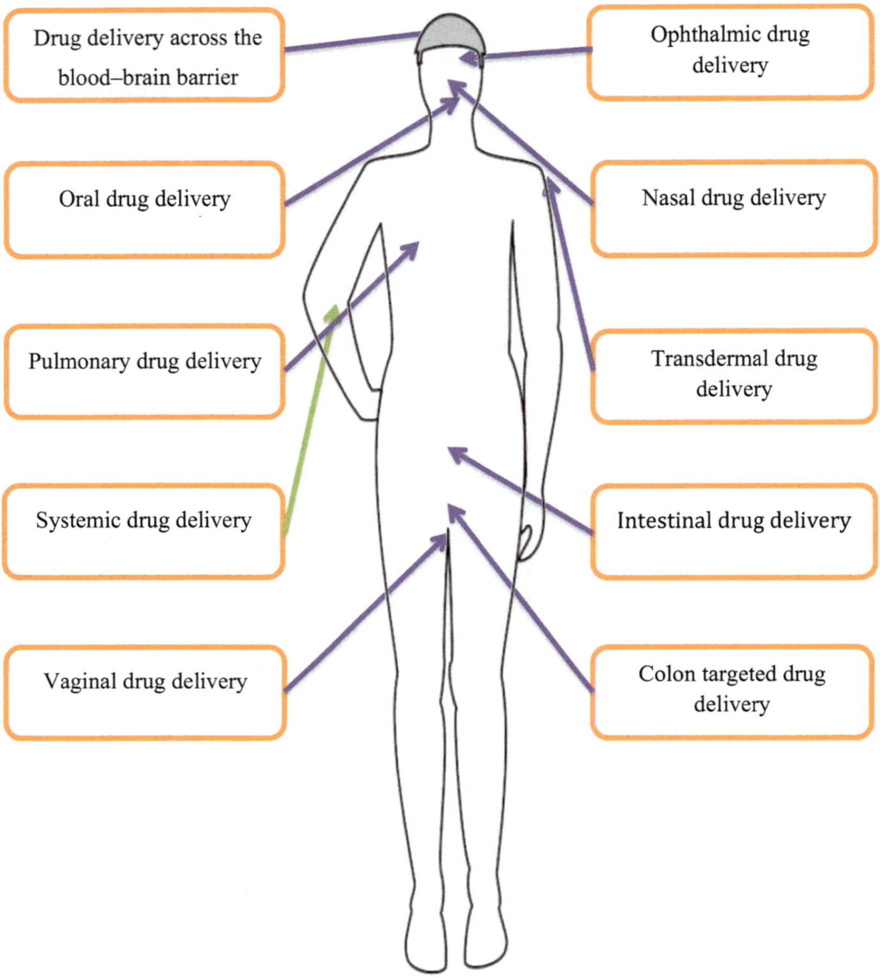

Fig. 1 Examples of nano-biomaterials and their delivery methods for usage in humans

delivery of organic remedies for the treatment of malignancy along with numerous other diseases, it offers a number of interests. In light of their numerous unique properties, including their ability to induce tumor-suppressing senescence and function as antimicrobials, biomaterials have been extensively explored in the context of treating illnesses. Caffeine and curcumin and have been linked to apoptosis [15], whilst cinnamaldehyde, curcumin, carvacrol, and eugenol have been linked to antimicrobial properties. Using nanotechnology allowed for the enhancement of various qualities, including targeting, bioavailability, and controlled release. For example, the bioactive ingredient in Nigella sativa called thymoquinone is examined after already being enclosed in a lipid nanomaterial. Its bioavailability increased six fold following encapsulation compared to free thymoquinone, protecting the gastrointestinal tract

[16]. Also, it improved the natural product's pharmacokinetic properties, improving medicinal value.

Gold/silver, inorganic, organic, and polymeric nanotechnology, including lipid vesicles are frequently taken into consideration when developing target-reach drug delivery systems. Certain drugs that have reduced solubility and uptake are combined with these nanoparticles [14]. However, these nanotechnologies' efficacies as drug delivery vehicles vary depending on their morphological and additional fundamental chemical and biophysical properties. As an instance, these nanomaterials between 10 and 1000 nm in size possess properties that render them efficient delivery systems [7]. A combination of their excellent biological action and being biodegradable many synthetic polymers, including poly(lactic-co-glycolic acid), polyethylene glycol, poly-l-lactic acid, and polyvinyl alcohol, as well as biopolymers like chitosan and alginates, are frequently used for the nanotechnology of biological substances [8]. Nanostructures and nanocapsules, two subtypes of biopolymeric nanomaterials that provide efficient drug delivery vehicles, can be distinguished. The delivery of targeted drugs can also benefit greatly from the usage of small phospholipids and lipid nanostructures like liposomes and micelles.

The biochemical and biophysical properties of the targeted drugs used for therapy are the main determinants of the choice of the best delivery based on nanotechnology [8]. However, issues like the toxicity displayed by nanoparticles can really be disregarded when thinking about the utilization of nanomedicine. To reduce toxicity risks, nanoparticles are now routinely employed in conjunction with natural goods. A widely regarded green chemistry strategy is the creation of drug-loaded nanoparticles since it lowers the quantity of potentially dangerous substances used in biosynthetic activity. Thus, using emerald nanoparticles to carry medications can lessen their side effects [12]. Further enhancing the biocompatibility of such nano-biomaterials are modifications to the nanostructures' shape, size, wettability, and surface properties. Hence, the targeted and site-specific administration of medications made possible by nanotechnology has several advantages in the treatment of chronic human illnesses. However, the lack of knowledge surrounding the toxicity of nanomaterials is a serious worry and clearly needs additional research in order to enhance safety while increasing the efficacy of these drugs. Therefore, careful design may be beneficial in overcoming the problems associated with the use of these nanoparticles.

Taking into account the aforementioned information, the review's objectives are to discuss various nano-based systems for drug delivery, notable implementations of nanomedicines based on biomaterials, and the specific sites, bioavailability, and regulated release of nanomedicine, as well as additional difficulties related to the use of nano-biomaterials in pharmaceuticals.

2 A Clinical Necessity of Nano-Biomaterials for Controlled Drug Delivery

The inherent issues with traditional dosage distribution techniques lead to the requirement of substances for controlled medication release. Administration of drugs often requires numerous, repetitive dosages, resulting in highly variable medication levels in the body during the treatment course. After delivery, medication levels rise to therapeutic doses, but occasionally hazardous side effects develop whenever the quantity exceeds the upper limit of what is safe [17]. As a consequence of degradation, metabolism, and transportation far from the targeted therapy, these approaches also quickly reduce drug levels to levels that are no more helpful [17]. Together, this causes materials and medicine waste and raises patient risk owing to probable serious adverse reactions and decreased treatment efficacy. Strategies for reducing the rate of release were created to solve these problems. These "sustained release" methods included the required treatment in capsule form that was typically taken orally, however some of them were also designed for parenteral use [18]. Drug release was slowed down by using cellulose coverings that slowly dissolve, adding chemicals that make drugs less soluble, compressing tablets, and using emulsion and suspensions [18] that are all contained in capsules. Nevertheless patient to patient variation, environmental influences, and the need for repeated dosing continued to have a significant impact on controlled release preparations [18].

The optimum controlled medication release method offers a number of benefits over extended release. These delivery substances preserve drug release within the therapeutic range and prevent the inefficiency of the drug concentration fluctuations of conventional formulations by releasing pharmaceuticals at rates that do not fluctuate over duration (zero-order release). Controlled release substances have the advantage of lowering the total quantity of medication required to attain treatment effectiveness since they prevent "peaks and valleys" and stay within the therapeutic range. These substances would help increase patient compliance, which would in developed countries is just 50%, by lowering the total amount of dosages necessary [19]. Such substances can also be administered or implanted precisely into a particular sick tissue, decreasing off-target adverse effects and boosting efficacy. Release of drugs can be regulated across extended therapeutic windows (i.e., days to years). In order to prevent off-target consequences, sustained release methods must improve the delivery of medications to particular tissues and cells inside the system [20]. Targeted delivery techniques are used for precise preservation and absorption by sick tissues and cells in order to increase tissue specificity [21]. In this method, delivery components are coupled to ligands which attach to surface proteins or receptors that are abundantly expressed in sick tissues and cells. Optimal controlled release polymers should also shield pharmaceuticals against quick bodily clearance and/or deterioration.

Engineering, physical science, biology, and clinical professionals must all work together in a comprehensive approach to create suitable nano-biomaterials for controlled drug release [22]. Design criteria involve: I including enough drug in the host matrix to accomplish extended-release characteristics necessary for efficacy of treatment; (ii) shielding therapies from degradation in vivo whilst also preserving bioactivity; and (iii) ensuring dependable release over the course of the medication regime, tend to range from days to years [23]. In order to prevent unpleasantness for the patient before and after administration, the substances themselves and associated decomposition products ought to be nontoxic and biodegradable inside the body. During the design process, it is also necessary to consider the cost of a certain material-drug composition because of the expense for synthesis and/or manufacturing of material.

3 Basic Principles for Creating Drugs Using Nanotechnology

Nanomaterials, such as biocompatible nanorobots and nanoparticles, are used for a number of purposes in the field of medicine known as nanomedicine, including diagnostic tests, medication administration, actuation, and sensing in a living cell [24]. Particularly less solubility materials including different problems with medicaments delivery, similarly poor oral bioaccessibility, decreased quantity to diffuse through surface, required higher injectable dosages, and noxious effects prior to standard vaccination technique. But what if you could enhance the medication delivery system while getting rid of all of these drawbacks by using nanotechnology techniques?

Drug creation at the nanoscale scale is the most futuristic method in the realm of nano-medicine applications since it allows for the modification of properties such solubility, diffusivity, drug release patterns, bioavailability, and immunogenicity. With reduced toxicity, less negative effects, enhanced drug distribution, and a higher pharmaceutical life lifetime, more efficient and simple administration techniques may also be developed. The methods created for medication delivery are either aimed at a targeted location or built for a regulated release of medicinal compounds there. Self-assembly, in which specified patterns or themes organically arise from constructing components, is a component of its evolution [25]. Additionally, they must overcome challenges such being opsonized or contained by the mononuclear phagocyte network.

Nanostructures have the ability to actively or passively transport drugs. In the latter, medications are predominantly integrated into the internal line of the composition via the hydrophobic impact. Every time the different parts of the nanomaterials are guided to specific regions, the proper quantity of the medicaments is released since it is contained at a low concentration in a hydrophobic environment [25]. Contrarily, the drugs that are to be released are already connected to the transporter nanostructure material in the later, allowing for straightforward dispersion. The time of release is critical in this technique because, if done improperly, if done correctly, the drug's biological action and effectiveness will increase but it won't reach the intended spot and will swiftly dissociate by the carrier [25]. Tailored drug delivery, whether can be active or passive and uses nanostructures as the drug delivery networks, is another crucial aspect of medicine administration. In targeted delivery, drugs are delivered by means of peptides, antibodies, and other compounds that bind to receptor sites made in the target area. Through preferences, which are influenced by alterable similar as pH, molecular site, temperature, and structure as they circulate through the circulation, the generated drug delivery vehicle molecule is carried to the target region in passive targeting. The organism's main targets are receptors on cellular membranes, lipids in cellular membranes, and proteins or antigens on certain substrates [26]. Today, the bulk of drug delivery methods made available by nanotechnology are focused on curing and treating cancer.

4 Nano-Biomaterials Delivery Parameters in Biological Systems

A number of considerations must be made when developing a system for delivering pharmaceuticals using nanoparticles, including the size, surface chemistry, and surface charge of the particulate matter, which will obviously vary based on the medications being administered and the kind of cancer being treated [27]. The biological interactions that take place within a person's or animal's body from the moment the nanobiomaterial is injected until it reaches the region of drug delivery must be thoroughly understood in order to design the nanoparticles for effective drug delivery. Once nanomaterials have made their way into the system of the body, the two crucial processes in drug delivery are the development of the nanobiomaterials system at the tumour site and entry into malignant cells to allow the medicine to escape into the appropriate cellular compartment.

The immune system of the human being, including the opsonization, mononuclear phagocytes, and complement activation, as well as elimination through the liver and kidneys, will regard the nano-biomaterials as a foreign entity upon introduction into the systemic circulation [28]. One benefit of adopting a nanoparticulate technology is that it can ultimately be eliminated by the host; however, the particles must circulate for a sufficient amount of time to recognize the tumour site and aggregate there in order to deliver the medicine. Hence, efficient drug administration requires striking a compromise between the host's speed of nano-biomaterials clearance and the system's characteristics that lead to tumor-specific retention. Size, surface characteristics, and the functions on the drug-encapsulating nano-biomaterial's surface are only a few of the variables that affect this preferred deposition [29].

Nanoparticles naturally have the capacity to collect at tumour locations. The increased permeation and persistence phenomenon, a physiological consequence of the tumour microenvironment, is the cause of this targeted delivery. As opposed to the strong endothelial connections found in healthy vasculature, the blood arteries that surround the tumour microenvironment typically have more leaks, allowing the nanoparticles to penetrate and aggregate at the tumour site. The build-up is made worse by inadequate lymph drainage in the vicinity of the tumour [30]. The distances among blood vessels vary depending on the kind of tumour and are typically around 100 and 800 nm. Nano-biomaterials of a diameter range of 30–200 nm had the highest tumour survival potential, according to experimental studies using several tumour models [31]. Also, it was shown that round nano-biomaterilas, as opposed to rod- or bar-shaped ones, might conformance into the tumour interstitium better successfully due to their ability to retain laminar flow characteristics [32].

Active targeting, particularly includes the nano-biomaterials molecularly recognizing the tumour, can solve passive targeting's several disadvantages, including unequal distribution throughout the solid tumour and trapping within the tumour [33]. It is possible to attach tumor-specific ligands to an exterior of the nanomaterial, allowing them to establish highly-affine interactions with receptors expressed on the tumour surface. These ligands can include antibodies or other compounds. In order to effectively internalize the nanomaterial into the cell and deliver the medicine, this also promotes receptor-mediated cellular uptake. Multipurpose nano-biomaterials are defined as nano-biomaterial that possess certain chemical identification signals; this topic will be covered in more detail in subsequent chapters [34].

Therefore, active and passive targeting cooperate to deliver the drug to the cancer. Therefore, active targeting increases the drug's retention and absorption by the cancer cells, whilst passive targeting ensures the concentration of the nano-particles from the bloodstream into the tumour milieu.

5 Nano-Biomaterials in Drug Delivery System

5.1 Alginate

Alginate is a different biopolymeric substance that is utilized for medication delivery. As an anionic bioadhesion polymer containing terminal carboxyl groups, such biopolymer has stronger mucoadhesive characteristics in comparison to cationic and neutrality polymers [35]. In order to reduce blood glucose levels and increase insulin levels in the blood, Patil and Devarajan [36] produced insulin-containing alginate nanoparticulates using nicotinamide as a penetration agent, resulting in a hypoglycemic effect. In the case of nicotinamide, sublingually administered nanoparticles (5 IU/kg) showed excellent pharmacological accessibility (> 100%) and bioavailability (>80%). NPs were shown to be potential transporters of insulin via the sublingual pathway in the streptozotocin-agravated diabetes-induced mouse by getting a pharmacologic significant power of 20.2% with a bio-accessibility of 24.1% in comparison to the subcutaneously at 1 IU/kg [36].

To specifically regulate the macro cell lung carcinomas, Roman et al. [37] created alginate microcapsules with the epidermal growth factor attached to its external portion. The nanomaterials also included a load of the cancer-causing chemical cisplatin. EGF considerably improved carrier system sensitivity and accelerated the rate of cell mortality in the H460-lung tumor cell when compared to the unbound drug.

5.2 *Chitosan*

Chitosan can be utilized to function at the constrictive epithelial connections since it has muco-adhesive qualities. As a result, sustained drug delivery systems targeting a range of epithelia, having the oral, nasal, intestinal, and pulmonary epithelia, are routinely made using chitosan-based nanoparticles. By administering chitosan nanoparticles covered with hyaluronic acid orally, Jain and Jain [38] looked at the 5-fluorouracil (5-FU) release pattern. In tests simulating the passage from either the gastro to the colonic, the release kinetics of 5-FU were found to be guarded from release in the small intestine and gastro. Also, the relative high active ingredient would've been capable of prolonging the duration of exposure, improving the antitumor activity and reducing cytotoxic effects in colon cancer therapy.

5.3 *Cellulose*

The main purpose of using cellulose and its variants in pharmaceutical drug administration is to vary the medications' gelation and solubility, which in turn affects how quickly they are released from the body [39]. By conjugating calcium alginate pellets with 5-fluoroacyl (5-FU) enriched carboxymethylcellulose (CMC), Agarwal et al. [40] successfully created a medication-targeted method that is specific to the colon. In the setting that mimicked a colonic milieu, the pellets with reduced CMC ratios displayed more edema and muco-adhesiveness. The 5-FU contained in the beads was released by the intestinal enzymes to 90% of its original concentration.

5.4 *Xanthan Gum*

The bacterium Xanthomonas campestris produces xanthan gum (XG), a high-molecular-mass heteropolysaccharide. A potent bioadhesive, it is a polyanionic polysaccharide. Since xanthan gum is believed to be non-irritating and non-toxic, it is commonly used as a medicinal addition [41]. Huang et al. [42] developed intravenous hydrogels with potent angiogenic activity made of aldehyde-ragulated xanthan and carboxymethyl-regulated chitosan to improve abdominal wall regeneration. The digestive system and open sores were among the tissues where the hydrogel showed releasing properties the most. The VEGF-containing hydrogel increased both angiogenesis activity and abdominal wall regeneration.

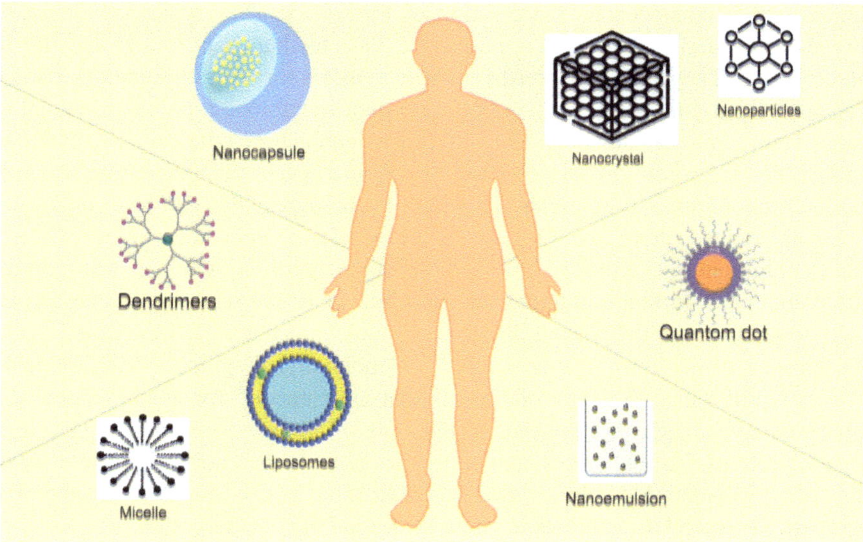

Fig. 2 An example of the nanocarriers utilised in sophisticated drug delivery systems

6 Nanotechnology-Based Medication Delivery Methods

6.1 Polymeric Micelles

Amphiphilic block copolymers are used to create polymeric micelles, which coalesce into a centred-shell configuration in watery solutions. The aquaphillic shell may stabilize the hydrophobic core, which may be loaded with hydrophobic medicines (such as docetaxel, camptothecin, and paclitaxel), making the entire system soluble in water. The Enhanced permeability and retention (EPR) effect allows polymeric micelles, which typically involve a size of less than 100 nm and a limited dissemination to slow rapid renal clearance, to accumulate in cancer tissues. Their polymeric shell also protects them from accidental interaction with biological components. Since their inner core design allows for the absorption of this type of medication, these nanostructures (Fig. 2) hold great potential for the delivery of hydrophobic drugs, which will boost durability and bioavailability [43, 44].

6.2 Liposomes

Alec Bangham discovered them for the first time in 1960. Liposomes, one of the most thoroughly researched medicament delivery carrier systems, are employed in the health and beauty industries to transport a range of chemicals. Drug delivery has long been enhanced by the creation of liposomes. They typically consist of phospholipids and steroids and are spherical vesicles between 50 and 450 nm in diameter [45]. They are considered as superior vehicles for drug administration since it is easy to put drugs into them and because their membrane topology identical to that of membranes of cells [45]. They have also demonstrated the ability to stabilize therapeutic molecules, enhance biodistribution, and permit the use of both aqueous and hydrophobic pharmaceuticals. They are also biodegradable and biocompatible.

For liposomes, there are four groups: (1) Ordinary liposomes: they feature a water-containing core wrapped in a cholesterol/phospholipid which can be neutral, cationic/anionic. In that condition, the lipid bilayer or the water-filled gap can be filled, respectively, with hydrophilic or hydrophobic molecules. (2) PEGylated types: The surfaces of the liposomes are coated with polyethylene glycol (PEG) to create a steric equilibrium. (3) Type of ligands that are intended for attachment to the outermost layer of the liposome or to the ends of already connected PEG chains include proteins, carbohydrates, and antibodies and (4) theranostic liposome form: this type of liposome combines the preceding three and frequently consists of a nanoparticle as well as elements for treatment, imaging, and targeting [46].

6.3 Dendrimers

Dendrimers are three-dimensional, established, monodisperse formations that are strongly branched. These molecular structures would be great drug delivery vehicles because of their globular form and ease because the outside may be customized in a regulated manner. There are two ways to create dendrimers synthetically: The initial one is a separate path, where the dendrimer forms in its centre before expanding outward, while the subsequent one is a converging path that begins on the dendrimer's exterior [47]. Drug is mostly transported by dendrimers using one of two methods: either the in vivo breakdown of the medication's dendrimer's covalent attachment due to the presence of appropriate enzymes or favourable conditions that could break down the bonds, or the discharge of the pharmaceutical as a result of environmental changes such as temperature, pH, etc. [48]. For transdermal, oral, pulmonary, ophthalmic, and selective administration of medicines, dendrimers have been created [49]. It was demonstrated that doxorubicin-folate linked poly-l-lysine dendrimers raised the amount of doxorubicin in the tumour by 121.5-fold following 24 h when compared to permitted doxorubicin, according to Jain et al.'s [50] description of the pharmaceutical carrier system for preventing cancer drug release, pH reliant pharmaceutical release, antiangiogenic, target particularity, and anticancer prospective.

6.4 Nanocrystals

Pure solid pharmaceutical nanocrystals have a size between 1000 and 2000 nm. The majority of the time, they are stabilized by polymeric steric stabilizers or surfactants and are entirely medication lacking any carriers molecules connected to it. Typically, adding a surfactant agent referred to as nano-suspension will improve a nanocrystals dispersion in a problematic liquid media. The predominant medium for dispersion in this scenario is water, but it might also be any other aquatic or non-aqueous medium-sized, such as liquefied polyethylene glycol and oils [51]. Nanocrystals display unique properties by overcoming problems such as increased maximum solubility, increased dissolving rate, and enhanced glueyness to surface/ cell membranes. Chitosan microparticles and cinaciguat nanocrystals were combined by Ni et al. [52] to transport the hydrophobic medication to the lungs. The swelling and muco-adhesive properties of the polymer were used in the creation of the nanoparticles to maintain the release of drugs. They found that inhalation efficiency could be diminished in illness-related conditions, thus more study is needed to show the technique's more substantial advantages.

6.5 Nanoparticles

These essential pharmacological chemicals can be adsorbed or bonded to surfaces and are found in submicron-sized polymer colloidal particles. Nanoparticles can be guided to precise locations thanks to surface modifications that enable biochemical contact with certain receptors on cells of interest [53]. The capacity of nanoparticles to deliver medications to the intended region despite navigating a variety of physiological barriers, including the blood–brain barrier, is an additional essential function of nanoparticles. This intravenous injection approach coats the drug-loaded nanoparticles using polysorbates to help them cross the blood–brain barrier [54]. This makes brain scapegoating conceivable. In order to create the needed nanoparticles coupled to epithelial growth factor antibodies, Acharya et al. [55] created immunonanoparticles with improved efficacy compared to MCF-7 breast malignant cell line by adding a centre localization pattern to the surfaces of the nanoparticles, which precisely directs the medication to the centre of breast tumour cells.

6.6 Metallic Nanoparticles

The use of nanoparticles of metal in several fields of medicine, including biosensors, bioimaging, hyperthermia, target/sustained delivery of drugs, and photoablation treatment, has gained popularity recently [56]. Additionally, these nanoparticles have been altered and functionalized with certain functional groups that enable them to link to antibodies, medicines, and additional ligands, giving these structures enhanced viability for use in healthcare research [57]. While gold, iron, silver, and copper are the most commonly investigated metallic nanoparticles, growing curiosity has been shown in other types of metallic nanoparticles, including, gadolinium, titanium oxide, platinum, zinc oxide, selenium, cerium dioxide, and palladium as well [57].

6.7 Quantum Dots (QDs)

Luminescent inorganic nanoparticles of semiconductors with dimensions of 2–10 nm are known as quantum dots [58]. They can precisely manipulate their absorption and emission properties because of their superb size and shape management. They have been widely investigated for photographic use in real-world environments [58] and are robust for months without degradation or modification. To facilitate precision targeting, particular ligands have been introduced to QDs with the intention of identifying cancer cells [59]. They will thus undoubtedly be chosen as long-term, extremely sensitive, and multiple contrasting testing methods employed for in vivo cancer identification and evaluation [60]. According to Li et al.'s [61] investigation into the glutathione-carried release of vital plasmid DNA using energy from pragmative CdTe quantum dots, similar QDs may be employed to specifically and clearly discharge load in living cells [61].

7 Future of Medication Delivery and Biomedicine Using Nanostructures

One of the most fascinating fields of research nowadays is nano-biomedicine. In the past 20 years, 1500 applications for patents are currently pending in this subject, and multiple clinical investigations have also been finished [62]. As mentioned in the various sections above, cancer seems to be the most prominent example of an illness that's where non-medical innovation has benefited in both detection and therapy. The application of nano-biomedicine and nano-drug techniques for delivery is unquestionably an expanding area that is going to be the coming field of investigation and advancement for a while in the years to come. This is because it delivers the right amount of drugs to the influenced cells, including the cancer/tumor cells, without hindering the bodily processes of cells that are healthy.

The sizes of the many nanoparticle examples shown in this message range from those that are correctly expressed by nanometers to those that have measurements in sub-micrometers (more than 100 nm). The next phase of research would concentrate on biomaterials with enhanced uniformity medicaments loading, and release capabilities. The use of metal-based nanoparticles for medicinal applications has advanced significantly in the years addressed by the present research. Future research into the utilization of such metals, having gold and silver, for both therapeutics and diagnostics may increase the usage of nano-biomedicines. Gold nanoparticles, which seem to be readily absorbed in delicate cancer cells and render the tumour susceptible to radiation-carried thermal treatment (for instance, in the near-infrared range), are among the primary sources of fascination in this research. The true influence of nano-biomedicine and nano-drug methods of administration on the medical field, especially in the treatment and detection of cancer, is still quite limited, notwithstanding the widespread knowledge of their future potential. This is due to the topic being a young one in scientific findings, with just two decades of actual study, and the fact that many important, basic characteristics are still unexplained. The primary focus of the upcoming study will be on the fundamental molecular indicators of diseased tissues, especially the ones that allow for precision targeting despite compromising regular functioning of cells. Eventually, the utilization of nano-biomedicine will proceed alongside our growing comprehension of illness at the atomic level or those that represent a nanomaterial-subcellular scale corresponding biomarker detection thereby opening up new avenues for therapy and diagnosis. Therefore, developing nanomedicine uses in the years to come will require knowledge of the molecular fingerprints of illness. For more widespread applications of nano-biomedicine beyond the ones we have discussed in this chapter using popular nanostructures, more research would be required.

Pharmaceutical activity in tissues/cellular erect, technical advancements for analysing these events, and the idea of regulated administration of specific medications at the afflicted regions are still distant from its full potential. The majority of investigation in the subject of nano-biomedicine is concentrated on biomaterial and compositional studies, which appear to represent the infancy of biomedicine applications. Valuable knowledge that might be employed in medical treatment and diagnostic studies will be obtained through time- and money-consuming transdisciplinary research and experiments on animals. A growing worldwide movement is the pursuit of more precise medical diagnosis and therapies, and the potential of nano-biomedicine and nano-drug delivery systems seems bright.

The creation of nanostructures that serve as tissue diagnostic and repair mechanisms with complete control from outside characteristics has garnered a lot of attention. This is still hypothetical research that hasn't yet materialized but that humans could accomplish in a few years. However, just as for their benefits, any drawbacks of nanomedicine need to be carefully examined for both people and the environment in its entirety. As a result, a careful analysis of the possible short-term or long-term negative impacts of new nanomaterials on humans and the natural world is necessary. Nano-biomedicines' affordable cost adds another area for research that needs more contributions as they gain in popularity. The regulation of nano-biomedicines, which was covered in greater detail in the chapter above, will advance along with novel applications for these substances.

8 Conclusion

The most recent advances in nano-biomedicine are highlighted in the present chapter, including innovative methods for diagnosis as well as technological advances in the delivery of both new and old drugs. There are several reported nano-dimensional entities, particularly nanostructures. These molecules can be utilized in practical systems for pharmacological activation, sensing, precise delivery to targets, and diagnostics. Initially, the main application of nanotechnology was to enhance the bioavailability, controlled administration, and dissolution of pharmaceuticals. The application of nanotechnology to boost the therapeutic value of by now-known natural bioactive chemicals has become common practise, yet the creation of nanopharmaceuticals is plagued in ambiguity and searching for compounds with pharmacological activity naturally occurring is not as common compared to it had been fifty years ago. The medical efficacy of these organic compounds has been greatly enhanced by the use of nanocarriers created with solid lipid nanoparticles, micelles, liposomes as crystal nanoparticles, and dendrimers along with gold, cadmium sulphide, silver, and titanium dioxide polymeric nanoparticles.

Natural, new biomaterials continue to be in demand due to their biodegradability, biocompatibility, accessibility, potential to regenerate, and low toxicity. Beyond merely acknowledging such proteins and polysaccharides as organic biopolymers, one of the most cutting-edge research topics at the moment is on enhancing their resilience in their presence of an alive platform and production circumstances. Also widely used are polymer nanoparticles (nanocapsules and nanospheres) produced via emulsion polymerization, solvent evaporation, and surfactant-free emulsion polymerization. In the past several years, as nano-biomedicine has expanded, a lot of focus has been placed on the interconnected of therapy and identification (theranostic), that employs tumours as a disease model.

Examples include the employing of cathepsin B to identify propagate, fluorogenic protein devices connected to glycol-based chitosan the nanoparticles, photodynamic colon cancer recognition via folic and alginate acid-mixed chitosan nanoparticles, the use of iron oxide covered biopolymeric (hyaluronic acid) substances in carcinoma treatment, and dextran. Since the 1990s, a staggering number of nanotechnology-based products after approval of FDA as study projects have been developed, which include polymer synthetic particles, liposome compositions, nanocrystals, micellar particles, protein's nanostructures, and many more. When paired with medications or biological agents, these things are frequently used. Even though nano-biomedicine has already significantly changed how we locate and use medications in biological systems, the main focus of subsequent studies will be on evaluating the safety/toxicity of nano-biomedicines and their approval procedures. We can now identify illnesses and, in certain cases, connect a diagnosis to a treatment. Advances in nano-biomedicine are to blame for this.

Acknowledgements Mr. Tadbeer Ahemad (ShriRam College of Pharmacy, Banmore, Morena) and Prof. Kirti Tiwari were especially helpful to the writer (Department of Mathematics, Bal Gopal Academy, Morena, Madhya Pradesh, India).

Funding There was no funding for this project from any source.

References

1. Swamy, M.K., Sinniah, U.R.: Patchouli (Pogostemon cablin Benth.): botany, agrotechnology and biotechnological aspects. Ind Crops Prod. **87**, 161–76 (2016)
2. Mohanty, S.K., Swamy, M.K., Sinniah, U.R., Anuradha, M.: Leptadenia reticulata (Retz.) Wight & Arn. (Jivanti): botanical, agronomical, phytochemical, pharmacological, and biotechnological aspects. Molecules **22**, 1019 (2017)
3. Rodrigues, T., Reker, D., Schneider, P., Schneider, G.: Counting on natural products for drug design. Nat. Chem. **8**, 531 (2016)
4. Siddiqui, A.A., Iram, F., Siddiqui, S., Sahu, K.: Role of natural products in drug discovery process. Int. J. Drug Dev. Res. **6**(2), 172–204 (2014)
5. Beutler, J.A.: Natural products as a foundation for drug discovery. Curr. Prot. Pharmacol. **46**(1), 9–11 (2009)
6. Thilakarathna, S.H., Rupasinghe, H.: Flavonoid bioavailability and attempts for bioavailability enhancement. Nutrients **5**, 3367–3387 (2013)
7. Bonifacio, B.V., da Silva, P.B., dos Santos Ramos, M.A., Negri, K.M.S., Bauab, T.M., Chorilli, M.: Nanotechnology-based drug delivery systems and herbal medicines: a review. Int. J. Nanomed **9**, 1 (2014)
8. Watkins, R., Wu, L., Zhang, C., Davis, R.M., Xu, B.: Natural product-based nanomedicine: recent advances and issues. Int. J. Nanomed. **10**, 6055 (2015)
9. Martinho, N., Damgé, C., Reis, C.P.: Recent advances in drug delivery systems. J. Biomater. Nanobiotechnol. **2**, 510 (2011)
10. Jahangirian, H., Lemraski, E.G., Webster, T.J., Rafiee-Moghaddam, R., Abdollahi, Y.A.P.: Review of drug delivery systems based on nanotechnology and green chemistry: green nanomedicine. Int. J. Nanomed. **12**, 2957 (2017)
11. Sharma, P., Tailang, M.: In vivo study of orodispersible tablet of primaquine. Int. J. Pharm. Sci. Res. **9**(8), 3506–3510 (2018)
12. Sharma, P., Bhargava, S., Parashar, D., Mangal, A.: Formulation and evaluation of nanoparticles containing cyclophosphamide. WJPR **7**(1), 785–798 (2017)
13. Park, S.H., Oh, S.G., Mun, J.Y., Han, S.S.: Loading of gold nanoparticles inside the DPPC bilayers of liposome and their effects on membrane fluidities. Colloid Surf. B **48**, 112–118 (2006)
14. Mirza, A.Z., Siddiqui, F.A.: Nanomedicine and drug delivery: a mini review. Int. Nano Lett. **4**, 94 (2014)
15. Wang, N., Feng, Y.: Elaborating the role of natural products-induced autophagy in cancer treatment: achievements and artifacts in the state of the art. BioMed. Res. Int. **2015**, 934207 (2015)
16. Abdelwahab, S.I., Sheikh, B.Y., Taha, M.M.E., How, C.W., Abdullah, R., Yagoub, U., El-Sunousi, R., Eid, E.E.: Thymoquinone-loaded nanostructured lipid carriers: preparation, gastroprotection, *in vitro* toxicity, and pharmacokinetic properties after extravascular administration. Int. J. Nanomed. **8**, 2163 (2013)
17. Schneider, C., Langer, R., Loveday, D., Hair, D.: Applications of ethylene vinyl acetate copolymers (EVA) in drug delivery systems. J. Controlled Release **262**, 284 (2017)
18. Langer, R.S., Peppas, N.A.: Present and future applications of biomaterials in controlled drug delivery systems. Biomaterials **2**(4), 201–214 (1981)
19. Sharma, P., Jain, V., Tailang, M.: Selection and role of polymers for designing of a drug carrier. Drug Carriers. IntechOpen, London (2022). https://doi.org/10.5772/intechopen.103125
20. Sharma, P., Jain, V., Tailang, M.: Advancement of nanocarrier-based engineering for specific drug delivery for cancer therapy. In: Targeted cancer therapy in biomedical engineering. Singapore: Springer Nature Singapore, pp. 465–486 (2023)

21. Kamaly, N., Xiao, Z., Valencia, P.M., Radovic-Moreno, A.F., Farokhzad, O.C.: Targeted polymeric therapeutic nanoparticles: design, development and clinical translation. Chem. Soc. Rev. **41**(7), 2971–3010 (2012)
22. Joshi, R., Raje, S., Akram, W., Garud, N.: Particle engineering of fenofibrate for advanced drug delivery system. FJPS **5**, 1–1 (2019)
23. Saadeh, Y., Vyas, D.: Nanorobotic applications in medicine: current proposals and designs. Am. J. Robot. Surg. **1**, 4–11 (2014)
24. Sharma, P., Tailang, M.: Opportunities and challenges to develop transdermal patch with nanocarrier. WJPR **6**(15), 392–400 (2017)
25. Lu, H., Wang, J., Wang, T., Zhong, J., Bao, Y., Hao, H.: Recent progress on nanostructures for drug delivery applications. J. Nanomater. **2016**, 20 (2016)
26. Kumari, A., Kumar, V., Yadav, S.: Nanotechnology: a tool to enhance therapeutic values of natural plant products. Trends Med. Res. **7**, 34–42 (2012)
27. Pankaj, S., Mukul, T.: PH dependent release potential of natural polymers in sustained release of ornidazole from colon targeted delivery system. Magnesium **10**(10), 10 (2019)
28. Nie, S.: Understanding and overcoming major barriers in cancer nanomedicine. Nanomedicine (Lond.) **5**, 523–528 (2010)
29. Sharma, P., Tailang, M.: Design, optimization and evaluation of buccal drug delivery system of propranolol for hypertension treatment. Int. J. Pharm. Sci. Res. **11**(1), 301–311 (2020)
30. Maeda, H., Nakamura, H., Fang, J.: The EPR effect for macromolecular drug delivery to solid tumors: improvement of tumor uptake, lowering of systemic toxicity, and distinct tumor imaging *in vivo*. Adv. Drug Deliv. Rev. **65**, 71–79 (2013)
31. Jain, R.K., Stylianopoulos, T.: Delivering nanomedicine to solid tumors. Nat. Rev. Clin. Oncol. **7**, 653–664 (2010)
32. Sharma, P., Tailang, M.: Design, optimization, and evaluation of hydrogel of primaquine loaded nanoemulsion for malaria therapy. FJPS **6**, 1–1 (2020)
33. Bertrand, N., Wu, J., Xu, X., Kamaly, N., Farokhzad, O.C.: Cancer nanotechnology: the impact of passive and active targeting in the era of modern cancer biology. Adv. Drug Deliv. Rev. **66**, 2–25 (2014)
34. Bao, G., Mitragotri, S., Tong, S.: Multifunctional nanoparticles for drug delivery and molecular imaging. Annu. Rev. Biomed. Eng. **15**, 253–282 (2013)
35. Sosnik, A.: Alginate particles as platform for drug delivery by the oral route: state-of-the-art. ISRN Pharm. **2014**, 926157 (2014)
36. Patil, N.H., Devarajan, P.V.: Insulin-loaded alginic acid nanoparticles for sublingual delivery. Drug Deliv. **23**, 429–436 (2016)
37. Roman, J.V., Galan, M.A., del Valle, E.M.M.: Preparation and preliminary evaluation of alginate crosslinked microcapsules as potential drug delivery system (DDS) for human lung cancer therapy. Biomed. Phys. Eng. Expr. **22**, 035015 (2016)
38. Jain, A., Jain, S.K.: Optimization of chitosan nanoparticles for colon tumors using experimental design methodology. Artif. Cells Nanomed. Biotechnol. **44**, 1917–1926 (2016)
39. Sun, B., Zhang, M., Shen, J., He, Z., Fatehi, P., Ni, Y.: Applications of cellulosebased materials in sustained drug delivery systems. Curr. Med. Chem. **26**(14), 2485–2501 (2017). https://doi.org/10.2174/0929867324666170705143308
40. Agarwal, T., Narayana, S.G.H., Pal, K., Pramanik, K., Giri, S., Banerjee, I.: Calcium alginate-carboxymethyl cellulose beads for colon-targeted drug delivery. Int. J. Biol. Macromol. **75**, 409–417 (2015)
41. Goswami, S., Naik, S.: Natural gums and its pharmaceutical application. J. Sci. Innov. Res. **3**, 112–121 (2014)
42. Huang, J., Deng, Y., Ren, J., Chen, G., Wang, G., Wang, F., Wu, X.: Novel in situ forming hydrogel based on xanthan and chitosan re-gelifying in liquids for local drug delivery. Carbohydr. Polym. **186**, 54–63 (2018)
43. Miyata, K., Christie, R.J., Kataoka, K.: Polymeric micelles for nano-scale drug delivery. React. Funct. Polym. **71**, 227–234 (2011)

44. Xu, W., Ling, P., Zhang, T.: Polymeric micelles, a promising drug delivery system to enhance bioavailability of poorly water-soluble drugs. J Drug Deliv. **2013**, 340315 (2013)
45. Bozzuto, G., Molinari, A.: Liposomes as nanomedical devices. Int. J. Nanomed. **10**, 975 (2015)
46. Sercombe, L., Veerati, T., Moheimani, F., Wu, S.Y., Sood, A.K., Hua, S.: Advances and challenges of liposome assisted drug delivery. Front Pharm **6**, 286 (2015)
47. Madaan, K., Kumar, S., Poonia, N., Lather, V., Pandita, D.: Dendrimers in drug delivery and targeting: drug-dendrimer interactions and toxicity issues. J. Pharm. Bioallied Sci. **6**, 139 (2014)
48. Tripathy, S., Das, M.: Dendrimers and their applications as novel drug delivery carriers. J. Appl. Pharm. Sci. **3**, 142–149 (2013)
49. Kesharwani, P., Jain, K., Jain, N.K.: Dendrimer as nanocarrier for drug delivery. Progr. Polym. Sci. **39**, 268–307 (2014)
50. Jain, K., Gupta, U., Jain, N.K.: Dendronized nanoconjugates of lysine and folate for treatment of cancer. Eur. J. Pharm. Biopharm. **87**, 500–509 (2014)
51. Du, J., Li, X., Zhao, H., Zhou, Y., Wang, L., Tian, S., Wang, Y.: Nanosuspensions of poorly water-soluble drugs prepared by bottom-up technologies. Int. J. Pharm. **495**, 738–749 (2015)
52. Ni, R., Zhao, J., Liu, Q., Liang, Z., Muenster, U., Mao, S.: Nanocrystals embedded in chitosan-based respirable swellable microparticles as dry powder for sustained pulmonary drug delivery. Eur. J. Pharm. Sci. **99**, 137–146 (2017)
53. Misra, R., Sahoo, S.K.: Intracellular trafficking of nuclear localization signal conjugated nanoparticles for cancer therapy. Eur. J. Pharm. Sci. **39**, 152–163 (2010)
54. Lanjhiyana, S.K.: Polysaccharides based novel and controlled released multiparticulate systems for colon-specific delivery: Contemporary scenario and future prospects. AJP **14**(4) (2020)
55. Acharya, S., et al.: Targeted epidermal growth factor receptor nanoparticle bioconjugates for breast cancer therapy. Biomaterials **30**, 5737–5750 (2009)
56. McNamara, K., Tofail, S.A.: Nanoparticles in biomedical applications. Adv. Phys. **2**, 54–88 (2017)
57. Kudr, J., et al.: Magnetic nanoparticles: from design and synthesis to real world applications. Nanomaterials **7**, 243 (2017)
58. Cai, W., Chen, X.: Nanoplatforms for targeted molecular imaging in living subjects. Small **3**(11), 1840–1854 (2007)
59. Kim, K.Y.: Nanotechnology platforms and physiological challenges for cancer therapeutics. Nanomed.: Nanotechnol. Biol. Med. **3**(2), 103–110 (2007)
60. Morrow Jr.K.J., Bawa, R., Wei, C.: Recent advances in basic and clinical nanomedicine. Med. Clin. North Am. **91**(5), 805–843 (2007)
61. Li, D., Li, G.P., Guo, W., Li, P., Wang, E., Wang, J.: Glutathione-mediated release of functional plasmid DNA from positively charged quantum dots. Biomaterials **29**(18), 2776–2782 (2008)
62. Pandit, A., Zeugolis, D.I.: Twenty-five years of nano-bio-materials: have we revolutionized healthcare? Future Med. **11**(9), 985–987 (2016)

Dr. Pankaj Sharma completed his B. Pharm and M. Pharm from RGPV University Bhopal in 2010 and 2012, respectively. He earned a Ph.D. in the year 2020. He has a total experience of 11 years in teaching and research. Presently, he is working as a Professor at the Department of Pharmaceutics, ShriRam College of Pharmacy, Banmore, Morena. His research interests include development of transdermal preparations, nanoemulsions, tablets, and other carrier mediated systems for effective delivery of drugs. He is actively involved in formulating nanocarrier for effective delivery of drugs for treatment of Malaria, Arthritis, etc. Dr. Sharma has received many awards; a few that stand out are the Young Scientist Award from the Indian Pharmaceutical Education Society, the Fellowship from the All India Council for Technical Education (AICTE), New Delhi, as well as several prizes for the best conference papers presented. Dr. Sharma is a graduate of the Indian Pharmacy Association and Jiwaji University in Gwalior, India. Dr. Pankaj Sharma spoke as a representative or an invited speaker at several conferences, seminars, symposiums, and workshops on his research and teaching experiences. He participates actively in a number of university, department, and institute committees. Prof. Sharma has guided 20 M. Pharm students under his supervision. In national and international publications, he has produced 26 peer-reviewed papers in the field of pharmaceutical sciences, along with 10 book chapters. Also, he reviews articles for a number of international scientific publications. He belongs to numerous Indian scientific organisations as a member or life member.

Advancement in Biomaterials in the Form of Implants

Riya Shivgotra, Bindu Soni, Manjot Kaur, and Shubham Thakur

Abstract Biomaterials are materials utilized to replace a human body part that serves the same function as the original or restore the function of damaged or degenerated tissues and organs. Implants such as sutures, bone plates, joint replacements, ligaments, vascular grafts, heart valves, intraocular lenses, and dental implants have overtaken alternative therapy approaches and entered the mainstream of dental care over the past ten years. One of the primary requirements for a biomaterial to be suitable for medical applications or implantation is that it should be nontoxic, not cause any immune response, and be chemically stable, biocompatible, and well-tolerated by the human body. Different materials like metals, ceramics, and polymers, are used for the manufacturing of implants, depending on whether a permanent or temporary implant is needed. However, they exhibit a number of disadvantages, including toxicity, a lack of mechanical stability, and processing complexity. Numerous biomaterials have been discovered, but most implanted biomaterials now in use cause either acute or chronic inflammatory reactions inside the body. Nanotechnology has aided in developing an entirely novel implant material with enhanced efficacy, low cost, and a significant surface-to-volume ratio. Modifying the surfaces of present implants to reduce the body's reaction and enhance the natural healing of wounds is also one strategy for developing such healing biomaterials. This chapter covers the advancements in biomaterials used for implants, including various techniques for physically and chemically modifying the surfaces, as well as the use of 3D printing and nanotechnology.

Keywords Biomedical implants · Biocompatibility · Inflammation · Nanotechnology · Surface modified biomaterials

R. Shivgotra · B. Soni · M. Kaur · S. Thakur (✉)
Department of Pharmaceutical Sciences, Guru Nanak Dev University, Amritsar, Punjab 143005, India
e-mail: shubhamdthakur@gmail.com; shubhamdpharma.rsh@gndu.ac.in

R. Malviya and S. Sundram (eds.), *Engineered Biomaterials*, Engineering Materials,
https://doi.org/10.1007/978-981-99-6698-1_10

1 Introduction

Contributions from scientists and engineers are resulting in significant new solutions for critical medical issues in a scientific world that is changing rapidly. Diabetes, osteoporosis, asthma, cardiac issues, cancer, and other critical disorders are no longer solely treated with conventional pharmaceutical formulations [1]. Novel possibilities for treating and curing disease are constantly being opened up and it is now understood that a drug's therapeutic value and potency are not directly linked together; rather, they are linked with the process of developing drug formulation and absorption into the body. As the biotechnology sector advances with the emergence of novel categories of biopharmaceuticals, there is a crucial need to enhance our fundamental understanding of the influence of drug delivery methods on safety and efficacy. Additionally, exploring innovative delivery methods is essential to address the unique challenges presented by these new biopharmaceuticals. Despite advances in the field, delivering drugs effectively still presents a significant challenge, owing to our incomplete knowledge of the biological obstacles that impede the delivery of drugs. Because of these unresolved necessities and constraints, significant research efforts have been directed toward designing, implementing, and translating biomaterials for drug delivery [2]. The utility and significance of metallic materials and their composites have been considered since the early nineteenth century. Over the last 2–3 centuries, a lot of advancement has been made in the synthetic materials field to gain insight into how biomaterial interacts with the human body and their effective use [3]. Surgical repair is frequently required as a consequence of trauma, degradation, and diseases. Biomaterials are synthetic substitutes that can replace various skeletal components, including the joints of the knees, hips, fingers, and elbows, vertebrae, teeth. These substances are designed to perform the same functions as the living materials they replace [4].

A biomaterial is generally defined as "a nonviable material used in a medical device, intended to interact with biological systems." If we remove the phrase "used in a medical device," this definition becomes more comprehensive of the extensive range of applications that involve the integration of biology with synthetic and altered natural materials. The different biological functions of human tissue are replaced or repaired using biomaterials, developed by humans and constantly exposed to bodily fluids that facilitate daily human activities [3]. Biomaterials are materials, whether natural or synthetic, that come into contact with biological fluids, blood, and tissue. Their design aims to ensure that they do not have any negative impact on the living organism or its components while serving various purposes such as prosthetic, therapeutic, and diagnostic applications. Distinguishing between the terms "Biomaterial" and "Biological material" is crucial. A biological material refers to a material that has been created by a biological system, such as the epidermis or an artery. In contrast, biomaterials are materials that have been engineered to interact with biological systems, often used in medical or scientific applications [5].

The term "bioimplants" refers to engineered medical devices developed to replace biological structural human body parts that are malfunctioning or damaged and support the given host [6]. In the past ten years, biomaterials have attracted greater attention as an option for improving and even saving the lives of numerous patients. In the United States alone, over 13 million medical devices are implanted each year [7]. The need for mechanical implantable devices is growing daily as a result of abrupt shifts in the global population's age distribution. The proportion of people over the age of 60 in the population is dramatically growing, which has increased consumer interest in artificial implants. By 2050, there is expected to be a 39% increase in the demand for medical implants [4]. The medical industry has extensively explored and created a diverse range of implants and devices for numerous applications within the human body. The main objective of these devices is to safeguard human lives. This encompasses both the technologies that augment the performance of natural human organs, like pacemakers, as well as artificial devices that offer physical assistance, such as knee replacements and synthetic blood vessels [8]. Figure 1 depicts the ideal properties of the biomaterials that biomaterials should possess for their effective use in various medical implant applications within the human body.

Considering all the progress made so far in material and biological research, we have come a long way since the early Egyptians and South-Central American cultures made use of implants. In the modern era, the quality and as well as availability of implantable material have significantly improved, leading to the development and widespread use of this promising therapeutic modality. Tooth implants have a long

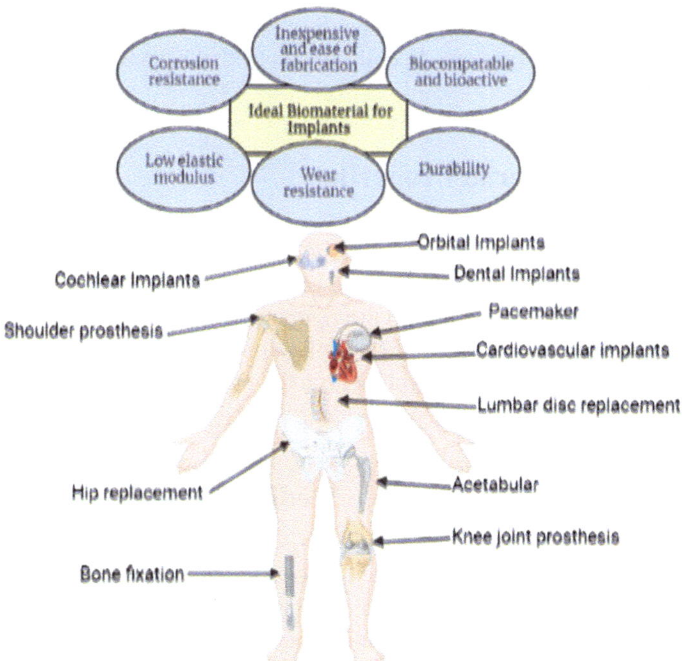

Fig. 1 Desirable properties of biomaterials for effective use in medical implants within the human body

history, dating back to ancient China and Egypt where they were made of stone and ivory. During the sixteenth and seventeenth centuries, dental implants frequently utilized gold and ivory as their primary materials. In the early twentieth century, implants made up of metals such as gold, lead, tantalum, stainless steel, and cobalt alloy were introduced as options for dental implant materials [9].

Biomedical implants, in general, involve a wide range of medical therapies for various health issues, such as medical devices that can be implanted in the cardiovascular system including stents, vascular grafts, heart valves, defibrillators, pacemakers, and other related devices. Neural implants are another type of implantable medical device, which can be used in different parts of the nervous system [10]. The "foreignness" of these biomaterials is a major obstacle for the current implant technology which should be overcome to enhance host tolerance, achieve the required release of drugs either in a specific location or over a particular time period, and minimize reactivity to foreign bodies [11]. As immunologists continue to gain insight into immune and foreign body reactions, the production of biocompatible materials with excellent performance will become progressively significant for the manufacturing of devices that are implanted for the controlled release of drugs, cell-based treatments, implantable detectors, and tissue engineering [2].

In this chapter, the biological characteristics of different biomaterials utilized in implant manufacturing—which involves metals, polymeric materials, and titanium alloys—are reviewed. With the objective to enhance the properties of metal implants, the chapter also discusses the essential novel advancement in techniques like surface modifications, 3D printing, and nanotechnology, utilized in their fabrication. Applications of medical implants in daily life have also been discussed.

2 Implant Biomaterial Properties

To create a biomaterial that is both safe and dependable for use in implants, it is necessary to identify an appropriate blend of mechanical, chemical, physical, and biological features that can withstand the rigors of long-term use without requiring further revision surgeries. Table 1 outlines key properties that a biomaterial must possess in order to function effectively within the human body [9, 12, 13].

3 Classification of Implantable Medical Material

The material utilized to produce implants possesses distinctive characteristics that make them the most suitable choice for the intended use. The three main criteria regarding selecting materials are (a) biocompatibility with the human body or the ability to integrate with the body without having an adverse effect, (b) mechanical properties comparable with the bones and skull of a human, and (c) affordability.

Table 1 Key properties to consider for selecting biomaterials in implant applications

Bulk properties	
Modulus of elasticity	The modulus of elasticity is a term used to describe the stiffness or ability of a material to resist deformation when subjected to stress. High Young's modulus materials are less ductile and more rigid. It is important to select implants made from materials that have an elastic modulus similar to bone (18 GPa) in order to produce a consistent load distribution across the implant and reduce relative motion at the implant-bone contact
Biocompatibility	The material's capability to not elicit a significant adverse reaction or immune response upon implantation in the body
Radiopacity	The material's ability to show up on X-rays or other imaging modalities is important for postoperative monitoring and assessment
Compressive, shear, and tensile strength	For applications involving bearing loads, it is crucial for the material to possess the capacity to endure mechanical stresses and loads without experiencing deformation or breakage
Yield strength, fatigue strength	Yield strength denotes the amount of stress a material can endure before it begins to deform plastically, while fatigue strength refers to the amount of cyclic loading a material can withstand before it fails due to fatigue crack propagation
Hardness and toughness	Hardness refers to the determination of a material's ability to withstand scratching or indentation. A harder material will be more resistant to deformation or wear, which is important for load-bearing applications. However, materials that are too hard can also be brittle and prone to fracture. Toughness is an estimation of a material's capacity to resist cracking or breaking under stress. However, materials that are too tough may be more difficult to machine or shape
Ductility	Ductility refers to a material's capability to undergo deformation under tensile stress without fracturing. This characteristic is particularly crucial to consider for biomaterials utilized in implants, especially in load-bearing scenarios like orthopedic implants
Surface properties	
Surface tension and surface energy	When considering biomaterials used in implants, surface tension plays a role in determining the material's wettability, which refers to its capacity to form a uniform and stable interface with liquids such as blood or other bodily fluids. Higher surface tensions result in lower wettability and decreased ability to interact with liquids Surface energy can influence the material's ability to adhere to tissues or to be recognized by the body's immune system. Materials with higher surface energies tend to be more biocompatible and promote better tissue integration

(continued)

Table 1 (continued)

Bulk properties	
Surface roughness	The surface roughness of a biomaterial can impact its biocompatibility, osseointegration, and wear resistance, among other factors. By enlarging the implant's surface area in contact with the bone and modifying the surface's roughness, the adherence of cells to the bone is improved, thereby influencing the reaction of cells and tissues. Different elements, including roughness, texture, and the orientation of abnormalities, have been used to categorize implant surfaces
Chemical properties	
Pitting corrosion	When an implant containing small holes or pits on the material's surface makes contact with a solution, the metal ions in the area dissolve, eliminating their positive charges in combination with chlorine ions resulting in pitting corrosion causing tiny pits or holes to develop on the material's surface., which can compromise its mechanical characteristics and increase the risk of implant failure
Crevices Corrosion	This happens at the junction connecting an implant screw with the bone. Crevice corrosion takes place by the positively charged environment produced when metallic ions dissolve
Galvanic corrosion	When two metals that are not compatible with one another come into touch with an electrolyte, such as physiological fluids, galvanic corrosion occurs
Electrochemical corrosion	It occurs when a material is exposed to an electrolyte (such as body fluids) and undergoes oxidation and reduction reactions that result in the discharge of metal ions and the deterioration of the material over a period of time

Additionally, it must be resistant to wear and corrosion in the human body's environment. The efficiency of materials employed for implants will be determined by these characteristics [9]. Several materials that can be employed as biomaterials for implant applications have been produced over the past few years considering the criteria outlined above. However, it might be hard for a single metallic substance to have all the required characteristics. The three main categories of biomaterials employed in biomedical purposes are ceramics, polymers, and metallic systems (Fig. 2) [6].

3.1 Ceramics

Among the materials used to design biomaterials, ceramics are one of the types. Examples of such materials include zirconium oxide, aluminum oxide, and bioactive glass which have higher yield strength, greater bioactivity, and osteointegration with bone, hydroxyapatite (HAp), along with different ceramics consisting of silica and calcium [5]. The inertness of ceramics within the body, their effortless molding into a range of forms and porosities, significant strength during compression, and

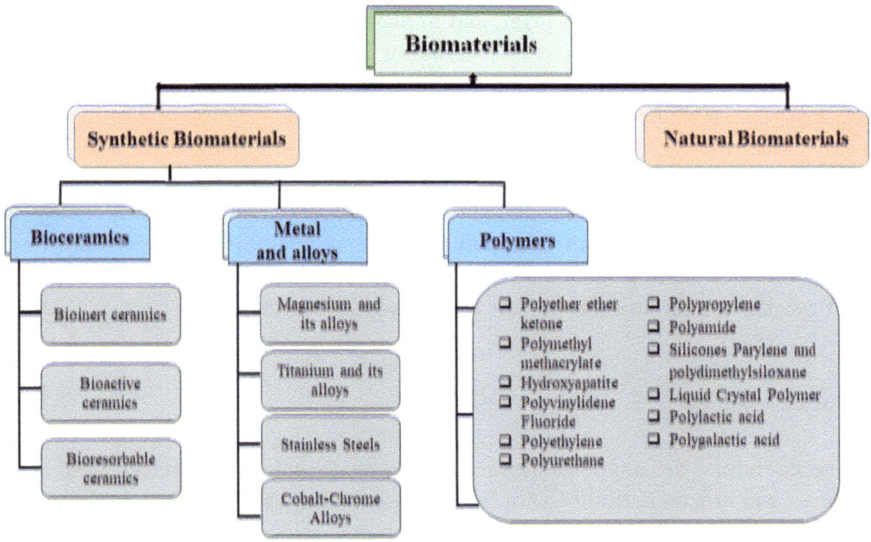

Fig. 2 A schematic representation of different types of biomaterials used in biomedical implants

superior wear properties all served as the rationale for their usage. Because of their superior robustness, resistance against biodegradation, and low electrical and thermal conductivity with a variety of diverse elastic properties, ceramics were first used for surgical implant devices [9]. Ceramics are also utilized for orbital and middle ear implants, bone grafts, dental and orthopedic implants, heart valves, coatings to make metallic implants more biocompatible, artificial knees, hip prostheses, bone transplants, and parts of the musculoskeletal system. For instance, higher Ca/P ratios in hydroxyapatite-based ceramics are preferred because of their chemical similarities between the hard tissues of teeth and bone. Ceramics are being used for bioimplant applications more frequently as a result of these desirable characteristics [6]. Ceramics are chemically inert, but because of their poor ductility and underlying brittleness, they must be handled carefully and replaced and their use has been restricted as a result [14].

3.1.1 Bioceramics

Ceramics are used to formulate biomaterials but are less frequently used than either metals or polymers. But because of their ideal biocompatibility and ability to integrate with bone, in addition to the fact that these materials are highly comparable to the mineral constituents of bones, bioceramics of phosphates are frequently employed for developing optimal biomaterials [5]. Bio-ceramics are biomaterials utilized as a cementing agent, supportive layer, and implant for repairing and replacing weakened skeletal system components. Bioceramics comprise about 50% of all biomaterial implants currently in use [15]. Bio-ceramics are a filler material used in orthopedic surgery to detect bone abnormalities. According to reports, "around 2 million bone grafting procedures, 280000 fractures of the hips, 700000 spinal tissues, and 250000 wrist fractures bone replacements have been performed throughout all over the world every year." Consequently, the demand for bio-ceramics is rising. Depending on the way they interact with the host bone, bio-ceramics may be divided into three groups: bio-inert ceramics, like alumina and zirconia, bioactive ceramics, including hydroxyapatite, and bioresorbable or biodegradable ceramics [16].

Bioinert Ceramics

Bioinert ceramics are a type of bioceramic materials that are characterized by their inertness and biocompatibility with living tissues. These ceramics are composed of materials such as alumina, zirconia, and other metal oxides, which do not react with biological systems and are not easily degraded by the body. Bioinert ceramics have numerous potential uses in the medical field because of their biocompatibility along with their ability to resist corrosion and wear and are commonly used in orthopedic implants as well as in bone grafts and other medical devices [17].

Aluminum oxide, commonly referred to as alumina, is a bioinert ceramic material extensively used for the production of orthopedic implants including hip and knee replacements [18]. Alumina is the best suitable material that can be used for implants that will be exposed to continuous stress and movement because of its superior mechanical properties, biocompatibility, and exceptionally high level of toughness and hardness. The ability of alumina to facilitate bone growth and attachment, which plays a crucial role in the implant's persistent efficacy, is the material's primary advantages when used in implants. Alumina implants also have a low rate of wear, which reduces the number of debris discharged into the body and lowers the chance of inflammation, implant loosening, and other challenges [6, 19]. Zirconia, commonly called zirconium oxide, is the other bioinert ceramic material widely employed in orthopedic implants, particularly dental implants. it can tolerate both higher stresses and high temperatures. Zirconia ceramics offer several advantages, such as high mechanical durability, biocompatibility, aesthetic appeal, and resistance to wear and chemical degradation. However, there are also some limitations associated with zirconia implants. They may be more prone to fractures since they are more brittle than conventional metal implants. Zirconia implants are also more expensive and require a more sophisticated manufacturing procedure in comparison to metal implants [17, 20]. Carbon in various forms, including diamond, glass oil, and graphite is biocompatible and does not trigger an immunological reaction. Regarding durability and elastic modulus, carbon is fragile and very similar to bones. As a result, it is biocompatible and appropriate to be used in coating medical equipment. Investigations and reports have been made on the usage of carbon in the manufacture of orthopedic and cardiovascular implants. As a bone covering for implants, pyrolyzed carbons are employed. Although it was initially employed for dental implants, the material's inadequate strength eventually led to its discontinuation. To enhance the rigidity and longevity of implants of knee joints, they tend to be reinforced with carbon fibers, while the implants that are subjected to bear heavy loads may utilize coatings of diamond to improve their lubricating properties [17, 21].

Bio-active Ceramics

Ceramics that are capable of interacting with living tissues and contribute to the process of healing and regeneration are referred to as bioactive ceramics. Because of their biocompatibility and capacity to adhere to bone tissue, these ceramics are frequently employed in medical applications such as bone grafts and dental implants. Materials like Hap, tricalcium phosphate, and bioactive glass are examples of bioactive ceramics. Bioactive ceramics may promote new bone formation and integration when used in medical implants, improving results and lowering the chance of implant failure [17]. Due to its remarkable Osseointegration and biocompatibility, HA attracted considerable amounts of interest [22]. Bioactive glass is composed of a silicone oxide network, along with certain network-modifying oxides such as magnesium oxide, and potassium oxides. The behavior of bio-glass ceramics, which contain approximately 60% silica is similar to that of inert ceramics [23]. It is possible to develop bioactive glasses in a variety of shapes, including microspheres and porous implants. Animals with bioactive glass have been found to have various bone formations. These possess advantageous properties, such as the tendency to chemically connect to the tissue, promote osseointegration, and exhibit resistance to bio-fouling. In maxillofacial surgery, dentistry, and orthopedics, bioactive silicate glasses are frequently utilized. While such materials are unsuitable for implants that are exposed to heavy loads due to their poor mechanical strength, they are often utilized to coat such implants for added benefits [24]. Glass ceramics possess acceptable thermal, mechanical, scratch, and abrasion resistance qualities. which makes them well-suited for use in orthopedic applications. The production process involves the use of metal alkoxides gel, which is heated at high temperatures to create amorphous glass. This glass is then subjected to further heating to induce crystal formation, and the resulting crystals are used in implants [6, 17].

Bioresorbable Ceramics

Bioresorbable ceramics are a type of material that can be absorbed by the body over time, eventually breaking down into natural components that can be safely metabolized or excreted. Bioresorbable ceramics demonstrate minimal reactivity with the tissues of the host body. Owing to their superior biocompatibility and chemical interactions, bioresorbable ceramics are frequently utilized in the fields of dental and orthopedic applications. Compared to conventional non-bioresorbable materials, these types of materials provide various advantages, including lower risk of long-term issues, better healing results, and the opportunity to avoid the need to have another surgery to remove the implant. However, bioresorbable ceramics have drawbacks as well, including decreased mechanical strength when compared to non-bioresorbable materials and possible health issues brought on by their degradation products [25].

3.2 Metals and Alloys

In the nineteenth century, during the Industrial Revolution, metal implants became very popular. There was an enormous demand for body tissue and bone regeneration during World War 1, which triggered a surge in the utilization of metal implants [26]. Several polymers along with composites are now been developed in order to meet the need for implants and advancements in technology, several different polymers and composites have now been created. need for implants and technological advancement. Due to their mechanical strength and biocompatibility, metals are still predominantly used as implantable materials. The inherent properties of these metals have been further enhanced via alloys. Only a few metals and metal alloys have been used in medical implants, including stainless steel (316L), titanium and alloys (Cp-Ti, Ti6Al4V), aluminum alloys, zirconium-niobium, and tungsten heavy alloys [5, 27]. Despite the development of several metals and alloys, some of them are harmful to humans and do have not much strength, making them incompatible with their use. Metal implants are used most frequently in the medical industry and are often employed as substitutes for hip, knee, plate, and dental implants as well as pins, dental materials, and screws. Also, numerous medical products made from metal, including craniofacial plates, and screws, components for artificial hearts, pacemakers, clips, valves, balloon catheters, bone fixation devices, and dental implants have been made possible by the field of biomaterials' rapid growth and development [28].

Biological conditions and implant functions determine the kind of metal that is employed in biomedical implants. The majority of implants, from cardiovascular to otorhinology, are composed of 316L-type stainless steel. Cobalt-Chromium-Molybdenum alloys are employed when the implant necessitates a high level of resistance to wear, such as in prosthetic joints [5]. Compared to ceramics and polymeric materials, the superior tensile strength and degradation limit of metals give manufacturers the ability for developing implants that can withstand high mechanical stress. Metals, however, possess weaker strengths and modulus of elasticity than ceramics and are more susceptible to failure with strain [13].

When a biomaterial of metal-based gets implanted in the biological medium, the biomaterial's surface may alter and disintegrate, generating certain by-products like ions and debris. Interactions within the surface of the metallic implant and cells/tissues take place as a result of this releasing process. The majority of alloys allow metal ions to enter the plasma. The chance of ion accumulation in organs like the liver and spleen, which subsequently form particulates and interfere with these organs' normal function, is increased when there is an excessive discharge of ions in the blood [29]. When this condition continues to occur, it causes cytotoxicity and eventually organ failure. Also, the human body does not fully accept metal-based materials, and inadequate implant attachment hinders tissue growth, leading to discomfort or significant pain at the implant location. Metallic implants have a larger risk of infection than ceramic implants, and they heal more slowly as well. These properties have encouraged current researchers to place the greatest emphasis on recognizing the characteristics of metallic surfaces with the objective to produce materials that are biocompatible [5]. Despite several shortcomings, metallic implants should be preferred because of their superior mechanical durability, low cost, and corrosion resistance. Platinum and gold, which are chemically inert, can be utilized as bioimplants and do not show corrosion in situ but they are very costly. Consequently, owing to their superior mechanical strength and improved biocompatibility, Titanium and Magnesium-based alloys are now frequently utilized in biomedical industries [30, 31].

3.2.1 Magnesium and Its Alloys

Magnesium (Mg) is a highly abundant element in nature, constituting 2.7% of the Earth's crust. Additionally, minerals such as dolomite and magnesite also contain magnesium [32]. Human cortical bone and magnesium have remarkably similar mechanical characteristics. Due to its remarkable characteristics including lower density, a modulus of elasticity that is comparable with bones, a low weight, biocompatibility, and superior mechanical characteristics, Mg along with its alloys are attracting a lot of interest in the domain of biodegradable alloys. Mg-based implants dissolve in the body since Mg corrodes more quickly in physiological environments that contain chloride. It has become a biocompatible and biodegradable material for use in implants which also reduces the need for additional surgery [33]. Additionally, it demonstrates minimized toxicity to the human body and also a stress-shielding effect.

In an in vivo analysis using rats, Kraus et al. [34] evaluated the impact of Mg-based implants made of Mg on the growth plates. In the right femoral bone of rats, an Mg implant was inserted. After that, the contralateral femoral bones had been then inserted inside their bodies. Using 5% Zinc (Zn), 0.25% Calcium (Ca), and 0.15% Manganese (Mn), an Mg implant was alloyed. The findings demonstrated that the chosen alloy had superior mechanical and yield strength as well as homogeneous, moderate deterioration, and minor gas leakage It, therefore meets the requirements to support application in orthopedics whereby attachment near the growth plate is needed. By means of an in vivo pig study, Naujokat et al. [35] evaluated the process of osteosynthesis of biodegradable magnesium plates. Pig bones were substituted by magnesium implants during the experiment. According to the outcomes, implants have no impact on the process of how bones heal and are not associated with any adverse effects on adjoining bones. As a result, it had been determined that implants based on Mg are permissible for internal bone fixation due to their biocompatibility with pig body tissues.

The major disadvantage of Mg and Mg-based alloys was the speed at which they corrode under physiological conditions. Since it has a high electrode potential (2.3 V), and because of this Mg is extremely prone to corrosion in both aqueous environments and bodily fluids. The reason for the high rate of erosion is that certain impurities, such as iron, copper, and nickel, accumulate in Mg during casting along with refining and speed up corrosion while they exist in concentrations over the permissible level. For Iron, Nickel, and Copper in Mg, the impurity standards were 35–50 ppm, 20–50 ppm, and 100–300 ppm, respectively. No electrochemically active cathodic sites have developed If the above elements exist within these limits, which would accelerate corrosion [36]. Because of the rapid in vivo deterioration, speedy corrosion leads to the frequent emission of byproducts like as hydrogen gases which indicates that there is a need for surface modifications [6]. It has been explored to alloy with different elements to combat rapid corrosion. The corrosion-resistant and mechanical characteristics of Mg, for instance, have been improved by the inclusion of elements like calcium, zinc, silver, aluminum, zirconium, yttrium, and neodymium. Ca–Mg, Zn–Mg, and Ca–Zn–Mg are common examples and because of this, they become perfect for replacing bone. The microstructure can be modified in accordance with mechanical properties similar to the types of bone, by carefully selecting an appropriate element and its composition [37].

3.2.2 Titanium

As a result of its high biocompatibility, resistance to corrosion, and physical and mechanical qualities, its growth as an emerging biomaterial for application in dental and medical applications has significantly accelerated in recent years [14]. Wilhelm Gregor made the first known discovery of titanium (Ti) in 1789. The aerospace and defense sectors were the first to employ Ti industrially, but this was only about 60 years ago. However, toward the end of the 1970s, the recognition of their biocompatibility resulted in the overwhelming need for Ti and Ti-based alloys in applications in biology and medicine [14, 38]. Due to its low electrical conductivity, resulting in the development of a thin passive oxide layer, it can be considered biocompatible. The implant is shielded from corrosion via this oxide layer. Comparing Ti's (around 110 GPa) young modulus to stainless steels and cobalt-chrome alloys (approximately 180 GPa and 210 GPa respectively), Ti's young modulus is closer to the bone. Exceptional tensile strength, excellent resistance to corrosion, significant rigidity, resistance to fractures, and biocompatibility are just some of the characteristics of Ti-based alloys that make them a good choice for developing bone plates, knee and hip joints, screws for fixing fractures, cardiovascular, orthopedic and dental implants [6, 39, 40].

Because pure Ti is susceptible to abrasive and adhesive wear, its usage in implanted devices is limited, as Olmedo et al. [41] reported in one of their studies. In this research, they examined the local effects of corrosion using Ti-based implants in rats. To get confirmation of pitting corrosion, they retain the implant for one minute first in a cell of NaCl electrolyte. These were then cleaned before being inserted into rats' bodies. According to the results, bone apposition was decreased by artificial pitting corrosion. It was determined that corrosion leads Ti to have an adverse impact on the body as hazardous metal ions were discharged into the rat's bodies, hence pure titanium shouldn't be used when developing durable medical implants. Wear debris that accumulates between implant surfaces can lead to bone degeneration in addition to the compromise stability of implants in long term. Because of this, it is used primarily as femoral stems and tibial trays in hip joint replacements, not as an articulating component [42]. Thus, Titanium represents one of those commonly utilized metals for the fabrication of medical implants used in the medical field, although extensive investigations are required to improve its durability against wear and tear. Further investigations may concentrate to enhance Ti's wear resistance via metal coupling. Considering metals can be hazardous, polymeric coating material is a more effective solution in order to enhancing their properties [4].

The most common titanium alloy types used in biomedicine are alpha (Ti–6Al–4V), near-α, α–β, and metastable β (Ti–6Al–7Nb). As a consequence of their exceptionally high modulus of elasticity values, these alloys, despite being widely employed as biometallic implants, have issues with shielding stress across the implant-tissue junction. Vanadium and aluminum compounds present in the physiological environment induce the release of poisonous aluminum and vanadium, which can have detrimental effects on health. In order to substitute Vanadium and aluminum in the alloy, a lot of attention has been given to alloys that comprise Zirconium, Niobium, Tantalum, or Molybdenum [6].

3.2.3 Stainless Steels

The group of ferrous alloys referred to as stainless steel (SS) is composed of iron (70%), nickel (10–14%), chromium (16–18%), and carbon (1%) [4]. Stainless steel is typically used for the medical implant used to replace hips, knees, or shoulders temporarily. After cardiac surgery, SS is typically used to make guidewires and sensor wires implanted into our bodies [43]. Chromium provides a thin, passive oxide coating that shields the implant surface from corrosion. Its mechanical characteristics are enhanced by the inclusion of carbon, primarily the toughness for fractures, corrosion resistance toward the corrosion, and implants' tribological function [44, 45]. In India, SS grades 304 and 316L are the most preferred implantable materials for the medical field because they are more affordable, have a wider range of resources available to them, are reliable, and are simpler to fabricate than Ti- and Cobalt-based alloys. AISI type 316L stainless steel is the most prevalent SS grade that is suitable for implant applications. They are a suitable material for orthopedic implants due to their load-bearing potential. However, because of a pitting corrosion attack with the ejection of ions (nickel and chromium), resulting in allergic reactions near the implant area, about 90% of SS implants of 316L grade lose their properties. As a result, adding a small percentage of molybdenum (2–4 weight %) increases the 316L stainless steel grade's ability to resist corrosion and also strengthens it [6, 46].

Due to its better characteristics and reduced toxicity, nickel (Ni)-based SS proved to be a superior alloy. Implants made of SS are transparent to X-rays; therefore, noble metals (gold or platinum) coatings, are used to augment their radiopacity. The mechanical properties of the implant shouldn't be compromised throughout this procedure, though [47]. Salahinejad et al. [48] evaluated and developed SS free of Ni considering it is hazardous to humans. With nitrogen in place of Ni, he increased steel strength and structural stability. In comparison to conventional Ni-based steel, the results showed excellent strength, superior wear, corrosion, and biocompatibility resistance.

Regardless of possessing numerous advantages, SS is less frequently used in biomedical implants because of the reason of corrosion in an environment that contains chloride, which can lead to metal ions like Ni and chromium to be released that are hazardous to humans. Furthermore, because it is prone to deformation, its ductility (flexibility) is restricted. Researchers in the future can therefore concentrate on boosting the corrosion resistance against corrosion while simultaneously working to increase the ductility of stainless steel.

3.2.4 Cobalt-Chrome Alloys

In the first decade of the twentieth century, this alloy was first utilized as a hip replacement implant material [26, 49]. Cobalt (Co) based alloys with chromium (Cr) (27–30%), molybdenum (Mo) (5–7%), and a small amount of manganese, silicon (1%), iron (0.75%), and Ni (0.5%) are regarded as among the best materials for implant applications. The other components include nitrogen, carbon, tungsten, phosphorus, and sulfur boron [50]. Co-based alloys are employed in applications for bioimplants, especially for developing surgical implants for the hip, knee, shoulder, and broken bone surfaces, in accordance with in vivo as well as in vitro studies because they exhibit greater biocompatibility, excellent mechanical, corrosion, and wear properties [51]. Due to their remarkable combination of strength and flexibility, Co–Cr–Mo alloys are among the most prevalent types of Co alloy. The alloy mentioned above has better elastic modulus, density, and hardness when compared to other metallic implants, making it an optimal choice for the implant process. Also, as these alloys retain their original qualities for a relatively long period after implantation, this alloy is typically employed in permanent implant fixation operations. Artificial ankles and knees are made of a Co–Cr–Mo alloy and ultra-high molecular weight polyethylene [52].

Implantation of Co–Cr alloy and Ti discs was performed in the subcutaneous layer of mice. in an in vivo study executed by one of the studies. Following skin closure, the implanted discs were immediately injected with a suspension comprising *S. aureus*. In accordance with the results, Co–Cr alloy discs are more bacterially resistant than Ti-made discs because they had fewer viable *S. aureus* compared to Ti discs. Ni, Mo, and Cr are a few other significant alloying components for Co-based alloys [53]. When these elements leak from the metal surface of Co alloys into bodily fluids during corrosion, it has been shown that they are harmful to humans as they can cause skin diseases as well as damage to the liver, blood cells, kidneys, and lungs [54]. The mechanical strength and resistance to corrosion of Co–Cr–Mo are improved through the inclusion of Ni, but the utilization of this alloy in bioimplants is constrained due to nickel's cytotoxicity. The tissue/implant interface experiences a stress-shielding phenomenon due to the substantially greater elastic modulus (200–250 GPa) and ultimate tensile strength (400–1000 GPa) of Co-based alloys, which are ten times greater than those of human bone. Surface modifications of Co–Cr–Mo alloys during plasma treatment with plasma increase their rigidity, wear, and corrosion resistance but due to their inferior tensile and frictional characteristics, they are still not suggested for joint fixtures [55, 56].

Because of their inferior frictional properties and corrosion behavior, Co-based alloys are not preferred materials for load bearing and joint surfaces. In comparison with the other Ti or other alloys, manufacturing Co-Cr alloys is more expensive and tedious as well. By employing titanium or zirconium-based coatings or bioceramics coatings on a material's surface, future studies can concentrate on improving the osteointegration of Co–Cr alloys. The development of Ni-free Co–Cr alloy should be the main area of study to lessen the toxicity while making them safe for people to use [4].

Other metals and alloys (Titanium, iridium, gold, platinum, and palladium have also been employed for developing dental implant devices. Zirconium, hafnium, and tungsten-based devices have recently been put through testing. These reactive metal groups and their alloys have been claimed to offer certain valuable benefits. Metals having relatively low strength values, such as gold, platinum, and palladium, have a limited range of applications. Due to their nobility and availability, these metals—especially gold—remain employed as implant materials in surgery [6].

3.3 Polymers

Since the 1980s, medical implants' properties have been modified by coating synthetic polymeric materials on them or using them as implantable materials [4]. Small repeating unit monomers (homopolymers and copolymers) are the building blocks of polymers, which are divided into two distinct categories referred to as biodegradable and non-biodegradable. Chitosan, Polyacetal, polycaprolactone, alginate, and polylactide are typical examples of biodegradable polymers, while polyethylene terephthalate, polymethylmethacrylate, polypropylene, polytetrafluoroethylene, etc., are examples of nonbiodegradable polymers [6].

The primary benefit of employing polymers is that they can conveniently have their composition modified, allowing for the modification of their physical characteristics in accordance with the application. In addition to pacemakers, implants made of polymers are commonly utilized to replace kidneys, skin, heart valves, bone, contact lenses, and artificial blood arteries. They can also be utilized to serve as a binding agent for sealing off the junction within the two materials, or as a protective covering on its own. In comparison to metals and ceramics, polymers exhibit lower strengths and elastic moduli. They lack adherence to biological tissue and are less mechanically and physically stable than metals. Due to biological factors, the polymers also deteriorate in the physiological environment. Because of this, they are not typically utilized in load-bearing implants [4]. Chitosan stands out among biodegradable polymers for its superior biodegradability, biocompatibility, antibacterial activity, and wound-healing properties. Primarily because of their inexpensive cost meanwhile ensuring appropriate mechanical characteristics, polymeric implants are highly attractive as bioimplants. Some biodegradable and non-biodegradable polymers that are commonly utilized and available commercially are mentioned below with brief explanations [8].

3.3.1 Polyether Ether Ketone

For instance, Polyether ether ketone (PEEK), is a semicrystalline thermoplastic polymer with ketone and ether functional groups. It exhibits higher compressive strength, i.e., 80%, and improved fatigue properties compared to pure PEEK due to the additional ketone group because it comprises 20% titanium oxide (TiO_2) particles in addition to the additional ketone group. PEEK is more mechanically strong than many metals, provides stability over high temperatures, and is tolerant to chemical and physical deterioration [57]. The mechanical similarities between PEEK and human cortical bone are one of the key advantages of employing PEEK is used in medical implants as a material. In addition, PEEK exhibits a low density equal to bone and Young's modulus of roughly 4GPa, it is nearly as similar to bone's 16 GPa [58]. The effect of coating PEEK over the Ti–6AI–4V structures surface with regard to corrosion was examined in one or more investigations. The findings indicated that the coating of PEEK materials ultimately boosted the material's corrosion resistance property because they induced the formation of voids and gaps between the materials and also prevented PEEK materials from conducting electrochemical reactions. considering its mechanical and physical qualities, it was also studied and determined that PEEK material will be potentially employed in cranial implants, total joint replacement, and trauma [59]. Kevlar, carbon fiber, glass fiber, and other bioceramic coatings are examples of additives that can be used to strengthen PEEK polymer. While the main disadvantage of utilizing PEEK polymer in the medical implant is its less reactivity with the tissue surrounding it, which makes it incompatible with the human body, and also it is significantly more costly in comparison to the conventional materials utilized in medical implants. Prospective studies may concentrate on enhancing the biocompatibility of PEEK polymer through adding bioactive ceramic fillers, like HAp, and also with surface modification techniques [60].

3.3.2 Polymethyl Methacrylate

In applications where great strength is not required, polymethyl methacrylate (PMMA), is a polymer made from synthetic materials that is non-biodegradable, lightweight, and an affordable substitute. Due to the absence of hazardous subunits like bisphenol, which are mostly present in other polycarbonates, it also exhibits less adverse impacts on human health PMMA, a non-biodegradable polymer that is typically used in applications requiring long-lasting durable implants. As a result, PMMA is among the best biomaterials for covering implants in the body. PMMA cement's main drawback is that it doesn't break down. Additionally, the material's high temperature of curing can result in necrosis of the tissue around it, which is a major concern [61, 62].

3.3.3 Hydroxyapatite

One of the primary minerals found in human bone is HAp of which 70% is made up of bones; the remaining 30% is collagen and water. The physical and chemical characteristics of synthetic HAp are comparable with bone because HAp constitutes the majority of bone [63]. It is referred as the most biocompatible material employed in bone replacement currently on the market, with great bioactivity, osteoconductivity, and less solubility within fluids. In order to improve their biocompatibility and facilitate bone regeneration, HAp is primarily applied as a surface coating to a variety of titanium or stainless steel-based implants. Additionally, it can be utilized as cement between permanent implants or for making small, less loaded medical implants. Its limited use in long-term medical implants is due to its poor mechanical strength and weak tensile strength [64].

3.3.4 Polyvinylidene Fluoride

The polymer known as Polyvinylidene Fluoride (PVDF) is used frequently in the medical sector and has undergone extensive characterization by numerous researchers across the world. Due to its nonreactivity, it is a suitable material for surgical meshes and sutures, and because of its piezoelectric properties, it is also a good choice for wound healing and can serve as a substrate for sensors [65]. However, considering its limitations including its incapability to make uniform films and poor adherence with other materials, pure PVDF film is quite difficult to obtain in biomedical devices used as packing films. In an effort to develop composites with the benefits of both materials, PVDF has been explored in combination with other materials. PVDF fabrication process has to optimize more so that it can be employed in nanoscale devices, which would be extremely advantageous for the industry given the challenges of thin film manufacturing [8].

3.3.5 Polyethylene

Polyethylene (PE) can be divided into groups based on their molecular weight, such as low-density polyethylene and high-density polyethylene, which have various applications depending on their properties. When PE was used successfully for total hip replacement, it was discovered that ceramic-polystyrene couplings performed better than conventional ceramic-ceramic couplings in terms of fracture rates and audible component-related noise [8]. Porous high-density polyethylene has good elasticity, remarkable biocompatibility, and potent anti-infective characteristics, as discovered by various studies employed for rhinoplasty surgery [66]. Furthermore, a study has been carried out on PE-related materials' surfaces to enhance many different characteristics for a number of applications [67].

3.3.6 Polyurethane

Several implants have made use of Polyurethane (PU), which is also simple to modify in accordance with various biomedical uses. On the other hand, PU may be degraded with chemical exposures in vivo, which causes the material to deteriorate. If managed properly, such type of degradation can be employed to aid in the development of new tissues [68]. Additionally, it was discovered that PU had lesser water permeability, which could be reduced even more by adding isopropyl myristate in a very concentration. Capsular contracture is quite infrequent with PU breast implants. Due to its excellent surface, thermomechanical, and biocompatible capabilities, thermoplastic PU also exhibits significant potential if combined with polydimethylsiloxane for application in implants; however, the material is still quite new, and only a few studies have examined its characteristics [69].

3.3.7 Polypropylene

Like PE, Polypropylene (PP) is a thermoplastic polymer that may be adjusted as per its density and divided into its copolymer and homopolymer components, with the primary difference being the material's strength. Numerous research recently explored its use in other body areas like implant-based breast reconstruction. There is significant debate over the usage of PP materials in this application, however [8]. According to Zheng et al. [70] using these implants causes an inflammatory reaction that makes the healing process less effective. PP meshes should still be utilized, according to Moalli et al. [71] who observed that these inflammatory reactions were necessary for the body's natural healing processes to take place. Additionally, PP and titanium have been coupled to develop a mesh with a smaller capsular contracture, an important concern in implant-based breast reconstruction [72]. PP is considered to be an ideal material, however, due to biocompatibility challenges, it cannot be used as a biomedical implant. To improve the material's biocompatibility before application in the human body, additional research should be conducted regarding surface treatments [8].

3.3.8 Polyamide

In both their natural and synthetic forms, polyamides (PA) are macromolecules that include repeating units connected with amide bonds. Nylon is the usually widely used type of PA in medical implants along with devices [73]. Although it's commonly used as a fiber material in composites to enhance their mechanical strength and in the production of sutures and dentures, its usage as a material for packaging films is infrequent [74]. In a recent study on microbiological contamination, nylon had been evaluated among several other materials and emerged with the lowest contamination level. A 3D printing technique may be used to efficiently develop nylon and certain of its composites, such as glass fiber nylon [8].

3.3.9 Silicones Parylene and Polydimethylsiloxane

Silicones are inert materials that are employed in a number of forms and applications. Silicone implants have found applications in laryngeal surgeries to address various problems, including incomplete glottis closure and vocal impairment caused by unilateral vocal fold paralysis. The vocal function of patients has improved, according to studies that analyze these implants Additionally, silicone serves as an encapsulant material in implants [75]. Silicone has been discovered to be more appropriate for extended encapsulation within the body because of its superiority to epoxy resin and PU coatings smoother topography and lower surface energy [76]. Additionally, the polymer itself is shielded from cells and molecules by these properties. The two silicone derivatives parylene and polydimethylsiloxane (PDMS) are frequently used in biomedical implants. In implanted neurological prostheses, parylene is commonly employed as a packaging material; parylene C is the most widely used of its various forms. In addition to pacemakers, blood pumps, breast prostheses, shunts, cochlear implants, esophagus replacements, and packaging for implantable electrical devices and sensors, PDMS is another widely recognized silicone derivative [68, 77]. For the treatment of ocular posterior segment disorders such as diabetic retinopathy, an implanted MEMS device that has been designed utilizing PDMS for direct on-demand drug administration to a human eye was recently developed [8]. Despite their widespread use concerns regarding silicone implant failure, in general, have been addressed.

3.3.10 Liquid Crystal Polymer

Liquid crystal polymers (LCPs) are frequently utilized in microwave frequency electronics because of their high impact strength and Young's modulus. In recent studies, the potential of LCPs as biomaterials for a wide range of implants and devices, including retinal and brain prosthetic implants, has garnered significant interest. In one study, it was discovered the usage of LCP packaging for retinal implants had an impact on pixel density and that the device's elevated pixel resolution allowed individuals to regain their ability to read and recognize faces. Additionally, LCPs have been employed to develop a flexible electrode array in rats for in vivo research on brain functioning and neurological disorders [8, 78]. In their study, Kim et al. [79] showcased the remarkable MRI compatibility of the material in cochlear implantable devices. Their findings suggest that the use of LCP packages resulted in reduced cochlear device size and that they hold promise for future exploration of the auditory perception mechanism.

3.3.11 Polylactic Acid

Polylactic acid (PLA) is a type of thermoplastic polymer that has the unique properties of being both biodegradable and compostable. It is made from renewable resources like corn starch, cassava roots, or sugarcane. It belongs to the family of polyesters and has various applications, including implants, medical devices, 3D printing, packaging, and textiles. As a widely used bioresorbable polymer, PLA has been studied for nearly 50 years [17]. The first reports of PLA's use as sutures and rods for treating dog jaw fractures date back to the 1960s, however, a Swedish scientist by the name of Scheele made the discovery of PLA in 1780. PLA implants are particularly useful because they can be designed to degrade over time, allowing the patient's tissues to take over the function of the implant. The two types of PLA that are most frequently used in the medical field are L-PLA which is hydrolysis resistant (mainly crystalline) and DL-PLA which is hydrolysis sensitive (primarily amorphous) [80].

3.3.12 Polygalactic Acid

Polygalactic acid (PGA) is a synthetic polymer synthesized from glycolic acid, which is produced by the cyclic acid diester's opening ring. Due to its strong crystallinity, poor solubility, and high melting point, PGA was initially utilized in the medical field for manufacturing biodegradable sutures. The primary advantage of PGA is that it is non-toxic, can be aggregated, and degrades naturally. Because PGA is hydrophilic, its breakdown rate is high. Usually, the strength of the implant drops to 50% after implantation and to 90% within 28 days. Because of this, therapies application of PLGA, the copolymer of PGA and PLA, is more common [17, 81].

Various biomaterials utilized in the production of implants with their applications, advantages, and disadvantages are shown in Table 2.

Table 2 An overview of biomaterials in the implant industry: exploring advantages, disadvantages and their applications

Biomaterial	Applications	Advantages	Disadvantages	References
Ceramics				
Bioinert ceramics (Alumina, zirconia, and titanium dioxide)	Orthopedic implants, such as joint replacements, dental implants, bone grafts cardiovascular implants, and orthodontic implants	Durable, long-lasting implants	Brittleness, difficulty in manufacturing, Implant loosening	[17–19]
Bio-active ceramics (Hydroxyapatite, tricalcium phosphate, and bioactive glass)	Dental implants, Orthopedic implants, Spinal implants, and maxillofacial implants	Bioactivity, biocompatibility, osteoconductivity, resorbability	Not be appropriate for load-bearing implants due to their weak mechanical durability, slow degradation, the potential for implant rejection, and variability in properties	[17, 22, 23]
Bioresorbable ceramics	Bone grafts, dental implants, orthopedic implants, drug delivery	Biocompatibility, osteoconductivity, gradual resorption, design flexibility, and reduced stress shielding, better healing results	Limited strength and availability, slow degradation rate, potential for foreign body reaction, and difficulty in removal	[25]
Metals and alloys				
Magnesium and its alloys	Orthopedic implants, cardiovascular implants, dental implants, drug delivery	Biodegradable, comparable to bones, light weight, biocompatibility, and superior mechanical properties	Susceptibility to corrosion, rapid degradation, as well lower mechanical strength	[33, 36, 37]

(continued)

Table 2 (continued)

Biomaterial	Applications	Advantages	Disadvantages	References
Titanium	Knee and hip joints, bone plates, screws for repairing fractures, artificial hearts, orthopedic, orthodontic, maxillofacial, cochlear, and dental implants	Biocompatibility, corrosion resistance, lightweight, strength, osseointegration	Cost, corrosion, allergic reactions, difficulty with imaging, difficulty with removal, implant fractures	[39, 40, 42]
Stainless steels	Orthopedic implants such as those used to replace hips, knees, or shoulders temporarily, dental implants, cardiovascular implants, surgical instruments	Biocompatibility, corrosion resistance, strength, and durability, Ease of sterilization, cost-effective excellent strength, reduced toxicity	Corrosion, allergies, MRI incompatibility, heavyweight, wear and tear	[44, 45, 48]
Cobalt-chrome alloys	Surgical implants for the hip, knee, shoulder, and shattered bone surfaces, dental implants, cardiovascular implants, neurological implants	Exhibit greater biocompatibility, excellent mechanical, corrosion, and wear properties better elastic modulus, density, and stiffness	Because of poor frictional qualities, not preferred materials for bearing and joint surfaces, allergic reactions, wear and tear, heavyweight, magnetic interference, difficulty in processing	[51, 52, 55]
Polymers				
Polyether ether ketone	Commonly utilized in orthopedic and dental implants, spinal implants, and Cardiovascular implants, along with this used for the development of coatings on metals	Biocompatibility, low density, radiolucent, corrosion-resistant, high strength, and stiffness	Poor bonding to the bone, Wear debris, high-cost, difficulties in sterilization, limited long-term data	[57–59]

(continued)

Table 2 (continued)

Biomaterial	Applications	Advantages	Disadvantages	References
Polymethyl methacrylate	Bone cement, dental, ocular, and facial Implants, Bone tissue regeneration, and hip joint replacement	Biocompatibility, mechanical strength, ease of use, radiopacity, low cost, and longevity	Brittleness, Foreign body reaction, poor long-term stability, difficulty in removal, risk of infection	[61, 62]
Hydroxyapatite	Dental implants, orthopedic implants, maxillofacial implants, drug delivery systems	Biocompatibility, osteoconductivity, stability, longevity, radiopacity	Brittle, poor wear resistance, limited bonding strength, difficulty in fabrication, and not appropriate for load-bearing implants or those that require high mechanical strength	[63, 64]
Polyvinylidene fluoride	Orthopedic implants, cardiovascular implants, neural implants, dental implants	Biocompatibility, chemical resistance, strength and durability, flexibility, radiolucency, and low friction coefficient	Limited strength, difficult to shape, wear resistance, cost, and lack of studies on the safety and efficacy of these implants on the long term basis	[8, 65]
Polyethylene	Total joint replacement, spinal implants, dental implants, craniofacial implants, cardiovascular implants	Biocompatibility, wear resistance, low friction, ease of shaping, and low cost	Wear debris, Stress shielding because it has a lower modulus of elasticity than bone, degrade over time, allergic reactions, unsuitable for larger joints or patients with more severe joint damage	[8, 66]

(continued)

Table 2 (continued)

Biomaterial	Applications	Advantages	Disadvantages	References
Polyurethane	Breast implants, vascular implants such as stents, dental implants, joint replacements, and soft tissue implants such as hernia mesh	Biocompatibility, durability, flexibility, low friction, and low water absorption rate, which means that it is less likely to absorb bodily fluids and other liquids thus reducing the risk of bacterial growth and infection, reducing capsular contracture	Breakdown over time, some people may have an adverse reaction to polyurethane, leading to inflammation or other complications, difficulty with imaging, not available in as many shapes and sizes, more expensive than other types of implants	[68, 69]
Polypropylene	Polypropylene mesh is often used to repair hernias, Suture material, orthopedic implants, breast implants, facial implants	Durability, resistance to chemicals, low friction, lightweight, and easy to handle	Poor biocompatibility, degradation particularly in the presence of UV light and high temperatures, poor imaging, limited flexibility, allergic reactions	[8, 70, 71]
Polyamide	Orthopedic implants such used to repair fractures, deformities, and other musculoskeletal conditions, cardiovascular implants such as stents and heart valve replacements, dental implants, soft tissue implants	Biocompatibility, strength, flexibility, low friction, resistant to moisture and chemicals	Degradation, poor wear resistance, can undergo hydrolysis in the presence of water, which can weaken the material and potentially compromise the integrity of the implant, allergic reactions, difficult to sterilize	[8, 73, 74]

(continued)

Table 2 (continued)

Biomaterial	Applications	Advantages	Disadvantages	References
Silicones parylene and polydimethylsiloxane (PDMS)	Used in the fabrication of breast implants, pacemaker insulation, and catheters, silicone-based coatings are utilized to enhance the biocompatibility of other implant materials Parylene is commonly used to protect medical implants, including pacemakers and cochlear implants, from corrosion and other environmental factors Polydimethylsiloxane is used in the fabrication of implantable devices	Silicones are biocompatible and have good mechanical properties, such as flexibility. They are also easy to sterilize and have a long shelf life Parylene is biocompatible and chemically inert, like silicones, but it also has excellent barrier properties, which can protect the underlying substrate from damage or degradation Polydimethylsiloxane exhibits good mechanical properties, including flexibility and low friction and can be easily molded into different shapes and sizes	Silicones and PDMS are non-biodegradable Parylene, on the other hand, is biostable but not biodegradable Silicone and PDMS implants may experience wear and tear over time and are not suitable for all implant types	[68, 75, 76]
Liquid crystal polymer	Orthopedic, dental, cardiovascular and neurological implants	Biocompatibility, high strength, and stiffness, flexibility, durability, radiolucency	Difficulty in processing, limited mechanical properties, high cost, limited availability	[78, 79]

(continued)

Table 2 (continued)

Biomaterial	Applications	Advantages	Disadvantages	References
Polylactic acid	Orthopedic implants include plates for bone fixation, screws and pins utilized in the manufacturing of scaffolds for tissue engineering, drug delivery, dental implants including root canal posts and bone grafting materials	Biocompatibility, biodegradability, versatility, reduced inflammation	Limited temperature stability, cost, and limited range of applications because of its degradation and mechanical properties	[17, 80]
Polygalactic acid	Can be used to create bioresorbable implants to repair and regeneration of bone, dental implants, used to create biodegradable stents, drug delivery implants, and Tissue engineering	Biocompatibility, Biodegradability, and additionally molded into different shapes and sizes	Poor mechanical strength, as it degrades, can release acidic byproducts that can lead to inflammation and tissue damage	[17, 81]

4 Recent Advancements in Biomaterials for Biomedical Implants

The enormous requirement for biomedical implants in the biomedical industry will likely continue to grow in the future. Bioimplants have emerged as a potentially revolutionary therapy option for diseases such as blindness, neurological disorders, orthopedic issues, cardiovascular disease, deformity, and dental deformities [82]. In recent years, many implant biomaterials have evolved. There are usually two types of implant failure: mechanical and biological. The strain on the implant is influenced by our body position while walking, and activities such as jogging, leaping, or cycling can increase the strain. If the implant is subjected to excessive load, it may experience mechanical failure. Additionally, biological failure can occur when the body reacts to the foreign material, and its defense mechanisms interfere with the implant's function. The "foreign-ness" of these biomaterials must be overcome for the currently available implant technology to improve host tolerance, maintain the required temporal and/or spatial drug release profile, and decrease foreign body reactions. If the treatment fails, the patient may undergo further procedures and experience significant trauma, along with the discomfort of their body rejecting the treatment. One of the most important another aspect among each of the characteristics of a permanent implant is corrosion resistance. Due to the presence of water, sodium, chlorine, proteins, and amino acids in the human body, they produce an exceptionally corrosive environment. However, if the material is not sufficiently adaptable, it will gradually decay, inducing an emission of hazardous ions, that cause infection and dysfunction in the tissues of the body [4]. Nanotechnology, 3D printing, novel types of biomaterials, surface coatings, and modifications are being employed to address issues with current biomaterials used for implant technology (Fig. 3). The development of a new class of implant materials with enhanced effectiveness, cost-effectiveness, and high surface area-to-volume ratio owes much to the crucial role played by nanotechnology. Moreover, implant surface coating and modification using various techniques are desirable to improve their mechanical integrity.

4.1 Advancement in the Form Surface Coating and Modification of Biomaterials

A biomaterial's surface chemistry and topography are crucial factors that affect protein adsorption, cell interaction, and host reaction [83]. The effective integration of bioimplants with human body tissues is the main necessity for the implant process. The surface morphology and chemistry of biomedical implants impart more command of the biological responses to lifespan and performance [6]. The modification of the surface of biometallic materials is being proposed to attain the needed qualities with the aim to enhance implant success rates, bioperformance, biocompatibility, and osteoconductivity. When the surface is adequately modified, the general

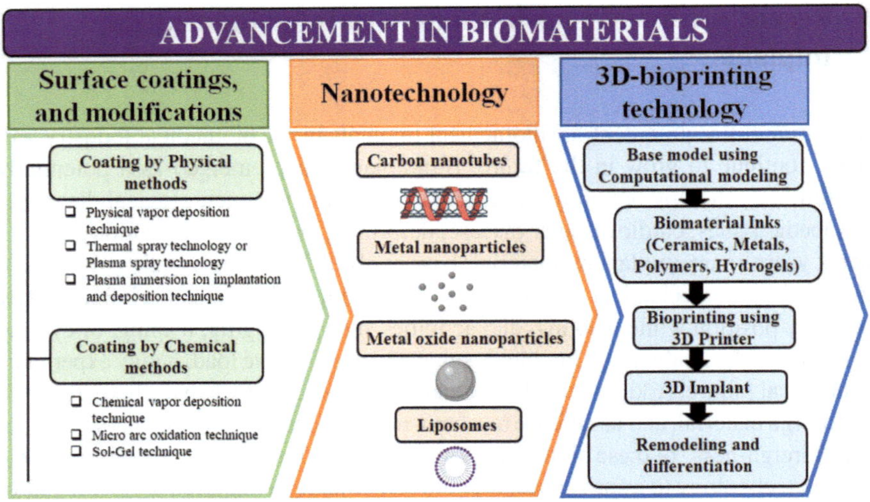

Fig. 3 Shows the advancement of biomaterials to address the challenges associated with existing biomaterials used in implant technology

functionality and attributes of the biomedical implant device won't be impacted for a while. The success rate for bioimplants can be considerably improved with the benefit of bio-integration and the load-bearing capacity of biomaterials [84, 85]. A similar response can be produced in vivo by materials comprised of polymeric, ceramic, or metallic bases that have various surface properties, ranging from being hard to soft or hydrophilic to hydrophobic. Non-specific protein adsorption is thought to be the reason for this observable phenomenon. On the other hand, every process in nature depends on unique molecular interactions, such as those between proteins and saccharides. The complex surfaces linked to biological recognition are effectively regulating biological responses, in contrast to relatively simple surface chemistries that organisms seem unresponsive to [83].

Nonfouling surfaces, also known as surfaces resistant to protein adsorption, and more recently surfaces that regulate protein adsorption, have both been the subject of extensive research [83]. Researchers have made efforts to improve implant bio-integration by altering the implant surface that eventually comes in contact with the body. The surface of the implants is being modified utilizing two different methods. Covering the metallic surface with organic or inorganic materials without altering the implant substrate is the first method [86]. The second method involves the employment of conversion coatings or surface-modified layers, in which a substrate's surface is chemically altered and a minor thickness increase is a consequence. The substrate components in this instance play a role in the development of conversion coatings [87].

Surface preparation through grinding and polishing is essential for conversion coating in order to increase surface roughness for improved mechanical interlocking of coatings. Surface modification through the application of an overlay coating is advised because this method is critical. In order to achieve the synergy of both characteristics, surface modification, and thin film deposition have recently been coupled [44]. For enhanced biocompatibility, corrosion resistance, antibacterial activity, and mechanical qualities, the surface of metallic implants is modified with a suitable coating material in the evolving biomedical implant industry. Although there are numerous ways to deposit bioactive surface coatings, a coating approach that is ideal for biomedical purposes has not yet been established. The coatings on implant materials are currently applied using one of several deposition techniques, including electrodeposition, electrophoretic deposition, sol-gel procedures, physical and chemical vapor deposition, and sol-gel procedures [6]. Physical processes involve exposing the surface to energetic charges or other physical resources such as flame or plasma. On the other hand, chemical methods entail immersing the surface in chemically active solutions, which results in the creation of a coating. The Physical method is regarded as the best among them because it offers precise stoichiometry, high density, excellent adhesion, and outstanding homogeneity when depositing metal or ceramic materials on the surface of implants [88].

4.1.1 Coating by Physical Methods

Physical Vapor Deposition Technique

Among the physical coating techniques, the Physical vapor deposition technique (PVD) is the most commonly used approach for coating the biomaterials over the metal implant surfaces. Using a number of physical processes, the material is first evaporated in this method. When the substance that has evaporated condenses over the surface of another metallic substrate, the presence of gases like nitrogen and argon in the vacuum chamber facilitates the formation of a coating [89]. The metal's surface topography is unaffected by the PVD method, which also exhibits good tribological properties. Because of its rapid deposition rate, PVD is capable of producing coatings with little processing time and basic equipment. Coatings having a thickness of less than 10 nm can also be produced using PVD [90].

Several studies came to the conclusion that PVD may be an effective and innovative method for coating the surface of biomaterials and is thus widely used in industries for producing medical implants [91]. Tantalum carbide was coated with Ti–6Al–4V using the PVD method in one study by Esmaeili et al. [92] and the results showed that the resulting film was uniform, smooth, and homogeneous. The layer was both physically and chemically stable, and it strengthened the material's hardness and resistance to corrosion. Despite this, PVD has several shortcomings. For instance, the difference in the thermal expansion of the coating and substrate causes the bonding force between them to weaken. Both the substrate's mechanical strength and wear resistance are not greatly increased by PVD. Undercuts are challenging to coat with PVD, and the process is also costly [93, 94]. Consequently, PVD is an innovative method with industrial uses for producing medical implants. Its use in the medical field is constrained by the unstable bond in between the substrate and coating surface.

Thermal Spray Technology or Plasma Spray Technology

In the biomedical industry, a plasma arc that has been generated by heating a gas to high temperatures is used as a heat source in the thermal spray technique known as "plasma spray technology" for producing surface coatings. Ionized states of matter, such as plasma, contain neutral particles like neutrons as well as charged particles like electrons and protons [95]. For coating biomaterials, the plasma spray process has multiple advantages. With good adhesion to the substrate, it may produce coatings that are homogeneous, and dense, and also enhance the performance and durability of biomaterials used for implants. Additionally, the coatings can be modified to have specific features including bioactivity, biocompatibility, and corrosion resistance A high-energy heat source is utilized during the coating process to melt and expedite the adherence of minute particles onto a substrate surface. Due to the process of heat transmission, these molten particles rapidly solidify and cool down on applying impact. Therefore, the buildup of these tiny particles produces a coating to form on the surface [96]. Producing coatings for implant surfaces using this method is among the most practical, simple, and reliable techniques available. Due to its high operating temperature, it is typically used for developing implants made of metal with coatings formed of apatite and its derivatives, including bioglass coating, calcium silicate coating, zirconia, and titanium coating, etc. [97]. Overall, the development of biomaterials for implants can be aided by plasma spray technology. This method has several disadvantages, including bioceramic disintegration during elevated plasma spraying temperatures, significant thermal residual stresses within the protective coating, and the existence of flaws that include gaps, cracks, and unmelted particles. The high temperature and gas environment have an impact on the substrate's thermal characteristics and crystallinity, which eventually have an impact on the osteogenic activity of the implant [4].

Plasma Immersion Ion Implantation and Deposition Technique

The Plasma Immersion Ion Implantation and Deposition (PIII&D) technique is a valuable method for surface modification that utilizes both the PVD approach and the Plasma Ion Implantation (PII) technique. Under this method, the substrate surface is continuously bombarded with the required material utilizing the evaporation process. When ions are injected into a surface, the metal phase, which usually acts as an anode, releases metal ions over the substrate [98]. The method is very versatile and allows for the simultaneous application of several processes, including surface etching and deposition. Furthermore, it has the additional advantage of handling items with irregular shapes and allowing for compositional control over coatings. This method enables the development of a transition layer on the substrate, that is deposited within the substrate and the film which progressively enhances the structural density and substrate adhesion. This process enhances the mechanical strength of implants and provides them with superior corrosion protection. These characteristics make PIII&D significantly better compared to PVD and other fabrication methods [99].

Zinc ions were applied to the titanium implant's surface using the PIII&D process by Liang et al. [100] and the results describe that the coating was thick and uniform throughout the implant's surface. The technique enhanced a material's resistance to corrosion, fatigue, and abrasion. The PIII&D technique was shown to be effective for processing materials of any shape and getting around various material restrictions. The PIII&D method was also explored by Sun et al. [101] who utilized it to deposit a titanium and oxygen film on the Ni and titanium alloy surface. The study results showed that the coating is uniform and dense and that PIII&D had no effect on the Ti-Ni's shape memory. This method enormously enhanced the coating's strength and resistance to corrosion. It was determined that layers produced utilizing the PIII&D process exhibit excellent wear properties and adhesion strength, making them suitable for coating metal surfaces. In order to coat biomaterials over the metal surface, PIII&D is one of the most popular methods and it is a desirable technique from an industrial standpoint. It needs a cheap setup, has the ability to alter various shapes, and enables large-scale mass production of medical implants. Future studies can concentrate on the deposition of various metal elements which further improve the characteristics of implants [4].

4.1.2 Coating by Chemical Methods

Chemical Vapor Deposition Technique

A thin film layer can be formed over the surface of a metal substrate using the coating technique known as chemical vapor deposition (CVD). A non-volatile substance is deposited onto the substrate surface in this procedure as a result of a chemical reaction that occurs within chemicals in the gaseous phase and the sample surface. The coating produced via this method is of high quality and has adjustable purity and density [102]. Inorganic compounds like carbon nanotubes, graphene, TiO_2, and other such materials are synthesized using this method in synthetic inorganic chemistry because the end product can be precisely controlled in terms of both quantity and quality [103]. The benefits of CVD include avoiding the line of sight, producing thick coating layers on substrate surfaces, and simultaneously co-depositing several materials. Several studies have found that CVD is a more advantageous choice than PVD and can be further improved to increase its resistance to corrosion [4]. Despite this, CVD use is not as popular as other techniques since it requires a costly and expensive to set up on a big scale, vapor deposition equipment, high reaction temperature, and hence has a low deposition rate. The exhaust gas and gas source employed in the procedure could have toxic consequences that would make the implant more toxic and dangerous for human health [104].

Micro Arc Oxidation Technique

A method for modifying the surface known as Micro Arc Oxidation (MAO) employs the anodizing process to modify the surface of metals and their alloys in the immediate extreme conditions of pressure and temperature produced by an arc discharge. Metal oxides and electrolyte components constitute the modified ceramic coating formed via MAO. The positive outcomes provided by MAO include its ease of use, minimal space requirements, high processing capacity, and environmental safety. Additionally, it has advantages for industry, such as cost-effectiveness and suitability for very large-scale production. High hydrophilicity in the coating produced through the MAO technique can resemble the process by which an implant interacts with its surrounding biological environment. The presence of metal ions enhances the implant's antibacterial properties as well [105].

According to the outcomes of a study by Cao et al. [106] the MAO method improved the adhesive strength of coating as the uniformity of PLC coating increased. Another study by Wang et al. [107] examined the characteristics of an MAO coating that was applied to magnesium in a two-step, current-decreasing mode. The results indicate that the two steps current decreasing mode improved the characteristics, contributed to energy conservation by lowering current consumption, and displayed the best corrosion resistance, maximum adhesion, and highest hardness. The MAO technique's main disadvantage is that the MAO electrolyzer has very high-power consumption and size limitations.

Sol-Gel Technique

It is the common technique to fabricate several oxide films is Sol-Gel. By using this process, hybrid organic-inorganic materials and inorganic materials can be synthesized at low temperatures. A gel is composed of a continuous solid phase that encloses a liquid phase, while a sol is a colloidal suspension comprising solid particles dispersed in a liquid. With this technique, the chemical reaction takes place in the solution rather than at the sample-to-gel interface [108]. Sol-gel technology offers several advantages, including the ease of fabrication, exceptional film uniformity, the ability to cover substrates of any size and over large areas, and low processing temperatures. To coat implants made of metal with a coating of hydroxyapatite, TiO_2, SiO_2, and various other films in order to form a 3D rigid network on the substrate surface, hydrolysis, and condensation reactions are used in this method. This method has several industrial uses for the reason that it is more rapid than other methods and uses a significantly lesser coating material [109]. The sol-gel technique, however, comes with some drawbacks, including the possibility of large volume shrinkage and disc cracking; the potential cost of the precursor used, which enhances the overall process cost; and the sensitivity of the membrane formed to heat treatment, resulting in film layer cracking. Due to these and other benefits, sol-gel technology has a promising future in the medical field. Further applications are restricted by its sensitivity to heat treatment. In the future, researchers may investigate its effect on thermal effects and further develop the method to improve its effectiveness and uses [4, 110].

4.2 Nanotechnology in Implant Technology

The properties of implants composed of biomaterials are further improved by a novel emerging use of nanotechnology in the implant field, yielding beneficial results. Recent years have observed an enormous rise in the impact of nanotechnology on the implant industry. Researchers are looking into the potential of using nanomaterials, particularly those with biologically inspired characteristics, to enhance the functionality of conventional implants [82]. Ceramics, composites, metals, and polymers are just a few of the many materials that have been produced as nanophase (100 nm particle size) components as a result of advancements in nanotechnology. Several of these materials exhibit improved osseointegration and new bone development capabilities. To facilitate osteoblast activity, enhance osteointegration, and address bone-related diseases, significant research has been done on nanomaterial monomers and nanocomposites in bone tissue engineering. Nanosized materials, including nanopillars, nanotubes, nanocubes, quantum dots, nanorods, nanoflowers, and metal–organic frameworks, are of great importance to take into consideration [111]. Nanotechnology permits advancements in methods of treatment, leading to more effective and long-lasting implants, preventable infections, and better bone and tendon healing.

The potential advantages of nanomedicine are now beginning to be grasped as a result of considerable fundamental scientific research efforts, particularly in orthopaedics. There is still more study to be done in order to fully understand the benefits and risks of this unique technology [82].

4.2.1 Metal Oxide Nanoparticles

Several medical devices and implants have been developed with metal oxide nanoparticles. Iron oxide's magnetic properties have been used for therapeutic and diagnostic reasons as contrast materials to provide magnetic resonance imaging, magnetic particle imaging, ultrasonic methods photoacoustic imaging, and magnetic particle hyperthermia. Zinc oxide (ZnO) has a valuable electrical structure for biological applications; for instance, cancer cells have been visualized using ZnO nanowires' inherent fluorescence [112]. Numerous biological uses exist for (TiO_2). For example, in bone-substituting materials, a thin layer of TiO_2 spontaneously forms on the top surface of metallic titanium, which encouraged the utilization of TiO_2 nanoparticles for the regeneration of bones. Due to its similar compatibility to titanium with hard tissues, zirconium oxide is currently utilized for dental implants [113].

4.2.2 Metal Nanoparticles

The absorption and scattering in the visible and near-infrared ranges have led to the usage of materials containing metal nanoparticles in scientific domains of sensing and diagnostics. To improve luminescence, gold nanoparticles can be added to the formulation of substrates or deposited on suitable substrates. The particle size and morphology affect their absorption and scattering capabilities, which in turn determines how this technology is used. Until the development of technologies based on nanomaterials, there weren't many effective detection approaches [114]. However, modern technologies based on nanotechnology facilitate the diagnosis of osteoporosis using a portable device. For instance, research has led to the development of a novel biochip that uses gold nanoparticles to identify an osteoporosis-related protein. It has been demonstrated that it can efficiently assess bone quality and precisely detect and characterize the rate of bone degradation. The assessment of the effectiveness of cancer treatment may also be aided by the use of fluorescent probes to identify nanoparticles [113, 114].

4.2.3 Carbon Nanotubes

The use of carbon nanotubes (CNTs) in various scientific fields has been prompted by their physical and chemical characteristics. Their application in nanobiotechnology has expanded as a result of surface modification studies performed on these particles and their molecular functionalization with biological molecules [115, 116]. Recently, researchers suggested employing a new type of tubular structure for bone regeneration called collagen-modified calcium carbonate nanotubes. The emphasis on using nanotechnology to increase the mechanical stability of ultra-high molecular weight polyethylene—the most often used polymer for packaging orthopedic implants—has developed primarily because of its wear resistance and higher biocompatibility [113]. The incorporation of carbon nanotubes into this polymer has demonstrated translational efficacy and can sometimes be employed as a liner for the acetabulum or an element of the tibia [82].

4.3 3D Bioprinting Technology

In recent years, the use of 3D printable biomaterials has significantly increased in the production of orthopedic implants. This is primarily due to the lightweight nature of the materials, minimal waste generated during production, porous structure that facilitates tissue growth, and the convenience of making patient-specific implants with complex topologies [17]. The biodegradable implants, which are 3D-printed and customized for each patient, are sustainable and dissolve naturally within the body. They offer superior healing properties compared to metal implants [117]. Challenges for modifying the structure of the production system are made possible by the development of 3D printing techniques. The current state of the art demonstrates a trend toward 3D printing in the production of complicated implants such as heart valves, blood arteries, tracheas, etc. [118]. The utilization of 3D printing for implants offers several benefits, such as rapid production, cost-effectiveness, precision in creating intricate porous structures, absorption of human bone cells, and the ability to facilitate bone integration, while eliminating the risk of surface coating detachment. Currently, several 3D-printed implants are available in the market, including knee joints, posterior lumbar interbody fusion devices, meniscus tissues, spinal implants, hip joints, and knee braces, among others. Implants can be 3D printed using different materials, such as metals, ceramics, and polymers [119].

The three categories of 3D printed biomaterials are similar to those used for traditional biomaterials: 3D printable metals, 3D printable polymers, and 3D printable ceramics. When the strength of the implant is an essential factor, such as in dental implants, orthopedic implants, and the repair of long bone fractures, metallic biomaterials that are 3D printable are used [17]. The most common 3D printing polymers utilized for manufacturing orthoses, implants, drug delivery systems, and orthopedic implants include PLA, PEEK, Collagen, Polyether, and Polyesters. Because Young's modulus of these materials is lower than or equivalent to that of bones and other body parts, implants made of these biomaterials offer superior durability and are biocompatible in comparison to those made of metal. The cortical long bones implants are made with 3D printable bioceramics. Orthodontic and orthopedic implants are developed from 3D printable biomaterials including HAp, bioglass, glass ceramics, zirconia, etc. Low fatigue life is a major drawback of the 3D-printed implant. The number of alternative stress cycles that a part can withstand before failing is measured as fatigue life. Materials can fracture at lower stress levels, even below their ultimate strength, if they are repeatedly subjected to alternate stresses over a long period of time. Every metal has been affected by this; hence the fatigue strength of 3D-printed implants must be examined [17, 119].

5 Conclusions and Future Prospects

Biomaterials used in implants have a promising future as they continue to advance in terms of their properties, applications, and performance. The application of implants has become essential in current medical practices and is widely used in people's daily lives. They assist millions of people worldwide to live longer and with improved quality of life and as a result, implant demand has surged over the past few years [83]. Patients prefer using implants more commonly as their first choice of treatment since they are becoming progressively more popular. Implants have overtaken different traditional treatment approaches and entered the mainstream of dental care over the past ten years [9]. Most body tissues found in the human body, including teeth, ligaments, bones, tendons, and others, have been effectively replaced by these biomaterials in the form of implants. Immune rejection is the biggest obstacle to adopting these biomaterials since, in the current context, permanent implants and bone replacement require biocompatibility along with the mechanical and biological properties of the biomaterial used. Many biomaterials have been identified until now, and because of their biocompatibility and biodegradability, these materials are widely used in biotherapy and medical research [3]. Different biomaterials are used for manufacturing of implants, including metals, ceramics, and polymers, depending on the requirement for either permanent or temporary implants. This article examines the use of several metals and their alloys, polymeric materials, and other materials employed for medical implants. To ensure biocompatibility and prevent corrosion in the body, metals are chosen with care. Because of its biocompatibility and surface hardness, titanium metal has multiple uses in the medical industry. The chapter discusses various metals and their alloys. Although these alloys were able to perform well as medical implants, there remains room for improvement in their nanotoxicity and manufacturing complexity. Polymeric materials are also employed as materials used for implantation typically as a surface coating material of metal implants to enhance their properties. The surfaces of implants can be fabricated and modified using a variety of methods, such as surface coatings and alterations to boost the mechanical strength of the materials. Even though these methods have been widely employed and have produced acceptable outcomes, there has yet opportunity for advancement in addressing their drawbacks. Nanotechnology and 3D printing have further facilitated the development of a novel type of material for implants with improved efficacy, affordability, and a substantial surface area-to-volume ratio. Designing materials that are biocompatible, biodegradable, and possess appropriate osseointegration properties, while simultaneously avoiding adverse effects on biological structures, is a significant challenge. Thus, focused and careful research is necessary in order to develop the most effective and ideal materials used for the production of implants. In conclusion, despite these challenges, future prospects of biomaterials used in implants are promising, and ongoing research and development will continue to improve their properties and performance. These advancements will lead to better outcomes for patients and enable the development of more advanced implantable devices.

References

1. Langer, R., Peppas, N.A.: Advances in biomaterials, drug delivery, and bionanotechnology. AIChE J. **49**(12), 2990–3006 (2003)
2. Fenton, O.S., Olafson, K.N., Pillai, P.S., Mitchell, M.J., Langer, R.: Advances in biomaterials for drug delivery. Adv. Mater. **30**(29), 1705328 (2018)
3. Bharadwaj, A.: An overview on biomaterials and its applications in medical science. In: IOP Conference Series: materials Science and Engineering, vol. 1116, no. 1, p. 012178. IOP Publishing
4. Gautam, S., Bhatnagar, D., Bansal, D., Batra, H., Goyal, N.: Recent advancements in nanomaterials for biomedical implants. Biomed. Eng. Adv. **13**, 100029 (2022)

5. Raghavendra, G.M., Varaprasad, K., Jayaramudu, T.: Biomaterials: design, development and biomedical applications. In: Nanotechnology Applications for Tissue Engineering, pp. 21–44. William Andrew Publishing (2015)
6. Amirtharaj Mosas, K.K., Chandrasekar, A.R., Dasan, A., Pakseresht, A., Galusek, D.: Recent advancements in materials and coatings for biomedical implants. Gels **8**(5), 323 (2022)
7. Wang, X.: Overview on biocompatibilities of implantable biomaterials. Lazinica, R. (ed.) Adv. Biomater. Sci. Biomed. Appl. Biomed. 111–155 (2013)
8. Teo, A.J., Mishra, A., Park, I., Kim, Y.J., Park, W.T., Yoon, Y.J.: Polymeric biomaterials for medical implants and devices. ACS Biomater. Sci. Eng. **2**(4), 454–472 (2016)
9. Saini, M., Singh, Y., Arora, P., Arora, V., Jain, K.: Implant biomaterials: a comprehensive review. World J. Clin. Cases: WJCC **3**(1), 52 (2015)
10. Arsiwala, A., Desai, P., Patravale, V.: Recent advances in micro/nanoscale biomedical implants. J. Control. Release **10**(189), 25–45 (2014)
11. Franz, S., Rammelt, S., Scharnweber, D., Simon, J.C.: Immune responses to implants—A review of the implications for the design of immunomodulatory biomaterials. Biomaterials **32**(28), 6692–6709 (2011)
12. Misch, C.E.: Prosthetic options in implant dentistry. In: Contemporary Implant Dentistry, vol. 3, pp. 105–126 (1999)
13. Deepashree, R., Devaki, V., Balu Kandhasamy, R.A.: Evolution of implant biomaterials: a literature review. J. Indian Acad. Dental Spec. Res. **4**(2) (2017)
14. Amarnath, G.S., Muddugangadhar, B.C., Tripathi, S., Dikshit, S., MS, D.: Biomaterials for dental implants: an overview. Int. J. Oral Implantol. Clin. Res. **2**(1), 13–24 (2011)
15. Yamamuro, T.: Bioceramics. In: Poitout, D.G. (ed.) Biomechanics and Biomaterials in Orthopedics, pp. 22–33. Springer, London (2004)
16. Elbadawi, M., Meredith, J., Hopkins, L., Reaney, I.: Progress in bioactive metal and ceramic implants for load-bearing application. Adv. Tech. Bone Regen. [Internet] **31**, 195–219 (2016)
17. Yadav, D., Garg, R.K., Ahlawat, A., Chhabra, D.: 3D printable biomaterials for orthopedic implants: solution for sustainable and circular economy. Resour. Policy **1**(68), 101767 (2020)
18. Raucci, M.G., Giugliano, D., Ambrosio, L.: Fundamental properties of bioceramics and biocomposites. In: Handbook of Bioceramics and Biocomposites, pp. 35–58. Springer, Cham (2016)
19. Thamaraiselvi, T., Rajeswari, S.: Biological evaluation of bioceramic materials-a review. Carbon **24**(31), 172 (2004)
20. Chevalier, J., Liens, A., Reveron, H., Zhang, F., Reynaud, P., Douillard, T., Preiss, L., Sergo, V., Lughi, V., Swain, M., Courtois, N.: Forty years after the promise of "ceramic steel?": zirconia-based composites with a metal-like mechanical behavior. J. Am. Ceram. Soc. **103**(3), 1482–1513 (2020)
21. Larose, J., Timms, D.: Hemocompatibility in mechanical circulatory support. In: Mechanical Circulatory Support: a Companion to Braunwald's Heart Disease, pp. 83–89. Elsevier (2020)
22. Dubok, V.A.: Bioceramics-yesterday, today, tomorrow. Powder Metall. Met. Ceram. **39**, 381–394 (2000)
23. Gupta, R., Kumar, A.: Bioactive materials for biomedical applications using sol-gel technology. Biomed. Mater. **3**(3), 034005 (2008)
24. Baino, F., Verné, E.: Glass-based coatings on biomedical implants: a state-of-the-art review. Biomed. Glasses **3**(1), 1–7 (2017)
25. Salinas, A.J., Esbrit, P., Vallet-Regí, M.: A tissue engineering approach based on the use of bioceramics for bone repair. Biomater. Sci. **1**(1), 40–51 (2013)
26. Gotman, I.: Characteristics of metals used in implants. J. Endourol. **11**(6), 383–389 (1997)
27. Sykaras, N., Iacopino, A.M., Marker, V.A., Triplett, R.G., Woody, R.D.: Implant materials, designs, and surface topographies: their effect on osseointegration. A literature review. Int. J. Oral Maxillofac. Implants **15**(5) (2000)
28. Niinomi, M.: Recent metallic materials for biomedical applications. Metall. Mater. Trans. A **33**, 477–486 (2002)

29. Cadosch, D., Chan, E., Gautschi, O.P., Filgueira, L.: Metal is not inert: role of metal ions released by biocorrosion in aseptic loosening—Current concepts. J. Biomed. Mater. Res. Part A: Off. J. Soc. Biomater. Jpn. Soc. Biomater. Austral. Soc. Biomater. Korean Soc. Biomater. **91**(4), 1252–1262 (2009)
30. Hermawan, H., Ramdan, D., Djuansjah, J.R.: Metals for biomedical applications. Biomed. Eng. Theory Appl. **29**(1), 411–430 (2011)
31. Hanawa, T.: Research and development of metals for medical devices based on clinical needs. Sci. Technol. Adv. Mater. (2012)
32. Bommala, V.K., Krishna, M.G., Rao, C.T.: Magnesium matrix composites for biomedical applications: a review. J. Magn. Alloys **7**(1), 72–79 (2019)
33. Harrison, R., Maradze, D., Lyons, S., Zheng, Y., Liu, Y.: Corrosion of magnesium and magnesium–calcium alloy in biologically-simulated environment. Progr. Nat. Sci.: Mater. Int. **24**(5), 539–546 (2014)
34. Kraus, T., Fischerauer, S., Treichler, S., Martinelli, E., Eichler, J., Myrissa, A., Zötsch, S., Uggowitzer, P.J., Löffler, J.F., Weinberg, A.M.: The influence of biodegradable magnesium implants on the growth plate. Acta Biomater. **15**(66), 109–117 (2018)
35. Naujokat, H., Seitz, J.M., Açil, Y., Damm, T., Möller, I., Gülses, A., Wiltfang, J.: Osteosynthesis of a cranio-osteoplasty with a biodegradable magnesium plate system in miniature pigs. Acta Biomater. **15**(62), 434–445 (2017)
36. Chakraborty Banerjee, P., Al-Saadi, S., Choudhary, L., Harandi, S.E., Singh, R.: Magnesium implants: prospects and challenges. Materials **12**(1), 136 (2019)
37. Jamel, M.M., Jamel, M.M., Lopez, H.F.: Designing advanced biomedical biodegradable Mg alloys: a review. Metals **12**(1), 85 (2022)
38. Sarraf, M., Rezvani Ghomi, E., Alipour, S., Ramakrishna, S., Liana, S.N.: A state-of-the-art review of the fabrication and characteristics of titanium and its alloys for biomedical applications. Bio-Des. Manuf. **26**, 1–25 (2021)
39. Khadija, G., Saleem, A., Akhtar, Z., Naqvi, Z., Gull, M., Masood, M., Mukhtar, S., Batool, M., Saleem, N., Rasheed, T., Nizam, N.: Short term exposure to titanium, aluminum and vanadium (Ti 6Al 4V) alloy powder drastically affects behavior and antioxidant metabolites in vital organs of male albino mice. Toxicol. Rep. **1**(5), 765–770 (2018)
40. Fellah, M., Labaïz, M., Assala, O., Dekhil, L., Taleb, A., Rezag, H., Iost, A.: Tribological behavior of Ti–6Al–4V and Ti–6Al–7Nb alloys for total hip prosthesis. Adv. Tribol. **21**, 2014 (2014)
41. Olmedo, D.G., Duffó, G., Cabrini, R.L., Guglielmotti, M.B.: Local effect of titanium implant corrosion: an experimental study in rats. Int. J. Oral Maxillofac. Surg. **37**(11), 1032–1038 (2008)
42. Kirmanidou, Y., Sidira, M., Drosou, M.E., Bennani, V., Bakopoulou, A., Tsouknidas, A., Michailidis, N., Michalakis, K.: New Ti-alloys and surface modifications to improve the mechanical properties and the biological response to orthopedic and dental implants: a review. Biomed. Res. Int. **14**, 2016 (2016)
43. Reclaru, L., Lerf, R., Eschler, P.Y., Meyer, J.M.: Corrosion behavior of a welded stainless-steel orthopedic implant. Biomaterials **22**(3), 269–279 (2001)
44. Bekmurzayeva, A., Duncanson, W.J., Azevedo, H.S., Kanayeva, D.: Surface modification of stainless steel for biomedical applications: revisiting a century-old material. Mater. Sci. Eng. C **1**(93), 1073–1089 (2018)
45. Godbole, N., Yadav, S., Ramachandran, M., Belemkar, S.: A review on surface treatment of stainless steel orthopedic implants. Int. J. Pharm. Sci. Rev. Res. **36**(1), 190–194 (2016)
46. Ma, L., Wiame, F., Maurice, V., Marcus, P.: Origin of nanoscale heterogeneity in the surface oxide film protecting stainless steel against corrosion. NPJ Mater. Degrad. **3**(1), 29 (2019)
47. Mahmoudi Hashemi, P., Borhani, E., Nourbakhsh, M.S.: A review on nanostructured stainless steel implants for biomedical application. Nanomed. J. **3**(4), 202–216 (2016)
48. Salahinejad, E., Hadianfard, M.J., Macdonald, D.D., Sharifi-Asl, S., Mozafari, M., Walker, K.J., Rad, A.T., Madihally, S.V., Tayebi, L.: In vitro electrochemical corrosion and cell viability studies on nickel-free stainless steel orthopedic implants. Plos One **8**(4), e61633 (2013)

49. Natiella, J.R., Armitage, J.E., Greene, Jr., G.W., Meenaghan, M.A.: Council on dental materials and devices. Current evaluation of dental implants. J. Am. Dental Assoc. **84**(6), 1358–1372 (1972)
50. Marti, A.: Cobalt-base alloys used in bone surgery. Injury **1**(31), D18-21 (2000)
51. Garcia-Mendez, M.C., Urrutia-Baca, V.H., Cuao-Moreu, C.A., Lorenzo-Bonet, E., Alvarez-Vera, M., Ortiz-Martinez, D.M., de la Garza-Ramos, M.A.: In vitro biocompatibility evaluation of a new Co–Cr–B alloy with potential biomedical application. Metals **11**(8), 1267 (2021)
52. González-Mora, V.A., Hoffmann, M., Stroosnijder, R., Espinar Escalona, E., Llamas Carrera, J.M., Fernández-Fairén, M., Gil Mur, F.J.: Influence of different CoCrMo counterfaces on wear in UHMWPE for artificial joints. J. Biomed. Sci. Eng. (2011)
53. Watanabe, K., Fukuzaki, S., Sugino, A., Benson, N., Metcalf, N., Nakamura, M., Matsumoto, M.: Cobalt–chromium alloy has superior antibacterial effect than titanium alloy: in vitro and in vivo studies. Spine **46**(17), E911 (2021)
54. Jaishankar, M., Tseten, T., Anbalagan, N., Mathew, B.B., Beeregowda, K.N.: Toxicity, mechanism and health effects of some heavy metals. Interdiscip. Toxicol. **7**(2), 60 (2014)
55. Guo, Z., Pang, X., Yan, Y., Gao, K., Volinsky, A.A., Zhang, T.Y.: CoCrMo alloy for orthopedic implant application enhanced corrosion and tribocorrosion properties by nitrogen ion implantation. Appl. Surf. Sci. **30**(347), 23–34 (2015)
56. Lourenço, M.L., Cardoso, G.C., Sousa, K.D., Donato, T.A., Pontes, F.M., Grandini, C.R.: Development of novel Ti–Mo–Mn alloys for biomedical applications. Sci. Rep. **10**(1), 6298 (2020)
57. Maddock, N.A., James, N.L., McKenzie, D.R., Patrick, J.F.: Technological advances for polymers in active implantable medical devices. Des. Manuf. Med. Dev. **1**, 239–272 (2012)
58. Kurtz, S.M., Devine, J.N.: PEEK biomaterials in trauma, orthopedic, and spinal implants. Biomaterials **28**(32), 4845–4869 (2007)
59. Costa, M.M., Dantas, T.A., Bartolomeu, F., Alves, N., Silva, F.S., Miranda, G., Toptan, F.: Corrosion behaviour of PEEK or β-TCP-impregnated Ti6Al4V SLM structures targeting biomedical applications. Trans. Nonferrous Metals Soc. China **29**(12), 2523–2533 (2019)
60. Panayotov, I.V., Orti, V., Cuisinier, F., Yachouh, J.: Polyetheretherketone (PEEK) for medical applications. J. Mater. Sci. Mater. Med. **27**, 1–1 (2016)
61. Li, C., Mason, J., Yakimicki, D.: Thermal characterization of PMMA-based bone cement curing. J. Mater. Sci. Mater. Med. **15**(1), 85–89 (2004)
62. Hacker, M.C., Krieghoff, J., Mikos, A.G.: Synthetic polymers. In: Principles of Regenerative Medicine, pp. 559–590. Academic Press (2019)
63. Mazumder, S., Nayak, A.K., Ara, T.J., Hasnain, M.S.: Hydroxyapatite composites for dentistry. Appl. Nanocompos. Mater. Dent. **1**, 123–143 (2019)
64. Kannan, M.B.: Hydroxyapatite coating on biodegradable magnesium and magnesium-based alloys. In: Hydroxyapatite (HAp) For Biomedical Applications, pp. 289–306. Woodhead Publishing (2015)
65. Klinge, U., Klosterhalfen, B., Öttinger, A.P., Junge, K., Schumpelick, V.: PVDF as a new polymer for the construction of surgical meshes. Biomaterials **23**(16), 3487–3493 (2002)
66. Zhou, J., Huang, X., Zheng, D., Li, H., Herrler, T., Li, Q.: Oriental nose elongation using an L-shaped polyethylene sheet implant for combined septal spreading and extension. Aesthetic Plast. Surg. **38**, 295–302 (2014)
67. Cools, P., Van Vrekhem, S., De Geyter, N., Morent, R.: The use of DBD plasma treatment and polymerization for the enhancement of biomedical UHMWPE. Thin Solid Films **1**(572), 251–259 (2014)
68. Rahimi, A., Mashak, A.: Review on rubbers in medicine: Natural, silicone and polyurethane rubbers. Plast. Rubber Compos. **42**(6), 223–230 (2013)
69. Pergal, M.V., Nestorov, J., Tovilović, G., Ostojić, S., Gođevac, D., Vasiljević-Radović, D., Djonlagić, J.: Structure and properties of thermoplastic polyurethanes based on poly (dimethylsiloxane): assessment of biocompatibility. J. Biomed. Mater. Res. Part A **102**(11), 3951–3964 (2014)

70. Zheng, F., Xu, L., Verbiest, L., Verbeken, E., De Ridder, D., Deprest, J.: Cytokine production following experimental implantation of xenogenic dermal collagen and polypropylene grafts in mice. Neurourol. Urodyn.: Off. J. Int. Cont. Soc. **26**(2), 280–289 (2007)
71. Moalli, P., Brown, B., Reitman, M.T., Nager, C.W.: Polypropylene mesh: evidence for lack of carcinogenicity. Int. Urogynecol. J. **25**, 573–576 (2014)
72. Bergmann, P.A., Becker, B., Mauss, K.L., Liodaki, M.E., Knobloch, J., Mailänder, P., Siemers, F.: Titanium-coated polypropylene mesh (TiLoop Bra®)—An effective prevention for capsular contracture? Eur. J. Plast. Surg. **37**, 339–346 (2014)
73. Nanni, F., Lamastra, F.R., Pisa, F., Gusmano, G.: Synthesis and characterization of poly (ε-caprolactone) reinforced with aligned hybrid electrospun PMMA/nano-Al2O3 fibre mats by film stacking. J. Mater. Sci. **46**, 6124–6130 (2011)
74. McMahon, R.E., Wang, L., Skoracki, R., Mathur, A.B.: Development of nanomaterials for bone repair and regeneration. J. Biomed. Mater. Res. B Appl. Biomater. **101**(2), 387–397 (2013)
75. Van Ardenne, N., Vanderwegen, J., Van Nuffelen, G., De Bodt, M., Van de Heyning, P.: Medialization thyroplasty: vocal outcome of silicone and titanium implant. Eur. Arch. Otorhinolaryngol. **268**, 101–107 (2011)
76. Kirsten, S., Uhlemann, J., Braunschweig, M., Wolter, K.J.: Packaging of electronic devices for long-term implantation. In: 2012 35th International Spring Seminar on Electronics Technology, pp. 123–127. IEEE (2012)
77. Qin, Y., Howlader, M.M., Deen, M.J., Haddara, Y.M., Selvaganapathy, P.R.: Polymer integration for packaging of implantable sensors. Sens. Actuat. B Chem. **31**(202), 758–778 (2014)
78. Sundaram, V., Sukumaran, V., Cato, M.E., Liu, F., Tummala, R., Weiland, J.D., Nasiatka, P.J., Tanguay, A.R.: High density electrical interconnections in liquid crystal polymer (LCP) substrates for retinal and neural prosthesis applications. In: 2011 IEEE 61st Electronic Components and Technology Conference (ECTC), pp. 1308–1313. IEEE (2011)
79. Kim, J.H., Min, K.S., An, S.K., Jeong, J.S., Jun, S.B., Cho, M.H., Son, Y.D., Cho, Z.H., Kim, S.J.: Magnetic resonance imaging compatibility of the polymer-based cochlear implant. Clin. Exp. Otorhinolaryngol. **5**(Suppl 1), S19–S23
80. Maurus, P.B., Kaeding, C.C.: Bioabsorbable implant material review. Oper. Tech. Sports Med. **12**(3), 158–160 (2004)
81. Suh, J.K., Matthew, H.W.: Application of chitosan-based polysaccharide biomaterials in cartilage tissue engineering: a review. Biomaterials **21**(24), 2589–2598 (2000)
82. Chen, M.Q.: Recent advances and perspective of nanotechnology-based implants for orthopedic applications. Front. Bioeng. Biotechnol. **10** (2022)
83. Ratner, B.D., Bryant, S.J.: Biomaterials: where we have been and where we are going. Annu. Rev. Biomed. Eng. **15**(6), 41–75 (2004)
84. Lam, M., Migonney, V., Falentin-Daudre, C.: Review of silicone surface modification techniques and coatings for antibacterial/antimicrobial applications to improve breast implant surfaces. Acta Biomater. **1**(121), 68–88 (2021)
85. Accioni, F., Vázquez, J., Merinero, M., Begines, B., Alcudia, A.: Latest trends in surface modification for dental implantology: innovative developments and analytical applications. Pharmaceutics **14**(2), 455 (2022)
86. Sikder, P., Ren, Y., Bhaduri, S.B.: Synthesis and evaluation of protective poly (lactic acid) and fluorine-doped hydroxyapatite—Based composite coatings on AZ31 magnesium alloy. J. Mater. Res. **34**(22), 3766–3776 (2019)
87. Metroke, T.L., Parkhill, R.L., Knobbe, E.T.: Passivation of metal alloys using sol-gel-derived materials—A review. Prog. Org. Coat. **41**(4), 233–238 (2001)
88. Campbell, A.A.: Bioceramics for implant coatings. Mater. Today **6**(11), 26–30 (2003)
89. Helmersson, U., Lattemann, M., Bohlmark, J., Ehiasarian, A.P., Gudmundsson, J.T.: Ionized physical vapor deposition (IPVD): a review of technology and applications. Thin Solid Films **513**(1–2), 1–24 (2006)

90. Liu, M.J., Zhang, M., Zhang, X.F., Li, G.R., Zhang, Q., Li, C.X., Li, C.J., Yang, G.J.: Transport and deposition behaviors of vapor coating materials in plasma spray-physical vapor deposition. Appl. Surf. Sci. **30**(486), 80–92 (2019)
91. Ozkucur, N., Wetzel, C., Hollstein, F., Richter, E., Funk, R.H., Monsees, T.K.: Physical vapor deposition of zirconium or titanium thin films on flexible polyurethane highly support adhesion and physiology of human endothelial cells. J. Biomed. Mater. Res. Part A: Off. J. Soc. Biomater. Jpn. Soc. Biomater. Austral. Soc. Biomater. Korean Soc. Biomater. **89**(1), 57–67 (2009)
92. Esmaeili, M.M., Mahmoodi, M., Imani, R.: Tantalum carbide coating on Ti–6Al–4V by electron beam physical vapor deposition method: study of corrosion and biocompatibility behavior. Int. J. Appl. Ceram. Technol. **14**(3), 374–382 (2017)
93. Qadir, M., Li, Y., Wen, C.: Ion-substituted calcium phosphate coatings by physical vapor deposition magnetron sputtering for biomedical applications: a review. Acta Biomater. **15**(89), 14–32 (2019)
94. Taran, A.V., Garkusha, I.E., Taran, V.S., Pyvovar, N.V., Muratov, R.M., Leonovych, A.V., Nikolaychuk, G.P., Baturin, A.A.: Structure of biocompatible nanocoatings obtained by physical vapor deposition on flexible polyurethane for medical applications. J. Adv. Microsc. Res. **13**(3), 313–319 (2018)
95. Vardelle, A., Moreau, C., Themelis, N.J., Chazelas, C.: A perspective on plasma spray technology. Plasma Chem. Plasma Process. **35**, 491–509 (2015)
96. Singh, H., Sidhu, B.S., Puri, D., Prakash, S.: Use of plasma spray technology for deposition of high temperature oxidation/corrosion resistant coatings—A review. Mater. Corros. **58**(2), 92–102 (2007)
97. Xue, T., Attarilar, S., Liu, S., Liu, J., Song, X., Li, L., Zhao, B., Tang, Y.: Surface modification techniques of titanium and its alloys to functionally optimize their biomedical properties: thematic review. Front. Bioeng. Biotechnol. **11**(8), 603072 (2020)
98. Anders, A.: Metal plasma immersion ion implantation and deposition: a review. Surf. Coat. Technol. **93**(2–3), 158–167 (1997)
99. Mändl, S., Manova, D.: Modification of metals by plasma immersion ion implantation. Surf. Coat. Technol. **15**(365), 83–93 (2019)
100. Liang, Y., Xu, J., Chen, J., Qi, M., Xie, X., Hu, M.: Zinc ion implantation-deposition technique improves the osteoblast biocompatibility of titanium surfaces. Mol. Med. Rep. **11**(6), 4225–4231 (2015)
101. Sun, T., Wang, L.P., Wang, M.: (Ti, O)/Ti and (Ti, O, N)/Ti composite coatings fabricated via PIIID for the medical application of NiTi shape memory alloy. J. Biomed. Mater. Res. B Appl. Biomater. **96**(2), 249–260 (2011)
102. Carlsson, J.O., Martin, P.M.: Chemical vapor deposition. In: Handbook of Deposition Technologies for Films and Coatings, pp. 314–363. William Andrew Publishing (2010)
103. Brunette, D.M., Tengvall, P., Textor, M., Thomsen, P., Thull, R., Grant, D.: Physical and chemical vapor deposition and plasma-assisted techniques for coating titanium. In: Titanium in Medicine: material Science, Surface Science, Engineering, Biological Responses and Medical Applications, pp. 283–341 (2001)
104. Stewart, C., Akhavan, B., Wise, S.G., Bilek, M.M.: A review of biomimetic surface functionalization for bone-integrating orthopedic implants: mechanisms, current approaches, and future directions. Prog. Mater Sci. **1**(106), 100588 (2019)
105. Li, L.H., Kong, Y.M., Kim, H.W., Kim, Y.W., Kim, H.E., Heo, S.J., Koak, J.Y.: Improved biological performance of Ti implants due to surface modification by micro-arc oxidation. Biomaterials **25**(14), 2867–2875 (2004)
106. Cao, X.Y., Tian, N., Dong, X., Cheng, C.K.: Implant coating manufactured by micro-arc oxidation and dip coating in resorbable polylactide for antimicrobial applications in orthopedics. Coatings **9**(5), 284 (2019)
107. Wang, Z.X., Zhang, J.W., Ye, F., Lv, W.G., Lu, S., Sun, L., Jiang, X.Z.: Properties of micro-arc oxidation coating fabricated on magnesium under two steps current-decreasing mode. Front. Mater. **18**(7), 261 (2020)

108. Feng, W., Mu-Sen, L., Yu-Peng, L., Yong-Xin, Q.: A simple sol-gel technique for preparing hydroxyapatite nanopowders. Mater. Lett. **59**(8–9), 916–919 (2005)
109. Castro, Y., Durán, A.: Control of degradation rate of Mg alloys using silica sol-gel coatings for biodegradable implant materials. J. Sol-Gel. Sci. Technol. **15**(90), 198–208 (2019)
110. Bollino, F., Catauro, M.: Sol-gel technology to prepare advanced coatings. Photoenergy Thin Film Mater. **25**, 321–378 (2019)
111. Zhang, L., Webster, T.J.: Nanotechnology and nanomaterials: promises for improved tissue regeneration. Nano Today **4**(1), 66–80 (2009)
112. Liu, T.Y., Liao, H.C., Lin, C.C., Hu, S.H., Chen, S.Y.: Biofunctional ZnO nanorod arrays grown on flexible substrates. Langmuir **22**(13), 5804–5809 (2006)
113. Ramos, A.P., Cruz, M.A., Tovani, C.B., Ciancaglini, P.: Biomedical applications of nanotechnology. Biophys. Rev. **9**(2), 79–89 (2017)
114. Bhattacharya, R., Mukherjee, P.: Biological properties of "naked" metal nanoparticles. Adv. Drug Deliv. Rev. **60**(11), 1289–1306 (2008)
115. Sharma, P., Kumar Mehra, N., Jain, K., Jain, N.K.: Biomedical applications of carbon nanotubes: a critical review. Curr. Drug Deliv. **13**(6), 796–817 (2016)
116. Yang, W., Thordarson, P., Gooding, J.J., Ringer, S.P., Braet, F.: Carbon nanotubes for biological and biomedical applications. Nanotechnology **18**(41), 412001 (2007)
117. Chia, H.N., Wu, B.M.: Recent advances in 3D printing of biomaterials. J. Biol. Eng. **9**(1), 1–4 (2015)
118. Javaid, M., Haleem, A.: Current status and challenges of additive manufacturing in orthopaedics: an overview. J. Clin. Orthop. Trauma **10**(2), 380–386 (2019)
119. Chunhua, S., Guangqing, S.: Application and development of 3D printing in medical field. Modern Mech. Eng. **10**(03), 25 (2020)

Smart Biomaterials in Drug Delivery Applications

S. Giridhar Reddy and H. C. Ananda Murthy

Abstract The controlled release of drugs is an essential aspect of drug delivery, ensuring optimal therapeutic effects while minimising side effects. Stimuli-responsive materials have emerged as a promising tool for developing controlled-release drug delivery systems. These substances can react to alterations in the surroundings like light, pH, and temperature and can dispense drugs in a regulated manner. In this section, the focus is on recent progressions in materials that are reactive to stimuli, such as hydrogels, nanoparticles, and liposomes, and their utilisation in the transportation of drugs. The challenges of developing stimuli-responsive drug delivery systems and potential strategies to overcome these obstacles are discussed. Overall, the development of stimuli-responsive materials has great potential in drug delivery, and further research is needed to exploit their capabilities fully.

Keywords Controlled drug delivery · Hydrogels · Biopolymers · Nanoparticles

1 Introduction

The conventional approaches to treating deadly diseases like cancer, coronary artery disease, and respiratory and abdominal infections face challenges such as invasive procedures for solid tumours, poor solubility of drugs, short drug circulation, drug resistance, non-specific targeting, and harmful side effects, both locally and systemically [1]. Millions of fresh incidents and fatalities are recorded worldwide due to cancer, among the primary causes of death [2, 3]. The incidence and mortality rates for 36 types of cancer in 185 countries are estimated by GLOBOCAN [3]. Hence, there is an urgent need for new and innovative treatment methods that can target

S. G. Reddy (✉)
Department of Chemistry, Amrita School of Engineering, Amrita Vishwa Vidyapeetham, Bengaluru 560035, India
e-mail: s_giri@blr.amrita.edu

H. C. A. Murthy
Department of Applied Chemistry, School of Natural Science, Adama Science and Technology University, Adama 1888, Ethiopia

R. Malviya and S. Sundram (eds.), *Engineered Biomaterials*, Engineering Materials,
https://doi.org/10.1007/978-981-99-6698-1_11

cancer cells precisely without harming healthy tissues and organs surrounding the tumour [4].

Recently drug delivery through nanotechnology has gained popularity due to the unique properties of nanoscale materials. These properties, including enhanced intracellular drug delivery and targeted subcellular structures, make it a promising field. Nanoscale materials can also overcome physiological barriers in living organisms, allowing them to reach previously inaccessible body regions [5].

Another emerging field is Supramolecular hydrogels which are versatile and rapidly developing nanostructures with potential applications in diverse fields, including nanofabrication, biosensing, catalysis, tissue engineering, and controlled drug delivery [6]. These hydrogels are formed by low-molecular-weight supramolecular hydrogelators that respond to a stimulus by forming a network of intertwined fibres, resulting in a solid-like material with a 3D structure and high-water content [7, 8].

Polymer micelles have emerged as a new area for drug delivery that can enhance treatment efficacy while minimising negative effects. These functional biomaterials rely on the self-assembly of amphiphilic polymers to create polymeric micelles, which can enclose hydrophobic drugs in their core via hydrophobic interactions [4, 8–12].

The release of medication can be managed by stimuli in the tumour microenvironment, such as enzymes, temperature, and pH [13, 14].

Another form of drug delivery system is transdermal drug delivery can bypass first-pass metabolism and achieve continuous medication release, but the protective skin layer (stratum corneum) poses a challenge [15]. Microneedles offer a user-friendly solution for effective transdermal drug delivery with high bioavailability [16]. These patches penetrate the stratum corneum and can deliver drugs painlessly and minimally invasively for skin vaccination and drug delivery [17–19].

Overall, controlled release is a critical aspect of drug delivery, where drugs are released at a controlled rate over time, ensuring maximum therapeutic efficacy while minimising unwanted side effects. Stimuli-responsive materials have gained significant attention recently as a promising tool for developing controlled-release drug delivery systems [20]. Such substances can react to environmental variations, including temperature changes, pH, magnetic or electric fields, and light, allowing them to dispense medication in a controlled manner [21]. The development of stimuli-responsive materials has revolutionised the field of drug delivery, as these materials can overcome many limitations of conventional drug delivery systems [22].

This chapter will briefly overview the recent advances in stimuli-responsive materials, their properties, and their potential applications in drug delivery [23].

Overall, stimuli-responsive materials have great potential to improve the efficacy and safety of drug delivery, and their continued development is expected to impact the field of medicine significantly. The various investigators, with their respective research areas on CDD, are influenced by different parameters such as pH, temperature, light, and magnetic or electric fields as shown in Fig. 1.

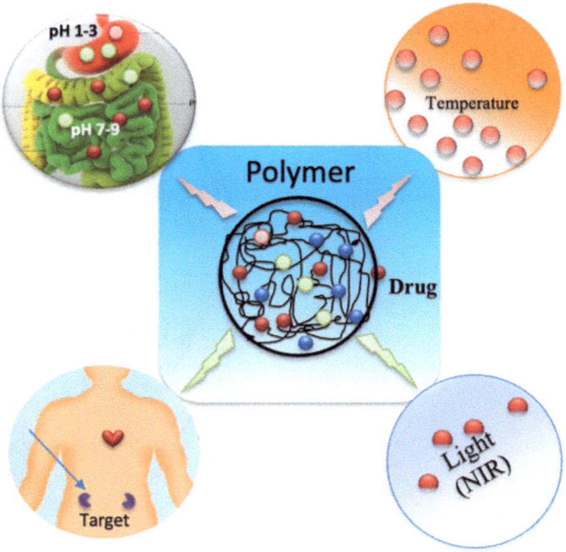

Fig. 1 Schematic representation of controlled release of drug delivery for stimuli-responsive for pH, temperature, light sensitivity (NIR) and magnetic

2 pH-Responsive Controlled Drug Delivery

pH is a widely studied environmental stimulus that can trigger drug release from responsive materials. These materials can swell or collapse in response to pH changes, leading to drug release. pH-responsive materials are particularly useful for targeted drug delivery [24], as different tissues and organs in the body can have varying pH levels. For instance, the stomach has a highly acidic pH (1–3), while the blood has a slightly alkaline pH (7.35–7.45).

Hydrogels and liposomes are two types of extensively studied pH-responsive materials. Hydrogels are 3D polymer networks that can swell or collapse in response to pH changes, allowing drug encapsulation and release. Liposomes are spherical structures made of a phospholipid bilayer that can encapsulate drugs and respond to pH changes for drug release [25].

Some latest articles published and their outcomes for pH-responsive stimuli are highlighted in Fig. 2.

2.1 pH-Responsive-Targeted Drug Delivery for Cancer Therapy

A new drug delivery system was developed and evaluated by Pilch et al. [26] for targeted cancer therapy. The system uses pH-responsive nanoparticles as carriers for an unsymmetrical bisacridine drug designed to target cancer cells selectively. It is demonstrated that the drug-loaded nanoparticles have a high drug-loading capacity.

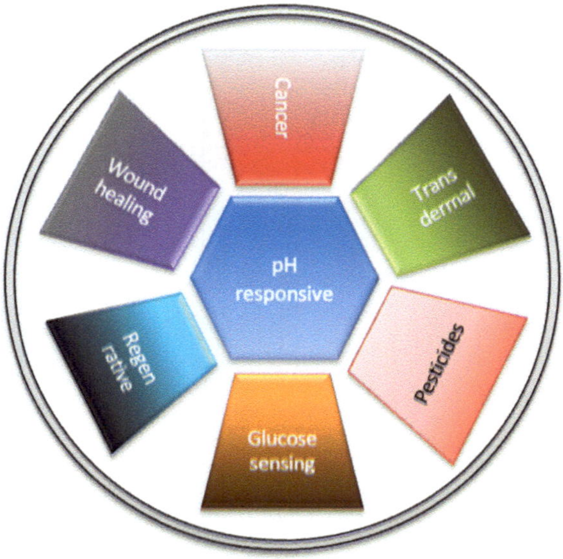

Fig. 2 Application of pH-responsive drug delivery in fields

The drug-loaded nanoparticles also exhibited significant cytotoxicity against cancer cells in vitro and were able to inhibit tumour growth in a mouse model of breast cancer. The study mentions the potential of pH-responsive drug delivery nanoplatforms as smart carriers for targeted cancer therapy.

The pH-responsive nanofiber buttresses were synthesised using the electrospinning technology by Altinbasak et al. [27]. These buttresses release drugs based on changes in pH and have controlled drug release over multiple days. It was effective against human cancer with enough tensile strength for surgical applications and could be used with a conventional surgical cutting stapler. This method holds great potential for drug delivery devices due to the large surface area, high drug payload capacity, and easy placement of the nanofiber buttresses.

A drug delivery system developed by the facile solvothermal method utilises hollow mesoporous structured $MnFe_2O_4$ nanospheres, as reported by Gao et al. [28]. This system is capable of pH-responsive release and is biocompatible. The study found that the nanospheres have a high capacity for drug loading, release drugs in response to changes in pH, and do not cause significant cell toxicity. The findings suggest that these nanospheres could be a promising drug delivery system for cancer treatment.

Son et al. [29] conducted experiments on mesoporous nanocarriers with fluorescence and pH-sensitive controlled release capabilities. The nanocarriers were synthesised using a facile method and were loaded with a natural compound called Dieckol. The release of Dieckol from the nanocarriers was controlled by the pH of the solution, with a faster release at lower pH levels. The fluorescence of the nanocarriers allowed for imaging and tracking of the carriers in cells. The study demonstrates the potential of these nanocarriers for targeted drug delivery and imaging applications.

Another hollow mesoporous structured $MnFe_2O_4$ nanospheres zeolitic-imidazolate framework-8 was developed by Anggraini et al. [30]. This system is capable of pH-responsive release and is biocompatible. The study found that the nanospheres have a high capacity for drug loading, release drugs in response to changes in pH, and do not cause significant cell toxicity. The findings suggest that these nanospheres could be a promising drug delivery system for cancer treatment.

The novel magnetic-luminescent nanomaterial, $CoFe_2O_4/mSiO_2–NH_2/$ Poly(MAA-co-IA)/MF_3 reported by Ghazimoradi et al. [31], which is designed to act as a carrier for the anti-cancer drugs methotrexate (MTX) and doxorubicin (DOX). The nanomaterial has a core/double-shell structure and can release drugs in a controlled manner that distinguishes between normal and cancerous tissues. The study shows that the nanocarrier is non-toxic to human cells and inhibits tumour growth in vitro. These results suggest that the nanomaterial can be used in cancer therapy as a drug delivery system for MTX and DOX.

The nanotherapeutics developed by Zhang et al. [32] can respond to changes in pH for treating ulcerative colitis (UC). However, current treatments target the entire colon, which leads to inadequate drug accumulation in inflamed colon lesions and causes side effects. To improve targeted delivery, the researchers used a core-shell nanoparticle design, coating nano-sized curcumin with two pH-sensitive materials, Eudragit® EPO and L100. The resulting CNs@EPO@L100 nanoparticles exhibited pH-responsive drug release behaviour, improved anti-inflammatory properties, and enhanced accumulation at the colon inflammation site. After oral administration, the nanoparticles significantly improved inflammatory symptoms in mice, offering a promising strategy for a precisely targeted therapy of UC.

A hydrogel made of Chitosan and Poly(acrylamide-maleic acid) that responds to changes in pH levels described by Akkaya et al. [33]. The hydrogel was loaded with Doxorubicin, an anti-cancer drug, and was found to have a high capacity for drug absorption. Tests showed that the hydrogel released more Doxorubicin at lower pH levels, which is desirable for cancer treatment. The hydrogel was also found to be biocompatible with breast cancer cells and effectively released the drug in a controlled manner. Additionally, the study used molecular docking and density functional theory analysis to investigate the interaction between Doxorubicin and the protein HER2, which is involved in cancer development.

Truong-Thi et al. [34] discuss developing a drug delivery system that uses mesoporous silica nanoparticles modified with heparin to improve the effectiveness of Platinum-based drugs in chemotherapy. The system was found to have a sustained and pH-responsive release of the drug and was shown to be biocompatible with MCF-7 cells. The results suggest that this system has the potential to overcome the limitations of toxicity and resistance associated with Platinum-based drugs.

The nanocarrier system for delivering daunorubicin (DNR) to breast cancer cells is reported by Sheykhisarem et al. [35]. The system consisted of Bi_2MoO_6 nanoparticles, NH_2-graphene oxide (GO), and Polyethylene glycol (PEG), and it was pH-sensitive and biocompatible. The nanocarrier system released 86.85% of the drug in acidic pH and only 15% in biological pH, indicating pH-sensitive behaviour. The system had high drug loading content and encapsulation efficiency and did not negatively affect healthy cells but reduced the survival rate of cancer cells. The nanocarrier system showed excellent blood compatibility, making it a promising candidate for anticancer nanocarrier systems.

Gou et al. [36] discuss a drug delivery system that utilises chiral mesoporous silica nanoparticles (CMSN) that have been modified with chitosan (CS) to deliver drugs to tumours in a targeted manner. The CS-CMSN system demonstrated the ability to effectively contain and release doxorubicin (DOX) in response to changes in pH levels. Both in vitro and in vivo experiments showed that the DOX@CS-CMSN system had better uptake in tumour cells and was more efficient at killing them while remaining safe for healthy cells. These findings suggest that the CS-CMSN nanocarrier system is a promising approach for targeted therapy in treating tumours.

The drug delivery system that responds to changes in pH levels for photodynamic therapy can treat tumours through chemotherapy reported by Zhang et al. [37]. The system is created using water-soluble conjugated polymers attached to the anticancer drug doxorubicin through acid-labile imine and acyl hydrazone bonds. The resulting PFE-DOX-2 system can release drugs in response to pH changes and produce reactive oxygen species when exposed to light, thus enhancing its therapeutic effectiveness. The system also permits the monitoring of drug release by observing changes in the fluorescence of the conjugated polymers. This research presents a promising approach to treating cancer in vivo using a multifunctional drug delivery system that responds to stimuli.

The research explored by Rao et al. [38] how palmyra palm kernel (PPK) behaves as a gel and its possible usage as a carrier for delivering drugs to treat colon cancer. The PPK hydrogels displayed impressive swelling properties when exposed to pH, and they effectively encapsulated 5-Fluorouracil (5-FU) with an efficiency rate of up to 62%. Molecularly dispersed 5-FU was found in the PPK hydrogel matrix, and in laboratory tests, the drug was gradually released over 12 h. The study indicates that PPK hydrogels carrying chemotherapeutic drugs can be a treatment alternative for colon cancer.

Efficient chemotherapy was reported by Heragh et al. [39] for synthesising cross-linked chitosan/laponite RD nanoparticles that co-encapsulate doxorubicin and saffron extract. The nanocarriers had a high level of encapsulation for both DOX and SAF and displayed a release behaviour sensitive to pH. The release of DOX and SAF was regulated by the cross-linked CS content, with the release rate increasing in acidic conditions. Additionally, the nanocarriers demonstrated antibacterial properties and high toxicity against cancer cells. The research indicates that using nanotechnology to combine natural antioxidants with synthetic anticancer agents can enhance the effectiveness of chemotherapy while minimising side effects.

The drug delivery by nanoassemblies that release doxorubicin in acidic conditions is typically found in cancerous tissues while exhibiting a lower release rate in normal physiological conditions, as reported by Teixeira et al. [40] discuss the creation of. The nanoassemblies' hybrid composition allowed for high encapsulation efficiency and drug-loading content, and they maintained their colloidal stability for at least four weeks. The DOX-loaded nanoassemblies were shown to have higher cytotoxicity in cancer cell lines compared to free DOX and lower cytotoxicity in normal cell lines. The results suggest that the nanoassemblies can be used in cancer therapy.

Li [41] describes the fabrication of pH-responsive nanoparticles using Polyamine-modified β-Cyclodextrin and Sodium dodecyl benzene sulfonate loaded with Doxorubicin and Celastrol. The nanoparticles were characterised using various techniques and demonstrated minimal release in acidic environments and effective release in basic environments. The release kinetics were studied using different models, and the nanoparticles were found to be highly toxic to tumour cells and low in cytotoxicity to normal epithelial cells. Flow cytometry analysis indicated that the nanoparticles could induce cell apoptosis and had great potential as an oral administration drug delivery system with sustained release and target specificity.

The process of creating hydrogels Batool et al. [42] that can sense pH, biodegrade and are compatible with living organisms. The hydrogels are made using Na-CMC/pectin Poly(methacrylic acid) and are loaded with the anti-cancer drug Cytarabine. Their properties, such as texture, morphology, loading efficiency, and drug release profile, were studied, and they were found to have effective drug encapsulation and continuous release for 24 h. In toxicity studies with rabbits, the hydrogels were safe and compatible with living organisms. In pharmacokinetic evaluation, they showed a significant increase in plasma half-life and AUC. The study's results suggest these hydrogels could be used as a controlled drug delivery system.

2.2 pH-Responsive Hydrogels/Nanoparticles in Wound Healing

The hydrogels that can hold two drugs discussed by Du et al. [43] have pH-responsive and antibacterial characteristics, which are suitable for dressing skin wounds. The hydrogels were loaded with two drugs (Gentamicin sulfate, and Lysozyme), an antibacterial agent and a growth factor. They were designed to be pH-responsive, releasing the drugs more rapidly in acidic environments like those found in infected wounds. The hydrogels were found to have good antibacterial activity and could promote wound healing in vitro. It was suggested that these hydrogels could be used in developing advanced wound dressings with improved efficacy.

Another hydrogel reported by Xie et al. [44] describes the development of a dressing loaded with microcarriers that release antibacterial and pro-vascularizing agents for wound healing. The dressing is pH-responsive and promotes the controlled release of vascular endothelial growth factor (VEGF) to aid in angiogenesis. The dressing effectively inhibits bacterial formation and accelerates wound healing, making it a promising treatment for infected wounds.

Pan et al. [45] focused on silica nanoparticles that are pH-sensitive and loaded with Chlorhexidine for the specific release of the drug on alkaline wounds. These nanoparticles were found to have antibacterial effects against both Gram-positive and -negative bacterial pathogens and exhibited no cytotoxicity. The nanoparticles' administration decreased the number of viable bacterial cells, suggesting they may be useful in treating chronic wound infections.

2.3 pH-Responsive Hydrogels for Regeneration of Bone Tissue

A pH-responsive drug carrier for treating bone infections and promoting bone tissue regeneration was developed by Zhang et al. [46]. The carrier comprises a borate glass core and a mesoporous hydroxyapatite HA shell (BG-HA), coated with glutaraldehyde-crosslinked chitosan (CS). Drug release is minimal at neutral pH, but in acidic environments, the CS swells and opens the pores on the HA shell, resulting in faster drug release. The carrier effectively eliminates bacteria like S. aureus and E. coli and has the potential to gain self-regulated drug release, control the acidic environment at bone infection sites, and promote the regeneration of bone tissue.

2.4 pH-Responsive Membrane for Glucose-Sensing

Patil [47] the study examined how permeable a pH-sensitive membrane made from polyimide and polyacrylic acid (PI-PAAc) was and how it could be turned into a glucose-sensitive membrane by attaching glucose oxidase (GOD) enzymes to it. The PI membranes were initially created using a photolithographic process, with PAAc being grafted onto the pores using a 248 nm KrF laser. The attached GOD molecules were responsive to glucose and could transform it into gluconic acid. The research suggests that these membranes can recognise and release corresponding solute amounts, making them a potential candidate for glucose-sensing purposes.

2.5 pH-Responsive Membrane for Transdermal Patches

Aashli et al. [48] the authors prepared transdermal films of Montelukast Sodium using the solvent casting method. The films were prepared using different concentrations of Polyvinyl alcohol (PVA), Polyethylene glycol (PEG), and Montelukast Sodium. The prepared films were characterised using FTIR, DSC, SEM, and XRD techniques. The drug release kinetics of Montelukast Sodium from the films were studied using the in-vitro release method. The prepared transdermal films showed improved chemical stability and extended drug release of Montelukast Sodium, making them an effective drug delivery system for transdermal administration.

2.6 pH-Responsive Hydrogels in Environmental Preservatives

The hydrogels that are pH-responsive using Carboxymethyl chitosan (CC), Sodium alginate (SA), and Carvacrol (CA) synthesised by Cheng et al. [49]. The hydrogels were formed through electrostatic interactions and hydrogen bonding, and increased CC content resulted in better thermostability. The hydrogels increased in size with higher pH, with CA release mainly controlled by Fickian diffusion. CA release increased with temperature and pH, allowing for pH stimulation and on-demand release. The hydrogels were stable for storage and biocompatible, providing a new solution for environmentally responsive preservatives and intelligent preservation.

2.7 pH-Responsive Hydrogels for Agricultural Chemicals Release

The biochar-supported γ-FeOOH nanoarrays (γ-FeOOH@BC) release agricultural chemicals in a controlled manner, as discussed by Wang et al. [50]. The improved charge of the material allows for accurate release through changes in environmental pH. The study found that QNC and PO43 were released at pH 7.8 and reached 88% at pH 11. The encapsulation efficiency is 75% even after five sensitivity tests, and the structures showed selective herbicidal activity against barnyard grass.

2.8 Graphene-Based pH-Responsive Drug Delivery Systems

An overview of graphene-based materials, including graphene oxide, reduced graphene oxide, and graphene quantum dots, and their potential applications in pH-responsive drug delivery systems was discussed by Banihashemian et al. [51]. The authors review different strategies for designing and preparing graphene-based drug delivery systems, such as chemical modification, functionalisation, and nanocomposite formation. The mechanisms of pH responsiveness in graphene-based drug delivery systems, such as protonation/deprotonation, electrostatic interactions, and hydrophobic/hydrophilic balance, are also discussed. Furthermore, the authors highlight the potential challenges of developing graphene-based pH-responsive drug delivery systems, including stability, biocompatibility, and toxicity.

2.9 pH-Responsive Drug Delivery by Nanocomposites/Blends

The production of halloysite nanotubes for a drug delivery system was reported by Lei et al. [52] for responsiveness to pH. The drug release can be regulated based on the pH of the environment surrounding the nanotubes by attaching pH-responsive polymers to their surface. They discovered that halloysite nanotubes' anisotropic structure enhances their potential to hold and release drugs compared to other nanoparticle types. The research suggests that this approach could have the potential for specific drug delivery and other biomedical applications.

Elham Saleh et al. reported a nanocomposite material that can act as a carrier for the controlled release of Piperine responsive to pH [53]. This material is made by combining bovine serum albumin and oxidised gum Arabic. The material was synthesised using a green approach and was stable and non-toxic. The controlled release of Piperine was demonstrated using in vitro experiments, and the results showed that the release rate was dependent on the pH of the solution. Also performed a molecular docking study to investigate the interactions between the nanocomposite material and Piperine. The study provides a promising approach for developing pH-responsive drug delivery systems using natural and biocompatible materials.

de Jesus Oliveira et al. [54] conducted a study where they developed pH-responsive phthalate cashew gum nanoparticles to enhance the delivery and effectiveness of anti-Trypanosoma cruzi drugs. The study emphasised the importance of utilising biodegradable and biocompatible materials for drug delivery systems. The nanoparticles were spherical, stable, and responsive to pH levels, allowing for sustained drug release. In vitro experiments indicated that drug-loaded nanoparticles had superior efficacy against Trypanosoma cruzi compared to free drugs. The results suggest that pH-responsive phthalate cashew gum nanoparticles could be a promising drug delivery system for treating Chagas disease.

Lee et al. [55], describes a novel approach for treating renal fibrosis, a condition characterised by the excessive accumulation of extracellular matrix proteins in the kidneys, leading to impaired kidney function. The approach involves using a pH-responsive nanocarrier to deliver nitric oxide (NO), a signalling molecule with potent anti-fibrotic effects, specifically to the fibrotic tissue in the kidneys. The nanocarrier is designed to release NO in response to the acidic environment characteristic of fibrotic tissue. The authors demonstrate that the pH-responsive nanocarrier effectively delivers NO to fibrotic kidney tissue in a mouse model of renal fibrosis, reducing fibrosis and improving kidney function. This approach has the potential to be a promising therapeutic strategy for the treatment of renal fibrosis.

The pH-responsive Sodium alginate (SA) and Lignosulphonic (LS) biodegradable polymeric blends with different proportions (80:20) were prepared by the solution casting technique to achieve the desired drug release rate. The prepared polymeric blends were characterised using FTIR. The drug release kinetics of Hydroxychloroquine sulphate (HCQ) from the polymeric blends were studied using the in-vitro release method. The SA/LS blends showed a controlled release of HCQ over an extended period [56]. Crosslinking affects the control drug release of Hydroxychloroquine sulphate (HCQ) drug using alginate beads. The ionotropic gelation method prepared the alginate beads and crosslinked them using different calcium chloride concentrations. The drug release kinetics of HCQ from the beads were studied using the in-vitro release method [57]. The authors concluded that the crosslinked alginate beads showed a controlled release of HCQ over an extended period.

3 Temperature-Responsive Drug Delivery

Temperature is another environmental stimulus that can be used to trigger the release of drugs from stimuli-responsive materials. Temperature-responsive materials can change their structure or solubility in response to temperature changes, resulting in drug release [22]. Temperature-responsive materials are particularly useful for localised drug delivery, as they can be triggered by changes in temperature at a specific site in the body.

One class of temperature-responsive materials that have been extensively studied is thermoresponsive hydrogels. These hydrogels can undergo a reversible phase transition from a swollen to a collapsed state in response to changes in temperature. Below a certain temperature, the hydrogel is in a swollen state and can encapsulate the drug. When the temperature is increased above a critical value, the hydrogel collapses, resulting in the release of the drug. Another class of temperature-responsive materials is liposomes. These materials can be designed to respond to changes in temperature, causing the release of the encapsulated drug.

In summary, temperature-responsive materials have emerged as a promising tool for developing controlled-release drug delivery systems. These materials can respond to changes in temperature to release drugs in a controlled manner and can be used for localised drug delivery. The development of temperature-responsive materials has opened up new possibilities for drug delivery. Further research in this area is expected to lead to the development of more effective and efficient drug delivery systems.

Some of the well-known polymers that release drugs with temperature application are furnished in Table 1.

These are a few examples of polymers commonly used in temperature-dependent drug delivery systems. Depending on the specific drug delivery application and desired temperature-responsive behaviour published by various researchers are provided below.

Table 1 List of well-known polymers that are temperature-responsive drug delivery

Polymer	Properties
• Poly(N-isopropylacrylamide) (PNIPAAm)	This is a well-known thermoresponsive polymer that is commonly used in temperature-dependent drug delivery systems due to its lower critical solution temperature (LCST) behaviour
• Polyethylene glycol (PEG)	It is a biocompatible polymer that can be used as a matrix material for drug delivery systems. It can be modified to exhibit temperature-responsive behaviour by introducing thermo-sensitive groups such as PNIPAAm
• Polymethacrylates	Is a pH-sensitive polymers that can also exhibit temperature-dependent behaviour when combined with other thermoresponsive polymers
• Chitosan	It is a biodegradable and biocompatible polymer that can prepare temperature-dependent drug delivery systems. It can be combined with other thermoresponsive polymers to enhance its temperature-responsive behaviour
• Poly(caprolactone) (PCL)	PCL is a biodegradable and biocompatible polymer that can prepare temperature-dependent drug delivery systems. It can be modified to exhibit temperature-responsive behaviour by incorporating thermoresponsive polymers

3.1 Temperature-Sensitive Drug Delivery for Cancer/Tumour Treatment

The nanocarrier is responsive to changes in temperature reported by Hemmatpour et al. [58]. The carrier is made from a natural and safe clay mineral, halloysite nanotubes, coated with a layer of Poly (N-isopropyl acrylamide) brushes. These brushes have a phase transition temperature of 32 °C, which means they can release drugs at high temperatures. The study shows that the nanocarrier is non-toxic and can successfully release drugs in response to temperature changes. Additionally, the team used a special type of microscopy to confirm that the nanocarrier can enter and interact with cancer cells. The results suggest that this nanocarrier could improve hyperthermia therapies by delivering drugs to cancer cells in response to temperature changes.

The hydrogels developed by Lin [59] respond to temperature and pH changes. The hydrogels are monomethoxy poly (ethylene glycol) copolymers and polypeptides with tertiary amine pendants that transform from α-helix to β-sheet at pH 7.4. The hydrogels can release doxorubicin quickly at acidic pH levels and slowly at neutral pH levels, making them suitable for controlled drug delivery. The hydrogels are also biocompatible in vivo.

Another hydrogel developed by Pourbadiei et al. [60] that responds to light and temperature is made of a copolymer of azobenzene derivative and N-isopropyl acrylamide (NIPAM). The hydrogel is created through host-guest interactions between the hydrophobic core of cyclodextrin and the azobenzene component. At low temperatures, the hydrogel can be injected through a syringe and become a gel after injection due to its LCST-based properties. The researchers explored the drug release behaviour under various light and temperature conditions. They also tested the efficacy of the Paclitaxel-loaded hydrogel in vitro against cancerous fibroblastic cells (A-431) and conducted in vivo studies using staining techniques.

The gold nanorods nanocomposite for synergistic breast cancer treatment was developed by Rawand et al. [61]. The nanocomposites are coated with a pH-temperature responsive polymer and Doxorubicin-loaded mesoporous silica for drug delivery. Upon NIR radiation, the nanocomposite generates heat to induce hyperthermia and trigger the release of Doxorubicin in response to high temperature/low pH at the tumour site. The nanocomposite demonstrated high efficiency in suppressing MDA-MB-231 cell proliferation in vitro and good capability in suppressing tumour growth in mice in vivo while exhibiting good biosafety.

The potential thermosensitive gels as carriers for tumour treatment in interventional therapy developed by Hu et al. [62] can extend the lives of cancer patients, especially those with primary liver cancer. These gels have biocompatibility and drug retention ability and can turn gel-like in response to temperature changes, allowing for rapid embolisation without the risk of ectopic embolisation. The authors reviewed recent advancements in embolisation materials, including chitosan, polyacrylamide and poloxamers, to provide insights into developing and implementing thermosensitive gels for cancer treatment.

Ruan et al. [63] address the issue of reversing drug resistance in cancer patients undergoing chemotherapy and suggest that delivering drugs to the mitochondria could be a solution. They developed a temperature-responsive drug delivery system that targets the mitochondria and overcomes doxorubicin resistance in lung cancer. The nanocarrier prevents drug efflux and allows drug accumulation in resistant tumours, enhancing cytotoxicity and reversal of drug resistance in mice. This is the first study to demonstrate the use of temperature-responsive drug delivery to target the mitochondria and reverse drug resistance in cancer.

The drug delivery system was designed for treating lung cancer using temperature-responsive micelles and hyaluronic acid biopolymers reported by Xu et al. [64]. The system is created using layer-by-layer self-assembly technology, which enables efficient drug loading and a controllable release mode based on environmental stimuli. The study investigates the responsive behaviour of the films to temperature and ionic strength and examines the influence of each component on the drug release profile's response to environmental triggers. The morphology integrity of the films is retained after several temperature-triggered cycles. Additionally, the study examines the system's intermolecular interaction and molecular motions to determine its delivery efficiency. The development of these LBL films offers innovative ideas for modifying traditional drug delivery materials and future functional material design.

The drug delivery system that responds to temperature and pH using a poly(N-isopropyl acrylamide)-co-poly(acrylamide) copolymer and a melamine cross-linker discussed by Thirupathi et al. [65]. The hydrogel system they created had phase transition properties sensitive to temperature. The drug release efficiency was sensitive to pH when tested with curcumin under different temperatures and pH conditions. Under a combination of pH 5.0 and 45 °C temperature stimuli, the system displayed a high drug-loading capacity and almost complete drug release within 8 h.

3.2 Temperature-Sensitive Drug Release Behaviours Using the Artificial Neural Network Technique

The pH and temperature-responsive hydrogel for drug delivery systems developed by Balan et al. [66] offer several advantages, such as high drug loading capacity, biocompatibility, and controlled drug release. The study found the hydrogel had varying fracture strengths and DOX loading capacities. Their drug release behaviour was influenced by time, pH, temperature and physical crosslinking degree of the hydrogel. The study successfully modelled the complex drug release behaviours using the artificial neural network (ANN) technique.

An interpenetrating polymer network (IPN) hydrogel that responds to pH and temperature was synthesised by Boztepe et al. [67]. They focussed on creating a model for predicting drug-release behaviour in hydrogels that respond to environmental changes, such as pH and temperature. This can be challenging due to the many variables in the body. The hydrogel was loaded with the drug doxorubicin (DOX), and its release was observed over time, considering pH and temperature changes. The study employed artificial neural networks (ANNs), least squares support vector machines (LS-SVM), and support vector regression (SVR) methodologies to model the experimental release data, with the ANN model proving to be the most effective at predicting the behaviour of the IPN hydrogel in releasing DOX. The performance of the models was assessed using statistical measures.

Computer simulations were explored by Sakai et al. [68] for the impact of temperature on Cyclophosphamide (CP) discharge from β-Cyclodextrin (β-CD). The researchers discovered that CP was released from β-CD at 127 °C and that the β-CD/CP complex displayed less rigidity at elevated temperatures. These results may aid in developing temperature-sensitive drug delivery systems that employ β-CD nanocarriers.

3.3 Temperature-Sensitive Drug Release-Copolymer as a Gatekeeper

Mesoporous silica drug delivery method reported by Thirupathi et al. [69]. They utilised a copolymer (PNIPAm-PAAm) that acted as a gatekeeper, reacting to changes in temperature and pH, to cover the surface. This system regulates drug delivery under various pH and temperature conditions. The system's biocompatibility was confirmed using MTT assay and cellular internalisation research. The research implies that the

MS@PNIPAm-PAAm NPs they produced could serve as a drug delivery mechanism where continuous drug release is required at elevated temperatures.

3.4 Temperature-Sensitive Drug Release from Microsphere

Microspheres made of poly(methyl methacrylate) (PMMA) and poly(methyl methacrylate-co-N-vinyl caprolactam) (P(MMA-NVCL) by oil-in-water emulsion method. It was photo polymerised under green LED irradiation to load erythromycin (EM), as reported by Yu et al. [70]. The microspheres were characterised using various techniques and were observed to have well-controlled release behaviour for up to 12 h. EM concentration primarily controlled the release mechanism, but hydrogen bond rupture and microsphere relaxation also contributed. The P(MMA-NVCL)-EM microspheres were temperature-sensitive and had a higher EM loading content than PMMA-EM microspheres.

Nanospheres of Chitosan/hydroxypropyl cellulose containing Flurbiprofen drugs were synthesised as reported by Işıklan et al. [71]. The nanospheres were found to have a temperature-responsive feature, and their release profiles depended on various factors. The study also demonstrated the biocompatibility of the nanospheres. The results suggest the nanospheres can be used as a controlled drug release carrier.

3.5 Temperature-Sensitive Drug Release in Regenerative Medicine

The drug delivery for stem cells in regenerative medicine is developed by Huang et al. [72]. A temperature-sensitive hydrogel using a combination of two materials, methylcellulose (MC) and polyvinyl alcohol (PVA) synthesised, to address the problem of low cell survival and retention in cell delivery therapy. Creating physical hydrogen bonds between MC and PVA chains produced porous hydrogels with a high water retention capacity, stable modulus, fast gelation, and injectability. These hydrogels are highly transparent to light. The team found that the hydrogels were non-toxic to rabbit adipose-derived stem cells and could be utilised as a delivery method for stem cells in regenerative medicine.

3.6 Temperature-Sensitive Drug Release from Hybrid Hydrogels

Hybrid hydrogels are developed by Emam et al. [73] with stimuli-responsive properties by using Succinylated cellulose nanocrystals (Su-CNC) and Poly(N-isopropylacrylamide) (PNIPAm) via free radical polymerisation. The resulting hydrogels exhibited pH and temperature responsiveness, with hydrophobicity and shrinkage occurring at 35 °C and higher temperatures. At pH 6, the hydrogels displayed significant equilibrium swelling ratios, and the Su-CNC/PNIPAm 2 hydrogel demonstrated potential for in vitro Famotidine release due to its pH responsiveness.

3.7 Temperature-Sensitive Drug Release from Nanocomposite

Uniform temperature-sensitive hollow mesoporous silica nanoparticles (HMSN@P(NIPAM-co-NHMA)) with high drug loading capacity were created by Zhu et al. [74] for drug delivery purposes. The nanoparticles had P(NIPAM-co-NHMA) polymer attached to their surface. They were tested using Puerarin (PUE) as a drug model, demonstrating excellent loading efficiency and temperature-sensitive release. The researchers also discovered that HMSN and HMSN@P(NIPAM-co-NHMA) were highly stable and biocompatible.

3.8 Temperature-Sensitive Drug Release in Ocular Treatments

In their research, Ow et al. [75] examine the impact of ocular diseases on patients and the healthcare system, and stress that nearly half of all vision problems can be prevented or treated. One possible solution could be to use temperature-sensitive hydrogels that are safe for the body and can be transformed into a gel by changing the temperature. Recent advancements in temperature-responsive polymers show potential for treating various eye conditions such as cataracts, dry eyes, glaucoma, retinal detachment, and age-related macular degeneration. These developments may have applications in various ocular treatments, such as vitreous substitutes for retinal surgery, topical eye drops, and lenses that can reduce inflammation and discomfort. A review article highlights the potential of these advances in treating ocular diseases.

3.9 Temperature-Sensitive Drug Release from Nanocomposites

A temperature-sensitive nanocarrier for hyperthermia therapies using halloysite nanotubes and poly(N-isopropyl acrylamide) brushes were reported by Hemmatpour et al. [76]. The nanocarrier was synthesised through a mussel-inspired dopamine polymerisation and surface-initiated atom transfer radical polymerisation. The resulting material has a polymer layer surrounding the nanotubes, and drug release studies demonstrated a temperature-responsive release behaviour due to the Poly(N-isopropyl acrylamide) brushes.

The drug delivery system is intelligent and functional by using silica nanoparticles with radially porous structures and functionalised ligands encapsulated with polymer developed by Choi et al. [77]. This system has a high capacity for drug loading and can respond to changes in pH and temperature. The researchers could load ibuprofen up to 270 wt.% with high efficiency and achieve pH and temperature-responsive drug release by using amine functionalisation and a temperature-responsive agarose gel surface coating. In vitro tests showed that the nanoparticles did not show toxicity on fibroblast cells, suggesting they have potential as nanomedicine for drug delivery applications.

3.10 Temperature-Responsive Block Copolymer

Kotsuchibashi et al. [78] explained two kinds of temperature-sensitive polymers depending on their characteristics: lower critical solution temperature (LCST) and upper critical solution temperature (UCST). The authors highlighted that block copolymers possessing dual-temperature-sensitive properties are predicted to show complicated states and structures. Recent advancements have produced block copolymers with multi-temperature-sensitive properties with potential applications in drug delivery, sensors, model proteins and memory storage. The review paper is split into three parts, concentrating on dual-temperature-responsive block copolymers, dual-temperature-responsive hydrogels and multi-temperature-responsive block copolymers.

3.11 Temperature Responsive Aerogels

Liu et al. [79] have developed a new drug delivery system, which uses temperature-responsive aerogels made from Hydroxypropyl methylcellulose (HPMC)-grafted N-isopropyl acrylamide (NIPAM). The researchers conducted several analyses to determine the structure and morphology of the aerogels and tested the release behaviour of a model drug, 5-Fluorouracil, from the aerogels. The study found that the release of the drug was controlled and sustained and dependent on temperature. The researchers also used models such as first-order kinetic, Higuchi, and Korsmyer-Peppas to analyse the sustained-release mechanism. The results suggest that HPMC-NIPAM aerogels could be a promising drug delivery system.

3.12 Temperature-Responsive Hybrid Nanoparticles

Hajebi et al. [80] describe the preparation of hybrid nanoparticles that are sensitive to temperature. The nanoparticles were created using N-Isopropyl acrylamide and vinyl-modified silica nanoparticles, with varying cross-linking densities achieved by altering the N, N-Methylene bisacrylamide percentage. The silica cores were then hydrolysed to form hollow poly(N-isopropyl acrylamide) nanogels. The researchers studied the sensitivity of these nanogels to temperature using UV-vis spectroscopy and dynamic light scattering. They also examined the release of the drug doxorubicin from the nanogels at different temperatures and found that the Korsmeyer-Peppas model provided the best fit for the release data.

3.13 Temperature-Sensitive Hydrogel Loaded with Micelles

Zhao and colleagues [81] developed a novel hydrogel loaded with micelles that respond to changes in pH and temperature and can release hydrophobic drugs like indomethacin in a controlled manner. The hydrogel was created using tremella polysaccharide, carboxymethyl cellulose, and a nonionic surfactant called decyl polyglucoside, and was prepared by crosslinking N-isopropyl acrylamide through radical polymerisation. The hydrogel was subjected to various tests to characterise its properties, and the researchers studied its drug release behaviour, which was found to be influenced by changes in pH and temperature. The study suggests that hydrogel has potential applications in biomedical fields such as drug delivery and tissue engineering.

4 Light-Responsive Controlled Drug Delivery

Light is another environmental stimulus that can be used to trigger the release of drugs from stimuli-responsive materials. Light-responsive materials can be designed to respond to specific wavelengths of light, resulting in the release of drugs. Light-responsive materials are particularly useful for localised drug delivery, as they can be triggered by light at a specific site in the body.

One class of light-responsive materials that have been extensively studied is photoresponsive hydrogels. These hydrogels can undergo a reversible structural change in response to light, releasing the drug. The release mechanism can vary depending on the type of photoresponsive material used. For example, some photoresponsive materials undergo a photochemical reaction, while others undergo a photothermal or photoisomerisation reaction. Another class of light-responsive materials is nanoparticles, which can be designed to release drugs in response to light.

Light-responsive materials are an efficient tool for developing controlled-release drug delivery systems. These materials can respond to specific wavelengths of light to release drugs in a controlled manner and can be used for localised drug delivery. The development of light-responsive materials has opened up new possibilities for drug delivery. Further research in this area is expected to lead to the development of more promising and efficient drug delivery systems.

4.1 Light-Responsive CDD for Cancer Therapy

A small device made of DNA developed by Liu et al. [82] can deliver drugs in a controlled manner to treat tumours. They modified GC-rich DNA duplexes with a light-sensitive molecule to construct the device. They placed them on the surface of nanoparticles that can convert near-infrared light into ultraviolet light. They then added the chemotherapy drug doxorubicin to the DNA duplexes, where its fluorescent signal was suppressed. When the nanoparticles were exposed to near-infrared light, they emitted UV light, which caused the light-sensitive molecule to break and release the doxorubicin in a controlled manner while restoring the fluorescent signal. The nanodevice can selectively induce apoptosis in tumour cells through light activation and allows in situ drug release monitoring. This approach has the potential for remote-controlled drug delivery.

Lipid nanocapsules reported by Brion et al. [83] are sensitive to light and can release an anti-tumour drug using photolysis. The researchers used photon upconverting nanoparticles (LNC-UCs) connected to a photocleavable linker based on coumarin. This linker can efficiently and quantitatively release the drug by upconversion luminescence-assisted photolysis when excited by a deep-red wavelength of light. The nanocapsules are stable without light and do not release the drug, making them suitable for nanomedicine applications. This technology allows for the precise control of drug release or activation by light, making it ideal for the targeted treatment of cancer and other diseases.

A hydrogel-based system for chemodynamic therapy (CDT) and photothermal therapy (PTT) to treat tumours was designed by Liu et al. [84]. The system utilises a composite nanomaterial called Cu-Hemin-Au loaded into agarose hydrogels, which are injected intratumorally and stay in the tumour for an extended period. When exposed to near-infrared light, Cu-Hemin-Au acts as a photothermal agent, generating heat, which causes the hydrogel to soften and release the Cu-Hemin-Au to achieve photothermal therapy (PTT). The Au nanoparticles in Cu-Hemin-Au also generate hydrogen peroxide, which reacts with Cu-Hemin-Au to produce reactive oxygen species for CDT. The hydrogel system achieves good PTT/CDT synergy and shows promising results in inhibiting tumour growth in mice. This hydrogel system may inspire the development of new hydrogels for antitumor therapy.

The drug delivery system that responds to near-infrared (NIR-II) light using Polyethylene glycol (PEG)-modified hollow Cu_xS nanoparticles developed by Zhou et al. [85]. The nanoparticles were designed to release drugs at a high penetration depth, making them ideal for chemo-photothermal therapy. They demonstrated high drug loading capacity and stimuli-responsive drug release triggered by NIR-II laser irradiation. The study found that the nanoparticles achieved a 98.5% reduction in tumour size through synergistic chemo-photothermal therapy. These findings suggest that CuxS-PEG nanoparticles have the potential as a comprehensive NIR-II light-responsive nanocarrier-based drug delivery system for various biomedical applications.

A gold nanorods (AuNRs) coated with mesoporous silica nanoshells (MSNs) and two gatekeepers, Lauric acid (LA) and Tannic acid (TA), for pH- and near-infrared (NIR) light-controlled release of drugs reported by Park et al. [86]. The pH-sensitive TA layer swells or shrinks based on pH changes, enabling pH-responsive drug release. In contrast, the thermosensitive LA layer undergoes a solid–liquid phase transition due to heat generated by NIR light in AuNRs, enabling NIR light-controlled drug release. The combination of these gatekeepers allows for on–off drug release switching in acidic environments under NIR light without premature drug leakage. The system has demonstrated effective anticancer performance.

In their study, Ray et al. [87] suggested that a combination of therapies could be an effective approach to enhance the effectiveness of treatment and decrease negative side effects. They created a new technology, which involved a dual-arm acridine photocage that releases two amino and carboxylic acids at the same time. This technology created a dual-drug delivery system that responds to visible light and generates single-component fluorescent organic nanoparticles. These nanoparticles were used to carry two anticancer drugs (chlorambucil and valproic acid) and were tested on HeLa cell lines in vitro. The results showed that the nanoparticles had better anticancer effectiveness and allowed for real-time fluorescence tracking of drug release. The flat structure of the photocage facilitated its insertion into DNA, thereby increasing the drugs' cancer-killing potential.

Another gold nanostructure as a drug carrier for tunable optical properties against cancer treatments was reported by Zafar et al. [88]. Stimuli-assisted drug delivery systems, particularly those that are externally responsive, have received attention from researchers as they provide more controlled drug delivery. Light-responsive DDSs, particularly those that respond to near-infrared light, are potential candidates due to their proficiency and spatiotemporal control. The article summarises various shapes and structures of gold nanoparticles and how they can enhance the effectiveness of drug delivery systems by exploiting the surface plasmon resonance phenomenon.

Folic acid-functionalized polydopamine nanoparticles (NPs) for the combined photothermal and chemotherapy treatment of breast cancer reported by Fan et al. [89]. These NPs respond to tumour acidity and near-infrared light (NIR), allowing targeted drug delivery and enhanced cellular uptake. In vitro and in vivo studies demonstrate that these NPs improve drug accumulation, penetration, and retention, leading to higher therapeutic efficacy and better survival rate in breast cancer-bearing mice. The article concludes that these NPs have the potential as a useful nanoscale vector for enhanced cancer therapy.

4.2 Light Sensitive-CDD by Hyperbranched Polymer

A light-sensitive polymer named Azo-HPG consists of hyperbranched Polyglycerol as the shell and Azobenzene as the core discussed by Rafiee et al. [90]. Anionic ring-opening polymerisation was used to create the polymer, and its structure was analysed using various techniques. The article emphasises the utilisation of azo compounds for designing photo-responsive compounds and their possible applications in medicine and industry. The polymer's potential as a drug delivery system was explored under different pH levels and UV irradiation, indicating that it could be used as a drug carrier.

4.3 Light Sensitive-CDD to Bacterial Infections

A near-infra-red (NIR) responsive nanocomposite to combat bacterial infections has become increasingly difficult to treat due to the rise of multidrug-resistant bacteria [91]. The nanocomposite comprises unzipped carbon nanotubes (uCNTs) modified with mussel-inspired polydopamine (PDA) coating, improving biocompatibility and NIR responsiveness. The uCNT@PDA nanocomposite exhibits excellent photothermal properties, with a temperature increase of 40 °C upon exposure to 808 nm NIR light. Additionally, they created a nanofiber mat using PCL/uCNT@PDA that has sturdy mechanical properties and can generate controlled photothermal heat. This mat was effective in killing both gram-positive and gram-negative bacteria, and it also released curcumin when exposed to near-infrared light.

Hydrogel is made from natural polymers sensitive to near-infrared (NIR) light and can be used for controlled drug delivery [92]. The hydrogel was made by combining Dextran, Graphene oxide, and Laponite and was loaded with Ciprofloxacin as a test drug. The hydrogel heated up when exposed to NIR light, triggering drug release. Experiments showed that the drug release remained controlled even when simulated at different tissue thicknesses. The hydrogel also showed good antibacterial properties and compatibility with blood. The study suggests that NIR light-responsive materials may have potential applications in antimicrobial therapy.

4.4 Light Sensitive-CDD for Periodontal Restoration

A new technique for treating periodontitis by developing a nano-drug delivery system that responds to near-infrared (NIR) light and uses carvacrol [93]. The system comprises upconversion nanoparticles (UCNPs), mesoporous silica ($mSiO_2$), and hydrophobic CA, which exhibits specific antibacterial, anti-inflammatory, and immunomodulatory properties when exposed to 808 nm NIR light. The system's unique features accelerate and provide a non-invasive treatment option for periodontitis. Additionally, the study shows that the immune microenvironment is remodelled through various classic inflammatory immune-related signalling pathways. Therefore, this herbal monomer-based light-responsive nano-drug delivery system shows potential as an effective and reliable treatment strategy for various deep-tissue diseases.

4.5 Light Sensitive-CDD on Skin Patch

A silica-polyvinylpyrrolidone (SiO_2-PVP) composite hydrogel was developed that responds to light and releases drugs locally [94]. The hydrogel was made using partially hydrolysed Tetraethyloxysilane and PVP, which provided flexibility and deformation ability to the SiO_2 framework. The hydrogel exhibited strong mechanical properties and could stretch up to 160% and swell up to 600%. When exposed to UV light and higher temperatures, the hydrogel exhibited a quicker release of drugs and could even release drugs through artificial skin. It was also found to be safe for use on the skin, indicating its potential use as a skin patch.

4.6 Light Sensitive-CDD for Wound Healing

A multifunctional hydrogel for treating chronic diabetic wounds consists of a bilayer structure that is covalently crosslinked, with the lower layer being thermoresponsive and the upper layer highly stretchable [95]. The hydrogel is embedded with peptide-functionalized gold nanorods (AuNRs) in each layer. Upon near-infrared (NIR) stimulation, the AuNRs are released in a two-stage sequential manner. The lower layer releases antibacterial peptide-functionalized AuNRs to control bacterial infections. The upper layer releases pro-angiogenic peptide-functionalized AuNRs to promote angiogenesis and collagen deposition during the subsequent healing phases.

4.7 Light Sensitive-CDD for Brain Diseases

Xue et al. [96] discuss the importance of exploring new materials for diagnosing and treating brain diseases. Near-infrared (NIR) light-responsive materials are particularly interesting due to their high spatial resolution and strong penetration into the skull. The article summarises recent advances in developing NIR light-responsive materials for diagnosing and treating various brain diseases, including approaches targeting the blood–brain barrier (BBB), imaging-guided therapies, and challenges and prospects for future research.

4.8 Light Sensitive-CDD by Photosensitive Hydrogels

Xing et al. [97] examine the potential benefits of using light-sensitive hydrogels for drug delivery. These hydrogels contain photosensitive components that enable drug release through photoisomerisation, photochemical, or photothermal reactions. Recent advances in materials science have created a wide range of photosensitisers that can respond to various light sources, including ultraviolet and near-infrared light, as well as up-conversion nanoparticles. Light-sensitive drug delivery systems have been employed in various applications, such as chemotherapy, immunotherapy, photodynamic therapy, gene therapy, and wound healing. The article provides an overview of current research, challenges, and prospects in light-responsive hydrogels for controlled drug delivery, emphasising the need to improve effectiveness, safety, patient compliance, and convenience.

4.9 Light Sensitive-CDD in Ophthalmology

Abdelmohsen et al. [98] discuss the potential of light-responsive biomaterials for precisely delivering therapeutic drugs and nucleic acids, focusing on their use in ophthalmology. Light-responsive drug delivery systems are considered non-invasive and have significant clinical potential. The article reviews various light-responsive polymers and their chemistry, highlighting their potential applications in ophthalmic drug delivery systems.

4.10 Light Sensitive-CDD in Pesticides

The article by Shan et al. [99] aims to enhance the efficacy of pesticide use while reducing negative impacts on both the environment and human health. The authors developed a biodegradable amphiphilic polymer via a simple and mild process. This polymer was then used to create a nanosized pesticide system that responds to light for controlled release of 2,4-dichloro phenoxy acetic acid (2,4-D). The system was found to release the pesticide stably in the absence of UV light and showed increasing release in the presence of UV light. The study demonstrated the system's efficacy in controlling weeds while reducing toxicity to non-target organisms.

4.11 Light Sensitive-CDD by Nanoparticles

Liu et al. [100] address the difficulties of targeted drug delivery and explore how light-responsive drug delivery systems can help overcome these challenges. They explain how photoremovable protecting groups, which are light-sensitive molecules that can be attached to a drug molecule to control its function and structure with light, have been used in recent applications of nanoparticle-based drug delivery. The authors also discuss various approaches for achieving long-wavelength light excitation and emphasise the importance of understanding these mechanisms for designing more effective and precise photoresponsive drug delivery systems.

4.12 Light Sensitive-CDD in Pulsatile Systems

Khalifa et al. [101] have written about the benefits of pulsatile drug delivery systems compared to conventional dosage forms. Pulsatile systems can administer precise amounts of medication to targeted locations at specific times, enhancing patient adherence and permitting customised treatment. The article concentrates on recent breakthroughs in externally controlled pulsatile release systems, such as electro-responsive, light-responsive, ultrasound-responsive, magnetically triggered, and wirelessly managed implantable systems. The present state of these technologies is explored, along with their possible future uses.

4.13 Light Responsive-CDD by Nanocarriers

A review article was reported by Tang et al. [102] on the development and production of nanocarriers for releasing drugs activated by near-infrared (NIR) light. The article concentrates on NIR light with a 780–1700 nm wavelength range. It can penetrate deep into tissues with minimal damage to cells, making it an ideal option for in vivo light-activated delivery systems. The review highlights three nanocarrier triggering mechanisms: chromophores' photoreactions, photothermal effects caused by inorganic or organic photothermal conversion agents, and photo-oxidation generated by photosensitisers. Additionally, the article discusses the challenges and prospects for creating NIR light-responsive nanocarriers.

Recent advancements in photo-responsive polymeric nanocarriers designed for drug delivery systems [103]. Traditional delivery systems have shortcomings linked to their limited efficacy and safety. Stimuli-responsive polymeric nanomaterials have gained attention in the biomedical sector due to their flexible behaviour in response to their environment. Light irradiation is particularly fascinating among the potential triggers for these materials because it can be precisely localised and non-contact. The review explores different photo-responsive polymeric nanocarriers, including nanoparticles, nanogels, micelles, nanofibers, dendrimers, and polymersomes, along with their drug release mechanisms. The authors conclude that photo-responsive polymeric nanocarriers have the potential to overcome the limitations of conventional drug delivery systems and enhance targeted drug delivery.

A new method for producing nanocarriers responsive to stimuli can be used for drug delivery and tissue engineering [104]. The approach involves using molecular motor-doped cholesteric liquid crystals (CLCs) and drugs to create microcapsules responsive to UV light. Under UV light, the molecular motors isomerise, inducing a change in the superstructure of the CLCs and accelerating drug release. The authors suggest this method provides new opportunities for designing and fabricating functional drug delivery systems.

5 Magnetic-Responsive Controlled Drug Delivery

Magnetic fields are another environmental stimulus that can be used to trigger the release of drugs from stimuli-responsive materials. Magnetic-responsive materials can be designed to respond to an external magnetic field, releasing drugs [105]. Magnetic-responsive materials are particularly useful for targeted drug delivery, as the magnetic field can guide the materials to a specific site in the body.

One class of magnetic-responsive materials that have been extensively studied in magnetic nanoparticles. A temperature-responsive material can be used to coat these nanoparticles, which react to temperature changes and cause the drug to be released [106]. Additionally, a pH-responsive material can be used to coat the nanoparticles, which react to pH changes and cause the drug to be released. Furthermore, an external magnetic field can trigger the release of the drug, allowing the nanoparticles to be guided to a particular location in the body [23].

Using magnetic-responsive materials has shown great potential in developing drug delivery systems that can be controlled and released in a targeted manner. By responding to an external magnetic field, these materials can be designed to release drugs in a controlled way. The emergence of magnetic-responsive materials has brought about new opportunities for drug delivery. As further studies are conducted in this area, more efficient and effective drug delivery systems are expected to be developed [107].

This chapter proposes the development of a soft, magnetically-responsive carrier for drug delivery. The researchers used a moulding process to create prototypes from a shape memory polymer called DiAPLEX MP-3510, which has a low transition temperature. The design was validated through simulations, and the thermal responsiveness of the prototypes was tested both ex-vivo and in a phantom. The results show that the design is viable and capable of delivering drugs to a targeted area in the large phantom intestine, utilising the rubbery modulus of the polymer and enhancing the recovery force through magnetic actuation [107].

Some of the articles published and their outcomes are furnished below.

5.1 Magnetic-Responsive CDD in Cancer Treatment

A review article on the latest uses of magnetic nanocarriers in treating tumours was reported by Zhu et al. [108]. Nanocarriers that include magnetic nanoparticles have been applied in different therapeutic fields, such as biocatalysis, MRI, magneto-thermal therapy (MHT), and photo-responsive therapy. Magnetic nanocarriers enable the efficient delivery of drugs to specific areas, and their position can be tracked to manage and continually assess the effectiveness of treatment. The article outlines multiple applications of magnetic nanocarriers in tumour treatment, including stimuli-responsive drug delivery, MHT, photoresponsive therapy, immunotherapy, gene therapy, and synergistic therapy. The authors also discuss the challenges associated with clinical application and the potential of magnetic nanocarriers in treating cancer.

The potential use of a novel degradable microgel for drug delivery and hyperthermia treatment was developed [109]. The microgel comprises Poly(N-isopropylacrylamide) cross-linked with N, N′-Bisacryloylcystine and contains superparamagnetic iron oxide nanoparticles. The microgel exhibits temperature sensitivity and stability across various conditions, and the presence of the nanoparticles allows for remote temperature control for improved drug release. The microgel loaded with the anticancer drug doxorubicin demonstrated enhanced drug release at higher temperatures and in the presence of glutathione. In vitro studies also showed comparable cytotoxicity to free doxorubicin against cancer cells but less toxicity to healthy cells.

A novel thermoresponsive magnetic hydrogel microparticle system for potential biomedical applications was reported by Durkut et al. [110]. The system comprises magnetic nanoparticles combined with Poly(N-vinylcaprolactam)-g-galactosylated chitosan hybrid hydrogel. The microparticles were synthesised using the suspension crosslinking method and characterised using various analyses. The resulting microparticles were biocompatible, temperature- and magnetic-responsive, and had enhanced mechanical stability. The system can be used as a controlled release system, in situ bioactive agent carrier, and anti-cancer therapy.

A dual-stimuli responsive system was developed for cancer treatment by incorporating Fe_3O_4 magnetic nanoparticles and Poly(N-isopropyl acrylamide) (PNIPAAm) microgels into electrospun polymeric fibres [111]. The study showed that adding microgels and Fe_3O_4 NPs to electrospun fibres can enhance Young's modulus. It also indicated that Fe_3O_4 NPs reduce the membranes' swelling ratio, as observed in swelling assays. Additionally, magnetic hyperthermia assays demonstrated that a higher concentration of NPs leads to better heating performance, and the composite membrane containing DMSA-coated NPs exhibited the highest temperature variation. The study also confirmed that the composite membranes are biocompatible and support long-term cell viability, indicating their potential for cancer treatment, particularly for solid tumours that are anatomically reachable.

The magnetic vortex Fe_3O_4@PVP@DOX nanostructures for accurate cancer therapy was developed by Wang et al. [112]. These nanostructures possess magneto-triggered on-demand hyperthermia, magnetic-responsive drug delivery, and MRI T2 signal enhancement. In vitro experiments yielded positive outcomes, while in vivo experiments inhibited tumour growth without adverse effects. The nanostructures present a novel method for developing stimuli-responsive theranostic platforms to achieve precision medicine for cancer.

Mazidi et al. [113] summarise recent developments in biomaterials that can be implanted to provide targeted drug delivery to cancerous tissues. The article explores systems that respond to redox, enzymes and pH changes, allowing for drug-controlled release. Additionally, the authors discuss systems regulated by near-infrared, ultrasound, electric, and magnetic stimuli, enabling drugs to be delivered as needed. The paper also covers the preparation process, implantable materials' physical and chemical properties, and the models used to predict drug release. The article highlights how implantable drug delivery systems that respond to specific stimuli can improve cancer treatment.

Nanocomposite material was synthesised using magnetic nanoparticles coated with silica and modified with Polyethylene glycol (PEG) conjugate [114]. This material allows for the high-capacity loading of the anti-cancer drug Doxorubicin. The researchers found that this material exhibited high cytotoxicity against tumour cells and could release the drug when exposed to an alternating magnetic field (AMF), increasing its effectiveness.

Magnetic thermosensitive hydrogels are synthesised using microfluidic technology to achieve targeted delivery of anticancer drugs that are both hydrophobic and hydrophilic [115]. They incorporated Fe_3O_4 nanoparticles into the shell layer of double emulsions, which enabled thermal effects and controlled release of the drugs under a high-frequency alternating magnetic field. The hydrogels showed effective cytotoxicity in inhibiting cancer cells, and the researchers deemed PNIPAM a safe carrier for drug delivery systems.

5.2 *Magnetic-Responsive CDD in Wound Healing*

Wound dressings are explored could effectively solve the problem of quick drug loss and weak healing effects in patients who have undergone bladder tumour resection [116]. To tackle this problem, they developed cellulose nanofibers wound dressing that reacts to pH and near-infrared light, magnetic switching Fe_3O_4 nanoparticles, and temperature switching Pluronic®F-127. This dressing can be attached to the tissue site, release drugs using an external magnetic field, and maintain a 3D network for an extended time. Furthermore, the dressing exhibited strong antibacterial properties, eliminated biofilms, and killed T24 tumour cells. It also stimulated wound healing through photothermal, photodynamic, and chemotherapy mechanisms. This wound dressing is appropriate for bladder postoperative infected wound healing and overcomes the issue of fast drug loss due to cyclical urination.

Li et al. [117] discuss how wound healing is a complicated process that requires suitable microenvironments to be successful. These microenvironments consist of various biological, chemical, and physical factors that are influenced by internal and external factors. However, delivering drugs to wounds is challenging due to the need for blood supply, persistent inflammation and other factors. Using stimuli-responsive biomaterials can provide accurate drug delivery and release by responding to various factors from wound microenvironments or external sources. The article discusses recent developments in stimuli-responsive biomaterials to regulate microenvironments during wound healing. Various biomaterials that respond to wound and physical microenvironments and new drug carriers are highlighted.

A hydrogel was developed based on 2D MXene that uses photo and magnetic-responsive drug delivery to treat chronic wounds [118]. This system includes magnetic colloids wrapped with MXene and dual-network hydrogels, which enable multiple response capabilities and controlled drug delivery. They tested this system on infected wounds in a rat model and found it has potential for clinical wound healing and other biomedical applications.

5.3 Magnetic-Responsive Targeted Drug Delivery

The nanocomposite combines magnetic hydroxyapatite and a metal–organic framework with a biocompatible Polyethene glycol coating designed for specific drug delivery [119]. The nanocomposite had a substantial drug-loading capacity and high encapsulation efficiency. The nanocomposite demonstrated low toxicity and high effectiveness against cancer cells. Additionally, it exhibited excellent antioxidant activity, making it a promising candidate for targeted drug delivery applications.

The Fe_3O_4@ZIF-8, a magnetic metal–organic framework for drug release applications, particularly for Norfloxacin, owing to its high specific surface area and inherent biodegradability examined by Cai et al. [120]. The study investigated the release behaviour of Norfloxacin and its antibacterial effectiveness at different pH levels. The findings showed that Fe_3O_4@ZIF-8 demonstrates pH-sensitive properties, and the release process followed the Bhaskar model. Moreover, the composite exhibited good biocompatibility and antibacterial activity against Escherichia coli. Due to its magnetic properties and pH-responsive behaviour, Fe_3O_4@ZIF-8 holds potential as a targeted drug delivery system.

The combined magnetic and polymeric materials were synthesised to achieve a unique magnetically responsive behaviour controlled remotely with magnetic fields [121]. Using magnetically responsive polymeric and elastomeric composites, drug release can be precisely controlled through magnetic actuation. Different magnetic fillers, polymers, and fabrication methods can create materials with integrated functions such as shape morphing and heat generation. The article comprehensively discusses the mechanism, challenges, components, fabrication techniques, properties, emerging advancements, applications, and prospects of using magnetically modulated and functionalised polymers and elastomers for drug delivery.

The magnetic polymer hybrids responsive to external stimuli (RMPHs) combine the advantages of organic and inorganic materials, where magnetic nanoparticles provide magnetic properties [122]. At the same time, the polymeric component is responsible for responding to changes in pH, temperature, redox reactions, or irradiation. RMPHs are attractive for various applications, such as catalysis, biotechnology, imaging, and cancer therapy, due to the ease of modification of the polymeric materials. The review explores different synthetic approaches to create RMPHs of different shapes, including nanometric core-shell structures, nano gels, microgels, and membranes, focusing on their potential targeted applications.

The potential use of mesoporous silica nanoparticles (MSNs), as Mohanan reported, drug delivery vehicles, Mohanan et al. [123]. It highlighted their structural stability, ability to load various drugs, and tunable morphologies/sizes. The surface silanol groups on MSNs allow for functionalisation with drugs, imaging agents, and targeting molecules, making them a popular platform for drug delivery. The article specifically focuses on developing stimuli-responsive silanol conjugates as a novel approach to delivering therapeutic drugs to a specific location on demand, using either endogenous or exogenous stimuli such as pH, redox potential, temperature, and hypoxia. The authors also address the challenges in understanding the role of silanols and overcoming limitations in synthesising stimuli-responsive mesoporous silica-based drug delivery systems.

Recent advancements in using magnetic hydrogels made with polysaccharides have various biomedical applications [124]. These hydrogels are considered intelligent platforms because they can be controlled remotely via a magnetic field and possess excellent biocompatibility and biodegradability. The author discusses the methods of synthesising these hydrogels and their synergistic properties when combined with magnetic nanoparticles. The paper emphasises the importance of these polysaccharide-based magnetic hydrogels in fields such as targeted drug delivery, tissue regeneration, and hyperthermia therapy. The article aims to encourage the development of new composite biomaterials that can help overcome challenges in treating different diseases.

5.4 *Magnetic-Responsive in Smart Nanomedicine*

As a review reported by Armenia et al. [125], smart nanomedicine can enhance drug delivery using smart nanomaterials that respond to internal and external stimuli. To increase responsiveness to external stimuli, photonic and magnetic nanoparticles are utilised. The authors highlight recent progress in integrating these materials within different carriers to improve efficacy, stability, and toxicity. They also discuss the current regulatory hurdles and the importance of standardising these materials.

A new hydrogel composite of ferrogel, responsive to magnetic fields by adding Fe_3O_4 nanoparticles to a pre-existing hydrogel, was temperature-responsive [126]. The study then investigated how the composition of the ferrogel affected its response to alternating magnetic fields. The researchers also used artificial neural networks to predict how the ferrogel would behave in different conditions accurately. These findings have important implications for developing smart materials for use in the biomedical field.

5.5 Magnetic-Responsive CDD in Osteomyelitis

The device to treat bone infections called osteomyelitis was reported by Shademani et al. [127], which includes a magnetic sponge capable of on-demand drug delivery. This device employs a combination of Silver nitrate and Gentamicin to combat drug-resistant pathogens effectively. The drug delivery system can release drugs for seven days and is easy to fabricate, allowing for control of the release profile.

5.6 Magnetic-Responsive CDD in Combination Therapy

The drawbacks of traditional antibacterial therapies and the rise of multidrug-resistant microorganisms were reported [128]. Nanoparticle-mediated therapeutics as a potential solution for targeted drug delivery and stimulus-responsive antibiotic delivery as a promising approach for site-specific drug delivery. The article also emphasises the advancements and outlooks of stimulus-responsive combination therapy for eliminating multidrug-resistant strains and biofilms.

5.7 Magnetic-Responsive CDD Via Magnetoelectric Nanoparticles

The benefits of magnetoelectric core-shell nanoparticles reported by Casillas-Popova et al. [129] as drug carriers due to their magnetic and electric properties. The study evaluated the performance of previously synthesised magnetoelectric nanoparticles with Methotrexate as a model drug. Mathematical modelling was used to understand the adsorption and release mechanisms. The study also investigated the impact of nanoparticle stability and temperature on methotrexate adsorption and release. The nanoparticles showed sensitivity to temperature and magnetic fields, with drug release achieved only under magnetic stimulation. The study suggests these nanoparticles have good in vitro performance and are expected to have suitable in vivo behaviour.

Magnetic/pH dual-sensitive hydrogels are synthesised by physically incorporating superparamagnetic iron oxide nanoparticles (Fe_3O_4) into Dextran hydrogels through Schiff base reactions [130]. The hydrogels displayed magnetic and pH-responsive behaviour and could release Doxorubicin in a controlled manner through a combination of diffusion, swelling, and erosion processes. The hydrogel/Fe_3O_4 composites have potential applications in various fields, including drug delivery.

6 Conclusion

Stimuli-responsive materials have shown great potential in developing controlled-release drug delivery systems. These materials can react to external stimuli, such as light, pH, and temperature, and dispense drugs in a regulated manner, providing optimal therapeutic effects while minimising side effects. Recent progressions in stimuli-responsive materials, including hydrogels, nanoparticles, and liposomes, have been discussed, and their utilisation in drug delivery has been highlighted. However, challenges remain in developing these systems, including stability, biocompatibility, and selectivity. Therefore, further research is required to overcome these challenges and fully exploit the capabilities of stimuli-responsive materials in drug delivery. Overall, using stimuli-responsive materials holds great promise for developing more effective and targeted drug delivery systems in the future.

References

1. Niculescu, A.G., Grumezescu, A.M.: Novel tumor-targeting nanoparticles for cancer treatment—A review. Int. J. Mol. Sci. **23**, 5253 (2022)
2. Sung, H., Ferlay, J., Siegel, R.L., Laversanne, M., Soerjomataram, I., Jemal, A., Bray, F.: Glob. Cancer Stat. (2020)
3. GLOBOCAN Estimates of Incidence and Mortality Worldwide for 36 Cancers in 185 Countries. CA Cancer J. Clin. **71**, 209–249 (2021)
4. Wang, S., Chen, Y., Guo, J., Huang, Q.: Liposomes for tumor targeted therapy: a review. Int. J. Mol. Sci. **24**, 2643 (2023)
5. Banerjee, R.: Nanotechnology in drug delivery: present status and a glimpse into the future. Ther. Deliv. **9**(4), 231 (2018)
6. Fleming, S., Ulijn, R.V.: Design of nanostructures based on aromatic peptide amphiphiles. Chem. Soc. Rev. **43**, 8150–8177 (2014)
7. Lin, G.G., Scott, J.G.: Peptide-directed self-assembly of hydrogels. Acta Biomater. **100**, 130–134 (2012)
8. Jonker, A.M., Löwik, D.W.P.M., Van Hest, J.C.M.: Peptide- and protein-based hydrogels. Chem. Mater. **24**, 759–773 (2012)
9. Luk, B.T., Zhang, L.: Current advances in polymer-based nanotheranostics for cancer treatment and diagnosis. ACS Appl. Mat. Interf. **6**, 21859–21873 (2014)

10. Zhang, L., Zhu, D., Dong, X., Sun, H., Song, C., Wang, C., Kong, D.: Folate-modified lipid–polymer hybrid nanoparticles for targeted paclitaxel delivery. Int. J. Nanomed. **10**, 2101–2114 (2015)
11. Ahmad, Z., Shah, A., Siddiq, M., Kraatz, H.-B.: Polymeric micelles as drug delivery vehicles. RSC Adv. **4**, 17028–17038 (2014)
12. Lin, Y.-K., Yu, Y.-C., Wang, S.-W., Lee, R.-S.: Temperature, ultrasound and redox triple-responsive poly(N-isopropylacrylamide) block copolymer: synthesis, characterization and controlled release. RSC Adv. **7**(43), 212–226 (2017)
13. Huang, X., Liao, W., Zhang, G., Kang, S., Zhang, C.Y.: PH-sensitive micelles self-assembled from polymer brush (PAE-g-cholesterol)-b-PEG-b-(PAE-g-cholesterol) for anticancer drug delivery and controlled release. Int. J. Nanomed. **12**, 2215–2226 (2017)
14. Zhang, C.Y., Wu, W.S., Yao, N., Zhao, B., Zhang, L.J.: PH-sensitive amphiphilic copolymer brush Chol-g-P(HEMA-co-DEAEMA)-b-PPEGMA: synthesis and self-assembled micelles for controlled anti-cancer drug release. RSC Adv. **4**(40), 232–240 (2014)
15. Dharadhar, S., Majumdar, A., Dhoble, S., Patravale, V.: Microneedles for transdermal drug delivery: a systematic review. Drug Dev. Ind. Pharm. **45**, 188–201 (2019)
16. Prausnitz, M.R.: Engineering microneedle patches for vaccination and drug delivery to skin. Annu. Rev. Chem. Biomol. Eng. **8**, 177–200 (2017)
17. Ye, Y., Yu, J., Wen, D., Kahkoska, A.R., Gu, Z.: Polymeric microneedles for transdermal protein delivery. Adv. Drug Deliv. Rev. **127**, 106–118 (2018)
18. Kim, J.Y., Han, M.R., Kim, Y.H., Shin, S.W., Nam, S.Y., Park, J.H.: Tip-loaded dissolving microneedles for transdermal delivery of donepezil hydrochloride for treatment of Alzheimer's disease. Eur. J. Pharm. Biopharm. **105**, 148–155 (2016)
19. Ma, G., Wu, C.: Microneedle, bio-microneedle and bio-inspired microneedle: a review. J. Control Release **251**, 11–23 (2017); Prausnitz, M., Langer, R.: Transdermal drug delivery. Nat. Biotechnol. **26**, 1261–1268 (2008)
20. Yao, J., Feng, J., Chen, J.: External-stimuli responsive systems for cancer theranostic. Asian J. Pharm. Sci. **11**(5), 585–595 (2016)
21. Yang, G., Liu, J., Wu, Y., Feng, L., Liu, Z.: Near-infrared-light responsive nanoscale drug delivery systems for cancer treatment. Coord. Chem. Rev. **320–321**, 100–117 (2016)
22. Kamaly, N., Yameen, B., Wu, J., Farokhzad, O.C.: Degradable controlled-release polymers and polymeric nanoparticles: mechanisms of controlling drug release. Chem. Rev. (2016)
23. www.ncbi.nlm.nih.gov
24. Li, L., Yang, W.-W., Xu, D.-G.: Stimuli-responsive nanoscale drug delivery systems for cancer therapy. J. Drug Target. (2019)
25. Gutiérrez, L., Stepien, G., Gutiérrez, L., Pérez-Hernández, M., Pardo, J., Pardo, J., Grazú, V., de la Fuente, J.M.: Nanotechnology in Drug Discovery and Development. Elsevier (2017)
26. Pilch, J., Potęga, A., Kowalczyk, A., Kasprzak, A., Kowalik, P., Bujak, P., Paluszkiewicz, E., Augustin, E., Nowicka, A.M.: PH-responsive drug delivery nanoplatforms as smart carriers of unsymmetrical bisacridines for targeted cancer therapy. Pharmaceutics **15**(1), 201 (2023)
27. Altinbasak, I., Kocak, S., Colby, A.H., Alp, Y., Sanyal, R., Grinstaff, M.W., Sanyal, A.: Biomater. Sci. **11**, 813–821 (2023)
28. Gao, J., Si, F., Wang, F., Li, Y., Wang, G., Zhao, J., Ma, Y., Yu, R., Li, Y., Jin, C., Li, D.: Hollow mesoporous structured MnFe2O4 nanospheres: a biocompatible drug delivery system with pH-responsive release for potential application in cancer treatment. Solid State Sci. **135**, 107066 (2023)
29. Son, M.J., Kim, T., Lee, S.-W.: Facile synthesis of fluorescent mesoporous nanocarriers with pH-sensitive controlled release of naturally derived dieckol. In: Colloids and Surfaces A: physicochemical and Engineering Aspects, Part A, vol. 657, p. 130535 (2023)
30. Anggraini, S.A., Prasetija, K.A., Yuliana, M., Wijaya, C.J., Bundjaja, V., Angkawijaya, A.E., Jiang, Y.-F., Putro, J.N., Hartono, S.B., Ayucitra, A., Santoso, S.P., Ismadji, S., Soetaredjo, F.E.: PH-responsive hollow core zeolitic-imidazolate framework-8 as an effective drug carrier of 5-fluorouracil. Mater. Today Chem. **27**, 101277 (2023)

31. Ghazimoradi, M., Tarlani, A., Alemi, A., Hamishehkar, H., Ghorbani, M.: "pH-responsive, magnetic-luminescent core/shell carriers for co-delivery of anticancer drugs (MTX & DOX) for breast cancer treatment. J. Alloy. Compd. **936**, 168257 (2023)
32. Zhang, G., Han, W., Zhao, P., Wang, Z., Li, M., Sui, X., Liu, Y., Tian, B., He, Z., Fu, Q.: Nanoscale **15**, 1937–1946 (2023)
33. Akkaya, B., Akkaya, R.: Sümeyye Idil Celikkaya, Nurgül Sarıaydin, Kayode Yomi Raheem, Doxorubucin loaded pH-responsive chitosan-poly(acrylamide-maleic acid) composite hydrogel for anticancer targeting. J. Mol. Struct. **1274**(1), 134536 (2023)
34. Truong-Thi, N.-H., Nguyen, N.H., Nguyen, D.T.D., Tang, T.N., Nguyen, T.H., Nguyen, D.H.: pH-responsive delivery of Platinum-based drugs through the surface modification of heparin on mesoporous silica nanoparticles. Euro. Polym. J. **185**, 111818 (2023)
35. Sheykhisarem, R., Dehghani, H.: In vitro biocompatibility evaluations of pH-sensitive Bi2MoO6/NH2-GO conjugated polyethylene glycol for release of daunorubicin in cancer therapy. Colloids Surf. B **221**, 113006 (2023)
36. Gou, K., Xin, W., Lv, J., Ma, Z., Yang, J., Zhao, L., Cheng, Y., Chen, X., Zeng, R., Li, H.: A pH-responsive chiral mesoporous silica nanoparticles for delivery of doxorubicin in tumour-targeted therapy. Colloids Surf. B **221**, 113027 (2023)
37. Zhang, C., Yuan, Q., Zhang, Z., Tang, Y.: A pH-responsive drug delivery system based on conjugated polymer for effective synergistic chemo-/photodynamic therapy. Molecules **28**(1), 399 (2023)
38. Rao, K.-M., Rao, K.S.V.K., Palem, R.-R., Uthappa, U.-T., Ha, C.-S., Han, S.-S.: PH sensitive drug delivery behavior of palmyra palm kernel hydrogel of chemotherapeutic agent. Gels **9**(1), 38 (2023)
39. Heragh, B.K., Taherinezhad, H., Mahdavinia, G.R., Javanshir, S., Labib, P., Ghasemsolb, S.: pH-responsive co-delivery of doxorubicin and saffron via cross-linked chitosan/laponite RD nanoparticles for enhanced-chemotherapy. Mater. Today Commun. **34**, 104956 (2023)
40. Teixeira, P.V., Adega, F., Martins-Lopes, P., Machado, R., Lopes, C.M., Lúcio, M.: PH-responsive hybrid nanoassemblies for cancer treatment: formulation development, optimization, and in vitro therapeutic performance. Pharmaceutics **15**(2), 326 (2023)
41. Li, B.-L., Zhang, J., Jin, W., Chen, X.-Y., Yang, J.-M., Chi, S.-M., Ruan, Q., Zhao, Y.: Oral administration of pH-responsive polyamine-modified cyclodextrin nanoparticles for controlled release of anti-tumour drugs. React. Funct. Polym. **172**, 105175 (2022)
42. Batool, N., Sarfraz, R.M., Mahmood, A., Rehman, U., Zaman, M., Akbar, S., Almasri, D.M., Gad, H.A.: Development and evaluation of cellulose derivative and pectin based swellable pH responsive hydrogel network for controlled delivery of cytarabine. Gels **9**(1), 60 (2023)
43. Du, M., Jin, J., Zhou, F., Chen, J., Jiang, W.: Dual drug-loaded hydrogels with pH-responsive and antibacterial activity for skin wound dressing. Colloids Surf. B: Biointerfaces **222**, 113063 (2023)
44. Xie, X., Lei, H., Fan, D.: Antibacterial hydrogel with pH-responsive microcarriers of slow-release VEGF for bacterial infected wounds repair. J. Mater. Sci. Technol. **144**, 198–212 (2023)
45. Pan, F., Giovannini, G., Zhang, S., Altenried, S., Zuber, F., Chen, Q., Boesel, L.F., Ren, Q.: PH-responsive silica nanoparticles for the treatment of skin wound infections. Acta Biomater. **145**, 172–184 (2022)
46. Zhang, R., Ding, J., Lu, X., Yao, A., Wang, D.: pH-responsive drug release and antibacterial activity of chitosan-coated core/shell borate glass-hydroxyapatite microspheres. Ceramics Int. **49**(3), 5161–5168 (2023)
47. Patil, R.S., Narayanan, A., Tantisuwanno, C., Sancaktar, E.: Immobilization of glucose oxidase on pH-responsive polyimide-polyacrylic acid smart membranes fabricated using 248 nm KrF excimer laser for drug delivery, **13**(1), 11 (2023)
48. Reddy, S.G., Kumar, B.S., Prashanthi, K., Murthy, H.A.: Fabricating transdermal film formulations of montelukast sodium with improved chemical stability and extended drug release. Heliyon **9**(3), e14469 (2023)

49. Cheng, M., Cui, Y., Guo, Y., Zhao, P., Wang, J., Zhang, R., Wang, X.: Design of carboxymethyl chitosan-reinforced pH-responsive hydrogels for on-demand release of carvacrol and simulation of release kinetics. Food Chem. **405**(Part A), 134856 (2023)
50. Wang, Y., Peng, Z., Yang, Y., Li, Z., Wen, Y., Liu, M., Li, S., Su, L., Zhou, Z., Zhu, Y., Zhou, N.: Auricularia auricula biochar supported γ-FeOOH nanoarrays for electrostatic self-assembly and pH-responsive controlled release of herbicide and fertilizer. Chem. Eng. J. **437**(Part 1), 134984 (2022)
51. Banihashemian, A.R., Kowsari-Esfahan, S., Ebrahimzadeh, M., Jafari, M., Akbari, A., Karimi-Maleh, H.: PH-responsive drug delivery using graphene-based materials. Nanoscale **13**(3), 1801–1823 (2021). https://doi.org/10.1039/D0NR06675J
52. Lei, X., Zhou, Y., Liu, X., Kong, L., Liao, L., Li, Y., Liu, M., Tian, L., Rao, W., Lv, G.: Effective pH-responsive nanocarrier based on the anisotropic surfaces of halloysite nanotubes for controlled drug release. Appl. Clay Sci. **232**, 106799 (2023). ISSN:0169-1317
53. Jalali, E.S., Shojaosadati, S.A., Hamedi, S.: Green synthesis of bovine serum albumin/oxidized gum Arabic nanocomposite as pH-responsive carrier for controlled release of piperine and the molecular docking study. Int. J. Biol. Macromol. **225**, 51–62 (2023). ISSN:0141-8130
54. de Jesus Oliveira, A.C., Silva, E.B., de Oliveira, T.C., Ribeiro, F.D.O.S., Nadvorny, D., de Freitas Oliveira, J.W., Borrego-Sánchez, A., da Franca Rodrigues, K.A., Silva, M.S., Rolim-Neto, P.J., Viseras, C.: pH-responsive phthalate cashew gum nanoparticles for improving drugs delivery and anti-Trypanosoma cruzi efficacy. Int. J. Bio. Mol. **230**, 123272 (2023)
55. Lee, T.-Y., Lu, H.-H., Cheng, H.-T., Huang, H.-C., Tsai, Y.-J., Chang, I.-H., Tu, C.-P., Chung, C.-W., Lu, T.-T., Peng, C.-H., Chen, Y.: Delivery of nitric oxide with a pH-responsive nanocarrier for the treatment of renal fibrosis. J. Controll. Release **354**, 417–428 (2023)
56. Giridhar Reddy, S.: Controlled release studies of Hydroxychloroquine sulphate (HCQ) drug-using Biodegradable polymeric Sodium alginate and Lignosulphonic acid Blends. Rasayan J. Chem. **4**, 2209–2215 (2021)
57. Giridhar Reddy, S.: Effect of crosslinking on control drug release of hydroxychloroquine sulphate drug-using alginate beads. Iran. J. Mater. Sci. Eng. **19**(2) (2022)
58. Hemmatpour, H., Haddadi-Asl, V., Burgers, T.C., Yan, F., Stuart, M.C., Reker-Smit, C., Vlijm, R., Salvati, A., Rudolf, P.: Nanoscale **15**, 2402–2416 (2023)
59. Lin, Z., Ding, J., Chen, X., He, C.: pH- and temperature-responsive hydrogels based on tertiary amine-modified polypeptides for stimuli-responsive drug delivery (2023)
60. Pourbadiei, B., Adlsadabad, S.Y., Rahbariasr, N., Pourjavadi, A.: Synthesis and characterization of dual light/temperature-responsive supramolecular injectable hydrogel based on host-guest interaction between azobenzene and starch-grafted β-cyclodextrin: Melanoma therapy with paclitaxel. Carbohydr. Polym. 120667 (2023)
61. Mustafa, R.A., Ran, M., Wang, Y., Yan, J., Zhang, Y., Rosenholm, J.M., Zhang, H.: A pH/temperature responsive nanocomposite for chemo-photothermal synergistic cancer therapy. Smart Mater. Med. **4**, 199–211 (2023)
62. Hu, X.-E., Shi, Y.-R., Zhu, X., Tian, K.-W., Xu, X.-L.: Temperature-responsive hydrogel for tumor embolization therapy. J. Drug Deliv. Sci. Technol. **80**, 104107 (2023)
63. Ruan, L., Chen, J., Du, C., Lu, H., Zhang, J., Cai, X., Dou, R., Lin, W., Chai, Z., Nie, G., Hu, Y. Mitochondrial temperature-responsive drug delivery reverses drug resistance in lung cancer. Bioact. Mater. **13**, 191–199 (2022)
64. Xu, L., Wang, H., Chu, Z., Cai, L., Shi, H., Zhu, C., Pan, D., Pan, J., Fei, X., Lei, Y.: Temperature-responsive multilayer films of micelle-based composites for controlled release of a third-generation EGFR inhibitor. ACS Appl. Polym. Mater. **2**(2), 741–750 (2020)
65. Thirupathi, K., Phan, T.T.V., Santhamoorthy, M., Ramkumar, V., Kim, S.-C.: PH and thermoresponsive PNIPAm-co-polyacrylamide hydrogel for dual stimuli-responsive controlled drug delivery. Polymers **15**(1), 167 (2023)
66. Balan, K.E., Boztepe, C., Künkül, A.: Modeling the effect of physical crosslinking degree of pH and temperature responsive poly(NIPAAm-co-VSA)/alginate IPN hydrogels on drug release behavior. J. Drug Deliv. Sci. Technol. **75**, 103671 (2022)

67. Boztepe, C., Künkül, A., Yüceer, M.: Application of artificial intelligence in modeling of the doxorubicin release behavior of pH and temperature responsive poly(NIPAAm-co-AAc)-PEG IPN hydrogel. J. Drug Deliv. Sci. Technol. **57**, 101603 (2020)
68. Sakai, S., Hirano, Y., Kobayashi, Y., Arai, N.: Effect of temperature on the structure and drug-release behaviour of inclusion complex of β-cyclodextrin with cyclophosphamide: a molecular dynamics study. Soft Matter. (2023)
69. Thirupathi, K., Santhamoorthy, M., Radhakrishnan, S., Ulagesan, S., Nam, T.-J., Phan, T.T.V., Kim, S.-C.: Thermosensitive polymer-modified mesoporous silica for pH and temperature-responsive drug delivery. Pharmaceutics **15**(3), 795 (2023)
70. Yu, S., Xing, J.: Preparation of temperature-responsive PMMA-based microspheres encapsulating erythromycin in situ by emulsion photopolymerization. J. Drug Deliv. Sci. Technol. **81**, 104256 (2023)
71. Işıklan, N., Erol, Ü.H.: Design and evaluation of temperature-responsive chitosan/hydroxypropyl cellulose blend nanospheres for sustainable flurbiprofen release. Int. J. Biol. Macromol. **159**, 751–762 (2020)
72. Huang, Z., Xiao, X., Jiang, X., Yang, S., Niu, C., Yang, Y., Yang, L., Li, C., Feng, L.: Preparation and evaluation of a temperature-responsive methylcellulose/polyvinyl alcohol hydrogel for stem cell encapsulation. Polym. Test. **119**, 107936 (2023)
73. Emam, H.E., Shaheen, T.I.: Design of a dual pH and temperature responsive hydrogel based on esterified cellulose nanocrystals for potential drug release. Carbohydr. Polym. **278**, 118925 (2022)
74. Zhu, Y., Zhang, M., Wei, S., Wang, B., He, J., Qiu, X.: Temperature-responsive P(NIPAM-co-NHMA)-grafted organic-inorganic hybrid hollow mesoporous silica nanoparticles for controlled drug delivery. J. Drug Deliv. Sci. Technol. **70**, 103197 (2022)
75. Ow, V., Loh, X.J.: Recent developments of temperature-responsive polymers for ophthalmic applications. J. Polym. Sci. (2022)
76. Hemmatpour, H., Haddadi-Asl, V., Burgers, T.Q., Yan, F., Stuart, M.C.A., Reker-Smit, C., Vlijm, R., Salvati, A., Rudolf, P.: Temperature-responsive and biocompatible nanocarriers based on clay nanotubes for controlled anti-cancer drug release. Nanoscale **15**, 2402–2416 (2023)
77. Choi, Y., Kim, J., Yu, S., Hong, S.: pH- and temperature-responsive radially porous silica nanoparticles with high-capacity drug loading for controlled drug delivery. Nanotechnology **31**, 335103 (2020)
78. Kotsuchibashi, Y.: Recent advances in multi-temperature-responsive polymeric materials. Polym. J. **52**, 681–689 (2020)
79. Liu, Z., Zhang, S., He, B.: Temperature-responsive hydroxypropyl methylcellulose-N-isopropylacrylamide aerogels for drug delivery systems. Cellulose **27**, 9493–9504
80. Hajebi, S., Abdollahi, A., Roghani-Mamaqani, H., Salami-Kalajahi, M.: Temperature-responsive poly(N-Isopropylacrylamide) Nanogels: the role of hollow cavities and different shell cross-linking densities on doxorubicin loading and release. Langmuir **36**(10), 2683–2694 (2020)
81. Zhao, H., Li, Y.: A novel pH/temperature-responsive hydrogel based on tremella polysaccharide and poly(N-isopropylacrylamide). Colloids Surf. A **586**, 124270 (2020)
82. Liu, Q., Cheng, H.-B., Ma, R., Yu, M., Huang, Y., Li, L., Zhao, J.: A DNA-based nanodevice for near-infrared light-controlled drug release and bioimaging. Nano Today **48**, 101747 (2023)
83. Brion, A., Chaud, J., Léonard, J., Bolze, F., Chassaing, S., Frisch, B., Heurtault, B., Kichler, A., Specht, A.: Red light-responsive upconverting nanoparticles for quantitative and controlled release of a coumarin-based prodrug. Adv. Healthc. Mater. **12**, 2201474 (2023)
84. Liu, Z., Chen, H., Huang, C., Huang, Q.: A light-responsive injectable hydrogel with remodeling tumor microenvironment for light-activated chemodynamic therapy. Macromol. Biosci. **23**, 2200329 (2023)
85. Zhou, T., Xie, S., Zhou, C., Chen, Y., Li, H., Liu, P., Jiang, R., Hang, L., Jiang, G.: All-in-one second near-infrared light-responsive drug delivery system for synergistic chemo-photothermal therapy. ACS Appl. Bio Mater. **5**(8), 3841–3849 (2022)

86. Park, J.H., Sung, K.E., Kim, K.H., Kim, J.R., Kim, J., Moon, G.D., Hyun, D.C.: Dual gate-keeping and reversible on-off switching drug release for anti-cancer therapy with pH- and NIR light-responsive mesoporous silica-coated gold nanorods. J. Ind. Eng. Chem. **106**, 233–242 (2002)
87. Ray, S., Banerjee, S., Singh, A.K., Ojha, M., Mondal, A., Singh, N.D.P.: Visible light-responsive delivery of two anticancer drugs using single-component fluorescent organic nanoparticles. ACS Appl. Nano Mater. **5**(5), 7512–7520 (2022)
88. Zafar, M., Ijaz, M., Iqbal, T.: Efficient Au nanostructures for NIR-responsive controlled drug delivery systems. Chem. Pap. **75**, 2277–2293 (2021)
89. Fan, R., Chen, C., Hou, H., Chuan, D., Mu, M., Liu, Z., Liang, R., Guo, G., Xu, J.: Tumor acidity and near-infrared light responsive dual drug delivery polydopamine-based nanoparticles for chemo-photothermal therapy. Adv. Funct. Mater. **31**, 2009733 (2021)
90. Rafiee, Z., Bodaghi, A., Omidi, S.: Fabrication of a photo- and pH-sensitive micelle by self-assembly of azobenzene polyglycerol for anticancer drug delivery. Monatsh. Chem. **154**, 259–265 (2023)
91. Patil, T.V., Dutta, S.D., Patel, D.K., Ganguly, K., Lim, K.-T.: Electrospinning near infra-red light-responsive unzipped CNT/PDA nanofibrous membrane for enhanced antibacterial effect and rapid drug release. Appl. Surf. Sci. **612**, 155949 (2023)
92. Luo, J., Ma, Z., Yang, F., Wu, T., Wen, S., Zhang, J., Huang, L., Deng, S., Tan, S.: Fabrication of laponite-reinforced dextran-based hydrogels for NIR-responsive controlled drug release. ACS Biomater. Sci. Eng. **8**(4), 1554–1565 (2022)
93. Hu, D., Zhang, C., Sun, C., et al.: Carvacrol combined with NIR light-responsive nano-drug delivery system with specific anti-bacteria, anti-inflammation, and immunomodulation for periodontitis. Nano Res. (2023)
94. Zhu, W., Zhao, Y., Wu, Z., Lv, F., Zhang, Y., Guo, S.: Application of UV responsive SiO2/PVP composite hydrogels as intelligent controlled drug release patches. Polymer **264**, 125535 (2023)
95. Huang, X., Xu, L., Yu, X., Li, Y., Huang, Z., Xu, R., Zeng, W., Zhang, Z., Li, W., Deng, F.: Near-infrared light-responsive multifunctional hydrogel releasing peptide-functionalized gold nanorods sequentially for diabetic wound healing. J. Colloid Interface Sci. **639** (2023)
96. Xue, D., Wang, Y., Zhang, H.: Advances of NIR light responsive materials for diagnosis and treatment of brain diseases. Adv. Opt. Mater. 2202888 (2023)
97. Xing, Y., Zeng, B., Yang, W.: Light responsive hydrogels for controlled drug delivery. Front Bioeng Biotechnol. **16**(10), 1075670 (2022)
98. Abdelmohsen, H.A.M., Copeland, N.A., Hardy, J.G.: Light-responsive biomaterials for ocular drug delivery. Drug Deliv. Transl. Res. (2022)
99. Shan, P., Lu, Y., Lu, W., Yin, X., Liu, H., Li, D., Lian, X., Wang, W., Li, Z., Li, Z.: Biodegradable and light-responsive polymeric nanoparticles for environmentally safe herbicide delivery. ACS Appl. Mater. Interfaces **14**(38), 43759–43770 (2022)
100. Liu, J., Kang, W., Wang, W.: Photocleavage-based photoresponsive drug delivery. Photochem. Photobiol. **98**, 288–302 (2022)
101. Khalifa, A.Z., Zyad, H., Mohammed, H., Ihsan, K., Alrawi, L., Abdullah, M., Akram, O.: Recent advances in remotely controlled pulsatile drug delivery systems. J. Adv. Pharm. Technol. Res. **13**(2), 77–82 (2022)
102. Tang, Y., Wang, G.: NIR light-responsive nanocarriers for controlled release. J. Photochem. Photobiol. C: Photochem. Rev. **47**, 100420 (2021)
103. Giménez, V.M.M., Arya, G., Zucchi, I.A., Galante, M.J., Manucha, W.: Soft Matter. (38), 2021 (2021)
104. Huang, R., Lan, R., Shen, C., Zhang, Z., Wang, Z., Bao, J., Wang, Z., Zhang, L., Hu, W., Yu, Z., Zhu, S.: Remotely controlling drug release by light-responsive cholesteric liquid crystal microcapsules triggered by molecular motors. ACS Appl. Mater. Interfaces **13**(49), 59221–59230 (2021)
105. Yang, L., Guo, C., Jia, L., Xie, K., Shou, Q., Liu, H.: Fabrication of biocompatible temperature- and pH-responsive magnetic nanoparticles and their reversible agglomeration in aqueous Milieu. Ind. Eng. Chem. Res. (2010)

106. Rakshit, S., Sivasankar, S.: Cross-linking of a charged polysaccharide using polyions as electrostatic staples. Soft Matter. (2011)
107. Heunis, C.M., Wang, Z., de Vente, G., Misra, S., Venkiteswaran, V.K.: A magnetic bio-inspired soft carrier as a temperature-controlled gastrointestinal drug delivery system. Macromol. Biosci. 2200559 (2023)
108. Zhu, J., Wang, J., Li, Y.: Recent advances in magnetic nanocarriers for tumor treatment. Biomed. Pharmacother. **159**, 114227 (2023)
109. Dagdelen, S., Mackiewicz, M., Osial, M., et al.: Redox-responsive degradable microgel modified with superparamagnetic nanoparticles exhibiting controlled, hyperthermia-enhanced drug release. J. Mater. Sci. **58**, 4094–4114 (2023)
110. Durkut, S.: Fe_3O_4 magnetic nanoparticles-loaded thermoresponsive poly(N-vinylcaprolactam)-g-galactosylated chitosan microparticles: investigation of physico-chemical, morphological and magnetic properties. J. Macromol. Sci. Part A, 181–191 (2023)
111. Gonçalves, A., Simões, B.T., Almeida, F.V., Fernandes, S.N., Valente, M., Vieira, T., Henriques, C., Borges, J.P., Soares, P.I.P.: Engineering dual-stimuli responsive poly(vinyl alcohol) nanofibrous membranes for cancer treatment by magnetic hyperthermia. Biomater. Adv. **145**, 213275 (2023)
112. Wang, X., Qi, Y., Hu, Z., et al.: Fe_3O_4@PVP@DOX magnetic vortex hybrid nanostructures with magnetic-responsive heating and controlled drug delivery functions for precise medicine of cancers. Adv. Compos. Hybrid Mater. **5**, 1786–1798 (2022)
113. Mazidi, Z., Javanmardi, S., Naghib, S.M., Mohammadpour, Z.: Smart stimuli-responsive implantable drug delivery systems for programmed and on-demand cancer treatment: an overview on the emerging materials. Chem. Eng. J. **433**(Part 1), 134569 (2022)
114. Demin, A.M., Vakhrushev, A.V., Pershina, A.G., Valova, M.S., Efimova, L.V., Syomchina, A.A., Uimin, M.A., Minin, A.S., Levit, G.L., Krasnov, V.P., Charushin, V.N.: Magnetic-responsive doxorubicin-containing materials based on Fe_3O_4 nanoparticles with a SiO_2/PEG shell and study of their effects on cancer cell lines. Int. J. Mol. Sci. **23**(16), 9093 (2022)
115. Chen, Z., Song, S., Ma, J., Da Ling, S., Wang, Y.D., Kong, T.T., Xu, J.H.: Fabrication of magnetic core/shell hydrogels via microfluidics for controlled drug delivery. Chem. Eng. Sci. **248**(Part B), 117216 (2022)
116. Dong, D., Chen, R., Jia, J., Zhao, C., Chen, Z., Lu, Q., Sun, Y., Huang, W., Wang, C., Li, Y., He, H.: Tailoring and application of a multi-responsive cellulose nanofibre-based 3D nanonetwork wound dressing. Carbohydr. Polym. **305**, 120542 (2023)
117. Li, H., Li, B., Lv, D., Li, W., Lu, Y., Luo, G.: Biomaterials releasing drug responsively to promote wound healing via regulation of pathological microenvironment. Adv. Drug Deliv. Rev. 114778 (2023)
118. Yang, X., Zhang, C.Q., Deng, D.W., Gu, Y.Q., Wang, H., Zhong, Q.F.: Multiple stimuli-responsive MXene-based hydrogel as intelligent drug delivery carriers for deep chronic wound healing. Small **18**, 2104368 (2022)
119. Zeyni, V., Karimi, S., Namazi, H.: Surface PEGylation of ZIF-8 metal-organic framework based on magnetic hydroxyapatite as a pH/magnetic targeting responsive system for anticancer drug delivery. Microporous Mesoporous Mater. **354**, 112544 (2023)
120. Cai, W., Zhang, W., Chen, Z.: Magnetic Fe3O4@ZIF-8 nanoparticles as a drug release vehicle: pH-sensitive release of norfloxacin and its antibacterial activity. Colloids Surf. B **223**, 113170 (2023)
121. Shrivastava, P., Vishwakarma, N., Gautam, L., Vyas, S.P.: Chapter 5—Magnetically responsive polymeric gels and elastomeric system(s) for drug delivery. In: Smart Polymeric Nano-Constructs in Drug Delivery, pp. 129–150. Academic Press (2023)
122. Nguyen, T.P.T., Ménager, C., Rieger, J., Coumes, F.: Rational design of stimuli-responsive magnetic polymer hybrid (nano)materials. Polym. Int. (2023)
123. Mohanan, S., Guan, X., Liang, M., Karakoti, A., Vinu, A.: Stimuli-responsive Silica Silanol conjugates: strategic nanoarchitectonics in targeted drug delivery. Small 2301113 (2023)

124. Fragal, E.H., Fragal, V.H., Silva, E.P., Paulino, A.T., da Silva Filho, E.C., Mauricio, M.R., Silva, R., Rubira, A.F., Muniz, E.C.: Magnetic-responsive polysaccharide hydrogels as smart biomaterials: synthesis, properties, and biomedical applications. Carbohydr. Polym. **292**, 119665 (2022)
125. Armenia, I., Ayllón, C.C., Herrero, B.T., Bussolari, F., Alfranca, G., Grazú, V., de la Fuente, J.M.: Photonic and magnetic materials for on-demand local drug delivery. Adv. Drug Deliv. Rev. **191**, 114584 (2022)
126. Boztepe, C., Daskin, M., Erdogan, A.: Synthesis of magnetic responsive poly(NIPAAm-co-VSA)/Fe3O4 IPN ferrogels and modeling their deswelling and heating behaviors under AMF by using artificial neural networks. React. Funct. Polym. **173**, 105219 (2022)
127. Shademani, A., Jackson, J.K., Thompson, C.J., Chiao, M.: Controlled and localized antibiotics delivery using magnetic-responsive beads for synergistic treatment of orthopedic infection. J. Biomed. Mater. Res. **110**(5), 1036–1051 (2022)
128. Bag, N., Bardhan, S., Roy, S., Roy, J.: D Mondal nanoparticle-mediated stimulus-responsive antibacterial therapy. Biomater. Sci. **11**, 1994–2019 (2023)
129. Casillas-Popova, N., Bernad-Bernad, M.J., Gracia-Mora, J.: Modeling of adsorption and release kinetics of methotrexate from thermo/magnetic responsive CoFe2O4–BaTiO3, CoFe2O4–Bi4Ti3O12 and Fe3O4–BaTiO3 core-shell magnetoelectric nanoparticles functionalized with PNIPAm. J. Drug Deliv. Sci. Technol. **68**, 103121 (2022)
130. Zeng, N., He, L., Jiang, L., Shan, S., Su, H.: Synthesis of magnetic/pH dual responsive dextran hydrogels as stimuli-sensitive drug carriers, Carbohydr. Res. **520**, 108632 (2022)

Advanced Tissue Engineering with Novel Engineered Biomaterials

Azadeh Izadyari Aghmiuni and Aref Gholami

Abstract Nowadays, the concept of regenerative medicine is known via matrix properties that can mimic natural tissue functions. In this field, the therapeutic successes in implantation and tissue regeneration depend on the type of biomaterials applied in the fabrication of polymer substrates or substitutes. Such materials lead to the design of intelligent/smart substrates, the promotion of specific bioactive signals, the improvement of cell-to-cell interaction, and control of the designed micro-environment functions to respond to the cellular behavior during the process of tissue repair. These scaffolds, alone or incorporated with bioactive drugs, cells, and biomolecules, can be promising in the medical and pharmaceutical fields. Accordingly, this chapter aims to introduce novel biomaterials applied in tissue engineering and focuses on their main advance and limitations.

Keywords Regenerative medicine · Biomaterials · Smart substrates · Tissue engineering · Natural polymers

A. Izadyari Aghmiuni (✉)
Department of Nanobiotechnology, Pasteur Institute of Iran, 1316943551 Tehran, Iran
e-mail: azadeh.izadyari@gmail.com

Department of Tissue Engineering and Applied Cell Science, School of Advanced Technologies in Medicine, Shahid Beheshti University of Medical Sciences, 19839-63113 Tehran, Iran

A. Gholami
Faculty of Engineering, Isfahan Technical and Vocational University (Mohajer), 81645-315 Isfahan, Iran

R. Malviya and S. Sundram (eds.), *Engineered Biomaterials*, Engineering Materials,
https://doi.org/10.1007/978-981-99-6698-1_12

1 Introduction

Tissue engineering involves biological principles and approaches along with polymer and materials sciences, and engineering techniques to design tissue substitutes for mimicking the structural and physiological functions of native tissues and consequently regenerating injured or diseased tissues [1]. However, this field is continuously developing through the input of different scientific areas such as nanotechnology, cell therapy, phytochemistry, electronics, etc., which can lead to the precise control of cell functions (proliferation, differentiation, migration). Although, nowadays, tissue engineering has evident growing impacts in therapeutic and research fields, known as a novel and improved therapy; however, it is still struggling with the issues of *in-vivo* transfer. In this field, novel methods like 3D and 4D bio-printing, microstructures (organs on a chip), and/or a combination of several methods can provide promising perspectives [2–5].

To fully understand the diagnostic therapeutic potential of this technique, it is crucial to possess a rationally designed network based on biomaterials. On this matter, there are research advances related to new biomaterials and their copolymers or composites that can control the interactions of cell/biomaterial networks and cell/cell and improve cell signaling pathways [6–8]. Hence, this chapter focuses on reviewing novel engineered biomaterials, biomimetic materials, and some of the most exciting studies regarding the use of these materials to apply in regenerative medicine and tissue engineering, along with their technological approaches. In the following, some of the new topics on tissue engineering have been reviewed including personalized regenerative medicine, 3D and 4D bio-printing, and engineered microstructure models for diagnostic or therapeutic strategies.

2 Novel Biomimetic Materials

Recently, many studies have been made on the use of engineered biomaterials/ biomimetic materials and their interactions with biological environments. Indeed, these materials can play a crucial role in cellular infiltration within an implanted matrix at the local implant sites and then form new tissues on the biomaterial-based matrix. In this matter, biomaterial itself can act as a mechanical substrate or pharmacological/therapeutic agent to improve biological signals and transfer these signals to cells for tissue regeneration [9–12]. Generally, biomimetic materials can be applied to increase tissue regeneration, however, having the engineered scaffold is essential for supplying mechanical support, preventing scar formation, and successful regeneration of tissue. These biomaterials can provide cell adhesion factors, proliferation, induction, differentiation, and migration and lead to the modification of cellular responses during tissue regeneration, via the incorporation of survival factors and creating covalently or non-covalently bonds with cells [13, 14].

In this regard, the recognition of biomaterials and their functions for the improvement of cellular adhesion sites, the spatial distribution of ligands, selection of suitable cells and their spatial distribution can play a pivotal role in designing biomimetic materials and bio-matrices [15–17]. The transformation of active materials from one state to another state is one of the attractive features of biomaterials that occurs in the presence of biological systems or response to some of the external stimuli like temperature, chemical composition, light, etc., and leads to new noninvasive surgical procedures [2, 18, 19]. Moreover, the incorporation of DNAs into biomaterials to control cellular responses and stimulate in situ production of growth factors, as well as the design of micro-fabricated bio-matrices by micro-fabrication techniques to improve cellular adhesion behavior, delivery of oxygen and nutrients, etc., are other applications of these materials [2, 19–23].

Based on the studies, the modification of biomaterial surface can also be a suitable way to fabricate a biomimetic matrix, and the use of bioactive molecules, especially ECM biomolecules can be a suitable option to achieve this aim. In this field, proteins like heparin, fibronectin, elastin, etc., are considered as the protein covers on these materials which can promote adhesion and growth of cells; and increase elastic modulus of artificial ECMs, and improve osteoblastic mineralization and osteoinduction [24–26]. The use of synthetic peptide sequences and short-chain peptides are other methods for surface modification of biomaterials that can be led to improved adsorption to the biomaterial surface and increased cell access to all surfaces. Such modified biomaterials can be used in the network of polymer matrices as potentially artificial ECMs to guide the formation of new tissues [24, 27].

Although, the common peptides in this field are Arg-Gly-Asp (RGDs) derived from fibronectin and laminin, however, other peptides like Tyr-Ile-Gly-Ser-Arg, Arg-Glu-Asp-Val, and Ile-Lys-Val-Ala-Val have also developed for modifying engineered matrices [24, 28, 29].

There are well-known techniques such as coupling with functional groups or bi-functional crosslinking that are used for ensuring the covalent binding between the peptide and the surfaces of the biomaterial and its immobilization and provide flexible movement in biological environments [24] Other techniques in this matter include the following [24, 30–33]:

1. Photochemical immobilization: when the polymer matrices lack functional groups to form the coupling reactions.
2. Limiting non-specific adsorption of protein and the reactive chains.
3. Bulk modification of biomaterial: in this technique, cell-signaling peptide is incorporated into the bio-matrix structure to exist the recognition sites in the bulk of the biomaterials.

Some of the suitable techniques for modifying the surface of biomaterials or biomimetic materials to mimic biological environments and achieve microenvironments similar to native ECMs have been indicated in Fig. 1.

Generally, biomimetic means the study and assessment of naturally made models, systems, processes, and elements, and then mimicking and copying their structures and functions. Rose petals, lotus leaves, beehives, butterfly wings, spider silk, water

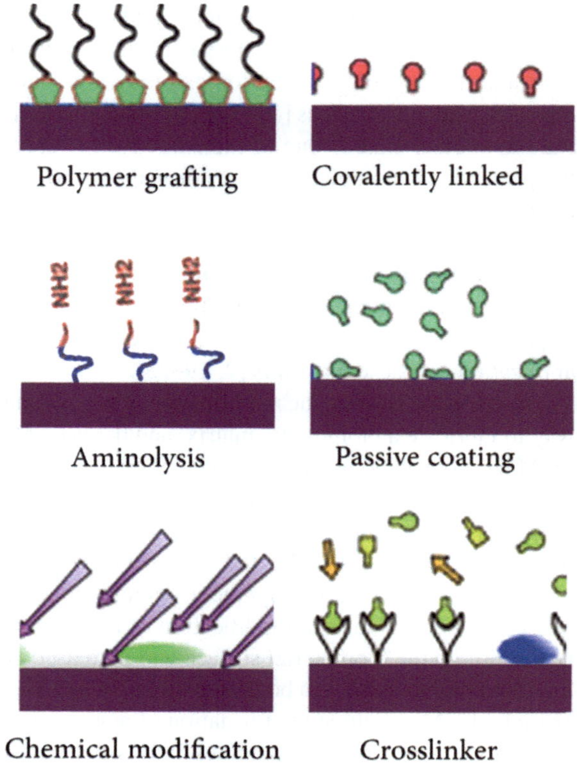

Fig. 1 Techniques for modifying the surface

strider foot, etc., are examples in this field that possess unique micro/nanostructures with hydrophobic nature. Common techniques for designing such structures include chemical deposition, electro-deposition, and hydrothermal reaction which are used for designing biomimetic matrices or biomaterial structures in various industries. Some studies in this field have been listed in Table 1.

Based on the reports, the photo-functionalization method by ultraviolet radiation is known as another method in the design of biomaterial/biomimetic matrices to treat diseases and regenerate tissue. This technique provides the physicochemical modifications of the bio-matrix surface and leads to the improvement of its biological properties [46]. Notably, given that direct exposure to ultraviolet light, after implantation of biomatrix, can be very harmful, hence photo-functionalization with this radiation should carry out before implantation [47]. Such a process can increase cellular adhesion and the adsorption of proteins, and help *in-vitro* osteogenic differentiation and mineralization [48, 49]. In this field, Dini et al. indicated that the UV-mediated photo-functionalization process can improve the adsorption of proteins in titanium-treated discs [50]. The authors stated that photo-functionalization of the PEO coating leads to the improvement of the biological activities of dental and bone implants by changing the physicochemical properties of titanium (Ti). Assessment of the human gingival fibroblast cells confirmed the non-cytotoxic effects of these discs. Hence, it seems

Table 1 Some of the common methods for designing biomimetic matrices and biomaterial structures

The aim	Techniques	Description	Refs.
Development of Ti-based alloys in dentistry	Electrochemical technique	Creation of Zr–Nb alloys-based microstructures for decreasing corrosion behavior and without any toxic effects	[34]
	Electrochemical technique	Improvement of binary Ti-Ag alloys-based surface characterization and cell responses	[35]
	Spark plasma sintering (SPS) technique	Creation of microstructures to improve mechanical properties and decrease Ti-Mn alloys-related cytotoxicity	[36]
	Casting technique	Improvement of mechanical property and deformation behavior of Ti–Cr-based alloys	[37]
Lotus-leaf-like surface	Simple electrochemical reaction	Superhydrophobic metal microstructure surface obtained via the non-electric chemical plating of the Cu onto the steel sheet	[38]
Superelastic alloy	Thermal process	Ti–Zr-based biomedical alloy containing Mo–Sn to improve superelastic properties via regulating Mo and Sn contents	[39]
Oil–water separation, improvement of the Cu corrosion resistance, stain repellency, sound absorption, and mechanochemical robustness	Electrodeposition	The design of the Cu-coated superhydrophobic copper foil in a zigzag manner	[40]
Oil–water separation, and the creation of superhydrophobic and superoleophilic surfaces	Electrochemical technique	The design of biomimetic graphene on the stainless steel surfaces inspired by rose petals, butterfly wings, nacre, and honeycomb	[41]
The control of the corrosion behavior of the alloy	Melting method	Design of Ti-6Al-4V-xZr-based microstructures and the passivation of Ti-6Al-4V-xZr alloy via increasing Zr content	[42]

(continued)

Table 1 (continued)

The aim	Techniques	Description	Refs.
Superhydrophobic films	Electrodeposition process	The design of corrosion-resistant superhydrophobic surfaces along with excellent chemical stability and self-cleaning effects	[43]
Superhydrophobic structures	Electrochemical co-deposition	The design of graphene oxide-myristic acid-based superhydrophobic structure (nickel-reduced) to increase corrosion resistance	[44]
Sheet-like superhydrophobic surfaces	Etching-heat combination treatment	The design of superhydrophobic sheets based on Cu to decrease corrosion	[45]

that such a bio-functional coating can act as a promising approach for biomedical applications.

Nagay et al. demonstrated that non-thermal plasma can improve surface and physicochemical properties of the soft liner surfaces based on the hexamethyl disiloxane, and lead to the inhabitation of C. albicans biofilm [51]. Based on their reports, these treated surfaces can provide candida-related stomatitis protection, and antimicrobial effects (decrease in fungal colonization) and act as a suitable candidate for medical applications. Cordeiro et al. also indicated that Ti-Nb-Zr-Ta alloys treated with plasma electrolytic oxidation (PEO) can create porous surfaces and crystalline structure along with an increase in electrochemical stability, hardness, roughness, and surface free energy [52]. Finally, the authors stated that such alloys can be suitable alternatives for fabricating dental implants, due to their nontoxic and elastic modulus similar to the bone.

In the following, Prosdent et al. indicated that some specimens of tissue conditioners such as visco-gels possess high water sorption and solubility [53]. Moreover, some other tissue conditioners such as dentu-soft and coe-soft (as novel biomimetic materials) can provide suitable functions to fabricate denture relining. Furthermore, the effects of the coating through the biomimetic technique on titanium substrates were assessed by Coelho et al. [54]. This technique consists of the immersion of titanium substrate into the solution of simulated body fluid with conditions similar to the blood plasma, to form the slow and well-organized nucleation and promote apatite to be deposited. Based on their results, the proposed biomimetic coating in this research can be effective to achieve hydroxyapatite and carbonated hydroxyapatite, as well as, enhance the osseointegration process of the implanted bones. Moreover, titanium treated with the method possesses a higher resistance to corrosion than pure titanium. Hence, they stated that titanium substrates coated with biomimetic materials can be a suitable candidate for bone implantation applications.

Given that the macroscopic structure of biomimetic materials has led to the limitation in the mechanical functions of the product biomaterial or matrix, the use of hierarchical architecture consisting of a microscopic cellular structure can remove the mentioned limitation. Several studies in this field illustrate that hierarchical structures of bone improve mechanical properties. Barth et al. stated that X-ray irradiation on the hierarchical structures of bone can progressively degrade the mechanical functions of human cortical bone (strength, fracture resistance, ductility, and toughness) [55, 56]. This degradation increases with the promotion of X-ray irradiation. Moreover, the characterization of hierarchical structures of bone in the conditions of aging, disease, and treatment was assessed by Milovanovic et al. for understanding the architectural modifications and mechanical integrity [57]. Based on the results, young and healthy bones possess the highest resistance to mechanical loading. These bones also indicated better mineralization along with preservation of osteocyte-lacunar properties, but aging bones and osteoporosis lead to alterations in bone material properties like increased osteocyte apoptosis and decreased mechanical competence. The authors stated that structural or compositional change in the bone material leads to the maintenance or decay of bones in disease and health conditions.

Furthermore, deformation and fracture in bones were assessed by Zimmermann et al., at the strain rate of 10^{-5}–10^{-1} (1/s) [58]. The results indicated that bone toughness decreased with increasing strain rate; so that higher strain rate led to lower intrinsic bone matrix toughness and consequently fracture induced by a loss in toughness in the matrix. In continuing, Peng et al. designed an ultra-light biomimetic material via the 3D printing technique [59]. Based on their reports, the designed biomimetic possessed an ultrahigh elasticity and stability, at the compressive strain of 95%, along with remarkable stiffness and resilience.

Chitosan-based flexible robust structures are considered another promising candidate in bio-friendly catalyst applications. Based on the reports of Tseng et al., TiO_2-attached biomimetic chitosan films can be effective in the photo-catalytic reduction of carbon dioxide [60]. Notably, the sol–gel process along with heat treatment was used for preparing these films. Moreover, the use of the nano-casting method provided the duplication of structures of natural leaves on these films. It was also found that hydroxyl and amine groups can fasten the TiO2 nanoparticles on the chitosan films. Adding glutaraldehyde leads to also the improvement of the film denseness, the adsorption of carbon dioxide, stability, and hydrophobicity of the surfaces. The authors stated such films as eco-friendly photocatalysts can be suitable candidates to produce pure chemical fuels via decreasing CO_2.

Generally, modified biomimetic surfaces can act as transport barriers for decreasing water/molecules/ions loss inside the cell. Likewise, the uptake of liquids/molecules can also decrease via these surfaces [61]. Selenium is another material that led to the inactivation of carcinogens and inhibition of cell damage from free radicals. Hence the use of this material in the modified biomimetic structures such as microgels and matrices, etc. not only can control and remove free radicals but also prevents dysfunctions of the cellular metabolism caused by conventional selenium-containing compounds [62].

Nowadays, selenium-based materials have developed in bio-medical applications owing to lower toxicity, higher reversibility, sensitivity, and programmability. However, the processing severe conditions leads to weak behavior in enzymes based on this material [63]. Based on the reports, the materials based on carbon and graphene can be effective in the inhibition of free radicals [64]. The research of Cao et al. is one of these studies performed to fabricate biomimetic catalysts [65]. In this study, a selenium-modified carbon nitride nanosheet was prepared that can be effective in the inhibition of free radicals. It was also found that such catalysts can provide an organic conjugated polymer based on selenium with non-toxic and biocompatible properties to apply biomedicine.

Other studies in the field of materials indicated, although polypropylene-based synthetic materials can play an important role in medical applications for repairing abdominal wall defects [66, 67], however, the rigidity of implantation and side effects of this material (chronic inflammation along with pain and visceral adhesion) are considered as the main problems [68]. On this matter, biomimetic/biological materials due to biocompatibility, biodegradability, and suitable absorbing nature are better candidates for regenerative medicine applications [8, 69, 70]. In this regard, the small intestinal sub-mucosa is known as the native collagen-rich extracellular matrix consisting of fibroblasts and growth factors and can lead to the relocation of cells and regeneration of defective tissue [71]. However, poor mechanical properties, insufficient neovascularization, and high biodegradation rate have limited the use of this bio-matrix in clinical treatments [72]. Cao et al. indicated that the control of ratios of elastin and collagen in the native matrices can be a suitable factor for decreasing/treating hernia [73]. In this study, they synthesized biomimetic composite nano-fibers based on the small intestinal sub-mucosa via principles of bio-mimicry. Their results demonstrated a lower biodegradation rate compared to pure small intestinal sub-mucosa. Moreover, it is known that the designed nano-composites possess antibacterial rates of ~997% and ~98% against E. coli and S. aureus, respectively; and can better maintain the stability of structures and micro-mechanical properties than that of the pure small intestinal sub-mucosa, as well. Based on in-vivo studies, these nano-composites can provide tissue regeneration without serious inflammatory reactions and act as a promising option for repairing the abdominal wall repair.

Nowadays, the more suitable techniques and architectures to regenerate tissues have led to the development of biomimetic matrices (substrates, scaffolds, hydrogels, etc.) feigning the structure and function of ECMs [74]. In this field, the techniques such as electrospinning, rapid prototyping, freeze drying, and sacrificial templates (like colloidal crystals, ice crystals, and fibers) play an important role in fabricating 3D biomimetic matrices [75–78].

Many studies are going on to explore new biomimetic materials and to design bio-matrices with better properties for applications of biomedicine; some of them are listed in Table 2.

Table 2 Some of the new biomimetic materials and bio-matrices along with their applications in biomedicine

Matrix	Technique	Description	Applications	Refs.
Porous silk-based biomaterials	Sacrificial templates of supra-colloidal assemblies (urea templating method)	This study was carried out to create hierarchically composite material and generate porous biomaterials with mimicking tissue properties. To this end, urea self-assembles were performed by the interaction of hydrogen bonding into crystalline supra-colloidal assemblies for forming macroscopic pores in the structure of the polymer scaffold. Based on the results, solvent interactions can improve the scope of polymers. Moreover, solvent and polymer-urea interactions can affect the morphology of supra-colloidal crystals and the pores of biomaterials and provide a potential for regenerating various tissues in which cellular alignments are observed	Skin, bone, muscle, and nerve tissue engineering	[79]
β-tricalcium phosphate scaffolds	Novel template-casting technique	This study indicates that the novel template-casting technique leads to completely interconnected substrates which their architecture and chemistry can fully manipulate via varying the casting material and template The processing consists of 4 step 1- β-TCP slurry preparation, 2- Casting and shaping into a paraffin beads-based template 3- Solidifying and drying 4- Sintering Based on the results, the designed scaffold possesses interconnected macro-pores along with solid struts that can be effective in eye tissue engineering for mimicking the reticular network Scaffolds also indicate high mechanical strength (~9 MPa) with ~79% porosity and suitable biocompatibility for the proliferation of Human mesenchymal cells	Eye tissue engineering	[80]

(continued)

Table 2 (continued)

Matrix	Technique	Description	Applications	Refs.
3D nanofibrous 58S-bioglass scaffold	Sacrificial template method	This study was performed to design mimicking scaffold of natural extracellular matrices (ECMs). To this end, natural 3D nano-fibrous bacterial cellulose was applied during the process, as the sacrificial template. Based on the results, designed scaffolds possessed morphological properties similar to nano-fibrous bacterial cellulose. The scaffolds also possess ~75% porosity and pores 39.4 nm and 60 μm, along with suitable biocompatibility with mouse osteoblast cells	Bone tissue engineering to imitate the structure and function of ECMs	[81]
Biomimetic structures	Direct homogenization techniques	This study was for the biomimetic design of cellular structures to regenerate micro-structures with cuttlebone properties. Based on the results, the unit cell with a similar topology to the cuttlebone unit cell was achieved	Design of cellular biomimetic structures based on characterization of cuttlebone	[82]
Meso-porous wollastonite particles	Creation of bone defect with a dental drill	This study aimed to assess the role of bioactive mesoporous wollastonite particles. To this end, after creating bone defects and filling the hole with the mentioned particles, the results illustrated the promotion of bone formation after 2 and 4 weeks. Based on the X-ray result, a high-dense and radio-opacity image indicates the closure of the created hole. Moreover, histological analysis demonstrated that mesoporous wollastonite particles-based treatment can improve the deposition of collagen in the bone defect areas	Bone tissue engineering	[83]

(continued)

Table 2 (continued)

Matrix	Technique	Description	Applications	Refs.
Nano-structured hydroxyapatite-bredigite scaffolds	Space holder technique	Given the weak mechanical properties of 3D pure hydroxyapatite scaffolds, this study aimed to design hydroxyapatite-based scaffolds containing bredigite nano-powder. It was found that the increased bredigite content led to a decrease in the micropore size of the scaffold (220–310 μm) along with porosity of ~63–76%. In contrast, the increase in bredigite content enhanced the compression strength and modulus of the scaffold. Based on the result, these scaffolds possess superior bioactivity and biodegradability than pure hydroxyapatite scaffolds so that hydroxyapatite scaffold containing 15 wt.% bredigite indicated better proliferation of cells compared with pure hydroxyapatite scaffold	Hard tissue engineering	[84]
Polycaprolactone substrates	Selective laser sintering	This study is aimed to prepare porous PCL scaffolds. Accordingly, the design of such substrates with a porous architecture leads to the improvement of mechanical properties. Based on the finite element analysis results, the mechanical properties of designed substrates can be computationally predicted. Accordingly, yield strength and compressive modulus were reported at 2.0–3.2 MPa and 52–67 MPa, respectively. Moreover, subcutaneously implantation of scaffolds containing bone morphogenetic protein-7 demonstrated excellent biological properties for tissue in-growth, so that, bone generation *in-vivo* was confirmed by histological study. This study illustrated that the integration of the computational design of matrices with free-form fabrication technique proves useful for the construction of engineered substrates	Bone and cartilage tissue engineering	[85]

(continued)

Table 2 (continued)

Matrix	Technique	Description	Applications	Refs.
Baghdadite-based coatings or filler materials	Direct solid-state synthesis routes	In this study, the aim is the preparation of dense Ca_3ZrSi_2O9 (baghdadite) bulk ceramics and the assessment of their mechanical properties. To this end, baghdadite was heated at 1350–1450 °C for 3 h. Accordingly, samples sintered at 1400 °C possessed better mechanical properties (bending strength: 98 ± 16 MPa, fracture toughness:1.3 ± 0.1 MPa $m^{0.5}$, and hardness: 7.9 ± 0.2 GPa)	Bone tissue engineering	[86]
Biomimetic nylon 6-baghdadite nano-composite scaffolds	Sacrificial template	This study was carried out to simulate the main properties in native tissue architecture and develop novel biomimetic custom-made matrices for tissue regeneration. To this end, the cuttlefish bone-based biomimetic scaffolds by combining nylon-6, and baghdadite along with sacrificial template cuttlefish bone, were introduced for biomedical applications. Based on the results, the combination of cuttlefish bone improves the mechanical properties of the scaffold. Such that, the mechanical analysis indicated ~ twofold and threefold increase in compressive modulus and strength, respectively. Moreover, the combination of cuttlefish bone nanoparticles and the nylon-6 matrix promotes the degradation rate of the substrate and its bioactivity in simulated body fluids. Such new nano-composite scaffolds with open pores (153–253 μm, depending on the cuttlefish bone contents) can be a promising candidate to use in regenerative medicine	Hard tissue engineering	[87]

(continued)

Table 2 (continued)

Matrix	Technique	Description	Applications	Refs.
Mussel-inspired ε-poly-l-lysine hydrogels	Horseradish peroxidase cross-linking	This study aimed to form an intimate assembly of hydrogels to prevent wound infections. Based on the results, the distribution of biomimetic catechol–lysine residue in polymer hydrogel leads to an increase in adhesion properties of wet tissues such as exuding ulcers, as well as the improvement of the capacity of *in-vivo* hemostatic and acceleration of wound repair. These hydrogels can also provide anti-infection properties	Skin tissue engineering	[88]
Polythiophene/ poly-isocyanide hybrid hydrogels	600 nm light shows high for increasing antibacterial activities	This study aimed to fabricate a biomimetic network based on conjugated hybrid hydrogels. Accordingly, the PIC[a] polymer acts as the scaffold for trapping and aligning the PMNT backbone into an ordered conformation. Moreover, the PMNT/PIC hybrid indicates a higher reactive oxygen species level compared to PMNT only, providing better photodynamic anti-microbial effects against various pathogenic bacteria	Wound healing	[89]

[a]Tri(ethylene glycol)-functionalized polyisocyanides

3 Herbal Materials-Based New Matrices

Given that the uncontrolled differentiation of cells can increase the risk of tumorigenicity, the use of therapeutic strategies for the complete and irreversible differentiation of these cells can play a crucial role in targeted differentiation. These strategies consist of the use of bio-materials and/or herbal bioactive materials for designing polymer matrices and providing suitable physicochemical/mechanical properties to differentiate stem cells and regenerate tissue [90, 91]. Such biomaterials play a key role in providing the microenvironments for the differentiation of cells and controlling cellular functions and interactions by creating an ideal matrix.

The studies of Izadyari Aghmiuni et al. indicate that herbal bioactive materials possess high therapeutic and regenerative effects [92, 93]. However, some of their forms such as mucilage, the extract/essential oil, gel, etc. cannot be applied to damaged tissues directly. This research team demonstrated that the use of these biomaterials in the structure of engineered matrices can turn the matrix into a smart biological substrate [92]. They stated that some of the herbal materials such as quince seed mucilage are known as polysaccharides that are composed of xylose and glucuronic acid. These polysaccharides can swell in water owing to the existence of hydrophilic groups, unlike chitosan which is suspended in an acidic solvent. Based on their reports, quince seed mucilage-based substrates can provide a better porous network than chitosan-based matrices. These substrates can also support the proliferation of fibroblast cells and provide higher water absorption capacity for applications of skin tissue engineering. Moreover, the authors indicated that the combination of this mucilage with PEG increased the transduction of mechanical signals and improved the biological signals for inducting stem cell differentiation into dermal tissue.

According to the reports of Tahmasebi et al., nano-fibrous scaffolds of poly(3-hydroxybutyrate-co-3-hydroxy valerate blended with aloe vera gel can be also effective in bone tissue engineering applications [94]. Indeed, an increase in the biocompatibility of these scaffolds is related to aloe vera contents. Likewise, they indicated that these hybrid scaffolds possess better alkaline phosphatase activities and mineralization, and high expression of bone-related genes/proteins compared with aloe vera-free nano-fibrous. It is also known that aloe vera gel possesses osteoinductive potential and can be used in the network of bio-implants. Moreover, Suganya et al. indicated that electrospun mesh of aloe vera can be an effective transdermal therapeutic strategy. Accordingly, the fibers of polycaprolactone containing aloe vera powder (10 wt %) fabricated and stated these meshes can provide better hydrophilic properties, more suitable mechanical properties (tensile strength:6.28 MPa, elastic modulus similar to dermal tissue) than polycaprolactone-collagen fibers. These electrospun fibers also lead to more desirable cell proliferation, better secretion of collagen, and a high level of expression of F-actin [95].

Another study is about an aloe vera-based scaffold that possesses wide applications in biomedicine. The popularity of this herbal polysaccharide is inducted by its bioactive components such as anthraquinones, anthrones, vitamins, etc., and their properties (anti-inflammatory, anti-microbial, and anti-oxidant), in tissue repair [6, 96]. Based on the reports of Kallyanashis Paul et al., aloe vera-based hydrogel plays a crucial role in reducing maternal birth injuries [97]. They stated that aloe vera-alginate hydrogel can act as an immediate treatment by stem cell delivery and regeneration of damaged tissues. Moreover, local injection of such hydrogels with/without stem cells can be suitable candidates for birth injury repair and preventing pelvic organ prolapse via improving smooth muscle and elastin contents.

Nettle (*Urtica dioica* L.) is another herb that can provide osteogenic differentiation when blended with other biomaterials. On this matter, Zadegan et al. reported that silk fibroin-nettle nano-fibers can improve water uptake on the scaffold, as well as cellular attachment and proliferation compared to silk fibroin nano-fibers [98]. Moreover, nettle-based nano-fibers can express both early and late markers of osteoblast differentiation. Furthermore, Ghiyasi et al. stated that *Urtica dioica*-based scaffolds containing ZnO-nanoparticles increase the antibacterial activities of these electrospun scaffolds via the creation of a synergy effect. Accordingly, the incorporation of Urtica dioica/ZnO-nanoparticles to poly(caprolactone) scaffolds increases the tensile strength up to 2.54 MPa, improves water uptake ability, and promotes the proliferation of fibroblast cells [99].

Rochette et al. also illustrated that apple extract (*Malus sp.*) and tutin due to the existence of polyphenolic compounds can be used in applications of dermal tissue engineering [100].

They assessed the effect of two topical formulations consisting of 1.25% of apple extract and 0.75% rutin on dermal models, (ex-vivo and 3D-engineered skin). Based on the results, these two formulas possess a protecting role against solar ultraviolet B (UVB) radiation and can inhibit radiation-induced metalloproteinase formation. Moreover, the studies by Hajiali et al. demonstrated that nano-fibrous dressing based on lavender essential oil provides high efficacy to treat UVB-induced dermal burns [101]. These dressings are also able to reduce and control the inflammatory responses induced by UVB radiation.

Another study relates to the induction of stem cell differentiation and proliferation by animal fats. According to the report of Pilehvar Soltanahmadi et al., emu oil not only increases the proliferation and differentiation of adipose tissue-derived stem cells into keratinocytes but also leads to the improvement of cell adherence and cytoprotection [102]. Finally, they stated that emu oil-based electrospun nano-fibrous can be a suitable option for designing wound dressings and/or engineered bio-matrices containing cells for dermal tissue regeneration.

Generally, research on biomaterials illustrates that herb-derived materials not only promote the proliferation and differentiation of stem cells but also can inhibit the proliferation of cancer cells that are influenced by the doses of the stimulant compounds. It means that specific doses of herbal biomaterials can promote or inhibit cell proliferation and induce its differentiation into the targeted cell. In this field, Zhang et al. demonstrated that 1–100 μg/ml concentrations of naringin derived from citrus can increase H-BM-MSCs[1] growth and their osteogenic differentiation [103].

[1] Human bone mesenchymal stem cell.

The studies of Yu et al. also show that naringin in 50μg/ml concentration activates the Notch signaling pathways and stimulates osteogenic differentiation. While higher concentrations than 100μg/ml suppress the rate of cell proliferation [104]. Based on the reports of Zhang, et al., naringin can also induce osteogenic activities and differentiation of canine bone marrow stromal cells at 10–6 mol/l concentrations [105]. Based on their reports, this range of concentration can lead to an increase in cellular proliferation and calcium nodules. In this regard, Beom Su Kim et al. also showed that the extract of brown algae Laminaria japonica (fucoidan) in 0.1–10μg/ml concentrations led to JNK- and ERK-dependent BMP2 –Smad 1/5/8 signaling and induction of osteoblast differentiation [106].

Fei Li et al. also stated that some of the phenylethanoid glycosides such as Echinacoside isolated from Cistanches Herba stem (as a novel biomaterial) can increase cell bioactivities and mineralization of osteoblastic [107]. To this end, the level of osteoprotegerin, osteocalcin, and collagen secretion on the cell culture was assessed. The results indicated that concentrations of 0.01–10 nmol/l can increase cell proliferation, osteocalcin levels, and collagen I content, as well as, lead to the promotion of the mineralization of osteoblastic. They reported that such glycosides possess a stimulatory effect on the formation of osteoblastic bone and can potentially be effective against osteoporosis.

Moreover, MaríaSatué et al. reported that flavonoids not only can stimulate stem cell differentiation into osteoblast but inhibit osteoclastogenesis in RAW 264.7 cells, as well [108]. Based on the results, doses greater than 100 μM of diosmetin, galangin, and chrysin flavonoids as well as 500 μM doses of taxifolin flavonoid possess toxic effects on cells. While, quercitrin with safe doses of 200 and 500 μM, and taxifolin in safe doses of 100 and 200 μM can induce osteocalcin mRNAs and bone sialoprotein expression and lead to higher osteocalcin levels.

The study of Kim et al. indicates Herba *Siegesbeckia*-extracted kirenol can promote osteoblast differentiation by activating the BMPs and signaling pathway in cells [109]. This natural diterpenoid can also promote mineralization, ALP activities, osteopontin, collagen content, and expression of OPG/RANKL. Choi et al. also stated that *Magnolia Officinalis*-isolated Honokiol can lead to the stimulation of osteoblast MC3T3-E1 cell functions and reduce/inhibit the production of bone-resorbing mediators [110]. They reported that such phenolic compounds can be effective in natural therapies for osteoporosis. Likewise, Hui-HuiXiao et al. indicated that Vanillic acid derived by *Sambucus williamsii* Hance can increase estrogen-like activities of cells [111]. The results of this study illustrate that these phenolic acids can stimulate cellular proliferation and ALP activity and promotes Runx2, osteocalcin, and the ratio of OPG-RANKL[2] mRNA expression.

Phytoestrogens and puerarin derived from *Pueraria Mirifica* are considered the other herbal biomaterials that can increase the growth of cells and the expression of osteoblastic differentiation [112]. On this matter, Tiyasatkulkovit et al. demonstrated that puerarin can improve the mRNA expression of ALP, decrease the expression level of RANKL, and induce bone gain via increasing osteoblast differentiation in

[2] Osteoprotegerin-receptor activator of nuclear factor kB ligand.

rat osteoblast-like cells and suppressing osteoclast functions. Accordingly, puerarin can induce the differentiation of osteoblast rather than the proliferation of osteoblast in an estrogen receptor-dependent manner.

Generally, the studies on biomaterials and the design of novel bio-matrices continue, and biomaterials-based approaches demonstrate their potential to develop tissue-engineered substitutes (such as hydrogels, substrates, or scaffolds) and regenerate damaged tissues.

3.1 New Approaches in Tissue Engineering

Tissue engineering develops alternative approaches for decreasing the risks of disease transfer of allografts and the lack of availability of auto-grafts [113, 114]. The strategy of this method includes the regeneration of the guided tissues via modeling human diseases, pharmaceutical screening tools, designed nano/micro-tissues, etc. In this field, natural polymers as scaffolding biomaterials play an important role; so that they can provide organ physiology-like structures and lead to the imitation of tissue functions.

Generally, tissue engineering creates a personalized medicine concept to a new level that can be used for tuning the suitable approaches to meeting the patient's requirement. Integration of technological advances like micro/nano-fabrications and technologies of cell therapy can be also effective in this regard. Personalized regenerative medicine, 3D and 4D bio-printing, and engineered microstructure models are considered as new topics in diagnostic or therapeutic strategies. On this matter, personalized regenerative medicine not only benefits patients due to the following treatment but also provides lower healthcare and social costs. Moreover, the use of 3D or 4D printing technology along with an imaging approach can play an impact on assembling and directing the 3D/4D systems for the design of clinical sizes, and mechanical and structural integrity of tissue-mimicking structures (Fig. 2).

Some of the biomaterials used in bio-printing techniques along with their applications include the following [115]:

12.4.1 Gold, titanium, platinum, steel, etc.: for applications of orthopedic implants, screws, pin

12.4.2 Ceramics and carbon compounds: For applications of bioactive orthopedic implant

12.4.3 Calcium phosphate salt, titanium oxide, aluminum oxide, and, glass: For applications of dental implants and artificial hearing aids

12.4.4 Poly (methyl methacrylate), polycarbonates, polyurethanes, poly(caprolactone): for applications of tissue-engineered scaffolds, drug delivery systems, and Prostheses

12.4.5 Carbon fibers, dental cement composites, types of bone cement, polyethylene with ultra-high molecular weight: for applications of dental fillings, orthopedic, rubber catheters, gloves, and implants

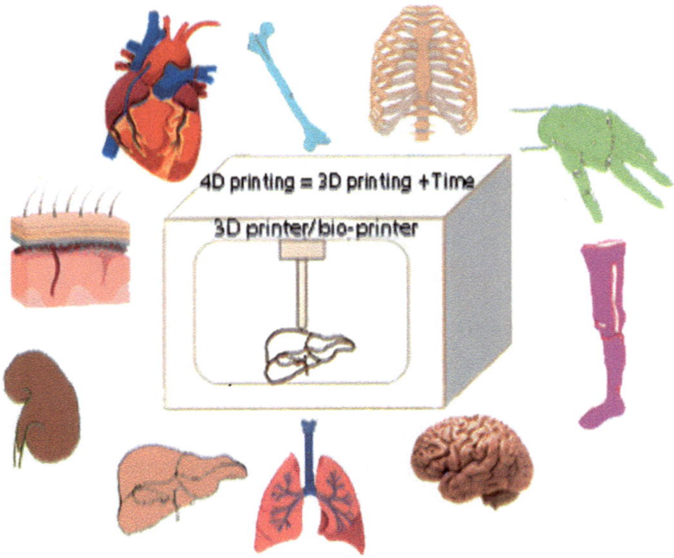

Fig. 2 The 3D and 4D-printed materials in biomedical applications

Types of 3D printing methods have been indicated in Table 3.

Generally, among the 3D printing/bio-printing techniques, micro-extrusion, inkjet, and poly-jet are considered as the most common methods applied in the biomedical. In this field, many biomaterials can be used in 3D bio-printing/printing techniques. A large variety of biopolymers (collagen, gelatin, alginate, etc.), and synthetic polymers (polyglycolic acid, polyvinyl alcohol, polyethylene glycol, polycaprolactone, etc.), are known as more common materials in this field [128–136]. Some of the 3D bio-printing/printing applications in biomedicine have been listed in Table 4.

4 Transforming 3D to 4D Printing

4D bio-printing is known as one of the additive processing techniques that possess excellent potential to manufacture living tissues and alter tissue mechanical properties [146]. Although 3D bio-printing techniques are widely used for the design of biological networks such as bones, skin, heart tissues, etc., however, 3D techniques possess a main disadvantage; so that, they can just assess the printed object's starting status. Indeed, natural regeneration entails the creation of a complex 3D network along with a micro-architecture similar to extracellular matrix structure, and tissue repair. While the majority of the 3D structures created by 3D printing techniques cannot simulate such networks because [135, 146, 147]. The 4D bio-printing can be used

Table 3 Types of 3D printing/bio-printing methods

Method	Description	Figures	Refs.
Inkjet	This type of printer operates via a collection of ink droplets at a specific region on the substrate. Thermal, piezoelectric, or electromagnetic forces are applied for expelling droplets from the reservoir nozzle. Although, these forces produce severe regional conditions, however, the transient nature of the pressure permits the cell to remain viable with low stress	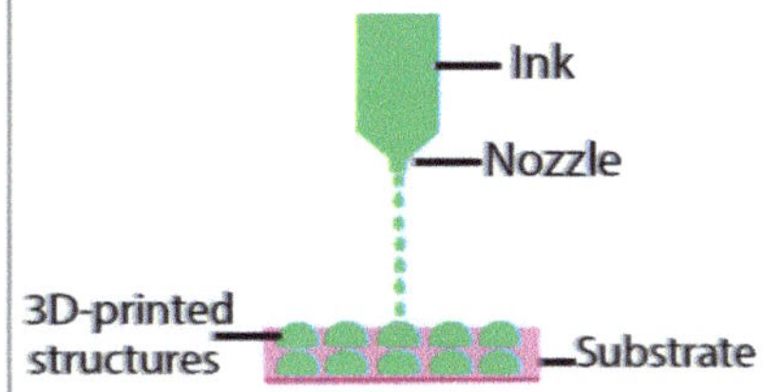	[116–122]
Polyjet	This method is similar to inkjet, except for photo-polymer liquids are sprayed onto the build platforms and cured by an ultraviolet beam. Benefits of this method include the creation of smooth and detailed prototypes and the achievement of complex shapes	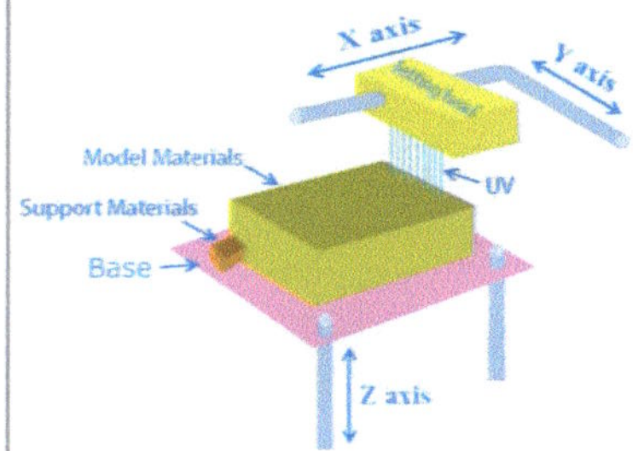	[115]

(continued)

Table 3 (continued)

Method	Description	Figures	Refs.
Micro-extrusion	This method is an additive manufacturing process that rotates around the accumulation of substances in successive layers for forming the ideal 3D structures. Micro-extrusion bio-printers possess a broader range of viscous bio-inks and are known as the most common and low-cost printing technique. This system includes pulley systems that can move along the x, y, and z axes, and possesses a temperature-controlled material processing/ dispensing system, a light source for illumination of the deposition areas, or activation of the photo-initiator	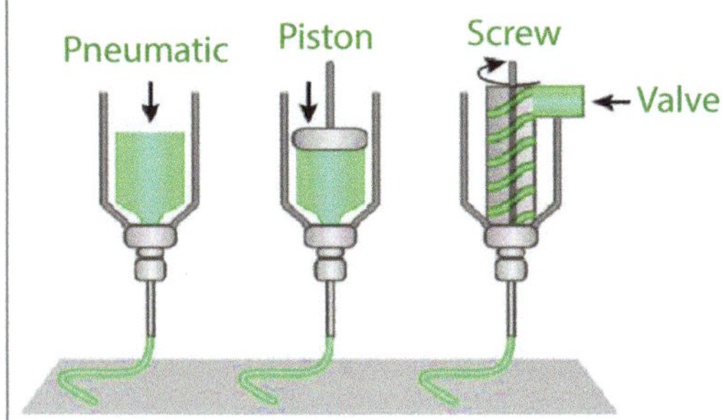	[117, 119, 120, 122, 123]
Laser-assisted 3D bio-printing	This method identified as laser-based 3D bio-printers is a droplet-based system and was primarily developed for transferring metals according to the principle of laser excitation. Nowadays, this method is used for biological materials like DNAs, peptides, and cells, as well as tissue engineering applications. Laser-based 3D bio-printing includes a laser energy-absorbing layer along with a pulsed laser beam and a focusing system. Indeed, the principle of operation entails using a high-energy pulsed laser to a donor slide coated with the bio-ink. High gelling speed and expensive processing are known as disadvantages of this method	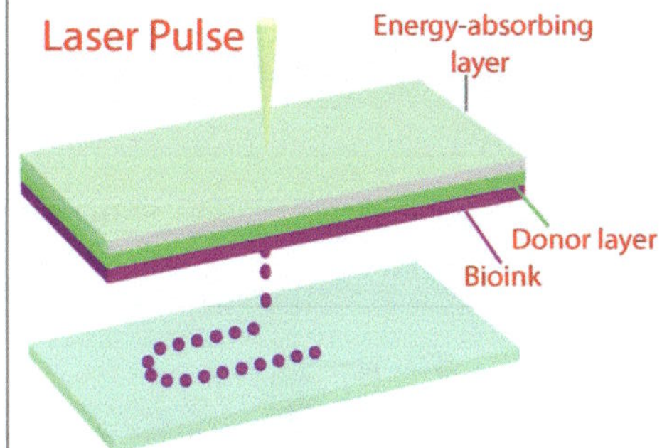	[119, 121, 122, 124, 125]

(continued)

Table 3 (continued)

Method	Description	Figures	Refs.
Selective laser sintering	This method is applied in producing strong and functional parts and includes a high-power laser beam for forming objects by fusing powdered materials	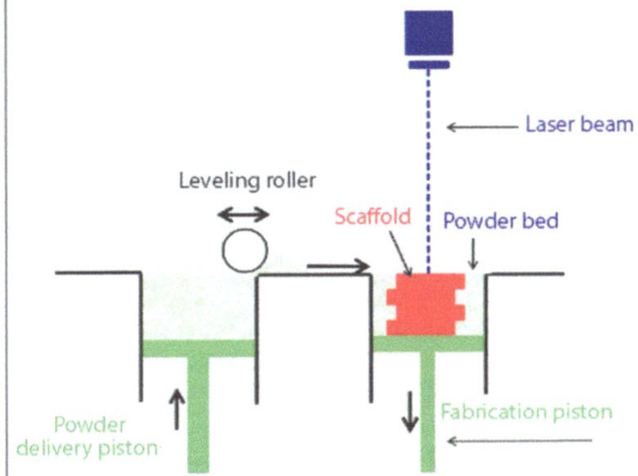	[126]

(continued)

Table 3 (continued)

Method	Description	Figures	Refs.
Electron Beam Manufacturing	This method is similar to selective laser sintering, except for a high-power electron beam is applied for fusing the powdered particles. The used materials in this method include titanium and cobalt-chrome alloys	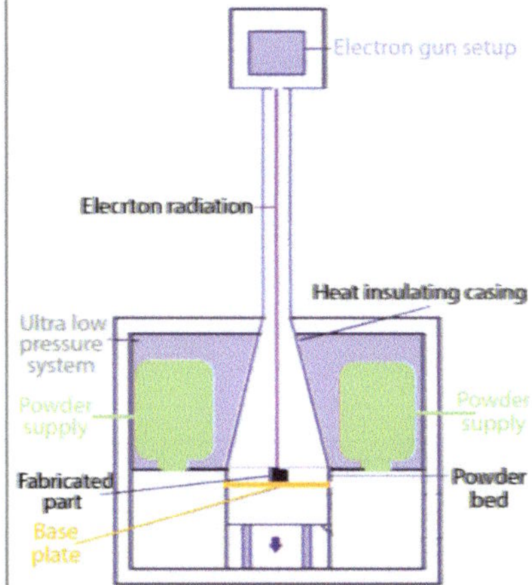	[115]

(continued)

Table 3 (continued)

Method	Description	Figures	Refs.
Stereolithography	The device utilized a moving helium-cadmium laser for producing an ultraviolet light spot on top of the polymer vat and an optical scanning system is applied to manage the movement of this spot	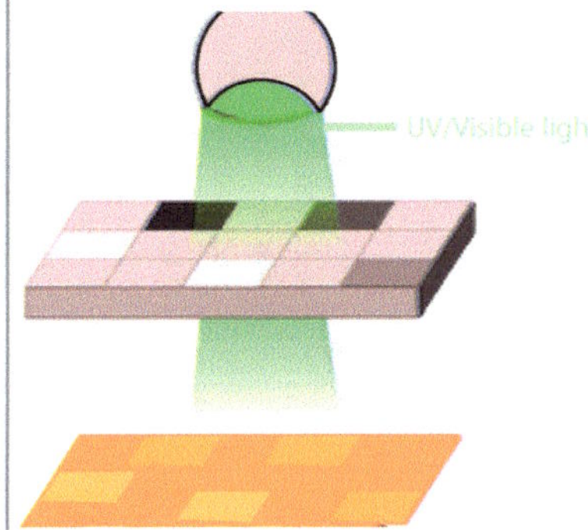	[125]
Continuous Liquid Interface Production	This method has many similarities with stereolithography and falls under the process of polymerization; so that, ultraviolet beams are passed by a transparent window to form build platforms for raising upwards holding the printed objects	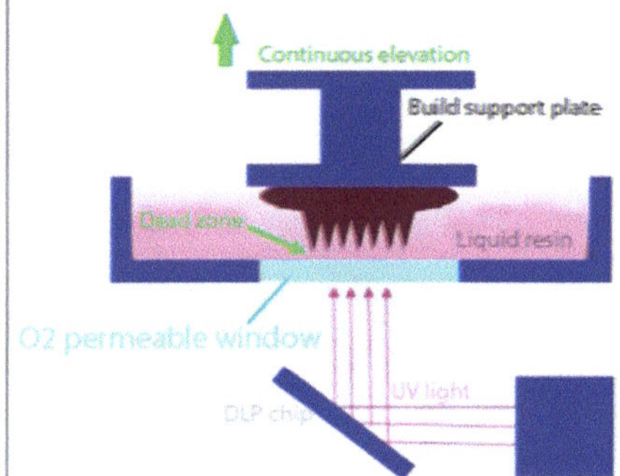	[127]

Table 4 The 3D bio-printing/printing applications in biomedicine

Type of 3D printing	Materials	The type of cell or test model	Fabrication process/ Function	Refs.
Heating jacket	Polycaprolactone/ tricalcium phosphate scaffolds coated with bd-ECM)	Adipose-derived stem cells	Polycaprolactone dried at 105 °C for 24 h, tricalcium phosphate formed with a particle size of <100 nm, polycaprolactone/ tricalcium phosphate scaffolds manufactured by a 3D printing technique, and finally the porcine bone dissolved into an acidic pepsin solution to be formed bdECM gels on designed substrate/To increase osteoconductivity and osteoinductivity	[137]
The MHDS 3D printing system with 4 extrusion heads	Polycaprolactone/ β-tricalcium phosphate Membranes	Fibroblasts and preosteoblasts	Membrane fabricated by melting polycaprolactone/ β-tricalcium phosphate and maintained at 110 °C to dispense material. Melted material dispensed from the nozzle (110 °C, 500 kPa), fabrication of layer-by-layer membrane, and formation of 3D-multilayer mesh-type structures/To facilitate bone regeneration	[138]
Bio-printing	Nanocellulose-based bioink cross-linked by alginate	Human chondrocytes	Printing at 40 kPa pressure and printing speed of 5 mm/s/Cartilage tissue engineering	[130]

(continued)

Table 4 (continued)

Type of 3D printing	Materials	The type of cell or test model	Fabrication process/ Function	Refs.
Extrusion	Polycaprolactone-printed 3D constructs laden with estrogen or progesterone	Hormones-derived induced pluripotent stem cells (estrogen and/or progesterone)	PCL thermoplastic pellets applied to load hormones, hormones-coated pellets extruded as filaments/For obstetric and gynecologic applications such as intrauterine devices	[139]
Hot melt extrusion	Oleo-gum-resins disks treated by nano-oxides of TiO_2, P25, Cu_2O, and MoO_3	Staphylococcus aureus, pseudomonas aeruginosa, escherichia coli, and candida albicans	3D-printed disks into model geometries and the disk-diffusion methodology was applied to assess antimicrobial and antifungal activities of studied materials against clinical isolates/To study the potential of 3D-printed bio-actives	[140]
Inkjet	Phosphoric acid-polyvinyl alcohol scaffold with a hydroxyapatite/ β-tricalcium phosphate	Rabbit bone marrow stromal cells	Scaffold printed by ZPrinter at 0.1 mm powder thickness/ Bone tissue engineering and medical implants	[141]
UV cross-linked	Methacrylated hyaluronic acid hydrogel	Human bone marrow-derived mesenchymal stromal cells	Methacrylated hyaluronic acid dissolved in an alpha minimum essential medium concentration, scaffold bio-printed by the dispensing system, scaffold UV cross-linked at 1800 mJ/Bone regenerative medicine	[142]

(continued)

Table 4 (continued)

Type of 3D printing	Materials	The type of cell or test model	Fabrication process/ Function	Refs.
Fused deposition modeling	Biphasic calcium phosphate scaffolds reinforced with zirconia	Human mesenchymal stem cells	Fabrication of the 3D slurry foam with blending agents, the removal of blending materials at high temperature, extruding at 600kPa pressure and 100 mm/min printing speed/Bone tissue engineering	[143]
Polyjet	Rigid acrylic resin	Preoperative planning in anatomical resection of hepatocellular carcinoma	Patient-specific intrahepatic vessel model/To understand the positional relationships between tumors and vessel, as well as operator and assistant during operation	[144]
Inkjet-based 3D printing	Calcium Sulfate Scaffolds	Osteoblast-like sarcoma cells	The scaffolds treated with heat, heat-treated scaffolds at 300 °C possessed almost adequate strength but severe toxicity, by contrast, the modified scaffold at 500–1000°C was non-toxic but possessed insufficient mechanical strength/ hard tissue engineering	[145]

for producing 3D bioactive structures having the capability of robust orientational change to adapt to novel favored stimulations over time [146, 148, 149].

The 4D-printed structures can alter over time in response to different stimuli and adjust to the native niches of fault fields, opening up novel avenues for tissue engineering, especially hard tissue [146, 150, 151]. Generally, the shape-conversion capabilities of 4D-printed hard tissue structures such as bones can be effective in healing individualized bone disorders [146, 152]. In this field, cells, growth factors, inorganic particles, etc. can act as suitable supports to improve mechanical properties, increase alkaline enzymatic activities, and form calcium in osteoblast cells [146, 153, 154]. Moreover, the formation of composite scaffolds can increase rates of shape recovery potential and be applied for regenerating small bones.

5 Summary and Conclusions

The development of novel engineered biomaterials or biomimetic matrices requires that the designed biomaterials should be not only biologically inert but also interact with biological environments and respond to them. These biomaterials must provide biological active signals and lead to an increase in cellular adhesion, proliferation, differentiation, and migration or apoptosis. In this field, the modification or treatment of biomaterials with biomolecules and herbal active molecules can play an important role in designing biomimetic novel bio-matrices to induce the cellular response and form new tissues. The surface and bulk modifications/changes of biomaterials with peptides can also provide to modulate cell functions via the alteration in concentrations or distribution on the biomaterial or bio-matrices. Hence, biomimetic/biomaterials-based matrices can mimic natural tissue functions and alone or incorporated with bioactive drugs, cells, and biomolecules provide a promising approach for regenerating tissues. Moreover, the strategies of bio-matrix fabrication for biomedical applications such as the regeneration of the guided tissues via modeling human diseases, pharmaceutical screening tools, designed nano/micro-tissues, etc. can allow new bio-surgical and therapeutic approaches to repair and replace damaged/diseased tissues like regeneration of broken bones, dermal wounds, articular cartilage disorders, and infarcted cardiac muscle, etc.

References

1. Gomes, M.E., Rodrigues, M.T., Domingues, R.M.A., et al.: Tissue engineering and regenerative medicine: new trends and directions—a year in review. Tissue Eng. Part B Rev. **23**, 211–224 (2017)
2. Izadyari Aghmiuni, A., Heidari Keshel, S.: Eye-on-a-chip. In: Principles of Human Organs-on-Chips, pp. 315–370 (2023)
3. Chen, L.-J., Raut, B., Kaji, H.: On-Chip Disease Models of the Human Retina, 2nd edn. Elsevier Inc. Epub ahead of print (2019). https://doi.org/10.1016/b978-0-12-813671-3.00012-8
4. Izadyari Aghmiuni, A., Heidari Keshel, S., Sefat, F., et al.: Fabrication of 3D hybrid scaffold by combination technique of electrospinning-like and freeze-drying to create mechanotransduction signals and mimic extracellular matrix function of skin. Mater. Sci. Eng. C **120**, 111752 (2021)
5. Ashammakhi, N., Ahadian, S., Zengjie, F., et al.: Advances and future perspectives in 4D bioprinting. Biotechnol. J. **13**, 1–12 (2018)
6. Darzi, S., Paul, K., Leitan, S., et al.: Immunobiology and application of aloe vera-based scaffolds in tissue engineering. Int. J. Mol. Sci. **22**, 1–19 (2021)

7. Wang, S., Meng, S., Zhou, X., et al.: PH-responsive and mucoadhesive nanoparticles for enhanced oral insulin delivery: the effect of hyaluronic acid with different molecular weights. Pharmaceutics **15**, 820 (2023)
8. Aghmiuni, A.I., Baei, M.S., Keshel, S.H., et al.: Design of novel 3D-scaffold as a potential material to induct epidermal-dermal keratinocytes of human-adipose-derived stem cells and promote fibroblast cells proliferation for skin regeneration. Fibers Polym. **21**, 33–44 (2020)
9. Brovold, M., Almeida, J.I., Pla-Palacín, I., et al.: Naturally-derived biomaterials for tissue engineering applications (2020). Epub ahead of print. https://doi.org/10.1007/978-981-13-0947-2_23
10. Ghorbani, M., Ramezani, S., Rashidi, M.: Fabrication of honey-loaded ethylcellulose/gum tragacanth nanofibers as an effective antibacterial wound dressing. Colloids Surfaces A Physicochem. Eng. Asp. **621**, 126615 (2021)
11. Zhong, S.P., Zhang, Y.Z., Lim, C.T.: Tissue scaffolds for skin wound healing and dermal reconstruction. Wiley Interdiscip. Rev.: Nanomedicine Nanobiotechnology **2**, 510–525 (2010)
12. Li, Q., Niu, Y., Xing, P., et al.: Bioactive polysaccharides from natural resources including Chinese medicinal herbs on tissue repair. Chinese Med. (United Kingdom) **13**, 1–11 (2018)
13. Wang, X., Ali, M.S., Lacerda, C.M.R.: A three-dimensional collagen-elastin scaffold for heart valve tissue engineering. Bioengineering 5. Epub ahead of print (2018). https://doi.org/10.3390/bioengineering5030069
14. Theocharis, A.D., Skandalis, S.S., Gialeli, C., et al.: Extracellular matrix structure. Adv. Drug Deliv. Rev. 1–76 (2015)
15. Ghiasi, M., Kalhor, N., Tabatabaei Qomi, R., et al.: The effects of synthetic and natural scaffolds on viability and proliferation of adipose-derived stem cells. Front Life Sci. **9**, 32–43 (2016)
16. Zahedi, E., Esmaeili, A., Eslahi, N., et al.: Fabrication and Characterization of Core-Shell Electrospun Fibrous Mats Containing Medicinal Herbs for Wound Healing and Skin Tissue Engineering. Mar. Drugs **17**, 27 (2019)
17. Stratton, S., Shelke, N.B., Hoshino, K., et al.: Bioactive polymeric scaffolds for tissue engineering. Bioact Mater **1**, 93–108 (2016)
18. Pires, P.C., Mascarenhas-melo F, Pedrosa K, et al. Polymer-based biomaterials for pharamaceutical and biomedical applications : a focus on topical drug administration. Eur. Polym. J. 187. Epub ahead of print 2023. https://doi.org/10.1016/j.eurpolymj.2023.111868
19. Huang, G., Li, F., Zhao, X., et al.: Functional and biomimetic materials for engineering of the three-dimensional cell microenvironment. Chem. Rev. **117**, 12764–12850 (2017)
20. Howard, D., Buttery, L.D., Shakesheff, K.M., et al.: Tissue engineering: strategies, stem cells and scaffolds. J. Anat. **213**, 66–72 (2008)
21. Ezhilarasu, H., Ramalingam, R., Dhand, C., et al.: Biocompatible aloe vera and tetracycline hydrochloride loaded hybrid nanofibrous scaffolds for skin tissue engineering. Int. J. Mol. Sci. **20**, 5174 (2019)
22. Thi Hiep, N., Chan Khon, H., Dai Hai, N., et al.: Biocompatibility of PCL/PLGA-BCP porous scaffold for bone tissue engineering applications. J. Biomater. Sci. Polym. Ed. **28**, 864–878 (2017)
23. Loh, Q.L., Choong, C.: Three-dimensional scaffolds for tissue engineering applications: role of porosity and pore size. Tissue Eng. Part B Rev. **19**, 485–502 (2013)
24. Shin, H., Jo, S., Mikos, A.G.: Biomimetic materials for tissue engineering. Biomaterials **24**, 4353–4364 (2003)
25. Yan, S., Zhang, Q., Wang, J., et al.: Silk fibroin/chondroitin sulfate/hyaluronic acid ternary scaffolds for dermal tissue reconstruction. Acta Biomater. **9**, 6771–6782 (2013)
26. Sengupta, P., Prasad, B.L.V.: Surface modification of polymeric scaffolds for tissue engineering applications. Regen. Eng. Transl. Med. **4**, 75–91 (2018)
27. Daamen, W.F., Veerkamp, J.H., Van Hest, J.C.M., et al.: Elastin as a biomaterial for tissue engineering. Biomaterials **28**, 4378–4398 (2007)
28. Hern, D.L., Hubbell, J.A.: Incorporation of adhesion peptides into nonadhesive hydrogels useful for tissue resurfacing. J. Biomed. Mater. Res. **39**, 266–276 (1998)

29. Tang X, Thankappan, S.K., Lee P, et al. Polymeric biomaterials in tissue engineering and regenerative medicine. In: Natural and Synthetic Biomedical Polymers. Elsevier, pp. 351–371 (2014)
30. Zhang, J., Sheng, X., Jiang, L.: The dewetting properties of lotus leaves. Langmuir **25**, 1371–1376 (2009)
31. Bixler, G.D., Bhushan, B.: Bioinspired rice leaf and butterfly wing surface structures combining shark skin and lotus effects. Soft Matter **8**, 11271–11284 (2012)
32. Panchanathan, D., Rajappan, A., Varanasi, K.K., et al.: Plastron regeneration on submerged superhydrophobic surfaces using in situ gas generation by chemical reaction. ACS Appl. Mater. Interfaces **10**, 33684–33692 (2018)
33. Feng, L., Zhang, Y., Xi, J., et al.: Petal effect: a superhydrophobic state with high adhesive force. Langmuir **24**, 4114–4119 (2008)
34. Zhou, F.Y., Wang, B.L., Qiu, K.J., et al.: Microstructure, corrosion behavior and cytotoxicity of Zr-Nb alloys for biomedical application. Mater. Sci. Eng. C **32**, 851–857 (2012)
35. Zhang, B.B., Qiu, K.J., Wang, B.L., et al.: Surface characterization and cell response of binary Ti-Ag alloys with CP Ti as material control. J. Mater. Sci. Technol. **28**, 779–784 (2012)
36. Zhang, F., Weidmann, A., Nebe, B.J., et al. Preparation of TiMn alloy by mechanical alloying and spark plasma sintering for biomedical applications. J. Phys. Conf. Ser. 144. Epub ahead of print 2009. https://doi.org/10.1088/1742-6596/144/1/012007
37. Ho, W.F., Chiang, T.Y., Wu, S.C., et al.: Mechanical properties and deformation behavior of cast binary Ti-Cr alloys. J. Alloys Compd. **468**, 533–538 (2009)
38. Han, J.T., Jang, Y., Lee, D.Y., et al.: Fabrication of a bionic superhydrophobic metal surface by sulfur-induced morphological development. J. Mater. Chem. **15**, 3089–3092 (2005)
39. Ijaz, M.F., Kim, H.Y., Hosoda, H., et al.: Superelastic properties of biomedical (Ti-Zr)-Mo-Sn alloys. Mater. Sci. Eng. C **48**, 11–20 (2015)
40. Huang, Z., Quan, Y., Mao, J., et al.: Multifunctional superhydrophobic composite materials with remarkable mechanochemical, r.obustness., stain, r.epellency., oil-water separation and sound-absorption properties. Chem. Eng. J. **358**, 1610–1619 (2019)
41. Tang, W., Sun, D., Liu, S., et al.: One step electrochemical fabricating of the biomimetic graphene skins with superhydrophobicity and superoleophilicity for highly efficient oil-water separation. Sep. Purif. Technol. **236**, 116293 (2020)
42. Xia, C., Zhang, Z., Feng, Z., et al.: Effect of zirconium content on the microstructure and corrosion behavior of Ti-6Al-4V-xZr alloys. Corros. Sci. **112**, 687–695 (2016)
43. Fan, Y., He, Y., Luo, P., et al.: A facile electrodeposition process to fabricate corrosion-resistant superhydrophobic surface on carbon steel. Appl. Surf. Sci. **368**, 435–442 (2016)
44. Jena, G., Thinaharan, C., George, R.P., et al.: Robust nickel-reduced graphene oxide-myristic acid superhydrophobic coating on carbon steel using electrochemical codeposition and its corrosion resistance. Surf. Coatings Technol. **397**, 125942 (2020)
45. Zeng, Y., Qin, Z., Hua, Q., et al.: Sheet-like superhydrophobic surfaces fabricated on copper as a barrier to corrosion in a simulated marine system. Surf. Coatings Technol. **362**, 62–71 (2019)
46. Ogawa, T.: Ultraviolet photofunctionalization of titanium implants. Oral Craniofacial Tissue Eng. 2
47. Cortat, B., Garcia, C.C.M., Quinet, A., et al.: The relative roles of DNA damage induced by UVA irradiation in human cells. Photochem. Photobiol. Sci. **12**, 1483–1495 (2013)
48. Aita, H., Hori, N., Takeuchi, M., et al.: The effect of ultraviolet functionalization of titanium on integration with bone. Biomaterials **30**, 1015–1025 (2009)
49. Att, W., Hori, N., Iwasa, F., et al.: The effect of UV-photofunctionalization on the time-related bioactivity of titanium and chromium–cobalt alloys. Biomaterials **30**, 4268–4276 (2009)
50. Dini, C., Nagay, B.E., Cordeiro, J.M., et al.: UV-photofunctionalization of a biomimetic coating for dental implants application. Mater. Sci. Eng. C **110**, 110657 (2020)
51. Nagay, B.E., Bitencourt, S.B., Commar, B.C., et al.: Antimicrobial and protective effects of non-thermal plasma treatments on the performance of a resinous liner. Arch. Oral Biol. **117**, 104822 (2020)

52. Cordeiro, J.M., Nagay, B.E., Ribeiro, A.L.R., et al.: Functionalization of an experimental Ti-Nb-Zr-Ta alloy with a biomimetic coating produced by plasma electrolytic oxidation. J. Alloys Compd. **770**, 1038–1048 (2019)
53. Maciel, J.G., Sugio, C.Y.C., de, C.ampos., C.G., et al.: Determining acceptable limits for water sorption and solubility of interim denture resilient liners. J. Prosthet. Dent. **121**, 311–316 (2019)
54. Coelho, M.FC., De Sousa, L.L., Ferreira, C.C., et al.: Biomimetic coating on titanium: evaluation of bioactivity and corrosion. Mater. Res. Express 6. Epub ahead of print (2019). https://doi.org/10.1088/2053-1591/ab67f1
55. Barth, H.D., Zimmermann, E.A., Schaible, E., et al.: Characterization of the effects of x-ray irradiation on the hierarchical structure and mechanical properties of human cortical bone. Biomaterials **32**, 8892–8904 (2011)
56. Barth, H.D., Launey, M.E., MacDowell, A.A., et al.: On the effect of X-ray irradiation on the deformation and fracture behavior of human cortical bone. Bone **46**, 1475–1485 (2010)
57. Milovanovic, P., Zimmermann, E.A., Riedel, C., et al.: Multi-level characterization of human femoral cortices and their underlying osteocyte network reveal trends in quality of young, aged, osteoporotic and antiresorptive-treated bone. Biomaterials **45**, 46–55 (2015)
58. Zimmermann, E.A., Gludovatz, B., Schaible, E., et al.: Fracture resistance of human cortical bone across multiple length-scales at physiological strain rates. Biomaterials **35**, 5472–5481 (2014)
59. Peng, M., Wen, Z., Xie, L., et al.: 3D printing of ultralight biomimetic hierarchical graphene materials with exceptional stiffness and resilience. Adv. Mater. **31**, 1902930 (2019)
60. Tseng, I.-H., Liu, Z.-C., Chang, P.-Y.: Bio-friendly titania-grafted chitosan film with biomimetic surface structure for photocatalytic application. Carbohydr. Polym. **230**, 115584 (2020)
61. Barthlott, W., Mail, M., Bhushan, B., et al.: Plant surfaces: structures and functions for biomimetic innovations. Nano-Micro Lett. **9**, 1–40 (2017)
62. Xu, Q., He, C., Xiao, C., et al.: Reactive oxygen species (ROS) responsive polymers for biomedical applications. Macromol. Biosci. **16**, 635–646 (2016)
63. Xia, J., Li, T., Lu, C., et al.: Selenium-containing polymers: perspectives toward diverse applications in both adaptive and biomedical materials. Macromolecules **51**, 7435–7455 (2018)
64. Tian, J., Liu, Q., Asiri, A.M., et al.: Ultrathin graphitic carbon nitride nanosheets: a novel peroxidase mimetic, Fe doping-mediated catalytic performance enhancement and application to rapid, highly sensitive optical detection of glucose. Nanoscale **5**, 11604–11609 (2013)
65. Cao, X.-N., Lian, S., Tong, Y., et al.: Fluorescent Se-modified carbon nitride nanosheets as biomimetic catalases for free-radical scavenging. Chem. Commun. **56**, 916–919 (2020)
66. Zhao, X., Chen, Z., Liu, Y., et al.: Silk fibroin microparticles with hollow mesoporous silica nanocarriers encapsulation for abdominal wall repair. Adv. Healthc. Mater. **7**, 1801005 (2018)
67. Zhang, W., Li, Y., Jiang, D., et al.: Promotion of hernia repair with high-strength, flexible, and bioresorbable silk fibroin mesh in a large abdominal hernia model. ACS Biomater. Sci. Eng. **4**, 2067–2080 (2017)
68. Skrobot, J., Zair, L., Ostrowski, M., et al.: New injectable elastomeric biomaterials for hernia repair and their biocompatibility. Biomaterials **75**, 182–192 (2016)
69. Song, Z., Peng, Z., Liu, Z., et al.: Reconstruction of abdominal wall musculofascial defects with small intestinal submucosa scaffolds seeded with tenocytes in rats. Tissue Eng. Part A **19**, 1543–1553 (2013)
70. Izadyari Aghmiuni, A., Heidari Keshel, S., Sefat, F., et al.: Quince seed mucilage-based scaffold as a smart biological substrate to mimic mechanobiological behavior of skin and promote fibroblasts proliferation and h-ASCs differentiation into keratinocytes. Int. J. Biol. Macromol.. Epub ahead of print October (2019). https://doi.org/10.1016/j.ijbiomac.2019.10.008
71. Da, L., Gong, M., Chen, A., et al.: Composite elastomeric polyurethane scaffolds incorporating small intestinal submucosa for soft tissue engineering. Acta Biomater. **59**, 45–57 (2017)

72. Jiang, W., Zhang, J., Lv, X., et al.: Use of small intestinal submucosal and acellular dermal matrix grafts in giant omphaloceles in neonates and a rabbit abdominal wall defect model. J. Pediatr. Surg. **51**, 368–373 (2016)
73. Cao, G., Wang, C., Fan, Y., et al.: Biomimetic SIS-based biocomposites with improved biodegradability, antibacterial activity and angiogenesis for abdominal wall repair. Mater. Sci. Eng. C **109**, 110538 (2020)
74. Kim, T.G., Shin, H., Lim, D.W.: Biomimetic scaffolds for tissue engineering. Adv. Funct. Mater. **22**, 2446–2468 (2012)
75. Kharaziha, M., Fathi, M.H., Edris, H.: Development of novel aligned nanofibrous composite membranes for guided bone regeneration. J. Mech. Behav. Biomed. Mater. **24**, 9–20 (2013)
76. Naghieh, S., Foroozmehr, E., Badrossamay, M., et al.: Combinational processing of 3D printing and electrospinning of hierarchical poly (lactic acid)/gelatin-forsterite scaffolds as a biocomposite: Mechanical and biological assessment. Mater. Des. **133**, 128–135 (2017)
77. Kharaziha, M., Nikkhah, M.: Spatial patterning of stem cells to engineer microvascular networks. Microscale Technol. Cell Eng. 143–166 (2016)
78. Li, S., Abdel-Wahab, A., Silberschmidt, V.V.: Analysis of fracture processes in cortical bone tissue. Eng. Fract. Mech. **110**, 448–458 (2013)
79. Hardy, J.G., Ghezzi, C.E., Saballos, R.J., et al.: Supracolloidal assemblies as sacrificial templates for porous silk-based biomaterials. Int. J. Mol. Sci. **16**, 20511–20522 (2015)
80. Liu, Y., Kim, J.H., Young, D., et al.: Novel template-casting technique for fabricating β-tricalcium phosphate scaffolds with high interconnectivity and mechanical strength and in vitro cell responses. J. Biomed. Mater. Res. Part A **92**, 997–1006 (2010)
81. Luo, H., Zhang, Y., Li, G., et al.: Sacrificial template method for the synthesis of three-dimensional nanofibrous 58S bioglass scaffold and its in vitro bioactivity and cell responses. J. Biomater. Appl. **32**, 265–275 (2017)
82. Cadman, J., Zhou, S., Chen, Y., et al. Characterization of cuttlebone for a biomimetic design of cellular structures. In: Acta Mechanica Sinica/Lixue Xuebao, pp. 27–35 (2010)
83. Saravanan, S., Selvamurugan, N.: Bioactive mesoporous wollastonite particles for bone tissue engineering. J. Tissue Eng. 7. Epub ahead of print (2016). https://doi.org/10.1177/2041731416680319
84. Eilbagi, M., Emadi, R., Raeissi, K., et al.: Mechanical and cytotoxicity evaluation of nanostructured hydroxyapatite-bredigite scaffolds for bone regeneration. Mater. Sci. Eng. C **68**, 603–612 (2016)
85. Williams, J.M., Adewunmi, A., Schek, R.M., et al.: Bone tissue engineering using polycaprolactone scaffolds fabricated via selective laser sintering. Biomaterials **26**, 4817–4827 (2005)
86. Schumacher, T.C., Volkmann, E., Yilmaz, R., et al.: Mechanical evaluation of calcium-zirconium-silicate (baghdadite) obtained by a direct solid-state synthesis route. J. Mech. Behav. Biomed. Mater. **34**, 294–301 (2014)
87. Abbasian, V., Emadi, R., Kharaziha, M.: Biomimetic nylon 6-baghdadite nanocomposite scaffold for bone tissue engineering. Mater. Sci. Eng. C **109**, 110549 (2020)
88. Wang, R., Li, J., Chen, W., et al.: A biomimetic mussel-inspired ε-poly-l-lysine hydrogel with robust tissue-anchor and anti-infection capacity. Adv. Funct. Mater. 27. Epub ahead of print (2017). https://doi.org/10.1002/adfm.201604894
89. Yuan, H., Zhan, Y., Rowan, A.E., et al.: Biomimetic networks with enhanced photodynamic antimicrobial activity from conjugated polythiophene/polyisocyanide hybrid hydrogels. Angew. Chemie Int. Ed. **59**, 2720–2724 (2020)
90. Izadyari Aghmiuni, A., Heidari Keshel, S.: The role of the extracellular matrix (ECM) and ECM-like polymeric substrates in health and disease. In: Berhardt, L.V. (ed.) Advances in Medicine and Biology. NOVA Medicine and Health, pp. 145–175 (2022)
91. Aghmiuni, A.I., Ghadi, A., Azmoun, E., et al.: Electrospun polymeric substrates for tissue engineering: viewpoints on fabrication, application, and challenges. In: Electrospinning—Material Technology of the Future. InTech, pp. 1–26 (2022)

92. Izadyari Aghmiuni, A., Heidari Keshel, S., Sefat, F., et al.: Quince seed mucilage-based scaffold as a smart biological substrate to mimic mechanobiological behavior of skin and promote fibroblasts proliferation and h-ASCs differentiation into keratinocytes. Int. J. Biol. Macromol. **142**, 668–679 (2020)
93. Izadyari Aghmiuni, A., Akbarzadeh Khiavi, A.: Medicinal plants to calm and treat psoriasis disease. In: El-Shemy, H.A. (ed.) Aromatic and Medicinal Plants—Back to Nature, pp. 1–28. InTech, Rijeka
94. Tahmasebi, A., Shapouri Moghadam, A., Enderami, S.E., et al.: Aloe vera-derived gel-blended PHBV nanofibrous scaffold for bone tissue engineering. ASAIO J. **66**, 966–973 (2020)
95. Suganya, S., Venugopal, J., Agnes Mary, S., et al.: Aloe vera incorporated biomimetic nanofibrous scaffold: a regenerative approach for skin tissue engineering. Iran. Polym. J. **23**, 237–248 (2014)
96. Rahman, S., Carter, P., Bhattarai, N.: Aloe vera for tissue engineering applications. J. Funct. Biomater. **8**, 6 (2017)
97. Paul, K., Darzi, S., Del Borgo, M.P., et al.: Vaginal delivery of tissue engineered endometrial mesenchymal stem/stromal cells in an aloe vera-alginate hydrogel alleviates maternal simulated birth injury. Appl. Mater. Today **22**, 100890 (2021)
98. Zadegan, S., Nourmohammadi, J., Vahidi, B., et al.: An investigation into osteogenic differentiation effects of silk fibroin-nettle (Urtica dioica L.) nanofibers. Int. J. Biol. Macromol. **133**, 795–803 (2019)
99. Ghiyasi, Y., Salahi, E., Esfahani, H.: Synergy effect of Urtica dioica and ZnO NPs on microstructure, antibacterial activity and cytotoxicity of electrospun PCL scaffold for wound dressing application. Mater. Today Commun. **26**, 102163 (2021)
100. Martins, R.M., Alves, G.D.A.D., Martins, S.D.S., et al. Apple extract (Malus sp.) and rutin as photochemopreventive agents: evaluation of ultraviolet B-induced alterations on skin biopsies and tissue-engineered skin. Rejuvenation Res. **23**, 465–475 (2020)
101. Hajiali, H., Summa, M., Russo, D., et al.: Alginate–lavender nanofibers with antibacterial and anti-inflammatory activity to effectively promote burn healing. J. Mater. Chem. B **4**, 1686–1695 (2016)
102. Pilehvar-Soltanahmadi, Y., Nouri, M., Martino, M.M., et al.: Cytoprotection, proliferation and epidermal differentiation of adipose tissue-derived stem cells on emu oil based electrospun nanofibrous mat. Exp. Cell Res. **357**, 192–201 (2017)
103. Peng-Zhang, Dai, K.R., Yan, S.G., et al.: Effects of naringin on the proliferation and osteogenic differentiation of human bone mesenchymal stem cell. Eur. J. Pharmacol. **607**, 1–5 (2009)
104. Yu, G.Y., Zheng, G.Z., Chang, B., et al.: Naringin stimulates osteogenic differentiation of rat bone marrow stromal cells via activation of the notch signaling pathway. Stem Cells Int. Epub ahead of print (2016). https://doi.org/10.1155/2016/7130653
105. Zhang, J., Xu, D.P., Shang, J., et al.: Effects of naringin on the proliferation and osteogenic differentiation of canine bone marrow stromal cells in vitro. J. Hard Tissue Biol. **30**, 73–78 (2021)
106. Kim, B.S., Kang, H.J., Park, J.Y., et al.: Fucoidan promotes osteoblast differentiation via JNK- and ERK-dependent BMP2-Smad 1/5/8 signaling in human mesenchymal stem cells. Exp. Mol. Med. **47**, 1–9 (2015)
107. Li, F., Yang, Y., Zhu, P., et al.: Echinacoside promotes bone regeneration by increasing OPG/RANKL ratio in MC3T3-E1 cells. Fitoterapia **83**, 1443–1450 (2012)
108. Satué, M., Arriero, M.D.M., Monjo, M., et al.: Quercitrin and taxifolin stimulate osteoblast differentiation in MC3T3-E1 cells and inhibit osteoclastogenesis in RAW 264.7 cells. Biochem. Pharmacol. **86**, 1476–1486 (2013)
109. Kim, M.B., Song, Y., Hwang, J.K.: Kirenol stimulates osteoblast differentiation through activation of the BMP and Wnt/β-catenin signaling pathways in MC3T3-E1 cells. Fitoterapia **98**, 59–65 (2014)
110. Choi, E.M.: Honokiol isolated from Magnolia officinalis stimulates osteoblast function and inhibits the release of bone-resorbing mediators. Int. Immunopharmacol. **11**, 1541–1545 (2011)

111. Xiao, H.H., Gao, Q.G., Zhang, Y., et al.: Vanillic acid exerts oestrogen-like activities in osteoblast-like UMR 106 cells through MAP kinase (MEK/ERK)-mediated ER signaling pathway. J. Steroid Biochem. Mol. Biol. **144**, 382–391 (2014)
112. Tiyasatkulkovit, W., Charoenphandhu, N., Wongdee, K., et al.: Upregulation of osteoblastic differentiation marker mRNA expression in osteoblast-like UMR106 cells by puerarin and phytoestrogens from Pueraria mirifica. Phytomedicine **19**, 1147–1155 (2012)
113. Delloye, C., Cornu, O., Druez, V., et al.: Bone allografts: what they can offer and what they cannot. J. Bone Joint Surg. Br. **89**, 574–580 (2007)
114. Seiler, J.G., 3rd., Johnson, J.: Iliac crest autogenous bone grafting: donor site complications. J. South. Orthop. Assoc. **9**, 91–97 (2000)
115. Tappa K, Jammalamadaka U. Novel biomaterials used in medical 3D printing techniques. J. Funct. Biomater. 9. Epub ahead of print (2018). https://doi.org/10.3390/jfb9010017
116. Xu, T., Jin, J., Gregory, C., et al.: Inkjet printing of viable mammalian cells. Biomaterials **26**, 93–99 (2005)
117. Seyedmahmoud, R., Messler, M.J., Loboa, E.G.: 3D bioprinting technologies for tissue engineering: a mini review. HSOA J. Stem Cells Res. Dev. Ther. 6
118. Ong, C.S., Yesantharao, P., Huang, C.Y., et al.: 3D bioprinting using stem cells. Pediatr. Res. **83**, 223–231 (2018)
119. Akkuş, H., Günaydin, B., Ustundag, C.: DOku mühendisliğinde 3 boyutlu biyo-baski için biyofonksiyonel mürekkepler. Int. J. 3D Print Technol. Digit. Ind. **4** 285–294 (2020)
120. Hacıoglu, A., Yılmazer, H., Ustundag, C.B.: 3D printing for tissue engineering applications. Politek. Derg. **21**, 221–227 (2018)
121. Tasnim, N., De la Vega, L., Anil Kumar, S., et al.: 3D bioprinting stem cell derived tissues. Cell. Mol. Bioeng. **11**, 219–240 (2018)
122. Vurat, M.T.: Üç-boyutlu biyobasım için yeni kompozit biyomürekkep geliştirilmesi ve karakterizasyonu
123. Cui, X., Boland, T., DD'Lima, D., et al.: Thermal inkjet printing in tissue engineering and regenerative medicine. Recent Pat. Drug Deliv. Formul. **6**, 149–155 (2012)
124. Murphy, S.V., Atala, A.: 3D bioprinting of tissues and organs. Nat. Biotechnol. **32**, 773–785 (2014)
125. Yilmaz, B., Tahmasebifar, A., Baran, E.T.: Bioprinting technologies in tissue engineering. Curr. Appl. Pharm. Biotechnol. 279–319 (2020)
126. Liu, T., Guessasma, S., Zhu, J., et al.: Microstructural defects induced by stereolithography and related compressive behaviour of polymers. J. Mater. Process. Technol. **251**, 37–46 (2018)
127. Ligon, S.C., Liska, R., Stampfl, J., et al.: Polymers for 3D printing and customized additive manufacturing. Chem. Rev. **117**, 10212–10290 (2017)
128. Rhee, S., Puetzer, J.L., Mason, B.N., et al.: 3D bioprinting of spatially heterogeneous collagen constructs for cartilage tissue engineering. ACS Biomater. Sci. Eng. **2**, 1800–1805 (2016)
129. Laronda, M.M., Rutz, A.L., Xiao, S., et al.: A bioprosthetic ovary created using 3D printed microporous scaffolds restores ovarian function in sterilized mice. Nat. Commun. **8**, 15261 (2017)
130. Markstedt, K., Mantas, A., Tournier, I., et al.: 3D bioprinting human chondrocytes with nanocellulose–alginate bioink for cartilage tissue engineering applications. Biomacromol **16**, 1489–1496 (2015)
131. Nguyen, D., Hägg, D.A., Forsman, A., et al.: Cartilage tissue engineering by the 3D bioprinting of iPS cells in a nanocellulose/alginate bioink. Sci. Rep. **7**, 1–10 (2017)
132. Tan, Z., Parisi, C., Di Silvio, L., et al.: Cryogenic 3D printing of super soft hydrogels. Sci. Rep. **7**, 16293 (2017)
133. Bee, S.-L., Hamid, Z.A.A., Mariatti, M., et al.: Approaches to improve therapeutic efficacy of biodegradable PLA/PLGA microspheres: a review. Polym. Rev. **58**, 495–536 (2018)
134. Fontes, A.B., Marcomini, R.F.: 3D bioprinting: a review of materials, processes and bioink properties. J. Eng. Exact Sci. **6**, 617–639 (2020)
135. Yu, J., Park, S.A., Kim, W.D., et al.: Current advances in 3D bioprinting technology and its applications for tissue engineering. Polymers (Basel) **12**, 2958 (2020)

136. Aduba, D.C., Jr., Margaretta, E.D., Marnot, A.E.C., et al.: Vat photopolymerization 3D printing of acid-cleavable PEG-methacrylate networks for biomaterial applications. Mater. Today Commun. **19**, 204–211 (2019)
137. Lee, J.S., Park, T.H., Ryu, J.Y., et al.: Osteogenesis of 3d-printed pcl/tcp/bdecm scaffold using adipose-derived stem cells aggregates; an experimental study in the canine mandible. Int. J. Mol. Sci. **22**. Epub ahead of print (2021). https://doi.org/10.3390/ijms22115409
138. Shim, J.H., Won, J.Y., Park, J.H., et al.: Effects of 3D-printed polycaprolactone/β-tricalcium phosphate membranes on guided bone regeneration. Int. J. Mol. Sci. **18**. Epub ahead of print (2017). https://doi.org/10.3390/ijms18050899
139. Tappa, K., Jammalamadaka, U., Ballard, D.H., et al.: Medication eluting devices for the field of OBGYN (MEDOBGYN): 3D printed biodegradable hormone eluting constructs, a proof of concept study. PLoS ONE **12**, 1–17 (2017)
140. Horst, D.J., Tebcherani, S.M., Kubaski, E.T., et al.: Bioactive potential of 3D-printed oleo-gum-resin disks: B. papyrifera, C. myrrha, and S. benzoin loading Nanooxides—TiO_2, P25, Cu_2O, and MoO_3. Bioinorg. Chem. Appl. (2017). Epub ahead of print. https://doi.org/10.1155/2017/6398167
141. Wang, Y., Wang, K., Li, X., et al.: 3D fabrication and characterization of phosphoric acid scaffold with a HA/β-TCP weight ratio of 60:40 for bone tissue engineering applications. PLoS ONE **12**, 1–17 (2017)
142. Poldervaart, M.T., Goversen, B., De Ruijter, M., et al.: 3D bioprinting of methacrylated hyaluronic acid (MeHA) hydrogel with intrinsic osteogenicity. PLoS ONE **12**, 1–15 (2017)
143. Sa, M.W., Nguyen, B.N.B., Moriarty, R.A., et al.: Fabrication and evaluation of 3D printed BCP scaffolds reinforced with ZrO2 for bone tissue applications. Biotechnol. Bioeng. **115**, 989–999 (2018)
144. Kuroda, S., Kobayashi, T., Ohdan, H.: 3D printing model of the intrahepatic vessels for navigation during anatomical resection of hepatocellular carcinoma. Int. J. Surg. Case Rep. **41**, 219–222 (2017)
145. Asadi-Eydivand, M., Solati-Hashjin, M., Shafiei, S.S., et al.: Structure, properties, and in vitro behavior of heat-treated calcium sulfate scaffolds fabricated by 3D printing. PLoS ONE **11**, e0151216 (2016)
146. Esworthy, T.J., Miao, S., Lee, S.-J., et al.: Advanced 4D-bioprinting technologies for brain tissue modeling and study. Int. J. Smart Nano Mater. **10**, 177–204 (2019)
147. Arslan-Yildiz, A., El, A.R., Chen, P., et al.: Towards artificial tissue models: past, present, and future of 3D bioprinting. Biofabrication **8**, 14103 (2016)
148. Gao, B., Yang, Q., Zhao, X., et al.: 4D bioprinting for biomedical applications. Trends Biotechnol. **34**, 746–756 (2016)
149. Li, Y.-C., Zhang, Y.S., Akpek, A., et al.: 4D bioprinting: the next-generation technology for biofabrication enabled by stimuli-responsive materials. Biofabrication **9**, 12001 (2016)
150. Saravanan, S., Vimalraj, S., Thanikaivelan, P., et al.: A review on injectable chitosan/beta glycerophosphate hydrogels for bone tissue regeneration. Int. J. Biol. Macromol. **121**, 38–54 (2019)
151. Graham, S., Marina, P.F., Blencowe, A.: Thermoresponsive polysaccharides and their thermoreversible physical hydrogel networks. Carbohydr. Polym. **207**, 143–159 (2019)
152. Suo, H., Zhang, D., Yin, J., et al.: Interpenetrating polymer network hydrogels composed of chitosan and photocrosslinkable gelatin with enhanced mechanical properties for tissue engineering. Mater. Sci. Eng. C **92**, 612–620 (2018)
153. Yang, J., van Lith, R., Baler, K., et al.: A thermoresponsive biodegradable polymer with intrinsic antioxidant properties. Biomacromol **15**, 3942–3952 (2014)
154. Chen, J.-P., Tsai, M.-J., Liao, H.-T.: Incorporation of biphasic calcium phosphate microparticles in injectable thermoresponsive hydrogel modulates bone cell proliferation and differentiation. Colloids Surfaces B Biointerfaces **110**, 120–129 (2013)

Azadeh Izadyari Aghmiuni has a Ph.D. in Chemical Engineering-Tissue Engineering. Now, she is working on the stem cells, biological macromolecules & natural/synthetic polymers for construction of engineered scaffolds and smart biological substrates to regenerate hard and soft tissues. In this field, she has one patent entitled "Hybrid nano-substrate containing collagen to regenerate skin wounds, IR101120". Moreover, she investigates on the application of the herbal active ingredient and targeted drug delivery (TDD) method in tissue engineering and regenerative medicine to treat diseases.

Aref Gholami has a B.Sc. in Electrical Engineering with Power Technology orientation. He has a lot of experience in the field of design of medical electronic devices; and now, is working in this field.

An Insight into Collagen-Based Nano Biomaterials for Drug Delivery Applications

Amit Kumar Verma

Abstract In modern medicine, drug delivery is a broad field of research for the evolution of novel materials or carrier systems for effective therapeutic delivery of drugs as the controlled drug delivery is a challenge on the basis of solubility, bioavailability, cytotoxicity along with pharmacokinetic parameters.

Protein-based drug delivery systems (DDS) have shown promising results due to structural support, cell-targeted delivery, bioavailability, biocompatibility and non-immunogenicity, etc. Collagen as an important extracellular matrix component has attracted drug delivery-based research in recent years. Collagen based-hydrogels/composites/biofilms are excellent objects for drug delivery, tissue engineering, wound dressings and gene therapeutics etc. due to high encapsulating capacity, mechanically strong swollen structural network and efficient mass transfer properties. Some of the applications of collagen are the formation of microspheres and microneedles for drug delivery, formulation of nanoparticles (NPs) for gene delivery, development of pellets and tablets for protein delivery, formation of gels and combination with liposomes for sustained drug delivery, cancer treatment and collagen shields in ophthalmology.

DDS based on NPs display enhanced efficacy of drugs and improve the drug's half-life, hydrophobic drug solubility and controlled/sustained drug release in the infected body regions. Stimuli-responsive NPs regulate drug biodistribution and reduce drug toxicity. Protein-based nanocomposites can be prepared through various physical and chemical methods like desolvation, emulsification, phase separation, electrospray, electrospinning and milling, etc. These methods have their operating ease and difficulties for the production of the desired quality of nanomaterials/composites.

Current Polymeric NPs systems are sensitive to stimuli such as temperature, light, pH, oxidizing/reducing agents, magnetic fields and enzymes which increases efficiency and specificity for various applications. Collagen with NPs results in stabilization of the nanoparticles and helps with entrapment of the drug, to attain steady and regulated drug release for ideal therapeutic reactions. Collagen NPs have

A. K. Verma (✉)
Department of Biosciences, Srinivasa Ramanujan Block, Jamia Millia Islamia, Jamia Nagar, New Delhi 110 025, India
e-mail: averma@jmi.ac.in

R. Malviya and S. Sundram (eds.), *Engineered Biomaterials*, Engineering Materials,
https://doi.org/10.1007/978-981-99-6698-1_13

advantage over other natural and synthetic polymeric NPs due to biocompatibility, biodegradability, low antigenicity, high contact surface and reduced toxicity.

Significant advancements have been achieved using collagen-based nano-DDS to deliver biomolecules with better efficacy at targeted sites. In spite of the substantial progress, collagen is still affected by low mechanical strength and high rate of degradability, which is a serious concern during clinical trials of targeting intracellular molecules like genes, drugs and growth factors, etc. In future, collagen-based nano-DDS will be the key player for the delivery of desired drugs/biomolecules at specific target for different medical conditions.

Keywords Drug Delivery Systems (DDS) · Collagen · Drug · Biomolecules · Nanoparticles (NPs)

1 Introduction

Drug delivery systems (DDS) are the important, impressive and effective tools for disease treatment, management and prevention in the field of modern medicine. The DDS comprises the introduction and delivery of active biomolecule or pharmaceutical at targeted location in the body with improvised efficiency, biocompatibility and safety. Further, DDS involves the augmented therapeutic performance via increased absorption, prolonged or sustained release at specific target site with minimal cytotoxicity and effectively removal from the body without causing adverse physiological conditions [1–3]. Guided drug delivery is the challenge, because the bioactive molecule has to survive through various cellular or tissue level barriers during circulation before binding to a particular target in body. In recent years, the use of nanoscale protein-based polymers has been sufficiently progressed for delivery of drugs and vaccines due to the ability to cross biological barriers and to reach at the target site at molecular level [4].

Nanoparticles (NPs) are microscopic structures with typical size range of 1–100 nm in diameter have gained the attention of scientists due to their small size, high surface-to-volume ratio, stability, self-assembly behaviour and mutual interactions at fluid interfaces, make them the suitable candidate for delivering the variety of drugs of active biomolecules in biomedicine [5, 6].

The crucial biological molecules- proteins and peptides are the most important objects of research in the areas of nanomedicine. These biomolecules based nanoconstructs are the centre of attraction in the field of nanomedicine, nanobiotechnology, toxicology and immunology to understand the interactions with other molecules in body for therapeutic and diagnostic purposes [7, 8]. The nature and organic composition of protein-based nanocomposites have the instinctive affinity to interact with living cells or tissues and they are often not considered as foreign entity and are readily eliminated from the body through macrophages. The typical property of protein nanomaterial is inherent uniform size and homogenous size distribution throughout the material. These materials can be modified biologically, chemically or physically for providing desired physical and biological qualities for efficient drug delivery [9, 10].

Collagen is unique, most abundant and major extracellular structural protein of extracellular matrix (ECM) and plays crucial role to the structural integrity of tissues/organs and cellular growth in vertebrates and other organisms [11–13]. Though having poor mechanical strength and rapid degradation property, collagen can form stable, mechanically strong and biocompatible nanocomposites with natural biopolymer like chitosan, alginate, cellulose as well as synthetic polymer such as polyamide, polyvinyl alcohol, poly ε-caprolactone (PCL) and poly (lactic-co-glycolic acid), etc., in different proportion using the chemical crosslinkers [14, 15]. Collagen can be utilized for DDS through variety of structures and shapes such as nanoparticles, nanofilms, protein cages, hydrogels and minirods with other protein polymer-based composites [16].

2 Collagen-Based Nano-Biomaterial Formulations

Nanotechnology is the modern branch of science and engineering deals with the technological developments of materials at the nanoscale for various physical, chemical, industrial and medical applications [17, 18]. This technology envisages the construction and characterization of engineered materials by rearranging at atomic, molecular and supramolecular scales at 1–100 nm size in one or more dimensions. Based on shapes, the nanomaterials could be of zero dimension in NPs, one dimension in nanorods, nanosheets in two dimension and three dimensions in nanotubes, nanowires or multi-nanolayers and dendrimers [19, 20]. In present scenario, nano biomaterials are utilized in different biomedical applications such as drug delivery, biosensing, bioimaging, antibacterial and anti-cancer activity, medical diagnosis, tissue engineering/regeneration, wound healing, medical equipment, cosmetics, food packaging industry and environmental remedy, etc. [21].

Nanocollagen (NC) is the typical regular collagen downsized at the nanoscale dimension of 1–100 nm, having desirable qualities of NPs like high surface area to volume ratio and biomimetic, bioavailability, low antigenicity, optimal penetration potential at particular location of collagen, making the collagen as very effective biomolecule for drug delivery and wound healing applications in medical sciences [21]. Collagen nanofibers are cylindrical network of fibres with less than 1000 nm of external diameter and more than 50 ratio of length and width [22]. Wang et al. [23] designed the NC fibre with uniform networking and structure with attributes of elasticity and resistance to break, making it suitable for tissue engineering application. The NC fibres could withstand 500 mN tensile load with 50 nN load resolution, which later augmented the load transfer efficiency.

Protein-based NPs preparation and drug encapsulation are performed in optimal environment to avoid the toxic solvents or organic chemicals as residue after the synthesis. NC can be generally produced by chemical, physical and self-assembly methods [18, 24]. The chemical methods are emulsification and coacervation or co-electrolyte complexation; electro-spraying and nano-spray drying are physical methods and desolvation is the self-assembly method [17]. Apart from these methods, others are milling, phase separation, interfacial polymerization and polymer chain collapse [25]. Each method of preparation of protein/collagen-based NPs has its own pros and cons, which are summarized in Table 1.

2.1 Emulsification or Solvent Extraction

Nanocollagen can be formulated by nanoemulsion solution along with drug followed by mechanical sheering. A nanoemulsion solution is the mixture of two immiscible liquids and categorized in two classes: oil-in-water (O/W) phase, where dispersive phase is oil and continuous phase is water, water-in-oil (W/O) in reverse order. Nanoemulsion droplets differ from regular/normal emulsion drop. The dimension of nanoemulsion drops is 20–200 nm while that of normal emulsion drop is around 1μm [26]. In this method, collagen in aqueous phase and hydrophilic surfactant in water is mixed with organic phase of lipophilic surfactant, oil and solvent miscible in water under normal room temperature conditions through mechanical agitation to generate homogeneous emulsion. The generated emulsion is combined with heated oil in drop wise fashion to form the NC emulsion particles [27] (Fig. 1). The nanoscale size of nanoemulsion drops allow the homogeneous dispersion and diffusion of active biomolecules or compounds through the skin barrier [28]. Nanoemulsion preparation required only 3–10% of surfactant, which resulted in better penetration power of drugs due to large surface area and low surface tension of the nanoemulsion solution, while on the other hand microemulsions required more than 20% of surfactant [29, 30].

Nanoemulsion can be formulated with low-energy or high-energy methods or the amalgamation of both methods. High-energy method involves technique like high pressure homogenizers, microfluidizers and ultrasonicators to generate high magnitude of sheer forces to rupture the oil and water phases to create the oil droplets. In contrast, low-energy method utilizes the intrinsic physicochemical attributes of surfactant and bulking agent in the system for emulsification [31].

In present day applications, nanoemulsion formulations are applied for drug delivery in medicine, food and cosmetic industries. Nanoemulsions are the common delivery module for the sustained release of active pharmaceuticals ingredients (APIs) and nutraceuticals into deeper skin layers [32]. Nanocollagen emulsions are produced at larger extent mainly in the area of cosmetic industry and drug delivery due to the potential to blend the cosmetic care active compounds while increasing their durability and absorption rate [33].

Table 1 Pros and cons of protein/collagen-based NPs formulation preparations. (Table adapted from open source, https://doi.org/10.3390/app112311369)

Process	Advantages	Disadvantages	References
Emulsification/ solvent extraction	Easy method, equipment required simple, high flexibility, stability, selectivity and encapsulation potential of NPs, controllable shape, size of NPs and secondary structure of protein, zeta potential in limit	Large sized NPs produced, thermodynamic instability needs more amount of surfactant and stabilizers or organic solvents may lead to toxicity	[63, 64]
Complex coacervation or polyelectrolyte complexation	highly stable and small sized NPs, NPs shape and size controllable with experiment conditions, protein or peptide like sensitive drug can be mixed	Challenge for scale up	[65]
Phase separation	Homogenous NPs formed, specific instrument not involved, particle size controlled by altering polymer amount	Particle size restricted to 50–500 nm, small scale production, involved organic solvents may be toxic	[37, 66]
Electrospinning	Flexibility and easy insertion of drug, NPs etc., high porosity, large surface-to-volume ratio, swelling capacity, controllable, gaseous permeation, economical and simple	Use of organic solvents, limited control of pore, structure	[41, 67]
Nanospray drying	Experimentally easy method, cost-effective, easy encapsulation of hydrophilic drugs, reduced degradation of heat-sensitive samples, NPs size can be restricted	Difficult incorporation of hydrophobic drugs, confined to small scale production, decreased encapsulation potential, high energy required	[68, 69]
Electrospray deposition	Formulation of dry and highly stable NPS, convenient synthesis due to one-step method, efficient drug loading capacity, scalable for industrial purposes, high yield and reproducibility, versatile method	Parameters like drying, sheer stress of nozzle produce thermal degradation of macromolecules	[47, 70]
Milling	Simple and cost-effective process for large scale fabrication of NPs	Cooling required during procedure due to heat generation. Coarse NPs produced, reduced control for NPs shape	[50, 71]
Self assembly	Small NPs with enhanced encapsulating efficiency can be formed through this highly stable process, simplicity, versatility	Uncontrollable NPs size and shape, high possibility of protein strain	[52, 53]

(continued)

Table 1 (continued)

Process	Advantages	Disadvantages	References
Desolvation	Controllable size and shape of NPs, better encapsulating potential and highly stable process	Process applicable with proteins affected by desoluble process or diluted by transporter proteins	[17, 51]
Interfacial polymerization	Simple process, undesirable monomer purity	Long duration of the process	[56, 58]
Polymer chain collapse	NPs attributes controllable, great stability and improved spherical shaped NPs	Restricted NPs diameter (5–20 nm) and challenging control for side reactions	[72, 73]

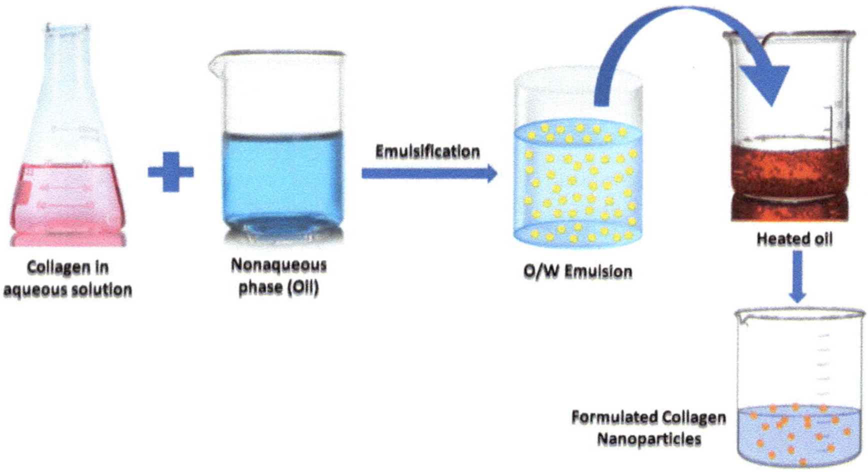

Fig. 1 Nanoemulsion—emulsion of drug and polymer solution is produced by homogenization followed by precipitation in nano droplets of encapsulated drug in polymeric matrix. (Image modified from open source, https://doi.org/10.3390/app112311369)

2.2 *Complex Coacervation or Polyelectrolyte Complexation*

Complex coacervation or polyelectrolyte complexation is the chemical nanoparticles fabrication process, where oppositely charged polyelectrolytes undergo liquid–liquid phase separation in an aqueous solution to create the polymer-dilute and polymer-dense (coacervate) phase (Fig. 2). The coacervates attributes can be modified by chirality of polyelectrolytes, pH, salt concentration, charge density, ionic strength and temperature [24].

Proteins can be made cationic and anionic by altering factors like pH due to their amphoteric nature with several charged functional groups. The charged protein

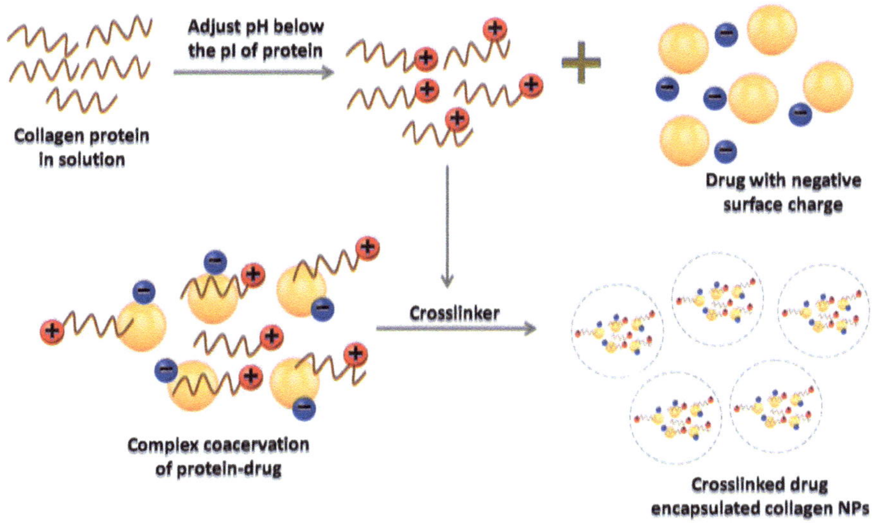

Fig. 2 Complex coacervation or polyelectrolyte complexation- alteration of pH to generate charged protein and interaction with charged drug followed by addition of crosslinker, produced the drug loaded collagen NPs. (Image modified from open source, https://doi.org/10.3390/app112311369)

electrostatically interact with polyelectrolytes in pH dependent manner and these interactions can be utilized for the formulation of stable and biocompatible NPs and coacervates for the sustained release of DNA for gene therapeutics application [34, 35]. Singh and Yethiraj [27] fabricated the NPs with native collagen and methylated collagen (MC) for transferring DNA into cells, using the complex coacervation approach. Native collagen/DNA-NPs due to low stability in serum aggregated at neutral pH and did not release DNA properly. While MC/DNA-NPs were more stable and small in size and released DNA in prolonged fashion up to 3 weeks. In another study, Truong-Le [36], encapsulated the DNA in gelatine NPs using complex coacervation approach. At pH 5, gelatin was positively charged and produced the complex coacervate with DNA and NPs sized from 200–700 nm with 25–30% loading capacity. Electrostatic and hydrophobic interactions along with hydrogen bonds play role in advanced protein-polymer synthesis.

2.3 *Phase Separation*

In the emulsion solvent evaporation for production of NPs, aqueous and organic phase plays crucial role. The advantage of the phase separation method is that it is a relatively simple procedure and needs minimum apparatus. In this method, the collagen polymer is dissolved in solution and phase separation is triggered, though

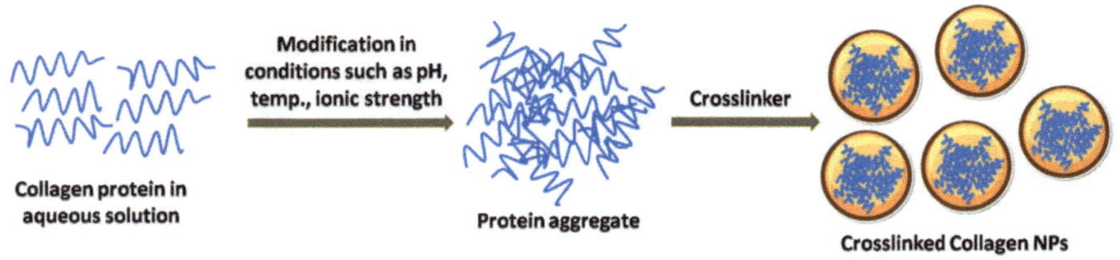

Fig. 3 Phase separation—By altering pH, temperature and ionic strength, collagen in aqueous solution form aggregates followed by addition of crosslinker, form the crosslinked collagen NPs through ultrasonication technique. (Image modified from open source, https://doi.org/10.3390/app112311369)

addition of surfactant to this protein solution. The resulting mixture is then ultrasonicated for proper homogenization followed by polymer droplets formation and isolation of solvent. The formulated NPs by the phase separation are having 50–500 nm of diameter (Fig. 3) [24]. Yoon et al. [37] fabricated the NPs of conjugated polymers through phase separation technique forming the lipid-incorporated NPs having the potential to form composites with inorganic/organic nanosubstances, cell specific ligands with polyethylene glycol for in vivo applications.

2.4 Electrospinning

Electrospinning is the process of production of nanocollagen fibres under the influence of electrostatic field in polymeric solution of collagen molecules (Fig. 4). In this technique a spinneret is charged at low current and high voltage (15–20 kV) followed by inclusion of droplets of polymeric solution resulting in highly charged surface and increase in length to form the cone known as Taylor cone. The conical shape is the resultant of electrostatic repulsion from coulambic forces of spinneret and charged surface of droplet. After reaching the threshold of electric field, electrostatic forces exceed the surface tension and elasticity of Taylor cone leading to stretching of cone for nanofibers generation. Finally, at the metallic collector, randomly interconnected dry nanofibers without solvent molecules are deposited [18]. The produced nanofibers through this method are cost effective and multifaceted, with high surface area-to-volume ratio and high porosity for applications such as cellular growth, tissue repair and regeneration and tissue engineering [38]. Through this technique, the fabricated collagen nanofibers have the advanced properties of response to exterior stimuli like modification in pH, exposure to light and magnetic field, retaining the shape memory and self-cleaning, etc. [39]. This process generates the interconnected network of fibres of nanoscale size, which are analogous to natural ECM morphology, so they promote the process of cellular proliferation [40]. Through this highly advantageous method, drugs, bioactive molecules, NPs, antimicrobial agents, growth factors and anti-inflammatory molecules can be inserted into the nanofibers

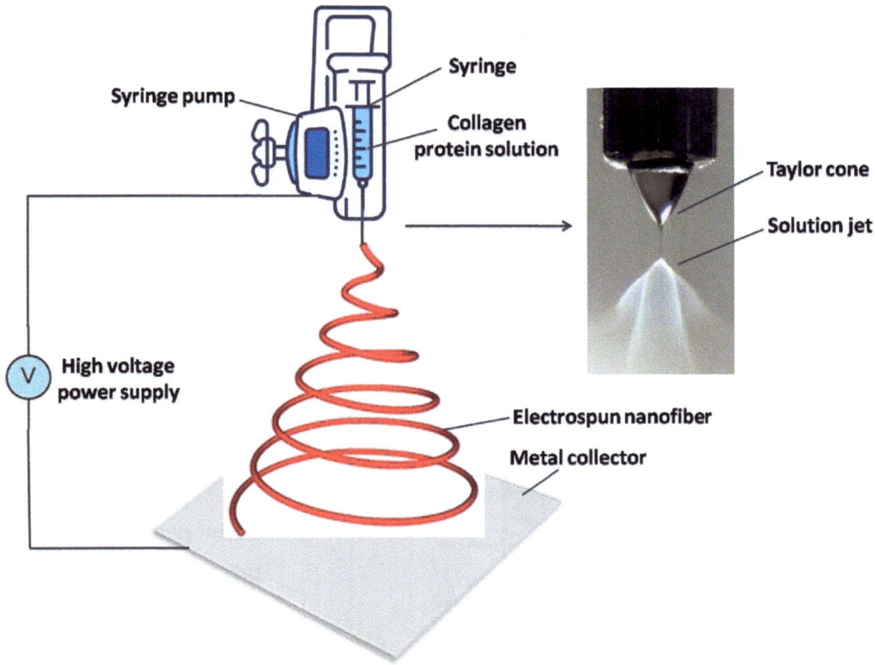

Fig. 4 Electrospinning—Collagen solution under influence of high electric field is passed through syringe, form Taylor cone followed by drying of solvent and generation of interconnected collagen nanofibers

[41]. Among all the methods for nanofibers formulations, electrospinning is mostly applied process due to economical and simplicity [42].

2.5 *Nanospray Drying*

This technique is the physical NPs fabrication process to produce the collagen NPs in liquid medium. The collagen solution along with hot CO_2 and nitrogen gas is sprayed through nozzle into the heated chamber to generate the hollow NPs, which are later collected at the bottom of the chamber under the influence of charged electrodes. Finally, the created NPs are frozen, lyophilized and crosslinked (Fig. 5). Liquid N_2 is used to avoid the denaturation of collagen molecules during the spraying in highly heated chamber [43]. This method is fast and economical to fabricate the collagen NPs at small scale and to encapsulate the hydrophilic drugs for delivery purposes. The addition of surfactant to the protein solution makes the spherical NPs and also stabilize them. The NPs size can be modified by restricting the nozzle size and rate of spraying [44, 45].

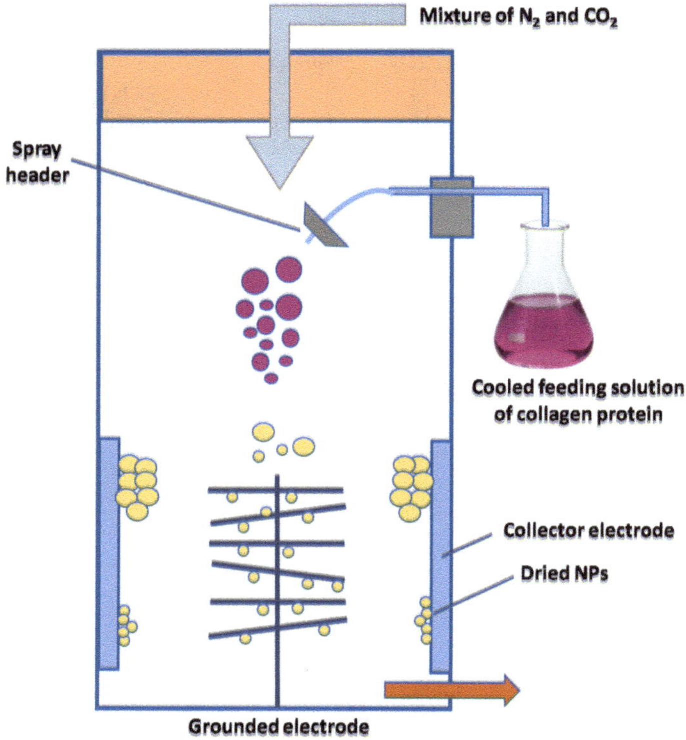

Fig. 5 Nanospray drying—at high temperature, stream of N_2, CO_2 along with cooled collagen is sprayed through nozzle in chamber and dried NPs are collected at the electrode. (Image modified from open source, https://doi.org/10.3390/app112311369)

2.6 Electrospray Deposition

This technique is the process of liquid atomization, where the high voltage is applied to the collagen solution and liquid jet stream is passed through nozzle to form the aerosolized drops of collagen NPs (Fig. 6). Drugs and nucleic acids can be efficiently incorporated into these generated NPs with high productivity [24, 46]. The operating conditions like applied voltage, operating distance, gauge diameter of needle and flow rate fluctuate according to different types of drug delivery modules. In the advanced technique of electrospray, coaxial spray of collagen in salt solution and drug is performed simultaneously to form the solid collagen NPs with encapsulated drug for delivery objectives [47]. Electrospray deposition method is economical, convenient to handle with efficient encapsulating capability. Solid and stable collagen NPs can be formulated to overcome the complications of reduced encapsulation potential and low rate of biocompatibility during drug delivery [48].

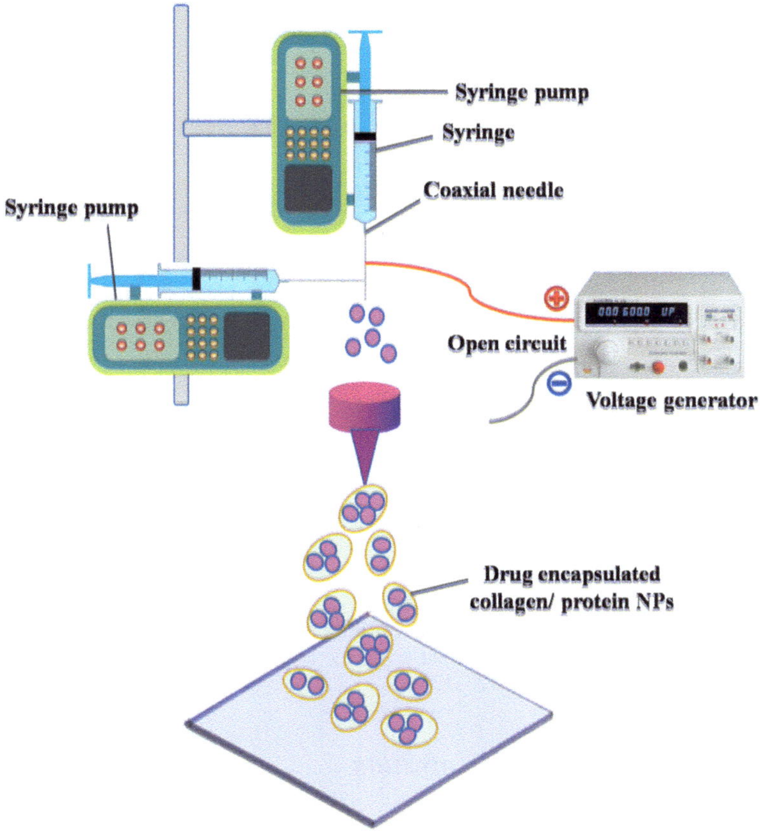

Fig. 6 Electrospray deposition—Liquid jet stream of drug and protein solution are passed through nozzle to generate the aerosolized droplets of drug encapsulated NPs at high temperature. (Image modified from open source, https://doi.org/10.3390/app112311369)

2.7 Milling

Milling is the mechanical process to fabricate the collagen NPs under the influence of high mechanical energy. The milling balls are utilized to generate mechanical collisions to produce disintegrated polymeric material into finer NPs (Fig. 7). During the milling, excess mechanical and kinetic energy generate the heat in the vessel; to avoid the heat milling vessel is cooled to reduce the breakdown of polymeric substance. Heat responsive protein or collagen material is milled under the control of liquid N_2 and cryogenic temperatures to counter the protein denaturation. Milling is the cost-effective technique to scaledown the particle size at large scale production of NPs [49, 50].

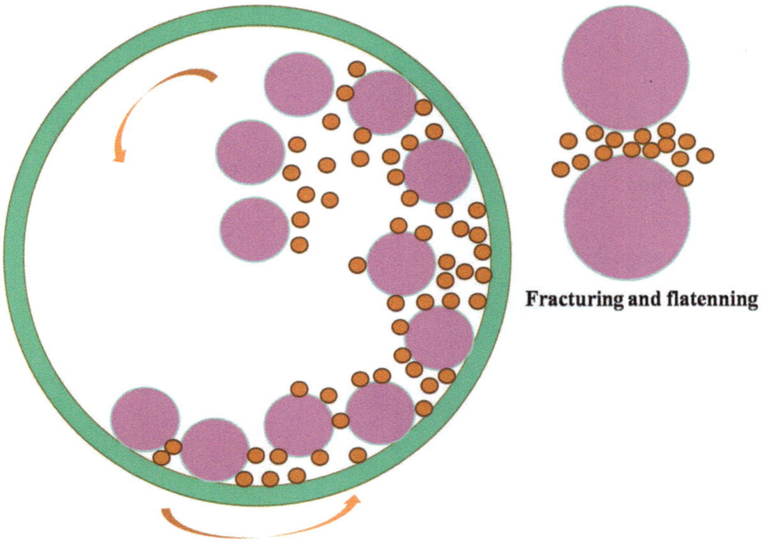

Fig. 7 Milling—Polymeric solution is mixed with milling balls under high mechanical and kinetic energy to produce the finer collagen NPs at cryogenic temperature. (Image modified from open source, https://doi.org/10.3390/app112311369)

2.8 *Self-Assembly and Desolvation*

When altered hydrophobic protein molecules are dissolved in aqueous solution, form the micelle NPs through self-assembly process. The hydrophobic pockets in these micelle NPs are the suitable place for encapsulation of active ingredients. In the self-assembly process, when the fibres of protein in solution at a critical solution temperature surpass the critical micelle concentration leads to the development of nanoscale composites. Later, through the process of solidification these nanoscale micelles become stable due to crosslinking between chains or fibres or sheets of protein molecules. By varying the self-assembly parameters of the collagen fabricates, sustained release of active molecules in the collagen-based wound dressings could be obtained to enhance the fibroblast production and expedited wound healing [51, 52]. Layer-by-layer (LbL) self-assembly process is conventional technique for fabricating multi-layer films and indulges the sequential adsorption of two or more building block that can complementarily combine with specific substrate at molecular level through electrostatic, hydrophobic interactions and hydrogen bonding. These interactions facilitate the deposition of alternate layers with opposite charged biomaterials and control the nanoscale characters like thickness, surface attributes and composition to the film [53, 54]. The main advantages of LbL self-assembly

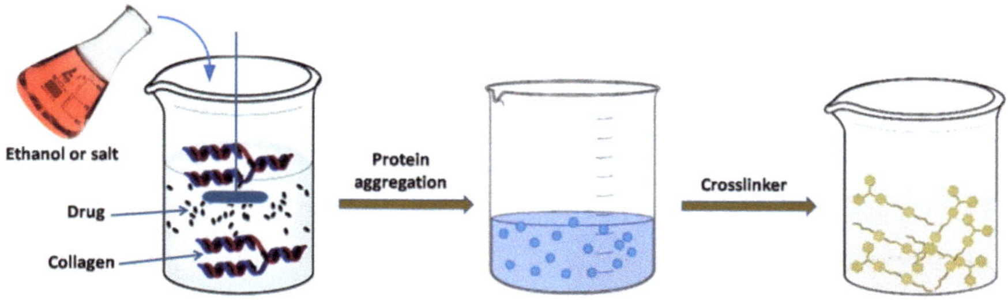

Fig. 8 Self-assembly or desolvation—collagen fibres solution with drug is dissolved by ethanol or salt exceeding critical micelle concentration (CMC) forming aggregates followed by addition of crosslinker to form the collagen-based NPs. In desolvation method, salt or alcohol is added to the collagen solution along with drug to form the nano droplets of encapsulated drug in polymeric matrix. (Image modified from open source, https://doi.org/10.3390/app112311369)

method is simplicity, versatility and lack of restrictions in view of choice of material or substrates [55]. The manipulation of surface physicochemical properties is possible by application of LbL assembly on certain surface.

Desolvation is the type of self-assembly method of nanocollagen formation by including the salt or alcohol to collagen solution with active biomolecule or drug as desolvation factor. The salt or alcohol modifies the conformation of collagen and reduces its solubility. Later, glutaraldehyde as crosslinking chemical is added after the critical amount of desolvation to the formulated aggregates of collagen solution following the NP generation [27] (Fig. 8). Through desolvation method, the size of the protein-based NPs can be customized with respect to parameters such as protein concentration, pH, temperature and addition of salt or alcohol. Small sized protein NPs can be generated by decreased protein concentration and increased pH [17]. Desolvation method is mainly applied for the formation of protein-based NPs.

2.9 *Interfacial Polymerization*

Polymeric membranes and particles can be fabricated through the technique of interfacial polymerization (Fig. 9). This method is a type of step-growth polymerization, where two immiscible solutions form the polymeric substances with particular topological and chemical attributes, anisotropic shapes, hollow structures or alternative surface chemistry at the interface region [56]. Here, the polymer is confined to interface region. Through this technique large scale production of ultrathin layers or membranes, hollow nanospheres and nanofibers is possible [57]. The category of synthesized polymeric materials is zero-dimension NPs, 1D nanofibers, 2D films and 3D composite membranes [58].

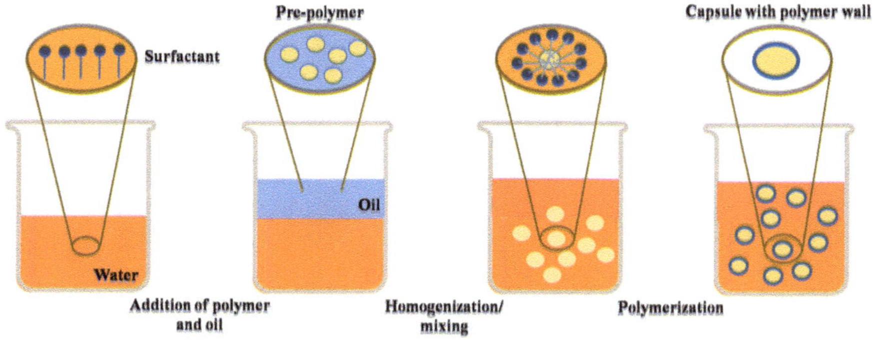

Fig. 9 Interfacial polymerization—two immiscible solutions like polymer and oil, form the polymeric material at the phase interface after polymerization. (Adapted from open source paper (https://www.researchgate.net/publication/267901674_A_Multifunctional_Smart_Coating_for_Automous_Corrosion_Control)

2.10 *Polymer Chain Collapse*

Polymer chain collapse technique can produce the single-chain polymer NPs (SCNP) with excellent stability in size range of 5–20 nm (Fig. 10). The NPs size can be controlled by polymer molecular weight and the extent of crosslinking [59]. The intramolecular polymer chain collapse method can generate the extremely small polymeric NPs of approximately 10nm in diameter, not possible by other methods of synthesis. Under the control of polymer precursor chain, the typical molecules of NPs can be generated [60]. Various types of SCNP fabrication processes are available on the basis of involved functional groups in the reactions [61]. Effective folding of single-chain polymer through single-chain nano objects can mimic and control the functionality and morphology of natural biomacromolecules such as enzymes, drug delivery modules and catalysts. NPs with systematically smaller and spherical than their homofunctional analogues can be produced by increasing the various types of chemical linker molecules through orthogonal chemistry methods [62].

3 Characteristic of Formulated Collagen-Based Nanomaterials or NPs

Protein-based nanomaterials are the emerging materials of research for drug delivery, medical devices, therapeutics, tissue engineering and other biomedicine areas due to small size, significant absorption, distribution in the human body, ability to cross the biological barriers and exhibiting negligible cytotoxicity [20]. When the size of the particle reaches at the nano level, most of the physical attributes of NPs are quite distinct form microparticles and bulk materials, providing more advanced options

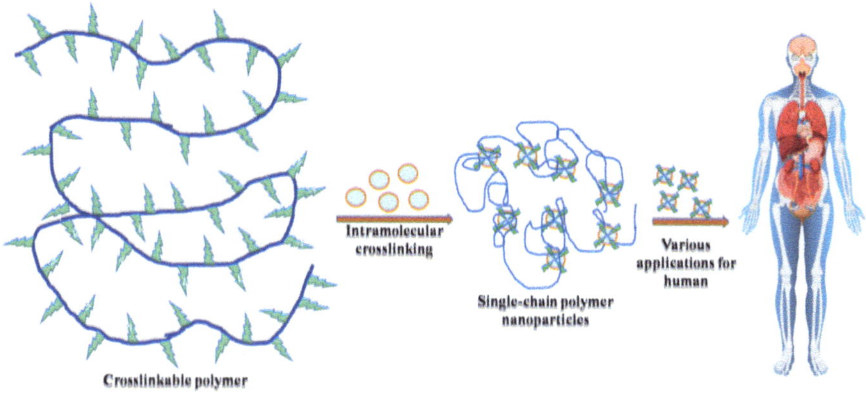

Fig. 10 Polymer chain collapse—different molecules are formulated to generate single-chain polymer NPs through polymer precursor chain via intermolecular crosslinking, which can be applied for several biomedical applications for human. (Image modified from open source, https://doi.org/10.3390/app112311369)

for encapsulation or adsorption in matrix as carriers of active ingredients for local or systemic delivery [74].

Removal of NPs from the body is dependent on various factors such as surface charge, hydrophobicity and size. The potential of NPs surface alteration can be determined through surface charge, density and hydrophilicity. The most preferred method about the surface charge of NPs is the measurement of zeta potential in aqueous solution. The common techniques for measuring the zeta potential are light scattering techniques such as dynamic light scattering (DLS) and laser Doppler electrophoresis [75]. Zeta potential predicts the stability of NPs and interaction with their surroundings and between the nanoparticles. Most water-soluble colloid systems are stabilized by electrostatic repulsion between NPs, to form the cohesive system as greater repulsion between NPs leads to reduction of distance between them [76]. The zeta potential also determines the location of charged active ingredient material inside the nanoparticles or adsorption on the NPs surface.

Collagen based NPs are the modern promising modules for controlled drug release due to unique features like easy to fabricate, high surface area-to-volume ratio and significant absorption capacity [77]. According to Shalaby et al. [78], high negative potential (-17.7 mV) of polydispersed collagen NPs increased long term stability, remarkable colloidal property and high dispersion due to negative-negative repulsion. The negative potential of NPs leads to electrostatic repulsion between particles to prevent the aggregation and also to prevent thrombosis [79]. Cationic NPs attached to the cell membranes due to electrostatic interactions. The size of formulated collagen NPs in the range of 5–200 nm measured by dynamic light scattering (DLS) method, is the promising drug delivery system through endocytosis into cells [80]. Collagen NPs infused gel had favourable effect on the healing of incision wounds in rabbits. Collagen regeneration and enhanced cellular proliferation at the wounded site exhibited the exception wound-healing capacity of collagen NPs [78]. In contrast to collagen microfibers, silver NPs (AgNPs) impregnated nanocollagen had better potential for rapid regeneration of wounded tissue [81]. In a report, Cardoso et al. [82], fabricated the AgNPs-type 1 collagen composite (AgNPcol) for antibacterial and cell viability potentials. All the formulations were analyzed on the basis of particle size, zeta potential and polydispersity index (PDI). The particle diameter and PDI were 64–81 nm and 0.40–0.77 of range, respectively. PDI value is the selecting parameter for low-polydispersity solution related to cell viability test. The positive zeta potential of AgNPcol expressed effective antibacterial activity against *Escherichia coli* and *Staphylococcus aureus*. The opposite surface charge and small size of NPs promote antibacterial activity.

Roshanbinfar et al. [83] synthesized the electrically conductive hydrogel of type I collagen and branched polyethyleneimine (bPEI)-caoted AuNPs for cardiac tissue generation by adding neonatal rat ventricular or human embryonic or induced pluripotent stem cell (hiPSC) derived cardiomyocytes. Along the collagen matrix, bPEI-AuNPs were uniformly distributed with distance in the magnitude of few hundred nanometers. A wide range of negatively charged biomedical drug, active ingredients and growth factors can be effectively loaded on the positively charged coating of AuNPs. During the cardiac cycle, high amount of forces are produced, that should be acceptable by engineered material to assure the no drug or cell content leakage. Production of hydrogel-based cardiac tissue material for large animal's mechanical stress is a tough task. The AuNPs escalated the hydrogel stiffness from 91 to 146 kPa and also expanded the electrical conductivity from 40 to 49–69 mS cm^{-1} for the slow and uniform release of incorporated drug. These hydrogels-based engineered tissues have the potential to reduce cardiac arrest and treatment of ailments of electrically sensitive tissues.

Terzopoulou et al. [84] designed the curcumin (Cur)-loaded chitosan (CS) NPs and incorporated into the hydrophilic and biocompatible collagen-based patch as topical preparation for psoriatic skin. Size of the NPs is very crucial parameter influencing the drug release profile and biological performance of NPs. NPs less than 200 nm in diameter are supposed to manage the drug discharge rate and penetrate the cells more adequately, while the zeta potential of greater than +30 mV contributed better stability and capacity to adhere the negatively charged biomembranes. The fabricated Cur-CS-NPs were expected to penetrate the upper layers of skin through injection. The collagen patches expressed the significant swelling ratio of upto ~1500% and after crosslinking with EDC (1-ethyl-3-(3-dimethylaminopropyl) carbodiimide hydrochloride)/ NHS (N-hydroxysuccinimide) the pore size decreased which leads to decreased hydrolysis rate. Release of curcumin within collagen patches generated the proliferative inhibitory effect on psoriatic human keratinocytes in vitro.

NP's introduction into the body may lead to rapid degradation and instability of NPs followed by elimination from the body through phagocytic processes [85]. The stability of NPs is depended on the composition of the composite and its interaction with surroundings. Lipid and polymer-based NPs are most sensitive to aggregation and instability [20]. In body, circulatory system factors like blood flow, excretion and phagocytic cells may decrease the NPs stability and delivery. NPs with diameter of

<10 nm usually removed through kidney in fast manner while NPs with >200 nm may initiate the complement system [86]. Several fabricated NPs utilize the coating of polyethylene glycol (PEG) to avert the fast excretion of NPs from body. PEGylation enhanced the circulation time though modification of NPs size and solubility and provide protection to NPs surface from degradation, secretion and clearance caused by enzymes and antibodies. At high concentration, PEG coated NPs may lead to anti-PEG antibodies could activate the clearance of PEGylated NPs [87].

3.1 Loading and Release of Active Ingredients/Drug for Delivery

Polymeric NPs for drug administration should have high drug loading capacity and the reduced volume of media. Active ingredients or pharmaceuticals can be incorporated in NPs through two approaches—first, encapsulate the drug during the synthesis of NPs and second, on the surface of NPs attach or insert drug through adsorption after the synthesis of NPs [88]. These alterations make the NPs quintessential for delivery applications and are suitable for delivery of various payloads of hydrophobic and hydrophilic materials along with cargos of different molecular weights like small molecules, bio-macromolecules, proteins and vaccines [89–92]. The loading efficiency and release kinetics of loaded pharmaceuticals on NPs can be accurately governed through modification of properties like composition, stability, response and surface charge of NPs [66, 93]. Nano-carriers are considered as promising therapeutics formulations due to high loading ability, significant intracellular drug delivery, effective cellular absorption and controlled release in cytosols with adequate biodegradation [78]. For the increment of biomolecule loading capability and cellular absorption potential, surface roughening of NPs has been formulated as another method [94].

Awad et al. [95] assessed the cytotoxicity effect of three combinations- collagen NPs, theophylline loaded with collagen NPs and free theophylline alone using MTT assay on NHSF cell line at the concentration of 180 μg/ml with double fold serial dilutions. At the lower doses, collagen NPs, theophylline loaded with collagen NPs and free theophylline had more than 97% cell viability. The cytotoxicity enhanced significantly by increasing the concentration of theophylline with 55.59% of viability at 180 μg/ml. Theophylline loaded with collagen NPs had higher cell viability of 78.53% and lower toxicity than at 180 μg/ml of theophylline alone. The order of cytotoxicity was collagen NPs < theophylline loaded with collagen NPs < theophylline for all the concentrations. Theophylline loaded with collagen NPs showed lower toxicity than free theophylline and slow release due to efficient encapsulation inside collagen NPs. The collagen NPs were least cytotoxic at all the tested concentrations due to natural biodegradable collagen and additives utilized for fabrication [96, 97]. The toxicity findings also expressed the biocompatibility, harmlessness of collagen NPs with normal cells making them suitable as drug carriers for medicine.

For the optimal delivery of active ingredients, drug release and polymer biodegradability are the crucial criteria. The release rates of NPs is based on desorption of the surface attached or adsorbed molecules, diffusion through NPs matrix, diffusion through the polymer wall, erosion of NPs matrix and diffusion process. The solubility, diffusion and biodegradation of NPs affect the discharge of drug in the system. The releasee of drug through protein-based polymeric NPs is accomplished via protein degradation, through pores, from polymer surfaces and electrical pulse-based delivery through magnetic field etc. [98]. The drug release potential of protein-based NPs can be controlled by modifying the geometry, coating thickness, porosity and crystallinity of the protein drug vehicles [99]. Spherical-shaped NPs with less than 200 nm diameter could cross the spleen to avoid the contact forces along the blood vessel walls and other flow barriers in body [100]. To obtain the high drug loading and emission, delivery vehicles can also be altered to attain the large pore volume and surface areas. A drug carrier's morphology, shape, diameter and thickness should be considered for effective tailored drug release potential [3]. Porosity of the NPs material is the preferred property to control the delivery of drug such as surface area, permeability, loading capacity, release rate and pharmacokinetics and is directly interconnected with pore size and pore interconnectivity. The increased pore percentage can accelerate the degradation and augmented drug release in polymer matrix [101].

At the high concentration or dose of pharmaceutical drug or active ingredients after systemic introduction in body may generate undesirable side effects. To achieve the similar efficiency or exceptional results, local delivery of therapeutic molecules at low dosage has been researched [102]. By regulating the degree of crosslinking in porous collagen-based scaffold, therapeutic substances can be encapsulated and their retention and release can also be controlled [103]. Instead of administering the growth factor directly, incorporation of growth factor loaded NPs in the crosslinked collagen scaffolds, the target molecules may be delivered via composite delivery system. For the controlled release of therapeutics, crosslinked collagen scaffold act as the vehicle for lodging the NP carriers. NPs create more opportunities for interaction with collagen, when they are directly or indirectly crosslinked with collagen. Metal oxide NPs utilized for crosslinking with collagen by directly binding to its side chain. The mechanical attributes of collagen-based scaffolds may be augments through metal NPs like ZnO, iron oxide, alumina oxide, etc. Through UV irradiation, radical oxygen species, promoted collagen crosslinking by metal oxide NPs and is incredible for formulation of collagen shields for controlled drug delivery for glaucoma treatment [104]. Choi et al. [105] developed the collagen constructs as inserts for wounded human gingival for optimization of the release profile of growth factor TGF-β1 and NPs of Au, TaO, dextran and ferritin. The released amount of biomolecule NPs was quantified through fluorescence intensity and X-ray scattering. The collagen hydrogel may be the promising system for sustained release of growth factor and imaging agents in human gingival to promote the regeneration of oral tissues.

4 Collagen-Based Nanocomposites for Delivery of Ingredients/Drug Through DDS

In the field of medicine, traditional drugs are the pivotal strategies to treat various ailments. The administration of classical drug in body at high dosage leads to mild or severe side effects, sometimes leads to life-threatening condition. The formulation of effective DDS for the transportation of natural or synthetic desired drug at target site with optimal amount has been a demanding clinical investigation and research for several years [106]. In the last decades, the growth of nanotechnology has unblocked the different innovative possibilities in biomedicine, specifically in the area of drug delivery. Different novel drug delivery vehicles at the micro or nano scale have been formulated and number of such delivery modules has been increasing tremendously to compensate for the need for targeted DDS [72].

Collagen is the most abundant extracellular protein found in connective tissues in mammals and can be used for DDS through various architectural types such as protein cages, nanoparticles, nanofilms, hydrogels and minirods with combination of different natural or synthetic polymers and crosslinking chemicals. Several methods such as emulsification, desolvation, electrospray, electrospinning, self-assembly, milling and phase separation, etc., are applied for the fabrication of collagen-based nanocomposites. Each preparatory process has various advantages and limitations. In Table 2, list of different collagen-based nanocomposites or nanoformulations, their method of preparation and loaded active ingredients or pharmaceuticals are mentioned.

Lin et al. [107] fabricated the biocompatible nanofiber membrane of protein collagen and zein by co-electrospinning method. Zein improvised the spinnability of collagen and generated nanofiber properties like fibre diameter, surface wettability, mechanical strength degradation potential and cell adhesiveness could be modified through alteration of collagen/zein ratio. The incorporated berberine drug in nanofiber membrane was assessed for controlled release and antibacterial activity and the berberine exhibited little effect on fibre morphology and cell viability. The wound healing process and histological experiments were conducted in vivo on the female Sprague–Dawley rats for the fabricated nanomembrane.

In a report by Karri et al. [108], a novel nanohybrid scaffold was fabricated through emulsification method using curcumin (CUR) impregnated in chitosan (CS) NPs and collagen scaffold for tissue regeneration application. The formulated CUR-CS-NPs were assessed on the basis of size, zeta potential, SEM, DSC, X-ray analyses and the novel nanohybird of CUR-CS-NPs-collagen scaffold was evaluated for water uptake, biocompatibility and sustained release parameters. The in vivo wound closure analysis exhibited the faster and significant contraction of wound treated by nanohybrid scaffold in contrast to control and placebo groups. The wound showed the complete epithelialization with thick tissue formed in nanohybrid treated scaffold than control and other group. The nanohybrid scaffold showed the promising potential for better wound healing capacity for diabetic wounds.

Table 2 List of various collagen-based nanocomposites, their method of preparation and loaded active ingredients. EDC-Hcl—3-ethyl carbodiimide-hydrochloride; MDA—malondialdehyde; chitosan—CS; NHS—N-hydroxysuccinimide; NLC—nanostructured lipid carrier; PCL—poly(ε-caprolactone); bPEI—branched polyethyleneimine; TGF-β1—transforming growth factor; FGF-2—fibroblast growth factor; PLGA—poly(lactic-co-glycolic acid); BMP-2—bone morphogenic protein; PLA—poly lactic acid; VEGF—vascular endothelial growth factor; PDGF—platelet derived growth factor; EGF—endothelial growth factor; PHA—polyhydroxy butyrate

Nanocomposite	Method of preparation	Loaded pharmaceutical	References
Collagen NPs	Nanoprecipitation method using non-solvent (ethanol)	Theophylline	[95]
Collagen NPs crosslinked with EDC-Hcl & MDA	Emulsification	Silymarin	[96]
Collagen patches-CS NPs crosslinked with Ion gelation method EDC/NHS/heparin	Ion gelation method	Curcumin	[84]
Collagen gel-CS NLC	Emulsification	SiRNA	[115]
Collagen hydrogel	Layer-by-layer self-assembly	Doxorubicin	[53]
Hydrolyzed collagen-PCL	Radical co-polymerization	Hydrocortisone	[114]
Collagen peptide-CS NPs	Ion gelation method	Doxorubicin	[116]
Collagen-bPEI-AuNP hydrogel	Desolvation	Phenylephrine, isoproterenol, doxorubicin	[83]
Collagen-CS dressing	Freeze drying	Ag NPs	[117]
Collagen type 1	Chemical synthesis	Ag NPs	[82]
Collagen-iron oxide NPs hydrogel	Freeze drying, coprecipitation	Fluorescein	[118]
Collagen-ferrtin NPs-TaO NPs hydrogel	Microemulsion	TGF-β1	[105]
Zein-collagen-PCl nanofiber scaffold	Electrospinning	Aloe vera	[110]
Collagen-graphene oxide sheet-hydrogel	Layer-by-layer self-assembly	FGF-2	[119]
Collagen-hydroxyapatite-PLGA scaffold	Emulsification (W/O/W method)	BMP-2, alendronate	[113]
Atelocollagen	Mixing	siRNA	[120]
Type 1 collagen hydrogel	Mixing	Royal jelly extracellular vesicles	[121]
Collagen peptide-chito oligosaccharides membrane	Electrospinning	Antibacterial activity	[122]
PLGA/collagen nanofibrous membrane	Electrospinning	Wound healing	[123]

(continued)

Table 2 (continued)

Nanocomposite	Method of preparation	Loaded pharmaceutical	References
Au-hydroxyapatite-collagen nanomaterial	Wet precipitation/ microwave assisted synthesis	Doxorubicin	[124]
Collagen-based 3D culture system-conjugated NPs	Self assembly	Coumarin	[109]
Collagen-hydrogel-nano/ microfibers	Electrospray	Rat pheochromocytoma cell cuture	[125]
PCL-collagen-nanobioglass nanocomposite	Electrospinning	Peripheral nerve regeneration	[126]
Collagen-chitosan membrane-chitosan NPs	Ion gelation, phase separation	Minocycline	[112]
Collagen-PLGA NPs asymmetric nanomembrane	Electrospinning	Aspirin, curcumin	[127]
Collagen-chitosan NPs	Emulsification	Curcumin	[108]
Ethyl cellulose-PLA-collagen nanofiber mat	Electrospinning	Silver sulfadiazine	[128]
Collagen-zein nanofiber membrane	Co-electrospinning	Berberine	[107]
PLGA-collagen scaffold nanomembrane	Electrospinning	Glucophage	[129]
Collagen-hyaluronic acid-gelatin NPs nanocomposite	Electrospinning	VEGF, PDGF, bFGF, EGF	[130]
PHA-gelatin-collagen nanofibre nanofiber composite	Electrospinning	Ostholamide	[131]
Collagen-polyamide nanofiber composite	Electrospinning	N-acetly cysteine	[132]
Gelatin-fish collagen-chitosan nanofiber bilayer scaffold	Electrospinning	*Lithospermi radix* extract	[133]
PCL-collagen-polyethylene oxide	Electrospinning	Doxycycline	[134]
PCL-collagen nanofibers	Electrospinning	Gentamicin sulphate	[135]

Targeted drug delivery and positive antitumor outcomes are the main challenges in cancer therapy. In view of this, Le et al. [109] developed the collagen-based 3D multicellular culture module with coumarin conjugated NPs by self-assembly method for the selection of new nanocarriers for drug delivery in preclinical research. The findings showed that 3-D cell colonies were successfully developed from 95-D, U87 and HTC116 cell lines after seven day culture in collagen matrix. The coumrarin-NPs were easily penetrated the matrix gel and targeted the tumor cells and the drug module was good for evaluating the therapeutic outcomes of drug transport in vivo or investigation in tumor biology for better DDS for cancer treatment.

Ghorbani et al. [110] formulated the nanofibers scaffold of zein-poly caprolactone (PCL)-collagen encapsulated zinc oxide (ZnO) NPs and aloe vera (Alv) through electrospinning method for wound healing potential evaluation. The nano-scaffolds were examined for the parameters- morphology, mechanical, thermal stability and hydrophilicity. The scaffolds with composition of ZnO (1%), Alv (8% wt) and zein/PCL (70:30 ratio) exhibited significant thermal stability and mechanical attributes. By reducing the zein/PCL ratio water contact angle of scaffolds could be enhanced. Cell culture experiments expressed good findings of cytocompatibility by nanoscaffolds. The cell adhesion property was also examined using fibroblast cells for 24 and 72 h. The nanoscaffolds sample exhibited better antibacterial activity against *S. aureus* and *E. coli* bacteria. These nanocomposites findings showed the promising material for wound healing.

Deepa et al. [111] biosynthesized the AgNPs using aqueous root extract from *Delphinium denudatum* (Dd) and further fabricated with collagen and doxycycline (DO) to assess the bactericidal property against the series of bacteria *S. aureus, Bacillus cereus, Pseudomonas aeruginosa, E. coli, Salmonella paratyphi, Shigella dysentriae* and *Vibrio cholerae*. The developed DdAgNPs was spherical shaped and in the size range of 40–85 nm. DdAgNPs alone showed antibacterial activity against only two organisms while in combination with collagen expressed strong inhibition against all the organisms. Doxycycline expressed no activity against *E. coli, S. dysentriae and V. cholerae* but combination with collagen exhibited antibacterial activity towards all the test organisms with increased zone of inhibition except *B. cereus.* Synergistic action of DdAgNPs, doxycycline and collagen resulted in improvised antibacterial activity.

Infections generated by the colonization of pathogens at the wounded site during bone tissue regeneration is considered as one of the main causes for the guided bone regeneration (GBR). Taking in view of GBR, Ma et al. [112] developed a novel asymmetric collagen/chitosan GBR membrane (CC) with minocycline (M)-loaded chitosan NPs by the process of ion gelation method. The morphology of chitosan NPs and membrane was analyzed by SEM and TEM and biocompatibility of CCM membranes were evaluated against MC3T3-E1 osteoblasts and L929 fibroblasts cells. The bacteriostatic rates of CCM membrane against *Porphyromonas gingivalis* and *Fusobacterium nucleatum* were 95.3% and 92.1%, respectively. The CCM membrane enhanced the osteoblasts and fibroblast growth. The CCM membrane in the rat models promoted angiogenesis and enhanced bone regeneration and could be better therapeutics for infection prevention and bone regeneration.

The successful restoration of bone defects caused by trauma, tumor and metabolic disorders is a clinical drawback in reconstructive surgery. Biomaterials are the critical parts of in situ bone tissue engineering due to contribution of 3-D scaffold, which facilitates ECM formation and cell growth. Lee et al. [113] constructed the collagen-nano hydroxyapatite (nHAP) scaffold (CHAS) incorporated with poly (lactic-co-glycolic acid) (PLGA) microspheres and two active biomolecules- bone morphogenic protein-2 (BMP-2) and alendronate (ALN) by emulsification method for the evaluation of synergistic effect on osteogenic regeneration. The release profile of BMP-2 and ALN encapsulated in PLGA microspheres showed sustained release upto 4 weeks and 2 weeks, respectively in contrast to BMP-2 and ALN without PLGA microspheres. Evaluation of accumulation and maturation of bone-like tissue in calvarial bone defect was performed by micro-computed tomography imaging at 2, 4 and 8 weeks after implantation in rat models. Enhanced bone regeneration was observed after 8 weeks post-implantation in rat with 8 mm critical-sized defect. The fabricated nano-composite could be the future biomaterial for cell-free tissue engineering.

Recently, conductive composites have been developed for in vitro and in vivo biomedical applications. The conductive composites like current-stimuli polymeric structures have been successfully introduced extra controlling DDSs equipped with switchable release profile. Nowadays, drug release from conductive polymeric materials by applying the required voltage at a precise duration time is available. The conductive polymers have broad applications in microelectronic industry, electrode production, photovoltaic devices, biotechnological systems such as nerve regeneration and drug delivery systems. Pourjavadi and Doroudian [114], developed a semi-conductive nanocomposite for electrically controlled drug delivery. Hydrolyzed collagen was altered with administration of polycaprolactone. The hydrogel was fabricated by radical co-polymerization of acrylic acid in the presence of crosslinker. The reaction parameters influencing the water absorbance of hydrogel were optimized through Taguchi method. In situ polymerization of aniline has been incorporated conductive nanofiber pathways throughout the hydrogel matrix. The nanocomposites were characterized through techniques such as ^{1}H NMR, TGA, AFM, SEM, FTIR, UV–vis cyclic voltametry and conductivity estimations. In vitro conductive stimuli drug release of hydrocortisone as model drug was evaluated. Any cytotoxicity was not observed for conductive and non-conductive hydrogels. According to findings, the fabricated nanoconstruct acts as precise externally controlled DDS that may be customized to meet the physiological processes requirement (Table 2).

5 Conclusion

Nanotechnology is a rapidly evolving multidisciplinary science technology that ensures the development of bio or synthetic polymers or in combinations with other chemicals to nanometer scale for various medical applications. In modern times, nanoparticles are among the most promising vehicle in drug delivery systems. In the field of biomedicine, protein-based nanoparticles have gained attention due to various advantages like economical, biocompatible, biodegradable, biomimetic, low immunogenicity, low cytotoxicity and existence of multifunctional chemical groups on them allow to carry large amount of drugs or bioactive molecules efficiently for targeted delivery.

Collagen, an abundant and major natural protein in the mammal's body for mechanical strength or support, is the attractive biopolymers for the delivery of therapeutic drugs, growth factors, hormones, proteins, gene and imaging probes in the field of DDSs and regenerative medicine, prosthetics and surgery. Generally the collagen-based materials sponges, hydrogels, membranes and scaffolds are common but emerging collagen-based nanocomposites is a relatively new technology and having challenges. Up to nanoscale size, high surface area-to-volume ratio and high mechanical strength, elasticity against mechanical, high capacity of adsorption, water dispersion ability to form clear colloidal solution make the collagen NPs as promising future biomaterial for DDSs at commercial level.

Several physical or chemical methods from easy to difficult operations such as emulsification, complex coacervation, phase separation, electrospinning, nanospray, layer-by-layer self-assembly, milling, freeze-drying are involved in the production of collagen-based nano-constructs. Each method has its own advantages or disadvantages for NPs production at large scale as mentioned in Table 1. Drug release efficiency and drug loading capacity of collagen nanocomposites can be modified or increased by controlling or maintaining the properties like surface charge, shape and size of the nanocomposites.

The innovative and formulated collagen-based nanomaterials offer a sustained and slow release of drug required for long term release profile. In case of sudden release of drug, the slow release material will fail in some anticancer therapy. The high swelling property and loading capacity of fabricated nanobiomaterial confront the fast release of drug. Therefore, in future, better nanobiomaterials are required for customized release of active biomolecules in response to stimuli of diseased state of the human or animal body or other in vitro systems, with having negligible cytotoxicity and biocompatibility.

References

1. Rosen, H., Abribat, T.: The rise and rise of drug delivery. Nat. Rev. Drug Discov. **4**, 381–385 (2005). https://doi.org/10.1038/nrd1721
2. Verma, D., Gulati, N., Kaul, S., Mukherjee, S., Nagaich, U.: Protein based nanostructures for drug delivery. J. Pharm. 9285854 (2018). https://doi.org/10.1155/2018/9285854
3. Jao, D., Xue, Y., Medina, J., Hu, X.: Protein-based drug-delivery materials. Materials (Basel) **10**(5), 517 (2017). https://doi.org/10.3390/ma10050517
4. Singh, R., Lillard, J.W.: Nanoparticle-based targeted drug delivery. Exp. Mol. Pathol. **86**(3), 215–223 (2009). https://doi.org/10.1016/j.yexmp.2008.12.004
5. Guo, D., Xie, G., Luo, J.: Mechanical properties of nanoparticles: basics and applications. J. Phys. D: Appl. Phys. **47**(1) (2014). https://doi.org/10.1088/0022-3727/47/1/013001
6. DeFrates, K., Markiewicz, T., Gallo, P., Rack, A., Weyhmiller, A., Jarmusik, B., et al.: Protein polymer-based nanoparticles: fabrication and medical applications. Int. J. Mol. Sci. **19**(6), 1717 (2018). https://doi.org/10.3390/ijms19061717
7. Cai, R., Chen, C.: The crown and the scepter: roles of the protein corona in nanomedicine. Adv. Mater. **31**(45), 1805740 (2019). https://doi.org/10.1002/adma.201805740
8. Ovais, M., Guo, M., Chen, C.: Tailoring nanomaterials for targeting tumor associated macrophages. Adv. Mater. **31**(19), 1808303 (2019). https://doi.org/10.1002/adma.201808303
9. Aljabali, A.A.A., Barclay, J.E., Lomonossoff, G.P., Evans, D.J.: Virus templated metallic nanoparticles. Nanoscale **2**(12), 2596–2600 (2010). https://doi.org/10.1039/C0NR00525H
10. Sainsbury, F., Saunders, K., Aljabali, A.A.A., Evans, D.J., Lomonossoff, G.P.: Peptide-controlled access to the interior surface of empty virus nanoparticles. ChemBioChem **12**(16), 2435–2440 (2011). https://doi.org/10.1002/cbic.201100482

11. Albu, M.G., Titorencu, I., Ghica, M.V.: Collagen-based drug delivery systems for tissue engineering. In: Pignatello, R. (ed.) Biomaterials Applications for Nanomedicine. IntechOpen 333–358 (2011). http://www.intechopen.com/books/biomaterialsapplications-for-anomedicine/collagen-based-drug-delivery-systems-for-tissue-engineering
12. Zhang, Y., Sun, T., Jiang, C.: Biomacromolecules as carriers in drug delivery and tissue engineering. Acta Pharm. Sin. B **8**(1), 34–50 (2018). https://doi.org/10.1016/j.apsb.2017.11.005
13. Ahmed, M., Verma, A.K., Patel, R.: Collagen extraction and recent biological activities of collagen peptides derived from sea-food waste: a review. Sustain. Chem. Pharm. **18**, 100315 (2020). https://doi.org/10.1016/j.scp.2020.100315
14. Daza, J.H.U., Righetto, G.M., Chaud, M.V., Martins, V.C.A., Camargo, I.L.B.C., Plepis, A.M.G.: PVA/anionic collagen membranes as drug carriers of ciprofloxacin hydrochloride with sustained antibacterial activity and potential use in the treatment of ulcerative keratitis. J. Biomater. Appl. **35**(3), 301–312 (2020). https://doi.org/10.1177/0885328220931733
15. Geanaliu-Nicolae, R.-E., Andronescu, E.: Blended natural support materials-collagen based hydrogels used in biomedicine. Materials **13**(24), 5641 (2020). https://doi.org/10.3390/ma13245641
16. Maham, A., Tang, Z., Wu, H., Wang, J., Lin, Y.: Protein-based nanomedicine platforms for drug delivery. Small **5**, 1706–1721 (2009). https://doi.org/10.1002/smll.200801602
17. Lo, S., Fauzi, M.B.: Current update of collagen nanomatrials-fabrication, characterisation and its applications: a review. Pharmaceutics **13**(3), 316 (2021). https://doi.org/10.3390/pharmaceutics13030316
18. Makkithaya, K.N., Nadumane, S., Zhuo, G.Y., Chakrabarty, S., Mazumdar N. Nanoparticle based collagen biomaterials for wound healing. In: Mazumdar, N. (ed.) Collagen biomaterials. IntechOpen (2022). https://doi.org/10.5772/intechopen.104851
19. Bhushan, B. (ed.): Springer Handbook of Nanotechnology. Springer, Berlin (2007)
20. Mitchell, M.J., Billingsley, M.M., Haley, R.M., Wechsler, M.E., Peppas, N.A., Langer, R.: Engineering precision nanoparticles for drug delivery. Nat. Rev. Drug Discov. **20**(2), 101–124 (2021). https://doi.org/10.1038/s41573-020-0090-8
21. Naskar, A., Kim, K.S.: Recent advances in nanomaterial-based wound-healing therapeutics. Pharmaceutics **12**(6), 499, 1–20 (2020). https://doi.org/10.3390/pharmaceutics12060499
22. Wilson, A.: The formation of dry, wet, spunlaid and other types of nonwovens. In: Applications of Nonwovens in Technical Textiles, pp. 3–17 (2010). https://doi.org/10.1533/9781845699741.1.3
23. Wang, X., Yan, Y., Yost, M.J., Fann, S.A., Dong, S., Li, X.: Nanomechanical characterization of micro/nanofiber reinforced type 1 collagens. J. Biomed. Mater. Res. Part A **83A**, 130–135 (2007). https://doi.org/10.1002/jbm.a.31207
24. Arun, A., Malrautu, P., Laha, A., Luo, H., Ramakrishna, S.: Collagen nanoparticles in drug delivery systems and tissue engineering. Appl. Sci. **11**(23), 11369 (2021). https://doi.org/10.3390/app112311369
25. Chak, V., Kumar, D., Visht, S.: A review on collagen based drug delivery systems. Int. J. Pharm. Teach. Pract. **4**, 811–820 (2013)
26. Musazzi, U.M., Franzè, S., Minghetti, P., Casiraghi, A.: Emulsion versus nanoemulsion: how much is the formulative shift critical for a cosmetic product? Drug Deliv. Transl. Res. **8**, 414–421 (2017). https://doi.org/10.1007/s13346-017-0390-7
27. Singh, A.N., Yethiraj, A.: Liquid–liquid phase separation as the second step of complex coacervation. J. Phys. Chem. B **125**(12), 3023–3031 (2021). https://doi.org/10.1021/acs.jpcb.0c07349
28. Singh, Y., Meher, J.G., Raval, K., Khan, F.A., Chaurasia, M., Jain, N.K., et al.: Nanoemulsion: concepts, development and applications in drug delivery. J. Control. Release **252**, 28–49 (2017). https://doi.org/10.1016/j.jconrel.2017.03.008
29. Tan, S.F., Masoumi, H.R., Karjiban, R.A., Stanslas, J., Kirby, B.P., Basri, M., et al.: Ultrasonic emulsification of parenteral valproic acid-loaded nanoemulsion with response surface methodology and evaluation of its stability. Ultrason. Sonochem. **29**, 299–308 (2016). https://doi.org/10.1016/j.ultsonch.2015.09.015

30. Dehghani, F., Farhadian, N., Golmohammadzadeh, S., Biriaee, A., Ebrahimi, A., Karimi, M.: Preparation, characterization and *in vivo* evaluation of microemulsions containing tamoxifen citrate anti-cancer drug. Eur. J. Pharm. Sci. **96**, 479–489 (2017). https://doi.org/10.1016/j.ejps.2016.09.033
31. Gurpreet, K., Singh, S.K.: Review of Nanoemulsion Formulation and Characterization Techniques. Indian J. Pharm. Sci. **80**, 781–789 (2018). https://doi.org/10.4172/pharmaceutical-sciences.1000422
32. Souto, E.B., Fernandes, A.R., Martin-Gomes, C., Coutino, T.E., Durazzo, A., Lucarini, M., et al.: Nanomaterials for skin delivery of cosmeceuticls and pharmaceuticals. Appl. Sci. **10**(5), 1594 (2020). https://doi.org/10.3390/app10051594
33. Katz, L.M., Dewan, K., Bronaugh, R.L.: Nanotechnology in cosmetics. Appl. Sci. **10**(5), 1594 (2020). https://doi.org/10.3390/app10051594
34. Feng, L., Zhu, C., Yuan, H., Liu, L., Lv, F., Wang, S.: Conjugated polymer nanoparticles: Preparation, properties, functionalization and biological applications. Chem. Soc. Rev. **42**, 6620–6633 (2013). https://doi.org/10.1039/C3CS60036J
35. Hong, S., Choi, D.W., Kim, H.N., Park, C.G., Lee, W.L., Park, H.H.: Protein-based nanoparticles as drug delivery systems. Pharmaceutics **12**(7), 604 (2020). https://doi.org/10.3390/pharmaceutics12070604
36. Truong-Le, V.L., August, J.T., Leong, K.W.: Controlled gene delivery by DNA–gelatin nanospheres. Hum. Gene Ther. **9**, 1709–1717 (1998). https://doi.org/10.1089/hum.1998.9.12-1709
37. Yoon, J., Kwag, J., Shin, T.J., Park, J., Lee, Y.M., Lee, Y., et al.: Nanoparticles of conjugated polymers prepared from phase-separated films of phospholipids and polymers for biomedical applications. Adv. Mater. **26**, 4559–4564 (2014). https://doi.org/10.1002/adma.201400906
38. Law, J.X., Liau, L.L., Saim, A., Yang, Y., Idrus, R.: Electrospun collagen nanofibers and their applications in skin tissue engineering. Tissue Eng. Regen. Med. **14**(6), 699–718 (2017). https://doi.org/10.1007/s13770-017-0075-9
39. Xue, J., Wu, T., Dai, Y., Xia, Y.: Electrospinning and electrospun nanofibers: methods, materials, and applications. Chem. Rev. **119**(8), 5298–5415 (2019). https://doi.org/10.1021/acs.chemrev.8b00593
40. Azimi, B., Maleki, H., Zavagna, L., De la Ossa, J.G., Linari, S., Lazzeri, A., Danti, S.: Bio-based electrospun fibers for wound healing. J. Funct. Biomater. **11**(3), 67 (2020). https://doi.org/10.3390/jfb11030067
41. Ghorbani, S., Eyni, H., Tiraihi, T., Asl, L.S., Soleimani, M., Atashi, A., et al.: Combined effects of 3D bone marrow stem cell-seeded wet-electrospun poly lactic acid scaffolds on full-thickness skin wound healing. Int. J. Polym. Mater. Polym. Biomater. **67**(15), 905–912 (2018). https://doi.org/10.1080/00914037.2017.1393681
42. Mbese, Z., Alven, S., Aderibigbe, B.A.: Collagen-based nanofibers for skin regeneration and wound dressing applications. Polymers (Basel) **13**(24), 4368 (2021). https://doi.org/10.3390/polym13244368
43. Haggag, Y.A., Faheem, A.M.: Evaluation of nano spray drying as a method for drying and formulation of therapeutic peptides and proteins. Front. Pharmacol. **6**, 140 (2015). https://doi.org/10.389/fphar.2015.00140
44. Lee, S.H., Heng, D., Ng, W.K., Chan, H.K., Tan, R.B.: Nano spray drying: a novel method for preparing protein nanoparticles for protein therapy. Int. J. Pharm. **403**, 192–200 (2011). https://doi.org/10.1016/j.ijpharm.2010.10.012
45. Gulfam, M., Kim, J.E., Lee, J.M., Ku, B., Chung, B.H.: Anticancer drug-loaded gliadin nanoparticles induce apoptosis in breast cancer cells. Langmuir **28**, 8216–8223 (2012). https://doi.org/10.1021/la300691n
46. Boda, S.K., Li, X., Xie, J.: Electrospraying an enabling technology for pharmaceutical and biomedical applications: a review. J. Aerosol Sci. **125**, 164–181 (2018). https://doi.org/10.1016/j.jaerosci.2018.04.002
47. Nagarajan, U., Kawakami, K., Zhang, S., Chandrasekaran, B., Nair, B.U.: Fabrication of solid collagen nanoparticles using electrospray deposition. Chem. Pharm. Bull. **62**, 422–428 (2014). https://doi.org/10.1248/cpb.c13-01004

48. Kempf, M., Miyamura, Y., Liu, P.Y., Chen, A.C., Nakamura, H., Shimizu, H., et al.: A denatured collagen microfiber scaffold seeded with human fibroblasts and keratinocytes for skin grafting. Biomaterials **32**, 4782–4792 (2011). https://doi.org/10.1016/j.biomaterials.2011.03.023
49. Mhetar, S.K., Ashok, N.A., Patil, R.L., Pawar, R.A., Patil, M.M., Shinde, H.T.: Cost effective ball milling machine for producing nanopowder. Int. Res. J. Eng. Technol. **4**, 330–334 (2017)
50. Kumar, M., Xiong, X., Wan, Z., Sun, Y., Tsang, D.C., Gupta, J., et al.: Ball milling as a mechanochemical technology for fabrication of novel biochar nanomaterials. Biores. Technol. **312**, 123613 (2020). https://doi.org/10.1016/j.biortech.2020.123613
51. Weber, C., Coester, C., Kreuter, J., Langer, K.: Desolvation process and surface characterisation of protein nanoparticles. Int. J. Pharm. **194**, 91–102 (2000). https://doi.org/10.1016/S0378-5173(99)00370-1
52. Kandamachira, A., Selvam, S., Marimuthu, N., Kalarical, S.J., Nishter, N.F.: Collagen-nanoparticle interactions: type I collagen stabilization using functionalized nanoparticles. Soft Mater. **13**, 59–65 (2014). https://doi.org/10.1080/1539445X.2015.1009550
53. Choi, D., Heo, J., Park, J.H., Jo, Y., Jeong, H., Chang, M., et al.: Nano-film coatings onto collagen hydrogels with desired drug release. J. Ind. Eng. Chem. **36925**, 326–333 (2016). https://doi.org/10.1016/j.jiec.2016.02.023
54. Liu, G., Li, L., Huo, D., Li, Y., Wu, Y., Zeng, L., et al.: A VEGF delivery system targeting MI improves angiogenesis and cardiac function based on the tropism of MSCs and layer-by-layer self-assembly. Biomaterials **127**, 117–131 (2017). https://doi.org/10.1016/j.biomaterials.2017.03.001
55. Jeong, H., Hep, J., Son, B., Choi, D., Park, T.H., Chang, M., et al.: Intrinsic hydrophobic cairnlike multilayer films for antibacterial effect with enhanced durability. ACS Appl. Mater. Interfaces **7**(47), 26117–26123 (2015). https://doi.org/10.1021/acsami.5b07613
56. Morgan, P.W.: Interfacial polymerization. In: Encylopedia of Polymer Science and Technology. Wiley, New York (2011). https://doi.org/10.1002/0471440264.pst168
57. Raaijmakers, M.J.T., Benes, N.E.: Current trends in interfacial polymerization chemistry. Prog. Polym. Sci. **63**, 86–142 (2016). https://doi.org/10.1016/j.progpolymsci.2016.06.004
58. Song, Y., Fan, J.B., Wang, S.: Recent progress in interfacial polymerization. Mater. Chem. Front. **1**, 1028–1040 (2017). https://doi.org/10.1039/C6QM00325G
59. Prasher, A., Loynd, C.M., Tuten, B.T., Frank, P.G., Chao, D., Berda, E.B.: Efficient fabrication of polymer nanoparticles via sonogashira cross-linking of linear polymers in dilute solution. J. Polym. Sci. Part A: Polym. Chem. **54**, 209–217 (2015). https://doi.org/10.1002/pola.27942
60. Kröger, A.P.P., Hamelmann, N.M., Juan, A., Lindhoud, S., Paulusse, J.M.J.: Biocompatible single-chain polymer nanoparticles for drug delivery-a dual approach. ACS Appl. Mater. Interfaces **10**, 30946–30951 (2018). https://doi.org/10.1021/acsami.8b07450
61. Hanlon, A.M., Chen, R., Rodriguez, K., Willis, C., Dickinson, J.G., Cashman, M., et al.: Scalable synthesis of single-chain nanoparticles under mild conditions. Macromolecules **50**, 2996–3003 (2017). https://doi.org/10.1021/acs.macromol.7b00497
62. Verso, F.L., Pomposo, J.A., Colmenero, J., Moreno, A.J.: Multi-orthogonal folding of single polymer chains into soft nanoparticles. Soft Matter **10**, 4813–4821 (2014). https://doi.org/10.1039/C4SM00459K
63. Wang, G., Uludag, H.: Recent developments in nanoparticle-based drug delivery and targeting systems with emphasis on protein-based nanoparticles. Expert Opin. Drug Deliv. **5**, 499–515 (2008). https://doi.org/10.1517/17425247.5.5.499
64. Komlosh, A., Weinstein, V., Loupe, P., Hasson, T., Timan, B., Konya, A., et al.: Physicochemical and biological examination of two glatiramer acetate products. Biomedicines **7**, 49 (2019). https://doi.org/10.3390/biomedicines7030049
65. Posadas, I., Monteagudo, S., Ceña, V.: Nanoparticles for brain-specific drug and genetic material delivery, imaging and diagnosis. Nanomedicine **11**, 833–849 (2016). https://doi.org/10.2217/nnm.16.15
66. Patra, J.K., Das, G., Fraceto, L.F., Campos, E.V.R., Rodriguez-Torres, M.D.P., Acosta-Torres, L.S., et al.: Nano based drug delivery systems: recent developments and future prospects. J. Nanobiotechnology **16**, 71 (2018). https://doi.org/10.1186/s12951-018-0392-8

67. Akhmetova, A., Heinz, A.: Electrospinning proteins for wound healing purposes: opportunities and challenges. Pharmaceutics **13**(1), 4 (2021). https://doi.org/10.3390/pharmaceutics13010004
68. Oliveira, A.M., Guimarães, K.L., Cerize, N.N., Tunussi, A.S., Poço JG. Nano spray drying as an innovative technology for encapsulating hydrophilic active pharmaceutical ingredients (API). J. Nanomedicine Nanotechnol. 4 (2013). https://doi.org/10.4172/2157-7439.1000186
69. Kim, M.T., Chen, Y., Marhoul, J., Jacobson, F.: Statistical modeling of the drug load distribution on trastuzumab emtansine (Kadcyla), a lysine-linked antibody drug conjugate. Bioconjug. Chem. **25**, 1223–1232 (2014). https://doi.org/10.1021/bc5000109
70. Bock, N., Woodruff, M.A., Hutmacher, D.W., Dargaville, T.R.T.R.: Electrospraying, a reproducible method for production of polymeric microspheres for biomedical applications. Polymers **3**, 131–149 (2011). https://doi.org/10.3390/polym3010131
71. Chelle, P., Yeung, C.H.T., Croteau, S.E., Lissick, J., Balasa, V., Ashburner, C., et al.: Development and validation of a population-pharmacokinetic model for Rurioctacog Alfa Pegol (Adynovate®): a report on behalf of the WAPPS-Hemo Investigators Ad Hoc Subgroup. Clin. Pharmacokinet. **59**, 245–256 (2020). https://doi.org/10.1007/s40262-019-00809-6
72. Lohcharoenkal, W., Wang, L., Chen, Y.C., Rojanasakul, Y.: Protein nanoparticles as drug delivery carriers for cancer therapy. Biomed. Res. Int. **180549**, 1–12 (2014). https://doi.org/10.1155/2014/180549
73. Zhang, X., Wang, D., Zhou, Z., Ma, Y.: Robust low-rank tensor recovery with rectification and alignment. IEEE Trans. Pattern Anal. Mach. Intell. **43**(1), 238–255 (2021). 10.1109/TPAMI.2019.29290
74. Garg, A., Visht, S., Sharma, P.K., Kumar, N.: Formulation, characterization and application on nanoparticle: a review. Der Pharm. Sin. **2**(2), 17–26 (2011)
75. Kaszuba, M., Corbett, J., Watson, F.M., Jones, A.: High-concentration zeta potential measurements using light-scattering techniques. Philos. Trans. R. Soc. A **2010**(368), 4439–4451 (1927). https://doi.org/10.1098/rsta.2010.0175
76. Mistry, A., Glud, S.Z., Kjems, J., Randel, J., Howard, K.A., Stolnik, S., et al.: Effect of physicochemical properties on intranasal nanoparticle transit into murine olfactory epithelium. J. Drug Target **17**, 543–552 (2009). https://doi.org/10.1080/10611860903055470
77. Nitta, S.K., Numata, K.: Biopolymer-based nanoparticles for drug/gene delivery and tissue engineering. Int. J. Mol. Sci. **14**, 1629–1654 (2013). https://doi.org/10.3390/ijms14011629
78. Shalaby, M., Ghareeb, A.Z., Khedra, S.M., Mostafa, H.M., Saeed, H., Hamouda, D.: Nanoparticles of bioactive natural collagen for wound healing: experimental approach (2023). https://doi.org/10.1101/2023.02.21.529363
79. Benyettou, F., Rezgui, R., Ravaux, F., Jaber, T., Blumer, K., Jouiad, M.: Synthesis of silver nanoparticles for the dual delivery of doxorubicin and alendronate to cancer cells. J. Mater. Chem. B **2015**(3), 7237–7245 (2015). https://doi.org/10.1039/C5TB00994D
80. Yildirimer, L., Thanh, N.T., Loizidou, M., Seifalian, A.M.: Toxicology and clinical potential of nanoparticles. Nano Today **6**, 585–607 (2011). https://doi.org/10.1016/j.nantod.2011.10.001
81. Subha, V., Kirubanandan, S., Ilangovan, R., Renganathan, S.: Silver nanoparticles impregnated nanocollagen as scaffold for soft tissue repair-synthesis, characterization, and *in vitro* investigation. Int. J. Med. Nano Res. **8**(1), 1–9 (2021). https://doi.org/10.23937/2378-3664.1410034
82. Cardoso, V.S., Quelemes, P.V., Amorin, A., Primo, F.L., Gobo, G.G., Tedesco A C. et al.: Collagen-based silver nanoparticles for biological applications: synthesis and characterization. J. Nanobiotechnology **12**, 36 (2014). https://doi.org/10.1186/s12951-014-0036-6
83. Roshanbinfar, K., Kolesnik-Gray, M., Angeloni, M., Schruefer, S., Fiedler, M., Schubert, D. et al.: Collagen hydrogel containing polyethylenimine-gold nanoparticles for drug release and enhanced beating properties of engineered cardiac tissues. Adv. Healthc. Mater. 2202408 (2023). https://doi.org/10.1002/adhm.202202408
84. Terzopoulou, Z., Michopoulou, A., Palamidi, A., Koliakou, E., Bikiaris, D.: Preparation and evaluation of collagen-based patches as curcumin carriers. Polymers (Basel) **12**(10), 2393 (2020). https://doi.org/10.3390/polym12102393

85. Mohanraj, V., Chen, Y.: Nanoparticles-a review. Trop. J. Pharm. Res. **5**, 561–573 (2006). https://doi.org/10.4314/tjpr.v5i1.14634
86. Hoshyar, N., Gray, S., Han, H., Bao, G.: The effect of nanoparticle size on *in vivo* pharmacokinetics and cellular interaction. Nanomedicine **11**(6), 673–692 (2016). https://doi.org/10.2217/nnm.16.5
87. McSweeney, M.D., Wessler, T., Price, L.S.L., Ciociola, E.C., Herity, L.B., Piscitelli, J.A., et al.: A minimal physiologically based pharmacokinetic model that predicts anti-PEG IgG-mediated clearance of PEGylated drugs in human and mouse. J. Control. Release **284**, 171–178 (2018). https://doi.org/10.1016/j.jconrel.2018.06.002
88. Breitenbach, M., Kamm, W., Hungere, K., Hund, H., Kissel, T.: Oral and nasal administration of tetanus toxoid loaded nanoparticles consisting of novel charged biodegradable polyesters for mucosal vaccination. In: Proceedings of the International Symposium on Controlled Release of Bioactive Materials, vol. 26, pp. 348–349 (1999)
89. Zhang, L., Beatty, A., Lu, L., Abdalrahman, A., Makris, T.M., Wang, G., et al.: Microfluidic-assisted polymer-protein assembly to fabricate homogeneous functional nanoparticles. Mater. Sci. Eng. C **111**, 110768 (2020). https://doi.org/10.1016/j.msec.2020.110768
90. Knight, F.C., Gilchuk, P., Kumar, A., Becker, K.W., Sevimli, S., Jacobson, M.E., et al.: Mucosal immunization with a pH-responsive nanoparticle vaccine induces protective $CD8^+$ lung-resident memory T cells. ACS Nano **13**(10), 10939–11096 (2019). https://doi.org/10.1021/acsnano.9b00326
91. Strand, M.S., Krasnick, B.A., Pan, H., Zhang, X., Bi, Y., Brooks, C., et al.: Precision delivery of RAS-inhibiting siRNA to KRAS driven cancer via peptide-based nanoparticles. Oncotarget **10**(46), 4761–4775 (2019). https://doi.org/10.18632/oncotarget.27109
92. Afsharzadeh, M., Hashemi, M., Mokhtarzadeh, A., Abnous, K., Ramezani, M.: Recent advances in co-delivery systems based on polymeric nanoparticle for cancer treatment. Artif. Cells Nanomedicine Biotechnol. **46**(6), 1095–1110 (2018). https://doi.org/10.1080/21691401.2017.1376675
93. Volpatti, L.R., Matranga, M.A., Cortinas, A.B., Delcassian, D., Daneil, K.B., Langer, R., Anderson, D.G. et al.: Glucose-responsive nanoparticles for rapid and extended self-regulated insulin delivery. ACS Nano **14**(1), 488–497 (2020). https://doi.org/10.1021/acsnano.9b06395
94. Niu, Y., Yu, M., Hartono, S.B., Yang, J., Xu, H., Zhang, H., et al.: Nanoparticles mimicking viral surface topography for enhanced cellular delivery. Adv. Mater. **25**, 6233–6237 (2013). https://doi.org/10.1002/adma.201302737
95. Awad, A., Shalaby, M., Batiha, G., Mady, R., Al-kuraishy, H., Shaheen, H.M.: Cytotoxicity effect assessment of theophylline loaded with collagen nanoparticles. Damanhour J. Veterinary Sci. **8**(2), 5–10 (2022). https://doi.org/10.21608/djvs.2022.140639.1073
96. Rathore, P., Arora, I., Rastogi, S., Akhtar, M., Singh, S., Samim, M.: Collagen–curcumin nanocomposites showing an enhanced neuroprotective effect against short term focal cerebral ischemia. RSC Adv. **10**(4), 2241–2253 (2020). https://doi.org/10.1039/C9RA08508D
97. Anwar, M.M., Shalaby, M.A., Saeed, H., Mostafa, H.M., Hamouda, D.G., Nounou, H.: Theophylline-encapsulated Nile Tilapia fish scale-based collagen nanoparticles effectively target the lungs of male Sprague–Dawley rats. Sci. Rep. **12**(1), 1–12 (2022).
98. Soppimath, K.S., Aminabhavi, T.M., Kulkarni, A.R., Rudzinski, W.E.: Biodegradable polymeric nanoparticles as drug delivery devices. J. Control. Release **70**, 1–20 (2001). https://doi.org/10.1016/S0168-3659(00)00339-4
99. Pritchard, E.M., Hu, X., Finley, V., Kuo, C.K., Kaplan, D.L.: Effect of silk protein processing on drug delivery from silk films. Macromol. Biosci. **13**, 311–320 (2013). https://doi.org/10.1002/mabi.201200323
100. Champion, J.A., Katare, Y.K., Mitragotri, S.: Particle shape: a new design parameter for micro- and nanoscale drug delivery carriers. J. Control. Release **121**, 3–9 (2007). https://doi.org/10.1016/j.jconrel.2007.03.022
101. Kundu, B., Soundrapandian, C., Nandi, S.K., Mukherjee, P., Dandapat, N., Roy, S., et al.: Development of new localized drug delivery system based on ceftriaxone-sulbactam composite drug impregnated porous hydroxyapatite: a systematic approach for *in vitro* and

in vivo animal trial. Pharm. Res. **27**, 1659–1676 (2010). https://doi.org/10.1007/s11095-010-0166-y
102. Tessmar, J.K., Göpferich, A.M.: Matrices and scaffolds for protein delivery in tissue engineering. Adv. Drug Deliv. Rev. **59**, 274–291 (2007). https://doi.org/10.1016/j.addr.2007.03.020
103. Solorio, L., Zwolinski, C., Lund, A.W., Farrell, M.J., Stegemann, J.P.: Gelatin microspheres crosslinked with genipin for local delivery of growth factors. J. Tissue Eng. Regen. Med. **4**, 514–523 (2010). https://doi.org/10.1002/term.267
104. Kukkar, P.M., Savkare, A.D., Bhavsar, M.R., Gholap, V.D.: A review on nanoparticle cross-linked collagen shield for sustained delivery of drug in glaucoma. Int. J. Pharm. Sci. Res. **8**, 2731–2739 (2017). https://doi.org/10.13040/IJPSR.0975-8232.8(7).2731-39
105. Choi, J., Park, H., Kim, T., Jeong, Y., Oh, M.H., Hyeon, T., et al.: Engineered collagen hydrogels for the sustained release of biomolecules and imaging agents: promoting the growth of human gingival cells. Int. J. Nanomed. **9**(1), 5189–5201 (2014). https://doi.org/10.2147/IJN.S71304
106. Verma, A.K.: Collagen-based biomaterial as drug delivery. In: Mazumder, N., Chakrabarty S. (eds.) Collagen Biomaterials. IntechOpen 1–40 (2022). https://doi.org/10.5772/intechopen.103063
107. Lin, J., Li, C., Zhao, Y., Hu, Z., Zhang, L.M.: Co-electrospun nanofibrous membranes of collagen and zein for wound healing. ACS Appl. Materi. Interfaces **4**(2), 1050–1057 (2012). https://doi.org/10.1021/am201669z
108. Karri, V.V.S.R., Kuppusamy, G., Taluri, S.V., Mannemala, S.S., Kollipara, R., Wadhwani, A.D., et al.: Curcumin loaded chitosan nanoparticles impregnated into collagen-alginate scaffolds for diabetic wound healing. Int. J. Biol. Macromol. **93**, 1519–1529 (2016). https://doi.org/10.1016/j.ijbiomac.2016.05.038
109. Le, V.M., Lang, M.D., Shi, W.B., Liu, J.W.: A collagen-based multicellular tumor spheroid model for evaluation of the efficiency of nanoparticle drug delivery. Artif. Cells Nanomedicine Biotechnol. **44**(2), 540–544 (2016). https://doi.org/10.3109/21691401.2014.968820
110. Ghorbani, M., Nezhad-Mokhtari, P., Ramazani, S.: Aloe vera-loaded nanofibrous scaffold based on zein/polycaprolactone/collagen for wound healing. Int. J. Biol. Macromol. **153**, 921–930 (2020). https://doi.org/10.1016/j.ijbiomac.2020.03.036
111. Deepa, C., Suresh, G., Devagi, P., Kowsalya, A., Kavitha, K., Ramesh, B.: Evaluation of the antibacterial activity of silver nanoparticles, doxycycline, collagen and their amalgamation – an *in vitro* study. Mater. Today: Proc. **43**(5), 3050–3053 (2021). https://doi.org/10.1016/j.matpr.2021.01.396
112. Ma, S., Adayi, A., Liu, Z., Li, M., Wu, M., Xiao, L., et al.: Asymmetric collagen/chitosan membrane containing minocycline-loaded chitosan nanoparticles for guided bone regeneration. Sci. Rep. **6**, 31822 (2016). https://doi.org/10.1038/srep31822
113. Lee, D., Wufuer, M., Kim, I., Choi, T.H., Kim, B.J., Jung, H.G., et al.: Sequential dual-drug delivery of BMP-2 and alendronate from hydroxyapatite-collagen scaffolds for enhanced bone regeneration. Sci. Rep. **11**(1), 746 (2021). https://doi.org/10.1038/s41598-020-80608-3
114. Pourjavadi, A., Doroudian, M.: Synthesis and characterization of semiconductive nanocomposite based on hydrolyzed collagen and *in vitro* electrically controlled drug release study. Polymer **76**, 287–294 (2015). https://doi.org/10.1016/j.polymer.2015.06.050
115. Tezgel, Ö., Distasio, N., Laghezza-Masci, V., Taddei, A.R., Szarpak-Jankowska, A., Auzély-Velty, R., et al.: Collagen scaffold-mediated delivery of NLC/siRNA as wound healing materials. J. Drug Deliv. Sci. Technol. **55**, 101421 (2020). https://doi.org/10.1016/j.jddst.2019.101421
116. Anandhakumar, S., Krishnamoorthy, G., Ramkumar, K.M., Raichur, A.M.: Preparation of collagen peptide functionalized chitosan nanoparticles by ionic gelation method: an effective carrier system for encapsulation and release of doxorubicin for cancer drug delivery. Mater. Sci. Eng. C **70**(1) (2017). https://doi.org/10.1016/j.msec.2016.09.003
117. Li, R., Xu, Z., Jiang, Q., Zheng, Y., Chen, Z., Chen, X.: Characterization and biological evaluation of a novel silver nanoparticle-loaded collagen-chitosan dressing. Regen. Biomater. **7**(4), 371–380 (2020). https://doi.org/10.1093/rb/rbaa008

118. Bettini, S., Bonfrate, V., Syrgiannis, Z., Sannino, A., Salvatore, L., Madaghiele, M., et al.: Biocompatible collagen paramagnetic scaffold for controlled drug release. Biomacromol **16**(9), 2599–2608 (2015). https://doi.org/10.1021/acs.biomac.5b00829
119. Choi, M., Chung, J.H., Cho, Y., Hong, B.Y., Hong, J.: Nano-film modification of collagen hydrogels for controlled growth factor release. Chem. Eng. Sci. **137**, 626–630 (2015). https://doi.org/10.1016/j.ces.2015.07.011
120. Minakuchi, Y., Takeshita, F., Kosaka, N., Sasaki, H., Yamamoto, Y., Kouno, M., et al.: Atelocollagen-mediated synthetic small interfering RNA delivery for effective gene silencing *in vitro* and *in vivo*. Nucleic Acids Res. **32**(13), e109 (2004). https://doi.org/10.1093/nar/gnh093
121. Ramírez, O.J., Alvarez, S., Contreras-Kallens, P., Barrera, N.P., Aguayo, S., Schuh, C.M.A.P.: Type I collagen hydrogels as a delivery matrix for royal jelly derived extracellular vesicles. Drug Deliv. **27**(1), 1308–1318 (2020). https://doi.org/10.1080/10717544.2020.1818880
122. Wang, Y., Zhang, C.L., Zhang, Q., Li, P.: Composite electrospun nanomembranes of fish scale collagen peptides/chito-oligosaccharides: antibacterial properties and potential for wound dressing. Int. J. Nanomed. **6**, 667–676 (2011). https://doi.org/10.2147/IJN.S17547
123. Liu, S.J., Kau, Y.C., Chou, C.Y., Chen, J.K., Wu, R.C., Yeh, W.L.: Electrospun PLGA/collagen nanofibrous membrane as early-stage wound dressing. J. Membr. Sci. **355**(1–2), 53–59 (2010). https://doi.org/10.1016/j.memsci.2010.03.012
124. Mondal, S., Hoang, G., Manivasagan, P., Moorthy, M.S., Phan, T.T.V., Kim, H.H., et al.: Rapid microwave-assisted synthesis of gold loaded hydroxyapatite collagen nano-bio materials for Drug Deliv. and tissue engineering application. Ceram. Int. **45**(3), 2977–2988 (2019). https://doi.org/10.1016/j.ceramint.2018.10.016
125. Wei, D., Sun, J., Yang, Y., Wu, C., Chen, S., Guo, Z., et al.: Cell alignment guided by nano/micro oriented collagen fibers and the synergistic vascularization for nervous cell functional expression. Mater. Today Chem. **8**, 85–95 (2018). https://doi.org/10.1016/j.mtchem.2018.03.001
126. Mohamadi, F., Ebrahimi-Barough, S., Nourani, M.R., Derakhshan, M.A., Goodarzi, V., Nazockdast, M.S., et al.: Electrospun nerve guide scaffold of poly (ε-caprolactone)/collagen/nanobioglass: an *in vitro* study in peripheral nerve tissue engineering. J. Biomed. Mater. Res. Part A **105A**, 1960–1972 (2017). https://doi.org/10.1002/jbm.a.36068
127. Ghavimi, M.A., Shahabadi, A.B., Jarolmasjed, S., Memar, M.Y., Dizaj, S.M., Sharifi, S.: Nanofibrous asymmetric collagen/curcumin membrane containing aspirin-loaded PLGA nanoparticles for guided bone regeneration. Sci. Rep. **10**, 18200 (2020). https://doi.org/10.1038/s41598-020-75454-2
128. Ahmadian, S., Ghorbani, M., Mahmoodzadeh, F.: Silver sulfadiazine-loaded electrospun ethyl cellulose/polylactic acid/collagen nanofibrous mats with antibacterial properties for wound healing. Int. J. Biol. Macromol. **162**, 1555–1565 (2020). https://doi.org/10.1016/j.ijbiomac.2020.08.059
129. Lee, C.H., Chang, S.H., Chen, W.J., Hung, K.C., Lin, Y.H., Liu, S.J., et al.: Augmentation of diabetic wound healing and enhancement of collagen content using nanofibrous glucophage-loaded collagen/PLGA scaffold membranes. J. Colloid Interface Sci. **439**, 88–97 (2015). https://doi.org/10.1016/j.jcis.2014.10.028
130. Lai, H.J., Kuan, C.h., Wu, H.C., Tsai, J.C., Chen, T.M., Hseish DJ et al. Tailored design of electrospun composite nanofibers with staged release of multiple angiogenic growth factors for chronic wound healing. Acta Biomater. **10**(10), 4156-4166 (2014). https://doi.org/10.1016/j.actbio.2014.05.001
131. Kandhasamy, S., Perumal, S., Madhan, B., Umamaheswari, N., Banday, J.A., Perumal, P.T., et al.: Synthesis and fabrication of collagen-coated ostholamide electrospun nanofiber scaffold for wound healing. ACS Appl. Mater. Interfaces **9**(10), 8556–8568 (2017). https://doi.org/10.1021/acsami.6b16488
132. Hou, J., Chen, L., Zhou, M., Li, J., Liu, J., Fang, H., et al.: Multi-layered polyamide/collagen scaffolds with topical sustained release of N-acetylcysteine for promoting wound healing. Int. J. Nanomed. **15**, 1349–1361 (2020). https://doi.org/10.2147/IJN.S232190

133. Yao, C.H., Chen, K.Y., Chen, Y.S., Li, S.J., Huang, C.H.: Lithospermi radix extract-containing bilayer nanofiber scaffold for promoting wound healing in a rat model. Mater. Sci. Eng. C **96**, 850–858 (2019). https://doi.org/10.1016/j.msec.2018.11.053
134. Tort, S., Acartürk, F., Beşikci, A.: Evaluation of three-layered doxycycline-collage loaded nanofiber wound dressing. Int. J. Pharm. **529**(1–2), 642–653 (2017). https://doi.org/10.1016/j.ijpharm.2017.07.027
135. Khodir, W.K.W.A., Razak, A.H.A., Ng, M.H., Guarino, V., Susanti, D.: Encapsulation and characterization of gentamicin sulfate in the collagen added electrospun nanofibers for skin regeneration. J. Funct. Biomater. **9**(2), 36 (2018). https://doi.org/10.3390/jfb9020036

Amit Kumar Verma since 16 years at Department of Biosciences, Jamia Millia Islamia earned his Master's degree from University of Pune and Ph.D. degree from University of Lucknow and started his career as teaching faculty at Jamia Millia Islamia. He has published several impacted research articles and supervised Ph.D. theses in the field of protein based biomaterials, parasitology, molecular biology and cancer biology. He is an active member of several science societies and committees and participated in many national and international conferences and seminars. Presently, he is actively involved in the research towards development of collagen based biomaterials for various biomedical applications.

Photo Responsive Material for 4D Printing in Tissue Engineering

Amisha, Shubham Thakur, and Amrinder Singh

Abstract Four-dimensional (4D) printing technology, which was facilitated by the inclusion of the time dimension to three-dimensional (3D) printing, has now risen as the next-generation approach to tissue engineering since it offers the fabrication of intricate, useful structures. Physiological attributes that are replicated by 3D-printed products are impressive, yet the biomedical devices developed employing 3D approach are inherently static and are therefore not intended to be deployed in unforeseen circumstances. In addition to possessing strong biodegradability and biocompatibility criteria, bioactive materials ought to adapt to the changing environment for better performance thus, 4D printing has now been devised as a remedy for the hurdle. Numerous biomedical applications for tissue engineering employ photo-sensitive biomaterials and can be activated by a variety of light wavelength ranges, including infrared (IR), and ultraviolet (UV). The impact of 4D printing is astounding in the biomedical sector, particularly in sectors like tissue engineering that enhances patient quality of life and strengthens our healthcare infrastructure. With the development of new photo-polymerized systems, these systems considerably lessen the detrimental effects on cellular proliferative and metabolic abilities. Thus, the convergence of light-responsive materials and medical applications of 4D printing techniques tend to be extremely promising. This chapter highlights the most recent application of photo responsive materials and 4D printing expertise in medical engineering, thereby offering an emerging application of 4D printing. Despite significant research into 4D printing, it is crucial to remember that further research is needed into implications and how this paradigm might be improved towards greater accuracy.

Amisha · A. Singh (✉)
Department of Pharmaceutics, ISF College of Pharmacy, Moga 142001, India
e-mail: amrinder.aulakh@yahoo.com

Amisha
e-mail: amisha.0718@gmail.com

S. Thakur (✉)
Department of Pharmaceutical Sciences, Guru Nanak Dev University, Amritsar 143005, India
e-mail: shubhamdthakur@gmail.com

R. Malviya and S. Sundram (eds.), *Engineered Biomaterials*, Engineering Materials,
https://doi.org/10.1007/978-981-99-6698-1_14

Keywords 4D printing · Biomaterials · Photo-responsive materials · Stimuli-responsive · Tissue engineering

Abbreviations

AM	Additive Manufacturing
3D	Three Dimensional
4D	Four Dimensional
4DP	Four Dimensional Printing
DLP	Digital Light Processing
FDM	Fused Deposition Modelling
SLS	Selective Laser Sintering
IJP	Inkjet Printing
SLA	Stereolithography
DIW	Direct Ink Writing
BTE	Bone Tissue Engineering
SMs	Stimuli-responsive Materials
SMMs	Shape Memory Materials
SCMs	Shape-Changing Materials
ABS	Acrylonitrile Butadiene Styrene
PLA	Polylactic Acid
PC	Polycarbonate
PA	Polyamide
PVB	Polyvinyl Butyral
PCL	Polycaprolactone
PS	Polystyrenes
DOD	Drop-On-Demand
NIR	Near Infrared
IR	Infrared
UV	Ultraviolet

1 Introduction

An escalating additive manufacturing (AM) technology acknowledged as three-dimensional (3D) printing was initially envisioned by Hull and his associates in 1986 [1]. This approach has been applied in a numerous assortment of therapeutic areas, along with the fabrication of organs, liver tissue, blood, arteries, bone, heart tissue, regenerative medicine, and drug delivery systems [2]. In the biological system, native tissues have a microenvironment that promotes their growth and governs the array of their biological functions. For better traits, the bioactive materials must be able to adapt to shifting conditions while also adhering to strict biocompatibility and biodegradability standards for improved features [3]. The capability to 3D print biomedical equipment to a patient's specifications adopting this technology's versatility enables the execution of specialized biomedical services. In spite of the fact that 3D-printed things can mimic remarkable physiological properties, biomedical devices made with this technology are intrinsically static and are not intended for dynamic environments [4]. To tackle this issue, 4D printing technique has been devised. Prof. Tibbits and Prof. Jerry originally laid out the concept of 4D printing in 2013, in which sophisticated components in a simulated environment alter their configuration over time [5]. A slashing manufacturing process, 4D printing approach is the upgraded form of 3D printing. The domain of tissue engineering has enthusiastically embraced this technology, and the advent of 4D bioprinting has been facilitated by the inclusion of time as the fourth dimension in 3D bioprinting technology. The morphology and functional modifications of printed structures over time are the primary techniques used in 4D bioprinting. These traits are necessary for the self-renewal of biosynthetic frameworks and long-term homeostasis [6]. Intricate biomaterials for 4D printing, also characterized as shape-morphing materials, distribute other elements or polymer composites by employing the layer-by-layer approach as previously employed in forging 3D objects. The choice of components and associated nature should be considered while choosing a bioprinter [7]. These sophisticated biomaterials undergo 4D printing in the vicinity of several external stimuli, such as moisture, pH, temperature, electric field, biomolecule, magnetic field, light, enzymes and cell adhesion force. In the modern day, numerous 4D printing technologies are adopted, namely digital light processing (DLP), fused deposition modelling (FDM), selective laser sintering (SLS), inkjet printing (IJP), stereolithography (SLA), micro-extrusion, and direct ink writing (DIW) [8, 9].

Recent events in modern society, especially within the realm of 4D printing, have paved the way for breakthroughs across a broad spectrum of disciplines. Numerous dynamic micro-environment tissue engineering fields, such as and fabrication of vascular stents and tissue, cardiac, muscular, vascularization, neuro, bone tissue engineering (BTE), have witnessed enormous advancements in 4D printing technology [10]. The discipline of tissue engineering strives to produce functional tissues that employs an assortment of cells, frameworks, biomaterials, and physiologically active components. Fabricating functional scaffolds that preserve and augment the functions of injured tissue is the prime objective of tissue engineering. It was first addressed in a meeting hosted by the National Science Foundation in the United States in the late 1980s [11]. 3D porous frameworks with the inclusion of time composed of biocompatible materials are frequently alluded to as tissue-engineered scaffolds. They serve a variety of purposes, including promoting targeted tissue repair and regeneration as well as proliferation and cell adhesion [12]. Human cells, physiological, mechanical and biochemical signals, as well as mechanical stabilization that is aided by the adoption of scaffolds—all collaborate together to promote healing [13]. The fabrication of these frameworks is feasible via metals, ceramics, polymer composites polymers, or fibres. Since polymers offer great chemical and biological properties, excellent degradability, and the propensity to imitate the endogenous extracellular matrix, that is vital for a biomaterial since it is used in the design and construction of scaffolds [14]. In the recent era, photo or light has been employed as an alternate treatment method for ailments like melanoma, depression, acne, skin problems or tissue engineering. It is an especially enticing source of energy as compared to other energy types, such as thermal or magnetic energy, which are frequently utilized as an external push to govern stimuli-responsive biomaterials [15]. Firstly, modern optical device enables light to be aimed at a particular region. Moreover, the duration or degree of exposure to light might be employed to modulate the stimulation intensity. Eventually, light might be segmented into a spectrum of wavelengths that can be individually modulated via distinct biomaterials to inspire new bio-inductive stimuli for upregulating interactions. Thus, contrasted to temperature or pH value fluctuation, light is a substantially less detrimental alternative for altering biomaterials in the physiological environment. As a corollary, light is a possible source of energy for constructing biomaterials that respond to stimuli for tissue engineering [16, 17].

So far, 4D bioprinting methodologies are still in its inception, and experts are continually learning about its mechanics and concepts. In this overview, we confer several types of stimuli-responsive materials (SMs), their shape-transformation mechanisms and their cell-inherent characteristics for 4D bioprinting and special emphasis is provided on photo responsive biomaterials as described in the literature at present [18]. A variety of approaches have additionally been explored to facilitate the functionalization and maturation of tissues or cells in 3D-printed structures over time. In addition, this chapter fuses the most recent research with potential therapeutic insights to prioritize on the development of tissue engineering 4D bioprinting and incorporate its sophisticated applications in tissue regeneration. Lastly, we confer about the foremost challenges that 4D bioprinting is facing as it develops and assess its potential future possibilities [19, 20].

2 Brief Assessment of 4D Bioprinting

In numerous biological applications, namely drug delivery systems, wound repair, and, tissue engineering, 4D bioprinting methodology has demonstrated incredible potential. While employing 3D bio printing, it is challenging to generate blood vessels and hollow structures. The 4D bioprinting method, which prints a flat biological scaffold that bends into the blood vessel in reaction to an external trigger, might be able to overcome this impediment [21]. Dynamic 3D-printed biological constructions are generated employing the blend of 4D bioprinting approach alongside novel stimuli-responsive materials. There has been a tremendous surge in research into the 4D bioprinting of smart polymers in the past few years [22]. There are primarily two sorts of SMs: Shape memory materials (SMMs) and shape-changing materials (SCMs). SMMs, construct printed parts of complex geometric structures via a shape-morphing effect and programming steps. SCMs contort as a consequence of environmental disparities, and after the cessation of external stimuli, these materials revert to their initial structure. In these materials, the deformation behaviour was

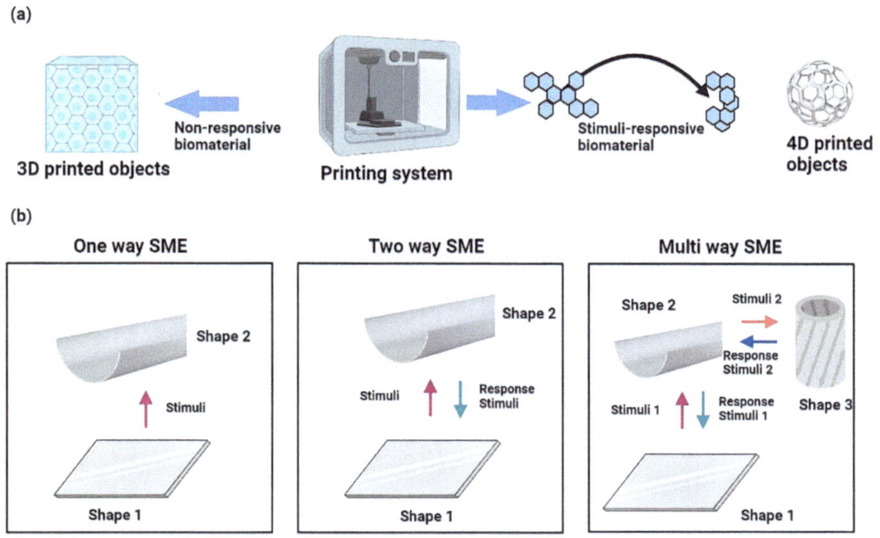

Fig. 1 **a** Instances of 3D and 4D printing. **b** Schematics of stimuli on different shape morphing effects

therefore predetermined [23, 24]. Self-assembly, self-actuating, self-healing, self-sensing, and shape memory, are five characteristic features of SMs discovered in 4D printed materials. External cues drive self-assembly to contour and merge into a pre-programmed configuration. External stimuli are recognized and evaluated through self-sensing behaviour. Despite having a self-actuating process, external impulses drive the actuation of automated materials. With the self-healing phenomena, structural damage is automatically repaired without outside assistance. Materials that have shape memory morph into pre-set geometry in response to external cues [25, 26]. As there are no substantial differences between 4D printing and 3D printing equipment, these two technologies are very compatible. The concept of additive manufacturing, which builds artefacts layer by layer from an array of materials, is the backbone for both 3D and 4D printing. A more refined iteration of 3D printing, 4D printing exploits stimuli-responsive substances to precisely produce complex, dynamic objects is further illustrated in Fig. 1. The advancement of 4D printing from the year of inception is illustrated in Fig. 2. The following is a brief explanation of how 4D printing technology is employed in tissue engineering and other biomedical applications and it is further elaborated in Table 1.

2.1 DLP

DLP printing has been demonstrated as a viable means of rapidly creating intricate 3D and 4D constructions since it leverages light as a crosslinking agent. This

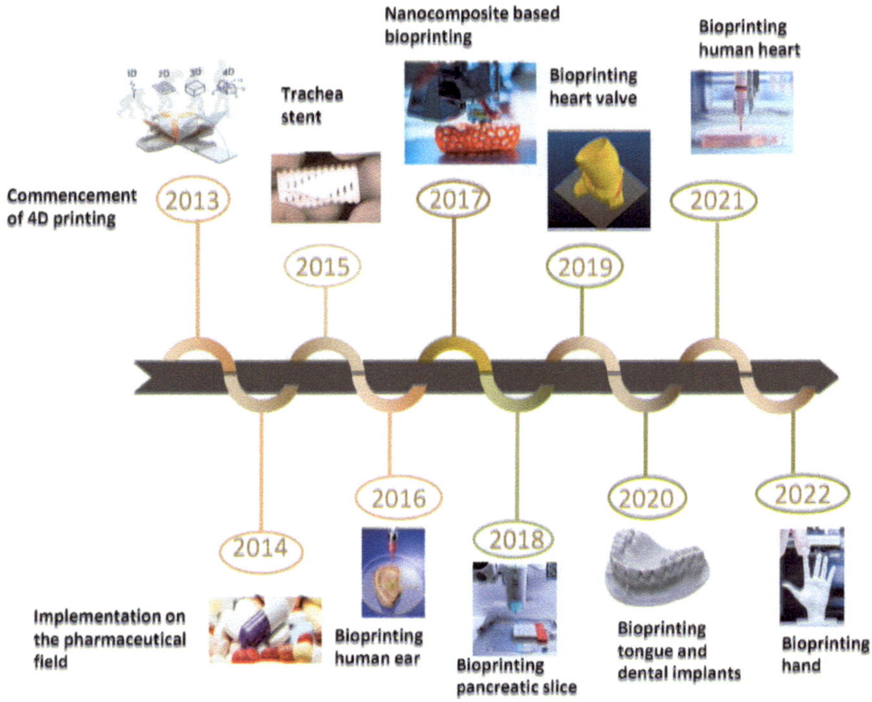

Fig. 2 Development of 4D printing year by year from the inception in various biomedical sector

approach provides an assortment of printing materials and could yield prints featuring improved print quality, print clarity, and print contour accuracy. In contrast to SLA, DLP cures liquid resin in a reservoir in approximately one second per layer adopting a digital light pattern. Since a DLP printer tends to make use of a surface-projection-based fabrication technique, it can print at high rates that are in the range of numerous times greater than what is feasible with SLA [27, 28]. Zhang et al. (2021) fabricated UV-curable and mechanically sturdy SMPs via DLP approach that has considerably increased the mechanical characteristics of SMP-relied 4D printing scaffolds, enabling their use in engineering applications like aerospace, intelligent furniture, and soft robots [29].

2.2 FDM

The fundamental printing principles employed by the FDM approach are quite similar to those used in SLA printing. The peculiarity in this process is that the substance used to print the object is dispensed from the nozzle. Each layer cools and hardens when it gets into the interface of the bottom layer. Generally defined, the FDM

Table 1 Current 4D printing technique for biomaterial

Printing technique	Advantage	Drawbacks	Materials employed	Applications
Digital light processing	High printing speed and resolution. Biocompatible with cells	Inadequate tensile attributes. Time for processing is quite high	Shape memory polymers, photo curable resins, Ceramics	Bone tissue engineering, Cardiac tissue engineering
Fused Deposition modelling	Needed no support framework. Fabricate an array of substances that are biocompatible. Designs porous scaffolds. High-density cells are transformed into tissues	Low spatial precision. High temperature demands to reduced cell viability	Shape memory polymers, Polymer composites	Bone tissue engineering, drug delivery
Selective laser sintering	Needed no support framework. Designs porous scaffolds	Highly porous and extreme temperature is required	Metals, polymers, ceramics	Bone regeneration and tissue engineering
Inkjet printing	Significant compatibility with bioactive compounds that have low viscosities. Producing bioink materials containing organisms. Nozzle without contact	Low spatial precision. High temperature demands to reduced cell viability	Shape memory polymers, hydrogel	Osteochondral tissue engineering, dental implants
Stereolithography	Smooth and fine surfaces are apparent in manufactured products. Incredibly precise method for generating complex frameworks. The loading of cells and biomolecules benefited from moderate conditions	Cytotoxic effect is induced by an uncured photo initiator. Biocompatibility and inefficient decomposition of photo-responsive materials. Printing numerous material layers is not possible Demands enormously potent UV rays to merge	Shape memory polymers, photo-curable resins	Dental implants and bone defect regeneration

(continued)

Table 1 (continued)

Printing technique	Advantage	Drawbacks	Materials employed	Applications
Micro extrusion	High printing speed and resolution. Tensile attribute is good	Low precision. Inadequate cell viability	Shape memory polymers, photo-curable resins	Soft tissue engineering
Direct Ink writing	Design multi-material constructions. Builds permeable scaffolds. Self-supporting filaments ought to be utilized	Necessitates post-processing. Inadequate print resolution and reproducibility	Shape memory polymers, photo-curable resins	Bone tissue regeneration and repair, Neural tissue engineering

process is applicable to composite materials and polymers. Thermoplastic polymers used in FDM need to be modified into a filament series before being ready for 3D printing [30]. Polymer-relied substances in a variety of forms, including polycarbonate (PC), polyamide (PA), polyvinyl butyral (PVB), polycaprolactone (PCL), and polystyrenes (PS) could be employed in FDM, but acrylonitrile butadiene styrene (ABS) and polylactic acid (PLA) are most frequently exploited. FDM is frequently adopted in 4DP owing to its less expensive and simple maintenance, and research is continually been carried out to enhance it and render it accessible to new biomaterials [31]. Miao et al. (2018) by merging the surface coating method (FCT) and FDM, examined whether topographical cues affected how human mesenchymal stem cells responded to skeletal muscle tissues. The stairway flaw was transformed into the ideal bioengineering approach. This work tackles the field of 4D printing by employing smart constructs that demonstrate structural fix and recovery procedures via a layer-by-layer coating and SMP [32].

2.3 SLS

A pretty recent 3DP technique called selective laser sintering (SLS) presents novel prospects for the manufacturing of pharmaceuticals. In order to develop incredibly intricate scaffolds, this approach precisely sinters powder particles in conjunction with a laser. Moreover, it prevents material waste and stimulates the recycling of feedstock since the unsintered powder remains accessible and could be employed again. Further, this demonstrates the technology's affordability; in a technological comparison, the expense of SLS was discovered to be less expensive in contrast to SLA, FDM, and injection moulding [33]. Wu et al. (2021) manufactured the ubiquitous robotics structure known as a gripper employing SLS-relied 4D printing of magnetism-responsive substances, and its propensity to deform when subjected to an exterior magnetic field was investigated. Further investigations into 4D printing driven by magnets would elaborate on the insights. Additional exploration is vital to enhance the gripper's framework for better integration. The amount of energy used while driving should also be considered [34].

2.4 Inkjet Printing

One of the oldest frequently utilized 4DP techniques was inkjet printing, often referred as drop-on-demand (DOD) printing. Since inkjet printing causes the printing head's ejection of photo resin droplets as a corollary of pressure and voltage pulses, it is referred to as DOD printing. UV-curable acrylate liquid is frequently included in inkjet printing fluids, since it quickly cures when exposed to UV light. On the building platform, with a second inkjet printing process, an inkjet printhead that may also move easily in 2D dispenses a liquid binder. The first pattern is lowered for the application of the second powder-liquid binder layer. Each step of the powder-inkjet printing process has to be repeated numerous times to produce the desired results [35]. The fabrication of 3D scaffolds continues to remain challenging as it's complicated to confine cells in 3D frameworks, and regulate cell–cell arrangement in 3D, so Cui et al. (2020) built a framework for printing in 4D via inkjet. This technology generates micropatterns that fold themselves into 3D. As a result, the standard 2D cell-seeding approach is complemented by a user-friendly platform provided by 4D printing and allows users to create a customized 3D cellularized scaffold. This research showed that 4D printing holds great promise for tissue engineering applications [36].

2.5 SLA

Another prevalent 4DP method is SLA, which employs light (laser) emitted at various frequencies to harden liquid resins and polymers. The technology utilizes lasers to layer by layer cures a liquid resin to fabricate a 3D framework. Simply stated, the reaction happens when a photo strikes the substance and the resin molecules link up to form a solid framework. After being dried, photopolymers transform into gels that cross-link to form solid polymers. In contrast to other methods, SLA could start designing with an elongation of up to 80%. It is frequently utilized to construct challenging and intricate buildings. Moreover, relevant materials for SLA printing have been discovered [37, 38]. Zhuo et al. (2022) supported the idea of the 4D printing by the usage of poly (N-vinyl caprolactam) an adaptable polymer for temperature, which is typically produced using radical-free polymerization. This led to the successful preparation of 4D prints adopting SLA 3D printer. Under changing temperatures, these prints demonstrated adaptive and reversible expansion/shrinkage characteristics [39].

2.6 Micro-Extrusion

Among the most popular forms of additive manufacturing processes, known as micro extrusion bioprinting, constructs layer-by-layer cell-laden frameworks by depositing cell-laden bio-inks in forging threads, filaments, or droplets via a tip or syringe. In order to dispense the biomaterial from the nozzle or syringe, micro extrusion uses force. A broad spectrum of substances with varying viscosities could be printed employing micro extrusion bioprinters. The viscosity and characteristics of the bio-ink have a significant impact on the printing's resolution and accuracy. However, additional factors should be considered, namely dispensing pressure or mechanical force, printing speed and distance. Micro extrusion bioprinting offers the benefit of allowing for the regulated fabrication of architectures with high cell densities in physiological circumstances [40, 41]. It is still challenging to execute 3D bio printing of living cellular entities featuring variability in cell types and extracellular matrices [ECMs) that are analogous to those of biological tissues. Thus, Wu et al. (2020) demonstrated how micro extrusion-based (ME) bioprinting techniques could substantially resolve this issue through the design of bio-ink materials. Ultimately, these findings demonstrated that complex structures comprising various cell types and ECMs might be bio-printed with potential applications in translational tissue engineering alongside basic biomedical research [42].

2.7 DIW

Viscoelastic inks are dispensed via needles adopting this extrusion-based technique at high pressure to generate dynamic, intelligent structures in DIW. Biodegradable scaffolds have been constructed using this approach for tissue engineering purposes. UV-assisted DIW was employed to print self-healing, highly flexible SMPs for wound healing applications and biomedical repair devices [43]. Weng et al. (2021) offered an approach for constructing self-morphing frameworks with high modulus (4.8 GPa) and substantial deformation. Employing composite inks with a high-volume proportion of fumed silica, solvent, short glass fibres, and photocurable polymer resin, the frameworks are printed utilising multi-material DIW. This technique for generating composite framework with programmable topologies and superior mechanical characteristics suggests possible uses in the transformation of lightweight structures into load-bearing structures [44].

3 Key Considerations About Photo Responsive Biomaterials

A light cue might be employed to alter the dimension or morphology of the structures utilizing photo-responsive materials, providing volume transformations, contraction, and bending. Externally transmitted optical signals could be captured by photo-responsive materials and translated into mechanical responses. Near-infrared (NIR), infrared (IR), and ultraviolet (UV) areas are among the light wavelength ranges that could activate photo-responsive biomaterials, that are widely used in biomedical applications including TE and controlled medication release [45, 46]. These substances encompass photosensitive nanomaterials, nitrobenzene, azobenzene, fulgide and stilbene-containing polymers, as well as their derivatives. Chromophores are incorporated into polymer resins to develop photo-responsive polymeric products. The fabrication of these materials is also feasible by incorporating photosensitive nanoparticles into polymers. Depending on the type of chromophore, the system's function might be either reversible or irreversible [47, 48].

The most prevalent reaction mechanisms for photo-responsive constituents, that have been extensively used to develop active 4D shape-changing frameworks, are photodegradation and photoisomerization of polymer chains. Further, real-time spatial and temporal control during hydrogel deterioration is provided through photodegradation of biomaterials [49]. By availing utilisation their photo-responsive properties, biomaterials could endure photodegradation, yielding dynamic hydrogel environments. Multiphoton lithography methodology with programmable 4D accredited for the simple creation of networks of microchannels with dimensions that are comparable to those of the natural human vascular systems. These photo-responsive 4D bio inks demonstrate the ability to imitate the dynamic features of extracellular matrix deterioration [50]. Rosenfled et al. (2021) present an innovative gelatin-methacryloyl (GelMA) hydrogel-derived from poly (ethylene glycol) methacrylate (PEGMA) that is biocompatible and naturally photodegradable and could be deployed to develop cells in 3D for at least 14 days. These hydrogels are quickly and easily segmented by exposure to UV light for 10 min, culminating in organized hydrogels featuring various shapes, sizes, and cell-filled hydrogel particles. On required, these structures could be fabricated arbitrarily. In addition, photopolymerization could substantially defer photodegradation, affording an opportunity to photopolymerization by adopting a photomask for the generation of hydrogel arrays [51]. Methodologies for photo responsive release of drug might augment therapeutic accumulation in light-sensitive areas. Current system's use of photocleavable groups often entails exposure to UV or blue light that hinders tissue penetration in detail and is detrimental to living organisms and healthy cells. Thus, Lv et al. (2020) proposed a straightforward response to these concerns by embedding the two components in polymeric micelles that are both biocompatible and biodegradable. The low-irradiance red photon-induced drug release system was designed with a dimension of about 40 nm with outstanding aqueous phase stability. The findings offer the first instance as well as a crucial reference for using photolysis resembling one-photon recreation in tissue engineering [52]. Watanabe et al. (2020) a method for producing hollow structures in photodegradable hydrogels produced in a microfluidic device is described. In order to create the desired hollow frameworks, an infrared femtosecond pulsed laser was used to analyse the hydrogel in a program-controlled manner. With the use of multi-photon excitation, the laser was used to cause photodegradation. After they added a microsphere suspension to the photodegradable hydrogel scaffolds, investigators were able to demonstrate that perfusable hollow structures could be constructed with specific designs applying the multi-photon excitation method [53]. Wu et al. (2022) created a tetra-ortho-methoxy-substituted Azo (mAzo) and CD-functionalized hyaluronic acid (HA) hydrogel that responds to red light. A biocompatible, 625 nm light-sensitive polymeric guest with improved hydrogen bonding and decreased photoisomerization was produced by combining red-shifted, photo-isomerized mAzo with HA. These hydrogels offer prospective benefits for tissue engineering, exclusively with regard to the restoration of functioning multi-tissue complexes [54]. Che et al. (2022) designed and prepared of a multi-sensitive hydrogel that responds to environmental factors that could be used to synthesize a biopolymer ink and can be employed for 3D bioengineering printing. While being subjected to visible light and body temperature, the N-succinyl hydroxybutyl methacrylated chitosan exhibits a sol–gel phase transition. The pH-responsive swelling of the optimized hydrogel efficiently delays shrinkage at a neutral pH. Moreover, the optimized hydrogel exhibits outstanding biocompatibility and biodegradability. These findings demonstrate the constructed hydrogel's tremendous potential to be implemented in an array of applications, encompassing delivery systems, wound healing, and tissue engineering [55].

A novel technique to generate methacrylated carboxymethyl chitosan (CMCS-MA) hydrogel, a shielding layer that can be agglomerated by visible light, was investigated by Xing et al. (2022). This new optimized hydrogel demonstrated quick light cure, high biocompatibility, and lysozyme degradation. Hence, the flexible and convenient biocompatible prepared hydrogels might have a promising use in the regeneration of periodontal tissue [56]. Using MHBC and soluble collagen, a simplistic photo/thermo dual curing composite hydrogel layout was brought on by Liu et al. (2021). Integration of temperature-induced sol–gel transition and contraction by M/C composite hydrogel, the optimum microstructure, customizable mechanical characteristics, enhanced cyto-compatibility, and biodegradability for three-dimensional cell culture. M/C composite hydrogel is a viable option for bio-ink that can be employed in in situ three-dimensional bioprinting [57]. Min et al. (2021) proposed the first hydrogel comprised of platelet lysate from human whole blood that could be methacrylate photopolymerized to produce a crosslinked hydrogel layer by layer, and then employed to construct cell-filled hydrogel frameworks. Efficient proliferation and cell dispersion at small concentrations of bio-ink optimizes biocompatibility and moldability. The 3D-printable PLMA bioinks could represent a viable method for designing tailored microenvironments incorporating materials derived from live organisms for the repair of diverse tissues [58].

The substantial light attenuation attributed to biological tissues, however, causes concern. Investigating the implementation of ultra-IR light, that penetrates more deeply into the tissue than UV radiation and less phototoxicity and absorption in live tissue might aid in resolving this challenge. Owing to their unanticipated cytotoxicity, several photo-initiators have limitations in tissue engineering applications. It has become feasible to generate new photo-polymerized systems with substantially reduced detrimental consequences on cellular metabolic activities and proliferative potential, paving the way to activities like cell encapsulation and the biofabrication of injectable hydrogels. Figure 3 further portrays the applicability of photo-responsive materials with their penetrability and ubiquitous optical attributes. The implementation of light responsive materials in biomedical field is represented in Table 2.

4 Applicability of 4D Printing in Biomedical Field

The advent of 4D printing methodology has enhanced 3D-printed constructions throughout time via enabling for shape and functional change. Substances for 4D printing ought to be incredibly printable [66]. Due to their outstanding shape memory

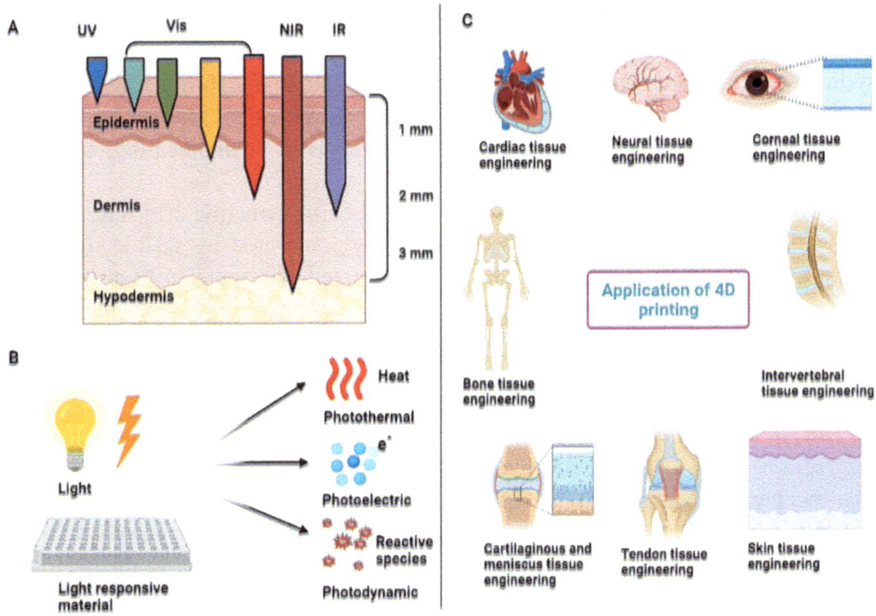

Fig. 3 **a** Schematics representation of penetrability of various light on different layers of skin. **b** Ubiquitous optical attributes. **c** applicability of photo-responsive materials

properties, biocompatibility and processing simplicity, SMPs have emerged as the most significant biomaterials with smart traits implemented for 4D bioprinting. which is depicted in Fig. 4. Researchers have recently devised many SMs, comprising biosensors and actuators, to be employed in biological applications [67]. Various aspects of 4D bioprinting that have recently been developed would be presented in this section. The implementation of 4D bioprinting technology has lately demonstrated tremendous potential for producing scaffolds as it's utilized to construct complex tissues and alter the shapes and functionalities of polymeric materials, that ensure the necessary cellular stresses [68]. Complex multilayer tissue constructs might be constructed employing innovative 4D printing techniques that could be beneficial for tissue engineering as well as other applications. Here, we'll concentrate on 4D tissue engineering and highlight several of the technology's existing tissue engineering applications particularly in various aspects of tissues, including bone, skin, nerve, vascular, and other tissues are elucidated, these applications show undeniable superiority in the repair of individualized tissue defects. 4D-printed structure's capabilities for self-remodelling and functional maturation could be used to create biologically complicated hierarchical structures that resemble native tissue [69, 70].

Table 2 Implementation of photo responsive stimuli at biomedical sectors

Applications	Study model	Inference	References
Cardiac tissue engineering	In vitro (hiPSC-CMs cells)	In this investigation, they developed 3D "spark cells" spheroids constructed up of human embryonic kidney cells that are ChR2-expressing and assessed their architecture as a measure of time and viable cells number. Then after, these "spark-cell" spheroids are employed to illustrate site-specific optical pacing of human stem cell-derived cardiomyocytes (hiPSC-CMs) employing non-localized light implementation	[59]
	In vitro (hiPSC-CMs)	For this analysis, highly aligned 4D NIR light-sensitive cardiac scaffolds with customizable curvature were generated using a DLP-relied printing technology. The 4D cardiac scaffolds might alter their shape on-demand to emulate and replicate the curved architecture of myocardial tissue for seamless integration because the curvature of the heart is variable across its surface	[60]
Brain model	In vitro (NSCs cloned from mouse neuroectoderm)	They describe a 4D printing framework that blends nanomaterials with the ubiquitous stimuli-responsive polymer to accomplish a remote-regulated, dynamic, on-time and positioned contour alteration for the 4D printed object. Implementing a NIR light-sensitive nanocomposite, 4D printed brain model was constructed to investigate the viability of photothermal cues and the ability to control the activity of neural stem cells	[61]
Bone tissue engineering	In vitro (mesenchymal stem cells)	In this investigation, they employed a bismuth sulfide/hydroxyapatite (BS/HAp) film to rapidly and repeatedly generate a photoelectric-responsive microenvironment around a transplant. Via modulating the in vivo photoelectric surroundings using NIR light, this research has developed a biological therapeutic technique that can accomplish in vitro distantly, reliably, and noninvasively directing cell differentiation behaviours	[62]

(continued)

Table 2 (continued)

Applications	Study model	Inference	References
Oesteoarthritis	In vitro (MC3T3-E1 cells)	A supramolecular interaction between azobenzene-modified mesoporous silica nanoparticles (bMSNs-AZO) and -cyclodextrin-modified poly(2-methacryloyloxyethyl phosphorylcholine) led to the development of visible light-responsive dual-functional biodegradable mesoporous silica nanoparticles with lubrication enhancement and drug delivery. After passing through the dermal tissue, visible light can efficiently cause azobenzene isomerization and hence cause medication release. Also, the hydration layer that CD-PMPC generated on the nanoparticles' surface was crucial in enhancing lubrication, which was advantageous for the treatment of osteoarthritis	[63]
Neural tissue engineering	In vitro (hNSCs)	Reduced graphene oxide nanomeshes (rGONMs), p-type semiconductors with a band-gap energy of less than 1 eV, have been constructed with the goal of stimulate the differentiation of human neural stem cells (hNSCs) into neurons using a NIR laser. The graphene layers—particularly the rGONMs—exhibited notable cell differentiation during NIR laser activation, including more cell elongation and higher differentiation of neurons than glia	[64]
	In vitro (CatCh-transfected PC12 cell)	They introduced a unique perspective to tissue engineering that blends optogenetics with bio-inspired neural scaffolds. To construct neuro-matrix interfaces, upconversion nanoparticles (UCNPs) were generated and merged with orientated fibrillar PCL membranes wrapped with collagen. The CatCh-transfected PC12 cells on these surfaces showed improved cell elongation and neurite extension, as well as elevated neurogenesis, upon NIR stimulation attributable to the outstanding bioactivity, directed fibrillation, and NIR-photoresponsivity	[65]

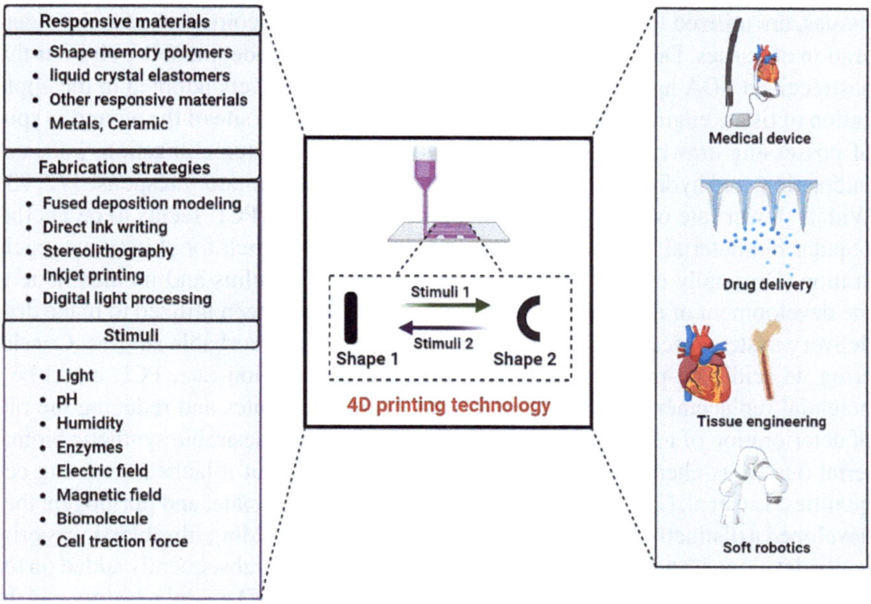

Fig. 4 Depiction of various aspects of 4D printing methodology, namely stimuli, stimuli-responsive material, fabrication technique with their applicability

4.1 Skin Tissue Engineering

The skin, the outermost and major organ of the human body, offers a physical barrier to shield individuals from hazards including diseases and microorganisms. Skin that has been wounded is consequently vulnerable to infection. While healing, uninfected wounds have a pH of 5.5–6.5, but infected wounds have a pH greater than 6.5. As a result, a condition brought on by bacteria is frequently considered prevalent when a wound's pH shifts from acidic to alkaline [71]. Because of fragile nature of skin tissue, hydrogels are frequently utilized to print skin analogues in 3D with respect to time. Skin constructs are constructed using a plethora of 3D printing techniques, including laser-assisted bioprinting as well as extrusion-based printing. To treat burns or persistent wounds, complete skin structures are also generated employing tissue engineering. A substantial amount of printing research involves materials made of collagen since it is a key component of natural skin. Nevertheless, collagen has a slow crosslinking rate and poor printability. Due to its antibacterial qualities and capacity to induce haemostasis, chitosan is favoured over collagen for applications involving wound healing. Furthermore, skin constructions can be printed using alginate, gelatine, GelMA, and fibrin. Synthetic polymers like polycaprolactone (PCL), polyethylene glycol, polyglycolic acid (PGA), polylactic acid (PLA), and related copolymers like polylactic-co-glycolic acids (PLGA) due to their higher compatibility with bodily tissues, are utilized for gathering matrices separately or as composites for the regeneration of tissues. Derived from rice and corn, PLA is a biodegradable polyester that has received FDA approval. PLA has indeed been extensively adopted in the application of tissue engineering scaffolds to transfer cells to the site of the wound in spite of possessing drawbacks such slow degradation rate, limited elongation, poor cell interaction, and hydrophobicity, which can cause an inflammatory response [72, 73]. With a slower rate of deterioration than PLA and PLGA, PCL seems to be another popular biomaterial that has been utilized in various research for skin tissue regeneration. Originally employed to generate biodegradable films and moulds. Due to the development of electrospinning technology, PCL has been utilized to make drug delivery systems, scaffolds for tissue regeneration, and absorbable sutures. Considering its acidic by-products and relatively slow degradation rate, PCL could be a potential replacement to strengthen the mechanical attributes and reducing the rate of deterioration of natural biomaterials [74, 75]. PEG is a desirable synthetic biomaterial due to its chemical and structural characteristics, but it lacks interacting cell qualities. Liu et al. (2022), employing gelatine, sodium alginate, and fibrinogen, they developed a distinctive bio-ink. A dermal layer design including fibroblasts was originally developed, and then laminin and keratinocytes were subsequently added on top of it by adjusting the ratio of the bio-components, inks the 3D model's layout, and the printing circumstances. Full-thickness skin tissue was generated through air–liquid interface (ALI) culturing adopting sterilized wire mesh. In addition to being physically similar to human skin, HE and immunofluorescence labelling revealed that the bio-printed skin also displayed particular markers relevant to epidermal transition and stratum corneum development [76].

4.2 Bone Tissue Engineering

A tissue with high mechanical strength and mineralization is bone. To better imitate bone tissue, it is crucial to strengthen the mechanical attributes of printed polymer. Numerous researches have been published in the various literatures that highlighted the assembly of printed scaffolds for BTE [77]. The polymers are blended with minerals, notably tricalcium phosphate (TCP) and hydroxyapatite (HAP), to strengthen the printed products. For bone tissue scaffolds created via 3D printing, the most prevalent polymer is PCL. And for several medicinal applications, PCL is an FDA-approved, biodegradable and biocompatible polymer. The fact that this polymer has a comparatively low Tm and Tg (60 °C) makes it a great product for fused deposition modelling among the key reasons for adopting it. PCL, however, lacks any cell-attractive moieties and is a hydrophobic polymer. To boost the hydrophilicity of 3D printed PCL scaffolds, it was incorporated with poly (propylene fumarate) (PPF) in a study [78]. The biological functions and microstructures, mechanical attributes, porosity, of 4D-printed SMPs scaffolds with collagen-dexamethasone (Col-Dex) and bioactive fillers (alendronate and hydroxyapatite) covering were thoroughly investigated in this research by Zhou et al. (2022). Under an alternating magnetic field, the SMP scaffolds developed in the study might be generated into temporary forms of tiny sizes and recovered to functioning shapes and sizes to treat bone inadequacies. Additionally, the new 4D printed scaffolds with the Col-Dex coating and bioactive fillers would offer a suitable method for enhanced bone tissue regeneration and repair of customized bone tissue [79]. In the proposed investigation, Hwangbo et al. (2021) successfully used 3D printing, leaching, and coating techniques to create a hierarchical scaffold of biomimetic collagen/hydroxyapatite with multiple microchannels in each framework (i.e., one-way shape morphing, 4D printing). In a posterolateral lumbar spinal fusion mouse model, this biomimetic microchannel framework significantly increased blood vessel ingrowth as well as novo bone formation. It also displayed remarkable promise for osteogenic activities in vitro. These in vivo and in vitro results point to the possibility of using the microchannel collagen/hydroxyapatite scaffold in place of bone grafts to facilitate effective fusion at high rates [80].

4.3 *Vascular Tissue Engineering*

Among the crucial components of tissue regeneration is vascularization, thus, 3D or 4D printing provides vascularization approaches and contributes their capabilities to generate vasculature and, as a result, vigorous vascularized constructions. Vascular regeneration, an extremely viable technique to regain the structure and functioning of affected organs and tissues by addressing the regeneration of blood vessels [81, 82]. The application of surgical bypass grafting, stents, angioplast, tissue-engineered vascular grafts, and non-biodegradable conduits have become prevalent as strategies for vascular regeneration. Vasculature rebuilding is involved in vascular regeneration to create biologically both structurally as well as functionally vessels from endothelial cells, stem cells, bioactive compounds, biomaterials, smooth muscle cells, and spheroids or similar cell aggregates [83]. Techniques for reconstructing vessels include the decellularization of existing vessels, the construction of vascular scaffolds, the development and production of self-constructing vascular grafts, and subsequent vessel maturation methods including the addition of culture additives and the use of vessel-reactors. Tissue engineering innovation is a great complement to these alternatives, namely autografts and allografts. This results from the rectifiable shortcomings of inadequate donors and immunological response [84]. Vascular tissue engineering utilizes one among two techniques: scaffold-based approaches or scaffold-free ways. Compared to scaffold-free approaches, a scaffold-based strategy may drive cell colonization and proliferation along the path. Recent trends have shown that 3D and 4D printing technology has advanced significantly. Because 3D or 4D printing technology is based on the distinct stacking concept, scaffolds with complicated characteristics, such as graded materials and multiscale porous structures, can be easily obtained. With the multidisciplinary technologies of 3D and 4D printing and scaffold-based tissue engineering, bio fabrication research is being conducted on a reasonable scale [85, 86]. Lin et al. (2020) adopted, two variants of shape memory, customized vascular stents with negative Poisson's ratio, and are manufactured via 4D printing. The architecture is optimized adopting the genetic technique. The mechanical characteristics of the stents are examined using axial compression, three-point bending, and radial compression tests. The finite element technique is additionally employed to explore the fluid–structure connection and stress distribution during the form process of recovery. The stents exhibit remarkable shape memory characteristics, and in vitro feasibility investigations show that they can rapidly broaden a mimicked constricted blood artery. For the management of vascular stenosis, 4D printed shape memory scaffolds with a negative Poisson's ratio framework are therefore very interesting [87]. Creating a T-shaped vascular bifurcation using 3D printed shape-changing layers and a mathematical model by Kitana et al. (2023) is made possible for the first time using a 4D bio fabrication-based methodology. Tubular constructions with different diameters might be created by carefully controlling the parameters (for example, crosslinking time). As a consequence, when immersed in water that has a few centimetres in diameter, the 3D-printed films show self-transformation into a T-junction. When the tubular T-junction is perfused with an aqueous liquid simulating blood flow via capillaries, it displays very little spillages with a peak flow velocity of 0.11 ms^{-1}. Moreover, human umbilical vein endothelial cells that have been planted on the inside of a simple T-junction exhibit exceptional growth characteristics and superb cell survival. The sizes that have been reached are comparable to those of natural blood vessels, which is still problematic to achieve using 3D bio fabrication. The formation of completely automatic, self-regulating vascular divergences as vascular transplants is facilitated through this method [88].

4.4 Muscle Tissue Engineering

The advancement of *ad hoc* approaches to stimulate functional tissue regeneration, following muscular disease or injuries, is an actual necessity [89, 90]. However, in cases of tumour ablations, severe trauma, or congenital muscular abnormalities, these impairments, recognized as volumetric muscle loss (VML), could lead to an insurmountable loss of muscle functionality and mass. Skeletal muscle restoration due to VML is still a hard and unresolved problem. However, a good biomimetic technique to get functioning muscular constructs has not yet been identified. Several tissue-engineered strategies for providing muscular grafts that truly rebuild and restore extremely enormous injuries have been formed. New techniques that can enable secure repair and regeneration of larger muscle tissue ought to be explored. The ideal biomaterial should have been able to fill the VML as well as the muscle basal lamina, stimulate cellular alignment, orientation, and maturation, and encourage cell and stem cell activity. This enables access to vascular and neural cells [91, 92]. The necessity of innovative muscle regeneration methods, such as the use of biomaterials that can transmit regenerative signals to either host muscle cells or transplanted cells, is brought on by the overstretched state of the body's normal muscle repair processes. Current research has focused on the utilisation of appropriate biomaterials that might be used as a prototype to direct tissue remodelling and eventually give cells the ideal microenvironment. These methods include using biomaterials solely as well as combining them with ex vivo-produced cells and exogenous growth agents [93, 94]. Although various approaches for the management of serious muscular injuries have been developed so far though, functional engineered muscles remain a critical clinical issue that must be addressed. Skeletal muscle's (SM) ability to regenerate is, in fact, insufficient for repairing significant damage, and SM reconstruction is still a challenging problem. In order to achieve this goal and encourage muscle regeneration, tissue-engineered muscles ought to offer a suitable biomimetic extracellular matrix (ECM) substitute that is defined by a myogenic milieu and an aligned/micro topographical structure. As a result, choosing the right materials and fabrication methods is essential to developing a successful therapeutic strategy. Tissue-specific decellularized ECM (dECM) is among the most potential material for assisting muscle repair and regeneration [95]. Technologically advanced bio fabrication strategies, like bioprinting, enables for the construction of frameworks laden with cells that could greatly speed up the healing process of tissue. Bioprinted cell-filled structures can imitate the extracellular matrix and offer a bioactive environment for cells to grow, distribute, and distinguish, leading in the skeletal muscle tissue's regeneration at the site of injury. In the analyses, Sonaye et al. (2022) developed bio-ink-capable alginate-gelatine composite inks. Then, they employed the inks inside an extrusion printing technique to create frameworks with customized contours for a potential VML therapy. Although the gelatine concentration was held constant at 6% w/v, the alginate concentration was changed between 4 and 12% w/v. The alginate-gelatine inks with 12 and 6% w/v alginate and gelatine respectively, according to rheological study, were the best for creating architecture with good structural fidelity and high resolution [96]. In this study, they investigated the function of topographical variables on modulating behaviours of human mesenchymal stem cells (hMSCs) onto skeletal muscle tissues via blending surface coating methodology with FDM. This staircase defect will then be translated into an efficient bioengineering strategy. This inclusive strategy provides to construct anisotropic scaffolds, contour-specific, multi-dimensional, employing distinct biomaterials. hMSC alignment is efficaciously aimed by 2D anisotropic frameworks, that were initially exhibited here with varying quantities of polycaprolactone. This is certainly relevant when a support loop is employed to fix the framework. The polymer solution from within FDM-printed sacrificial frameworks is interface encased to produce 3D anisotropic frameworks featuring thin wall characteristics. Subsequently, using these frameworks, seeded hMSCs are driven through a spinning bioreactor. Utilizing layer-by-layer coating, including an SMP, smart constructs demonstrating contour stabilization and recovery procedures are equipped, trying to bring this investigation into the domain of 4D printing. The topographical indications produced by FCT, based on immunofluorescence labelling and real-time quantitative polymerase chain reaction evaluation, significantly improve the interpretation of myogenic genes like desmin and myosin heavy chain-2. The fact that many organs and tissues, including skeletal muscle, contain highly coordinated and anisotropic extracellular matrix components leads researchers to the conclusion that this FCT method offers extensive application potentials in tissue engineering [32].

4.5 Nerve Tissue Engineering

The anisotropic alignment of the nerve fibres generates the directional (uniaxial) architecture of nerve tissue. Since a lesion, a damaged nerve's both distal and proximal ends are brought together using nerve aides. Moreover, they could be constructed employing 3D printing to create patient-specific constructions with a sophisticated internal layout. Products for nerve tissue engineering are frequently made using an AM approach, including ink jet printing, SLA, FDM, and bioprinting. To manufacture constructions for nerve applications in tissue engineering, polyurethane, collagen, fibrin, gellan gum, PLA, carboxymethyl chitosan, GelMA, agarose, and PEGDA were also employed [97, 98]. The constructs primarily contain neural stem cells (NSCs), although investigations on nerve regeneration have also used glial cells, astrocytes, bone marrow stem cells (BMSCs), retinal ganglion cells (RGCs), Schwann cells (SCs), and primary cortical neurons. Since the nervous system is comprised of various cell types, including neurons, SCs, and glial cells, numerous researches have focused on printing these types of cells to construct a complete nerve tissue. The direction of these cells is also crucial because of the anisotropic nature of nerve tissue. This issue can be successfully handled by bio printing which makes it possible to print cells on fibres or in tubes of hydrogel with a specified alignment. The 3D or 4D printed frameworks can also include biochemically significant molecules [6, 99]. Apsite et al. (2020) highlight the employment of the 4D bio fabrication approach to construct a synthetic nerve graft. Electrospinning was designed to create bilayer scaffolds that included randomly aligned methacrylated hyaluronic acid (HA-MA) fibres and uniaxially aligned polycaprolactone-poly (glycerol sebacate) (PCL-PGS) fibres meant for the incubation of PC-12 neuron cells. After a fibrous bilayer is immersed in an aqueous buffer, tubular structures immediately form. The width of these tubes might be altered by adjusting bilayer variables like the thickness of each layer, the medium counterion concentration, and the overall thickness. Over the course of four weeks of real-time deterioration, the scaffolds were designed to reveal a self-folding scroll-like structure with remarkable stability [100]. Low-level light treatment (LLLT) has been demonstrated to be effective in treating neurological disorders and degenerative nerve diseases. In light of this, Zhu et al. (2017) hypothesize that integrating LLLT with a 3D-printed neural scaffold would culminate in an innovative approach to managing neural degeneration. To accomplish this, they applied red laser light to stimulate neural stem cells on 3D printed scaffolds, and afterwards we investigated at how the cells responded in terms of differentiation and proliferation. Thus, they demonstrate that after 15 s of laser stimulation and one day of culture, the rate of proliferation of cell and the generation of intracellular reactive oxygen species (ROS) were both significantly elevated [99].

4.6 Other Tissue Engineering Applications

Applications for tissue engineering of the cornea, meniscus, cartilage, and urethra have used 4D or 3D printing. A significant contributor to chronic impairment is cartilage deterioration, which is typically treated with highly invasive treatments. The emergence of 3D bioprinting as a prospective tissue engineering methodology to address this problem. The structures created with this technique lack the dynamic reactivity of native tissues and are stagnant. Moreover, it was still difficult to fabricate curved or tubular structures, particularly while using soft tissues. The above issues might be addressed by 4D bioprinting, a recently developed, next-generation bio-fabrication technology [6]. The application of corneal transplantation is anticipated to be a viable approach for a myriad of corneal disease requirements. Although it has been used as an effective treatment for the majority of corneal problems, patients still have a number of difficulties since rejection is one unidentified hazard of corneal transplant material, and there aren't sufficient healthy donor corneas attainable. A lot of emphasis has been put on the creation of tissue-engineered scaffolds for corneal regeneration and healing utilizing corneal tissue engineering (CTE). The development of a substrate with equivalent mechanical and transmittance qualities is investigated by employing various methods in order to enhance corneal tissue regeneration. Due to their amazing spatial control, which results in a 3D corneal framework filled with cells, bio-printed scaffolds have recently attracted enough attention in this regard to simulate corneal structure [101]. In spite of the fact that tissue engineering is frequently utilized to create intricate, three-dimensional biocompatible architectures, researchers are currently working to expand the approach into the fourth dimension. This 4D is the gradual metamorphosis of three-dimensional materials, specifically the alteration of their composition, function, and/or shape in retort to certain external cues. The study explores the creation of a 4D biomaterial with an intrinsic stimulating function by Miotto et al. (2019). Contractile cells change the form and structure of tissues by acting as bio-actuators. Namely, curving collagen-based hydrogels under controlled, cell-driven conditions to produce cornea-shaped, curved stromal tissue analogues. The following is accomplished by leveraging a contraction-inhibiting peptide amphiphile to modify the activity of the bio-actuators in particular regions of the gels. The ability of the self-curved structures to promote the in vitro development of a corneal epithelium is then assessed, along with the gel stiffness, cell phenotypic, and cell and collagen fibril rearrangement. The results show that self-curved gels made using a 4D engineering technique have structural and mechanical properties that are much better than those of planar 3D scaffolds and more like native tissue. From this aspect, the study shows how versatile cell bio-actuators are for 4D tissue engineering applications [102].

SLS, ink-jet bioprinting, and extrusion-based bioprinting, for instance, have all been employed to cartilage tissue engineering. The printing polymers include polyurethane (PU), nanocellulose, collagen, alginate and agarose. The most often used cells in cartilage regeneration are chondrocytes [103]. The innovative 4D bioprinting technique described by Kalogeropoulou et al. (2020) relies on the multi-material extrusion of two hydrogel-based (bio) inks. This method enables the creation of bilayer scaffolds with an alginate and HA-Tyr hydrogel composite on top and a bottom component made of a HA-Tyr precursor. The varying swelling capacities of the two inks prompted the scaffolds to demonstrate self-bending while immersed in aqueous solutions. Rheological assessment of the inks was carried out to validate their suitability for extrusion printing. By adjusting the layer thickness, printing angle, and infill density, the resultant curvature might be controlled. Eventually, Alg/HA-Tyr ink was blended with human mesenchymal stem cells, and self-bending live scaffolds were bio-printed. For up to 14 days, the shape-shifting scaffolds could sustain a high cell viability, and they could continue to do so for up to 4 weeks [104].

A healthy knee joint requires an adequately functioning meniscus. This fibrocartilaginous tissue, which maintains proper biomechanics, is intricate and vital. Meniscus injuries, especially those to the interior section of the meniscus, seldom heal and typically worsen, leading to structural collapse, meniscus degeneration, and the beginning of osteoarthritis. Existing research in meniscal tissue engineering integrates cell-based therapies, polymeric biomaterials, 3D or 4D-printed frameworks and growth factors, to aid in the mending of meniscal anomalies. Today's orthopaedic sector have a huge opportunity to generate more precise implants attributable to 3D or 4D printing technology, which also accelerates up the production of meniscal structures and alters how orthopaedic surgeons approach therapies [105]. In their investigation, Li et al. (2021) offered a poly(-caprolactone) framework that has been modified with a poly (lactic-co-glycolic acid) microsphere that is loaded with the pharmaceutical kartogenin (KGN) to act as a drug delivery system prior to getting injected having the meniscus extracellular matrix (MECM). As a result, they recommended a strategy to enhance meniscus regeneration leveraging the 3D porous PCL/MECM frameworks that released KGN. The overall findings indicated that the PCL/MECM framework constructed had massively improved bioactivity and hydrophilicity. The introduction of MECM components appeared to aid cell adhesion and proliferation as well as preserve a promising capacity to promote cell migration, according to in vitro studies using mesenchymal stem cells generated from synovium. Moreover, PLGA microspheres loaded on scaffolds with KGN-incorporating revealed a prolonged release profile and enhanced SMSC's chondrogenic development over the course of a 14-day period [106].

Urethral strictures are indeed a prevalent and significant diagnosable disorder that can harm organs and adversely impact one's life. Experts and doctors are fascinated in urethral TE owing to the hunt for the best materials for urethral repair. A substantial number of preclinical research and tremendous advancement have been seen throughout the past few decades. On the other hand, urethral TE has thus far only achieved modest advancements in clinical practise. These analyses laid the groundwork for future research into the utilization of 3D and 4D bioprinting, urethral frameworks could be created that closely aligns the cellular activity and mechanical behaviour of native urethral tissue [107]. In this study, Gu et al. (2021) used a hydrophilic monomer that can be reconstituted with visible light is 2-hydroxyethyl methacrylate. The composite hydrogel's sodium alginate combined with a natural gel network equilibrium water content, mechanical and physical characteristics, and cytotoxicity were investigated, and an attempt was made to 3D print the hydrogel. Then, the viability of urethral tissue engineering with the hydrogel was assessed. The results demonstrated that the hydrogel made of poly (2-hydroxyethyl methacrylate) (p-HEMA) underwent visible-light curing at 405 nm and it was accomplished using the composite photo initiator system. The p-HEMA hydrogel's mechanical characteristics, biocompatibility, and other qualities were enhanced by the addition of sodium alginate [108]. Zhang et al. (2017) have constructed in vitro urethras with various polymer varieties and structural features adopting 3D bioprinting technology. The usage of PCL and PLCL polymers with a spiral scaffold design may be used to simulate the structure and mechanical properties of a rabbit's natural urethra, and a fibrin hydrogel containing cells may offer a better environment for cell development. Adopting an integrated bioprinting methodology, tubular constructs were fabricated, and smooth muscle and urothelial cells were uniformly distributed into the interior and exterior regions of the framework within the cell-laden hydrogel. The native urethra of the rabbit was mechanically comparable to the spiral PCL/PLCL (50:50) structure. When the bio-printed urethra's cell bioactivity was evaluated, it was discovered that SMCs and UCs still had more than 80% of their original viability 7 days after printing [109].

5 Current Hinderance and Future Outlook

The next era of tissue engineering techniques, referred as 4D bioprinting, that integrates "time" as the fourth dimension to 3D bio printing, is anticipated to facilitate the construction of complex entities featuring potentially controlled shapes and functionalities on request. With the emergence of stimuli-responsive biomaterials such as light responsive biomaterials and improved knowledge of tissue regeneration over the past several years, the biomedical enterprise and clinical applications have garnered 4D bioprinting innovation a significant amount of attention [110]. To illustrate, the domain of customized tissue regeneration offers tremendous application potential for 4D bioprinting innovation. For the 4D-printed implantation, the lesion sites would be precisely geometrized with a specified size and form. In the post-printing phase, the operational modification of the transplant would include characteristics that mirror biological processes, promoting tissue development and remodelling. Personalized tissue engineering now has new possibilities to design neo-tissue growth owing to recent breakthroughs in computational model systems. By developing self-growing frameworks, the transformational capabilities of 4D bioprinting could also help treat adolescent patients [111].

A Hi-tech, innovative technique named 4D bioprinting generates tailored elements of sophisticated component by interpreting digital medical imaging and demonstrates remarkable automation control. Via constructing and creating 3D cell-laden dynamical constructs, this technology is progressing and paving the way for novel approaches in biological research [112]. Exorbitant researchers recently shown how to construct several stimuli-responsive biopolymers utilizing 4D bioprinting and assessed their biomechanical qualities by taking in consideration various therapeutic applications notably tissue engineering, biosensors, implants, wound healing, actuators and DDSs [113]. The fields of expertise that ought to be investigated into in the near future are as follows: advances in the vascular tissue's biocompatibility and cohesion, necessitating empirical modelling, the research on innovative multi-stimuli materials, and the adoption of 4D bioprinting methodology in the market [114]. In order to hasten tissue regeneration and strengthen patient's life quality who seem to have organ anomalies, tissue engineering has become a massive breakthrough. It is regarded as a perfect substitute for organ transplantation. The use of various biomaterials and their mixtures, whether derived from natural sources (such gelatine, zein) or artificial sources (like polyurethanes PCL and PLGA), offers a wide range of choices. In general, natural polymers are preferred over synthetic ones for use. Specifically, the use of biopolymers generated from plants is strongly recommended over those derived from animals since they have fewer ethical concerns, are simpler to extract, and are less expensive. The prerequisites of the targeted tissue govern the precise biomaterial utilization, although the application of composite constituents provides the ability to precisely alter the attributes of the applied frameworks. Moreover, utilizing bioactive mineral fillers, such as titanium, niobium pentoxide, and silica, in tissue regeneration is a beneficial means of making utilization of their inherent mechanical and tissue-regenerating capabilities. It is absolutely amazing how sophisticated procedures have developed in the field of tissue engineering, where the use of stem cells has been deemed dependable because of their capacity to distinguish into diverse tissues. Frameworks can also be constructed in 3D and 4D leveraging modern, affordable technology. Although numerous researchers have done substantial research on tissue engineering, there are still limited therapeutic applications and therefore need additional research [14].

Contemplating that 4D printing is still a relatively fresh discipline, there remain numerous constraints and hurdles to be resolved. The implementation of 4D printing is still in its early inception, and the components used for fabricating it are still being investigated. In fact, there is currently not any printer available that is made exclusively for 4D printing. In anticipation of developing more highly precise medical devices, the related technology must be progressively developed. This criterion cannot be met, however, with the printing accuracy and material performance as they currently stand. The biological environment is also intricate, changing, and unique to each individual. In the hopes of bringing 4D printing to biomedicine, the printed objects need to be customized to the organism's microenvironment, for example via microfluidic systems. Microfluidic devices biomimetic milieu creates the ideal platform for cells to reach their biological potentials and produce a functional tissue [115]. Although many nanocomposites and advanced polymers can change their shape or function in response to stimulation, they frequently lack crucial characteristics, such as a certain mechanical strength, biocompatibility, no cytotoxicity, and the response stimulus strength that does not harm tissues. These characteristics are necessary for the materials used in bioengineering. As a result, a few dynamic materials satisfy the criteria listed above. Further, the vast majority of materials primarily react to one stimulation, and the large majority of these cues are exclusively temperature-related, which confines the spectrum of biological applications they can be utilized for. There might be the potential to adopt new devices for biomedical applications because to developments in functionalizing current monomers and polymers to strengthen their biocompatibility or render them printable [116]. The recently created and then used 4D system known as soybean oil epoxidized acrylate is an example of a new material. Rapid initiatives are required to investigate and create innovative multi-responsive hydrogels and SMPs. Further, the fabrication of versatile functional bioengineered tissues exhibiting remarkable biomechanical attributes can be accelerated by the augmentation of MNP nanoparticles, ceramic, and graphene into hydrogels and SMPs. The development of printed products is booming attributable to 4D bioprinting technology, which also holds the potential to completely transform TERM and DDS. However, it has a number of drawbacks, including the inability to forecast and design structural resilience into 4D-printed structures [117]. To address these difficulties, focused study on the empirical modelling of this technology is required. Given that 4D bioprinting has been implemented in applications involving tissue implantation, it is vital to create novel bio-ink substances that have dynamic behaviour, improved multi-stimuli sensitivity, cellular adherence, and optimal stiffening for in-vitro testing. Innovative bioprinters, structural designs and procedures are anticipated for 4D bioprinting expertise in addition to the bio-ink materials listed above [118]. The hurdles preventing the invention in 4DP from being commercialized provide an opportunity to examine strengths, weaknesses, opportunities, and risks. By overcoming each of these obstacles, 4DP of polymer-relied composites will broaden the range of biomedical applications it may be used for, including those for medical devices, cardiac, cartilaginous, bone, skin, neuro, DDSs, and individualized medicine. Further, the tremendous improvements in the contemporary iteration of 4D printing approach would help us better understand stimuli-response substances with their characteristics, and prospective biological attributes. In addition to the TERM applications, 4D bioprinting might potentially be employed to facilitate organ replacement in chronic disorders or fatal accidents [114]. Hence, it remains difficult to create printed things that can simultaneously respond to many stimuli and go through intricate shape-transformation processes. To actualize the manufacture of complicated self-transforming items, computer design advancements and intricate adaptable stimuli-responsive approaches must be incorporated. This will help to accelerate the scaling-up of fabrication and acknowledge the intricately foreordained regulation of 4DP in tissue engineering. Platform designed by MIT called Project Cyborg that allows for the simulation of self-assembling systems and the integration of architectures with programmable substances. The tracking of bodily motion has been done using biosensors and bio-actuators. However, more research is needed to offer accurate, least invasive control of the stimuli delivered to bio-printed structures. Yet, these programmable and stimuli-cues biomaterials could be expensive, and it is often expensive to produce such materials using advanced computer design tools. Production expansion should be feasible and affordable in the interim. As a result, there is still a trade-off between the excellence of 4D bio-printed constructions and their manufacturing viability [9].

In spite of these restrictions, 4D printing's prospects are encouraging. All groundbreaking discoveries and methods are first only visible at their exposed edges, while enormous "icebergs" lie concealed below the surface. Many only perceive the transition from 4 to 5G in the communication sector as an increase in speed; they are not aware of the economic potential it opens up. Moreover, the aforementioned restrictions would logically be lifted with the advancement of engineering technology. Instead of popularization, the early creation of individualized medical devices will help to address this constraint. As for adapting to the body, items might be made of a variety of dynamic materials to retort to an assortment of stimuli [119].

6 Conclusion

The 4D printing industry has made great strides during the past ten years. Since its inception, it has expanded its influence into a number of industrialized regions. In 4D printing, a shape-changing mechanism is often induced over time employing additive manufacturing processes and stimuli-responsive materials. Single-printed object might be exploited for numerous procedures, improving flexibility and adaptability. To put it another way, 4D printing represents a new advancement in 3D printing. The utilization of stimuli-responsive components in this approach is crucial for smart materials, and 3D printing technology enables dynamic printed framework. This trait demonstrates the tremendous future potential of 4D printing. So far it has been successfully applied to fields that seek for dynamic constructs, such as tissue engineering and organ transplants. As a corollary, the combination of tissue engineering and light-responsive biomaterials used in 4D printing led to significant improvements in our understanding of therapeutic processes based on the interest in how monochromatic light interacts with biological tissues. A comprehensive understanding of photodynamic mechanisms would be made possible by the science of tissue engineering which is currently undergoing rapid expansion. The state of the graphics in this domain of expertise could be directly improved with the help of this review chapter. Retrieving and exhibiting crucial data for the oversight of endeavours that seek to use these sophisticated protocols. There is no disputing about the immense potential of combining these principles, which are at the forefront of technology of knowledge and can aid in the creation of new biological assays used in a variety of clinical therapies. Moreover, 4D printing can be developed to create specialized prosthetics, tailored implants, and anatomical models using imaging techniques like CT scans and MRI. Although there have been numerous improvements, 4D printing is still in its initial stages. The aspects of 4D printing methodology, such as imaging techniques, additive manufacturing methods, stimulus-responsive materials and stimuli, thus need more exploration. Nevertheless, not all biomaterials are stimulus-responsive materials, and all stimulus-responsive materials might not be utilized for printing equipment. Sensitive materials should be employed for 4D printing. Moreover, often these materials only adapt to a unique stimulation. Thus, by searching into unique responsive materials and attempting to make the current responsive materials printable, this technology can advance. Notwithstanding these difficulties, the future of 4D printing technology is optimistic. As a result, the limelight could seek the development of larger-scale printers combining robotics, AM techniques, and multi-stimuli sensitive materials to print them, as well as more advancements on the nanoscale focusing on medical medication delivery deployment inside the body. In summation, 4D bioprinting shows promise as a means to establish the active and multiple-layered organization of tissue structure. Additionally, it affords the opportunity to fabricate functional, shape-programmed entities in a regulated manner. The development of 4D bioprinting will offer a potential method for meeting medical needs and examining its broad use in the biomedical industry. However, before the technology can be used to construct dynamic and hierarchical native cellular architectures, a few issues that coexist with promising application prospects must be resolved.

References

1. Wan, Z., Zhang, P., Liu, Y., Lv, L., Zhou, Y.: Four-dimensional bioprinting: Current developments and applications in bone tissue engineering. Acta Biomater. **101**, 26–42 (2020)
2. Top, N., Sahin, I., Gokce, H., Gokce, H.: Computer-aided design and additive manufacturing of bone scaffolds for tissue engineering: state of the art. J. Mater. Res. 1–21 (2021)
3. Jiang, D., Ning, F., Wang, Y.: Additive manufacturing of biodegradable iron-based particle reinforced polylactic acid composite scaffolds for tissue engineering. J. Mater. Process. Technol. **289**, 116952 (2021)
4. Santoni, S., Gugliandolo, S.G., Sponchioni, M., Moscatelli, D., Colosimo, B.M.: 3D bioprinting: current status and trends—a guide to the literature and industrial practice. Bio-Des Manuf. **5**(1), 14–42 (2022)
5. Chu, H., Yang, W., Sun, L., Cai, S., Yang, R., Liang, W, et al.: 4D printing: a review on recent progresses. Micromachines (Basel) **11**(9) (2020)
6. Tamay, D.G., Dursun Usal, T., Alagoz, A.S., Yucel, D., Hasirci, N., Hasirci, V.: 3D and 4D printing of polymers for tissue engineering applications. Front. Bioeng. Biotechnol. **7**, 164 (2019)
7. Khalid, M.Y., Arif, Z.U., Noroozi, R., Zolfagharian, A., Bodaghi, M.: 4D printing of shape memory polymer composites: a review on fabrication techniques, applications, and future perspectives. J. Manuf. Process. **81**, 759–797 (2022)
8. Noroozi, R., Zolfagharian, A., Fotouhi, M., Bodaghi, M.: 4D-printed shape memory polymer: modeling and fabrication. In: Smart Materials in Additive Manufacturing, Volume 2: 4D Printing Mechanics, Modeling, and Advanced Engineering Applications: Elsevier, pp. 195–228 (2022)
9. Arif, Z.U., Khalid, M.Y., Zolfagharian, A., Bodaghi, M.: 4D bioprinting of smart polymers for biomedical applications: recent progress, challenges, and future perspectives. React Funct Polym. 105374 (2022)
10. Wang, Y., Cui, H., Esworthy, T., Mei, D., Wang, Y., Zhang, L.G.: Emerging 4D printing strategies for next-generation tissue regeneration and medical devices. Adv. Mater. **34**(20), e2109198 (2022)
11. Donderwinkel, I., Tuan, R.S., Cameron, N.R., Frith, J.E.: Tendon tissue engineering: current progress towards an optimized tenogenic differentiation protocol for human stem cells. Acta Biomater. (2022)
12. Dhandayuthapani, B., Yoshida, Y., Maekawa, T., Kumar, D.S.: Polymeric scaffolds in tissue engineering application: a review. Int. J. Polym Sci. (2011)

13. Gao, J., Yu, X., Wang, X., He, Y., Ding, J.: Biomaterial-related cell microenvironment in tissue engineering and regenerative medicine. Engineering (2022)
14. Eldeeb, A.E., Salah, S., Elkasabgy, N.A.: Biomaterials for tissue engineering applications and current updates in the field: a comprehensive review. AAPS Pharm Sci. Tech. **23**(7), 267 (2022)
15. Lee, H.P., Gaharwar, A.K.: Light-responsive inorganic biomaterials for biomedical applications. Adv. Sci. (Weinh) **7**(17), 2000863 (2020)
16. Naskar, A., Kim, K.S.: Photo-stimuli-responsive CuS nanomaterials as cutting-edge platform materials for antibacterial applications. Pharmaceutics **14**(11) (2022).
17. Dolmans, D.E., Fukumura, D., Jain, R.K.: Photodynamic therapy for cancer. Nat. Rev. Cancer **3**(5), 380–387 (2003)
18. Bajpai, A., Baigent, A., Raghav, S., Bradaigh, C.O., Koutsos, V., Radacsi, N.: 4D printing: materials, technologies, and future applications in the biomedical field. Sustainabilit **12**(24), 10628 (2020)
19. Yang, G.H., Yeo, M., Koo, Y.W., Kim, G.H.: 4D bioprinting: technological advances in biofabrication. Macromol. Biosci. **19**(5), e1800441 (2019)
20. Amukarimi, S., Rezvani, Z., Eghtesadi, N., Mozafari, M.: Smart biomaterials: from 3D printing to 4D bioprinting. Methods **205**, 191–199 (2022)
21. Yang, Q., Gao, B., Xu, F.: Recent advances in 4D bioprinting. Biotechnol. J. **15**(1), e1900086 (2020)
22. Agarwal, T., Hann, S.Y., Chiesa, I., Cui, H., Celikkin, N., Micalizzi, S., et al.: 4D printing in biomedical applications: emerging trends and technologies. J. Mater Chem. B **9**(37), 7608–7632 (2021)
23. Haleem, A., Javaid, M., Singh, R.P., Suman, R.: Significant roles of 4D printing using smart materials in the field of manufacturing. Adv. Ind. Eng. Polym. Res. **4**(4), 301–311 (2021)
24. McLellan, K., Sun, Y.C., Naguib, H.: A review of 4D printing: materials, structures, and designs towards the printing of biomedical wearable devices. Bioprinting e00217 (2022)
25. Zhou, H., Dong, G., Gao, G., Du, R., Tang, X., Ma, Y., et al.: Hydrogel-based stimuli-responsive micromotors for biomedicine. Cyborg Bionic Syst. **2022**, 9852853 (2022)
26. Shen, Z., Chen, F., Zhu, X., Yong, K.T., Gu, G.: Stimuli-responsive functional materials for soft robotics. J. Mater. Chem. B (2020)
27. He, Y., Yu, R., Li, X., Zhang, M., Zhang, Y., Yang, X., et al.: Digital light processing 4D printing of transparent, strong, highly conductive hydrogels. ACS Appl. Mater. Interfaces **13**(30), 36286–36294 (2021)
28. Kang, X., Zhang, X.B., Gao, X.D., Hao, D.J., Li, T., Xu, Z.W.: Bioprinting for bone tissue engineering. Front. Bioeng. Biotechnol. **10**, 1036375 (2022)
29. Zhang, B., Li, H., Cheng, J., Ye, H., Sakhaei, A.H., Yuan, C., et al.: Mechanically robust and UV-curable shape-memory polymers for digital light processing based 4D printing. Adv. Mater. **33**(27), e2101298 (2021)
30. Gao, Q., Lee, J.S., Kim, B.S., Gao, G.: Three-dimensional printing of smart constructs using stimuli-responsive biomaterials: a future direction of precision medicine. Int. J. Bioprinting **9**(1), 638 (2023)
31. Miao, S., Cui, H., Nowicki, M., Lee, S.J., Almeida, J., Zhou, X., et al.: Photolithographic-stereolithographic-tandem fabrication of 4D smart scaffolds for improved stem cell cardiomyogenic differentiation. Biofabrication **10**(3), 035007 (2018)
32. Miao, S., Nowicki, M., Cui, H., Lee, S.J., Zhou, X., Mills, D.K., et al.: 4D anisotropic skeletal muscle tissue constructs fabricated by staircase effect strategy. Biofabrication **11**(3), 035030 (2019)
33. Fina, F., Goyanes, A., Madla, C.M., Awad, A., Trenfield, S.J., Kuek, J.M., et al.: 3D printing of drug-loaded gyroid lattices using selective laser sintering. Int. J. Pharm. **547**(1–2), 44–52 (2018)
34. Wu, H., Wang, O., Tian, Y., Wang, M., Su, B., Yan, C., et al.: Selective laser sintering-based 4D printing of magnetism-responsive grippers. ACS Appl. Mater. Interfaces **13**(11), 12679–12688 (2021)

35. Murthy, H., Thakur, N., Shankhwar, N.: Nickel-based inks for flexible electronics—a review on recent trends. J. Adv. Manuf. Syst. **21**(03), 591–624 (2022)
36. Cui, C., Kim, D.O., Pack, M.Y., Han, B., Han, L., Sun, Y., et al.: 4D printing of self-folding and cell-encapsulating 3D microstructures as scaffolds for tissue-engineering applications. Biofabrication **12**(4), 045018 (2020)
37. Quan, H., Zhang, T., Xu, H., Luo, S., Nie, J., Zhu, X.: Photo-curing 3D printing technique and its challenges. Bioact Mater. **5**(1), 110–115 (2020)
38. Deshmane, S., Kendre, P., Mahajan, H., Jain, S.: Stereolithography 3D printing technology in pharmaceuticals: a review. Drug Dev. Ind. Pharm. **47**(9), 1362–1372 (2021)
39. Zhuo, S., Geever, L.M., Halligan, E., Tie, B.S.H., Breheny, C.: A development of new material for 4D printing and the material properties comparison between the conventional and stereolithography polymerised NVCL hydrogels. J. Funct. Biomater. **13**(4) (2022)
40. Davoodi, E., Sarikhani, E., Montazerian, H., Ahadian, S., Costantini, M., Swieszkowski, W., et al.: Extrusion and microfluidic-based bioprinting to fabricate biomimetic tissues and organs. Adv. Mater. Technol. **5**(8) (2020)
41. Colosi, C., Costantini, M., Barbetta, A., Dentini, M.: Microfluidic bioprinting of heterogeneous 3D tissue constructs. Methods Mol. Biol. **1612**, 369–380 (2017)
42. Wu, Y., Wenger, A., Golzar, H., Tang, X.S.: 3D bioprinting of bicellular liver lobule-mimetic structures via microextrusion of cellulose nanocrystal-incorporated shear-thinning bioink. Sci. Rep. **10**(1), 20648 (2020)
43. Ahmed, A., Arya, S., Gupta, V., Furukawa, H., Khosla, A.: 4D printing: fundamentals, materials, applications and challenges. Polymer **228**, 123926 (2021)
44. Weng, S., Kuang, X., Zhang, Q., Hamel, C.M., Roach, D.J., Hu, N., et al.: 4D printing of glass fiber-regulated shape shifting structures with high stiffness. ACS Appl. Mater. Interfaces **13**(11), 12797–12804 (2021)
45. Luo, Y., Lin, X., Chen, B., Wei, X.: Cell-laden four-dimensional bioprinting using near-infrared-triggered shape-morphing alginate/polydopamine bioinks. Biofabrication **11**(4), 045019 (2019)
46. Cui, H., Miao, S., Esworthy, T., Lee, S.-j., Zhou, X., Hann, S.Y., et al.: A novel near-infrared light responsive 4D printed nanoarchitecture with dynamically and remotely controllable transformation **12**, 1381–1388 (2019)
47. Patdiya, J., Kandasubramanian, B.: Progress in 4D printing of stimuli responsive materials. Polym.-Plast. Technol. Mater. **60**(17), 1845–1883 (2021)
48. Shi, H., Wang, C., Ma, Z.: Stimuli-responsive biomaterials for cardiac tissue engineering and dynamic mechanobiology. APL Bioeng. **5**(1), 011506 (2021)
49. Brown, T.E., Anseth, K.S.: Spatiotemporal hydrogel biomaterials for regenerative medicine. Chem. Soc. Rev. **46**(21), 6532–6552 (2017)
50. Lim, K.S., Klotz, B.J., Lindberg, G.C., Melchels, F.P., Hooper, G.J., Malda, J., et al.: Visible light cross-linking of gelatin hydrogels offers an enhanced cell microenvironment with improved light penetration depth **19**(6), 1900098 (2019)
51. Rosenfeld, A., Gockler, T., Kuzina, M., Reischl, M., Schepers, U., Levkin, P.A.: Designing inherently photodegradable cell-adhesive hydrogels for 3D cell culture. Adv. Healthc. Mater. **10**(16), e2100632 (2021)
52. Lv, W., Long, K., Yang, Y., Chen, S., Zhan, C., Wang, W.: A red light-triggered drug release system based on one-photon upconversion-like photolysis. Adv. Healthc. Mater. **9**(21), e2001118 (2020)
53. Watanabe, U., Sugiura, S., Kakehata, M., Yanagawa, F., Takagi, T., Sumaru, K., et al.: Fabrication of hollow structures in photodegradable hydrogels using a multi-photon excitation process for blood vessel tissue engineering. Micromachines (Basel) **11**(7) (2020)
54. Wu, K., Wu, X., Zhang, Y., Chen, S., Qiao, Z., Wei, D., et al.: Semiconvertible hyaluronic hydrogel enabled red-light-responsive reversible mechanics, adhesion, and self-healing. Biomacromol **23**(3), 1030–1040 (2022)
55. Che, Q.T., Charoensri, K., Seo, J.W., Nguyen, M.H., Jang, G., Bae, H., et al.: Triple-conjugated photo-/temperature-/pH-sensitive chitosan with an intelligent response for bioengineering applications. Carbohydr. Polym. **298**, 120066 (2022)

56. Xing, X., Su, J., Liu, Y., Lin, H., Wang, Y., Cheng, H.: A novel visible light-curing chitosan-based hydrogel membrane for guided tissue regeneration. Colloids Surf. B Biointerfaces **218**, 112760 (2022)
57. Liu, Y., Luo, X., Wu, W., Zhang, A., Lu, B., Zhang, T., et al.: Dual cure (thermal/photo) composite hydrogel derived from chitosan/collagen for in situ 3D bioprinting. Int. J. Biol. Macromol. **182**, 689–700 (2021)
58. Min, S.J., Lee, J.S., Nah, H., Kim, S.H., Moon, H.J., Reis, R.L., et al.: Development of photo-crosslinkable platelet lysate-based hydrogels for 3D printing and tissue engineering. Biofabrication **13**(4) (2021)
59. Chua, C.J., Han, J.L., Li, W., Liu, W., Entcheva, E.: Integration of engineered "spark-cell" spheroids for optical pacing of cardiac tissue. Front. Bioeng. Biotechnol. **9**, 658594 (2021)
60. Wang, Y., Cui, H., Wang, Y., Xu, C., Esworthy, T.J., Hann, S.Y., et al.: 4D printed cardiac construct with aligned myofibers and adjustable curvature for myocardial regeneration. ACS Appl. Mater. Interfaces **13**(11), 12746–12758 (2021)
61. Cui, H., Miao, S., Esworthy, T., Lee, S.J., Zhou, X., Hann, S.Y., et al.: A novel near-infrared light responsive 4D printed nanoarchitecture with dynamically and remotely controllable transformation. Nano Res. **12**, 1381–1388 (2019)
62. Fu, J., Liu, X., Tan, L., Cui, Z., Zheng, Y., Liang, Y., et al.: Photoelectric-responsive extracellular matrix for bone engineering. ACS Nano **13**(11), 13581–13594 (2019)
63. Zhao, W., Wang, H., Wang, H., Han, Y., Zheng, Z., Liu, X., et al.: Light-responsive dual-functional biodegradable mesoporous silica nanoparticles with drug delivery and lubrication enhancement for the treatment of osteoarthritis. Nanoscale **13**(13), 6394–6399 (2021)
64. Akhavan, O., Ghaderi, E., Shirazian, S.A.: Near infrared laser stimulation of human neural stem cells into neurons on graphene nanomesh semiconductors. Colloids Surf. B Biointerfaces **126**, 313–321 (2015)
65. Wu, C., Su, B., Xin, N., Tang, J., Xiao, J., Luo, H., et al.: An upconversion nanoparticle-integrated fibrillar scaffold combined with a NIR-optogenetic strategy to regulate neural cell performance. J. Mater Chem. B **11**(2), 430–440 (2023)
66. Griffin, M., Castro, N., Bas, O., Saifzadeh, S., Butler, P., Hutmacher, D.W.: The current versatility of polyurethane three-dimensional printing for biomedical applications. Tissue Eng. Part B Rev. **26**(3), 272–283 (2020)
67. Vanaei, S., Parizi, M.S., Salemizadehparizi, F., Vanaei, H.R.: An overview on materials and techniques in 3D bioprinting toward biomedical application. Eng. Regen. **2**, 1–8 (2021)
68. Varma, M.V., Kandasubramanian, B., Ibrahim, S.M.: 3D printed scaffolds for biomedical applications. Mater. Chem. Phys. **255**, 123642 (2020)
69. Norotte, C., Marga, F.S., Niklason, L.E., Forgacs, G.: Scaffold-free vascular tissue engineering using bioprinting. Biomaterials **30**(30), 5910–5917 (2009)
70. Miri, A.K., Khalilpour, A., Cecen, B., Maharjan, S., Shin, S.R., Khademhosseini, A.: Multiscale bioprinting of vascularized models. Biomaterials **198**, 204–216 (2019)
71. Battisti, M., Vecchione, R., Casale, C., Pennacchio, F.A., Lettera, V., Jamaledin, R., et al.: Non-invasive production of multi-compartmental biodegradable polymer microneedles for controlled intradermal drug release of labile molecules. Front. Bioeng. Biotechnol. **7**, 296 (2019)
72. Sheikholeslam, M., Wright, M.E., Jeschke, M.G., Amini-Nik, S.: Biomaterials for skin substitutes. Adv. Healthc. Mater. **7**(5), 1700897 (2018)
73. Ferri, J.M., Jorda, J., Montanes, N., Fenollar, O., Balart, R.: Manufacturing and characterization of poly (lactic acid) composites with hydroxyapatite. J. Thermoplast. Compos. Mater. **31**(7), 865–881 (2018)
74. Mondal, D., Griffith, M., Venkatraman, S.S.: Polycaprolactone-based biomaterials for tissue engineering and drug delivery: Current scenario and challenges. Int. J. Polym. Mater. Polym. Biomater. **65**(5), 255–265 (2016)
75. Li, M., Chen, J., Shi, M., Zhang, H., Ma, P.X., Guo, B.: Electroactive anti-oxidant polyurethane elastomers with shape memory property as non-adherent wound dressing to enhance wound healing. Chem. Eng. J. **1**(375), 121999 (2019)

76. Liu, J., Zhou, Z., Zhang, M., Song, F., Feng, C., Liu, H.: Simple and robust 3D bioprinting of full-thickness human skin tissue. Bioengineered **13**(4), 10087–10097 (2022)
77. Wang, W., Caetano, G., Ambler, W.S., Blaker, J.J., Frade, M.A., Mandal, P., et al.: Enhancing the hydrophilicity and cell attachment of 3D printed PCL/graphene scaffolds for bone tissue engineering. Materials (Basel) **9**(12) (2016)
78. Lee, S.J., Lee, D., Yoon, T.R., Kim, H.K., Jo, H.H., Park, J.S., et al.: Surface modification of 3D-printed porous scaffolds via mussel-inspired polydopamine and effective immobilization of rhBMP-2 to promote osteogenic differentiation for bone tissue engineering. Acta Biomater. **40**, 182–191 (2016)
79. Zhou, W., Dong, X., He, Y., Zheng, W., Leng, J.: In-vitro and in-vivo studies of 4D printed shape memory scaffolds with bioactive fillers and coating for enhanced bone tissue regeneration. Smart Mater. Struct. **31**(10), 105002 (2022)
80. Hwangbo, H., Lee, H., Roh, E.J., Kim, W., Joshi, H.P., Kwon, S.Y., et al.: Bone tissue engineering via application of a collagen/hydroxyapatite 4D-printed biomimetic scaffold for spinal fusion. **8**(2), 021403 (2021)
81. Nurden, A.T.: Platelets, inflammation and tissue regeneration. Thromb. Haemost. **105**(Suppl 1), S13-33 (2011)
82. Gu, W., Hong, X., Potter, C., Qu, A., Xu, Q.: Mesenchymal stem cells and vascular regeneration. Microcirculation **24**(1) (2017)
83. Arslan-Yildiz, A., El Assal, R., Chen, P., Guven, S., Inci, F., Demirci, U.: Towards artificial tissue models: past, present, and future of 3D bioprinting. Biofabrication **8**(1), 014103 (2016)
84. Carrabba, M., Madeddu, P.: Current strategies for the manufacture of small size tissue engineering vascular grafts. Front. Bioeng. Biotechnol. **6**, 41 (2018)
85. Zhu, J., Wang, Y., Zhong, L., Pan, F., Wang, J.: Advances in tissue engineering of vasculature through three-dimensional bioprinting. Dev. Dyn. **250**(12), 1717–1738 (2021)
86. Zhang, J., Wehrle, E., Rubert, M., Muller, R.: 3D bioprinting of human tissues: biofabrication, bioinks, and bioreactors. Int. J. Mol. Sci. **22**(8) (2021)
87. Lin, C., Zhang, L., Liu, Y., Liu, L., Leng, J.: 4D printing of personalized shape memory polymer vascular stents with negative Poisson's ratio structure: a preliminary study. Sci. China Technol. Sci. **63**(4), 578–588 (2020Apr)
88. Kitana, W., Apsite, I., Hazur, J., Boccaccini, A.R., Ionov, L.J.A.M.T.: 4D biofabrication of T-shaped vascular bifurcation **8**(1), 2200429 (2023)
89. Wang, H.D., Lough, D.M., Kurlander, D.E., Lopez, J., Quan, A., Kumar, A.R.: Muscle-derived stem cell-enriched scaffolds are capable of enhanced healing of a murine volumetric muscle loss defect. Plast. Reconstr. Surg. **143**(2), 329e-e339 (2019)
90. Mase, V.J., Jr., Hsu, J.R., Wolf, S.E., Wenke, J.C., Baer, D.G., Owens, J., et al.: Clinical application of an acellular biologic scaffold for surgical repair of a large, traumatic quadriceps femoris muscle defect. Orthopedics **33**(7), 511 (2010)
91. Kwee, B.J., Mooney, D.J.: Biomaterials for skeletal muscle tissue engineering. Curr. Opin. Biotechnol. **47**, 16–22 (2017)
92. Mulbauer, G.D., Matthew, H.W.: Biomimetic scaffolds for skeletal muscle regeneration. Discoveries **7**, e90 (2019)
93. Downing, K., Prisby, R., Varanasi, V., Zhou, J., Pan, Z., Brotto, M.: Old and new biomarkers for volumetric muscle loss. Curr. Opin. Pharmacol. **59**, 61–69 (2021)
94. Gilbert-Honick, J., Grayson, W.: Vascularized and innervated skeletal muscle tissue engineering. Adv. Healthc. Mater. **9**(1), e1900626 (2020)
95. Baiguera, S., Del Gaudio, C., Di Nardo, P., Manzari, V., Carotenuto, F., Teodori, L.: 3D printing decellularized extracellular matrix to design biomimetic scaffolds for skeletal muscle tissue engineering. Biomed. Res. Int. **2020**, 2689701 (2020)
96. Sonaye, S.Y., Ertugral, E.G., Kothapalli, C.R., Sikder, P.: Extrusion 3D (bio)printing of alginate-gelatin-based composite scaffolds for skeletal muscle tissue engineering. Materials (Basel) **15**(22) (2022)
97. Layrolle, P., Payoux, P., Chavanas, S.: Message in a scaffold: natural biomaterials for three-dimensional (3D) bioprinting of human brain organoids. Biomolecules **13**(1) (2022)

98. Esworthy, T.J., Miao, S., Lee, S.J., Zhou, X., Cui, H., Zuo, Y.Y., et al.: Advanced 4D bioprinting technologies for brain tissue modeling and study. Int. J. Smart Nano Mater. **10**(3), 177–204 (2019)
99. Zhu, W., George, J.K., Sorger, V.J., Grace, Z.L.: 3D printing scaffold coupled with low level light therapy for neural tissue regeneration. Biofabrication **9**(2), 025002 (2017)
100. Apsite, I., Constante, G., Dulle, M., Vogt, L., Caspari, A., Boccaccini, A.R., et al.: 4D biofabrication of fibrous artificial nerve graft for neuron regeneration. Biofabrication **12**(3), 035027 (2020)
101. Orash Mahmoud Salehi, A., Heidari-Keshel, S., Poursamar, S.A., Zarrabi, A., Sefat, F., Mamidi, N., et al.: Bioprinted membranes for corneal tissue engineering: a review. Pharmaceutics **14**(12) (2022)
102. Miotto, M., Gouveia, R.M., Ionescu, A.M., Figueiredo, F., Hamley, I.W., Connon, C.J.: 4D corneal tissue engineering: achieving time-dependent tissue self-Curvature through localized control of cell actuators. Adv. Funct. Mater. **29**(8), 1807334 (2019)
103. Avila, H.M., Schwarz, S., Rotter, N., Gatenholm, P.: 3D bioprinting of human chondrocyte-laden nanocellulose hydrogels for patient-specific auricular cartilage regeneration. Bioprinting **1**, 22–35 (2016)
104. Kalogeropoulou, M.: 4D bioprinting for cartilage tissue engineering: a controlled shape transformation approach (2021)
105. Singh, R., Gupta, N., Vanathi, M., Tandon, R.: Corneal transplantation in the modern era. Indian J. Med. Res. **150**(1), 7–22 (2019)
106. Li, H., Liao, Z., Yang, Z., Gao, C., Fu, L., Li, P., et al.: 3d printed poly(epsilon-caprolactone)/meniscus extracellular matrix composite scaffold functionalized with kartogenin-releasing PLGA microspheres for meniscus tissue engineering. Front. Bioeng. Biotechnol. **9**, 662381 (2021)
107. Pastorek, D., Culenova, M., Csobonyeiova, M., Skuciova, V., Danisovic, L., Ziaran S.: Tissue engineering of the urethra: from bench to bedside. Biomedicines **9**(12) (2021)
108. Gu, X., Xu, Y., Li, S., Wang, Z., Meng, Q., Yu, J.: Preparation of a photocured biocompatible hydrogel for urethral tissue engineering. ACS Appl. Polym. Mater. **3**(7), 3519–3527 (2021)
109. Zhang, K., Fu, Q., Yoo, J., Chen, X., Chandra, P., Mo, X., et al.: 3D bioprinting of urethra with PCL/PLCL blend and dual autologous cells in fibrin hydrogel: An in vitro evaluation of biomimetic mechanical property and cell growth environment. Acta Biomater. **50**, 154–164 (2017)
110. Li, Y.C., Zhang, Y.S., Akpek, A., Shin, S.R., Khademhosseini, A.: 4D bioprinting: the next-generation technology for biofabrication enabled by stimuli-responsive materials. Biofabrication **9**(1), 012001 (2016)
111. Bryant, S.J., Vernerey, F.J.: Programmable hydrogels for cell encapsulation and neo-tissue growth to enable personalized tissue engineering. Adv. Healthc. Mater. **7**(1), 1700605 (2018)
112. Taylor, J.M., Luan, H., Lewis, J.A., Rogers, J.A., Nuzzo, R.G., Braun, P.V.: Biomimetic and biologically compliant soft architectures via 3D and 4D assembly methods: a perspective. Adv. Mater. **34**(16), e2108391 (2022)
113. Kamolz, L.P., Kotzbeck, P., Schintler, M., Spendel, S.: Skin regeneration, repair, and reconstruction: present and future. Eur. Surg. **54**(3), 163–169 (2022)
114. Guan, J., Jia, Y., Zhang, B., Zhao, G.: Application of 4D bioprinting in tissue engineering. Chin J. Tissue Eng. Res. **26**(3), 446 (2022)
115. Chiu, D.T., Demello, A.J., Di Carlo, D., Doyle, P.S., Hansen, C., Maceiczyk, R.M., et al.: Small but perfectly formed? Successes Challenges Oppor. Microfluid. Chem. Biol. Sci. **2**(2), 201–223 (2017)
116. Lee, J.Y., An, J., Chua, C.K.: Fundamentals and applications of 3D printing for novel materials. Appl. Mater. Today **7**, 120–133 (2017)
117. Miao, S., Zhu, W., Castro, N.J., Nowicki, M., Zhou, X., Cui, H., et al.: 4D printing smart biomedical scaffolds with novel soybean oil epoxidized acrylate. Sci. Rep. **6**, 27226 (2016)
118. Chan, B.Q.Y., Chong YT, Wang, S., Lee, C.J.J., Owh, C., Wang, F., et al.: Synergistic combination of 4D printing and electroless metallic plating for the fabrication of a highly conductive electrical device **430**, 132513 (2022)

119. Zhou, W., Qiao, Z., Nazarzadeh Zare, E., Huang, J., Zheng, X., Sun, X., et al.: 4D-printed dynamic materials in biomedical applications: chemistry, challenges, and their future perspectives in the clinical sector. J. Med. Chem. **63**(15), 8003–8024 (2020)

Amisha graduated in Pharmacy from Shalom Institute of Health and Allied Science, Prayagraj, Uttar Pradesh. She is currently pursuing post-graduation in Pharmaceutics discipline from ISF College of Pharmacy, Moga, Punjab. She had completed her SSC and HSC from her hometown in Muzaffarpur, Bihar in 2015 and 2017 respectively. She had one-month industrial training in production department, quality assurance, packing department and inventory control from pinnacle life science private limited in 2019. Presently, she has been engaged in research work for her M.Pharm thesis work on the liposomal gel for dermal application under the supervision of Dr. Amrinder Singh.

Shubham Thakur completed his undergraduate and postgraduate studies in pharmacy at Guru Nanak Dev University in Amritsar in 2016 and 2018 respectively. Presently, he serves as an Indian Council of Medical Research-Senior Research Fellow and is pursuing a Ph.D. in the Department of Pharmaceutical Sciences at Guru Nanak Dev University, Amritsar. His research focuses on the advancement of novel drug delivery systems, aiming to enhance the kinetics, efficacy, and safety of existing drugs. Specifically, his current project involves developing a safe oral formulation for pregnant women. Till now he has published 25 articles and 5 book chapters in Scopus indexed journals and books with h-index of 5.

Amrinder Singh graduated in Pharmacy from ISF College of Pharmacy, Moga, Punjab. He went for post-graduation in Pharmaceutics discipline to Teerthanker Mahaveer University, Moradabad and passed out in 2012. Post this, he taught in a Pharmacy college as a lecturer in Amritsar, Punjab, for about 4 years. Then, he successfully qualified GPAT in 2017 with 96 percentile. After that, he started Ph.D. under supervision of Professor Subheet Kumar Jain in Department of Pharmaceutical Sciences, Guru Nanak Dev University, Amritsar and completed in 2023. He got Senior Research Fellowship from Department of Science and Technology, New Delhi, Government of India to pursue his research work. He has been credited with 22 peer reviewed international scientific publications throughout his Doctorate journey. Presently, he has been engaged as an Associate Professor in Pharmaceutics at Chitkara College of Pharmacy, Rajpura, Punjab, India and is involved in various research, academic and administrative tasks.

Surface-Modified Biomaterials in Medical Device Development

Bindu Soni, Riya Shivgotra, Manjot Kaur, and Shubham Thakur

Abstract Biomaterials refer to materials used in direct interaction with biological systems. Biocompatible and surface-modifiable metallic, polymeric, and ceramic biomaterials are currently used in various applications, allowing for adjustments to improve their performance while maintaining their bulk properties. Surface modification is done to enhance biocompatibility and come into contact as a bioactive substance for certain applications, such as protein surface modification on ceramics, polymers, metals, and composites. Surface modification techniques include calcium phosphate deposition, covalent binding of poly (ethylene glycol) and poly (heparin), and plasma polymerization. Analyzing the surface chemistry, structure, morphology, and topography of biomaterials is crucial in surface modification to improve interactions with blood, fight infection, interact with soft tissues, repair and regenerate nerve cells, manage stem cell growth and differentiation, and interact better with bone. Biomedical devices that can replace or repair damaged tissues and organs depend heavily on the usage of substances that can interact with people's bodies without triggering negative reactions. Although joint replacement surgery is a standard procedure, the inserted biomaterial may not last for a long time. Factors such as unfavorable immune system responses, the development of biofilms, or issues with the implants' fabrication, biocompatibility, manufacturing processes, and their mechanical, chemical, or tri-biological processes may lead to their failure. Altering the surface of biomaterials can prevent these failures and improve the way the body responds to their implantation. Thus, the current chapter aims to show novel methodologies and applications of surface-modified biomaterials in the development of medical devices. It suggests novel studies on extending the lifespan of medical equipment and biomaterials.

Keywords Surface-modifiable · Implantable medical devices · Surgical implant · Implants fabrication

B. Soni · R. Shivgotra · M. Kaur · S. Thakur (✉)
Department of Pharmaceutical Sciences, Guru Nanak Dev University, Amritsar, Punjab 143005, India
e-mail: shubhamdpharma.rsh@gndu.ac.in

S. Thakur
ICMR, New Delhi, India

R. Malviya and S. Sundram (eds.), *Engineered Biomaterials*, Engineering Materials,
https://doi.org/10.1007/978-981-99-6698-1_15

Abbreviations

CVD	Chemical vapor deposition
EPD	Electrophoretic Deposition
Co-Cr	Cobalt-Chromium
MRI	Magnetic Resonance Imaging
CT	Computer Tomography
UHMWPE	Ultra-High Molecular Polyethylene
DDS	Drug Delivery System
PEG/PEO	Polyethylene glycol/oxide
PLA	Polylactic acid
PGA	Polyglycolic acid
TMC	Tri-Methyl Carbonate
PDS	Polydioxanone
PE	Polyethylene
PA	Polyamide
PMMA	Polymethyl methacrylate
PU	Polyurethane
PET	Polyethylene terephthalate
SR	Synthetic Rubber
PTFE	Polytetrafluoroethylene
PS	Polystyrene
PLA	Polylactic acid
PGA	Polyglycolide

1 Introduction

In various fields of medicine, there has been extensive use of biomaterials throughout a significant duration, and their perception has varied over time. Williams gave the first comprehensive explanation of biomaterials in 1987, describing them as "Non-living materials utilized in healthcare devices with a focus on interacting with biologically systems". These substances had to be risk-free, non-carcinogenic, chemically and physiologically stable, and mechanically robust enough to withstand repeated loads during their existence [1]. As advancements in the applications and compositions of biomaterials progressed, these definitions were modified to incorporate these developments [2, 3]. The following examples illustrate the evolution of biomaterial definitions.

- "Synthetic substance utilizes to modify a component of a living system or to interact closely with living tissue."
- "A pharmacologically and systemically inactive material intended for integration with or implantation among biological systems."

- "A nonviable substance employed in a medical apparatus meant to communicate with biological systems."
- "Materials that interact with biological fluids such as blood, tissues, and fluids have been developed that can be used in prosthetic devices, diagnostics, therapeutics, and storage without negatively impacting a living organism and its constituent parts. These materials come from both synthetically produced and naturally occurring substances."
- "Any chemical (apart from a medicine) or mixture of compounds, whether artificial or of natural origin, that can be utilized to treat, improve, or replace any tissue for any period of time, either in its overall terms or as a system component."

Any substance designed by the manufacturer for a medical purpose, whether used alone or in conjunction, including a piece of equipment, an implant, a machine, a material, reagent, software, or other related item, is referred to as a medical device. Medical devices have become valuable tools in disease diagnosis, therapy, prevention and screening, and palliative care and are widely used in various settings, including at home by individuals, in remote clinics by paramedical staff and clinicians, by dentists and opticians, and in advanced medical facilities by healthcare professionals [4]. Recent advancements in manufacturing techniques have made it possible to produce medical devices using different materials (composites, metals, polymers, and ceramics). However, living tissues are sensitive to material surfaces which initiates several immune reactions (inflammation, rejection, and infection).

Long time ago understood that a medical device's success depends on how well it interacts with the host tissues. Significant advances in material science have been made as a result of extensive research into this interface issue, with the goal of enhancing the interface in biological and medicinal applications [5]. The ability to modify a surface of a device in order to give it particular physicochemical properties has emerged as a workable way to produce desired biological reactions. One major benefit of surface modification is that it only affects the device surface, resulting in minor modifications to the devices' bulk physical and/or chemical properties and easier regulatory approval processes. Regardless of the materials' classification, numerous techniques have been established to alter their surfaces to maintain control biological reactions and enhance device performance.

These techniques are utilized to lessen protein adsorption, slow down osseointegration, increase electrical conductivity, and improve wear and resistance to corrosion [1]. The two major types of surface modifications are (i) physicochemical modifications, which involve chemical reactions (oxidation, reduction, salinization, acetylation), etching, mechanical roughening, and changes to the surface's atoms, compounds, molecules, or topography, and (ii) surface coatings (noncovalent, covalent, and thin-film coating), which use materials other than the underlying support that bind the biomolecules [6]. The techniques of surface modification are widely recognized as being easily adopted by existing researchers and medical device companies due to its exciting multi-disciplinary opportunities.

This chapter's main emphasis is on surface modification of biomaterials utilized with devices for medical purposes, since it is currently increasing popularity as a

Table 1 Selection of biomaterial in various therapies

S. no.	Therapy branch	Implants	Material	References
1	Dentistry	Dental implants	Calcium pyrophosphate, Cobalt chromium alloys, Polymethylmethacrylate, Gold, brushite, etc	[8, 9]
2	Orthopedic	Artificial joints, ligaments and tendons	Zirconia toughened alumina, cobalt chromium molybdenum alloy Polyetheretherketone, etc	[10, 11]
3	Cardiac	Heart valves, stents and grafts	Dissolvable zinc stents, ptfe (teflon), cobalt alloy (e.g. Stellite 21, Haynes), Lti pyrolytic carbon polyacetal, Polypropylene knitted (Dacron), etc	[12, 13]

flexible method of creating custom device interfaces. The chapter is structured to begin by going over the criteria for choosing biomaterials to have their surfaces modified and the techniques for doing so, then to look at how biomaterials and their surfaces interact, subsequently various approaches and thorough procedures for surface modification are described, along with legal and ethical concerns. The application of surface-modified biomaterials in medical devices is addressed in the final section, along with the current gaps and potential remedies.

2 Selection of Biomaterials for Surface Modification

Biomaterials have been chosen for surface modification in biomedical applications based on bulk characteristics such as non-toxicity, free from corrosion, minimal degradation, elastic modulus, and wear resistance. A functional group must be attached to the surface in order to change the surface's wettability, sealability, printability, dye uptake, glazing resistance, adherence to different materials, and interaction with the environment's biological organisms, among other biomaterial barrier qualities [7]. Due to the body's extremely sensitive chemical stability, implanted materials ought to possess bio-inert properties to avoid rusting and the unintended release of metallic ions. The selection of biomaterial in various therapies as shown in Table 1.

3 Provision for Biomaterial Surface Modification

Due to their higher affinity for a wider range of proteins than others, albumin, fibrinogen, IgG, fibronectin, and Von Willebrand factor make up the majority of plasma proteins found in utilized biomaterials. The process of protein adsorption and interaction can change the shape of the biomaterial that is produced. Especially fibrinogen, which becomes more adhesive when in contact with biomaterial or implant surfaces, can cause denaturation of biomaterials. In order to prevent neoepitope exposure after cell and tissue interactions and preserve the biomaterial for the duration of its intended duration of storage, it is crucial to change the biomaterial's surface.

Protein adhesion, cell disruption, and inflammatory response are all significantly influenced by the surface chemistry and structure of biomaterials. For instance, in vitro cell adsorption can be produced by changing the surface chemistry of a biomaterial. However, traditional chemistry is not a requirement for the in vivo mechanisms of externally applied biomaterials. Typically made of polymers, ceramics, metals (cobalt-chrome, titanium), and other materials, biomaterials have a variety of surface properties that affect how they react in vivo, ranging from hydrophilic to hydrophobic and tough to soft. In order to construct highly friendly biomaterial architectures, a variety of different techniques, including physical modifications, chemical alterations, and radiation, are employed to alter the surface of biomaterials.

Surface modification methods that significantly affect protein adsorption and biological responses in vitro include those that modify polymer chemistry, wetting capacity, and domain. Additionally, it has been demonstrated that protein adsorption and cellular responses are significantly influenced by the structure and design of biomaterials. Several topical experiments have made use of methods for changing surfaces that have been mentioned in the literature, such as plasma therapy, chemical graft modification, and self-assembled monolayer methods [14]. This instance of a foreign body reaction is typical of how cells in higher organisms react to artificial biomaterials that have been inserted.

4 Interaction of Biomaterial at the Surface

The method that tissues and the implant surface interact is always varying. Within the initial several seconds following implantation, dissolved ions, unbound biomolecules, and water are everywhere over the implant surface. Figure 1 illustrates how the bio fluid surrounding the wound changes in composition and a layer of biomolecules is adsorbing to the surface to begin the healing process. The adsorbed layer then controls how the cells behave once they reach the surface. Eventually, a tissue integration or fibrous capsule will form as the types of cells and their activity on the surface alter. The atomic, molecular, and higher-level physical textural properties operate as contact points for biological elements like proteins, cells, tissues, etc. At the same time, it's crucial to understand that the biomolecules react to the chemical activity of the implant surfaces.

The various bonding modalities connected to each of these biological units have an impact on how a surface is integrated hierarchically into its bonding environment. Chemical species that start distinct levels of bonding have an impact on adhesion qualities. A component experiences various chemical processes depending on the

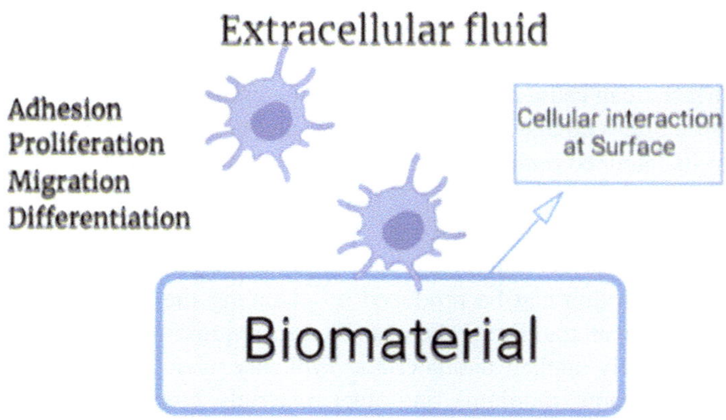

Fig. 1 Biomaterial interaction at the surface

surroundings, which makes it more difficult to understand the specific nature of the interactions [15, 16].

5 Different Techniques Used for Surface Modification

Surface modification involves using physical, chemical, and mechanical techniques to modify or coat a material's surface, resulting in desired properties that differ from the original material. In regenerative medicine and the engineering of tissues, activating the surface is essential to allow for the binding of adhesive biomolecules to scaffolds. Achieving high binding efficiency requires selecting an appropriate technique to conjugate biomolecules or biomaterials onto scaffold surfaces, considering factors such as the presence of reactive functional groups, hydrophobicity, and hydrophilicity. Recent years have seen significant attention given to chemical and physical methods due to their effectiveness in achieving high binding efficiency, while mechanical modification techniques are widely used because of their strong impact on cell adhesion and growth. The various surface modification techniques and different types of cell adhesion molecules that can be used to modify scaffold surface characteristics [17].

Surface modification of biomaterials is performed to create a particular physical and chemical environment that favorably influences the biological response in either soft tissue or hard tissue. Macro, micro, and even nano-scale elements should all be present in the physical environment that promote cell adhesion, proliferation, and migration in situations when tissue integration is required. The functionality of various devices, such as articulating surfaces or cardiovascular apparatus, might occasionally be negatively impacted by textured surfaces, it is crucial to note [18]. The different techniques of surface modification are shown in Fig. 2.

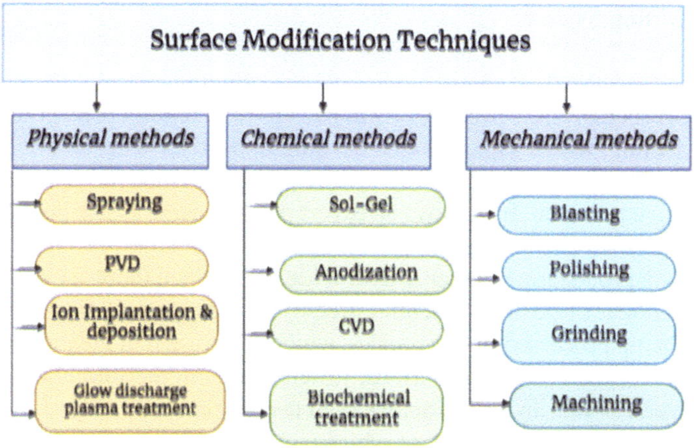

Fig. 2 Different techniques used for surface modification of Biomaterial

5.1 Mechanical Methods

5.1.1 Blasting

Blasting is a common method utilized for sterilizing surfaces, roughen surfaces, or make surfaces smooth. It involves applying harsh, abrasive particles to metal surfaces. In biomedical devices, such as orthopedic implants, the blasted surfaces must be biocompatible. Therefore, ceramic particles like titanium, alumina, or hydroxyapatite are commonly used for blasting [19, 20]. Blasting can produce rough surfaces that strengthen the bond between implants and bone, promoting osseointegration. The chemical makeup of the surface may change as a result of blasting, which could have an impact on biocompatibility.

Medical device surfaces need to be handled carefully to keep them biocompatible after treatment. In addition, nano-rough surfaces can be created by blasting, which are far more resistant to corrosion, wear, and corrosive wear than conventionally grained or sandblasted surfaces. A further approach for producing micron, submicron, and nanostructures on metal surfaces is laser ablation. It has some benefits, including the ability to use it on geometrically difficult implants and a minimal risk of contamination. On titanium implants, laser ablation can make nanoscale grooves that greatly improve osseointegration and increase the extent to which bone develops when the treated implants were in connect with it. Additionally, surfaces that have been ablated by lasers may be extremely hydrophobic and selectively inhibit bacterial adhesion.

5.1.2 Grinding System

The expected electrical grinding system and the grinding's electrochemical operations. In this process, an electrode is positioned above a metallic bond grinding wheel that is conducting, with a gap of around 0.3 mm, or nearly 1/6 the size of the surface of the wheel. The electrode is given a negative potential, while the grinding wheel is given a positive potential. It employs an expert pulse generator. The potential during the process electrolytically breaks down the conductive alkaline machining fluid used in the grinding procedure, producing hydroxide ions. When the work piece receives the proper amount of positive potential, free hydroxide ions (OH-) are attracted to the surface being processed by substances, which are present in the machining fluid. This emphasizes the ions to the surface, which results in the formation of a stable oxidized layer. For our investigation, we used test pieces made of type 316 stainless steel and the Ti6AI-4V alloy, each of which is frequently used as metallic biomaterials. A grinding wheel with a metal-resin-hybrid bond (#8000, diamond abrasives, with a diameter of approximately 2 microns) was used. [21].

5.2 *Chemical Methods*

5.2.1 Sol-Gel Method

For the purpose of developing biomaterials for drugs, the sol-gel method is a straightforward, wet chemical process that doesn't necessitate a high pH or high sintering temperature. The term "sol-gel coating" refers to the colloidal dispersion of solid particles (1–500 nm) in a liquid solution. The techniques of spraying, dipping, spinning, and doctor-blading can be used to apply a sol to a substrate. To create the top layer, the substrate gel is dried or calcined. By immersing a metal sample in a calcium and phosphorus gel at lower temperatures, calcium phosphate layers are created on the sample using this approach. The covering coating is calcined at 400–600 °C depending upon the material since the layer produced is porous and less dense [22].

5.2.2 Anodization

Anodizing is a process that is widely used to treat titanium and its alloys. It is similar to acid etching. In this procedure, the substance is dissolved in an ionic solution of $H2SO_4$, $H3PO_4$, or acetic acid while being exposed to anodic voltage. The process results in the surface of titanium and its alloys developing a thicker oxide layer, increasing their corrosion resistance. However, because this oxide layer is porous, sealing is frequently needed. On titanium substrates, titanium nanotubular structures can develop when fluoride-containing electrolytes are employed. The voltage, electrolyte concentration, and reaction time are some of the anodization parameters that significantly affect the shape and structure of these nanotubular layers.

Researchers have looked into how these anodized structures affect bacterial adherence, cell function, and bone formation. Studies have demonstrated that compared to normal flat titanium substrates, anodized surfaces with titanium nanotubes of 40–60 nm in diameter significantly increase adhesion and proliferation of bone marrow cells. Additionally, calcium deposition on anodized substrates is significantly higher than on conventional titanium substrates. *S. aureus* and *S. epidermidis* are two bacteria that are impacted by the diameter of the nanotubes' effects on bacterial adhesion to nanotubular titanium surfaces. The lowest bacterial adherence is seen on surfaces with 80 nm diameter nanotubes, which are less rigid than ordinary flat surfaces or nanotubular surfaces with tube diameters of 20, 40, and 60 nm.

5.2.3 Chemical Vapor Deposition (CVD)

Numerous commonly used formal techniques include CVD, low-pressure CVD, molecular beam epitaxy, and deposition of physical layer procedures like heat evaporation and laser ablation deposition. It is a technique used to deposit thin films onto substrates using gas-phase chemical processes. The more contemporary methods include sputtering, plasma atomic layer deposition, and plasma-enhanced CVD. In order to construct electronic devices that use electricity, magnetism, and photonics, including their utilization in biological contexts, these procedures are useful for depositing metal, semiconductor, and dielectric layers.

5.3 Mechanical Methods

5.3.1 Thermal Spray Technique

The thermal spray technology is rapidly evolving, with exciting developments focused on innovative uses of coatings. One area of particular interest is the application of coatings in cutting-edge energy generation processes, biomaterials, electronic-based features, and self-disinfecting surfaces enabled by photocatalysis, electrolysis, and various other applications. These advancements in thermal spray technique are driving progress in multiple directions and hold great promise for the future. [23].

5.3.2 Ion Implantation and Deposition

Ion implantation and deposition, a method that includes bombarding a substrate with high-energy ions to modify its surface properties. The deposition technique is a crucial step in micro- and nanofabrication. This process involves adding a layer of material to a bare silicon wafer to prepare it for subsequent lithography steps. Deposition can be performed using different methods, including plasma-based and non-plasma-based techniques.

5.3.3 Electrophoretic Deposition

The process of movement of charged particles in the electrolytic solution is necessary for the electrophoretic deposition (EPD) process, which has been intensively investigated. In an electric field, ceramic particles acquire charges whether they are in an aqueous or non-aqueous medium. When the initial surface oxide layer provides sufficient protection, depositing nanoparticles onto the surface of a material can enhance its resistance to wear and corrosion, particularly in the case of titanium implants. EPD system is used to achieve this, with a suspension containing uniformly dispersed powder particles in an aqueous or non-aqueous medium, as well as the appropriate anode and cathode electrodes. Hence, better deposition is achieved with a homogenous particle dispersion that possesses a suitable conductivity and medium dielectric constant [24].

6 Methods of Surface Modification of Biomaterial

Different methods with their advantages and disadvantages of biomaterials used in medical devices as shown in Table 2.

Table 2 Advantages and disadvantages of metals, ceramics, and polymers in medical devices

Different materials	Metals	Ceramics	Polymers	Composites
Advantages	Strength and Durability Biocompatibility Corrosion Resistance Ease of Fabrication Radiopacity	Biocompatibility Wear Resistance Corrosion Resistance Strength and Hardness Radiolucency	Biocompatibility Flexibility Lightweight Versatility Durability	Non-toxic Strength and durability Light weight Moldability Radiolucency
Disadvantages	Metal Allergies Thermal conductivity Costly	Lack of Flexibility Brittleness Rough surface finish	Limited temperature range Moisture absorption Biocompatibility	Limited material option Lack of standardization Limited shelf life Difficulty in manufacturing

6.1 Metallic Biomaterials

The widespread usage of metallic materials in biomedical technology is a result of their favorable mechanical and chemical characteristics. Stainless steels, cobalt-chrome, the metals titanium, and their alloys are among the most widely utilized. In load-bearing implant applications, these metals can be alloyed to tune attributes among the most often utilized metals are titanium and its alloys. to satisfy biological requirements. Surface modification techniques can be employed to enhance biocompatibility, reduce osseointegration time, and prevent corrosion without altering bulk properties.

Stainless steel (SS 316L), titanium alloy (Ti-6Al-4V), and cobalt-chromium (Co-Cr) alloy have attracted the most interest for orthopedic applications among the reported metallic biomaterials. It has been attempted to alter the bulk characteristics of metallic biomaterials to have mechanical properties similar to those of native bone in order to lessen stress shielding at the interface between tissue and metallic biomaterials. The recommended material for dental implants in dental applications is commercially pure titanium (CPTi). because of its better properties and the escalating cost of Pd. Titanium and its alloys have low Young's modulus, which is seen as a biomechanical advantage for replacing hard tissues. In cardiovascular implants, certain alloys, such as nickel-titanium alloy (Nitinol), have drawn increased interest in magnetic resonance imaging (MRI) due to their inert, robust, and non-magnetic qualities [25, 26].

6.1.1 Steel

Since stainless steel has a high level of corrosion resistance and strength, it is frequently utilized for biomedical implants. To keep the nickel content low, nickel-free stainless steel is used, and nitrogen is added to increase biocompatibility. Despite stainless steel's lower biocompatibility and corrosion resistance compared to titanium, temporary bone fracture therapies and implant studies frequently employ it since it is more affordable. Custom-sized stainless-steel implants can be created using 3D printing, and 3D dental implants can be created using liquid phase sintering.

Implants composed of pure metals, which had lesser corrosion resistance and mechanical strength, had been developed before stainless steel entered the biomedical business. The two types of stainless steel utilized in biomedical applications are standard stainless steel and Ni-free stainless steel, which is devoid of nickel. Stainless steel is widely used for bone fracture treatments, such as screws, nails, and fracture plates, and for fabricating durable implant trials. Compared to titanium, stainless steel has inferior biocompatibility, osseointegration, and corrosion resistance [27].

6.1.2 Titanium and Ti-Based Alloys

The Titanium and Ti-based alloys are well suited for usage with a high pace of loading than stainless steel for the reason that they have a higher strength-to-weight ratio. In comparison to steel, titanium implants have a significantly stronger bond with tissues because of the titanium dioxide layer's high dielectric constant, which forms quickly on the surface of bare titanium. It may increase the mechanical durability of titanium alloy by annealing, quenching, and thermal aging, and it can be further modified by alloying other elements like aluminum and niobium. For functionally demanding anatomically complex locations like cranio-maxillofacial surgery, 3D-printed titanium implants can be tailored.

Techniques for melting electron beams and directly sintering metal have been used to create customized titanium prostheses. To reduce stress shielding, porosity might be added to the implant's structure. To reduce stress shielding and encourage tissue regeneration or vascularization, porosity might be added to the implant's structure. The structure is rich in interconnected grooves and has 3D Ti-6Al-4V dental implants' Young's modulus gradient changes from 104 GPa at the metal inner core to 77 GPa for the very porous outer shell when gradient porosity is generated utilizing the selective remain sintering procedure [28, 29].

6.1.3 Cobalt-Based Biomaterials

Since Co-Cr alloys have a higher wear resistance compared to Ti alloys, they frequently used in prosthetic hip joints. Although they are less biocompatible and less capable of osseointegration than Ti, their higher elastic modulus and stiffness result in higher stress shielding compared to Mg, Ti, and Ti alloys. Ti is frequently utilized in clinical settings for components higher stress shielding compared to Mg, Ti, and Ti alloys, while Co-Cr is the preferred material for components that do not come into contact with the bone [30]. At the point where Co-Cr and Ti come into contact, metal corrosion and shredding remain serious problems.

6.1.4 Bio Implants Based on Tantalum

Tantalum has been studied for usage in a range of applications where outstanding durability against corrosion even under acidic conditions and biocompatibility in biomaterials are required. Tantalum's ability to resist corrosion is because a protective coating of natural, stable Ta_2O_5 forms over the implant surface. The functioning of porous tantalum to connect with bone makes it a promising material for use in artificial joints as a polymeric matrix or includes a coating on titanium and implants are constructed from stainless steel to improve osseointegration and prevention from corrosion [31]. Tantalum has a density of 16.6 g/cm^3 and an elastic modulus of more than 186 GPa Because they are significantly different from native cortical (12–18 GPa) and cancellous bone (0.1–0.5 GPa) in these respects, these features are detrimental when used in orthopedic implants.

It is challenging to produce this metal in large quantities due to its refractory properties, especially given its unusually high melting point of approximately 3017 °C. Due to Ta's high temperature conductivity, patients who undergo cranioplasty may also experience temperature-dependent headaches. To meet the growing need for orthopedic applications, metallic bioimplants have been developed using cutting-edge manufacturing techniques. An efficient method to create implants of the proper size may be discovered through a three-dimensional visualization of the patient and printing of the modified implant. Laser sintering is a technique that can be used to produce high-fidelity copies of pre-packaged implants using reasonably priced 316L stainless steel. Another use of liquid phase sintering is the fabrication of stainless-steel 3D dental implants.

Current Challenges with Metals

The utilization of 3D printing in numerous industries, including building, aerospace engineering, and clothing, has significantly advanced the discipline. The development of sophisticated multi-material scaffolds for tissue regeneration and the fulfillment of the needs of personalized medicine in the medical and healthcare sectors are all made possible by 3D printing, which has the potential to revolutionize tissue and organ engineering. [32]. It is clear that a key aspect of 3D printing is its capacity to produce implants that are exactly suited to the anatomy of the patient. This is especially true in surgeries to treat craniofacial ruptures or fractures, for children with dwarfism, or in cancer patients, when the implant is tailored to match the excised tissue and may lessen pressure exerted on the existing bone compared to a travail-customized implant.

However, combining these two approaches offers the most promise for personalized therapy [33]. The model is then used to provide data for the implant's high-accuracy quick prototyping, as well as to prescribe the implant's macroscopic features and give the material a desired structure. An example of this is the modification of porosity, in which the dimensions, orientation, size, and connectivity of the pores may be changed to promote the formation of capillaries, the ingrowth of tissues, and the supply of nutrients to support the developing tissues. Theoretically, printing many materials at once would allow for the construction of complex objects like tissues and organs utilizing a single technique. Metals, synthetic polymers, glass, ceramics, active molecules like proteins and factors, and living cells and organic materials can be found in one particular structure. [34].

The excessive structural rigidity of Co-Cr alloys can be mitigated by employing 3D printing to include nano- and micro-geometry into the bulk of the alloy. This lessens the stiffness differential between the alloy and bone and lowers the elastic modulus. He was able to effectively create his Co-Cr implants with the appropriate macro-geometry and interconnected pore architecture using his EBM, the preferred method for his 3D printing of Co-Cr alloys. After 26 weeks in adult sheep femurs, the implants were shown to have satisfactory overall bone-implant connection. The implant's surroundings gradually underwent tissue ingrowth and densification, and the generated bone's mineral crystallinity, apatite to collagen ratio, and carbonate to phosphate ratio differed between Co-Cr and Ti-6Al. It was in the range of 4 V implants [35].

Printing implants, both solid and mesh, is possible using EBM. Co-29Cr-6Mo alloy mesh-strut and foam-ligament microstructures both produced columnar-directed Cr23C6 precipitate architectures that were spaced about 2 m apart in the build direction to represent the directional solidification of solid cylindrical components during the assembly process in the melt pool. While similarly manufactured solid Ti-6Al-4V implants displayed acicular platelets in the - phase, mesh and foam implants largely displayed residual 0-martensitic phase. It displayed a fine cellular microstructure with enriched Mo and depleted Co grain boundaries when the CoCrMo implant was created using SLM. This provides the implant more corrosion resistance than a cast alloy by reducing carbide precipitation and the creation of the Martensitic phase on its surface. These characteristics, which also limit the discharge of metal ions into the peri-implant environment, lower the possibility of developing metallosis. Rates of ion release and corrosion were correlated with the number of laser melt pool borders [26].

There are many technological obstacles that are keeping 3D printing from developing. The presence of a vascular network poses the toughest challenge from a biological standpoint. An adequate supply of nutrients, correct gas exchange, and effective waste disposal are necessary for the growth of 3D tissues or organs during perfusion; these processes would not occur in the absence of such a network. As a result, the artificial organ's overall efficacy and cell longevity would be affected. However, there is still a significant obstacle to overcome before a system can successfully deliver nutrients, growth hormones, and oxygen to the cells without interfering with their metabolic activities [36].

Despite multiple attempts by researchers to develop vascular trees using computer models, it has not yet been possible to produce fractured veins with branched channels that are advantageous mechanically. For instance, electro-hydrodynamic inkjet printing (e-jetP) has the ability to generate prints with far higher resolution than traditional inkjet printing techniques. By utilizing an electric field to overcome surface tension, a metallic nanoparticle solution, such as Ag, Cu, Au, or Co, is evacuated as droplets in e-jetP [37]. Rapid layer-by-layer assembly with fine control over the form, dimension, and resolution of micro- and nano-scale features is made possible by accelerated solvent evaporation. Although there are currently few macroscopic implants that utilize this technology, these forms may be ideal for implantable electrodes and sensors [38].

Using Additive Manufacturing (AM) Surface Modifications on Metallic Biomaterials

For all metallic devices to perform their function efficiently and to endure for a long time, AM-based surface changes of metallic biomaterials are essential [39]. Passivating materials (Al, Cr, Ti) are added, as well as surface energy, topography, and crystalline structure to increase osseointegration, biocompatibility, and prevent corrosion without affecting the bulk properties of implants [40, 41]. Although including roughness to implant surfaces may reduce their mechanical properties, it can improve tissue integration [42].

It has been demonstrated that porous materials with purposeful topology promote adhesion and advantageous differentiation. Based on computer tomography (CT) images, AM is used to build biomimetic surface structures that have substantially identical surface characteristics to those found in living cells. The most prevalent technique for preventing corrosion is the incorporation of self-passivating materials (Ti, Cr, and AL). Implant of the lifespan is increased by the desirable wear qualities of some alloys, such as CoCrMo, as opposed to softer metals like titanium. But even these artificial limbs can become worn down gradually, releasing cobalt and chromium ions that can lead to metallosis and osteolysis. Degradation of UHMWPE, which leads to the creation of microparticles, might come from the incompatibility of hardness between the metal and the polymer.

6.2 Ceramic Biomaterials

Ceramics are utilized in various medical fields such as dentistry, orthopedics, and medication delivery. They are very helpful in the construction of bone scaffolds for the vital repair of fractures and other injuries in which the normal healing capacity of the bone is compromised. Ceramics are also used in dental procedures for enamel and root replacement, bone cement for minor fractures, ceramic-on-ceramic articulating surfaces for joints, and the hip, knee alignment, and shoulder area implants as coatings healing, and other drug release applications [43]. Ninety-nine percent of the calcium in the human body is found in bones and teeth, making it the fifth most common element in the body. The fact that many biomedical ceramics are calcium-based is therefore not surprising. Both bone and teeth are mostly made of ceramic materials, which have the right chemical makeup and crystallography for these uses.

Due to the similarities in chemical and crystal structure, as well as in micro- and nano topography, biomimetic devices are made that replicate the appearance and functionality of biological tissue, thereby preventing inflammation or rejection brought on by immune reactions to foreign objects [44]. The additional benefit of biomimicry is that it breaks down into ions and calcium phosphates the fact that the body is able to rapidly resorb or excrete. Whether used as a restorable material, degradation has a minor impact, and the rate of product absorption can be modified in accordance with the speed of healing of the tissue by altering the surface, the structure of the crystals, or the calcium phosphate phase. Calcium phosphate scaffolds have a high degree of elemental resemblance to actual bone in applications that come into touch with it, although they lack trace elements like Si^{4+}, Mg^{2+}, Sr^{2+}, Zn^{2+}, Cu^{2+}, and Fe^{2+}/Fe^{3+}, among others. Recent studies have concentrated on ionic elements that are naturally present in bone that can be added to bone scaffolds. Further improving the biomimetic features of ceramics in biomedical applications are the ionic' roles in angiogenesis, osteogenesis, and osteo induction [45]. Different ceramic materials used for surface modification of biomaterials are given below.

6.2.1 Alumina

Alumina is a bio-inert material with excellent immunological compatibility, good biocompatibility, and protection against rust. It is utilized in joint prosthesis, including the acetabulum and femoral heads in hip arthroplasty on worn surfaces due to its strong mechanical resistance to wear and polish ability. Two components are often polished together throughout the manufacturing process to achieve the best possible agreement between them. This lessens the production of harmful residues for the patient as well as friction between joint components. In order to attain lower wear values, it is also typical to attach an alumina femoral head with an ultra-high molecular polyethylene (UHMWPE) acetabular component.

6.2.2 Zirconia

With the use of proximal femur, zirconia was first examined as a biomaterial in the late 1960s. The combination of zirconia and yttrium, known as Tetragonal zirconia polycrystals, often known as TZP (tetragonal zirconia polycrystals), is the most common widespread after testing various chemical composition possibilities and is most frequently utilized in femoral heads [46]. It is one of the ceramic materials used in prosthetic equipment most frequently because of its mechanical and chemical qualities. It combines with oxygen and produces the biocompatible zirconia oxide (ZrO_2) as a result. Due to their great biocompatibility, good flexural strength, and fracture resistance, ZrO_2 implants are mostly employed in femoral heads for hip arthroplasty and dental implants.

6.2.3 Hydroxyapatite

Apatite crystals constitute the minerals found in bone tissues. The chief inorganic constituent responsible for the development of bones and teeth is a high hardness calcium phosphate salt, known as HA. Due to its chemical makeup and structural similarity with natural HA, HA ceramic exhibits exceptional biocompatibility.

6.2.4 Carbon

Carbon implants have an advantage over metallic implants in that they do not experience fatigue. However, because of their fragility and low tensile strength, they are not suitable for situations that necessitate support for substantial loads. Although it is legal to use it in blood-contact devices, it is still a good idea to double-check.

6.2.5 Bio-Glass and Vitro-Ceramic

A versatile and varied class of biomaterials, bio-glass has various capabilities depending on its composition. In order to generate a substance that might adhere to bone, Larry Hench originally developed it in 1969. A type of glass commonly known as "Bioglass" or "45S5" is composed of 46.1 mol% SiO_2, 24.4 mol% Na_2O, 26.9 mol% CaO, and 2.6 mol% P_2O_5. These biomaterials are used in prosthesis when highly hard tissues are required, such as in dentistry. Vitro-ceramic glass is ideal for application in dental implants. and bone development due to its strong mechanical qualities, good biocompatibility, bioactivity, and lack of toxicity. There currently exist a number of variations of the original 45S5 composition, each of which has a particular function or use as a biomaterial based on its chemical make-up.

Current Challenges with Ceramics

Because they perform better than metals and polymers in terms of compressive strength and corrosion resistance, additively produced ceramics are mostly used in dental applications. However, resorbable drug delivery systems are the main focus of surface modification of ceramics through AM. The low breakdown temperature of the additional material is one of the major difficulties, which makes it challenging to sinter an AM ceramic loaded with pharmaceuticals or biomolecules. The ceramic powder particles are held together by a binder, but they are nevertheless prone to bulk disintegration and the early release of loaded molecules.

Using AM Surface Modifications on Ceramics Biomaterials

In order to achieve consistent release profiles and ensure implant stability in vivo, surface modification techniques can be employed. One such approach is to introduce layers of polymeric coatings to the surface to prevent the sudden release of medication. Another technique involves coating ceramics with polymers and dosing them with desired medicine after calcination to regulate drug elution. Trace elements have also been utilized as additives or dopants to accelerate bone development and vascularization. When the drug delivery system (DDS) is subcutaneous or gastrointestinal, several binder formulations have been employed to create components with a bulk interior chemistry and a less soluble, polymer-enhanced outer surface chemistry. [47].

Mesoporous silica/bioglass scaffolds can be surface-modified to control the release profile of medications that have been adsorbed. The potential of surface-modified 3D-printed composite scaffolds was established in a study looking at bone replacement therapy for patients needing surgical bone removal due to osteoarticular pulmonary disease. These printed scaffolds are intriguing for further investigation in the area of bone treatment since they can compete with CaP scaffolds while maintaining a consistent drug release profile.

6.3 Polymeric Biomaterials

Polymers are extensively utilized as biomaterials in the field of biomedical engineering for various purposes, such as cartilage and bone repair, articulating surfaces for hips, knees, and shoulders, drug delivery systems, skin coatings for burn treatment, and optical applications. One of the most impressive qualities of polymers is their ability to biodegrade into components that can be processed by the body. Biodegradable polymers like polyethylene glycol/oxide (PEG/PEO), polylactic acid (PLA), polyglycolic acid (PGA), tri-methyl carbonate (TMC), and polydioxanone (PDS) are frequently employed for this purpose [48–50]. Medical applications utilize a variety of polymers, including polyethylene (PE), polyamide (PA), polymethyl methacrylate (PMMA), polyurethane (PU), polyethylene terephthalate (PET), synthetic rubber (SR), polytetrafluoroethylene (PTFE), polystyrene (PS), polylactic acid (PLA), and polyglycolide (PGA). The foundation for composites such as PEEK reinforced with glass or carbon or a variation like UHMWPE can be made from some of the above-mentioned polymers.

Latex, cellulose, gums, and starch are naturally occurring polymers, synthetic and semi-synthetic polymers with biodegradable or non-biodegradable polymers are the three categories of polymers now in use. Syringes and other disposable medical equipment, orthopedic uses, contact lens and corneas, sutures, cardiovascular devices, membranes, dental reconstruction, and adhesives are just a few of the medical fields where polymers are commonly used. It is also very important to consider the degree of crystallinity when choosing a polymer for medical purposes. The three states of polymers in this situation are amorphous, crystalline, and semi-crystalline [26].

6.3.1 Natural Polymers

Natural rubber latex is a colloidal system best recognized for its use in dental tools, condoms, and other medical supplies. Extensive research and documentation have been conducted on allergies to natural rubber latex. However, a recently developed manufacturing technique that avoids the use of carbamates or sulfur has resulted in a more biocompatible product. This development has opened up new possibilities for using the product as a delivery system for drugs, proteins, and nanoparticles, as well as for aiding in the controlled formation of bones and the healing of soft scar tissue [26].

6.3.2 Synthetic Polymers

Polyolefins are plastic resins that are produced from either propylene or ethylene, depending on the type of polyolefin. These plastics are hydrophobic and inert, meaning that they do not break down in the body. Polyethylene, or PE, is a type of polyolefin that comes in five different variants: high-density polyethylene (HDPE), low-density polyethylene (LDPE), linear low-density polyethylene (LLDPE), ultra-low-density polyethylene (ULDPE) and ultra-high-molecular weight polyethylene (UHMWPE). PE can be processed by various methods such as blowing, extrusion, and injection molding, utilized in sutures and netting and is physiologically inert. UHMWPE, which is one variant of PE, is commonly used for artificial joint surfaces because of its high impact strength, chemical stability, and good biocompatibility. However, the release of particles due to the friction of the components can cause reactions in the body that can lead to mechanical failure of the prostheses.

To resolve this issue, highly cross-linked polyethylene has been used, and the material is treated to gamma radiation to improve its biocompatibility and wear resistance. Antioxidants like vitamin E were added to the biomaterial to produce the second generation of cross-linked UHMWPE. Numerous medical fields, including orthopedics, cardiology, and neurology, use UHMWPE. Some examples of products made of UHMWPE are surgical cables for bone fractures, high-strength orthopedic sutures for soft tissue repair, catheters, stent grafts, heart valves, and disc replacements for spinal repair. Similar to polyethylene, polypropylene is a type of polyolefin that is biologically inert and used to manufacture nets and sutures.

6.3.3 PVC

It needs to be handled carefully when used in medical applications. The application of stabilizing and plasticizing agents causes the receptor to become poisonous in an unwanted way. It is typically utilized in catheters, blood bags, and tubes.

6.3.4 Silicones

These are hydrophobic, exhibit without additionally the application of plasticizers, biological stability and depending on their intended use, elicit completely different biological reactions in the host. It is frequently utilized in ophthalmological applications, breast implant encapsulation, and the induction of an inflammatory response in the synovial membrane in intra-articular implants. Hepatic cancers have been linked to silicone oil residues.

6.3.5 Acetal

It is a hard plastic substance made from formaldehyde that is highly durable and has a low coefficient of friction. This makes it a useful material for machining to create prototypes of medical devices and parts. Acetal is susceptible to the radiation used in sterilization, so it should be kept in mind that this could make the material brittle and easily break.

6.3.6 Polyamides

The most widely used synthetic polyamide used in medicinal uses is nylon. Due to its great tensile strength, it is utilized in suture lines. A composite material of nylon and PU with nylon strength and elasticity makes up the balloons of the catheters used in angioplasty. Recent studies have concentrated on synthetic polymers with assurance for use in soft and hard tissue regeneration, such as polycaprolactone (PCL), polypropylene fumarate (PPF), polyether ether ketone (PEEK), and functionalized polyurethanes [50, 51]. Furthermore, due to their attraction as naturally occurring biopolymers or their degrading properties, natural polymers including collagen and fibrin as well as polysaccharides like alginate, silk, hyaluronic acid, and chitosan have been investigated for biomedical uses.

Collagen, the most common protein and naturally occurring polymer in the ECM, is commonly used for surface modification. Cell mobility is supported and adhered to by collagen. The less biodegradable and even permanent polymers are polytetrafluoroethylene (PTFE) and poly (methyl methacrylate) (PMMA). PMMA, with its excellent optical properties, is used for contacts and intraocular lenses. However, foreign body reactions can occur, leading to the need for surface modification [52]. PMMA is used frequently in bone cement to anchor and secure implants. In particular, the difficulties of enhancing ocular compatibility and maintaining the duration of implantable ocular devices, the organ's shape, and the primary physiological roles of the component tissues influence the choice of biomaterials for ocular implants.

Current Challenges with Polymeric Biomaterials

Co-polymerization allows for greater customization of a polymer's chemical and mechanical properties, including biodegradability, toughness, and stiffness. To meet the requirements of each application, a different composition is used. Devices like sutures or bone screws, for instance, require stronger toughness or higher modulus, and include a rate of degradation which is comparable to the rate of tissue healing. Due to its excellent biocompatibility and ease of modifying its degradability, poly lactic-co-glycolic acid or PLGA), which degrades compared to formulations with a greater glycolide concentration, significantly more slowly, is one of the most often used polymers. Polymers frequently have high surface energies that make implants less wettable and without charges that promote cell and protein attachment. Its restricted bioactivity may hinder cell adhesion, prolonging the healing process, and perhaps causing the development of fibrotic tissue.

Using AM Surface Modifications on Polymeric Biomaterials

Polymeric biomaterials have limited biomedical applications due to the lack of surface characteristics. While AM techniques can modify hatch and strut dimensions, they minimal changes to surface energy. However, through nano- and micro-topography, crystallography and modifications to chemical composition including AM can be used to modify surface energy. For instance, type I collagen was found to promote more effective bone replacement than CaP or CaP/5% wt.% collagen scaffolds when it was added to the binder of a ceramic 3DP process [53]. Similarly, according to plasma or other chemical methods to attach functional groups can increase cell adhesion and proliferation, but are lengthy and require post-processing [54]. AM can create drug delivery systems with specific surface chemistries, topographies, and morphologies, which affect the rate of device resorption as well as the rate at which the drug dissolves into the tissue throughout it. The increased surface roughness can also increase the fixation of the implant [55].

6.4 Composite Biomaterials

A composite material is a material formed by combination of two or more resources with the aim of achieving properties that are greater than those of the individual constituents. The composite material consists of a base material, the reinforcement, and a matrix. For instance, carbon fibers can be added to a polymer to increase its hardness, mechanical strength, and fatigue resistance. Bioenergetics and bioactive ceramic materials are often used to produce composite materials that improve mechanical strength and bioactivity. For example, zirconia and hydroxyapatite (HA) can be used in combination to enhance HA's mechanical qualities while maintaining the ability to attach to bone tissue. Ceramic coatings are applied to metal-based composite implants to increase biocompatibility, strengthen the implant, and facilitate the growth of bone on its surface.

Ceramic reinforcements in polymer matrix implants, including HA, bio-glass, or calcium phosphates, can boost biocompatibility and raise the elastic modulus of the base substance, resulting in the mechanical properties of implants a closer resemblance to those of bone. This reduces the likelihood of stress shielding. Additionally, unlike metal implants, polymer-based composite materials have advantages over them, such as the lack of corrosion and toxicity brought on by the release of nickel or chromium metal ions, which could cause allergic reactions in patients. When compared to metal alloys and ceramic materials, polymer composites are suitable for diagnostic procedures like computerized tomography and magnetic resonance imaging (MRI), and do not obstruct radiographs [22].

6.4.1 Ceramic-Ceramic Composites

As comparison to HAP alone, biocomposites with a bioactive glass layer offer higher adhesion and mechanical strength while retaining the bioactivity of HAP. The Ti-6Al-4V with bioglass-apatite coatings have been the subject of numerous research, which have revealed better mechanical properties. Particularly, ZrO_2 and $Y2O_3$ composites on Ti-6Al-4V have shown enhancement on mechanical characteristics and increased bioactivity [42, 43]. It has also been discovered that bioceramic-free composites, like porcelain wollastonite, have improved mechanical characteristics. Also, an animal investigation using composites of HAP and bio-inert alumina demonstrated excellent osseointegration with bone [56, 57].

6.4.2 Polymer-Ceramic Composites

Metal oxides like ZnO, $Al2O_3$, TiO_3, and SiO_2 are examples of inorganic materials that can be coupled with novel, high-performance materials. Thanks to research on polymer (organic)-inorganic composites. To enhance adhesion and microstructural homogeneity, a homogeneous PEEK/bioactive glass composite coating was created for NiTi. The PEEK polymer, which was present in the composite, provided the material with good resistance to chemical and tribological effects as well as high strength, according to the researchers [58]. The different materials used for manufacturing biomaterials for medical devices are as shown in Fig. 3.

7 Sterilization of Medical Devices and Biomaterials

Sterilization is a significant development in the production of many medical devices, as most devices must be free of living organisms and pyrogens to ensure patient safety. However, the choice of sterilization method is often not given due consideration during product development, and it is sometimes assumed that the process will be straightforward. There are many sterilization methods available, including traditional techniques such as autoclaving and ethylene oxide (EO) sterilization, as well as newer technologies like gamma irradiation and sterilization utilizing gas plasma at low temperatures. Every technique has both advantages and disadvantages, and no one approach works for all kinds of medical equipment [59].

Sterilization can affect the physical and chemical characteristics of biomedical polymers, and can lead to the production of toxic degradation products. Some sterilization methods, such as autoclaving, may not be suitable for certain types of biomedical polymers due to changes in mechanical characteristics or the creation of degradation products. EO sterilization is widely used in industry due to its ability to kill bacteria, spores, and viruses at low temperatures, but it has drawbacks such as residual gas and toxicity. Gas plasma sterilization is effective at sterilizing many materials, but it uses oxidative chemicals that can damage some biomedical elastomers [60].

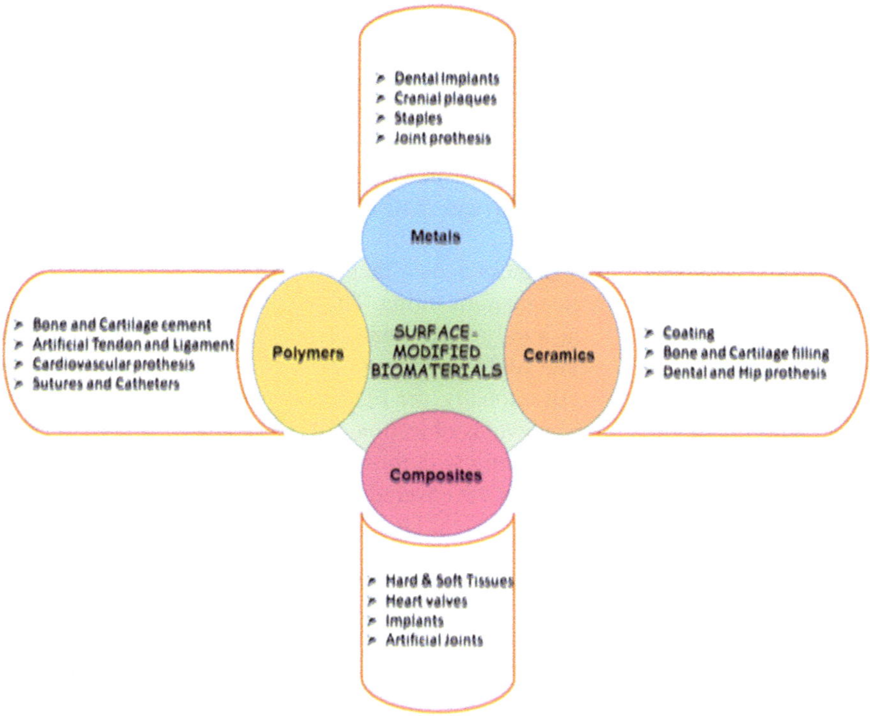

Fig. 3 Materials used in medical biomaterials

In conclusion, the choice of sterilization method is an important consideration in the development of medical devices. Each method has its own benefits and drawbacks, and it is essential to carefully consider the sterilization's effect on chemical and physical characteristics of biomedical materials [61].

8 Regulatory Issues Regarding Biomaterial in Medical Device

The advancements in additive manufacturing (AM) technology, particularly in creating nanoscale features and interacting with the human body, have opened up a range of possibilities [62, 63]. However, the increasing use of nanomaterials in AM also raises new concerns related to safety and regulation. The safety risks associated with nanomaterials are still not fully understood, in particular when these materials come into interaction with membranes and body fluids. Nanoparticles have been found to pose a specific risk owing to their extensive surface area, which can significantly alter the means by which chemical reactions react with their surrounding atmosphere.

Additionally, these particles can pass through defenses like the blood-brain barrier because they are small enough. Regulatory restrictions are required for AM materials that come into touch with the human body in order to address these risks. The FDA has recognized the value of personalized medicine and the potential of AM in the next years. Devices developed using AM methods were to be subject to a set of regulatory standards to be established by the FDA. In May 2016, the FDA released a drafting guidance for 3D-printed medical devices that included its most recent opinions on device design and testing [64, 65].

The primary obstacle the business is now facing is acquiring FDA certification for the materials used in additive manufacturing. There needs to be minimal variance in mechanical attributes from build to build and testing criteria need to be established. There are now just three ASTM standards in use for testing materials made via additive manufacturing. Additionally, the FDA is responsible for ensuring that patient-specific devices, also known as objection-of-care products, adhere to regulatory standards. These devices, along with those used for pharmaceutical and cell treatment produced via AM, are not yet covered by the drafted guidelines [25].

9 Future Prospectives

Today's implantable devices, although being the ultimate achievement of biomaterial science and the outcome of numerous years of research, struggle with the same challenges that have only permitted implants to be implemented as a support to non-surgical or grafting approaches. Despite being in its early stages, additive manufacturing has already caught the attention of biomaterials researchers as a fresh and potent tool for creating new devices and modifying the basic manner that we approach design. The future prospects of surface modification of biomaterials for medical devices are extremely promising. There is a growing emphasis on developing novel approaches for functionalization by grafting techniques, which can improve the performance of medical devices at the interfaces of biomacromolecules, cells, tissues, and biomaterials.

One of the key areas of focus for future research is the development of surface modification methods that can enhance the biocompatibility of medical devices [66–68]. This is especially crucial in the case of implantable medical devices, where poor biocompatibility can lead to complications such as implant rejection or infection. Novel approaches such as plasma-induced graft polymerization and photo-induced graft polymerization hold great promise in this regard. Another important area of future research is the development of surface modification methods that can improve the antimicrobial properties of medical devices. This is particularly important given the growing problem of antibiotic resistance, which makes it difficult to treat infections that arise in the context of medical device implantation.

However, more in-depth studies are necessary to address several issues in the future. Novel approaches such as ozone graft polymerization and radiation-induced graft polymerization may offer potential solutions in this regard. In addition to the creation of novel surface modification approaches, future research in this field is likely to focus on achieving greater precision and control over the grafting process. This may involve the use of advanced analytical techniques to better understand the mechanisms underlying surface modification, as well as the invention of new materials and scaffolds that are more amenable to surface modification.

Overall, the future of surface modification of biomaterials in medical devices with the continued innovation and investment in this field, it is likely that we will see significant advancements in the design and performance of medical devices, leading to better patient outcomes and improved quality of life for individuals around the world. As global attention on medical implants and devices increases, the technology behind surface modification of polymeric biomaterials is expected to extend to an industrial level, opening up a wider range of biomedical applications [69, 70]. With continuous progress in this field, it is reasonable to believe that various biomedical applications will finally be achieved. However, more in-depth studies are necessary to address several issues in the future.

10 Conclusion

In conclusion, while implantable devices have come a long way through years of research in biomaterial science, they still have drawbacks that limit their application in comparison with non-surgical or grafting techniques. However, the emerging technology of additive manufacturing (AM) has the capacity to revolutionize the way we think about design and develop novel devices. By allowing for direct control over both bulk and surface characteristics in the three-dimensional space, AM enables the use of materials that were previously unsuitable for certain applications. Additionally, surface modification methods such as nanometer-scale electrochemical anodizing have the potential to improve cellular interactions and protein adhesion in biomedical applications. With AM, we can look forward to personalized implants being created for each patient that are reliable, match their demands, and become the standard treatment for the many people throughout the world who are afflicted with illness and strategy that leverages.

References

1. Patel, N.R., Gohil, P.P.: A review on biomaterials: scope, applications & human anatomy significance. Int. J. Emerg. Technol. Adv. Eng. **2**(4), 91–101 (2012)
2. Ratner, B.D., Hoffman, A.S., Schoen, F.J., Lemons, J.E.: Biomaterials Science: An Introduction to Materials in Medicine. Elsevier (2004)
3. Kulinets, I.: Biomaterials and their applications in medicine. In: Regulatory Affairs for Biomaterials and Medical Devices, pp. 1–10. Woodhead Publishing (2015)
4. Martin, J.L., Norris, B.J., Murphy, E., Crowe, J.A.: Medical device development: the challenge for ergonomics. Appl. Ergon. **39**(3), 271–283 (2008)

5. Madalina Mihai, M., Maria Holban, A., Giurcaneanu, C., Gabriela Popa, L., Mihaela Oanea, R., Lazar, V., Carmen Chifiriuc, M., Popa, M., Ioan, P.M.: Microbial biofilms: impact on the pathogenesis of periodontitis, cystic fibrosis, chronic wounds and medical device-related infections. Curr. Top. Med. Chem. **15**(16), 1552–1576 (2015)
6. Agarwal, R., García, A.J.: Surface modification of biomaterials. In: Principles of Regenerative Medicine, pp. 651–660). cademic Press (2019)
7. Fabbri, P., Messori, M.: Surface modification of polymers: chemical, physical, and biological routes. In: Modification of Polymer Properties, pp. 109–130. William Andrew Publishing (2017)
8. Merola, M., Affatato, S.: Materials for hip prostheses: a review of wear and loading considerations. Materials **12**(3), 495 (2019)
9. Abraham, A.M., Venkatesan, S.: A review on application of biomaterials for medical and dental implants. Proc. Inst. Mech. Eng. Part L J. Mater. Des. Appl. **237**(2), 249–273 (2023)
10. Li, S., Wang, S., Liu, W., Zhang, C., Song, J.: Current strategies for enhancement of the bioactivity of artificial ligaments: a mini-review. J. Orthop. Transl. **1**(36), 205–215 (2022)
11. Milinkovic, D.D., Zimmermann, F., Balcarek, P.: Medial patellofemoral ligament reconstruction using nonresorbable sutures yields comparable outcomes to reconstruction with a pedicled quadriceps tendon autograft when performed in addition to bony risk factor correction. Knee Surg. Sports Traumatol. Arthrosc. **31**(1), 264–271 (2023)
12. Wang, Y., Li, G., Yang, L., Luo, R., Guo, G.: Development of innovative biomaterials and devices for the treatment of cardiovascular diseases. Adv. Mater. **34**(46), 2201971 (2022)
13. Bui, H.T., Khair, N., Yeats, B., Gooden, S., James, S.P., Dasi, L.P.: Transcatheter heart valves: a biomaterials perspective. Adv. Healthcare Mater. **10**(15), 2100115 (2021)
14. Raval, N., Kalyane, D., Maheshwari, R., Tekade, R.K.: Surface modifications of biomaterials and their implication on biocompatibility. In: Biomaterials and Bionanotechnology, pp. 639–674. Academic Press (2019)
15. Kurella, A., Dahotre, N.B.: Surface modification for bioimplants: the role of laser surface engineering. J. Biomater. Appl. **20**(1), 5 (2005)
16. Steckbeck, S.: Experimental Analysis of Interactions between Biomolecules and Inorganic Surfaces. Doctoral dissertation, Bremen, Universität Bremen, Diss. (2014)
17. Amani, H., Arzaghi, H., Bayandori, M., Dezfuli, A.S., Pazoki-Toroudi, H., Shafiee, A., Moradi, L.: Controlling cell behavior through the design of biomaterial surfaces: a focus on surface modification techniques. Adv. Mater. Interfaces **6**(13), 1900572 (2019)
18. Bose, S., Robertson, S.F., Bandyopadhyay, A.: Surface modification of biomaterials and biomedical devices using additive manufacturing. Acta Biomater. **15**(66), 6–22 (2018)
19. Brunette, D.M., Tengvall, P., Textor, M., Thomsen, P.: Titanium in Medicine: Material Science, Surface Science, Engineering, Biological Responses and Medical Applications. Springer, Berlin (2001)
20. Ahirwar, H., Zhou, Y., Mahapatra, C., Ramakrishna, S., Kumar, P., Nanda, H.S.: Materials for orthopedic bioimplants: modulating degradation and surface modification using integrated nanomaterials. Coatings **10**(3), 264 (2020)
21. Ohmori, H., Katahira, K., Nagata, J., Mizutani, M., Komotori, J.: Improvement of corrosion resistance in metallic biomaterials using a new electrical grinding technique. CIRP Ann. **51**(1), 491–494 (2002)
22. Callister, W.D.: Fundamentals of Materials Science and Engineering. Wiley, London (2000)
23. Priyadarshini, B., Rama, M., Chetan, V.U.: Bioactive coating as a surface modification technique for biocompatible metallic implants: a review. J. Asian Ceram. Soc. **7**(4), 397–406 (2019)
24. Gibson, I.R., Bonfield, W.: Preparation and characterization of magnesium/carbonate co-substituted hydroxyapatites. J. Mater. Sci. Mater. Med. **13**(7), 685–693 (2002)
25. Kulkarni, M., Mazare, A., Schmuki, P., Iglič, A., Seifalian, A.: Biomaterial surface modification of titanium and titanium alloys for medical applications. Nanomedicine **111**(615), 111 (2014)
26. Festas, A.J., Ramos, A., Davim, J.P.: Medical devices biomaterials—a review. Proc. Inst. Mech. Eng. Part L J. Mater. Des. Appl. **234**(1), 218–228 (2020)

27. Lara, N., Neto, A.F., da Mota, A.J., de Souza, A.E., Brito, A.G., Da Silva, A.C., Martinez, A.C., Nozaki, A.P., da Silva, A.A., de Vasconcelos Ferraz, A., Moraes, Â.M.: Engenharia de Materiais: materializando o futuro. Pimenta Cultural (2022)
28. Soni, R., Pande, S., Kumar, S., Salunkhe, S., Natu, H., Hussein, H.M.: Wear characterization of laser cladded Ti-Nb-Ta alloy for biomedical applications. Crystals **12**(12), 1716 (2022)
29. Su, Y., Luo, C., Zhang, Z., Hermawan, H., Zhu, D., Huang, J., Liang, Y., Li, G., Ren, L.: Bioinspired surface functionalization of metallic biomaterials. J. Mech. Behav. Biomed. Mater. **1**(77), 90–105 (2018)
30. Devgan, S., Sidhu, S.S.: Evolution of surface modification trends in bone related biomaterials: a review. Mater. Chem. Phys. **15**(233), 68–78 (2019)
31. Bandyopadhyay, A., Mitra, I., Shivaram, A., Dasgupta, N., Bose, S.: Direct comparison of additively manufactured porous titanium and tantalum implants towards *in vivo* osseointegration. Addit. Manuf. **1**(28), 259–266 (2019)
32. Hojjatzadeh, S.M., Parab, N.D., Yan, W., Guo, Q., Xiong, L., Zhao, C., Qu, M., Escano, L.I., Xiao, X., Fezzaa, K., Everhart, W.: Pore elimination mechanisms during 3D printing of metals. Nat. Commun. **10**(1), 3088 (2019)
33. Ni, J., Ling, H., Zhang, S., Wang, Z., Peng, Z., Benyshek, C., Zan, R., Miri, A.K., Li, Z., Zhang, X., Lee, J.: Three-dimensional printing of metals for biomedical applications. Mater. Today Bio. **1**(3), 100024 (2019)
34. Sing, S.L., Tey, C.F., Tan, J.H., Huang, S., Yeong, W.Y.: 3D printing of metals in rapid prototyping of biomaterials: techniques in additive manufacturing. In: Rapid Prototyping of Biomaterials, pp. 17–40. Woodhead Publishing (2020)
35. Liu, X., Chu, P.K., Ding, C.: Surface modification of titanium, titanium alloys, and related materials for biomedical applications. Mater. Sci. Eng. R. Rep. **47**(3–4), 49–121 (2004)
36. Mondy, W.L., Cameron, D., Timmermans, J.P., De Clerck, N., Sasov, A., Casteleyn, C., Piegl, L.A.: Computer-aided design of microvasculature systems for use in vascular scaffold production. Biofabrication **1**(3), 035002 (2009)
37. Setti, L., Fraleoni-Morgera, A., Ballarin, B., Filippini, A., Frascaro, D., Piana, C.: An amperometric glucose biosensor prototype fabricated by thermal inkjet printing. Biosens. Bioelectron. **20**(10), 2019–2026 (2005)
38. Farahani, R.D., Dubé, M., Therriault, D.: Three-dimensional printing of multifunctional nanocomposites: manufacturing techniques and applications. Adv. Mater. **28**(28), 5794–5821 (2016)
39. Stevens, M.M., George, J.H.: Exploring and engineering the cell surface interface. Science **310**(5751), 1135–1138 (2005)
40. Elfick, A.P., Green, S.M., Krikler, S., Unsworth, A.: The nature and dissemination of UHMWPE wear debris retrieved from periprosthetic tissue of THR. J. Biomed. Mater. Res. Part A Off. J. Soc. Biomater. Japan. Soc. Biomater. Aust. Soc. Biomater. Korean Soc. Biomater. **65**(1), 95–108 (2003)
41. Roy, M., Balla, V.K., Bandyopadhyay, A., Bose, S.: MgO-doped tantalum coating on Ti: microstructural study and biocompatibility evaluation. ACS Appl. Mater. Interfaces **4**(2), 577–580 (2012)
42. Bai, L., Gong, C., Chen, X., Sun, Y., Zhang, J., Cai, L., Zhu, S., Xie, S.Q.: Additive manufacturing of customized metallic orthopedic implants: materials, structures, and surface modifications. Metals **9**(9), 1004 (2019)
43. Darsell, J., Bose, S., Hosick, H.L., Bandyopadhyay, A.: From CT scan to ceramic bone graft. J. Am. Ceram. Soc. **86**(7), 1076–1080 (2003)
44. Bose, S., Fielding, G., Tarafder, S., Bandyopadhyay, A.: Understanding of dopant-induced osteogenesis and angiogenesis in calcium phosphate ceramics. Trends Biotechnol. **31**(10), 594–605 (2013)
45. Tarafder, S., Bose, S.: Polycaprolactone-coated 3D printed tricalcium phosphate scaffolds for bone tissue engineering: *in vitro* alendronate release behavior and local delivery effect on *in vivo* osteogenesis. ACS Appl. Mater. Interfaces **6**(13), 9955–9965 (2014)

46. Khaled, S.A., Burley, J.C., Alexander, M.R., Roberts, C.J.: Desktop 3D printing of controlled release pharmaceutical bilayer tablets. Int. J. Pharm. **461**(1–2), 105–111 (2014)
47. Song, S.W., Hidajat, K., Kawi, S.: Functionalized SBA-15 materials as carriers for controlled drug delivery: influence of surface properties on matrix–drug interactions. Langmuir **21**(21), 9568–9575 (2005)
48. Stallard, C.P., Solar, P., Biederman, H., Dowling, D.P.: Deposition of non-fouling PEO-like coatings using a low temperature atmospheric pressure plasma jet. Plasma Processes Polym. **13**(2), 241–252 (2016)
49. Gloria, A., Causa, F., Russo, T., Battista, E., Della Moglie, R., Zeppetelli, S., De Santis, R., Netti, P.A., Ambrosio, L.: Three-dimensional poly (ε-caprolactone) bioactive scaffolds with controlled structural and surface properties. Biomacromol **13**(11), 3510–3521 (2012)
50. Puppi, D., Chiellini, F., Piras, A.M., Chiellini, E.: Polymeric materials for bone and cartilage repair. Prog. Polym. Sci. **35**(4), 403–440 (2010)
51. Kurtz, S.M., Devine, J.N.: PEEK biomaterials in trauma, orthopedic, and spinal implants. Biomaterials **28**(32), 4845–4869 (2007)
52. Dawson, J.I., Wahl, D.A., Lanham, S.A., Kanczler, J.M., Czernuszka, J.T., Oreffo, R.O.: Development of specific collagen scaffolds to support the osteogenic and chondrogenic differentiation of human bone marrow stromal cells. Biomaterials **29**(21), 3105–3116 (2008)
53. Inzana, J.A., Olvera, D., Fuller, S.M., Kelly, J.P., Graeve, O.A., Schwarz, E.M., Kates, S.L., Awad, H.A.: 3D printing of composite calcium phosphate and collagen scaffolds for bone regeneration. Biomaterials **35**(13), 4026–4034 (2014)
54. Yu, D.G., Branford-White, C., Ma, Z.H., Zhu, L.M., Li, X.Y., Yang, X.L.: Novel drug delivery devices for providing linear release profiles fabricated by 3DP. Int. J. Pharm. **370**(1–2), 160–166 (2009)
55. Lee, K.Y., Mooney, D.J.: Alginate: properties and biomedical applications. Prog. Polym. Sci. **37**(1), 106–126 (2012)
56. Balamurugan, A., Balossier, G., Michel, J., Ferreira, J.M.: Electrochemical and structural evaluation of functionally graded bioglass-apatite composites electrophoretically deposited onto Ti6Al4V alloy. Electrochim. Acta **54**(4), 1192–1198 (2009)
57. Stojanovic, D., Jokic, B., Veljovic, D., Petrovic, R., Uskokovic, P.S., Janackovic, D.: Bioactive glass–apatite composite coating for titanium implant synthesized by electrophoretic deposition. J. Eur. Ceram. Soc. **27**(2–3), 1595–1599 (2007)
58. Boccaccini, A.R., Peters, C., Roether, J.A., Eifler, D., Misra, S.K., Minay, E.J.: Electrophoretic deposition of polyetheretherketone (PEEK) and PEEK/Bioglass® coatings on NiTi shape memory alloy wires. J. Mater. Sci. **41**, 8152–8159 (2006)
59. Lerouge, S., Guignot, C., Tabrizian, M., Ferrier, D., Yagoubi, N., Yahia, L.H.: Plasma-based sterilization: effect on surface and bulk properties and hydrolytic stability of reprocessed polyurethane electrophysiology catheters. J. Biomed. Mater. Res. **52**(4), 774–782 (2000)
60. Hooper, K.A., Cox, J.D., Kohn, J.: Comparison of the effect of ethylene oxide and γ-irradiation on selected tyrosine-derived polycarbonates and poly (L-lactic acid). J. Appl. Polym. Sci. **63**(11), 1499–1510 (1997)
61. Simmons, A.: Future trends for the sterilisation of biomaterials and medical devices. In: Sterilisation of Biomaterials and Medical Devices, pp. 310–320. Woodhead Publishing (2012)
62. Srivastava, M., Rathee, S., Maheshwari, S., Kundra, T.K.: Additive Manufacturing: Fundamentals and Advancements. CRC Press (2019)
63. Guddati, S., Kiran, A.S., Leavy, M., Ramakrishna, S.: Recent advancements in additive manufacturing technologies for porous material applications. Int. J. Adv. Manuf. Technol. **105**, 193–215 (2019)
64. Morrison, R.J., Kashlan, K.N., Flanangan, C.L., Wright, J.K., Green, G.E., Hollister, S.J., Weatherwax, K.J.: Regulatory considerations in the design and manufacturing of implantable 3D-printed medical devices. Clin. Transl. Sci. **8**(5), 594–600 (2015)
65. Schuh, J.C., Funk, K.A.: Compilation of international standards and regulatory guidance documents for evaluation of biomaterials, medical devices, and 3-D printed and regenerative medicine products. Toxicol. Pathol. **47**(3), 344–357 (2019)

66. Harawaza, K., Cousins, B., Roach, P., Fernandez, A.: Modification of the surface nanotopography of implant devices: a translational perspective. Mater. Today Bio. **1**(12), 100152 (2021)
67. Douglass, M., Garren, M., Devine, R., Mondal, A., Handa, H.: Bio-inspired hemocompatible surface modifications for biomedical applications. Prog. Mater. Sci. **17**, 100997 (2022)
68. Zhang, D., Chen, Q., Shi, C., Chen, M., Ma, K., Wan, J., Liu, R.: Dealing with the foreign-body response to implanted biomaterials: strategies and applications of new materials. Adv. Func. Mater. **31**(6), 2007226 (2021)
69. Punj, S., Singh, J., Singh, K.: Ceramic biomaterials: properties, state of the art and future prospectives. Ceram. Int. **47**(20), 28059–28074 (2021)
70. Alghamdi, H.S., Jansen, J.A.: The development and future of dental implants. Dent. Mater. J. **39**(2), 167–172 (2020)

B. Soni is presently a Ph.D. Scholar in the Department of Pharmaceutics, at Guru Nanak Dev University in Amritsar, Punjab. She completed her undergraduate and postgraduation at Punjab Technical University (PTU) in 2013 and 2016, Jalandhar. She completed her master's degree with distinction and got her name in the merit list of PTU. She has qualified for GPAT and has worked for six years as an Assistant Professor in the field of Pharmacy. Her research interests are centered around novel drug delivery systems, aiming to enhance drug bioavailability, as well as pharmacovigilance.

R. Shivgotra completed her bachelor's and master's degrees in pharmacy from Guru Nanak Dev University in Amritsar in 2020 and 2022, respectively. She qualified GPAT and secured the top rank in her postgraduate degree. She currently pursues her Ph.D. degree at the same university's Department of Pharmaceutical Sciences. Her primary area of interest is the development of novel drug delivery systems with the aim of enhancing the pharmacokinetics, therapeutic effectiveness, and safety of currently available drugs. Her current work focuses primarily on developing a safe and effective formulation to treat systemic fungal infections.

M. Kaur have done under-graduation and post-graduation degree in 2016 and 2018, respectively, as a university gold Medallist, from Guru Nanak Dev University, Amritsar, Punjab. Presently, Kaur is a Senior Research Fellow under INSPIRE program of the Department of Science and Technology New Delhi, India, and pursuing Ph.D. in Pharmaceutical Sciences at Department of Pharmaceutical Sciences, Guru Nanak Dev University, Amritsar, Punjab, India. The development of carrier systems for topical formulations with higher drug availability at the site of action is my field of work. Currently, Kaur is working on topical gel formulations with better antifungal activity. Till now, Kaur have published 11 articles and 2 book chapters in Scopus-indexed journals with an H-index of 4.

S. Thakur completed his undergraduate and postgraduate studies in pharmacy at Guru Nanak Dev University in Amritsar in 2016 and 2018 respectively. Presently, he serves as an Indian Council of Medical Research-Senior Research Fellow and is pursuing a Ph.D. in the Department of Pharmaceutical Sciences at Guru Nanak Dev University, Amritsar. His research focuses on the advancement of novel drug delivery systems, aiming to enhance the kinetics, efficacy, and safety of existing drugs. Specifically, his current project involves developing a safe oral formulation for pregnant women. Till now he has published 25 articles and 5 book chapters in Scopus-indexed journals and books with an h-index of 5.

Nanoporous Materials as Versatile Nanoplatforms for Drug Delivery Applications: Properties, Recent Progress, and Challenges

R. Abdel-Karim

Abstract Nanoporous materials (NPMs) are a class of materials in the quest for high-surface-area conductive and catalytic materials. Developing a broad spectrum of nanoporous materials has enhanced their extensive applications in catalysis, sensing, separation, and environmental, energy, and biomedical areas. Various types of pores like open, closed, transport, and blind pores in the porous solid allow them to adsorb drugs and release them in a more reproducible and predictable manner. According to the composition of the materials, the nanomaterials used in nano drug delivery systems (NDDSs) can be classified into organic, inorganic, and composite materials. In this book chapter, we present an overview of newly developed synthetic strategies for producing NPMs along with an in-depth discussion of the application of NPMs for drug delivery. The remarkable characteristics of NPFs are highlighted throughout this book chapter. Finally, challenges and future perspectives relating to nanoporous materials foams are concluded.

Keywords Nanotechnology; nano drug delivery systems NDDS · Drug loading · Drug release · Drug loading content · Drug loading efficiency · Mesoporous · Porous MOF · Porous organic carriers · Porous inorganic carriers

Abbreviations

NPMs	Nanoporous materials
NDDSs	Nano drug delivery systems
DSS	Drug delivery systems
DNCs	Drug nanocrystals
ADDCs	Amphiphilic drug-DCs
MBioFs	Metal-biomolecule frameworks
MS	Meso silica

R. Abdel-Karim (✉)
Department of Metallurgy, Cairo University, Giza, Egypt
e-mail: randaabdelkarim@gmail.com

R. Malviya and S. Sundram (eds.), *Engineered Biomaterials*, Engineering Materials,
https://doi.org/10.1007/978-981-99-6698-1_16

SCF	Supercritical fluid method
POPs	Porous Organic Carriers
MOFs	Metal-organic frameworks
HTCC	Chitosan chloride
NPCs	NanoPorous Carbon
OPL	Oil palm leaves
CaPNPs	Nanoporous Calcium Phosphate
ACP	Amorphous calcium phosphate
NPSi	Nanoporous Silicon
MPS	Mesoporous Silica Nanoparticles
CUR	Nano-encapsulated curcumin
PDT	Presumptive Drug Test
DMMA	3,4-Dimethoxy-N-methylamphetamine
NPG	Nanoporous Gold
MPA	Mercaptopropionic acid
MNs	Microneedles
PMMNs	Stainless-steel-based porous metallic microneedles
NAA	Nanoporous Alumina
TNT	Nanoporous Titana
MCNPs	Mesoporous carbon nanoparticles
EPR	Enhanced Permeability and Retention Effects

1 Introduction

The distinctive surface and structural characteristics of nanoporous materials NPMs include a large surface area, controllable pore sizes and pore geometries, and porous structures on the surface of the material [1]. These distinctive properties of nanoporous materials highlight their use in biomedical applications such as tissue engineering [2], implants [3], and targeted drug delivery [4].

NPMPs are favored in drug delivery systems (DSS) due to their superior bioavailability, solubility, and retention time. Drugs that have been encapsulated in nanoparticles are more effective than microparticles because of the nanoparticles' enhanced localization, especially after extravasation. When a medicine is encapsulated, it is shielded from a breakdown in the bloodstream and can therefore reach its destination in one piece. To prevent the medicine from being diluted too much along the way, nanoparticles can be adjusted to target a specific organ or tissue. The use of nanoparticles as drug carriers improves the drug's efficacy, which may reduce adverse effects and healthcare expenditures [5]. Materials engineering is the basis for a wide variety of reactive stimuli systems, including those that are affected by changes in pH [6], temperature [7], magnetic strength [8], ultrasound [9], and light [10].

DDS is a vehicle that is able to control the rate of drug release and can be directed to certain areas of the body by adding specific groups to the applied materials. The therapeutic effectiveness of a medicine may vary greatly depending on how it is administered. DDSs are made to maintain a therapeutic effect throughout the course of treatment, all the way to the intended site. The small carrier structure must meet the following criteria: (1) it must be non-toxic and biodegradable; (2) it must have a high loading capacity; (3) it must be able to prevent drugs decomposition; (4) By modifying pore openings with caps, it must seal the drug to shield the body from any side effects and avoid the early drug release before reaching the target-location; (5) it must be able to guide the small carrier structure toward the target-site [11].

The nano drug delivery systems NDDs are classified into four primary groups based on the fabrication technique of nanomedicines. The first type is based on inert carriers, such as inorganic porous carriers. The carrier containing drugs as part of, such as linear and branched PDCs and infinite coordination polymer (ICP) I-type nanomedicines is an example of the second type. Drug nanocrystals (DNCs), amphiphilic drug-DCs (ADDCs), metal-biomolecule frameworks (MBioFs), and ICP II-type, are some examples of the third type of nanomedicine, where no carrier is used [4].

The weight percentage of drug-loading content and drug-loading efficiency are calculated using Eqs. (1) and (2), respectively.

$$Drug\ Loading\ efficiency\ (wt\%) = \frac{Mass\ of\ loaded\ drug}{Initial\ mass\ of\ NPMs} \times 100 \quad (1)$$

$$Drug\ Loading\ efficiency\ (wt\%) = \frac{Mass\ of\ loaded\ drug}{Mass\ of\ total\ drug\ in\ feed} \times 100 \quad (2)$$

Drug-loading efficiency is a measure of how well medications in the feed are utilized during the production of nanomedicines, meanwhile, drug-loading content is a measure of the weight ratio of drugs to nanomedicines carriers. Drug-loading efficiency is affected by the process of drug-loading, and the weight of the drugs in the feed. The drug-loading content is dependent on the structure, as well as the physical and chemical characteristics of the drug carrier. Crystallization and covalent and coordinate bonding frequently result in high drug-loading efficiency, while physical and electrostatic adsorption often leads to a reduction of drug-loading efficiency during the drug-loading process [4].

NDDS has been investigated in a number of different porous nanomaterials, both organic and inorganic, as well as hybrids. Many synthesizing techniques such as electrochemical methods [12, 13], non-electrochemical methods [14, 15], sol-gel [16], microwave-based fabrication [17], combustion methods [18], ion beam irradiation [19], and pulsed laser approaches [20] are used in the production of NPMs.

This book chapter presents an overview of newly developed synthetic strategies for producing nanoporous materials NPMs along with an in-depth discussion of the application of NPMs for drug delivery. The remarkable characteristics of NPMs are discussed throughout this book chapter. Finally, challenges and future perspectives of nanoporous materials are included.

2 Drug Loading Techniques

As illustrated in Fig. 1, drugs have been loaded into mesoporous matrices using a variety of approaches, including the process of physical mixing, techniques that are solvent-based, melt operations, processes based on supercritical fluids, techniques using microwave irradiation, the technique of co-spraying drying, and the surfactant-assisted drug loading method [21].

2.1 The Solvent-Based Techniques

In accordance with Budiman et al. [22], the solvent evaporation approach was used to efficiently load drugs onto meso silica (MS) using chloroform as the solvent. Due to its high efficiency and well-known unit activities like filtering and drying, this approach is appealing.

2.2 The Co-spray Drying

It is one of the solvent-based techniques. Jaime-Escalante et al. [23] presented an overview of the application of the spry drying technique for drug loading. The type of precursors, solvents, aging time, drying temperature, and pressure can affect the features of drug-loaded particles.

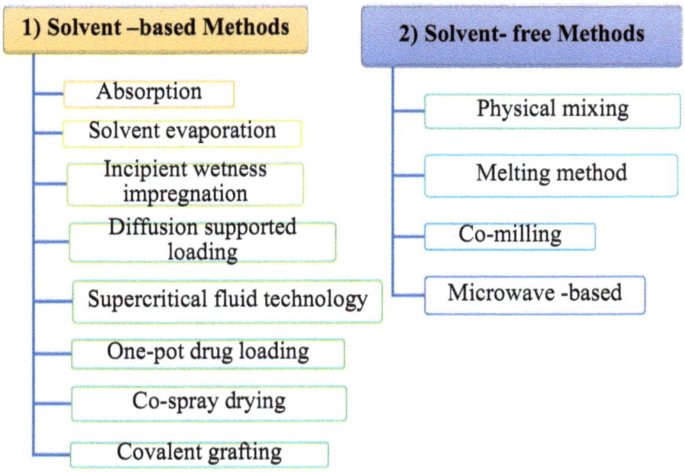

Fig. 1 Schematic illustration of different methods of drug loading into porous matrix

2.3 The Supercritical Fluid Method SCF

A review of the application of supercritical fluid technology for drug loading and delivery systems was presented by Chakravarty et al. [24]. SCF is non-toxic, inert, affordable, and ecologically benign. Many materials such as H_2O, N_2, Xe, SF_6, N_2O, C_2H_4, CHF_3, ethylene, propylene, propane, ammonia, n-pentane, and ethanol have been used for the SCF process, with CO_2 exhibiting the best behavior.

2.4 The Physical Mixing

It is one of the solvent-free techniques. The drug and mesoporous materials must be physically mixed to achieve a homogeneous dispersion. The adsorbent is added to the drug solution and stirred with a magnetic stirrer for an appropriate amount of time, followed by drying at 60 °C. This technique can be applied for preparing multiple drugs [21].

2.5 The Melting Technique

It is one of the solvent-free techniques. Melting the drug and the carrier together at high temperatures to create a physical mixture is the basis of the melting approach, which is both simple and solvent-free [21]. However, drug stability and thermal qualities are crucial and may restrict the process's usefulness.

2.6 Microwave-Based Methods

It is one of the non-solvent-based techniques. Using microwave heating and/or electromagnetic field could cause drying, polymeric cross-linkages, and drug-polymer interaction, and can affect the structure of drug crystallites. Microwave technology has the possibility of managing physicochemical characteristics and drug delivery profiles without the use of excessive heat, drawn-out procedures, or hazardous chemicals [25].

Table 1 Types and classifications of porous materials/carriers NPMs for drug delivery. Modified after Ahuja et al. [26]

Pore category	Pore aperture	Pore formation
1. Microporous	<2 nm	Mesoporous
2. Mesoporous	2> and <50 nm	This type of pore is formed due to the presence of some structural defects
3. Macroporous	>50 nm	The macroporous structure is a result of some structural defects such as cracks, fissures, and etching grooves

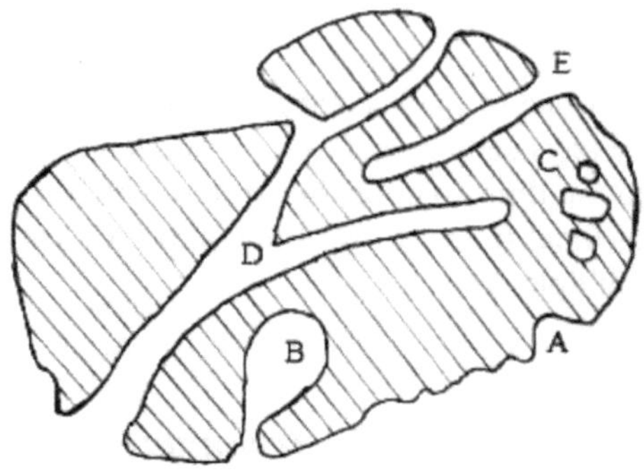

Fig. 2 Different forms of porous solid particles. A—Surface roughness, B—ink bottle porosity (blind pores), C—closed porosity type, D—transport porosity or porosity through the solid, and E—cylindrical blind type [26]

3 Properties and Characterization of Nanoporous Materials NPMs

As indicated in Table 1, and according to the International Union of Pure and Applied Chemistry, micropores are defined as pores with a diameter of 2 nm or less, mesopores span 2–50 nm, and macropores are defined as those greater than 50 nm in diameter [26, 27].

As illustrated in Fig. 2, the pores in the solid particles could be irregularly shaped and connected. The open pores are connected to the solid's external surface and can facilitate the drug's diffusion through the solid. On the other hand, a closed pore is an isolated vacancy within a solid and does not permit the diffusion of an adsorbate through the solid materials. Blind pores do not lead to other pores or surfaces and connect to the outside surface to the internal microporosity [26].

4 Classifications of Nanoporous Materials (NPMs) Applied for DDS

Below is a discussion of a number of organic, inorganic, and hybrid porous nanomaterials (Fig. 3) that have been investigated for drug delivery [28–35].

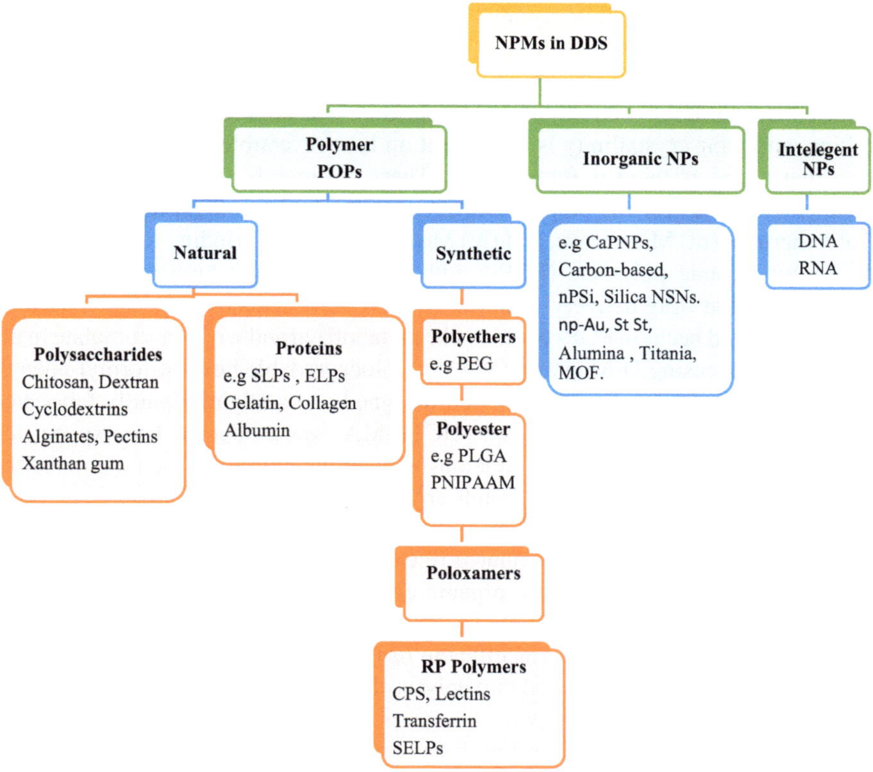

Fig. 3 Classification of porous nanoparticles NPMs in drug delivery systems DDS. *Abbreviations* SLPs: silk-like proteins type, ELPs: Elastin-like proteins type, PEG: Polyethylene glycol peloric type, PLGA: Poly (lactic-co-glycolic acid), PNIPAA<: Poly (N-isoppproplylacryla-mide), PF127: Pulronic F127, CPPs: Cell-penetrating peptides, SELPs: Silk-elestinlike protein type [28]

4.1 Porous Organic Carriers POPs

Porous organic substances are more frequently used in drug delivery systems due to their higher biocompatibility, lower cost, lower cytotoxicity, and lower immunogenicity than inorganic nanoparticles. This section reviews the main porous organic particles that are currently being explored and their most recent developments [30]. The morphologies of porous organic polymeric carriers include microspheres, porous fibers, porous microneedles, and hydrogel systems. As illustrated in Fig. 3, several synthetic polymers as well as many natural polymers, including chitosan, poly-sucrose, and alginate, have been used to create porous particles [31].

POPs are long-lasting porous substances produced by controlling the covalent bonding of stiff, multi-dimensional, organic building blocks using π-conjugation. POPs with interconnected gaps that are permeable to liquids, gases, and tiny molecules are created by the stiff building blocks. By adjusting the linker lengths, one may precisely control the pore volume, while functional groups serve as a foundation for further adjustments. POPs have an advantage over other materials because, like metal-organic frameworks MOFs, their constituent parts can be assembled precisely with enhanced atom economy. In addition, they can be strongly or weakly crystalline, and their crystallinity is dependent on how reversible the covalent connection that connects them is formed [32]. There are mainly four different types of synthetic methods for the preparation of crystalline polymers, (i) Ionothermal, (ii) Solvothermal, (iii) Microwave, and (iv) Mechanochemical grinding techniques [33].

Porous organic polymers (POPs) with low toxicity and high specific surface area are ideal as drug delivery vehicles. However, their potential applications were severely limited because POPs could not be metabolized and would accumulate in the body. By condensing D-Mannitol (MA) and biodegradable hexa (p-formyl phenyl) cyclotriphosphazene compounds, a well-designed and straightforwardly fabricated acetal-linked porous organic polymer, HCTP-MA, was produced. The oxygen-rich acetal links endowed the porous skeletons with strong polarity, which improved the drug loading capacity. Hydrogen bonds and van der Waals' forces helped the polyhydric drug loading into the porous skeleton. The pH could allow for the acetal gatekeepers' degradation and regulated release of cancer drugs [34]. According to Liu et al. [35] materials with an organic composition can have either natural or synthetic ingredients [35]. Sajid et al. [28] have investigated some natural as well as synthetic types of polymers as efficient drug-delivery tools for protein. Natural polymers were further subdivided into polysaccharide-based and protein-based polymers. Meanwhile, synthetic polymers were subgrouped into polyesters, polyethers, poloxamers, and recombinant protein-based polymers.

4.1.1 Natural Polymers

Chitosan, albumin, alginate, hydroxyapatite, and hyaluronic acid are popular natural polymers that are currently used as drug delivery systems [36]. Additionally, some natural polymers such as apoferritin DN-Apo [37], arginine, dextrin, polysaccharides, poly (glycolic acid), and poly (lactic acid) are involved in polymeric drug delivery systems [38]. Lv et al. [39] have encapsulated N-((2-hydroxy-3-trimethylammonium) propyl) anticancer drug on chitosan chloride (HTCC), with enhanced drug transportation across the intestinal barrier to the target site. Small interfering RNA (siRNA)—a class of nucleic acid-based drugs—was applied as a targeted therapy protocol in combination with polymeric, organic, and inorganic nanoparticles.

According to Suresh et al. [40] functionalized antibodies were added to gelatin nanoparticles. Collagen's cousin, gelatin, is another natural molecule with a practical application: lowering the body's immune response to drugs being administered intravenously. Gelatin was synthesized via a two-step desolvation procedure. Tris-glycine was used to quench excess glutaraldehyde. The obtained, porous gelatin hydrogels can provide high delivery efficiency with low toxicity due to the presence of the aldehyde groups.

4.1.2 Synthetic Polymers

POP morphologies, pore structures, and particular surface areas can be controlled using template techniques, template-free techniques, mechanical techniques, and control techniques such as electrospinning and 3D printing. The regulated release provided by POPs' dual action, molecular receptors, and stimuli-response to heat, pH, and light can positively affect cancer treatment. There are many factors that can affect the performance of polymeric DDS such as; the polymer assembly and manufacturing with specific porous structures, hydrophilicity, functionalization, and stimuli-responsiveness with the right choice of preparation techniques. Some polymers such as sodium polystyrene sulfonate and poly (fluoro acrylic acid) can serve as active drugs themselves [41].

According to Hao et al. [42], a triazine-based porous polymer containing thiophene units (2, 4, 6-tris (5-bromothiophene2-yl)-1, 3, 5-triazine [TBrTh]–1, 3, 5-benzene-triyltriboronic acid pinacol ester [BTBPE]–CTF) was synthesized using one-pot Suzuki cross-coupling reaction. This material exhibited good drug loading and releasing behavior due to π-π stacking as a result of the highly conjugated structure and abundant nitrogen sites in triazine rings. Bialik et al. [43] presented a review on the application of biodegradable synthetic polyesters as macromolecular carriers of antihypertensive drugs.

4.2 *Inorganic Porous Carriers*

This category of NPMPs for drug delivery includes carbon-based, metals, metal oxides, calcium phosphates, and metal-organic frameworks (Fig. 3).

4.2.1 Nanoporous Carbon NPCs

Nanostructured carbon is an important vehicle for the delivery of therapeutics for a range of disorders [44]. As illustrated in Fig. 4, carbon nanomaterials are classified as CNT, fullerenes, nanodiamonds, and graphite as well as nanoporous carbon. Nanoporous carbon can be synthesized using one of the following techniques; activation, self-activation, template method, or self-templating.

Yallappa et al. [45] extracted biocompatible NPC from Oil palm leaves (OPL) by pyrolysis at 600 °C in N_2 atmosphere, using a mixture of sulfuric and nitric acids (3:1 vol/ vol). NPC was combined with coumarin-6 (a fluorescence dye) for cellular imaging and drug delivery to cancer cells.

Ejsmont et al. [46] synthesized ordered mesoporous carbon drug carriers applying a soft-templating method altered with the use of nitrogen precursors and via a hard-templating method followed by chit with various morphologies. The loading capacity and drug release of carbon carriers toward losartan potassium (a drug used to treat high blood pressure) are affected by the specific surface areas and the amount of

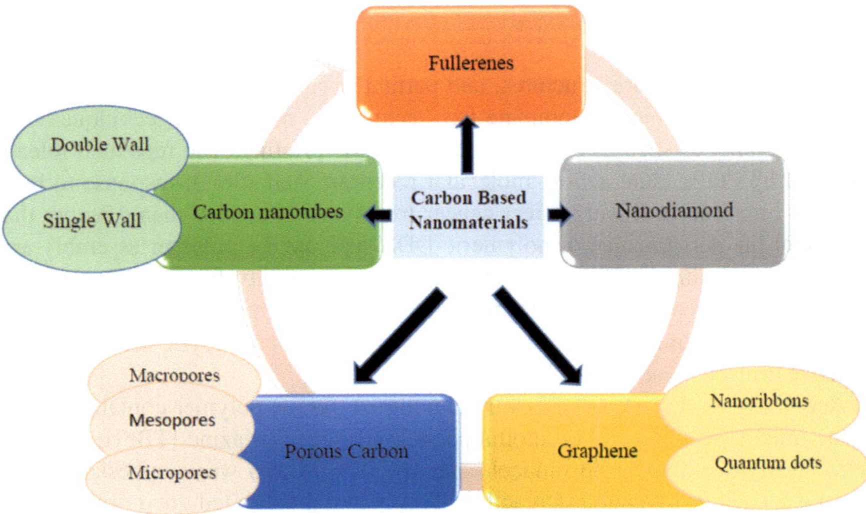

Fig. 4 Classification of carbon-based nanomaterials CBNs according to their shape and geometrical structure

acidic/basic oxygen-containing functional groups. Chen et al. [47] studied a drug delivery system of encapsulated protein/peptide molecules (Zadaxin) from CNTs controlled by electric fields.

Ine´s A´vila [48] the efficiency of porous carbon materials prepared by an aluminosilicate nano-templating technique and served as the carrier for sodium ibuprofen drug was investigated. The loading procedure conducted by either adsorption or evaporation under the effect of the pH of the release medium (1.8 and 7.4) was also studied.

The kinetics of drug release can be calculated using different mathematical relations, such as the Higuchi equation (Eq. 3), the mathematical model by Korsmeyer–Peppas (Eq. 4), or the first-order model (Eq. 5)

$$f = k_H.t^{1/2} \tag{3}$$

$$f = k_K.t^n \tag{4}$$

$$f = 1 - \exp(-k_F.t) \tag{5}$$

where f is the fraction of drug loading, k_H is the constant used in the Higuchi method, k_K is called the constant used in Korsmeyer–Peppas mathematical relation, n is the exponent, k_F is called the rate constant in the first order equation, and t is the time taken for drug release [48, 49].

4.2.2 Nanoporous Calcium Phosphate CaPNPs

Rout et al. [50] fabricated magnetic nanoporous amorphous calcium phosphate/ $CoFe_2O_4$ nanocomposite with high surface area for anticancer drug delivery and dual imaging. The surface modification of the composite was achieved by the addition of N-phosphonomethyl iminodiacetic acid (PMIDA) and the formation of an iminodiacetic acid group. Sun et al. [51] synthesized porous amorphous calcium phosphate (ACP) with a surface area of over 400 m^2/g-based on the Brunauer–Emmett–Teller BET method- by mixing phosphoric acid and methanol suspension containing amorphous calcium carbonate nanoparticles. The ACP synthesized showed cytocompatibility and high efficiency as a drug carrier for the bisphosphonate drug alendronate (AL) in vitro. The modified ACP showed long-term stability in the air, and water as well as in the α-MEM cell culture medium.

4.2.3 Nanoporous Silicon NPSi

NPSi is an excellent drug carrier based on its high porosity and surface area, biocompatibility, as well as biodegradability. For drug delivery systems, pore size is important because drug molecules must be suitable in order to pass through the pores. The quantity of medicine absorbed by pores is directly proportional to their porosity [52]. Their photoluminescent properties and simple silicon electrochemical anodization manufacturing are two of their many benefits [53]. Some review articles can provide significant studies about preparation technologies, properties, and application of nanoporous silicon for drug delivery [54–56].

Yang et al. [57] synthesized highly porous Si by electrochemical etching technique. Porous Si nanoparticles are safe and have low toxicity because they degrade into orthosilicic acid in the body. Colloidal instability, decreased blood circulation retention times, and unpredictable behavior in vitro and in vivo systems are some of the disadvantages of this biodegradable and biocompatible material. Hernández-Montelongo et al. [58] synthesized nPSi layers by electrochemical etching. The nanocomposites composed of oxidized nPSi (nPSi-Ox) microparticles cascade were treated with chitosan (CHI) and β-cyclodextrin (βCD) biopolymers. The as-received nPSi– βCD composites allowed a higher control in the drug release kinetic of florfenicol (FF) compared to nPSi-Ox microparticles. Perrone Donnorso et al. [59] fabricated water-soluble nanoporous silicon nanoparticles (NPS) with the anodizing approach. Annealed nPSi can be used to load both hydrophilic and hydrophobic drugs. They showed improved luminescent properties and can prevent aggregation in a water solution with a higher therapeutic index compared to traditional drugs.

4.2.4 Mesoporous Silica Nanoparticles (MPS)

In 1992, liquid crystal shaping was first used to synthesize the M41S family of silicate and alumina silicate-mesoporous materials MPS. Members of this family include MCM 41 (Mobil Composition of Matter 41) with two-dimensional arrays of cylindrical pores 2–5 nm diameter, MCM-48 with a cubic three-dimensional structure having well-recognized bi-continuous channels with 2–5 nm diameter, and MCM-50 with a laminar structure [60, 61].

Zhao et al. [62] fabricated SBA-15 and SBA-16 drug carriers characterized by thicker walls and larger pore sizes compared to MCM-X-type silica. The MCM-41 and SBA-15 are ordered MS materials [62]. The mesoporous silica drug carriers are classified based on their size, shape, and country of production, such as Santa Barbara Amorphous-1, Santa Barbara Amorphous-15, Santa Barbara Amorphous-16, Mobil Composition of Matter type No. 41, Fudan University-type mesoporous material-12, Korea Advanced Institute of Technology-6, and Korea Advanced Institute of Technology-5 type are examples. MS groups. MSU (Michigan State University), TUD (Technische Universiteit Delft), FSM (Folded-Sheet Mesoporous Material), as well as MPS (synthesized mesoporous silica), are also some categories of MS [63, 64].

Ahmidi et al. [65] presented a review of MPS nanoparticles as an important vehicle for anti-cancer. A normal MPS has a particle size of about 100 nm, a pore volume of about 0.9 cm^3/g, a pore volume surface area of about 900 m^2/g, and pore sizes of about 2 nm [66, 67]. Generally, the bioactive cargo is loaded into the extremely stable pores of silicon mesoporous materials. In a perfect world, cargo-loaded pores would be sealed off and discharged inside the cell. The capacity of MPS to withstand environmental stresses like changes in temperature and pH is a major benefit of these nanoparticles. Their large surface area and high pore volume enhance the loading of comparatively higher amounts of drugs. Meanwhile, their controllable and uniform pore size (3–50 nm) allows the functionalization of specific molecules onto the surface [68].

The mesoporous silica-based materials used for drug delivery systems have high biocompatibility, large surface areas, excellent compatibility with other materials, resistance to high temperatures, and chemical stability [69]. The drug release process using silica-based materials is dependent on the diffusion process, drug loading technique, the physicochemical properties of the drug, and the hydrophilicity of the platform [70].

Vallet-Regi et al. [71], manufactured two types of mesoporous silica (MCM-41s) with different pore sizes as drug carriers for ibuprofen (model drug). Ibuprofen (under static conditions) showed different release profiles based on the drug charging technique. The rate of the ibuprofen drug release decreased by using MCM 41 with finer pore size. Wang et al. [72] synthesized mesoporous nano silica hollow spheres (NSHS) with an average dimension of 190 nm and SSA of 330 $m^2\ g^{-1}$ using a microwave technique. These NSHS remained functional up to 700 °C when exposed to air.

Mesoporous materials based on silica are useful for many DDS, such as immediate drug delivery, targeted drug delivery, and stimuli-responsive systems. (CDDSs). The drug release rate from mesoporous silica can also be controlled by introducing suitable polymers or functional groups, such as CN, SH, NH_2, and Cl. Multifunctional DDS based on MSNs are currently under development to release anticancer medications on demand and in a targeted way while limiting the drug's premature release.

Manzano et al. [73] presented an overview of synthesizing, characteristics, and most recent applications of mesoporous silica for drug delivery. The siliation process, as well as the co-condensation process, are the techniques applied for functionalizing the surface of MPSs with the desired functional groups [74].

El-Ghannam et al. [75], investigated the efficiency of a porous silica–calcium phosphate nanocomposite (SCPC) as a drug delivery vehicle for 5-Fluorouracil (5-FU) anti-cancer in vitro and in vivo, with minimal side effects. Meihua et al. [76] prepared superior core-shell mesoporous silica MPS coated with a CaP-hyaluronic acid hybrid for the drug delivery of anticancer. An additional hyaluronic acid layer was applied in order to target CD44 over-expressed cancer cells.

Chen et al. [77] provided a review on nano-encapsulated curcumin (CUR) and its function in biomedical applications. It allows many pharmacological roles such as anti-inflammatory, anti-bacterial, anti-cancer, anti-Alzheimer, and anti-fungal. It has low bioavailability due to its limited absorption of water.

According to Ellahioui et al. [78] photosensitizers' anti-cancer efficiency can be improved through integration within MPSs, which decreases their tendency to aggregate. Recently, Presumptive Drug Test (PDT) using MPSs modified with ruthenium (III) compounds was attempted, along with the ensuing sensitization. The ruthenium complex makes tumor cells more sensitive to light, which then destroys them.

Liu et al. [79] took a different tack, employing the MPS as a synergistic inorganic nanohybrid instrument. Two distinct changes to the MPS surfaces turned them into "Janus" nanoparticles. Lung cancer cells' CD44 receptors were successfully targeted by the HA ligand on the modified Janus MSNs, and the acidic pH of A549 cells triggered charge reversal in the DMMA molecules- a psychoactive drug, leading to increased absorption.

Mesoporous silica nanoparticles (MSNs) can contain pharmaceuticals and bioactive drugs that can be released when a variety of stimulus-responsive molecular "gatekeepers" or "nano valves" are activated by redux stimuli-, pH, temperature, and magnetic field stimuli [80].

4.2.5 Nanoporous Gold (NPG)

Many important references are recommended for reviewing the production, characterization, and modification of NPG as a tool for drug delivery systems [81, 82]. As illustrated in Fig. 5, there are three common electrochemical techniques for the production of NPG on a solid support. (1) electrochemical etching, (2) electrodeposition, and (3) chemical and electrochemical dealloying [83].

Neupane et al. [84] developed NPG wires modified with thiolated β-cyclodextrin HS-β-CD drug delivery system containing DOX- anti-tumor drugs. The drug release from this implantable and biocompatible material was activated by pH- stimuli,

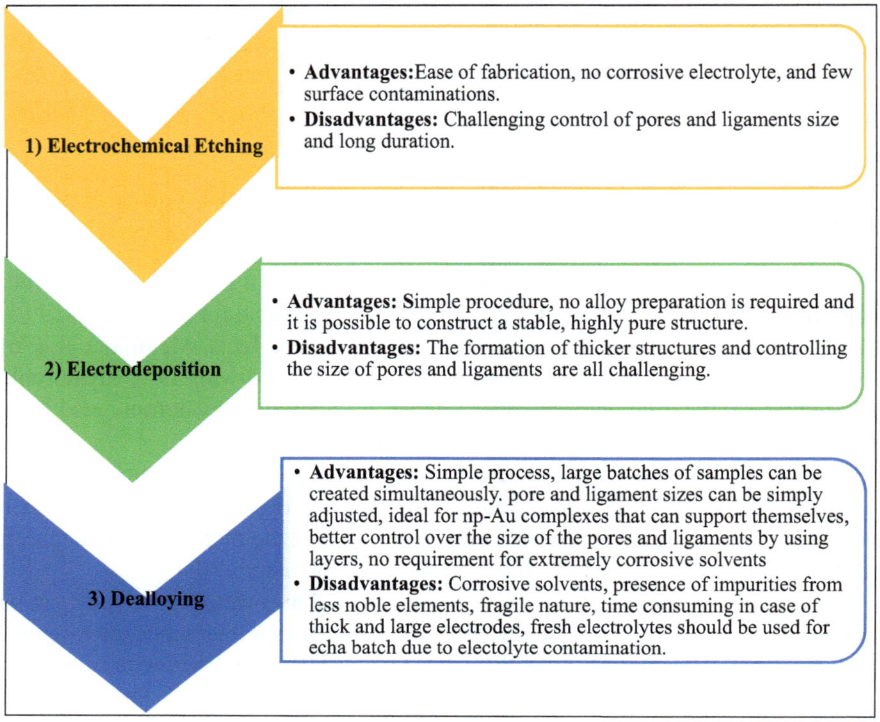

Fig. 5 Different approaches of nanoporous np-Au fabrication, advantages, and disadvantages of each process

showing a lower rate at the physiological condition and an accelerated release under acidic conditions. NPG materials are characterized by ease of size regulation, ease of manufacturing, mechanical and chemical stability, control of pore size, and ease of surface modification and drug loading. On the other hand, NPG is not biodegradable and should be removed after application.

The regulated porous architecture of np-Au is very important for the field of nanomedicine. The stability and efficiency of drug distribution are enhanced by capping groups on the surface of the porous gold carriers, and the mechanism of action is determined by the size of the carriers [85].

Kang et al. [86] synthesized NPG using templated pulsed *electrochemical deposition* technique for producing Ag–AgAu nano-segments followed by selective etching of the more active Ag phase. The dimercaptosuccinic acid (MSA) modified NPG showed the best performance of light-stimulated drug-releasing response for doxorubicin drug compared to those of the mercaptopropionic acid (MPA) modified NPG and the bare NPG.

4.2.6 Nanoporous Stainless Steel StSt

Bae et al. [87] synthesized nanoporous stainless steel by surface electropolishing at a steady voltage of 20 V and 10 A for 10 min in an acidic atmosphere (50% H_2SO_4). The drug loading and release rates of everolimus (EVL-immunosuppressive agent) were studied. The pre-polishing group had a greater drug storage rate (i.e., a lower drug release rate), and the post-polishing group had a higher overall amount of EVL on the surface (i.e., a higher drug storage rate).

Microneedles (MNs) are one of the biomedical tools that can facilitate transdermal drug delivery by piercing human skin and providing transport conduits across the stratum corneum with a thickness of 10–15 μm. Stainless steel-based porous metallic microneedles (PMMNs) are fabricated by hot embossing, debinding, and sintering processes using stainless steel powder, pore fillers, and binders [88].

4.2.7 Nanoporous Alumina NAA

Nanoporous anodic alumina (NAA) is an important inorganic porous material used recently for biomedical applications. A two-step anodization process is applied for synthesizing NAA templates with the superior ordering of nanopores. NAA is characterized by low conductivity, optical transparency, chemical inertness, and more recently, biological stability and biological compatibility [89]. Pourmadadi et al. [90] presented a review of the biomedical application of porous alumina with a focus on drug delivery systems. Properties such as biocompatibility and cytotoxicity as well as production techniques and novel strategies for improving the therapeutic efficiency of alumina-based nanocarriers have been reported. Synthesis methods for preparing alumina and alumina-containing composites include co-precipitation, hydrothermal, sol-gel, evaporation-induced self-assembly, as well as electrochemical anodizing. As illustrated in Fig. 6, in addition to the conventional anodization strategies mentioned above, there are several other types of modified anodization approaches that can be applied to obtain more complex structures [91]. Jeon et al. [92] created a hybrid system that combines NAA chips with electrically sensitive polymers, using electropolymerizing polypyrrole doped with dodecyl benzenesulfonate anion (PPy/dB) on top of NAA templates. This polymeric blend was subjected to a large volume change in order to accomplish pore mouth actuation by an external electrical stimulus.

4.2.8 Nanoporous Titana TNT

TNT-based implants are significant nanomaterials for regulated drug delivery. Numerous electrochemical techniques are used to create porous titanium nanotubes (TNTs). The drug delivery procedure can be controlled by many factors such as; the dimensions and charges of drug molecules, the dimensions of nanotubes, the charge and surface chemistry of nanotubes, the interfacial interaction between drug

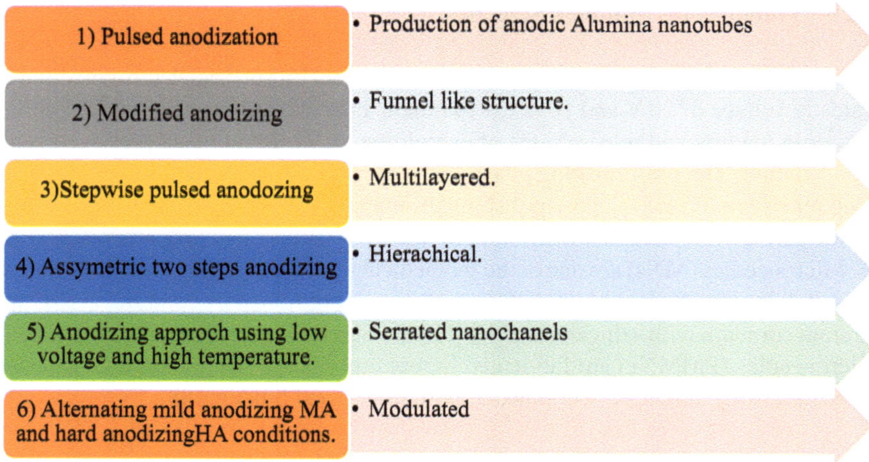

Fig. 6 Different anodization approaches used for fabrication of NAA with different morphology

molecules and nanotube surface, the rate of drug solubility, the diffusion coefficient, and finally the pH of the electrolyte [93].

The drug delivery system is a diffusion-controlled process, which can be studied by Fick's first law. The molar flux (J) of drug molecules is dependent on the concentration gradient (dc/dx), as presented by Fick's first law of diffusion presented in Eq. (6). According to Fick's second law presented in Eq. (7), the second derivative of the concentration change with distance determines how fast the concentration of a solution c is changed at a given position in space. It addresses how the gradient of concentration can alter over time and at any distance. When a drug release complies with Fick's law, it is referred to as Fickian diffusion; when it does not, it is considered as non-Fickian or anomalous diffusion [94].

$$J\alpha\frac{dc}{dx}orJ = D.\frac{dc}{dx} \tag{6}$$

$$\frac{dc}{dt} = D.\frac{d^2c}{dx^2} \tag{7}$$

where dc is the change in drug concentration in (g/cm^3), dx is the change in distance (cm), D is the diffusion constant in (cm^2/s), J flux in ($g.cm^{-2\,s-1}$), and dt is the time in (s).

Aw et al. [95] created TNT-based drug delivery systems with an extended-release capacity, preventing the breakdown of fragile drugs and proteins. Furthermore, polymeric carriers containing different drugs were loaded in the TNTs providing a time-controlled drug release system. Combined NAAs and TNTs were structures designed with sensing functions by applying external stimulating resources (thermal, magnetic, electromagnetic, ultrasonic, or mechanical activation) [89].

4.2.9 Metal–Organic Frameworks (MOFs)

The drug release process from MOFs is characterized by controlled, delayed liberation of the drugs via matrix breakdown, in contrast to other nanocarriers, which usually release the drug molecules in a burst. BioMIL-1 MOFs containing iron demonstrated up to 75% higher nicotinic acid loading and controlled drug delivery compared to native MOF structures. MOF-based delivery devices have been used to accomplish multimodal medicine administration and imaging by combining these approaches. Notably, MOFs can carry different types of drugs utilizing the noncovalent approach, including drugs with controlled drug release patterns and no burst effects, such as cisplatin and ibuprofen. Zeolite and mesoporous silica are useful for constructing robust frameworks because they do not trigger a burst-release reaction in MOF complexes [89].

Yang et al. [96] provided a review focusing on the stimuli and multi-stimuli-responsive. MOFs applied for drug delivery systems. MOFs-based DDSs are characterized by accurate pore size, adjustable composition and structure, controllable size, variable functionality, high loading capacity, and improved biocompatibility. Many metals such as Fe, Zn, Zr, Mn, Mg, and Cu are widely used for the manufacturing of MOFs, due to their reasonable toxicity as measured by oral lethal dose 50 (LD50). MOF-based materials are characterized by malleable structures that allow for a wide range of morphologies, compositions, sizes, and chemical properties, making them ideal as drug delivery systems. Finally, weak coordination connections allow MOFs to be biodegradable [97].

The chemical characteristics of MOFs can be modified by adding some inorganic clusters and/or organic ligands, which is not the case with other porous materials. Some metal-organic frameworks (MOFs) have lanthanide elements, for instance, fluoresce when exposed to ultraviolet light. Furthermore, the organic compounds can be modified either before or after synthesis to incorporate the desired functional groups [98]. Production techniques of MOFs include; microwave-assisted solvothermal synthesis, reverse-phase microemulsion synthesis, electrochemical synthesis, techniques of dry-gel conversion, preparation techniques using a microfluidics device, mechano-chemical or sonochemical synthesis, synthesis using step-by-step, and high-throughput methods. Figure 7 summarizes the benefits and drawbacks of the above-mentioned manufacturing techniques [99].

One of the most crucial and significant properties of MOFs' is their capacity to interact with biological systems in response to a variety of stimuli, including pH, temperature, light, magnetic field, pressure, glucose level, and numerous stimuli-responsive systems. This characteristic enhances the solubility of amorphous and poorly soluble medicines and allows for precise and sustained drug release [100].

According to Al Haydar et al. [101], different types of biocompatible MOFs such as Ca-MOF and Fe-MILs (53, 100, and 101) were synthesized using a solvothermal process and applied for FBP delivery. The percentage of FBP drug release was evaluated using the following Eq. (8);

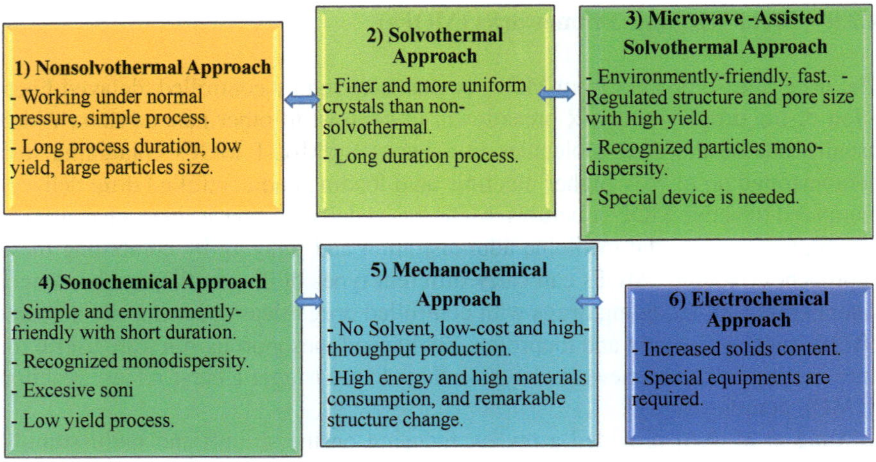

Fig. 7 Schematic illustration of MOFs manufacturing techniques and their characteristics

$$\%Release = \frac{Actual\ drug\ released\ at\ any\ time\ (mg)}{Amount\ of\ drug\ loaded\ within\ MOF\ (mg)} \tag{8}$$

Zhang et al. [102] synthesized mesoporous silica nanoparticles (MSN)-ZnO hybrid composite through electrostatic contact. The amine-functionalized mesopores were loaded with 0.02 mg/mg of indocyanine green (ICG). These QDs were combined with the drug erlotinib (ER) in order to target overexpressed EGFR receptors. EGFR is the epidermal growth factor receptor protein involved in cell signaling pathways that control cell division and survival. The hybrid's surface functionalization with N-(4-bromopthaloyl)-chitosan has positively affected the interaction with ER and increased its solubility.

According to Sun et al. [103] gold NRs-Capped and Ce6-doped mesoporous silica nanorods hybrid, were for photosensitive and hyperthermic anti-cancer treatment. The removal of gold nanorods from the MPS nanorods' surface under the effect of photothermal therapy (PTT) was followed by photodynamic therapy (PDT). Zhao et al. [104], synthesized a fluorescent mesoporous silica-carbon dot nanohybrid via a microwave-assisted solvothermal process.

5 Conclusions, Challenges, and Future Prospective

Nanoporous materials NPMs with controlled porous structure, microstructure, specific surface functionalization, and varied geometries are significant tools as drug delivery systems DDS. Different manufacturing technologies, including electrochemical methods, non-electrochemical methods, sol-gel, microwave-based fabrication, combustion methods, ion beam irradiation, and pulsed laser approaches, can easily synthesize nanoporous materials. Currently, research is shifted toward the use of several materials in combination to combine various therapeutic modalities and regulate drug release. In addition, researchers working in these areas should concentrate on the scalability of the synthesis methodologies.

Advanced DDS include biodegradable, natural or synthetic polymers, inorganic (metals, ceramics, carbon-based), and hybrid porous nanomaterials. For example, mesoporous carbon nanoparticles (MCNPs) still encounter the following unresolved issues in drug delivery systems: (1) standard technologies to produce MCNPs with the suitable size and hydrophilic surfaces for injection of medication; (2) inappropriateness of conventional oxidation methods used for the surface modification of MCNPs due to partial destruction of the carbonaceous framework and the mesostructure of MCNPs; and (3) the constrained availability of MCNPs [4]. In the case of NAA, more research should be focused on some industrial factors, such as the processing technique of the nanoporous alumina NAA powder, and the sintering temperature, rate, and time, which can affect the behavior of alumina [90].

- Some ideas are recommended for the future development of porous organic polymers POPs:
- Standard nPOP synthesis techniques need to be allocated in order to facilitate the clinical transition of anti-cancer drug delivery,
- The majority of POPs are made for passive targeting based on Enhanced Permeability and Retention (EPR) Effects; adding an active targeting technique will support drug presented at the location of cancer cells.
- More focus on the post-synthetic modification capability of POPs may enable liposome and biomolecule-modified nPOPs,
- Crystallinity is occasionally neglected in favor of shape, size, stability, and biodispersibility. However, a full investigation of how crystallinity affects cancer therapeutic effectiveness is still lacking.
- Studying the high toxicity of cancer medications has been a major worry [32].

Organometallic frameworks MOFs with their morphology, substantial surface area and porous structure, adaptable pore size, and simplicity of chemical functionalization are promising nanocarriers for DDSs. For synthesized MOFs, it is essential to thoroughly evaluate the biocompatibility, toxicity, and solubility of the chosen building blocks and metal nodes [96].

References

1. Hadden, M., Martinez-Martin, D., Yong, K.-T., Ramaswamy, Y., Singh, G.: Recent advancements in the fabrication of functional nanoporous materials and their biomedical applications. Materials **15**, 2111 (2022)
2. Maksoud F J, Velázquez de la Paz M F, Hann A J, Thanarak J, Reilly G C, Claeyssens F, Green N H and Zhang Y S. Porous Biomaterials for tissue engineering: a review. J. Mater. Chem. B; 2022, 10: 8111–8165

3. Benˇcin, M., Junkar, I., Vesel, A., Mozetiˇ, M., Igliˇ, A.: Nanoporous stainless steel materials for body implants—review of synthesizing procedures. Nanomaterials **12**, 2924 (2022)
4. Shen, S., Wu, Y., Liu, Y., Wu, D.: High drug-loading nanomedicines: progress, current status, and prospects. Int. J. Nanomed. **12**, 4085–4109 (2017)
5. Reisner, D E. Bionanotechnology II: Global Prospects. 1st ed: Taylor and Francis CRC Press; 2011. Chapter 11, Nanotechnology in biomaterials: nanoparticles as drug delivery systems; p. 217–233
6. Zhuo, S., Zhang, F., Yu, J., Zhang, X., Yang, G., Liu, X.: PH-sensitive biomaterials for drug delivery. Molecules **25**, 5649 (2020)
7. Mazzotta, E., Tavano, L., Muzzalupo, R.: Thermo-sensitive vesicles in controlled drug delivery for chemotherapy. Pharmaceutics **10**(3), 150 (2018)
8. Jiang M, Liu Q, Zhang Y, Wang H, Zhang J, Chen M, Yue Z, Wang Z, Wei X, Shi S, Wang M, Y Hou, Wang Z, Sheng F, Tian N, Wang Y . Construction of magnetic drug delivery system and its potential application in tumor theranostics. Biomedicine & Pharmacotherapy 2022; 154:113545.
9. Entzian, K., Aigner, A.: Drug delivery by ultrasound-responsive nanocarriers for cancer treatment. Pharmaceutics **13**(8), 1135 (2021)
10. Xiang, W., Chen, L.: In light-sensitive Drug Delivery System Nanoparticles. Mediate Oxidative Stress, Am J Transl Res. **12**(5), 1469–1480 (2020)
11. G. Wang, Silica Nanoporous Materials Adsorption and Release Study, Syracuse University, Dissertation, New York, USA, 2011.
12. Kharissova O V, Torres-Martínez L, Kharisov B I. Handbook of Nanomaterials and Nanocomposites for Energy and Environmental Applications. Springer International Publishing; 2021. Chapter 4, Nanoporous Metallic Foams for Energy Applications: Electrochemical Approaches for Synthesizing and Characterization; p.489–511
13. Abdelfatah A, Reda Y, Abdel-Karim R, El-Raghy S, Zohdy K M. Electrochemical characterization of electrodeposited Ni–Cu foams and their application as electrodes for supercapacitors. Frontiers in Mechanical Engineering 2020, 6, paper number 36.
14. Ahmad I. Adv Mater Appl: Micro to Nano Scale. One Central Press, 2017. Chapter 4, Electrochemical deposition of porous metallic foams for energy applications; p.78–91.
15. Ebothé J., Nanofabrication using Nanomaterials. One Central Press, 2016. Chapter 7, Fabrication of Nanoporous Alumina; p. 197–218.
16. Lei, Q., Guo, J., Noureddine, A., Wang, A., Wuttke, S., Brinker, C.J., Zhu, W.: Sol–gel–based advanced porous silica materials for biomedical applications. Adv. Funct. Mater. **30**, 1909539 (2020)
17. Głowniak, S., Szczęśniak, B., Choma, J., Jaroniec, M.: Advances in Microwave Synthesis of Nanoporous Materials. Adv. Mater. **33**, 2103477 (2021)
18. Aihara H, Zider J, Fanton G, and Duerig T. Combustion synthesis porous nitinol for biomedical applications. International Journal of Biomaterials 2019; Article ID 4307461: 11 pages.
19. Bai, Z., Zhang, L., Li, H., Liu, L.: Nanopore creation in graphene by ion beam irradiation: geometry, quality, and efficiency. ACS Appl. Mater. Interfaces **8**(37), 24803–24809 (2016)
20. Wanga, C., Huang, H., Zhang, Z., Zhang, L., Yan, J., Ren, L.: Formation, evolution and characterization of nanoporous structures on the Ti_6Al_4V surface induced by nanosecond pulse laser irradiation. Mater. Des. **223**, 111243 (2022)
21. Trzeciak, K., Chotera-Ouda, A., Bak-Sypien, I.I., Potrzebowski, M.J.: Mesoporous silica particles as drug delivery systems—The state of the art in loading methods and the recent progress in analytical techniques for monitoring these processes. Pharmaceutics **13**, 950 (2021)
22. Budiman, A., Aulifa, D.L.: Encapsulation of the drug into mesoporous silica by solvent evaporation: A comparative study of drug characterization in mesoporous silica with various molecular weights. Heliyon **7**, e086272 (2021)
23. Jaime-Escalante, B., Melgoza-Contreras, L.M., Leyva-Gómez, G., Mendoza-Muño, N.: Synthesis and drug loading improvements on mesoporous SBA-15 by spray drying. Drug Dev. Ind. Pharm. **47**(12), 1895–1903 (2021)

24. Chakravarty, P., Famili, A., Nagapudi, K., Al-Sayah, M.A.: Using supercritical fluid technology as a green alternative during the preparation of drug delivery systems. Pharmaceutics **11**, 629 (2019)
25. Won TW. Use of microwave in the processing of drug delivery systems, current drug delivery 2008; 5(2):77–84.
26. Ahuja, G., Pathaka, K.: Porous carriers for controlled/modulated drug delivery. Indian J. Pharm. Sci. **71**(6), 599–607 (2009)
27. Jibowu, T.: A Review on nanoporous metals. Front Nanosci Nanotech **2**(4), 165–168 (2016)
28. Sajid, M., Akash, H., Rehman, K., Chen, S.: Natural and synthetic polymers as drug carriers for delivery of therapeutic proteins. Polym. Rev. **55**(3), 371–406 (2015)
29. Ameen S, Akhtar M S and Shin H-S. Nanopores, IntechOpen; 2021. Chapter 2: Potential Application of Nanoporous Materials in Biomedical Field; 17 pages.
30. Radhakrishnan, D., Mohanan, S., Choi, G., Choy, J.-H., Tiburcius, S., Trinh, H.T., Bolan, S., Verrills, N., Tanwar, P., Karakoti, A., Vinu, A.: The Emergence of nanoporous materials in lung cancer therapy. Sci. Technol. Adv. Mater. **23**(1), 225–274 (2022)
31. Sayed E, Haj-Ahmad R, Ruparelia K, Arshad M S, Chang M-w and Ahmad Z. Porous inorganic drug delivery systems – a review. AAPS Pharm Sci Tech 2017; 18: 1507–1525.
32. Singh, N., Son, S., An, J., Kim, I., Choi, M., Kong, N., Tao, W., Kim, J.S.: Nanoscale porous organic polymers for drug delivery and advanced cancer theranostics. Chem. Soc. Rev. **50**, 12883–12896 (2021)
33. Zhang, S., Yang, Q., Wang, C., Luo, X., Kim, J., Wang, Z.: Porous organic frameworks: advanced materials in analytical chemistry. Adv. Sci. **5**, 1801116 (2018)
34. Xu, Z., Hu, L., Ming, J., Cui, X., Zhang, M., Dou, J., Zhang, W., Zhou, B.: Self-gated porous organic polymer as drug delivery system for pH stimuli-responsive controlled quercetin. Microporous Mesoporous Mater. **303**, 110259 (2020)
35. Liu, T., Liu, G.: Porous organic materials offer vast future opportunities. Nat. Commun. **11**, 4984 (2020)
36. Idrees H, Zohaib S, Zaidi J, Sabir A, Khan R U, Zhang X, and Hassan S-u. A review of biodegradable natural polymer-based nanoparticles for drug delivery applications. Nanomaterials 2020; 10: 1970.
37. Luo, Y., Wang, X., Du, D., et al.: Hyaluronic acid-conjugated apoferritin nanocages for lung cancer targeted drug delivery. Biomater Sci. **3**(10), 1386–1394 (2015)
38. Sung YK and Kim S W. Recent advances in polymeric drug delivery systems. Biomaterials Research. 2020; 24(12).
39. Lv, U.P.P., Wei, W., Yue, H., et al.: Porous quaternized chitosan nanoparticles containing paclitaxel nanocrystals improved therapeutic efficiency in non-small-cell lung cancer after oral administration. Biomacromol **12**(12), 4230–4239 (2011)
40. Suresh, D., Zambre, A., Mukherjee, S., et al.: Silencing AXL by covalent siRNA-gelatin-antibody nanoconjugate inactivates mTOR/EMT pathway and stimulates p53 for TKI sensitization in NSCLC. Nanomedicine **20**, 102007 (2019)
41. Tang Y, Varyambath A, Ding Y, Chen B, Huang X, Zhang Y, Yu D-g, Kim I and Song W. Porous organic polymers for drug delivery: hierarchical pore structures, variable morphologies, and biological properties. Biomater. Sci. 2022; 10: 5369.
42. Hao D, Wang D-Y, Dong B, Xi S-C, and Jiang G. Facile synthesis of a triazine-based porous organic polymer containing thiophene units for effective loading and releasing of temozolomide. e-Polymers 2022; 22: 664–675.
43. Bialik M, Kuras M, Sobczak M & Oledzka E. Biodegradable synthetic polyesters in the technology of controlled dosage forms of antihypertensive drugs – the overview, expert opinion on drug delivery 2019; 16 (9): 953–967.
44. Xu, G., Liu, S., Niu, H., Lv, W., Wu, R.: Functionalized mesoporous carbon nanoparticles for targeted chemo-photothermal therapy of cancer cells under near-infrared irradiation. RSC Adv. **4**(64), 33986–33997 (2014)
45. Yallappa, S., Abdul Mana, S.A., Hegde, G.: Synthesis of a biocompatible nanoporous carbon and its conjugation with florescent dye for cellular imaging and targeted drug delivery to cancer cells. New Carbon Mater. **33**(2), 162–172 (2018)

46. Ejsmont A, Stasiłowicz-Krzemie ´n A, Ludowicz D, Cielecka-Piontek J and Goscianska J. Synthesis and characterization of nanoporous carbon carriers for losartan potassium delivery. Materials 2021; 14: 7345.
47. Chen, Q., Liang, L., Zhang, Z., Wang, Q.: Release of an encapsulated peptide from carbon nanotubes driven by electric fields: A molecular dynamics study. ACS Omega **6**, 27485–27490 (2021)
48. Ine´s A´vila M, Alonso-Morales N, Baeza J´ A, Rodrı´guez J J and Gilarranz M A. High load drug release systems based on carbon porous nanocapsule carriers. Ibuprofen case study, J. Mater. Chem. B 2020; 8: 5293.
49. M P Paarakh, P A Jose, C M Setty, G V Peter. Christopher G. Release kinetics– concepts and applications. International Journal of Pharmacy Research & Technology 2018; 8: 12–20.
50. Rout, S.R., Behera, B., Maiti, T.K., Mohapatra, S.: Multifunctional magnetic calcium phosphate nanoparticles for targeted platin delivery. Dalton Trans. **41**, 10777–10783 (2012)
51. Sun, R., Åhlén, M., Tai, C.-W., Bajnóczi, É.G., de Kleijne, F., Ferraz, N., Persson, I., Strømme, M., Cheung, O.: Highly porous amorphous calcium phosphate for drug delivery and biomedical applications. Nanomaterials **10**, 20 (2020)
52. Zhang, R., Hua, M., Liu, H., Li, J.: How to design nanoporous silica nanoparticles in regulating drug delivery: Surface modification and porous control. Mater. Sci. Eng. B **263**, 114835 (2021)
53. Anglina, E.J., Chengb, L., Freeman, W.R., Sailor, M.J.: Porous silicon in drug delivery devices and materials. Adv. Drug Deliv. Rev. **60**(11), 1266–1277 (2008)
54. Kuntyi P, Zozulya G, and Shepida M. Porous silicon formation by electrochemical etching. Advances in Materials Science and Engineering 2022; Article ID 1482877, 15 pages.
55. Jung, D., Cho, S.G., Moon, T., Sohn, H.: Fabrication and characterization of porous silicon nanowires. Electron. Mater. Lett. **12**, 17–23 (2016)
56. Lai Y, Thompson J R, and Dasog M, Metallothermic reduction of silica nanoparticles to porous silicon for drug delivery using new and existing reductants, Chem. Eur. J. 2018; 24,:7913 – 7920.
57. Yang, X., Xi, F., Chen, X., et al.: Porous silicon fabrication and surface cracking behavior research based on anodic electrochemical etching. Fuel Cells. **21**(1), 52–57 (2021)
58. Hernández-Montelongo, J., Oria, L., Cárdenas, A.B., Benito, N., Romero-Sáez, M., Recio-Sánchez, G.: Nanoporous silicon composite as potential system for sustained delivery of florfenicol drug. Phys. Status Solidi B **255**, 1700626 (2018)
59. Perrone Donnorso, M., Miele, E., De Angelis, F., La Rocca, R., Limongi, T., Zanacchi, F.C., Marras, S., Brescia, R., Di Fabrizio, E.: Nanoporous silicon nanoparticles for drug delivery applications. Microelectron. Eng. **98**, 626–629 (2012)
60. Zienkiewicz-Strzalka, M., Pikus, S., Skibinska, M., Blachnio, M., Derylo-Marczewska, A.: The structure of ordered mesoporous materials synthesized from aluminum phyllosilicate clay (Bentonite). Molecules **28**, 2561 (2023)
61. Hoffmann, F., Cornelius, M., Morell, J., Fröba, M.: Silica-based mesoporous organic–inorganic hybrid Materials. Angew. Chem. Int. Ed. **45**, 3216–3251 (2006)
62. Zhao, D., Huo, Q., Feng, J., Chmelka, B.F., Stucky, G.D.: Nonionic triblock and star diblock copolymer and oligomeric surfactant syntheses of highly ordered hydrothermally stable mesoporous silica structures. J. Am. Chem. Soc. **120**, 6024–6036 (1998)
63. Slowing, I.I., Viveroescoto, J., Wu, C., Lin, V.: Mesoporous silica nanoparticles as controlled release drug delivery and gene transfection carriers. Adv. Drug Deliv. Rev. **60**, 1278–1288 (2008)
64. Huang, X., Li, L., Liu, T., et al.: The shape effect of mesoporous silica nanoparticles on biodistribution, clearance, and biocompatibility in vivo. ACS Nano **5**(7), 5390–5399 (2011)
65. Ahmadi, F., Sodagar-Taleghani, A., Ebrahimnejad, P., Moghaddam, S.P.H., Ebrahimnejad, F., Asare-Addo, K., Nokhodchi, A.: A Review on the latest developments of mesoporous silica nanoparticles as a promising platform for the diagnosis and treatment of cancer. Int. J. Pharm. **625**, 122099 (2022)
66. Kaushik A, Jayant R D, Nair M. Advances in personalized nanotherapeutics; Springer; 2017. Chapter 5, Nanomaterials for drug delivery; p. 57–77.

67. Narayan, R., Nayak, U.Y., Raichur, A.M., Garg, S.: Review mesoporous silica nanoparticles: A comprehensive review on synthesis and recent advances. Pharmaceutics **10**, 118 (2018)
68. Song, S., Hidajat, K.: Functionalized SBA-15 materials as carriers for controlled drug delivery: influence of surface properties on matrix-drug interactions. Langmuir **21**, 9568–9575 (2005)
69. Heikkilä, T., Salonen, J., Tuura, J., Kumar, N., Salmi, T., Murzin, D.Y., et al.: Evaluation of mesoporous TCPSi, MCM-41, SBA-15, and TUD-1 materials as API carriers for oral drug delivery. Drug Deliv. **14**, 337–347 (2007)
70. Horcajada, P., Ramila, A., Perez-Pariente, J., Vallet-Regí, M.: Influence of pore size of MCM-41 matrices on drug delivery rate. Microporous Mesoporous Mater. **68**, 105–109 (2004)
71. Vallet-Regi, M., Ramila, A., Del Real, R., Pérez-Pariente, J.: A new property of MCM-41: drug delivery system. Chem. Mater. **13**, 308–311 (2001)
72. Wang, Y., Zhao, Q., Han, N., Bai, L., Li, J., Liu, J., Che, E., Hu, L., Zhang, Q., Jiang, T., Wang, S.: Mesoporous silica nanoparticles for drug delivery: current insights. Nanomedicine **11**, 313–327 (2015)
73. Manzano, M., Vallet-Reg, M.: Mesoporous silica nanoparticles for drug delivery. Adv. Funct. Mater. **30**, 1902634 (2020)
74. Issa, A.A., Luyt, A.S.: Kinetics of alkoxysilanes and organoalkoxysilanes polymerization: A review. Polym (Basel) **11**(3), 537 (2019)
75. El Ghannam, A., Ricci, K., Malkawi, A., Jahed, K., Vedantham, K., Wyan, H., Allen, L.D., Dréau, D.: A ceramic-based anticancer drug delivery system to treat breast cancer. J. Mater. Sci. Mater. Med. **21**, 2701–2710 (2010)
76. Meihua, Y., Siddharth, J., Peter, T., Jiezhong, C., Wenyi, G., Chengzhong, Y.: Hyaluronic acid modified mesoporous silica nanoparticles for targeted drug delivery to CD44-overexpressing cancer cells. Nanomaterials **5**, 178–183 (2013)
77. Chen, Y., Lu, Y., Lee, R.J., et al.: Nano encapsulated curcumin: and its potential for biomedical applications. Int. J. Nanomedicine **15**, 3099–3120 (2020)
78. Ellahioui, Y., Patra, M., Mari, C., et al.: Mesoporous silica nanoparticles functionalized with a photoactive ruthenium (ii) complex: Exploring the formulation of a metal-based photodynamic therapy photosensitizer. Dalton Trans. **48**(18), 5940–5951 (2019)
79. Liu, Y., Dai, R., Wei, Q., et al.: Dual-functionalized janus mesoporous silica nanoparticles with active targeting and charge reversal for synergistic tumor-targeting therapy. ACS Appl. Mater. Interfaces **11**(47), 44582–44592 (2019)
80. Karimi, M., Mirshekari, H., Aliakbari, M., Sahandi-Zangabad, P., Hamblin, M.R.: Smart mesoporous silica nanoparticles for controlled-release drug delivery. Nanotechnol. Rev. Nanotechnol. Rev. **5**(2), 195–207 (2016)
81. Sondhi, P., Lingden, D., Bhattarai, J.K., Demchenko, A.V., Stine, K.J.: Applications of nanoporous gold in therapy, drug delivery, and diagnostics. Metals **13**, 78 (2023)
82. Shuangjue, X., Wang, W.X., Xu, P.: Nanoporous gold: A review and potentials in biotechnological and biomedical applications. Nano Select. **2**, 1437–1458 (2021)
83. Bhattarai, J.K., Neupane, D., Nepal, B., Mikhaylov, V., Demchenko, A.V., Stine, K.J.: Preparation, modification, characterization, and biosensing application of nanoporous gold using electrochemical techniques. Nanomaterials **8**, 171 (2018)
84. Neupane, D., Bhattarai, J.K., Demchenko, A.V., Stine, K.J.: A pH-ensitive thiolated β-cyclodextrin-modified nanoporous gold for controlled release of doxorubicin. J Drug Deliv Sci Technol. **60**, 101985 (2020)
85. Xiao, S., Wang, S., Wang, X., Xu, P.: Nanoporous gold: A review and potentials in biotechnological and biomedical applications. Nano Select. **2**, 1437–1458 (2021)
86. Kang, T.Y., Park, K., Kwon, S.H., Chae, W.-S.: Surface-engineered nanoporous gold nanoparticles for light-triggered drug release. Opt. Mater. **106**, 109985 (2020)
87. Bae, I., Lim, K.-S., Park, J.-K., Song, J.H., Oh, S.-H., Kim, J.-W., Zhang, Z., Park, C., Koh, J.-T.: Evaluation of cellular response and drug delivery efficiency of nanoporous stainless steel. Material Biomaterials Research **25**, 30 (2021)
88. Bao, L., Park, J., Bonfante, G., Kim, B.: Recent advances in porous microneedles: materials, fabrication, and transdermal applications. Drug Deliv. Transl. Res. **12**, 395–414 (2022)

89. Sezer A D. IntechOpen; 2016. Chapter 3, Smart drug delivery strategies based on porous nanostructure materials in smart drug delivery system; p. 63–90.
90. Pourmadadi, M., Farokh, A., Rahmani, E., Shamsabadipour, A., Eshaghi, M.M., Rahdar, A., Fernando, L., Ferreira, R.: Porous alumina as potential nanostructures for drug delivery applications, synthesis, and characteristics. Journal of Drug Delivery Science and Technology **77**, 103877 (2022)
91. Kapruwan P, Ferré-Borrull J, and Marsal L F. Nanoporous anodic alumina platforms for drug delivery applications: recent advances and perspective. Adv. Mater. Interfaces 2020; 2001133.
92. Jeon, G., Yang, S.Y., Byun, J., Kim, J.K.: Electrically actuatable smart nanoporous membrane for pulsatile drug release. Nano Letter **11**(3), 1284–1288 (2011)
93. Orosz, K.E., Gupta, S., Hassink, M., Abdel-Rahman, M., Moldovan, L., Davidoff, F.H., Moldovan, N.I.: Delivery of antiangiogenic and antioxidant drugs of ophthalmic interest through a nanoporous inorganic filter. Molecule Vision **10**(68), 555–565 (2004)
94. Losic, D., Aw, M.S., Santos, A., Gulati, K., Bariana, M.: Titania nanotube arrays for local drug delivery: recent advances and perspectives. Expert Opin. Drug Deliv. **12**(1), 103–127 (2015)
95. Aw, M.S., Kurian, M., Losic, D.: Non-eroding drug-releasing implants with ordered nanoporous and nanotubular structures: concepts for controlling drug release. Biomaterials Science **2**, 10–34 (2014)
96. Yang, J., Wang, H., Liu, J., Ding, M., Xie, X., Yang, X., Peng, Y., Zhou, S., Ouyang, R., Miaoa, Y.: Recent advances in nanosized metal organic frameworks for drug delivery and tumor therapy. RSC Adv. **11**, 3241 (2021)
97. Wu, M.-X., Yan, Y.-W.: Metal–organic framework (MOF)-based drug/cargo delivery and cancer therapy. Adv. Mater. **29**, 1606134 (2017)
98. He, S., Wua, L., Li, X., Sun, H., Xiong, T., Liu, J., Huang, C., Xua, H., Sunf, H., Chene, W., Gref, R., Zhang, J.: Metal-organic frameworks for advanced drug delivery. Acta Pharmaceutica Sinica B **11**(8), 2362–2395 (2021)
99. Sun Y, Zheng L, Yang Y, Qian X, Fu T, Li X, Yang Z, Yan H, Cui C, Tan W. Metal–organic framework nanocarriers for drug delivery in biomedical applications. Nano-Micro Lett. 2020; 12:103, 29 pages.
100. Maranescu, B., Visa, A.: Applications of metal-organic frameworks as drug delivery systems. Int. J. Mol. Sci. **23**, 4458 (2022)
101. Al Haydar, M., Abid, H.R., Sunderland, B., Wang, S.: Metal-organic frameworks as a drug delivery system for flurbiprofen. Drug Des. Dev. Ther. **11**, 2685–2695 (2017)
102. Zhang, Y.Y., Zhang, L., Lin, X.W., et al.: Dual-responsive nanosystem for precise molecular subtyping and resistant reversal of EGFR targeted therapy. Chem. Eng. J. **372**, 483–495 (2019)
103. Sun, Q., You, Q., Pang, X., et al.: A photoresponsive and rod-shape nanocarrier: single wavelength of light-triggered photothermal and photodynamic therapy based on AuNRs-capped & Ce6-doped mesoporous silica nanorods. Biomaterials **122**, 188–200 (2017)
104. Zhao, S., Sun, S., Jiang, K., et al.: In situ synthesis of fluorescent mesoporous silica–carbon dot nanohybrids featuring folate receptor-overexpressing cancer cell targeting and drug delivery. Nano-Micro Lett. **11**(1), 32 (2019)

R. Abdel-Karim currently a Professor at the department of Metallurgy- Faculty of Engineering, Cairo University. She got her Ph.D. from RWTH-Aachen University in 1994. She has the following academic products; over 100 published technical papers and 10 book chapters in the field of: Corrosion and Electrochemistry, Nanomaterials and Nanotechnology. She is the supervisor of more than 40 master's and Ph.D. thesis in the field of metallurgy, materials selection, as well as different applications of nanotechnology such as; fuel cells, supercapacitors, li-ions batteries, sensors. Her research interests include the development of advanced materials for renewable energy systems, electrochemical synthesis and characterization of multifunctional materials; for H2 generation.

Development of Cardiovascular Biomaterials From Collagenous Tissues

Gowrav Baradwaj, Kshitija Aherkar, R. Mythreyi, T. S. Gopenath, and Kanthesh M. Basalingappa

Abstract Cardiovascular diseases are a worldwide issue, accounting for a significant number of deaths each year. Various types of prostheses made of synthetic textile fibers were initially used to treat cardiovascular diseases but were later abandoned. This is because it was found that the degradation of the implantation causes some of the polymers to lose mechanical strength. These polymers would also pollute the environment significantly. It became crucial to create a biomaterial that would be stable for a longer time while generating little to no pollution as a result. As collagen defines the majority of tissues, it is extensively used in cardiac tissue engineering and regeneration. Collagen networks are typically highly organized three-dimensional structures that entrap other materials. This chapter will discuss various methods for using collagen-based biomaterials in treating cardiovascular diseases, both existing and in development.

Keywords Cardiovascular diseases · Tissue engineering · Collagen · Biomaterials

Abbreviations

AF	Angiogenic factor
BM-MSC	Bone marrow derived mesenchymal stem cell
CABG	Coronary Artery Bypass Surgery

Supplementary Information The online version contains supplementary material available at https://doi.org/10.1007/978-981-99-6698-1_17.

G. Baradwaj · K. Aherkar · R. Mythreyi · K. M. Basalingappa (✉)
Department of Molecular Biology, School of Life Sciences, JSS Academy of Higher Education and Research, Mysuru, Karnataka, India
e-mail: kantheshmb@jssuni.edu.in

T. S. Gopenath
Department of Biotechnology and Bioinformatics, School of Life Sciences, JSS Academy of Higher Education and Research, Mysuru, Karnataka 570015, India
e-mail: gopenath@jssuni.edu.in

R. Malviya and S. Sundram (eds.), *Engineered Biomaterials*, Engineering Materials,
https://doi.org/10.1007/978-981-99-6698-1_17

CB	Cardiovascular Biomaterials
CVD	Cardiovascular disease
DDS	Drug Delivery Systems
ECM	Extracellular Matrix
GAGs	Glycosaminoglycans
MI	Myocardial Infarction
PCL	Polycaprolactone
PEG	Polyethylene glycol
PET	Polyethylene terephthalate
PHEMA	Poly (2-hydroxyethyl methacrylate)
PNIPAAM	Poly (N-isopropyl acrylamide)
SM	Stem cell
TE	Tissue Engineering
VEGF	Vascular Endothelial Growth Factor
3D	3 Dimensional

1 Introduction

A multitude of illnesses affecting the heart and blood vessels are referred to as "cardiovascular diseases" (CVDs) [1, 2]. They consist of cerebrovascular disease, coronary heart disease, congenital heart disease, pulmonary embolism, deep vein thrombosis, etc. [2–6]. Myocardial infractions and strokes commonly occur suddenly. A blockage that halts the flow of blood to the heart or brain, respectively, is the main cause of the acute episodes. The most prevalent cause of this is fatty buildup on the inside surfaces of blood arteries that transport blood to the brain or heart [3, 4]. Strokes may be caused by blood clots or hemorrhages from a brain blood vessel. The word "biomaterial" refers to any material that interacts with biological systems. They can be created in the lab using metallic, polymer, ceramic, or composite materials, or they can be natural or synthetic materials derived from nature [5–7]. To enhance or replace a natural function, biomaterials are commonly used in medical devices [3–5]. Examples include hip replacements, heart valves, and supplies frequently used in surgery and dentistry. Biomaterials play a crucial part in modern medicine, helping patients recover from illness or damage by restoring function [6]. In therapeutic settings, biomaterials are employed to sustain, enhance, or replace damaged tissue or bioactivity. Ancient Egyptians were the first to use biomaterials; they did it by using sutures made of animal sinew [2, 5, 6]. The study of biomaterials nowadays is influenced by a variety of scientific fields, including engineering, medical science, biology, physics, and chemistry [5]. The growth of field has been significant over the past decade as a consequence of developments in TE, regenerative medicine, and other fields. Implants for hearing loss, dental implants, prosthetic joints, ligaments, tendons, heart valves, stents, and grafts are all examples of medical implants [7–9]. Using dissolvable bandages, stitches, clamps, and staples to close wounds can all hasten the healing of human tissues [10]. Cells, bioactive substances, and biomaterial scaffolds have all been used to regenerate human tissues [3, 8, 10]. One example is a lab-grown human bladder, and another is a hydrogel that regenerates bone [8–11]. Additionally, biocompatible molecular probes and nanoparticles are employed to help cancer molecular imaging and therapy [12]. Biosensors are used to quantify and communicate information about the presence and amount of certain chemicals. Sensors that gauge brain activity and blood glucose levels are two examples [10–12]. Researchers are looking towards drug delivery systems that either apply or transport drugs to a medical goal [5–8]. Examples include vascular stents coated with drugs and implanted chemotherapy wafers for cancer patients [8–10]. The cardiovascular system is one of the most significant applications for biomaterials. Cardiovascular biomaterials are anticipated to account for most of the market for biomaterials in 2014, with a value of around $20.7 billion [10–12]. Cardiovascular biomaterials (CB) must mix with the surrounding environment and be compatible with blood to be employed [5–8]. When utilizing CB, two important considerations should be given equal weight: the material's physical and mechanical qualities, including its strength, deformation, fatigue, creep, resistance to friction and wear, flow characteristics, pressure loss, and other designed features [7–9]. There should also be consideration for biocompatibility, which describes how a substance interacts with tissue [8]. To analyze and evaluate these features, various in vivo and in vitro experiments must be carried out [7–10].

1.1 Synthetic Methods and Their Issues with Environment

Traditional tissue-engineering frameworks will not be applicable to the myocardium and the treatment of conditions like myocardial infarction; instead, the concept of an implanted biological substance that transmits information and perhaps cells is significantly more important [13]. The overwhelming majority of injected cells die as a result of environmental stress brought on by the absence of a dimensionally flexible biomimetic microenvironment, as well as direct exposure to free radicals, oxygen stress, and inflammatory cytokines for any cells administered into the myocardium [14, 15]. Since hydrogels are the most likely choice, this necessitates the development of specialized biomaterials that can successfully transport cells to the myocardium [12, 16–18]. Natural and synthetic hydrogels were investigated. In addition, while poly (ethylene glycol) (PEG), poly (2-hydroxyethyl methacrylate) (PHEMA), and poly (N-isopropyl acrylamide) (PNIPAAM) have some applicable characteristics, they are unlikely to satisfy all the criteria in the absence of additional biological functionalization and activity [14]. Due to the fluid nature of the heart's setting, strong elastomeric hydrogels should also provide ideal tissue compliance, and conductive materials could allow electrical signal transmission during cardiac action [14, 17, 18].

Again, collagen and fibrin are of special relevance as natural ECM components that could offer a firmer basis for building the necessary hydrogels [9–12]. Given that it is challenging to give such hydrogels the essential stiffnesses, more advanced molecular engineering techniques may be required [13–15]. The development of "small molecular hydrogels" is dependent on peptides like ^{D}FEFK^{D}FEFKYRGD, which in turn has been proven to provide a scaffold for mesenchymal stem cells altered with hepatocyte growth factor to maintain cardiac rhythm, post infarction while reducing ventricular remodeling in an animal model [17, 18]. It is important to investigate this approach even if these are obviously quite different from the previously established cellular engineering scaffolds [15–18].

1.2 Development of Biomaterials From Collagenous Tissue

Collagen is perhaps the most widely utilized organic polymer for uses in tissue engineering since it is present in the ECM of almost all human tissues [11–14]. Since the first analysis of the relationship between cells and acquired collagen was conducted in the early twentieth century, collagen has been employed as a biomaterial [15–18]. Collagen is used for biological applications because of a variety of characteristics that make it a great material: Benefits include biodegradability, poor antigenicity, and excellent biocompatibility. Cell receptors which can recognize a particular sequence of peptide inside collagen molecules help to further increase cell adhesion [12, 13, 15, 16]. As an added bonus, collagen, one of the most common and well-preserved proteins in vertebrates, may be extracted from a variety of sources. Bovine tendons and skin are frequently used in extraction of collagen, which makes it an affordable option for scaffold-based tissue engineering [19–22]. Since the 1970s, the application of collagen as a framework for the construction of a DDS has piqued the interest of numerous investigators throughout the world for multiple applications, like regeneration of bone, ocular, heart, and brain medicine [18–20]. Various role of collagen in cardiovascular medicines is shown in Fig. 1.

As a result of complicated design of cardiac structure, such as heart valves, tissue engineering solutions for heart disorders primarily rely on implantation and colonization of the acellular matrix. However, the effectiveness of xenogenic cellular heart valves remains a concern because of their high immunogenic potential and proclivity to calcify [23–25]. This immunologic problem prompted the creation of readily accessible human acellular scaffolding for heart and vascular reconstructive surgery, such as Cryolife®, which provides tissue replacement spanning from valves in the heart to vascular conduits [25, 26]. Collagen-based biomaterials are crucial for regenerative medicine and transplantation. Collagen is still the protein of priority for production of biomaterials due to its low immunogenicity and great biocompatibility [23]. It can be taken from a variety of tissues and combined with other molecules like stem cells, angiogenic factors, or grafted during CABG surgery [19, 23, 25]. In the laboratory, it can be used to replace decellularized ECM for fundamental investigations or as a tissue substitution material for medical purposes [27–30]. The majority of current research is directed at improving the mechanical strength, biodegradability,

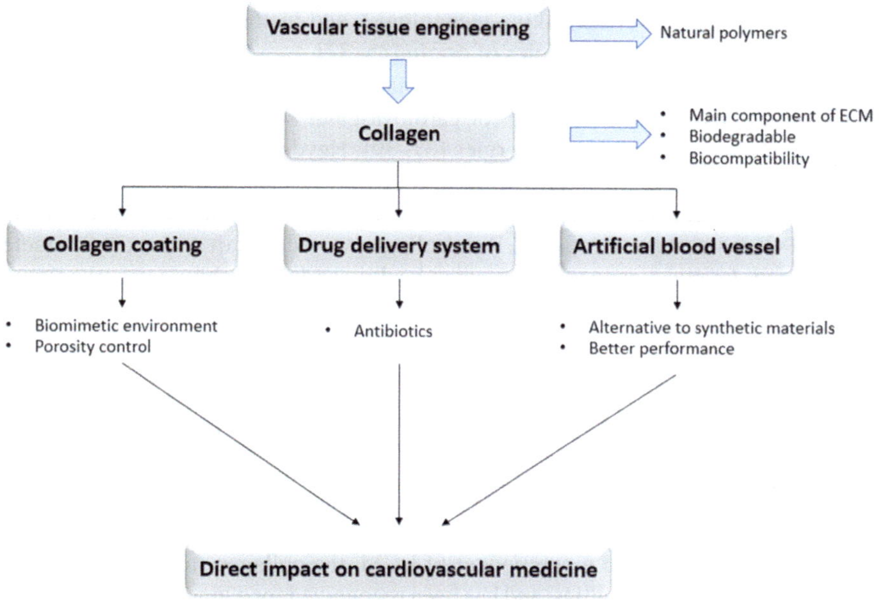

Fig. 1 Role of Collagen in Cardiovascular medicines

or delivery features of collagen-based composites for medical purposes [31–34]. Effect of collagen scaffold with various materials is shown in Fig. 2.

Collagen-based biomaterials can be created using two basic processes. The first is a decellularized collagen, also known as a matrix, that preserves its original cell shape and ECM structure [30–32]. The second method creates a useful scaffold by purifying, extracting, and polymerizing collagen [33–35]. Both procedures can

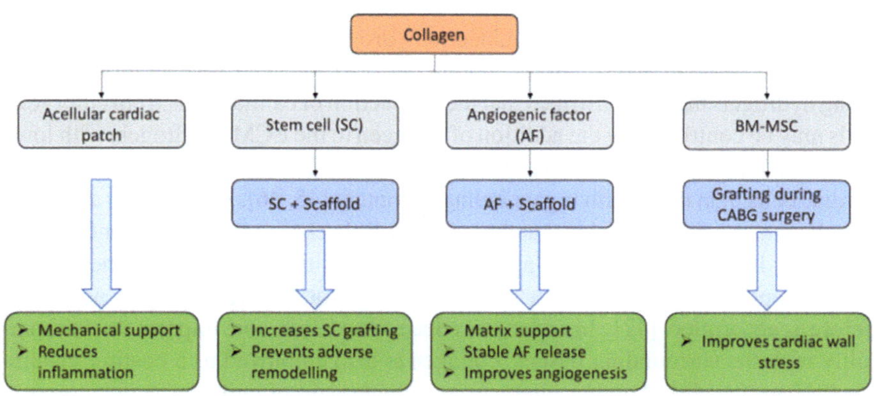

Fig. 2 Effect of collagen scaffold with various materials

go through a range of cross-linking technologies and methods that are suitable for different tissue sources and species of origin [35–37].

Collagen can be utilized as a vascular implant (or stent), a component of artificial heart valves, and an outer coating for cardiovascular stents since it is "hypoallergenic," or less prone to induce allergies [37–39]. Heydarkhan-Hagvall et al. developed a 3D electrospinning hybrid scaffold for cardiovascular tissue engineering that utilizes collagen/elastin/polycaprolactone (PCL) [37, 38]. Recently, a functional, bilayer PCL/silica nanofibrous and PCL/collagen vascular graft with a tiny diameter was created. Its makeup encourages quick healing and delays blood vessel aging or replacement by native blood vessels [38–40].

1.3 Advantages and Disadvantages of Collagenous Tissue

Although collagen has been widely used in the synthesis of biomaterials in recent years, it comes with a lot of benefits and drawbacks. As the primary protein in all connective tissues and a component of ECM, the collagen family is one of the most prevalent protein families, and has been recognized for a long time as having the potential to serve as the foundation for biomaterials, either as natural, unaltered tissue grafts or as manufactured goods for cardiac repair [41]. It has been proposed that therapy using an injectable collagen-based matrix might mediate apoptosis, vascularization, and regeneration to offer functional recovery in models of ischemia, necrosis, and/or infarction. The adhesion of cells to scaffold and grow has been facilitated by the use of them as matrices in conjunction with various anchoring materials [42, 43]. Collagen matrix has been shown to mediate several healing pathways and stop the progression of cardiac decompensation after MI. The healing of the myocardium is also observed to be significantly impacted by the timing of the collagen hydrogel application. Timing of collagen matrix hydrogel distribution impacts its ability to inhibit adverse remodeling, minimize scar size, and enhance infarcted heart performance [44]. Collagen is typically added to hydrogels formed of fibrin to boost their mechanical strength and to attract endothelial cells for heart regeneration. A study using hydrogels has demonstrated that the production of cardiac cells from embryonic cells may be controlled by the addition of collagen to the ECM. Hydrogels with lower concentrations of collagen in the ECM are less capable of differentiating hESC into cardiac cells than those with higher collagen content [45, 46]. Another key advantage of collagen as a biomaterial is that there is very little chance that an immune reaction will arise. In addition to this collagen has the capability to interact with the host and either support new host tissue or act as a scaffold for the production of new tissue before reabsorption [47]. In order to reduce blood loss and improve biocompatibility with the surrounding tissues, collagen is also employed as a coating material for PET grafts [48]. Another interesting characteristic of collagen is its composite formation with other polymers. Increased angiogenesis and thickness of ventricular wall were seen after a VEGF-collagen patch was put onto a heart with a damaged right ventricular free wall. When given with a fibrin glue hydrogel rather than alone, intramyocardially delivered BM-MSC in a rodent Myocardial Infarction model lived longer. The result was an upgrade in Cardiac function, and recovery was associated with a smaller scar. These results suggested that growth factor immobilization on a collagen patch enhances heart healing through the encouragement of cell recruitment and proliferation [49, 50]. The strong thrombogenicity of collagen has led to its use as a hemostatic agent. One of the key players in the coagulation cascade is collagen. Numerous collagen-based hemostats are available or being tested in clinical settings for a variety of surgical procedures. A sealant made of cow collagen and bovine thrombin is one of these, used for spinal and cardiovascular surgery [51].

Despite the above said advantages, there exist certain disadvantages of using collagen tissue as a biomaterial for treating cardiovascular diseases. Low immunogenicity, a porous structure, high permeability, biocompatibility, and biodegradability are only a few advantages of collagen. Although collagen may be obtained and purified from a number of sources, type I collagen is difficult to get since the pure form is costly. Collagen scaffolds lose their structural stability and mechanical strength after being hydrated, which limits their application in certain tissues [52]. Collagen's biodegradability is an intriguing quality that can be both a benefit and a drawback. The treatment is rendered worthless if the collagen breaks down before the treated tissue has healed entirely. Bovine spongiform encephalopathy and transmissible spongiform encephalopathy are risks connected with biomaterials made from collagen taken from cattle. When used as a biomaterial alone, collagen has a poor mechanical strength. However, when mixed with other polymers, such as GAG and chitosan, their impacts increased their mechanical strength while simultaneously lowering the porosity of the resultant scaffold [53]. Collagen isolated from different sources has variable crosslinking density, fiber size, impurities, and porosity. Comparing Endothelial Progenitor Cells embedded in blood vessel-derived dECM to collagen hydrogel, they showed greater capacity for proliferation and improved creation of vascular networks during the healing of cardiovascular injury [54]. Another problem is the brittleness and temperature sensitivity of collagen scaffolds, whether crosslinked or decellularized. As a result, these biomaterials are usually unable to withstand physiological burst pressures [55, 56]. Another drawback is that the scaffolds cannot be sterilized using normal high-heat techniques. The molecular makeup of the collagen scaffold is altered by alternative sterilizing techniques using low-dose gamma irradiation, which also reduces its mechanical and enzymatic resilience [57–59]. The potential for producing a substrate for arrhythmia is yet another significant worry with the use of biomaterials. ECM-based biomaterials also have limited control over the kinetics of their degradation. Purified non-crosslinked collagen hydrogel, for instance, survived only for a few days after injection into the heart, yet, after seven weeks, crosslinked high-density collagen scaffolds in pig muscle only partially degraded [60–62]. Despite the generally low immunogenicity of ECM-based biomaterials, it has been demonstrated that leftover nucleic acids and cell membrane epitopes may cause an unfavorable immune reaction after implantation [63, 64]. Progressive collagen degradation in swine valves has also been shown to result in significant ultrastructural, pre- and post-implantation alterations. This process appears to be accelerated by the removal of the endothelium and acid muco polysaccharides during commercial processing of swine valves, which allows plasma proteins to diffuse into the valvular structure [65, 66].

1.4 Challenges and Future Scope

Despite the enormous discoveries and improvements made in the field of cardiac tissue engineering, a lot more research is still needed to effectively treat cardiovascular illnesses, which are a significant global concern. One such field is the application of biomaterials. Biomaterials are widely employed in the disciplines of regenerative medicine and tissue engineering. While there are number of different biomaterials accessible today that are utilized to treat different heart problems, these materials do offer a risk to the body, and the rate of cardiac muscle regeneration is not always 100%. Much research is presently being conducted in this area to reduce problems from the employment of these biomaterials and improve the effectiveness of the therapy. Finding the appropriate biomaterial-cell type combination for tissue regeneration is the aim of biomaterials and tissue engineering. The danger of producing a substrate for arrhythmia is the main worry with employing biomaterials to treat heart illness. Although this topic hasn't received much attention in the literature, a recent study did discover that the biomaterials' spreading characteristics have an impact on how much the electrophysiological alterations are altered [67]. We need to identify beneficial cells that stimulate myogenesis, reroute apoptotic mechanisms, and restart the dormant-cell process in order to treat heart failure. More research is required on the type of biomaterial, dose, delivery technique, and timing. It is crucial to get through these obstacles and increase our knowledge of injectable cardiac tissue engineering from every angle [68]. Another difficulty is to limit the immune response to avoid the encapsulation of biomaterials or tissue constructions since doing so hinders transplanted cells from completely integrating with their original environment. As the heart is the body's main source of bioelectricity, creating conductive polymers that can help cardiomyocytes beat in unison will improve communication between transplanted tissues and the native myocardium [69, 70]. This is crucial because electrical field stimulation improves electrical signal propagation, cell polarization, and protein structure [71, 72].

Despite these difficulties, research in biomaterials and tissue engineering has achieved significant strides over the past two decades and holds out a great amount of hope. The end goal is to partially or fully regenerate an organ using biomaterial scaffolds in conjunction with the proper cell types. In the end, this will lessen the need for organ transplants, treat present irreparable illnesses, and enhance quality of life. More study is required in this field, but injectable cardiac tissue engineering combined with previously existing medications may reduce mortality and enhance quality of life in MI patients [72].

2 Conclusion

The extraction and modification of collagen for the creation of biomaterials for cardiac tissue engineering are the main topics of current research in this area. Biomaterials based on collagen are a great option for transplantation and regenerative medicine. Collagen continues to be the protein of choice for creating biomaterials due to its great biocompatibility and low immunogenicity. It may be taken out of many tissues and mixed with other substances. It can be used to replace decellularized ECM in the lab for basic research or as a tissue replacement for medicinal applications. Collagen can also be used in conjunction with other polymers to improve cell adherence and proliferation of cardiac tissues. Other naturally occurring polymers can also be used to treat cardiovascular diseases. In order to improve collagen-based composites' mechanical strength, biodegradability, or transport properties for medical applications, the majority of current research is concentrated in this area. Further study attempts to incorporate natural polymers into more challenging procedures and formulations of biodegradable, injectable, and mechanically stronger components for restoring the performance of diverse organs.

References

1. Nabel, E.G.: Cardiovascular disease. N. Engl. J. Med. **349**(1), 60–72 (2003)
2. U.S. Preventive Services Task Force, Curry, S.J., Krist, A.H., Owens, D.K., Barry, M.J., Caughey, A.B., Davidson, K.W., Doubeni, C.A., Epling, J.W., Kemper, A.R., Kubik, M., Landefeld, C.S., Mangione, C.M., Silverstein, M., Simon, MA., Tseng, C.W., Wong, J.B.: Risk assessment for cardiovascular disease with nontraditional risk factors: US Preventive Services Task Force recommendation statement. JAMA **320**(3), 272–280 (2018)
3. Lopez, E.O., Ballard, B.D., Jan, A.: Cardiovascular Disease. InStatPearls [Internet]. StatPearls Publishing (2022)
4. Libby, P., Ridker, P.M., Hansson, G.K.: Progress and challenges in translating the biology of atherosclerosis. Nature **473**(7347), 317–325 (2011)
5. Davies, M.J., Woolf, N., Rowles, P.M., Pepper, J.: Morphology of the endothelium over atherosclerotic plaques in human coronary arteries. Br. Heart J. **60**(6), 459–464 (1988)
6. Yusuf, S., Hawken, S., Ounpuu, S., Dans, T., Avezum, A., Lanas, F., McQueen, M., Budaj, A., Pais, P., Varigos, J., Lisheng, L.: INTERHEART Study Investigators. Effect of potentially modifiable risk factors associated with myocardial infarction in 52 countries (the INTERHEART study): Case-control study. Lancet **364**(9438), 937–52 (2004)
7. Lam, M.T., Wu, J.C.: Biomaterial applications in cardiovascular tissue repair and regeneration. Expert Rev. Cardiovasc. Ther. **10**(8), 1039–1049 (2012)
8. Badylak, S.F.: The extracellular matrix as a biologic scaffold material. Biomaterials **28**(25), 3587–3593 (2007)
9. Boyd, W., Johnson, W., Sultan, P., Deering, T., Matheny, R.: Pericardial reconstruction using an extracellular matrix implant correlates with reduced risk of postoperative atrial fibrillation in coronary artery bypass surgery patients. InHeart Surg. Forum **13**(5), E311 (2010)
10. Venugopal, J.R., Prabhakaran, M.P., Mukherjee, S., Ravichandran, R., Dan, K., Ramakrishna, S.: Biomaterial strategies for alleviation of myocardial infarction. J. R. Soc. Interface **9**(66), 1–9 (2012)

11. Liu, Z., Wang, H., Wang, Y., et al.: The influence of chitosan hydrogel on stem cell engraftment, survival and homing in the ischemic myocardial microenvironment. Biomaterials **33**(11), 3093–3106 (2012)
12. Spadaccio, C., Chello, M., Trombetta, M., Rainer, A., Toyoda, Y., Genovese, J.A.: Drug releasing systems in cardiovascular tissue engineering. J. Cell. Mol. Med. **13**(3), 422–439 (2009)
13. Bella, J., Hulmes, D.J.: Fibrillar collagens. Fibrous proteins: Structures and mechanisms. 457–90 (2017)
14. Boccafoschi, F., Habermehl, J., Vesentini, S., Mantovani, D.: Biological performances of collagen-based scaffolds for vascular tissue engineering. Biomaterials **26**(35), 7410–7417 (2005)
15. Bou-Gharios, G., Ponticos, M., Rajkumar, V., Abraham, D.: Extra-cellular matrix in vascular networks. Cell Prolif. **37**(3), 207–220 (2004)
16. Furthmayr, H., Timpl, R.: Immunochemistry of collagens and procollagens. Int. Rev. Connect. Tissue Res. **1**(7), 61–99 (1976)
17. Copes, F., Pien, N., Van Vlierberghe, S., Boccafoschi, F., Mantovani, D.: Collagen-based tissue engineering strategies for vascular medicine. Front. Bioeng. Biotechnol. **12**(7), 166 (2019)
18. Williams, D.F.: Challenges with the development of biomaterials for sustainable tissue engineering. Front. Bioeng. Biotechnol. **31**(7), 127 (2019)
19. Hussey, G.S., Dziki, J.L., Badylak, S.F.: Extracellular matrix-based materials for regenerative medicine. Nat. Rev. Mater. **3**(7), 159–173 (2018)
20. Vunjak-Novakovic, G.: Tissue engineering of the heart: An evolving paradigm. J. Thorac. Cardiovasc. Surg. **153**(3), 593–595 (2017)
21. Saul, J.M., Williams, D.F.: Hydrogels in regenerative medicine. In: Handbook of Polymer Applications in Medicine and Medical Devices, pp. 279–302. William Andrew Publishing (2011)
22. Parenteau-Bareil, R., Gauvin, R., Berthod, F.: Collagen-based biomaterials for tissue engineering applications. Materials **3**(3), 1863–1887 (2010)
23. Prockop, D.J., Kivirikko, K.I.: Collagens: Molecular biology, diseases, and potentials for therapy. Annu. Rev. Biochem. **64**, 403–434 (1995)
24. Exposito, J.Y., Cluzel, C., Garrone, R., Lethias, C.: Evolution of collagens. Anat. Rec. **268**, 302–316 (2002)
25. Miller, E.J.: Structural studies on cartilage collagen employing limited cleavage and solubilization with pepsin. Biochemistry **11**, 4903–4909 (1972)
26. Stachowiak, A.N., Irvine, D.J.: Inverse opal hydrogel-collagen composite scaffolds as a supportive microenvironment for immune cell migration. J. Biomed. Mater. Res. A **85**, 815–828 (2008)
27. Joyce, K., Fabra, G.T., Bozkurt, Y., Pandit, A.: Bioactive potential of natural biomaterials: Identification, retention and assessment of biological properties. Signal Transduct. Target. Ther. **6**(1), 122 (2021)
28. Lin, K., Zhang, D., Macedo, M.H., Cui, W., Sarmento, B., Shen, G.: Advanced collagen-based biomaterials for regenerative biomedicine. Adv. Funct. Mater. **29**(3), 1804943 (2019)
29. Zhu, J., Li, Z., Zou, Y., Lu, G., Ronca, A., D'Amora, U., Liang, J., Fan, Y., Zhang, X., Sun, Y.: Advanced application of collagen-based biomaterials in tissue repair and restoration. J. Leather Sci. Eng. **4**(1), 30 (2022)
30. Sorushanova, A., Delgado, L.M., Wu, Z., Shologu, N., Kshirsagar, A., Raghunath, R., Mullen, A.M., Bayon, Y., Pandit, A., Raghunath, M., Zeugolis, D.I.: The collagen suprafamily: From biosynthesis to advanced biomaterial development. Adv. Mater. **31**(1), 1801651 (2019)
31. Chen, H., Xue, L., Gong, G.: Collagen-based materials in reproductive medicine and engineered reproductive tissues. J. Leather Sci. Eng. **4**(3), 1–15 (2022)
32. Ramshaw, J.A., Werkmeister, J.A., Glattauer, V.: Collagen-based biomaterials. Biotechnol. Genet. Eng. Rev. **13**, 335–382 (1996)
33. Kwon, H., Brown, W.E., Lee, C.A., Wang, D., Paschos, N., Hu, J.C., Athanasiou, K.A.: Surgical and tissue engineering strategies for articular cartilage and meniscus repair. Nat. Rev. Rheumatol. **15**(9), 550–570 (2019)

34. Salvatore, L., Gallo, N., Natali, M.L., Campa, L., Lunetti, P., Madaghiele, M., Blasi, F.S., Corallo, A., Capobianco, L., Sannino, A.: Marine collagen and its derivatives: Versatile and sustainable bio-resources for healthcare. Mater. Sci. Eng. C **113**, 110963 (2020)
35. Liu, W., Lin, H., Zhao, P., Xing, L., Li, J., Wang, Z., Ju, S., Shi, X., Liu, Y., Deng, G., Gao, G., Sun, L., Zhang, X.: A regulatory perspective on recombinant collagen-based medical devices. Bioact Mater. **12**, 198–202 (2022)
36. Yang, Y., Campbell Ritchie, A., Everitt, N.M.: Recombinant human collagen/chitosan-based soft hydrogels as biomaterials for soft tissue engineering. Mater. Sci. Eng. C **121**, 111846 (2021)
37. Gelse, K., Poschl, E., Aigner, T.: Collagens–structure, function, and biosynthesis. Adv. Drug Deliv. Rev. **55**(12), 1531–1546 (2003)
38. Church, R.L., Pfeiffer, S.E., Tanzer, M.L.: Collagen biosynthesis: Synthesis and secretion of a high molecular weight collagen precursor (procollagen). Proc. Natl. Acad. Sci. U.S.A. **68**(11), 2638–2642 (1971)
39. An, B., Lin, Y.S., Brodsky, B.: Collagen interactions: Drug design and delivery. Adv. Drug Deliv. Rev. **97**, 69–84 (2016)
40. Delgado, L.M., Fuller, K., Zeugolis, D.I.: Collagen cross-linking: Biophysical, biochemical, and biological response analysis. Tissue Eng. Part A **23**(19–20), 1064–1077 (2017)
41. R.H., A, N.T., P., R.G., G., S.E., R., H.G., T.: A collagen mesh prosthesis for wound repair and hernia reinforcement. InSurgical Forum **13**, 29–31 (1962)
42. Burdick, J.A., Mauck, R.L., Gorman III, J.H., Gorman R.C.: Acellular biomaterials: An evolving alternative to cell-based therapies. Sci. Transl. Med. **5**(176):176ps4 (2013)
43. Johnson, T.D., Christman, K.L.: Injectable hydrogel therapies and their delivery strategies for treating myocardial infarction. Expert Opin. Drug Deliv. **10**(1), 59–72 (2013)
44. Blackburn, N.J., Sofrenovic, T., Kuraitis, D., Ahmadi, A., McNeill, B., Deng, C., Rayner, K.J., Zhong, Z., Ruel, M., Suuronen, E.J.: Timing underpins the benefits associated with injectable collagen biomaterial therapy for the treatment of myocardial infarction. Biomaterials **1**(39), 182–192 (2015)
45. Chaudhuri, R., Ramachandran, M., Moharil, P., Harumalani, M., Jaiswal, A.K.: Biomaterials and cells for cardiac tissue engineering: Current choices. Mater. Sci. Eng., C **1**(79), 950–957 (2017)
46. Duan, Y., Liu, Z., O'Neill, J., Wan, L.Q., Freytes, D.O., Vunjak-Novakovic, G.: Hybrid gel composed of native heart matrix and collagen induces cardiac differentiation of human embryonic stem cells without supplemental growth factors. J. Cardiovasc. Transl. Res. **4**, 605–615 (2011)
47. Ramshaw, J.A., Werkmeister, J.A., Glattauer, V.: Collagen-based biomaterials. Biotechnol. Genet. Eng. Rev. **13**(1), 335–382 (1996)
48. Jaganathan, S.K., Supriyanto, E., Murugesan, S., Balaji, A., Asokan, M.K.: Biomaterials in cardiovascular research: Applications and clinical implications. Biomed. Res. Int. **8**, 2014 (2014)
49. Miyagi, Y., Chiu, L.L., Cimini, M., Weisel, R.D., Radisic, M., Li, R.K.: Biodegradable collagen patch with covalently immobilized VEGF for myocardial repair. Biomaterials **32**(5), 1280–1290 (2011)
50. Frederick, J.R., Fitzpatrick III, J.R., McCormick, R.C., Harris, D.A., Kim, A.Y., Muenzer, J.R., Marotta, N., Smith, M.J., Cohen, J.E., Hiesinger, W., Atluri, P.: Stromal cell-derived factor-1α activation of tissue-engineered endothelial progenitor cell matrix enhances ventricular function after myocardial infarction by inducing neovasculogenesis. Circulation **122**(11_suppl_1), S107–17 (2010)
51. Nair, L.S., Laurencin, C.T.: Biodegradable polymers as biomaterials. Prog. Polym. Sci. **32**(8–9), 762–798 (2007)
52. Dong, C., Lv, Y.: Application of collagen scaffold in tissue engineering: Recent advances and new perspectives. Polymers **8**(2), 42 (2016)
53. Ullah, S., Chen, X.: Fabrication, applications and challenges of natural biomaterials in tissue engineering. Appl. Mater. Today **1**(20), 100656 (2020)

54. Ning, L.J., Zhang, Y.J., Zhang, Y., Qing, Q., Jiang, Y.L., Yang, J.L., Luo, J.C., Qin, T.W.: The utilization of decellularized tendon slices to provide an inductive microenvironment for the proliferation and tenogenic differentiation of stem cells. Biomaterials **1**(52), 539–550 (2015)
55. Li, Q., Williams, C.G., Sun, D.D., Wang, J., Leong, K., Elisseeff, J.H.: Photocrosslinkable polysaccharides based on chondroitin sulfate. J. Biomed. Mater. Res. Part A: Off. J. Soc. Biomater., Jpn. Soc. Biomater., Aust. Soc. Biomater. Korean Soc. Biomater. **68**(1), 28–33 (2004)
56. Nettles, D.L., Vail, T.P., Morgan, M.T., Grinstaff, M.W., Setton, L.A.: Photocrosslinkable hyaluronan as a scaffold for articular cartilage repair. Ann. Biomed. Eng. **32**, 391–397 (2004)
57. Johnson, K.A., Rogers, G.J., Roe, S.C., Howlett, C.R., Clayton, M.K., Milthorpe, B.K., Schindhelm, K.: Nitrous acid pretreatment of tendon xenografts cross-linked with glutaraldehyde and sterilized with gamma irradiation. Biomaterials **20**(11), 1003–1015 (1999)
58. Roe, S.C., Milthorpe, B.K., True, K., Rogers, G.J., Schindhelm, K.: The effect of gamma irradiation on a xenograft tendon bioprosthesis. Clin. Mater. **9**(3–4), 149–154 (1992)
59. Cheung, D.T., Perelman, N., Tong, D., Nimni, M.E.: The effect of γ-irradiation on collagen molecules, isolated α-chains, and crosslinked native fibers. J. Biomed. Mater. Res. **24**(5), 581–589 (1990)
60. Huang, N.F., Yu, J., Sievers, R., Li, S., Lee, R.J.: Injectable biopolymers enhance angiogenesis after myocardial infarction. Tissue Eng. **11**(11–12), 1860–1866 (2005)
61. S. Zaitseva, T., Yang, G., Dionyssiou, D., Zamani, M., Sawamura, S., Yakubov, E., Ferguson, J., Hallett, R.L., Fleischmann, D., Paukshto, M.V., Huang, N.F.: Delivery of hepatocyte growth factor mRNA from nanofibrillar scaffolds in a pig model of peripheral arterial disease. Regen. Med. **15**(6), 1761–73 (2020)
62. Khanna, A., Zamani, M., Huang, N.F.: Extracellular matrix-based biomaterials for cardiovascular tissue engineering. J. Cardiovasc. Dev. Dis. **8**(11), 137 (2021)
63. Neve, A., Cantatore, F.P., Maruotti, N., Corrado, A., Ribatti, D.: Extracellular matrix modulates angiogenesis in physiological and pathological conditions. BioMed Res. Int. (2014)
64. Ahmed, M., Ffrench-Constant, C.: Extracellular matrix regulation of stem cell behavior. Curr. Stem Cell Rep. **2**, 197–206 (2016)
65. Ferrans, V.J., Boyce, S.W., Billingham, M.E., Jones, M., Ishihara, T., Roberts, W.C.: Calcific deposits in porcine bioprostheses: Structure and pathogenesis. Am. J. Cardiol. **46**(5), 721–734 (1980)
66. Bruck, S.D.: Problems and challenges of biomaterials in cardiovascular applications: A status report. Biomater. Med. Devices Artif. Organs **11**(4), 271–280 (1983)
67. Suarez, S.L., Rane, A.A., Muñoz, A., Wright, A.T., Zhang, S.X., Braden, R.L., Almutairi, A., McCulloch, A.D., Christman, K.L.: Intramyocardial injection of hydrogel with high interstitial spread does not impact action potential propagation. Acta Biomater. **15**(26), 13–22 (2015)
68. Chachques, J.C.: Cellular cardiac regenerative therapy in which patients? Expert Rev. Cardiovasc. Ther. **7**(8), 911–919 (2009)
69. Griffith, L.G., Naughton, G.: Tissue engineering—current challenges and expanding opportunities. Science **295**(5557), 1009–14 (2002)
70. Tandon, N., Cannizzaro, C., Chao, P.H., Maidhof, R., Marsano, A., Au, H.T., Radisic, M., Vunjak-Novakovic, G.: Electrical stimulation systems for cardiac tissue engineering. Nat. Protoc. **4**(2), 155–173 (2009)
71. Radisic, M., Park, H., Shing, H., Consi, T., Schoen, F.J., Langer, R., Freed, L.E., Vunjak-Novakovic, G.: Functional assembly of engineered myocardium by electrical stimulation of cardiac myocytes cultured on scaffolds. Proc. Natl. Acad. Sci. **101**(52), 18129–18134 (2004)
72. Cui, Z., Yang, B., Li, R.K.: Application of biomaterials in cardiac repair and regeneration. Engineering **2**(1), 141–148 (2016)

Gowrav Baradwaj is a postgraduate student under Dr. Kanthesh BM, an Associate Professor and Coordinator of Division of Molecular Biology, in JSS Academy of Higher Education & Research. He is currently pursuing his master's in Molecular biology. He has completed his undergraduate in NITTE University Centre for Science Education and Research, Mangaluru in Biomedical Science Hons. After completion of postgraduate studies, Gowrav wants to pursue his PhD and continue with research in the field of Stem cell research and cancer.

Kshitija Aherkar is a postgraduate student at JSS Academy of Higher Education and Research studying under Dr. Kanthesh BM, an Associate Professor and Division Coordinator of Molecular Biology, in JSS Academy of Higher Education & and Research. She is currently pursuing studies on her master's degree in Molecular Biology. She completed her undergraduate studies at St Xavier's College, Mumbai with a dual major in Microbiology and Biochemistry. Kshitija aspires to pursue her Ph.D. after finishing her Masters degree and conduct research in the fields of genetics and antibiotic resistance.

Ms. Mythreyi R is a Ph.D. Scholar under the guidance of Dr. Kanthesh M Basalingappa, an Associate Professor and Course Coordinator of Division of Molecular Biology, JSS Academy of Higher Education & and Research—Mysuru. Her area of Research Interest is currently aimed towards inhibition of Molecular signaling pathways involved in cancer and development of drugs with the help of Phyto-chemicals, and genetics. She graduated her UG with a triple major of Chemistry, Botany and Zoology in the year 2019 from Kuvempu University, Shankaraghatta, Shivamogga. She has been awarded with the Post graduate degree, an M.Sc. in Molecular biology from School of Life Sciences, JSS Academy of Higher Education and Research- Mysuru in the year 2021. The author is currently pursuing her Ph.D. under the Division of Molecular Biology in JSS Academy of Higher Education and Research—Mysuru. During her post graduation, the author has been awarded with Indian academy of Sciences Summer research fellowship-2020, worked on a survey of natural compounds with the capacity for inhibition of SARS-CoV2. The author has also carried out a project related to hematology of turner syndrome. The author has cleared GATE 2021 in 2 papers Biotechnology and Life Sciences. The author has a review publication during her post graduation. She is a budding researcher in the field of Molecular Biology, with a vision for invention and innovation.

Dr. Kanthesh M Basalingappa is an Associate Professor of Molecular Biology at the Division of Molecular Biology, JSS Academy of Higher Education & Research, Mysuru. India. His goal of research is to determine the role of RNA Binding proteins in tumor progression and metastasis. Post-transcriptional regulation of gene expression by RNA binding protein is a crucial mechanism in regulating the timing and the amount of expression of genes. Growing evidence indicate that the alteration of the expression and function of RNA binding proteins could potentially play a role in inflammation and cancer. Dr.Kanthesh BM did is PhD from University of Madras (2005), He also did postdoctoral research at the University Malaya, Kuala Lumpur, Malaysia (2007–2009); West Virginia University, Morgantown, USA (2090-2011) and University of Oklahoma Health Sciences, Oklahoma, USA (2011–2014). He received Malaysia Prestigious Bio-Malaysia Gold medal Award (2008). For his research area is Arbovirus infections, in that they done patented work on "Early detection of BK virus using molecular methods". He also Received Dr. Wilson Aruni's "Best Research Mentor and Teacher Gold medal Award" from the Indian Association of Applied Microbiology (IAAM) (2018). He has been engaged in teaching and research in Microbiology and Molecular Biology for the past 20 years. He has published over 80 original research papers, 15 book chapters, and 15 review articles. He is also Professional and Scientific Memberships in, American Association for Cancer Research (AACR), Life Member of Indian Association of Applied Microbiology (IAAM), Life Member of Indian Association of Biomedical Scientists (IABMS), Indian Association of Medical Microbiologist (IAMM). He Received Fellowship Award from Indian Association of Applied Microbiology (FIAAM). At present he is a having collaboration with Royal Research Foundation, a research institute in India. I appreciate your support.

Stimuli-Responsive Material in Controlled Release of Drug

Karan Trehan, Muskaan Saini, and Shubham Thakur

Abstract In the past decade, "Stimuli-responsive materials" have emerged as effective drug delivery systems (DDSs) that are expected to continue advancing. These materials react to various triggers like light, temperature, magnetic fields, electricity, pH, and redox potential to release drugs at specific locations, enhancing drug efficacy. They can be shaped and possess unique characteristics, such as mechanical, chemical, electrical, optical, and biological properties. The main advantage of these materials is their ability to provide targeted drug delivery, minimizing side effects and increasing effectiveness. Understanding the stimuli that trigger desired responses and designing polymer systems accordingly, as well as characterizing the resulting changes, is crucial. Smart polymers alter their shape, chemical structure, or mechanical properties in response to triggers, allowing precise drug delivery at specific times and locations within the body. Such "intelligent polymers" are gaining popularity in the pharmaceutical industry due to their ability to release drugs at targeted times and locations. Examples include chitosan, cellulose, PEGs, gelatin, albumin, and polyacrylic acid, which are used for various drug delivery routes. However, ensuring their safe and effective use in clinical applications requires considering factors like biocompatibility and drug stability. Smart polymers also hold potential in biotechnology and biosensors. Challenges remain, including environmental interactions and the complexity of their structures. Durability and long-term stability need improvement, but stimuli-responsive materials can be tailored to target specific cells or body parts in forms like pH-sensitive, temperature-sensitive, and magnetic-responsive materials. They can regulate drug release, protect drugs against degradation, and enhance bioavailability, promising transformative advancements in drug delivery and treatment. This chapter provides a summary of stimuli-responsive materials' role in controlled drug release, highlighting different material types, required stimuli, advantages, and disadvantages in developing novel DDSs.

Keywords Stimuli-responsive materials · Smart materials · Chitosan · Bioavailability · Controlled release

K. Trehan · M. Saini · S. Thakur (✉)
Department of Pharmaceutical Sciences, Guru Nanak Dev University, Amritsar, Punjab 143005, India
e-mail: shubhamdthakur@gmail.com; shubhamdpharma.rsh@gndu.ac.in

R. Malviya and S. Sundram (eds.), *Engineered Biomaterials*, Engineering Materials,
https://doi.org/10.1007/978-981-99-6698-1_18

1 Introduction

The effective and precise delivery of drugs to diseased tissues is essential for treating numerous disorders, such as cancer, cardiovascular, and inflammatory diseases [1]. However, traditional drug administration methods often lead to non-specific targeting of different tissues and organs, requiring large amounts of drugs and causing poor therapeutic efficacy and adverse effects to normal tissues [2]. To tackle these obstacles, molecular prodrugs have been developed to increase the targeting of affected cells and tissues and enhancement of treatment effectiveness [3]. However, designing and preparing prodrugs can be a complex process, requiring special drug modifications for individual drugs, making it difficult to establish a standardized method for prodrug development [4]. To tackle these challenges, an alternative approach involves utilizing nanocarrier systems based on nanoscale materials to deliver medications to particularly affected tissues [5]. Nanoparticles used to transport drugs have numerous advantages, including safeguarding therapeutic drugs against degradation, allowing the administration of drugs at low dosages, prolonging the time they stay in the body, and boosting their effectiveness in treating diseases while reducing their impact on healthy tissues [6]. Applying specialized groups on the surface of nanoparticles shields the therapeutic cargo from harsh conditions in the body, ensuring it stays intact. Furthermore, these modified nanoparticles can be accurately directed to affected cells and tissues, leading to better treatment outcomes with negligible adverse effects [7].

Achieving spatiotemporal regulation over the targeted drug delivery is critical to enhancing treatment efficiency once nanocarriers have accumulated in pathological areas via passive or active targeting [8]. Due to the swift advances in nanobiotechnology, artificial biomaterials that react to stimuli have become increasingly popular because of their possible application in medical and pharmaceutical fields, such as medical transportation, biosensing, and tissue adhesion. These materials exhibit quick fluctuations in structure, solubility, or polarity on utilizing external or internal triggers, such as temperature, light, magnetic fields, pH, or enzymes [9]. By making use of these materials as carriers for drugs on a nanoscale, it becomes possible to transport medications with precision to specific locations in a safe and controlled way. This method helps to prevent harmful side effects that may arise from harming healthy tissues.

The idea of SRDD relies on the ability of biological systems to react to different stimuli. For instance, drugs can be released as a reaction to changes in pH levels, such as when the stomach environment changes from acidic to alkaline. Bacteria also react to fluctuations in environmental factors, such as temperature and pH. The development of DDSs that respond to stimuli has become feasible thanks to advancements in nanotechnology, materials science, and chemistry. These systems typically consist of a stimulus-sensitive component, a drug, and a carrier matrix. The stimulus-sensitive component, which can be a polymer, liposome, or nanoparticle, is designed to react to a specific stimulus to trigger drug release. The drug is usually

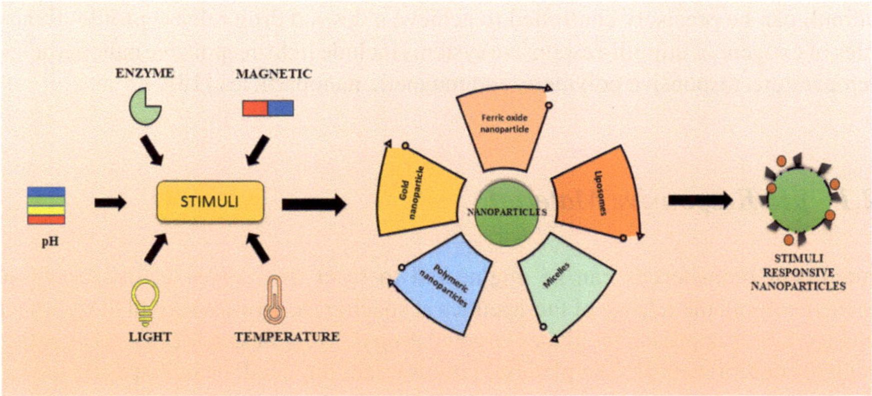

Fig. 1 Various forms of stimuli and stimuli-responsive material

encapsulated or bound to the carrier matrix, which can be made from natural or synthetic materials. Figure 1 illustrates the various forms of stimuli and stimuli-responsive materials.

SRDDSs have several advantages, one of which is their ability for localized drug administration, such as to tumors or inflamed tissues. For instance, pH-responsive systems can target tumors since they have a lower pH compared to healthy tissues. Enzyme-responsive systems, on the other hand, can target inflamed tissues where there is a high concentration of enzymes [10]. Another benefit of stimuli-responsive DDS is the potential to improve drug efficacy while reducing toxicity. By releasing drugs in a controlled manner, these systems can maintain therapeutic drug levels for longer periods, decreasing the need for frequent dosing. Moreover, they can decrease drug toxicity by minimizing off-target effects and improving drug selectivity [11]. Despite the promising potential of stimuli-responsive DDS, there are still several challenges that need to be addressed. A significant obstacle is the insufficient perception of the complex interactions between the drug, carrier matrix, and stimulus-sensitive component. Additionally, the stability and biocompatibility of these systems need to be optimized to ensure their safe and effective use in vivo.

2 Types of Stimuli-Responsive Material

SRDDSs are categorized into two main types: endogenous and exogenous systems. Endogenous SRDDSs utilize biological signals present in the target tissue or organ, such as pH, enzymes, and redox reactions. These systems are highly specific and can selectively target diseased tissues with higher levels of these biological signals. pH-responsive and enzyme-responsive nanocarriers are instances of endogenous stimuli-responsive systems [12]. Exogenous SRDDSs are triggered by external triggers such as light, temperature, and magnetic fields. These systems are highly tuneable, as the stimuli can be precisely controlled to achieve a desired drug release profile. Examples of exogenous stimuli-responsive systems include light-responsive nanoparticles, temperature-responsive polymers, and magnetic nanoparticles [13].

2.1 pH-Responsive Material

Intelligent biomaterials can be engineered to react to changes in their environment, enabling the release of therapeutics at specific locations where pH levels have changed. This approach facilitates targeted drug delivery in specific organs or pathological conditions [14]. The pH levels in diseased, inflamed, or infected tissues can decrease due to irregular metabolism or abnormal blood vessel growth, resulting in oxygen and nutrient deprivation, and a shift to glycolytic metabolism [15]. In sub endosomal and lysosomal organelles within cells have a pH range of 5.5–6.8 and 4.5–5.5, respectively [16]. Changes in pH can impact crucial crosslinking mechanisms for injectable hydrocolloids and self-regenerating substances. The addition or removal of protons from acidic or basic groups creates unique interactions between a drug and a substrate, resulting in a specific release profile that can be beneficial in targeting a particular tissue or cell [17].

The acidic microenvironment in solid tumors presents an opportunity for developing DDSs that are responsive to pH changes. This technology can also be applied to other diseases such as gastric ulcers through the development of smart delivery systems [18]. These materials have been developed and studied in various forms, including hydrogels, nanoparticles, beads, hollow particles, and three-dimensional porous structures [19]. There are three categories of these polymers: pH-ionizable responsive polymers, acid-cleavable polymers, and carbon dioxide (CO_2) responsive polymers. The category of pH-ionizable responsive polymers is grouped into three subtypes: anionic polymers, cationic polymers, and zwitterionic polymers. Table 1 shows the various pH-responsive polymers and their actuation responses.

Poly (acrylic acid) (PAAc) is a pH-responsive anionic polymer that can change its solvation state and shape when pH exceeds its distinctive pKa. This property makes anionic polymers suitable for developing pH sensors. PAAc also contains carboxylic acid groups that can be easily modified chemically post-polymerization [19].

Table 1 Various pH-responsive polymers and their actuation response

Polymer	Properties	Actuation response
pAAc	Anionic	pH variation
pDMAEMA	Cationic	pH variation
pNIPAm	Neutral	pH, ionic strength, temperature
p(HEA-co-HEMA-co-PSPMA)	Anionic	pH, ionic strength
p(HEA-co-DMA-co-dopamine)	Neutral	pH, ionoprinting, catechol-metal coordination chemistry

In contrast to anionic polymers, cationic polymers usually contain amine-containing functional groups such as amine, amidine, and pyridine. These groups become positively charged when they are protonated, with the exception of quaternary amines which are permanently positively charged. Cationic polymers such as PDMAEMA, PDADMAC, and certain natural polymers are widely used in various because of their positive charge [20]. PAMAM is a usually used dendrimer that has a distinct chemical structure and nanometer-sized width. It possesses many advantages over PEI, such as its ability to dissolve in water, be biodegradable and biocompatible, making it suitable for in situ biosensing applications [8]. PDADMAC, a positively charged polyelectrolyte, has been utilized for actuation purposes. It adheres strongly to pNIPAm-co-AAc microgels via electrostatic forces, causing the microgels to tightly bind to the gold-coated surface. In particular, in conditions of high humidity, the PDADMAC layer expands, maintaining the device in an unbent state. Conversely, under low humidity conditions, the PDADMAC layer loses moisture, resulting in the bending of the device [21]. PDMAEMA is another cationic polymer that has unique temperature and pH sensitivities. It has tertiary amino groups and responds to changes in temperature and pH, collapsing when the solution pH is raised to 7.5 or when the temp. goes above 40 °C. When PDMAEMA is added to an opal template, it causes the photonic crystal to change color by swelling and deswelling in reaction to alterations in the pH and temperature of the solution [22]. Zwitterionic polymers contain both anionic and cationic groups, whereas anionic and cationic polymers have only one type of charged group. Zwitterionic polymers can have anionic and cationic groups positioned either on the side chains of different comonomer units, which is referred to as polyampholyte, or on the same monomer unit, known as polybetaine [8]. Zwitterionic polymer surfaces are ideal for biosensing applications due to their high hydrophilicity, charge, and hydration layer that prevents biofouling, and can help overcome the limitations of current implanted continuous glucose monitors, which exhibit low reliability and signal noise, likely caused by the immune response to the wound healing process [23].

Polyacids, also referred to as pH-responsive acidic polymers, contain acidic functional groups within their structure. The degree of hydrophilicity in aqueous environments is adjusted by changes in the external pH, which is dependent on the quantity of negatively charged groups on the polymer chain. These polymers are categorized based on their functional groups [24]. At high pH levels, the carboxylic acid groups in the polymer undergo deprotonation, releasing H+ ions, and resulting in an augmentation of the quantity of negatively charged moieties on the polymer chain. The pH level at which the acid becomes ionized is determined by its dissociation constant (ka), which varies as per the polymer's composition and molecular weight and differs from that of monoacids [25]. Polymers containing sulfonic acid are considered strong polyelectrolyte hydrogels due to their extensive ionization. These polymers have a sulfonate group (SO_3H) attached to their structure and have pKa values ranging from 2 to 3. To address the challenges associated with sulfonic acid polymers, a novel group of pH-sensitive polymers was developed. These polymers feature a sulfonamide group attached to aniline. The sulfonyl group in the structure acts as an electron-withdrawing element from the nitrogen atom of the amine, causing the hydrogen atom bound to it to become ionized [24].

Le et al. developed a hydrogel actuator with anisotropic properties by combining pAAm and pAAc. They achieved this by immersing a pAAm hydrogel in a solution of AAc monomers and polymerizing them using photopolymerization. The uneven distribution of pAAc within the hydrogel resulted from imperfect penetration of UV light, leading to asymmetrical swelling and shrinking responses to pH changes [26]. Similarly, Shang et al. fabricated a composite actuator using patterned pNIPAm/pAAc that could undergo controlled shape changes in response to pH, ionic strength, and temperature variations. The actuator was created by immersing a pAAc hydrogel in a NIPAm/water solution and exposing it to UV light through a mask. At pH 11, the hydrogel bent towards the pNIPAm-rich side [27]. Huang et al. conducted a study where a digital projector was utilized to dynamically and spatially control light for triggering polymerization. This resulted in varying conversion and crosslinking density within a p(HEA-co-HEMA-co-PSPMA) hydrogel. The hydrogel was 3D printed and exhibited a non-uniform swelling ratio and mechanical properties, leading to the transformation into a specific shape upon swelling [28].

Actuators based on cationic polymers have also been explored, albeit not as extensively as anion-responsive actuators. Huang et al. demonstrated that a bilayer hydrogel actuator made of pDMAEMA/p(DMAEMA-co-AAm) becomes increasingly curved as the solution pH decreases [29]. This behavior is attributed to the pDMAEMA layer swelling significantly at lower pH because of amine protonation, in contrast to the p(DMAEMA-co-AAm) layer [30]. Furthermore, Lee et al. demonstrated the utilization of varying crosslinking densities in the hydrogel network to develop actuators responsive to pH changes. They employed the ionoprinting technique along with catechol-metal coordination chemistry to manufacture a pH-responsive actuator. By locally imprinting Fe3+ ions into poly (N-hydroxyethyl acrylamide-co-dopamine methacrylamide) (p(HEA-co-DMA)) through the application of an electrical potential via an iron electrode, they successfully created the pH-responsive actuator [31].

2.2 *Magnetic Field Responsive Material*

Magnetic stimulation offers several advantages for drug delivery, such as precise targeting of the delivery system, monitoring of concentration and distribution, and control over the rate of drug release [32]. Iron oxide nanoparticles (IONPs) are FDA-approved inorganic nanomaterials that respond to external magnetic fields. They are biocompatible, low in toxicity, and easy to synthesize, making them highly attractive for numerous studies. Due to their superparamagnetic properties, Iron oxide nanoparticles can function as imaging enhancers for MRI and allow for drug release to be controlled remotely using a magnetic field [33]. Permanent magnets are positioned around the substance to make a magnetic field gradient, which holds or eliminates the superparamagnetic iron oxide nanoparticles from the bloodstream. When subjected to an AMF, IONPs produce heat via Brownian losses, Neel relaxation, or both. This thermal energy is employed to enhance drug diffusion, induce changes in shape, or accelerate drug release by breaking heat-sensitive covalent bonds [32].

Research has investigated the use of magnetic-responsive polymeric nanoparticles as a way to remotely control drug release and combine cancer diagnosis and treatment. Nanotechnology is promising for delivering therapeutics and diagnostic techniques to mitigate the negative consequences of conventional cancer treatments and detect diseases immediately [34]. A nano-vehicle was created using multi-block polyurethanes to achieve precise tumor theranostics, with polymers containing removable PEG and biodegradable PCL flexible components, disulfide linkages that respond to reduction, and alkynyl groups that can be clicked to attach targeting ligands after conjugation. The breaking of the PEG layer acted as a switch, activating the targeted ligands via click chemistry in an extracellular acidic environment, which led to the deterioration of the nanoparticle core and discharge of the payload within cancer cells. These polymeric materials demonstrated excellent MRI contrast effects [35], MRI that is guided by magnetic fields, and the utilization of multiple techniques to target drugs to the tumor, leading to a pronounced suppression of cancer with minimal side effects. Other investigations have employed FA-conjugated bovine serum albumin (BSA) [36], and a theranostic micellar DDS created using PDMA-b-PCL micelles that delivered SN-38, ultrasmall SPIONs, and siRNA targeting VEGF, acting as a negative MRI contrast agent in T2-weighted imaging [34].

2.3 Electro-Responsive Material

Recently, there has been significant attention in electro-responsive actuation because of the simplicity of introducing and regulating electricity in materials through the magnitude and waveform of the applied voltage [37]. Polyaniline, polythiophene, polypyrrole, and their derivatives, which are conducting polymers, have found extensive use in diverse applications like neural prostheses, and controlled drug delivery [38]. These polymers can be employed in developing wireless dermal or implantable electronic delivery devices that use external electric fields to achieve on-demand drug release by adjusting the field's intensity [39]. By utilizing advanced devices that provide precise control over voltage, it is feasible to achieve an accurate degree of authority over drug release [40]. Figure 2 illustrates the mechanism of electro-responsive material DDS.

Although electro-responsive systems offer many advantages, the need for invasive surgery to implant electronic delivery devices is a significant drawback. Furthermore, the shallow penetration depth and potential tissue injury at higher voltages could pose a significant obstacle to these systems [41]. One solution to this problem is an electro-responsive DDS that releases drugs through carriers that dissolve because of chemical reactions or changes in pH values brought about due to an electric potential [42]. Poly(pyrrole) (PPy), a conductive polymer, has been widely utilized

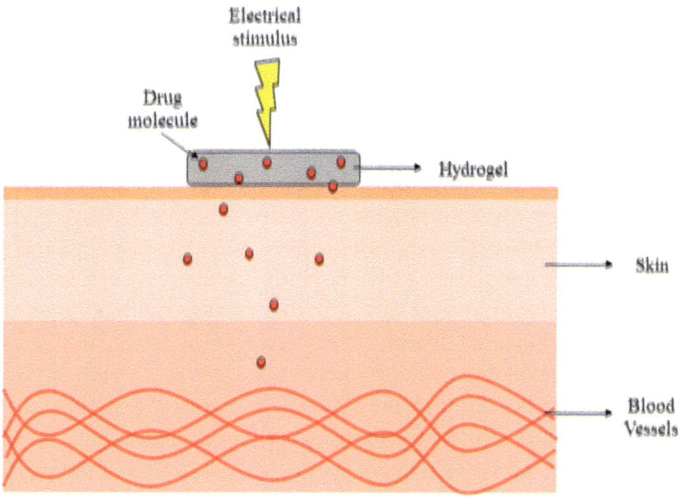

Fig. 2 Mechanism of electric responsive DDS

for controlled drug delivery activated by electrostimulation because of its excellent biocompatibility. A direct release method was developed for PPy, which can release a versatile payload option regardless of their molecular weight and polarity. Biotin was introduced as a stimulus to commence the release of biotin and the connected payload during electrical stimulation by reducing the PPy backbone [43]. This system can deliver various compounds, biomolecules, and drugs, with nerve growth factor (NGF) serving as an experimental system. NGF belongs to the neurotrophin family, which affects neuronal development and fate in the central and peripheral nervous systems [41].

Intrinsic Conductive Polymers (ICPs) are a class of polymers that typically contain alternating s and p bonds, and their electrical characteristics arise from p-conjugated systems. Among all the ICPs, Polyaniline (PANI) is considered the most economically viable, due to cost-effectiveness, high electrical conductivity, ease of processing, and good thermal stability [44]. PANI exhibits three different oxidation states of polyaniline, i.e., fully reduced (leucoemeraldine), half oxidized (emeraldine), and fully oxidized (pernigraniline), which exhibit unique electrical characteristics and colors. PANI has applications in gas sensors, particularly for detecting hydrogen chloride, hydrogen sulfide, and ammonia gases, as it can be oxidized in acidic solutions [8].

PANI's conductivity increases when it shifts from the emeraldine salt to the emeraldine base form. Acids are employed as stimulus in this process, providing enough protons with strong ionizing and conducting properties. PANI-based polymers' chemical composition, synthetic approach, and molecular weight all affect their conductivity [45]. PANI is also researched for possible utilization in glucose, DNA sensors, as well as in actuator production. A study explored electro-responsive actuators made of camphorsulfonic acid-doped PANI microfibers. These microfibers converted electrical energy into thermal energy (called the electrothermal effect), causing them to expand and bend when exposed to an electrical current. Applying a low voltage (less than 0.5 V) and an external magnetic field enabled the actuator to swing up to 9000 times per minute, within the range of insect wing flapping speed (between 1000 and 15,000 swings per minute) [8].

Ge et al. have reported on the initial successful in vivo application of drug delivery activated by an external electric field. They utilized conductive polymer nanoparticles that were encapsulated in thermo-responsive block copolymers, which can change from a liquid to a hydrogel state. The release of negatively charged fluorescein and positively charged daunorubicin from the conductive nanoparticles was ascribed to redox reactions induced by the electric field. This finding suggests the possibility of non-invasively initiating drug delivery by manipulating the characteristics of the electric field and selectively releasing co-loaded drug molecules [46]. Yan et al. have devised a system of voltage-sensitive vesicles that can undergo reversible self-assembly. This system includes a cyclodextrin-capped polystyrene chain and a ferrocenyl group-containing polyethylene oxide homopolymer. The self-assembled structure can be dissociated by applying an oxidizing voltage to expel the charged ferrocenyl cations from the cyclodextrin cavity.

On the other hand, a reducing voltage can revert the ferrocenyl functionality to its neutral state and allow for self-assembly. Yan et al. have introduced a system of electrical-responsive vesicles. This system consists of a polystyrene chain with a cyclodextrin motif at the end and a polyethylene oxide homopolymer containing a ferrocenyl group. When an oxidizing voltage is applied, the charged ferrocenyl cations are ejected from the cyclodextrin cavity, resulting in the dissociation of the self-assembled structure. The release of Rhodamine B was dependent on the applied voltage [47]. Chang and colleagues have developed a novel system of voltage-responsive reversible self-assembly using double hydrophilic block copolymers. The self-organized polyethylene glycol-poly(acrylic acid) (PEG113-b-PAA30) block copolymers into micelles by adding a cationic ferrocenyl surfactant. Oxidation of the ferrocenyl group into ferrocenium cations disrupts the micelles because of enhanced water solubility. The system demonstrated electrically induced disassembly, and the release of rhodamine 6G increased with raising oxidative voltage via electrostatic attraction [48].

Another investigation by Kim et al. introduced liposomes that can be triggered by an electric field by attaching random copolymers onto the surface of phosphatidylcholine liposomes. The imposition of an electric field ionizes the carboxyl groups of the attached chain, which in turn causes shear stress on the liposomal membrane, resulting in the release of the entrapped molecules. The liberation of calcein depended on the carboxyl content in the liposomes and the imposed electrical potential. Moreover, H+ generated by water electrolysis was also able to cause membrane compression, providing an alternative pathway to membrane disruption [49]. Qu et al. developed an injectable hydrogel by combining chitosan-g-polyaniline with oxidized dextran. They discovered that applying a voltage to the hydrogel resulted in the pulsatile release of both hydrophilic and lipophilic agents, which returned to a passive release rate when the voltage was switched off.

In a separate study, Liu et al. created a voltage-responsive hybrid hydrogel by incorporating reduced graphene oxide into a poly (vinyl alcohol) matrix. The study examined the release of the pain reliever, lidocaine hydrochloride, from hydrogels comprising varying concentrations of reduced graphene oxide (rGO) by applying an oxidative voltage (2–15 V). When no voltage was employed, the release of lidocaine decreased as the rGO content increased. However, when a voltage was applied, the release rate increased significantly with higher rGO content [50]. Servant et al. developed electro-responsive hydrogels using multi-walled carbon nanotubes (MWNT) and graphene as conductive components. The hydrogels were created by polymerizing methacrylic acid and N,N′-methylene bisacrylamide in the existence of either MWNTs or graphene. When stimulated with electrical voltage, the hydrogels underwent deswelling, causing water and loaded substances to be expelled. Comparing the two types of hydrogels in vitro and in vivo, the study found that the graphene-containing gel released more radiolabelled sucrose ([14C]-sucrose) in a pulsatile manner when stimulated with an electrical impulse [51, 52]. Table 2 shows the electro-responsive drug release in vitro/in vivo.

2.4 Light-Responsive Material

Using light as a trigger for drug delivery provides precise control over when and where drugs are released [53]. Light-responsive materials have advantages over traditional DDSs, one being their potential to focus on specific sites using different wavelengths of light [54]. The characteristics of light, like frequency, charge, strength, and length can be modified to attain precise drug discharge [55]. Therapeutic drugs loaded onto nanocarriers can selectively home in diseased tissues which use either passive or active binding and internalization pathways. Once the nanocarriers accumulate in diseased tissues and cells, radiation using light can be applied to the affected area with minimal impact on healthy tissues, and the light exposure can be managed in both spatial and temporal aspects.

Light-responsive drug delivery nanocarriers can be divided into three categories which depend on the wavelength of the light they respond to, namely ultraviolet (UV) light (200–400 nm), visible (Vis) light (400–700 nm), and near-infrared (NIR) light (700–1000 nm) [56]. Among these, UV light is preferred because of its ability to activate a broad range of light-responsive materials and trigger the release of therapeutic drugs through various reactions such as cleavage, isomerization, rearrangement, or cross-linking. Nevertheless, the utilization of UV radiation has certain disadvantages, such as inadequate tissue penetration and elevated phototoxicity, which can induce harm to the tissues, making it inefficient in treating deep-seated ailments. Hence, drug-delivery nanosystems activated by UV light have restricted usefulness in clinical applications [57].

In contrast to UV light, NIR and certain visible light wavelengths in the range of 600–1000 nm are more suitable for light-responsive materials since they can penetrate deeper into tissues. This is because they are less affected by natural pigments like

Table 2 Electro-responsive drug release in vitro/in vivo

Reference	Method	Materials	Key findings
Ge et al. [47]	External electric field triggered drug release	Conductive polymer, poly(pyrrole), and block copolymers consisting of a combination of d,l-lactic acid and glycolic acid linked to polyethylene oxide, with additional units of d,l-lactic acid and glycolic acid attached to the other end of the polyethylene oxide chain	1. Remote activation of drug release by means of an electrical force 2. The controlled discharge of drug molecules that are loaded together can be accomplished by modifying the characteristics of the electric field, specifically by using reduction or oxidation
Yan et al. [48]	Voltage-responsive vesicles	A polymer made of polystyrene with a ˇ-cyclodextrin (ˇ-CD) design at one end and a homopolymer of poly (ethylene oxide) with a ferrocenyl group at the other end	1. Reversible self-assembly of voltage-responsive vesicles 2. Controlled release of encapsulated molecules (Rhodamine B) through voltage variation
Chang et al. [49]	Electro-responsive reversible self-assembly system	A double hydrophilic block copolymer called Poly(ethylene glycol)-b-poly(acrylic acid) (PEG113-b-PAA30) and together with a positively charged surfactant containing ferrocenyl, known as (11-ferrocenylundecyl) trimethylammonium bromide	1. Electrically induced disassembly process through the reversible electrochemical activity of the ferrocene group 2. Controlled release of encapsulated molecules (rhodamine 6 G) through electrostatic attraction and increasing oxidative voltage

(continued)

Table 2 (continued)

Reference	Method	Materials	Key findings
Kim et al. [50]	Electro-responsive liposomes	The surface of phosphatidylcholine liposomes was coated with random copolymers made up of poly(hydroxyethyl acrylate-co-hexadecyl acrylate-co-carboxyethyl acrylate)	1. Carboxyl functionalities of the anchored chain ionize and induce mechanical force on the liposomal membrane upon utilization of an electric field 2. Release of encapsulated molecules (calcein), this effect is due to both the voltage applied and the amount of carboxyl groups present in the liposomes
Liu et al. [53]	Electrically triggered device for drug release controlled remotely	A glass substrate featuring an electrode made of indium tin oxide (ITO), as well as layers of polyvinyl alcohol and poly(3-hexylthiophene)	1. Low oxidative voltage ($\approx$1 V) results in the conductive polymer switching from being hydrophobic to hydrophilic 2. Release of encapsulated molecules through increased penetrability resulting from this shift voltage of 10 V for 5 min

blood, water, melanin, and other biomolecules [58]. A study by Juzenas et al. has shown that the depth of skin penetration of incident light in rats varies based on the wavelength, with NIR light having a much deep penetration depth than UV light. Specifically, at $\lambda = 705$ nm, the tissue permeation depth is 7.5 ± 0.5 mm, at $\lambda = 633$ nm it is 6.3 ± 0.5 mm, and at $\lambda = 408$ nm it is only 1.0 ± 0.02 mm [59]. As a result, NIR light has become a promising candidate for photo-responsive drug delivery as it can penetrate deeper into tissues while minimizing tissue damage.

Numerous light-responsive nanocarriers have been evaluated for cancer theranostics and drug delivery. These comprise a wide range of materials, such as polyplexes, polyion complex vesicles (PICsomes), nanoparticles, upconverting nanoparticles (UCNPs), nano gels, polymeric micelles, liposomes, nanorods [60], nano rattles [61]. Light-responsive nanocarriers exhibit various activities in response to light, including altering the conformation of specific molecules [62], cleaving light-sensitive chemical bonds for nanocarrier dissociation [63], activation of drug delivery from nanocarriers in areas affected by the disease [64], enabling light-activated imaging or imaging-guided therapy [65], generating singlet oxygen (O_2^1) for photodynamic therapy (PDT) [66], and while photothermal therapy (PTT) employs photothermal effects to destroy tumors [67]. In addition, light-responsive materials can alter the hydrophilicity-hydrophobicity balance or change the structure of light-sensitive nanocarriers, resulting in their formation or assembly in reaction to light [60]. In addition to their precise control and specificity, light-responsive materials also offer several other advantages over traditional DDSs. They are biocompatible, biodegradable, and can be easily synthesized using convenient and economical means. These properties make light-responsive materials a highly promising approach for drug delivery.

Light-responsive materials can be synthesized using numerous methods, like polymerization, crosslinking, and chemical modification. The preference for the preparation method relies on the particular characteristics of the material needed for drug delivery, such as mechanical strength, degradation rate, and responsiveness to light [53]. A photo-isomerizable linker refers to a photo-responsive group or linker that can undergo a reversible transition between two structures or form a photochromic group upon exposure to light [68]. Azobenzene and spiropyran are frequently utilized linkers. Upon exposure to light, azobenzene can switch from its trans-state to a metastable cis-form, with isomerization taking place in as little as microseconds or as long as sub-nanoseconds, as cited in reference [69]. There are different types of azobenzene chromophores classified into three categories, each with distinct properties for the cis-to-trans isomerization thermal relaxation and absorption wavelength for trans-to-cis isomerization. Pseudo-stilbenes have a fast thermal conversion rate and absorb light at longer wavelengths. Amino-azobenzenes have moderate lifetimes and a slight shift toward longer wavelengths in their trans-absorption band. Classical azobenzene, on the other hand, absorbs light in the UV-violet range and can retain its cis configuration for days in the absence of light. Three-layered azobenzene has been utilized for photo-controlled drug release systems like vesicle-like aggregates and hydrogel systems [70].

Spiropyran is a hydrophobic isomer that does not carry an electric charge. It can undergo reversible isomerization to the hydrophilic merocyanine form when exposed to light [71, 72], this change in form from hydrophobic to hydrophilic causes spiropyran-containing nanostructures to become unstable and leads to drug release [73]. Spiropyran has been employed as a starter for creating amphiphilic hyperbranched polyglycerols. The transformation of spiropyran from its non-polar, hydrophobic form to the zwitterionic merocyanine form, which is hydrophilic, can be induced by light irradiation. This eventually leads to the dismantling of the polymeric micelles. Dithienylethene is another reversible photo-responsive linker that can be used [74]. Conversely, a photo-sensitive bond that can induce permanent detachment of a medication transporter and accelerate the liberation of the confined medication when exposed to light is referred to as a photo-labile bond. O-nitrobenzyl and coumarin-4-ylmethyl are a couple of instances of these bonds [75]. Generally, linkers that use O-nitrobenzyl as a base are broken down at wavelengths between 300 nm and 365 nm. The speed of the breakdown can differ, lasting from a couple of min. to numerous hours, depending on the wavelength to which it is exposed. Coumarin-4-ylmethyl-based linkers, on the other hand, are generally cleaved with UV light at 254 nm. These linkers have been incorporated into diverse DDSs, such as nanoparticles and microspheres [70]. New research has indicated that photo-switching nanocarriers have promise for loading a range of bioactive substances and releasing them upon exposure to UV–Vis light, including drugs like paclitaxel, docetaxel, and doxorubicin, and for treating cancer. Nonetheless, the limited wavelength range of UV–Vis light might constrain their application. As a result, nanocarriers that respond to NIR light have been created to regulate drug delivery and enable deeper tissue penetration [76]. For example, polymeric micelles incorporated with IR-780 can respond to NIR for doxorubicin release [77].

Nanocarriers that are responsive to light can assist in the efficient delivery of bioactive agents like genes, photosensitizers, & anticancer drugs into cells [60]. These carriers can be engineered to respond to light and release their cargo within cells. For instance, PICsomes incorporating Al (III) phthalocyanine chloride disulfonic acid (AlPcS2a) can be activated by laser irradiation, leading to the escape of the drug from the endosomes and resulting in higher photocytotoxicity compared to AlPcS2a alone. A type of polyplex micelle with three layers that responds to light has been created using polycationic polymers. These micelles can condense pDNA and carry dendrimer phthalocyanine, which can enable successful gene transfection throughout the body via photochemical internalization (PCI), which is activated by light and allows for escape from the endosomal/lysosomal pathway [78]. Furthermore, light-responsive nanocarriers have the potential to be simulated for tumor therapy and theranostics guided by imaging. By combining light-triggered generation of ROS and photothermal effects, these nanocarriers can lead to tumor ablation or be used in conjunction with other bioactive agents for multimodal cancer theranostics. Additionally, light-responsive nanocarriers have illustrated high efficacy in treating cancers that are resistant to multiple drugs (MDR) [60].

2.5 Enzyme-Responsive Material

Enzyme-responsive nanomaterials offer a promising avenue for enhancing on-demand delivery, efficient cellular internalization, and tumor accumulation of imaging contrast or therapeutic agents. The combination of the particular enzyme stimuli with the intriguing characteristics of nanomaterials presents numerous advantages over other stimuli types. Enzymes are natural substances that do not require external triggers such as light, ultrasound, or magnetic fields, which makes them inherently safe and compatible with living organisms. Moreover, enzymatic reactions typically occur swiftly and efficiently, taking place in gentle reaction conditions such as mild temperatures, pH levels, and aqueous environments. Enzymes further demonstrate remarkable selectivity and specificity toward their substrates, facilitating controlled chemical reactions and biological responses [79].

The enzyme classes that are frequently cited as triggers are proteases [80], esterases, phosphatases, kinases, and oxidoreductases. Proteases have the ability to degrade proteins or peptides, while oxidoreductases can catalyze the movement of electrons initiated by a reducing agent to an oxidizing agent. Kinases regulate protein activity via phosphorylation, and phosphatases mediate the opposite action of dephosphorylation with an enhanced comprehension of enzyme functions, enzymes are becoming increasingly important as biomarkers for the diagnosis or prognosis of tumors. For example, diagnostic probes can be developed by attaching enzyme-sensitive substrates to imaging reporters, which can regain their signals, such as fluorescence or photoacoustic signals, from "off" to "on" states after enzymatic substrate processing. Enzyme-sensitive prodrugs are also being explored as an important application to attain specific activation of drugs in target tissues while minimizing harmful adverse effects [81]. Figure 3 illustrates the mechanism of enzyme-responsive DDS.

Several recent studies have highlighted the possibility of enzyme-responsive DDSs for cancer treatment. HA (Hyaluronic acid) is a naturally occurring compound that specifically targets tumor cells that overexpress CD44. HA can be broken down by HAase (hyaluronidases) present in the extracellular region of tumors. HA nano gels have been created for delivering DOX (doxorubicin) to CD44-overexpressing tumor cells. These nano gels were made by copolymerizing methacrylated HA and di (ethylene glycol) diacrylate (DEGDA) in an aqueous solution [82]. The use of both lipase and hyaluronidase resulted in an increased buildup of DOX in the tumor area and notable inhibition of tumor growth. In another study, a DDS was developed that consisted of a liposomal center and a cross-linked HA shell, which was programmed to sequentially and specifically deliver tumor necrosis factor-related

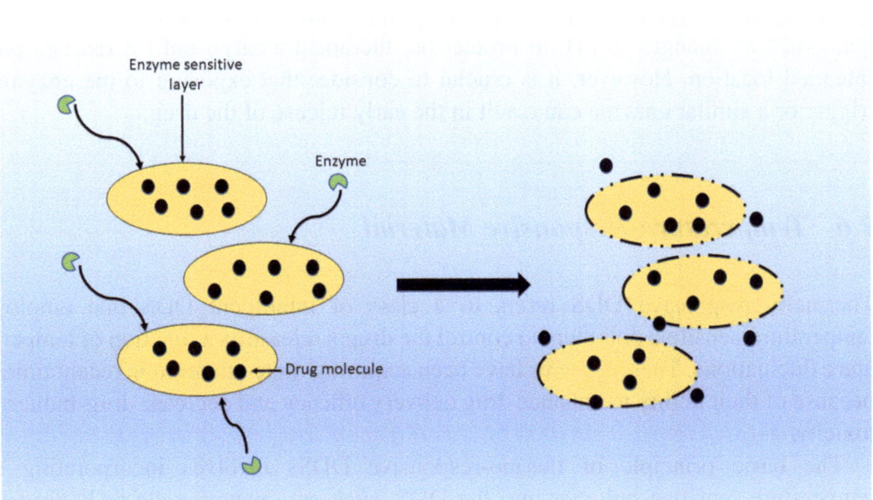

Fig. 3 Mechanism of enzyme-responsive DDS

apoptosis-inducing ligand (TRAIL) and DOX to the site of the tumor [83]. In the tumor microenvironment, the HAase enzyme rapidly degraded the HA shell surrounding the liposomes, leading to the quick release of TRAIL outside of the cells. Afterwards, the liposomes were internalized, leading to the integration of TRAIL and DOX. The co-delivery system of TRAIL and DOX showcased a remarkable IC50 value of 83 ng/mL against MDA-MB-231 cells. This demonstrated a potency that was 5.9 times greater than that of DOX used alone in the therapy [85].

In the study Zhang et al., constructed a PEGylated system that responds to Cathepsin B, an enzyme commonly found in the tumor microenvironment, releasing gemcitabine in their study [84]. The quantity of gemcitabine discharged highly relied on the existence or nonexistence of Cathepsin B [84]. Van Hove et al. created a PEG hydrogel that included pro-angiogenic peptides that were responsive to MMP2, an enzyme that can be utilized in cancer treatment or to reduce inflammation [33]. The hydrogel successfully released the peptides at the desired sites without any nonspecific degradation [33]. Lee et al. focused on the stabilization of insulin at increased temperatures by preparing a trehalose-based hydrogel that could trigger insulin release in response to glucose [85]. Their finding suggested that the trehalose hydrogel, which responded to glucose, effectively combatted heating stress, preserving over 50% of the cargo that was loaded at high temperatures [85]. The trehalose glycopolymers in the hydrogel served as active stabilizers for proteins, such as insulin.

Enzyme-responsive systems offer various applications in a broad spectrum of diseases and medical conditions. One of the key benefits of these systems is their selectivity and natural capacity to react to internal triggers, leveraging the pathological or physiological microenvironment. This specificity can help minimize toxicity in healthy tissues and cells. However, one major drawback is the potential for drug release prior to arriving at the desired destination. To overcome this issue, enzyme-triggered biomaterials often possess additional properties that respond to other triggers, such as changes in pH, to protect the therapeutic cargo until it reaches the intended location. However, it is crucial to consider that exposure to the enzyme trigger or a similar enzyme can result in the early release of the drug.

2.6 Temperature-Responsive Material

Thermally responsive DDS refers to a class of intelligent DDS that employ temperature-sensitive polymers to control the drug's release as a function of temperature fluctuations. These systems have been acquiring more attention in recent times because of their ability to enhance drug delivery efficacy and decrease drug-induced toxicity.

The basic principle of thermo-responsive DDSs involves incorporating a temperature-sensitive polymer into the DDS, such as a nanoparticle or hydrogel. Thermo-responsive polymers are capable of undergoing significant alterations in their physicochemical properties when exposed to temperature, particularly in aqueous solutions, which is preferred over the employment of organic solvents in biological settings. Thermo-responsive polymers are divided into two groups based on their behavior: those with a lower critical solution temperature (LCST) and those with an upper critical solution temperature (UCST) [86]. Polymers with an LCST can disintegrate in water or organic solutions when the temperature is below a particular level. However, when the temperature goes above that threshold, they start to resist water because of intensified hydrophobic interactions among the molecules. This feature has been utilized in various disciplines, including creating models of proteins, initiating self-assembly, regulating protein activity, transporting drugs, and producing sensors [87].

Uchiyama et al. prepared a polymer that reacts to changes in temperature and contains fluorescent components capable of detecting temperature variations in organelles within living cells through changes in fluorescence intensity resulting from dehydration [88]. Perera et al. developed created hydrogels based on gelatin that was crosslinked with temperature-responsive copolymers containing thiol groups. The purpose was to explore how the stiffness of the hydrogel could activate cardiac fibroblasts in a controlled manner. The researchers achieved this by using L-cysteine to soften the hydrogel, which increased the thiol-disulfide exchange reactions and reduced the areas of cultured cells [89]. Chen et al. developed a method of incorporating gold nanoparticles into temperature-responsive star-shaped copolymers, which led to increased solubility in water above the LCST [90].

Polymers that exhibit a UCST type of temperature responsiveness demonstrate solubility in solvents above the critical temperature but become insoluble below it due to specific interactions like electrostatic interactions and hydrogen bonding. These polymers have found numerous applications in the development of assembled nanomaterials, sensors, and protein separations [91]. One illustration is the work of Pslanisamy et al., which developed temperature-responsive films using a layer-by-layer approach and tannic acid was used in conjunction with copolymers that respond to temperature changes similar to the UCST mechanism. The thickness of the resulting films varied depending on factors. These films remained sensitive to temperature changes even after undergoing 55 cycles of heating and cooling [92]. Jana et al. conducted a study on copolymers containing tryptophan, which exhibited varying temperature-responsive properties depending on the surrounding conditions. The characteristics observed encompassed solvent-induced UCST, ion-induced LCST, and pH-induced UCST. These occurrences arise from a delicate balance between interactions involving the alignment of polar molecules and the bonding between hydrogen atoms. Additionally, the combination of ion pairs with small molecules and interactions involving ions contribute to these phenomena [93]. Additionally, Zhang et al. have created poly(amido thioether)s that display sensitivity to variations in temperature, pH, CO_2, and oxidation levels, contingent on their hydroxyl/diethylamino ratios. These polymers exhibit both LCST and UCST properties in aqueous solutions, with fOH values <0.84 and >0.89, respectively. Interestingly, polymers with fOH values between 0.84 and 0.89 display dual-temperature responses, demonstrating both LCST and UCST behaviors [94].

Temperature-responsive polymers have numerous possible biomedical applications, including DDSs, cell sheet technologies, sensors, protein separations, and assembled nanomaterials. Temperature-sensitive polymers can be utilized to fabricate materials that exhibit "smart" behavior and alter their properties upon exposure to changes in temperature. One major application of temperature-responsive polymers is in DDSs. It is possible to engineer drug release systems that are temperature-responsive and can be programmed to release drugs at precise temperature thresholds or respond to variations in temperature. For example, LCST polymers can be used to create drug carriers that release drugs at specific temperatures. This is useful for targeted drug delivery, as the polymer will only deliver the medication to the intended site. Another application of temperature-responsive polymers is in cell sheet technologies. These technologies use temperature-responsive polymers to create cell sheets that can be detached from a culture dish without the use of enzymes or mechanical stress. This is beneficial for creating tissue engineering constructs and regenerative medicine. Overall, temperature-responsive polymers have many potential applications in various biomedical fields. These polymers can be tailored to specific applications, making them versatile and useful tools for biomedical research and development [95]. Table 3 shows the application of light-responsive material [87].

Table 3 Functions of thermo-responsive material [87]

Application	Description
Drug delivery	Thermo-responsive materials can be employed to produce DDSs that respond to changes in temperature. They can encase drugs and release them at specific temperatures or target specific tissues
Tissue engineering	These can be used to create scaffolds for tissue engineering. They have the ability to provide physical reinforcement and imitate the biological extracellular matrix present in bodily tissues
Biosensors	One could utilize these to generate biological sensors that detect alterations in temperature. They can be combined with biological molecules to detect specific molecules or cells
Self-assembly	These have the ability to spontaneously arrange themselves into various formations, such as tiny spheres, small sacs, and extremely small particles. These formations hold promise for use in delivering medication, creating artificial tissue, and developing devices to detect biological molecules
Chromatography	These materials can be used in chromatography to separate biomolecules based on temperature. This method is called temperature-responsive chromatography
Adsorption materials	These can be employed as adsorption materials that selectively adsorb particular molecules based on temperature. They can be used in wastewater treatment and purification processes

2.7 Ultrasound Responsive Material

Scientists have been highly interested in polymeric materials that respond to ultrasound for many years. These smart materials are more useful than other responsive materials, like those that respond to UV, heat, or pH, because they can deliver drugs more effectively and precisely target treatment without being too invasive. Mechanical waves Having frequencies that are 20 kHz or higher are known as ultrasound waves, which can be directed and transmitted through certain mediums. Ultrasound waves have been employed in numerous clinical scenarios, providing a solution that is both non-invasive and cost-efficient. These applications encompass a range of fields, such as live imaging, physical therapy, beauty and skincare, and the culinary sector [96]. Materials that are made up of polymers and are capable of reacting to ultrasound comprise various forms [97]. These particular drug delivery mechanisms, which react to ultrasound, have the capability to carry various types of medications, like small molecule drugs, DNA, and proteins [98]. These vehicles that respond to ultrasound have a wide range of potential uses, including, tumor treatment, disrupting the blood–brain barrier, fighting infectious diseases, delivering drugs through the skin, and thrombolysis [96]. It is thought that the mechanism is linked to the creation of both thermal and non-thermal impacts through the energy released by ultrasound radiation. The process of transforming sound energy into heat energy generates heat within the tissue that comes into contact with the ultrasound waves, which is referred to as the thermal effect. This can disrupt the cell membrane and lead to higher permeability of blood vessels [99].

The use of hyperthermia as a treatment method has resulted in noteworthy advancements in the therapy of tumors. The non-thermal effect is another mechanism that could be responsible for ultrasound-mediated drug delivery, primarily linked to cavitation. This phenomenon can arise in pre-existing microbubbles or cavitation nuclei [96]. The creation of effective nanosystems/nanoplatforms that can respond to ultrasound signals and have specific composition, structure, and features is essential for developing theranostic applications that are activated by ultrasound. This typically involves utilizing the underlying material and synthetic chemistry to design these nanosystems. At present, nanoparticles being investigated for their responsiveness to ultrasound fall into three categories: organic, inorganic, or a combination of both known as organic–inorganic hybrid nanosystems. As a result, the methods used to create these ultrasound-responsive nanosystems differ greatly from one another. This indicates that the fundamental material chemistry also differs among these adaptable nanosystems. To provide an example, when creating nanoparticles that can change phases in response to ultrasound for therapeutic and diagnostic purposes, perfluorocarbon (PFC) is commonly used. In contrast, porphyrin-based nanomedicine is commonly employed in the treatment of tumors that exhibit a favorable response to ultrasound [100].

3 Challenges and Future Perspective

Although Substantial headway has been achieved in the creation of DDSs that can respond to stimuli, there are still several obstacles that must be overcome in order to make them more practical for clinical use. A key challenge is the capability to achieve precise control over the release of drugs. The release kinetics of the drug should be precisely controlled to ensure that the therapeutic dose is delivered at the right time and site. Several factors, such as the dimensions and configuration of the nanoparticles, the choice of polymer, and the type of stimulus, can influence drug release kinetics. Thus, further studies are needed to optimize these parameters and develop efficient and controlled DDSs.

A further difficulty in DDSs is ensuring that the nanoparticles used are both biocompatible and biodegradable. If the nanoparticles are toxic, they can lead to undesired side effects that can limit their practical application in medicine. Therefore, it is crucial to use materials that are safe for use in humans and that can break down naturally. Numerous studies have investigated the use of natural polymers such as chitosan, alginate, and cellulose for this purpose [101]. Additionally, ensuring the reliability of SRDDS (Super Reliable Data Storage) during storage and transportation poses an additional obstacle. Factors such as temperature, pH, and shear stress can affect the stability of nanoparticles. Hence, it is essential to develop DDSs that can keep their integrity and effectiveness during storage and transportation.

Although several in vitro and in vivo studies have proven the potential of these systems, their clinical efficacy and safety remain to be established. Therefore, it is necessary to conduct extensive preclinical and clinical trials to validate the efficacy, safety, and feasibility of these systems for clinical use [102, 103]. Despite the challenges, SRDDS holds tremendous possibilities for the treatment of several diseases. In the future, several developments are expected in this field, including developing new stimuli-responsive materials, optimizing drug release kinetics, and validating clinical efficacy and safety. Furthermore, the integration of advanced technologies, such as artificial intelligence and machine learning, can significantly improve the development and optimization of SRDDS.

4 Conclusion

Stimuli-responsive materials have brought about a ground-breaking transformation in the realm of drug delivery and biomedical applications. These materials' advancement has facilitated the precise and controlled discharge of drugs and other bioactive substances, opening up new horizons for treating a wide range of diseases and disorders. The design and development of stimuli-responsive materials have been challenging tasks, requiring the integration of several scientific fields, including chemistry, physics, and biology. However, substantial progress has been made in this area, with various stimuli-responsive materials being developed and studied extensively.

Polymer blends have emerged as a popular approach due to their easy fabrication and manipulation of device properties. Interpolymer complexes, interpenetrating polymer networks, and block copolymers have also been recognized as effective stimuli-responsive materials. The use of various synthetic routes and architectures has further enhanced the performance and functionality of these materials. Temperature-sensitive, pH-sensitive, and electric and magnetic field-responsive materials have been extensively studied for biomedical applications. These materials respond to various stimuli and exhibit a desired response, making them useful for controlled drug release. Controlled DDSs offer unique advantages over conventional therapy, including improved efficacy, reduced toxicity, and enhanced patient compliance. Responsive materials that can react to specific stimuli have found extensive use in diverse biomedical fields such as cancer therapy, gene therapy, tissue engineering, and diagnosis. For instance, temperature-sensitive gels have been utilized to deliver drugs precisely to tumors, while pH-sensitive systems have been employed for gene delivery purposes. Additionally, materials responsive to electric and magnetic fields have demonstrated potential in both diagnosing and treating various diseases.

Despite the considerable advancements achieved in the field of stimuli-responsive materials, there remain several challenges that require attention. These challenges encompass optimizing material properties, establishing scalable and cost-effective synthesis methods, and identifying appropriate target diseases and patient populations. Future research endeavors in this area are anticipated to focus on advancing state-of-the-art materials that exhibit intelligent responsiveness to multiple stimuli, resulting in improved specificity and sensitivity. Furthermore, the integration of advanced imaging techniques like magnetic resonance imaging and positron emission tomography with stimuli-responsive materials is expected to enable real-time monitoring and visualization of drug release and treatment efficacy. To summarize, stimuli-responsive materials hold great promise for the progress of advanced DDSs and biomedical applications. By incorporating diverse scientific disciplines and employing advanced synthetic methodologies and structures, scientists have successfully developed materials capable of responding to various stimuli and eliciting desired reactions. Further investigation in this area is anticipated to produce innovative materials that could significantly transform the diagnosis and treatment of various diseases and disorders.

References

1. Rosenblum, D., Joshi, N., Tao, W., Karp, J.M., Peer, D.: Progress and challenges towards targeted delivery of cancer therapeutics. Nat. Commun. **9**(1), 1410 (2018)
2. Srinivasarao, M., Low, P.S.: Ligand-targeted drug delivery. Chem. Rev. **117**(19), 12133–12164 (2017)
3. Xiao, M., Sun, W., Fan, J., Cao, J., Li, Y., Shao, K., Li, M., Li, X., Kang, Y., Zhang, W., Long, S.: Aminopeptidase-N-activated theranostic prodrug for NIR tracking of local tumor chemotherapy. Adv. Func. Mater. **28**(47), 1805128 (2018)

4. Liu, J., Zhang, R., Xu, Z.P.: Nanoparticle-based nanomedicines to promote cancer immunotherapy: Recent advances and future directions. Small **15**(32), 1900262 (2019)
5. Qiu, M., Singh, A., Wang, D., Qu, J., Swihart, M., Zhang, H., Prasad, P.N.: Biocompatible and biodegradable inorganic nanostructures for nanomedicine: Silicon and black phosphorus. Nano Today **1**(25), 135–155 (2019)
6. Chamundeeswari, M., Jeslin, J., Verma, M.L.: Nanocarriers for drug delivery applications. Environ. Chem. Lett. **17**(2), 849–865 (2019)
7. Ahmed, A., Sarwar, S., Hu, Y., Munir, M.U., Nisar, M.F., Ikram, F., Asif, A., Rahman, S.U., Chaudhry, A.A., Rehman, I.U.: Surface-modified polymeric nanoparticles for drug delivery to cancer cells. Expert Opin. Drug Deliv. **18**(1), 1–24 (2021)
8. Hu, L., Zhang, Q., Li, X., Serpe, M.J.: Stimuli-responsive polymers for sensing and actuation. Mater. Horiz. **6**(9), 1774–1793 (2019)
9. Das, S.S., Bharadwaj, P., Bilal, M., Barani, M., Rahdar, A., Taboada, P., Bungau, S., Kyzas, G.Z.: Stimuli-responsive polymeric nanocarriers for drug delivery, imaging, and theragnosis. Polymers **12**(6), 1397 (2020)
10. Ding, C., Tong, L., Feng, J., Fu, J.: Recent advances in stimuli-responsive release function DDSs for tumor treatment. Molecules **21**(12), 1715 (2016)
11. Li, L., Yang, W.W., Xu, D.G.: Stimuli-responsive nanoscale DDSs for cancer therapy. J. Drug Target. **27**(4), 423–433 (2019)
12. Raza, A., Rasheed, T., Nabeel, F., Hayat, U., Bilal, M., Iqbal, H.M.: Endogenous and exogenous SRDDS for programmed site-specific release. Molecules **24**(6), 1117 (2019)
13. Cook, A.B., Decuzzi, P.: Harnessing endogenous stimuli for responsive materials in theranostics. ACS Nano **15**(2), 2068–2098 (2021)
14. Li, S., Hu, K., Cao, W., Sun, Y., Sheng, W., Li, F., Wu, Y., Liang, X.J.: PH-responsive biocompatible fluorescent polymer nanoparticles based on phenylboronic acid for intracellular imaging and drug delivery. Nanoscale **6**(22), 13701–13709 (2014)
15. Municoy, S., Álvarez Echazú, M.I., Antezana, P.E., Galdopórpora, J.M., Olivetti, C., Mebert, A.M., Foglia, M.L., Tuttolomondo, M.V., Alvarez, G.S., Hardy, J.G., Desimone, M.F.: Stimuli-responsive materials for tissue engineering and drug delivery. Int. J. Mol. Sci. **21**(13), 4724 (2020)
16. Hu, Y.B., Dammer, E.B., Ren, R.J., Wang, G.: The endosomal-lysosomal system: From acidification and cargo sorting to neurodegeneration. Transl. Neurodegener. **4**, 1 (2015)
17. Wang, L., Neumann, M., Fu, T., Li, W., Cheng, X., Su, B.L.: Porous and responsive hydrogels for cell therapy. Curr. Opin. Colloid Interface Sci. **1**(38), 135–157 (2018)
18. Aycan, D., Alemdar, N.: Development of pH-responsive chitosan-based hydrogel modified with bone ash for controlled release of amoxicillin. Carbohyd. Polym. **15**(184), 401–407 (2018)
19. Ouyang, L., Sun, Z., Wang, D., Qiao, Y., Zhu, H., Ma, X., Liu, X.: Smart release of doxorubicin loaded on polyetheretherketone (PEEK) surface with 3D porous structure. Colloids Surf., B **1**(163), 175–183 (2018)
20. Bahadır, E.B., Sezgintürk, M.K.: Poly (amidoamine)(PAMAM): An emerging material for electrochemical bio (sensing) applications. Talanta **1**(148), 427–438 (2016)
21. Li, W., Feng, R., Wang, R., Li, D., Jiang, W., Liu, H., Guo, Z., Serpe, M.J., Hu, L.: Polyelectrolyte-based physical adhesive hydrogels with excellent mechanical properties for biomedical applications. J. Mater. Chem. B **6**(29), 4799–4807 (2018)
22. Zhao, W., Quan, M., Cao, Z., Zhang, Y., Wen, J., Pan, D., Dong, Z., Yang, Z., Wang, D., Cao, H., He, W.: Visual multi-triggered sensor based on inverse opal hydrogel. Colloids Surf., A **5**(554), 93–99 (2018)
23. Xie, X., Doloff, J.C., Yesilyurt, V., Sadraei, A., McGarrigle, J.J., Omami, M., Veiseh, O., Farah, S., Isa, D., Ghani, S., Joshi, I.: Reduction of measurement noise in a continuous glucose monitor by coating the sensor with a zwitterionic polymer. Nat. Biomed. Eng. **2**(12), 894–906 (2018)
24. Aguilar, M.R., San Román, J., (eds.): Smart Polymers and Their Applications. Woodhead Publishing (2019)

25. Bazban-Shotorbani, S., Hasani-Sadrabadi, M.M., Karkhaneh, A., Serpooshan, V., Jacob, K.I., Moshaverinia, A., Mahmoudi, M.: Revisiting structure-property relationship of pH-responsive polymers for drug delivery applications. J. Control. Release **10**(253), 46–63 (2017)
26. Le, X.X., Zhang, Y.C., Lu, W., Wang, L., Zheng, J., Ali, I., Zhang, J.W., Huang, Y.J., Serpe, M.J., Yang, X.T., Fan, X.D.: A novel anisotropic hydrogel with integrated self-deformation and controllable shape memory effect. Macromol. Rapid Commun. **39**(9), 1800019 (2018)
27. Shang, J., Theato, P.: Smart composite hydrogel with pH-, ionic strength-and temperature-induced actuation. Soft Matter. **14**(41), 8401–8407 (2018)
28. Huang, L., Jiang, R., Wu, J., Song, J., Bai, H., Li, B., Zhao, Q., Xie, T.: Ultrafast digital printing toward 4D shape changing materials. Adv. Mater. **29**(7), 1605390 (2017)
29. Cheng, Y., Huang, C., Yang, D., Ren, K., Wei, J.: Bilayer hydrogel mixed composites that respond to multiple stimuli for environmental sensing and underwater actuation. J. Mater. Chem. B **6**(48), 8170–8179 (2018)
30. Burillo, G., Bucio, E., Arenas, E., Lopez, G.P.: Macromol. Mater. Eng. **292**, 214 (2007)
31. Lee, B.P., Konst, S.: Novel hydrogel actuator inspired by reversible mussel adhesive protein chemistry. Adv. Mater. **26**(21), 3415–3419 (2014)
32. Wells, C.M., Harris, M., Choi, L., Murali, V.P., Guerra, F.D., Jennings, J.A.: Stimuli-responsive drug release from smart polymers. J. Funct. Biomater. **10**(3), 34 (2019)
33. Wu, B.Y., Xu, Y.W., Le, X.X., Jian, Y.K., Lu, W., Zhang, J.W., Chen, T.: Smart hydrogel actuators assembled via dynamic boronic ester bonds. Acta Polym. Sin. **1**(50), 496–504 (2019)
34. Zhou, Q., Zhang, L., Yang, T., Wu, H.: Stimuli-responsive polymeric micelles for drug delivery and cancer therapy. Int. J. Nanomed. **13**, 2921 (2018)
35. Wei, H., Zhang, X., Cheng, C., Cheng, S.X., Zhuo, R.X.: Self-assembled, thermosensitive micelles of a star block copolymer based on PMMA and PNIPAAm for controlled drug delivery. Biomaterials **28**(1), 99–107 (2007)
36. Li, H., Yan, K., Shang, Y., Shrestha, L., Liao, R., Liu, F., Li, P., Xu, H., Xu, Z., Chu, P.K.: Folate-bovine serum albumin functionalized polymeric micelles loaded with superparamagnetic iron oxide nanoparticles for tumor targeting and magnetic resonance imaging. Acta Biomater. **15**(15), 117–126 (2015)
37. Hu, L., Wan, Y., Zhang, Q., Serpe, M.J.: Harnessing the power of stimuli-responsive polymers for actuation. Adv. Func. Mater. **30**(2), 1903471 (2020)
38. Qu, J., Zhao, X., Ma, P.X., Guo, B.: Injectable antibacterial conductive hydrogels with dual response to an electric field and pH for localized "smart" drug release. Acta Biomater. **1**(72), 55–69 (2018)
39. Kolosnjaj-Tabi, J., Gibot, L., Fourquaux, I., Golzio, M., Rols, M.P.: Electric field-responsive nanoparticles and electric fields: Physical, chemical, biological mechanisms and therapeutic prospects. Adv. Drug Deliv. Rev. **1**(138), 56–67 (2019)
40. Deng, Z., Guo, Y., Zhao, X., Ma, P.X., Guo, B.: Multifunctional stimuli-responsive hydrogels with self-healing, high conductivity, and rapid recovery through host–guest interactions. Chem. Mater. **30**(5), 1729–1742 (2018)
41. Zhang, A., Jung, K., Li, A., Liu, J., Boyer, C.: Recent advances in stimuli-responsive polymer systems for remotely controlled drug release. Prog. Polym. Sci. **1**(99), 101164 (2019)
42. Gao, S., Tang, G., Hua, D., Xiong, R., Han, J., Jiang, S., Zhang, Q., Huang, C.: Stimuli-responsive bio-based polymeric systems and their applications. J. Mater. Chem. B **7**(5), 709–729 (2019)
43. George, P.M., LaVan, D.A., Burdick, J.A., Chen, C.Y., Liang, E., Langer, R.: Electrically controlled drug delivery from biotin-doped conductive polypyrrole. Adv. Mater. **18**(5), 577–581 (2006)
44. Baker, C.O., Huang, X., Nelson, W., Kaner, R.B.: Polyaniline nanofibers: Broadening applications for conducting polymers. Chem. Soc. Rev. **46**(5), 1510–1525 (2017)
45. Arteshi, Y., Aghanejad, A., Davaran, S., Omidi, Y.: Semi self-doped electroconductive and biocompatible polyaniline/sulfonated β-cyclodextrin (PANI/SCD) inclusion complex with potential use in regenerative medicine. Int. J. Polym. Mater. Polym. Biomater. (2019)

46. Ge, J., Neofytou, E., Cahill, T.J., III., Beygui, R.E., Zare, R.N.: Drug release from electric-field-responsive nanoparticles. ACS Nano **6**(1), 227–233 (2012)
47. Yan, Q., Yuan, J., Cai, Z., Xin, Y., Kang, Y., Yin, Y.: Voltage-responsive vesicles based on orthogonal assembly of two homopolymers. J. Am. Chem. Soc. **132**(27), 9268–9270 (2010)
48. Chang, X., Cheng, Z., Ren, B., Dong, R., Peng, J., Fu, S., Tong, Z.: Voltage-responsive reversible self-assembly and controlled drug release of ferrocene-containing polymeric superamphiphiles. Soft Matter. **11**(38), 7494–7501 (2015)
49. Kim, J.A., Kim, J.C.: Temperature and electric field-triggerable liposomes incorporating poly (hydroxyethyl acrylate-co-hexadecyl acrylate-co-carboxyethyl acrylate). J. Ind. Eng. Chem. **25**(62), 383–391 (2018)
50. Liu, H.W., Hu, S.H., Chen, Y.W., Chen, S.Y.: Characterization and drug release behavior of highly responsive chip-like electrically modulated reduced graphene oxide–poly (vinyl alcohol) membranes. J. Mater. Chem. **22**(33), 17311–17320 (2012)
51. Servant, A., Methven, L., Williams, R.P., Kostarelos, K.: Electroresponsive polymer-carbon nanotube hydrogel hybrids for pulsatile drug delivery in vivo. Adv. Healthc. Mater. **2**(6) (2013)
52. Liu, S., Chen, X., Wang, Y., Li, L., Wang, G., Li, X., Chen, H., Guo, J., Lin, H., Lian, Q.Q., Ge, R.S.: A role of KIT receptor signaling for proliferation and differentiation of rat stem Leydig cells in vitro. Mol. Cell. Endocrinol. **15**(444), 1–8 (2017)
53. Olejniczak, J., Carling, C.J., Almutairi, A.: Photocontrolled release using one-photon absorption of visible or NIR light. J. Control. Release **10**(219), 18–30 (2015)
54. Rapp, T.L., DeForest, C.A.: Targeting drug delivery with light: A highly focused approach. Adv. Drug Deliv. Rev. **1**(171), 94–107 (2021)
55. Cheng, L., Wang, C., Feng, L., Yang, K., Liu, Z.: Functional nanomaterials for phototherapies of cancer. Chem. Rev. **114**(21), 10869–10939 (2014)
56. Silva, J.M., Silva, E., Reis, R.L.: Light-triggered release of photocaged therapeutics-Where are we now? J. Control. Release **28**(298), 154–176 (2019)
57. Zhao, W., Zhao, Y., Wang, Q., Liu, T., Sun, J., Zhang, R.: Remote light-responsive nanocarriers for controlled drug delivery: Advances and perspectives. Small **15**(45), 1903060 (2019)
58. Juzenas, P., Juzeniene, A., Kaalhus, O., Iani, V., Moan, J.: Noninvasive fluorescence excitation spectroscopy during application of 5-aminolevulinic acid in vivo. Photochem. Photobiol. Sci. **1**(10), 745–748 (2002)
59. Mi, P.: Stimuli-responsive nanocarriers for drug delivery, tumor imaging, therapy and theranostics. Theranostics. **10**(10), 4557 (2020)
60. Li, C., Zhang, Y., Li, Z., Mei, E., Lin, J., Li, F., Chen, C., Qing, X., Hou, L., Xiong, L., Hao, H.: Light-responsive biodegradable nanorattles for cancer theranostics. Adv. Mater. **30**(8), 1706150 (2018)
61. Zhao, Y.: Light-responsive block copolymer micelles. Macromolecules **45**(9), 3647–3657 (2012)
62. Yan, B., Boyer, J.C., Branda, N.R., Zhao, Y.: Near-infrared light-triggered dissociation of block copolymer micelles using upconverting nanoparticles. J. Am. Chem. Soc. **133**(49), 19714–19717 (2011)
63. Luo, D., Li, N., Carter, K.A., Lin, C., Geng, J., Shao, S., Huang, W.C., Qin, Y., Atilla-Gokcumen, G.E., Lovell, J.F.: Rapid light-triggered drug release in liposomes containing small amounts of unsaturated and porphyrin–phospholipids. Small **12**(22), 3039–3047 (2016)
64. Wang, S., Lin, J., Wang, Z., Zhou, Z., Bai, R., Lu, N., Liu, Y., Fu, X., Jacobson, O., Fan, W., Qu, J.: Core–satellite polydopamine–gadolinium-metallofullerene nanotheranostics for multimodal imaging guided combination cancer therapy. Adv. Mater. **29**(35), 1701013 (2017)
65. Fan, W., Lu, N., Xu, C., Liu, Y., Lin, J., Wang, S., Shen, Z., Yang, Z., Qu, J., Wang, T., Chen, S.: Enhanced afterglow performance of persistent luminescence implants for efficient repeatable photodynamic therapy. ACS Nano **11**(6), 5864–5872 (2017)
66. Wang, Z., Huang, P., Jacobson, O., Wang, Z., Liu, Y., Lin, L., Lin, J., Lu, N., Zhang, H., Tian, R., Niu, G.: Biomineralization-inspired synthesis of copper sulfide–ferritin nanocages as cancer theranostics. ACS Nano **10**(3), 3453–3460 (2016)

67. Zhu, F., Tan, S., Dhinakaran, M.K., Cheng, J., Li, H.: The light-driven macroscopic directional motion of a water droplet on an azobenzene–calix [4] arene modified surface. Chem. Commun. **56**(74), 10922–10925 (2020)
68. Goulet-Hanssens, A., Barrett, C.J.: Photo-control of biological systems with azobenzene polymers. J. Polym. Sci., Part A: Polym. Chem. **51**(14), 3058–3070 (2013)
69. Giménez, V.M., Arya, G., Zucchi, I.A., Galante, M.J., Manucha, W.: Photo-responsive polymeric nanocarriers for target-specific and controlled drug delivery. Soft Matter. **17**(38), 8577–8584 (2021)
70. Zhang, D., Shah, P.K., Culver, H.R., David, S.N., Stansbury, J.W., Yin, X., Bowman, C.N.: Photo-responsive liposomes composed of spiropyran-containing triazole-phosphatidylcholine: investigation of merocyanine-stacking effects on liposome–fiber assembly-transition. Soft Matter. **15**(18), 3740–3750 (2019)
71. Volarić, J., Szymanski, W., Simeth, N.A., Feringa, B.L.: Molecular photoswitches in aqueous environments. Chem. Soc. Rev. **50**(22), 12377–12449 (2021)
72. Movia, D., Prina-Mello, A., Volkov, Y., Giordani, S.: Determination of spiropyran cytotoxicity by high content screening and analysis for safe application in bionanosensing. Chem. Res. Toxicol. **23**(9), 1459–1466 (2010)
73. Tao, Y., Chan, H.F., Shi, B., Li, M., Leong, K.W.: Light: A magical tool for controlled drug delivery. Adv. Func. Mater. **30**(49), 2005029 (2020)
74. Razavi, B., Abdollahi, A., Roghani-Mamaqani, H., Salami-Kalajahi, M.: Light-and temperature-responsive micellar carriers prepared by spiropyran-initiated atom transfer polymerization: Investigation of photochromism kinetics, responsivities, and controlled release of doxorubicin. Polymer **20**(187), 122046 (2020)
75. Hansen, M.J., Velema, W.A., Lerch, M.M., Szymanski, W., Feringa, B.L.: Wavelength-selective cleavage of photoprotecting groups: strategies and applications in dynamic systems. Chem. Soc. Rev. **44**(11), 3358–3377 (2015)
76. Wang, S., Lin, J., Wang, T., Chen, X., Huang, P.: Recent advances in photoacoustic imaging for deep-tissue biomedical applications. Theranostics. **6**(13), 2394 (2016)
77. Wang, J., Liu, Y., Ma, Y., Sun, C., Tao, W., Wang, Y., Yang, X., Wang, J.: NIR-activated supersensitive drug release using nanoparticles with a flow core. Adv. Func. Mater. **26**(41), 7516–7525 (2016)
78. Yen, H.C., Cabral, H., Mi, P., Toh, K., Matsumoto, Y., Liu, X., Koori, H., Kim, A., Miyazaki, K., Miura, Y., Nishiyama, N.: Light-induced cytosolic activation of reduction-sensitive camptothecin-loaded polymeric micelles for spatiotemporally controlled in vivo chemotherapy. ACS Nano **8**(11), 11591–11602 (2014)
79. Pham, S.H., Choi, Y., Choi, J.: Stimuli-responsive nanomaterials for application in antitumor therapy and drug delivery. Pharmaceutics **12**(7), 630 (2020)
80. Mu, J., Lin, J., Huang, P., Chen, X.: Development of endogenous enzyme-responsive nanomaterials for theranostics. Chem. Soc. Rev. **47**(15), 5554–5573 (2018)
81. Yang, C., Wang, X., Yao, X., Zhang, Y., Wu, W., Jiang, X.: Hyaluronic acid nanogels with enzyme-sensitive cross-linking group for drug delivery. J. Control. Release **10**(205), 206–217 (2015)
82. Jiang, T., Mo, R., Bellotti, A., Zhou, J., Gu, Z.: Gel–liposome-mediated co-delivery of anti-cancer membrane-associated proteins and small-molecule drugs for enhanced therapeutic efficacy. Adv. Func. Mater. **24**(16), 2295–2304 (2014)
83. Zhu, Q., Chen, X., Xu, X., Zhang, Y., Zhang, C., Mo, R.: Tumor-specific self-degradable nanogels as potential carriers for systemic delivery of anticancer proteins. Adv. Func. Mater. **28**(17), 1707371 (2018)
84. Van Hove, A.H., Beltejar, M.J., Benoit, D.S.: Development and in vitro assessment of enzymatically-responsive poly (ethylene glycol) hydrogels for the delivery of therapeutic peptides. Biomaterials **35**(36), 9719–9730 (2014)
85. Lee, J., Ko, J.H., Mansfield, K.M., Nauka, P.C., Bat, E., Maynard, H.D.: Glucose-responsive trehalose hydrogel for insulin stabilization and delivery. Macromol. Biosci. **18**(5), 1700372 (2018)

86. Bordat, A., Boissenot, T., Nicolas, J., Tsapis, N.: Thermoresponsive polymer nanocarriers for biomedical applications. Adv. Drug Deliv. Rev. **1**(138), 167–192 (2019)
87. Qiao, S., Wang, H.: Temperature-responsive polymers: Synthesis, properties, and biomedical applications. Nano Res. **11**, 5400–5423 (2018)
88. Uchiyama, S., Gota, C., Tsuji, T., Inada, N.: Intracellular temperature measurements with fluorescent polymeric thermometers. Chem. Commun. **53**(80), 10976–10992 (2017)
89. Perera, M.M., Fischesser, D.M., Molkentin, J.D., Ayres, N.: Stiffness of thermoresponsive gelatin-based dynamic hydrogels affects fibroblast activation. Polym. Chem. **10**(46), 6360–6367 (2019)
90. Chen, Y., Wang, Z., Harn, Y.W., Pan, S., Li, Z., Lin, S., Peng, J., Zhang, G., Lin, Z.: Resolving optical and catalytic activities in thermoresponsive nanoparticles by permanent ligation with temperature-sensitive polymers. Angew. Chem. Int. Ed. **58**(34), 11910–11917 (2019)
91. Kotsuchibashi, Y.: Recent advances in multi-temperature-responsive polymeric materials. Polym. J. **52**(7), 681–689 (2020)
92. Palanisamy, A., Sukhishvili, S.A.: Swelling transitions in layer-by-layer assemblies of UCST block copolymer micelles. Macromolecules **51**(9), 3467–3476 (2018)
93. Jana, S., Anas, M., Maji, T., Banerjee, S., Mandal, T.K.: Tryptophan-based styryl homopolymer and polyzwitterions with solvent-induced UCST, ion-induced LCST and pH-induced UCST. Polym. Chem. **10**(4), 526–538 (2019)
94. Zhang, H., Zhang, J., Dai, W., Zhao, Y.: Facile synthesis of thermo-, pH-, CO_2-and oxidation-responsive poly (amido thioether) s with tunable LCST and UCST behaviors. Polym. Chem. **8**(37), 5749–5760 (2017)
95. Khutoryanskiy, V.V., Georgiou, T.K., (eds.): Temperature-Responsive Polymers: Chemistry, Properties, and Applications. John Wiley & Sons (2018)
96. Wei, P., Cornel, E.J., Du, J.: Ultrasound-responsive polymer-based DDSs. Drug Deliv. Transl. Res. **11**, 1323–1339 (2021)
97. Wei, P., Sun, M., Yang, B., Xiao, J., Du, J.: Ultrasound-responsive polymersomes capable of endosomal escape for efficient cancer therapy. J. Control. Release **10**(322), 81–94 (2020)
98. Kuo, J.H., Jan, M.S., Sung, K.C.: Evaluation of the stability of polymer-based plasmid DNA delivery systems after ultrasound exposure. Int. J. Pharm. **257**(1–2), 75–84 (2003)
99. Kamaly, N., Yameen, B., Wu, J., Farokhzad, O.C.: Degradable controlled-release polymers and polymeric nanoparticles: Mechanisms of controlling drug release. Chem. Rev. **116**(4), 2602–2663 (2016)
100. Cui, X., Han, X., Yu, L., Zhang, B., Chen, Y.: Intrinsic chemistry and design principle of ultrasound-responsive nanomedicine. Nano Today **1**(28), 100773 (2019)
101. Idrees, H., Zaidi, S.Z., Sabir, A., Khan, R.U., Zhang, X., Hassan, S.U.: A review of biodegradable natural polymer-based nanoparticles for drug delivery applications. Nanomaterials **10**(10), 1970 (2020)
102. Torchilin, V.P.: Multifunctional, stimuli-sensitive nanoparticulate systems for drug delivery. Nat. Rev. Drug Discovery **13**(11), 813–827 (2014)
103. Song, Y., Jing, H., Vong, L.B., Wang, J., Li, N.: Recent advances in targeted stimuli-responsive nano-based DDSs combating atherosclerosis. Chin. Chem. Lett. **33**(4), 1705–1717 (2022)

Karan Trehan a Pharmacy graduate from Guru Nanak Dev University, Amritsar in 2021, is currently pursuing his Master's in Pharmaceutics from the same university. His research focus lies in the development and characterization of formulations containing 7-methylxanthine, a promising new drug molecule. With a keen interest in drug development and innovation, Karan is dedicated to exploring the potential of 7-methylxanthine for improved therapeutic outcomes.

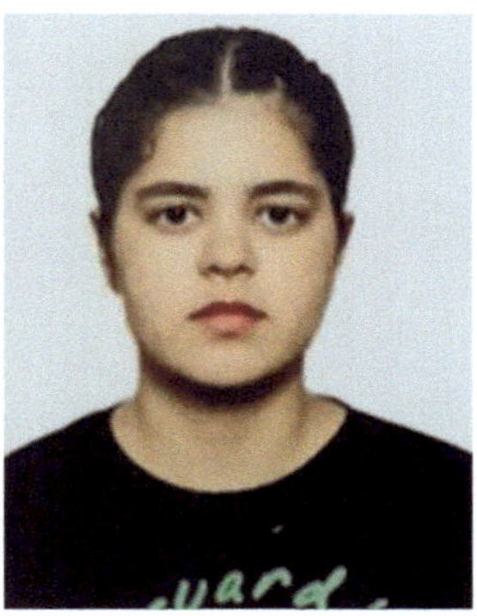

Muskaan Saini graduated in Pharmacy from Guru Nanak Dev University, Amritsar, in 2021. Currently, she is pursuing a Master's degree in Pharmaceutics from the same university. Her research focuses on the development and characterization of highly effective oral iron supplements, aiming to enhance their delivery and maximize their therapeutic impact.

Shubham Thakur completed his undergraduate and postgraduate studies in pharmacy at Guru Nanak Dev University in Amritsar in 2016 and 2018 respectively. Presently, he serves as an Indian Council of Medical Research-Senior Research Fellow and is pursuing a Ph.D. in the Department of Pharmaceutical Sciences at Guru Nanak Dev University, Amritsar. His research focuses on the advancement of novel drug delivery systems, aiming to enhance the kinetics, efficacy, and safety of existing drugs. Specifically, his current project involves developing a safe oral formulation for pregnant women. Till now he has published 25 articles and 5 book chapters in Scopus indexed journals and books with h-index of 5.

Gold Nanoparticles: Clinical Applications

Sheikdawood Parveen, T. Sathiyapriya, D. Tharani, S. U. Mohammed Riyaz, Rakshi Anuja Dinesh, Jayashree Shanmugam, K. Rajakumar, Dmitry Zherebtsov, Manikandan Dhayalan, and Antony Stalin

Abstract Global Cancer Therapy (GLOBOCAN) estimates that there are around 18 million new cancer cases each year. Various techniques have been used for cancer therapy. Innovation of the nano approach will be signed in to locate one of the chief concerns while using chemotherapy and radiation. Targeting the gold nanoparticle (GNP)-based systems to the tumor can increase local radiation dosage while also enhancing treatment through controlled releases of chemotherapeutics. In the last 20 years, more than 20 medicinal items based on nanotechnology have received clinical use accreditation. This review article's goal is to determine ways to employ GNP-based therapy systems more quickly in clinical settings while still reducing

S. Parveen
Department of Chemistry, Dr. Mahalingam College of Engineering and Technology, Pollachi 642003, India

T. Sathiyapriya
Department of Chemistry, Coimbatore Marine College, Coimbatore 641032, India

D. Tharani
Department of Science and Humanities, Nehru Institute of Engineering and Technology, Coimbatore 641105, India

S. U. Mohammed Riyaz
PG & Research Department of Biotechnology, Islamiah College (Autonomous), Vaniyambadi, Tamil Nadu 635752, India

R. A. Dinesh · J. Shanmugam
Department of Biotechnology, Sathyabama Institute of Science and Technology, Chennai 600119, India

K. Rajakumar · D. Zherebtsov
Nanotechnology Research and Education Centre South, Ural State University, Chelyabinsk 454080, Russia

M. Dhayalan (✉)
Department of Prosthodontics, Saveetha Dental College & Hospitals, Saveetha Institute of Medical and Technical Sciences (Saveetha University, TamilNadu, Chennai 600 077, India
e-mail: manikandandhayalan88@gmail.com

A. Stalin
Institute of Fundamental and Frontier Sciences, University of Electronic Science and Technology of China, Chengdu 610064, China

R. Malviya and S. Sundram (eds.), *Engineered Biomaterials*, Engineering Materials,
https://doi.org/10.1007/978-981-99-6698-1_19

normal tissue toxicity and improving treatment efficacy. Future cancer treatments will be improved by nanomedicines and have fewer adverse effects. Gold nanoparticles that have been altered and have regulated geometrical and optical characteristics have undergone substantial research and are employed in biomedicine. The application of gold nanoparticles in varied sizes, forms, and architectures in biomedicine will be covered in this study's summary of recent advancements and continuing research. The major focus of the paper is how gold nanoparticle conjugates are used in biological diagnostics and analytics.

Keywords Gold nanoparticles · Surface functionalization · Cytotoxicity · Transferrin

1 Introduction

The development of chemistry via the use of novel nanoparticles (NPs) and their notable applications in the nursing of certain diseases have captured the attention of people all over the globe [1]. With various sluggish practices in medical investigation and medication, nanotechnology is dynamic for the transmission of medicinal materials. Innumerable uses of NPs exist in the vast nanomedicine sector, including cancer treatment, research techniques, and therapeutic solutions [2]. Their distinctive physicochemical characteristics greatly facilitate the choice of NPs for cancer treatment. Due to their higher size-to-volume ratio, the NP units utilized in nanotechnology with sizes that span between 1.0 nm and 100 nm completely incorporate novel or inventive properties [3]. Metal NPs are the subject of investigation because they provide a cutting-edge technique for treating cancer locally. Over recent years, numerous distribution resources were proposed for the synthesizing of nanomaterials.

Numerous nanoparticles and nanomaterials have evolved from various metal elements thanks to contemporary advancements in nanotechnology and medical research. These materials are manufactured utilizing a variety of methods [4, 5]. The capability to influence nanoparticle structures unwraps up numerous opportunities to reconnoiter these nanoparticles for drug delivery and diagnostic purposes [6]. Due to their optical qualities, gold nanoparticles are used in imaging-based therapeutics and ultrasensitive detection. In this study, we emphasize brand-new ideas for investigating gold nanoparticles in cancer detection and therapy.

1.1 Gold Nanoparticles

One of the key metals that have been disclosed is gold. Gold has been sought after and studied for an extremely long time in several fields. In research studies employed in numerous medical domains, effectuated Gold Nanoparticles (GNPs) with systematized optical and geometrical features are emphasized [7]. The recognition of GNPs as an exceptional vehicle for delivering therapeutic medicines to their target cells. The efficiency of their discharge at the location is crucial for efficient therapy since drug particles conveyed have to be much petite in size if included in drug-related particles or big biomolecules, such as amino acids, nucleic acids, RNA, or DNA [8]. Due to their biocompatibility, inertness, and ability to be functionalized with a variety of other functional groups, GNPs typically possess a surface-to-volume ratio that is much greater. As a result, they can be extremely effective adjuvants in the medical field, enhancing immunogenicity while reducing toxicity and providing stable storage for medications and other drugs used in vaccinations [9].

2 Transferrin and Other Functionalized Materials to Influence at GNPs for Medical Applications

Since the harmfulness and absorption of NPs are closely linked to their surface composition, surface modification of nanoparticles is a crucial way to resolve or minimize these difficulties. The impacts and biokinetics of nanoparticles after systemic administration are dangerously influenced by the surface functionalization of nanoparticles and their resulting physicochemical characteristics [10–12]. A top-down method is used to confine gold nanoparticles in cellular membranes that resemble those of real red blood cells (RBCs). It is possible to observe in the gold nanoparticles the robust membrane layer that covers the biological traits of the donor cells [13]. GNP's ability to quench fluorescence was lessened by the addition of coumarin to the PEG spacer's surface [14]. Transferrin (TF), which can be accurately detected and taken up by TF receptors dynamically distributed on the surface of distinct tumor cells, is an ideal "ligand" for conjugating medicines [15]. As a result, researchers have started using the transferrin-receptor interface as a potentially effective method for the cellular absorption of medicines and genes [16–19]. Gold nanoparticles (GHMP) stabilized by sodium hexametaphosphate (HMP) were described by the researcher in an aqueous media. To create molecular probes with possible therapeutic applications, polyethylene glycol, and transferrin were used to modify the surface of Au-HMP nanoparticles. In vitro, cell survival assays employing naturally occurring G-HMP nanoparticles and their equivalents with modified surfaces demonstrate the nanoparticles' biocompatibility [20]. Gastric cancer is studied using a tailored fluorescent probe made of transferrin-linked cysteine-capped GNP [21].

3 Gene Transfection & Drug Delivery

Chemotherapy medications are often administered via traditional drug delivery methods, which spread the medication throughout the body while also directing some of it toward the tumor. This conundrum of negative consequences is avoided by battering medication delivery mechanisms, a procedure in which a precise active ingredient instead medication is supplied to a specific location in a regulated manner. The enhanced permeability and retention (EPR) effect is so-called because tumor tissue often contains drippy vasculature that facilitates the collection of nanoparticles. This method of submissive pointing makes advantage of the pathologic features of the tumor. The drawbacks of this strategy include inconsistent drug targeting and inadequate medication dispersion in cancer cells. Endocytosis and subsequent drug release are made possible by these ligands' interactions with the appropriate tumor cell receptors.

Gene therapy, the action or prevention of ailment by gene transmission, is observed by several as a probable revolution in medicine [22, 23]. GNPs in particular, function as elegant constituents for nucleic acid transfer applications [24, 25]. It can be invented in an accessible approach with low size dispersity [26, 27], functional diversity can be gladly accomplished by the formation of diverse functional monolayers, permitting multiple functional moieties such as nucleic acids and pointing agents to be located on the element surface [28]. Lastly, the cytotoxicity [29], biodistribution [30], and in vivo excretion properties [31] can be qualified by amendable particle size and surface functionality.

4 Cellular Detections

It may take several days to identify the cell pathogen, which eats up vital time for those with severe bacterial infections. Innovative platform technology is needed to enhance bacterial identification and detection in clinical practice. Aptamer-GNPs, a non-polymerase chain reaction (PCR)-based method for recognizing Staphylococcus aureus, one of the most important human pathogens that root more than 5 lakh infections globally, helped to accomplish the target. This method's quickness, great sensitivity, and cost are its main advantages. Thus, within 1.5 h it finds a single Staphylococcus aureus cell [32].

For more than a few decades, scientists have been enthralled by the potential applications of nanoparticles and gold colloids in a variety of disciplines, including biology, chemistry, engineering, and medicine. Using the Brust-Schiffrin method, monodispersed gold nanoparticles with diameters ranging fluctuate 1 to 3 nm were produced. Antibodies are now being developed in combination with gold nanorods and nanoshells for photothermal treatment to target cancer cells. The important class of materials known as golden nanoconjugates has already demonstrated promise in applications related to fundamental cell biology. Using golden nanoconjugates to transfer drugs to cells is a particularly successful strategy. Numerous separate research teams have looked into the toxicity of different kinds and sizes of gold-bearing nanoconjugates [33].

Breast cancer has emerged as the biggest danger facing women in the previous several decades. Precise molecular diagnosis of the illness and forecast of the tumor assists in the cancer decision-making process. Based on Förster resonant energy transfer (FRET), a functionalized double aptamers (DA-GNP) gold nanoprobe is created for the classification of breast cancer. Eventually, the fluorescent-labeled ER and HER2 a typical biomarker for breast cancer classification, particular aptamers are constrained to the surface of gold nanoparticles (PNG) and fluorescence is quenched. Breast cancer subtype-specific fluorescence is identified. The biomarker protein is coupled to the labeled fluorescent aptamer, which aids in the quantitative identification of various breast cancer subtypes [34].

The most critical part of cancer treatment is early detection which opens up safe and timely treatment procedures. Today, biological markers are widely used in the process of detecting cancer, which seems to be a reliable and adaptable method in turn increasing the patient's overall survival rate. Nanotechnology also contributes to the development of nanoscale biosensors that are highly sensitive and adaptable. This makes it easier to recognize biochemical and biophysical signals linked to health risks (communicable or non-communicable disorders like cancer) at the molecule or cell level. Aptamer-functionalized nanoparticles may be used in a wide variety of cutting-edge applications due to their distinctive molecular design, in addition to their usual role in biosensing. The adoption of this technology has increased during the last several years by cancer researchers due to how simple it is to create nanogold complexes. Surface mobility, its appealing and adaptable optical properties, and its superior cytocompatibility in clinical settings are additional attributes that make it more suitable. This has led to the effective use of these sensors in a variety of optical and electrochemical applications, which has aided in the creation of novel treatments for the eradication of this terrible illness [35].

5 Applications in Gene Transfection

The innovative materials known as gold nanoparticles have a broad range of uses in both technology and medicinal fields. Several types of research are being in process for coating DNA with gold particles and injecting them into plant embryos or plant cells this will enable the association of some genetic material into the cells and ensure some transformations in it. This will confirm that a portion of the genetic material will penetrate the cells and transform them.

Colloidal gold nanoparticles have been used for a variety of unique purposes throughout various fields [36]. JeongHoonByeon demonstrated the cytotoxicity and transfection properties of thiol-capped gold nanoparticles that function as a potential material for biomedical applications. Thiol-fused nanoparticles were cultured with human embryonic kidneys for 24 h to assess the survival of the kidney cells. The results demonstrated approximately 78% improved cell viability for fused particulate matter [37]. With a mortality rate of roughly 20%, lung cancer is one of the noncommunicable illnesses that sees an increase in cases every year. To solve the aforementioned issue lung cancer therapy uses an interfering RNA (siRNA) using Mini Circle DNA Transfection (McD) via Gold Conjugation [38].

To deliver genes, peptide-modified gold nanoparticles (GNPs) have been created using a simple and gentle technique. Four functional segments of a peptide with the order (CRGDKGPDC) GPLGLAGIIIGRRRRRRRR-NH_2 (CPIR28) have been created. The GNPs with diameters of 10 nm are efficiently created by a moderate one-pot reaction involving CPIR28, $HAuCl_4$, and $NaBH_4$ in an aqueous solution, and they contain CPIR28 molecules embedded on their surface. The R8 section is responsible for the GNPs' significant surface positive charge density. Electrostatic interactions with negatively charged DNA may result in condensed complexes, which have little cytotoxicity. In this system, CPIR28 serves as a superb stabilizer by grafting on the surface of the GNPs to prevent their aggregation and precipitation in addition to acting as a modifier to control the nucleation and growth of GNPs to control the shape and size. Additionally, the surface-fused peptide molecules have positive charges that allow them to connect with cells and aid in the effective delivery of genes [39].

Silicon nanowire arrays (SN), which reacted to reactive oxygen species (ROS) safely, were used to provide a platform for secure gene and drug transfection. The ROS-responsive SN is a novel, inventive, and secure platform for medicine and gene transfection because of the platform's moral biocompatibility.

6 Detection and Imaging

Inorganic nanoparticles are being studied more and more as therapeutic and diagnostic tools in the oncology field in recent years. With some potential for clinical trials, inorganic nanoparticles have shown efficiency in tumor photography and treatment both internally and externally. Applications for gold nanoparticles in the shapes of spheres, shells, rods, cages, etc., as well as photothermal treatment, have all been studied. Other applications include carbon nanotubes, iron oxide magnetic nanoparticles, ceramic nanoparticles, semiconductor fluorescent quantum dots, and nanoscale ceramics. Because of their exceptional x-ray attenuation capabilities, unique optical features, and surface modification properties, gold nanoparticle development and applications have a lot of potential. According to reports, gold nanoparticles show promise as CT contrast agents and can be used for medication administration, photothermal and photodynamic treatments, and radiation. Due to its high atomic number, gold has a significant X-ray absorption. Gold nanoparticles can also benefit greatly from the easily changed surface when acting as target CT contrast agents. Gold nanoparticles' CT imaging capabilities are said to depend on their size. The attenuation of X-rays in CT imaging increases when gold nanoparticle size decreases. Gold nanoparticles are availed as disparity agents in computed tomography imaging and ASX-ray-based radiotherapy (RT) agents because of their X-ray attenuation qualities. Gold nanoparticles also have the potential to be very effective in photothermal therapy (PTT) and photodynamic therapy (PDT), which both aim to kill cancerous tumor cells by converting light into heat or inducing the production of free radicals. This is a result of their plasmonic and optical characteristics. When employing PTT and PDT, the ability of gold nanoparticles to target tumor tissue precisely is advantageous. This tumor selectivity prevents healthy tissue from being amputated. This tumor-specificity may be affected by the increased permeability and retention (EPR) effect of GNPs and surface modification with active targeting ligands. Furthermore, gold nanoparticles are promising in the realm of drug administration due to their easily adjustable surface. Multifunctional drug delivery platforms can be created by embedding gold nanoparticles into big structures like liposomes and polymeric particles that can transport medicines or imaging agents. Theranostic delivery, which combines the payload of medications with diagnostic agents for delivery at the same time, is currently receiving a lot of attention in the biomedical sectors. Longer incubation times and larger concentrations of gold nanoparticles can effectively accumulate inside pancreatic cancer cells. The outcomes of this study will assist in the creation and improvement of therapeutic and diagnostic agents based on gold nanoparticles for X-ray Drug Delivery System applications. Building a therapeutic system out of functionalized gold nanoparticles that may particularly respond to the tumor environment is a novel approach. The healing method included employing a specific peptide substrate for the protease to bind doxorubicin to Gold nanoparticles through a thiol-Au bond. The rapid release of the anti-tumor medication (doxorubicin) from the functionalized gold nanoparticles, which may simultaneously limit tumor growth and enable fluorescence imaging, has been shown to occur when MMP-2 protease is overexpressed in tumor tissue and intracellular GSH.

7 Applications in Cells

Cell manipulation frequently takes advantage of interactions between nanoparticles and lasers. Methods for the precise delivery of medications to mammalian cells are crucial for molecular medicine. The last mediator in the vast majority of biological cascades is a protein. Despite several meticulous attempts, efficient protein transport into cells remains challenging. Due to the plasmonic effects of the laser on the gold nanoparticles, proteins may temporarily pass through the cellular membrane and enter the cell. With a 43% efficiency and excellent levels of cell survival, this technique effectively delivered green fluorescent protein. In some biological investigations, the delivery of caspase 3, a protein that regulates apoptosis, may also be seen and measured. The methodical technique makes the efficient transfection of around 10,000 cells per second possible, which is comparable to existing protein transfection techniques like microinjection. Additionally, gold nanoparticle-mediated laser transfection ensures a precise point in time of delivery, enabling a thorough temporal investigation of cellular pathways and protein trafficking. GNPs may be employed in disease diagnosis and treatment because cell death is essential for human health and is related to several major disorders. After entering the body and coming into contact with blood-borne human cells, GNPs specifically target organs and the immune system. Due to their stability and ease of decorating, GNPs are frequently employed in the treatment of diseases, particularly in the administration of medications and photothermal therapy with external stimulation. According to published research, GNPs were also found to cause a variety of cell deaths, mostly through intrinsic routes such as ROS generation, altered mitochondrial activity, and ER stress. The Surface chemistry, dispersion state, and also size, and shape of the Gold nanoparticles play important roles in determining cell death. Smaller GNPs have a higher propensity to induce necrosis. In comparison to hexagonal and spherical gold nanoparticles, gold nanorods are more likely to induce apoptosis. Apoptosis causes minimal to no inflammation in the tissues around the cell, hence changing a cancer cell's fate to undergo apoptotic death through GNP intervention may be beneficial. Furthermore, ligands decorated on GNPs can be targeted to boost the efficiency of cell death induced by targeted tissues and cells. Several various types of apoptosis might happen simultaneously and interact with each other through the various cell death pathways. Yet, the majority of research has solely addressed one particular form of cell death. Therefore, further study should focus on the ability of GNP features to predict various forms of cell death.

8 ADC-AntiBody Drug Conjugates

One of the main cancer treatment choices is chemotherapy [22]. Even though a range of chemotherapy drugs are often used in medical settings, there are still substantial challenges such as side effects and drug resistance [35]. There have been several attempts to make cytotoxic medications stronger, including the use of very potent substances like auristatin and maytansine as well as the mixing of many chemotherapeutic agents [7–19]. However, systemic adverse effects and a narrow therapeutic window limit their practical applicability. With the introduction of monoclonal antibodies, it is now possible to administer medicine to specific areas of the body utilizing their strong binding properties. Antibody—drug conjugates (ADCs), developed and produced over many decades by combining antibodies with cytotoxic medications, are based on this concept [10]. ADCs connect to tumor cell receptors with preference [33]. The receptor-ADC complex is then typically absorbed by the endocytosis pathway after that. Cytotoxic chemicals are released once the linker is split. As a consequence, these drugs have the potential to be cytotoxic in several ways, including by binding to DNA's minor groove or reacting with tubulin.

8.1 Clinical Studies Using Antibody–Drug Conjugates

The number of ADCs being tested in clinical trials for the treatment of solid tumors and hematologic malignancies has sharply increased during the last five years [2, 3]. A substantial category of anti-cancer drugs called ADCs has just emerged.

8.2 Gemtuzumabozogamicin

The FDA86 has authorized *Gemtuzumabozogamicin* (GO; Mylotarg®) as the first ADC. A cleavable hydrazone linker connects calicheamicin with a CD33 monoclonal antibody to form GO [4]. After GO was given accelerated approval in 2000, patients with acute myeloid leukemia (AML) deemed unable to undergo other cytotoxic chemotherapy were treated with it. This decision was made based on three phase II studies [32]. Hepatic venous-occlusive disease and a delayed return to normal hematological function were among the side effects of GO, with overall response rates (ORR) for GO ranging from 26 to 30%. The inclusion of GO during induction treatment was explored in one phase III study (NCT00085709) in individuals under the age of 61, but no appreciable advantage of GO was found. In addition, hazardous side events such as hepatotoxicity, infusion responses, and pulmonary toxicity were noted [26]. Due to these clinical outcomes, Pfizer voluntarily withdrew GO in 2010, following a dosage change (patients now get three doses of 3 mg/m^2 GO as an alternative to one dose of 9 mg/m^2 GO before) [14].

8.3 Brentuximabvedotin

A protease-cleavable dipeptide linker is used to conjugate MMAE with an anti-CD30 antibody to produce *brentuximabvedotin* (BV; Adcetris®), the second ADC to get FDA clearance [12]. The most recurrent effects were thrombocytopenia, anemia, neuropathy, and neutropenia [39]. Neuropathy was the most frequent side effect in patients getting BV57 therapy. In 2015, BV got complete approval as a consequence of the favorable findings of the phase III study (AETHERA, NCT01100502) which examined the use of BV as centralized therapy for Hodgkin's lymphoma [23].

8.4 Ado-Trastuzumabemtansine

The third ADC on the market, *ado-trastuzumabemtansine* (T-DM1; Kadcyla®), was released in 2013 [1]. The results of the phase III study (EMILIA,

NCT00829166)93,94 that T-DM1 was approved for were. The median progression-free survival (PFS) time in the T-DM1 arm was 9.6 months, compared to just 6.4 months in the active comparator (P0.001) [36]. The overall response rate (ORR) was 43.6% against 30.8% for the T-DM1, and overall survival (OS) was 30.9 months versus 25.1 months for the comparator (P 0.001). (P 0.001). In terms of the percentage of serious adverse events (15.5% vs 18.0%) 94 and the total incidence of adverse events (40.8%), the T-DM1 arm outperformed the comparator arm (57.0%).

8.5 Inotuzumabozogamicin

Isotuzumabozogamicin (InO; Besponsa®), the fourth approved ADC treatment to be marketed in 2017 [18], is the drug. An anti-CD22 antibody and a calicheamicin derivative make up InO, which is joined via a cleavable hydrazone linker [16]. Acute lymphocytic leukemia patients in this trial were given the option of receiving InO or a predetermined investigator's choice. In InO arm, full remission occurred at a rate of 80.7% vs 29.4% in the comparison arm (P 0.001). PFS in the InO arm was 5.0 months, whereas the comparative arm's PFS was just 1.7 months (P 0.001) [11]. Other phase III trials are still being conducted, such as post-induction chemotherapy and combination with frontline treatment (NCT03150693) (NCT03959085).

8.6 Polatuzumabvedotin-Piiq

One of most latest ADC to enter the market (in June 2019) is polatuzumabvedotin-piiq (Polivy®), which was produced via conjugating MMAE to an anti-CD79b antibody utilizing a protease cleavable dipeptide linker [9]. The results of the GO29365 phase Ib/II trial (NCT02257567) were used to decide the polatuzumabvedotin-expedited piiq's approval. In this trial, patients with large B-cell lymphoma do give the choice of receiving *polatuzumabvedotin* with *bendamustine* and *rituximab* (BR).

8.7 The Future of Antibody–Drug Conjugates

A fast-expanding area of cancer treatment is ADCs. Numerous ADC technologies that have been developed during the last ten years have opened up a wide range of design options for new ADCs. For instance, both solid and hematologic cancers have intriguing antigen targets [30]. Microtubule inhibitors, anthracyclines, and amatoxins are only a few of the very effective medications that have been found; they may be useful supplements to auristatin [25, 27, 29, 37]. ADCs' therapeutic window has been improved by the characterization of next-generation linkers [17, 20, 38]. Bispecific

ADCs with increased selectivity and potency are among the next directions, as are ADCs that can deliver various payload classes [21].

8.8 Peptide/DNA-Gold Nanoparticle Conjugates

Making peptide-drug conjugates (PDCs) is a useful method for precisely delivering deadly medicines to cancer cells. One barrier to the practical application of peptides is their very poor stability in the blood, liver, and kidneys. One method to solve this issue is to combine PDCs with gold nanoparticles (GNPs), which have shown promising physicochemical and safety qualities for drug delivery systems and seem to possess a greater plasma half-life than PDCs. It was feasible to isolate peptides that were exclusively taken up by A20 murine lymphoma cells by employing a phage library that included 7mer linear peptides. By using flow cytometry and confocal imaging, peptide specificity was determined.

We looked into the physicochemical properties of the coated particles employing electrophoresis, TEM, UV—VIS, and FTIR. It was determined how stable free and PDC-coated AuNP were. By biopanning the phage library, several new peptides that were incorporated into A20 cells were found. To create PDCs that included Bend, Melph, or Chlor, one of them (P4) was employed. Although all three PDCs had low half-lives of 10.6 to 15.4 min, they all only destroyed A20 target cells. The half-lives were increased to 21.0–22.3 h when these samples were coated with PEG–GNPs. PDC-PEG-GNPs' cytotoxicity prevented the target cells from surviving. Furthermore, although pre-incubation of free PDCs for 24 h substantially eliminated their cytotoxic activity, pre-incubation of PDC-PEGGNPs for 72 h sustained it. Peptide-drug integration has the potential to raise the target effectiveness of chemotherapeutic drugs despite their brief half-lives. It is simple to conjugate them into gold nanoparticles to get over this obstacle. This coupling also makes it possible to create slow-release versions of PDC-containing targeted drug delivery systems [15].

8.9 Toxicity

It is generally known that the use of GNPs in the biomedical area has opened up new possibilities for disease diagnosis and therapy. However, the extent to which they are hazardous once given to patients, either by intravenous injection or oral dose, determines how well they may be applied in the biomedical area, particularly in therapeutic and diagnostic modalities. Cells can become poisonous to any molecular compound at relatively large concentrations. Even therapeutic chemicals can cause an imbalance in cellular activity when exposed to excessive concentrations, which eventually results in harmful toxic consequences. What maximum concentration of GNPs may be provided without causing detrimental toxicity at the cellular level is therefore the correct issue to ask. How much GNP may be used for therapeutic and diagnostic purposes without jeopardizing cellular, mitochondrial, and genomic health, more specifically? The toxicity profile of GNPs is affected by some factors, in addition to the concertation effect, including size, shape, surface coating, surface charge, interaction with biological molecules, and others. Early research using a mouse model found that the toxicity of GNPs was influenced by their size [8]. Larger core-sized GNPs are less cytotoxic than smaller-sized GNPs, according to research. The effects of three distinct diameters were assessed in research employing human embryonic stem cells. The 1.5 nm core GNPs were found to be cytotoxic at concentrations of 0.1 gmL^{-1}, however, the 4 and 14 nm were not at concentrations of 10 gmL^{-1}, demonstrating that bigger GNPs are less hazardous to the cells [31]. Regarding the development of a database of various size-dependent (1.5–100 nm), It is necessary to take into account both in vitro and in vivo toxicity to completely comprehend the size-dependent toxicity profile. Steckiewicz et al. in vitro studies on the cytotoxicity of three various GNP shapes—star, rod, and spherical-on three different human cell lines osteosarcoma, human fetal osteoblast, and pancreatic duct cells were conducted to evaluate the toxicity of the form [34]. According to the study, compared to other shapes evaluated, star-shaped particles had the greatest damaging effects on cell lines. The range of the gold nano star's IC50 values in the three different cell types was 0.46–1.76 gmL^{-1}. In a separate study, the toxicity of gold nanoparticles such as hollow gold-silver nanoparticles, gold nanospheres, gold nanorods, gold nanocages, gold nanostars, and SiO_2-gold core–shell nanoparticles was investigated [24]. The extensive investigation examined these nanoparticles at concentrations ranging from 1.5 to 200 μmL^{-1}. The study used MCF7 breast cancer cells, and the results indicate that utilizing gold nanosphere and nanostar increased cell viability by more than 90% at a concentration of 12.5 μmL^{-1}. Contrary to the findings published by Steckiewicz et al who found that nanostars induced the highest toxicity in cells, the cytotoxicity of cells was significantly increased by the use of gold nanorods [34]. According to these latest findings, altering a particle's shape can be harmful to cells, particularly when a nanorod's high aspect ratio disrupts the lipid bilayer and causes cytotoxicity. Gold nanostars were discovered to raise the Bax proteins and reduce the BCL-2 protein, which supports apoptotic programmed cell death in addition to stiffening the lipid bilayer. The nanoparticles' hazardous activity is influenced by their surface coating and surface charges. Although many studies have found similarities in the toxicity profiles of anionic and cationic surface coatings, the cationic GNP has a two- to three-fold higher particle absorption rate, leading to high concentration-induced toxicity [5, 13]. Strong cationic ligand-coated GNP absorption is made possible by the cell membrane's abundance of anionic uptake transporters. The toxicity brought on by NPs is also influenced by the density of the surface ligand and its chemical interactions with different biological components floating in the circulation. Cells may suffer damage from the nanoparticle agglomeration that follows the formation of the protein corona. Given that GNPs are mostly found in the cellular compartment known as the endocytic compartment [28]. The bulk of circulating GNPs in the bloodstream are thought to be collected by the liver in animal models, according to biodistribution studies. These most recent discoveries suggest that primary hepatocytes or hepatic cell lines should be used for doing in vitro research. Numerous studies have been published that demonstrate how intracellular ROS may be produced by GNPs, although controlled application can result in potent antioxidants by limiting the formation of NF-B and the subsequent inflammatory reactions. In the first phases of cancer treatment, when the therapeutic objective is to accelerate cell death, the advantageous impact of GNPs' toxic activity may be helpful. Prior to using GNPs' risky behavior in cancer research, issues with protein corona formation and their efficient clearance from the body of treated patients must be properly evaluated and resolved.

8.10 Recent Trends and Future Perspectives

For applications in medicine, it is essential to create efficient procedures for functionalizing GNP with diverse substances that allow for stability in vivo and precise interaction with biological targets. The best-stabilizing agents nowadays are thought to be the thiolated derivatives of PEG and other compounds. PEG-coated particles in particular have the potential to circulate for extended periods and tend to be less susceptible to immune system cell components. By now, everyone is aware that GNP conjugates provide reliable labels for resolving bioimaging issues. Numerous two-photon luminescence manifestations of GNP, optical coherence tomography, acoustic tomography, etc. may all be performed using these conjugates. GNP conjugates have been used in analytical investigations employing both straightforward solid-phase or homophase approaches and cutting-edge experimental techniques (Dot analysis, immunochromatography). GNP-based plasmon photothermal laser treatment for cancer was first introduced in 2003 and has just begun the clinical approval stage. How rapidly the following significant issues can be overcome will determine the real clinical success of this technology: It is obvious that more, continuing study is required to fully understand the biodistribution and toxicity of GNP. This research should focus on three areas: (1) developing efficient techniques for non-optical heating or fiber-optic radiation delivery to malignancies inside the body; (2) enhancing contrast and uniformity of accumulation by conjugate delivery approaches; (3) emergent techniques for in situ photo thermolysis control. The creation of a coordinated program should be prioritized to ascertain the relationships between experimental factors (model, dosages, administration technique, research, etc.). Coordination is also required for the creation of particle standards and test procedures for the toxicity of nanomaterials.

8.11 Conclusion

The unique qualities of GNPs make it possible to use them in medical applications. Several synthetic approaches for creating particles of different sizes and shapes provide a broad toolbox for conjugates with enhanced affinity for specific cell receptors, effective cell internalization, extended circulation half-life, tumor permeability, and high biocompatibility. These characteristics of GNPs, along with their electrical and optical responsiveness, have made it possible to create strong platforms for developing new vaccines, effective gene therapy techniques, improved radiation procedures, and novel cancer treatment approaches. Even if gold is less hazardous, health complications might still occur due to the relatively slow rate of clearance from tissues and circulation. For this reason, methods that improve the precise targeting of sick cells must be standardized before GNPs are used regularly in humans.

References

1. Amiri-Kordestani, L., Blumenthal, G.M., Xu, Q.C., Zhang, L., Tang, S.W., Ha, L., Weinberg, W.C., Chi, B., Candau-Chacon, R., Hughes, P., Russell, A.M.: FDA Approval: Ado-Trastuzumab Emtansine for the Treatment of Patients with HER2-Positive Metastatic Breast CancerFDA Approval Summary for T-DM1 for HER2+ MBC. Clin. Cancer Res. **20**(17), 4436–4441 (2014)
2. Beck, A., Wurch, T., Bailly, C., Corvaia, N.: Strategies and challenges for the next generation of therapeutic antibodies. Nat. Rev. Immunol. **10**(5), 345–352 (2010)
3. Birrer, M.J., Moore, K.N., Betella, I., Bates, R.C.: Antibody-drug conjugate-based therapeutics: state of the science. JNCI: J. Natl. Cancer Inst. **111**(6), 538–49 (2019)
4. Bross, P.F., Beitz, J., Chen, G., Chen, X.H., Duffy, E., Kieffer, L., Roy, S., Sridhara, R., Rahman, A., Williams, G., Pazdur, R.: Approval summary: gemtuzumab ozogamicin in relapsed acute myeloid leukemia. Clin. Cancer Res. **7**(6), 1490–1496 (2001)
5. Canepa, E., Salassi, S., Simonelli, F., Ferrando, R., Rolandi, R., Lambruschini, C., Canepa, F., Dante, S., Relini, A., Rossi, G.: Non-disruptive uptake of anionic and cationic gold nanoparticles in neutral zwitterionic membranes. Sci. Rep. **11**(1), 1256 (2021)
6. Charbgoo, F., Nejabat, M., Abnous, K., Soltani, F., Taghdisi, S.M., Alibolandi, M., Shier, W.T., Steele, T.W., Ramezani, M.: Gold nanoparticle should understand protein corona for being a clinical nanomaterial. J. Control. Release **28**(272), 39–53 (2018)
7. Chari, R.V., Miller, M.L., Widdison, W.C.: Antibody–drug conjugates: an emerging concept in cancer therapy. Angew. Chem. Int. Ed. **53**(15), 3796–3827 (2014)
8. Chen, Y.S., Hung, Y.C., Liau, I., Huang, G.S.: Assessment of the in vivo toxicity of gold nanoparticles. Nanoscale Res. Lett. **4**, 858–864 (2009)
9. Deeks, E.D.: Polatuzumab vedotin: first global approval. Drugs **79**(13), 1467–1475 (2019)
10. Diamantis, N., Banerji, U.: Antibody-drug conjugates—an emerging class of cancer treatment. Br. J. Cancer **114**(4), 362–367 (2016)
11. DiJoseph, J.F., Armellino, D.C., Boghaert, E.R., Khandke, K., Dougher, M.M., Sridharan, L., Kunz, A., Hamann, P.R., Gorovits, B., Udata, C., Moran, J.K.: Antibody-targeted chemotherapy with CMC-544: a CD22-targeted immunoconjugate of calicheamicin for the treatment of B-lymphoid malignancies. Blood **103**(5), 1807–1814 (2004)
12. Doronina, S.O., Toki, B.E., Torgov, M.Y., Mendelsohn, B.A., Cerveny, C.G., Chace, D.F., DeBlanc, R.L., Gearing, R.P., Bovee, T.D., Siegall, C.B., Francisco, J.A.: Development of potent monoclonal antibody auristatin conjugates for cancer therapy. Nat. Biotechnol. **21**(7), 778–784 (2003)
13. Goodman, C.M., McCusker, C.D., Yilmaz, T., Rotello, V.M.: Toxicity of gold nanoparticles functionalized with cationic and anionic side chains. Bioconjug. Chem. **15**(4), 897–900 (2004)
14. Jen, E.Y., Ko, C.W., Lee, J.E., Del Valle, P.L., Aydanian, A., Jewell, C., Norsworthy, K.J., Przepiorka, D., Nie, L., Liu, J., Sheth, C.M.: FDA Approval: Gemtuzumab Ozogamicin for

the Treatment of Adults with Newly Diagnosed CD33-Positive Acute Myeloid LeukemiaFDA Approval Summary: Gemtuzumab Ozogamicin. Clin. Cancer Res. **24**(14), 3242–3246 (2018)
15. Kalimuthu, K., Lubin, B.C., Bazylevich, A., Gellerman, G., Shpilberg, O., Luboshits, G., Firer, M.A.: Gold nanoparticles stabilize peptide-drug-conjugates for sustained targeted drug delivery to cancer cells. Journal of nanobiotechnology. **16**(1), 1–3 (2018)
16. Kantarjian, H.M., DeAngelo, D.J., Stelljes, M., Martinelli, G., Liedtke, M., Stock, W., Gökbuget, N., O'Brien, S., Wang, K., Wang, T., Paccagnella, M.L.: Inotuzumab ozogamicin versus standard therapy for acute lymphoblastic leukemia. N. Engl. J. Med. **375**(8), 740–753 (2016)
17. Kovtun, Y.V., Audette, C.A., Mayo, M.F., Jones, G.E., Doherty, H., Maloney, E.K., Erickson, H.K., Sun, X., Wilhelm, S., Ab, O., Lai, K.C.: Antibody-maytansinoid conjugates designed to bypass multidrug resistance. Can. Res. **70**(6), 2528–2537 (2010)
18. Lamb, Y.N.: Inotuzumab ozogamicin: first global approval. Drugs **77**, 1603–1610 (2017)
19. Lambert, J.M., Berkenblit, A.: Antibody–drug conjugates for cancer treatment. Annu. Rev. Med. **29**(69), 191–207 (2018)
20. Lyon, R.P., Bovee, T.D., Doronina, S.O., Burke, P.J., Hunter, J.H., Neff-LaFord, H.D., Jonas, M., Anderson, M.E., Setter, J.R., Senter, P.D.: Reducing hydrophobicity of homogeneous antibody-drug conjugates improves pharmacokinetics and therapeutic index. Nat. Biotechnol. **33**(7), 733–735 (2015)
21. Maruani, A., Smith, M.E., Miranda, E., Chester, K.A., Chudasama, V., Caddick, S.: A plug-and-play approach to antibody-based therapeutics via a chemoselective dual click strategy. Nat. Commun. **6**(1), 6645 (2015)
22. Miller, D.R.: A tribute to Sidney Farber–the father of modern chemotherapy. Br. J. Haematol. **134**(1), 20–26 (2006)
23. Moskowitz, C.H., Nademanee, A., Masszi, T., Agura, E., Holowiecki, J., Abidi, M.H., Chen, A.I., Stiff, P., Gianni, A.M., Carella, A., Osmanov, D.: Brentuximab vedotin as consolidation therapy after autologous stem-cell transplantation in patients with Hodgkin's lymphoma at risk of relapse or progression (AETHERA): a randomised, double-blind, placebo-controlled, phase 3 trial. The Lancet. **385**(9980), 1853–1862 (2015)
24. Pakravan, A., Salehi, R., Mahkam, M.: Comparison study on the effect of gold nanoparticles shape in the forms of star, hallow, cage, rods, and Si-Au and Fe-Au core-shell on photothermal cancer treatment. Photodiagn. Photodyn. Ther. **1**(33), 102144 (2021)
25. Patterson, A.W., Peltier, H.M., Sasse, F., Ellman, J.A.: Design, synthesis, and biological properties of highly potent tubulysin D analogues. Chemistry–A Eur. Journal. **13**(34), 9534–41 (2007)
26. Petersdorf, S.H., Kopecky, K.J., Slovak, M., Willman, C., Nevill, T., Brandwein, J., Larson, R.A., Erba, H.P., Stiff, P.J., Stuart, R.K., Walter, R.B.: A phase 3 study of gemtuzumab ozogamicin during induction and postconsolidation therapy in younger patients with acute myeloid leukemia. Blood, The Journal of the American Society of Hematology. **121**(24), 4854–4860 (2013)
27. Quintieri, L., Geroni, C., Fantin, M., Battaglia, R., Rosato, A., Speed, W., Zanovello, P., Floreani, M.: Formation and antitumor activity of PNU-159682, a major metabolite of nemorubicin in human liver microsomes. Clin. Cancer Res. **11**(4), 1608–1617 (2005)
28. Saleh, H.M., Soliman, O.A., Elshazly, M.O., Raafat, A., Gohar, A.K., Salaheldin, T.A.: Acute hematologic, hepatologic, and nephrologic changes after intraperitoneal injections of 18 nm gold nanoparticles in hamsters. Int. J. Nanomed. **11**, 2505 (2016)
29. Sasse, F., Sieinmetz, H., Heil, J., Hoefle, G., Reichenbach, H.: Tubulysins, new cytostatic peptides from myxobacteria acting on microtubuli production, isolation, physico-chemical and biological properties. J. Antibiot. **53**(9), 879–85 (2000)
30. Senter, P.D.: Potent antibody drug conjugates for cancer therapy. Curr. Opin. Chem. Biol. **13**(3), 235–244 (2009)
31. Senut, M.C., Zhang, Y., Liu, F., Sen, A., Ruden, D.M., Mao, G.: Size-dependent toxicity of gold nanoparticles on human embryonic stem cells and their neural derivatives. Small **12**(5), 631–646 (2016)

32. Sievers, E.L.: Efficacy and safety of gemtuzumab ozogamicin in patients with CD33-positive acute myeloid leukaemia in first relapse. Expert Opin. Biol. Ther. **1**(5), 893–901 (2001)
33. Sievers, E.L., Senter, P.D.: Antibody-drug conjugates in cancer therapy. Annu. Rev. Med. **17**, 64 (2013)
34. Steckiewicz, K.P., Barcinska, E., Malankowska, A., Zauszkiewicz–Pawlak, A., Nowaczyk, G., Zaleska-Medynska, A., Inkielewicz-Stepniak, I.: Impact of gold nanoparticles shape on their cytotoxicity against human osteoblast and osteosarcoma in in vitro model. Evaluation of the safety of use and anti-cancer potential. J. Mater. Sci.: Mater. Medicine. **30**,1–5 (2019)
35. Tsuchikama, K., An, Z.: Antibody-drug conjugates: recent advances in conjugation and linker chemistries. Protein Cell **9**(1), 33–46 (2018)
36. Verma, S., Miles, D., Gianni, L., Krop, I.E., Welslau, M., Baselga, J., Pegram, M., Oh, D.Y., Diéras, V., Guardino, E., Fang, L.: Trastuzumab emtansine for HER2-positive advanced breast cancer. N. Engl. J. Med. **367**(19), 1783–1791 (2012)
37. Vetter, J.: Toxins of Amanita phalloides. Toxicon **36**(1), 13–24 (1998)
38. Widdison, W.C., Ponte, J.F., Coccia, J.A., Lanieri, L., Setiady, Y., Dong, L., Skaletskaya, A., Hong, E.E., Wu, R., Qiu, Q., Singh, R.: Development of anilino-maytansinoid ADCs that efficiently release cytotoxic metabolites in cancer cells and induce high levels of bystander killing. Bioconjug. Chem. **26**(11), 2261–2278 (2015)
39. Younes, A., Gopal, A.K., Smith, S.E., Ansell, S.M., Rosenblatt, J.D., Savage, K.J., Ramchandren, R., Bartlett, N.L., Cheson, B.D., De Vos, S., Forero-Torres, A.: Results of a pivotal phase II study of brentuximab vedotin for patients with relapsed or refractory Hodgkin's lymphoma. J. Clin. Oncol. **30**(18), 2183 (2012)

Biocidal Effect of Copper Contained in a Mineral Tailing on the Growth of *Shewanella Putrefaciens*

Hugo J. Marín-García, Ramiro Escudero-García, Carlos Cortés-Penagos, and Ricardo Morales-Estrella

Abstract Biotechnological applications in the mining area are considered competitive alternatives with respect to traditional processes for treating low-grade ores or tailings; even as a stage prior to leaching by cyanidation, by eliminating mineral species called cyanicides. In this sense, the use of the *Shewanella putrefaciens* bacterium to chemically interact with the copper tailings, which also contain 4.35 g of gold/ton, but that it is mostly encapsulated in iron oxides, makes its implementation feasible due to its iron-reducing capacity; nevertheless, the chemical interaction between the tailings and the liquid medium solubilizes the remaining copper in the ore, completely inhibiting the growth of *Shewanella putrefaciens* bacteria. In this research work, the maximum concentration of copper was determined, which does not stop the growth of the colony of this microorganism during the bioreduction of the iron species contained in the mineral. Experimental results show that the copper concentration in the ore must be less than 35 ppm, prior to contact with bioreducing bacteria to, for example, reduce oxidized iron species, prior to cyanide leaching of gold-bearing species. Therefore, to carry out the bioreduction of the iron present in the tailings, it will be necessary to first reduce the amount of copper in the ore to the concentration mentioned above, to prevent the copper content from decreasing by exhaustive chemical leaching (affecting the economy of the process), or genetically modifying the bacterium with a plasmid that provides greater resistance to copper.

Keywords *Shewanella putrefaciens* · Tailings · Copper · Bioleaching · Cyanidation

---

H. J. Marín-García · R. Escudero-García (✉) · R. Morales-Estrella
Institute of Research in Metallurgy and Materials, Morelia, Michoacán, México
e-mail: ramiro.escudero@umich.mx

H. J. Marín-García
e-mail: hugo.marin@umich.mx

R. Morales-Estrella
e-mail: ricardo.morales@umich.mx

C. Cortés-Penagos
Faculty of Chemical Pharmacobiology, Universidad Michoacana de San Nicolás de Hidalgo, Santiago Tapia 403, Col. Centro, C.P. 58000 Morelia, Michoacán, México

R. Malviya and S. Sundram (eds.), *Engineered Biomaterials*, Engineering Materials,
https://doi.org/10.1007/978-981-99-6698-1_20

1 Introduction

A very important characteristic of mining activity is the generation of large volumes of solid waste that can be potentially toxic to the environment, called tailings, which may still contain valuable metals and can therefore be considered for reprocessing [1, 2]; on the other hand, they generally present a quartz matrix, although it totally depends on the location of the deposit [3, 4]. The tailings with 0.26% w/w copper used in this work were generated in a mining operation to obtain copper from the leaching with sulfuric acid of the mineral known as malachite ($Cu_2(OH)_2CO_3$), which is totally soluble. In this acid at room temperature [5, 6], the main mineral matrix is made up of albite and quartz, while 68.72% of the gold present in the tailings (4.35 g/ton) is contained in the species soluble in aqua regia, mainly hematite [7]; because of this mineralogical characteristic, the use of the *Shewanella putrefaciens* bacterium in the bioreduction process of the hematite species was considered, since it is widely recognized for its ability to feed on oxidized iron species [8–11]. During the microbial interaction with a mineral, the dissolution of certain metallic species occurs, which can have a negative effect on the microorganism itself [12]; however, resistance to metals in general is a common phenotype in multiple microorganisms, and microbial communities are highly sensitive or adaptable to environmental changes [13, 14].

Several transition metals generate cations that have an important role as nutrients (or trace elements) in several biochemical reactions, although at high concentrations, these cations form non-specific compounds in the cell, causing toxic effects; Copper as a transition metal has the property of switching between its oxidation states from cuprous (Cu(I)) to cupric (Cu(II)), this reduction potential allows it to serve as an ideal biological cofactor, generally in aerobic organisms. Therefore, for there to be a physiological effect, the copper cation must enter the cell, but in high concentration it is extremely toxic [13, 15].

There are four bacterial protein channels for copper homeostasis in enterobacteria (Escherichia coli): cue, cus, Pco and Cop [13], although there are other bacterial genera that possess them (Acidithiobacillus sp, Shewanella sp, Pseudomonas sp) [16–18]. Most Gram (–) bacteria, such as *Shewanella putrefaciens*, use the cue and cus chromosomal systems, which detoxify and redistribute copper by changing charge and efflux, and are gradient regulated chemiosmotic agent that crosses the cytoplasmic membrane of the bacterium [13, 15]. The Pco and Cop systems refer to plasmid-mediated resistance to copper, although the latter is also found chromosomally and was detected in *Pseudomonas syringae*. These two systems provide greater resistance to copper, while the former focus on low or tolerable concentrations [13, 18, 19].

Although the resistance to copper by the *Shewanella putrefaciens* bacterium in mining applications has not been reported, this characteristic is relevant for biohydrometallurgical applications, as extensively described by Martínez-Bussenius et al. [16] in acidophilic bacteria. However, resistance to copper has been detected by *Shewanella putrefaciens* in strains of sea urchins and shellfish [17, 20], in isolates of sea urchins the copper resistance gene cusB and CopA were detected, while the strains isolated from shellfish only determined the maximum tolerable concentration (CMT) when cultivated in culture media added with copper sulfate (CuSO4), obtaining bacteria that grew in 100, 3000, and >5000 mg of copper per liter; therefore, the purpose of this work is to analyze the biocidal effect of copper contained in the tailings on the growth of *Shewanella putrefaciens* ATCC® 8071 for a specific application in biomining.

2 Experimental Procedure

2.1 Reactivation and Conservation of the Reference Strain

The *Shewanella putrefaciens* ATCC® 8071 bacterium was acquired through the National Collection of Microbial Strains and Cell Cultures belonging to CINVESTAV-IPN, which was kept in an internal working strain to avoid contamination and loss, and was later sown on marine agar [21], and TSI agar to verify its production of hydrogen sulfide. Once the bacteria grew on the marine agar, Gram staining was performed.

2.2 Bacterial Growth Curve and Preparation of Working Inocula

From the colony developed in the marine agar, an inoculum was taken with the bacteriological loop and suspended in marine broth and incubated at 35 ± 0.3 °C with 150 rpm of constant agitation for 72 h, aliquots were taken and inoculated into tubes with marine broth enriched with magnetite PFD 00-007-0322 ($FeO \cdot Fe_2O_3$), which were incubated under the same conditions mentioned above. The kinetics were carried out at 35 ± 0.3 °C with 150 rpm of constant agitation for 240 h, then the optical density of the biomass generated at 620 nm was read by the ultraviolet–visible light spectroscopy technique, taking as standards the scale of McFarland [22, 23]. Regarding the working inoculum, the previous conditions are repeated, but the incubation of the marine broths was for 48 h to ensure that the bacteria is in the exponential growth phase.

2.3 Effect of Copper from Copper Tailings on Bacterial Growth

Mixtures of marine broth with 5% solids (direct tailings and washed tailings) were prepared separately, and sterilized, they were subjected to 35 ± 0.3 °C with 150 rpm of constant agitation for 500 h to observe if the culture broth reacts with the tailings, later in the supernatant pH and copper in solution were determined by atomic absorption spectroscopy, while the solids were analyzed by scanning electron microscopy. The aforementioned liquids were subjected to sterilization and previously prepared inoculums were added to them, these samples were incubated aerobically at 35 ± 0.3 °C for 72 h, later, a 20 μL inoculum was taken and spread out with a glass rod. in marine agar, they were incubated aerobically at 35 ± 0.3 °C for 48 to 72 h to observe the development of CFU.

2.4 Determination of the Maximum Tolerable Concentration of Copper by the Bacteria

Culture media were prepared with marine agar enriched with copper sulfate hexahydrate with the following concentrations: 3.12, 6.35, 9.53, 12.71, 15.89, 19.06, 22.24, 25.42, 28.60, 31.77, 34.95, 38.13, 41.30, 44.48, 47.64., 76.26, 101.67, 127.10, 152.51, 177.93, 203.35 and 228.77 ppm of Cu to determine the Maximum Tolerable Concentration (CMT) of copper to which the bacterium *Shewanella putrefaciens* ATCC® 8071 can be subjected; in case of growth above 50.84 ppm of Cu, it is considered a strain resistant to this metal, but if it grows below it is considered sensitive [20, 24]. For the sowing of the culture media in an aerobic environment, a 20 μL inoculum was taken and sown in an extended way with a glass rod and after 72 h of incubation at 35 ± 0.3 °C they were observed in a colony counter.

3 Results and Discussion

3.1 Reactivation and Conservation of the Reference Strain

The *Shewanella putrefaciens* ATCC® 8071 strain sown on marine agar showed abundant growth under aerobic conditions (Fig. 1, section a), the cream-colored, slightly viscous, and concave colony together with the Gram bacilli (–), are characteristic of this bacterium (Section b) [25, 26]. Regarding subsection c, the formation of a black precipitate is observed in the TSI medium, indicating that the bacterium generates hydrogen sulfide and does not ferment sugars, by maintaining the red color of the indicator, which differentiates it from Enterobacteriaceae and the genus *Pseudomonas sp.* [27].

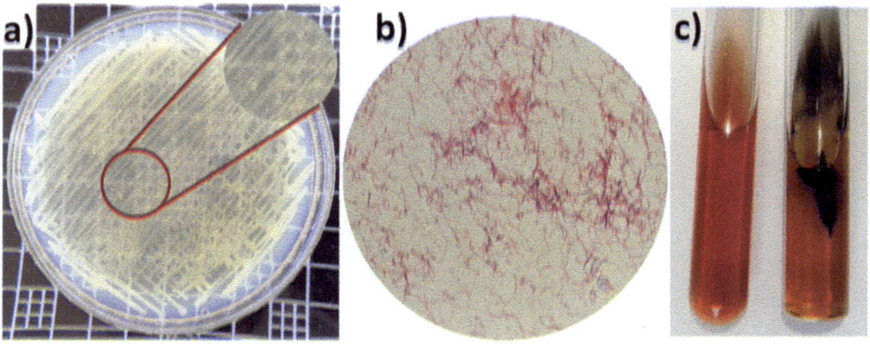

Fig. 1 **a** Growth in an aerobic environment of *Shewanella putrefaciens* ATCC® 8071 in marine agar; **b** Gram bacilli (–) of *Shewanella putrefaciens* ATCC® 8071; **c** Comparison of TSI medium before and after incubation

3.2 Bacterial Growth Curve and Preparation of Working Inocula

As previously mentioned, the *Shewanella putrefaciens* bacterium is a microorganism that bioreduces iron in an anaerobic environment; therefore, it was necessary to adapt it to the respiration of this metal and thus generate the first inoculum (Fig. 2, section a), since in the previous step its growth was in an aerobic environment. Regarding the kinetics of bacterial growth, this is shown in subsection b, in which it is observed that at 72 h of growth the inoculum presented a concentration of 1.11×10^9 CFU/ml, which when added to the vial is reduced to 9.20×10^8 CFU/ml, due to the greater volume of broth used, it is also observed that in the first 24 h the adaptation stage occurs and from 48 to 96 h it is in the exponential growth phase, During the next 48 h of growth, the stationary phase occurred, although it is not represented graphically, since the death phase was clearly observed after 144 h due to the drastic decrease in CFU/ml, a trend that was maintained until the end. at 240 h, coinciding with the control broth. On the other hand, it was possible to make a comparison between the inoculum and the 72-h sample, since there was a significant difference in the bacterial concentration, due to the fact that the bacteria grew in an aerobic environment in the inoculum; therefore, it required more time to develop, but once it enabled the mechanisms of iron respiration, its growth accelerated.

3.3 Effect of Copper Content of Tailings on Bacterial Growth

Figure 3 shows the dissolution of copper caused by chemical leaching between the tailings TH (Tailing Head), HTwas (Head Tailing washed), and the marine broth. The measured pH is close to neutrality and the characteristic yellow color is observed; after 500 h of agitation, the coloration of the marine broths turned green, being more

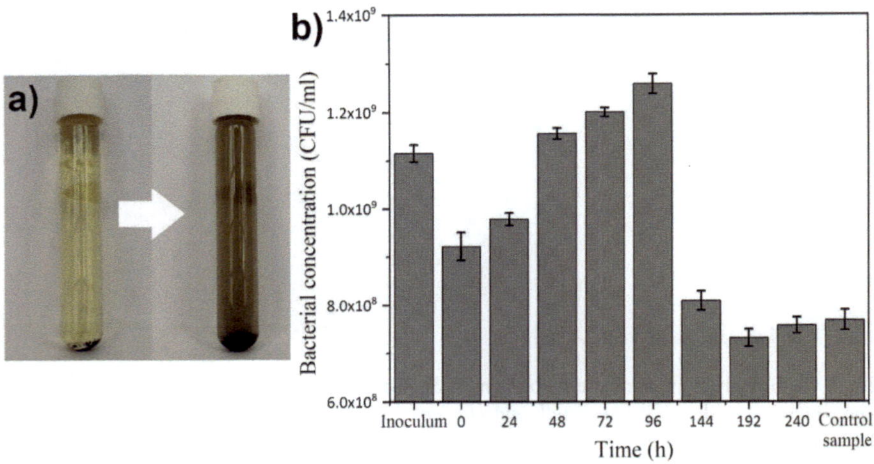

Fig. 2 **a** Inoculum with sedimented magnetite and during shaking; **b** Growth kinetics of *Shewanella putrefaciens* ATCC® 8071 in an anaerobic environment.

intense in the RSL than in the RL, mainly due to the excess of sulfuric acid that the RSL still contained; at this point the pH decreased to 5.63 and dissolved 120.74 ppm Cu, while the RL slightly decreased the pH to 6.05 and dissolved 67.36 ppm Cu. It is also possible to observe that the greatest chemical leaching of copper occurs during the first 250 h, since there is no significant difference with the longer leaching time, and the pH remains without significant changes.

Regarding the analysis by scanning electron microscopy, Fig. 4 shows the micrographs obtained with the electron backscattering technique, observing that in the TH before the interaction with seawater (section a), the particles present a cloudy layer on its surface associated with the remaining presence of sulfuric acid, since in the HT was it disappears and the quality of the image of the particles with higher atomic weight is improved (paragraph c) [7]. After 250 h of agitation with the marine broth, both the RSL and the RL present agglomerated particles made up of low atomic weight elements (from the marine broth), due to the homogeneity that the grays present in the backscattered electron technique; it was possible to highlight some particles or areas with brightness indicating elements of greater atomic weight (see sections b and d); nevertheless, with this result it is shown that the high concentration of salts, when dried on the tailings, generate agglomerates that can interfere with the efficiency of the next gold extraction process if a washing stage is not carried out that eliminates or considerably reduces said salts.

Figure 5, section a, shows the effect of the copper concentration on the growth of *Shewanella putrefaciens* ATCC® 8071, after incubation of the TH and HT was supernatants, plus the sterility control; the three presented null development of CFU, indicating that the concentrations of 59.73 and 120.73 ppm of Cu in the supernatants exceed the resistance mechanisms that the bacteria presents naturally, while the

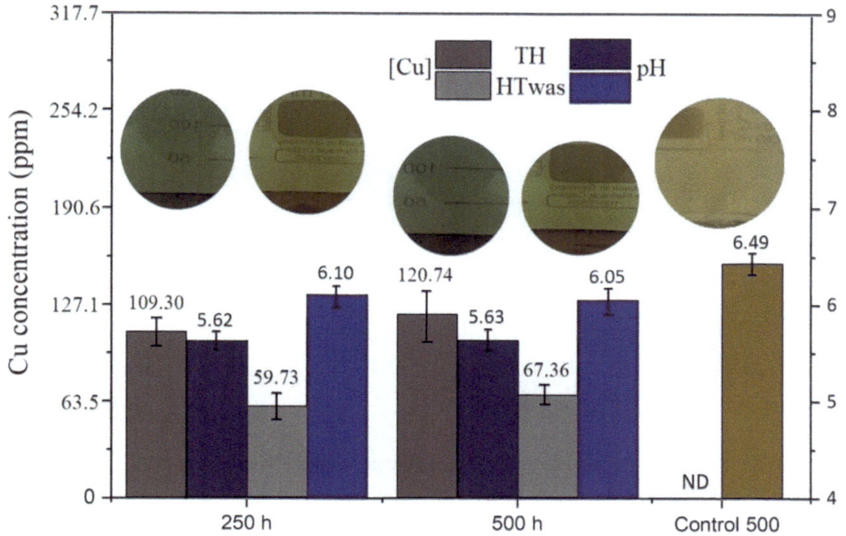

Fig. 3 Copper concentration and pH in the marine broth, after 250 and 500 h of agitation of the Tailing Head (TH), and the Head Tailing washed (HTwas)

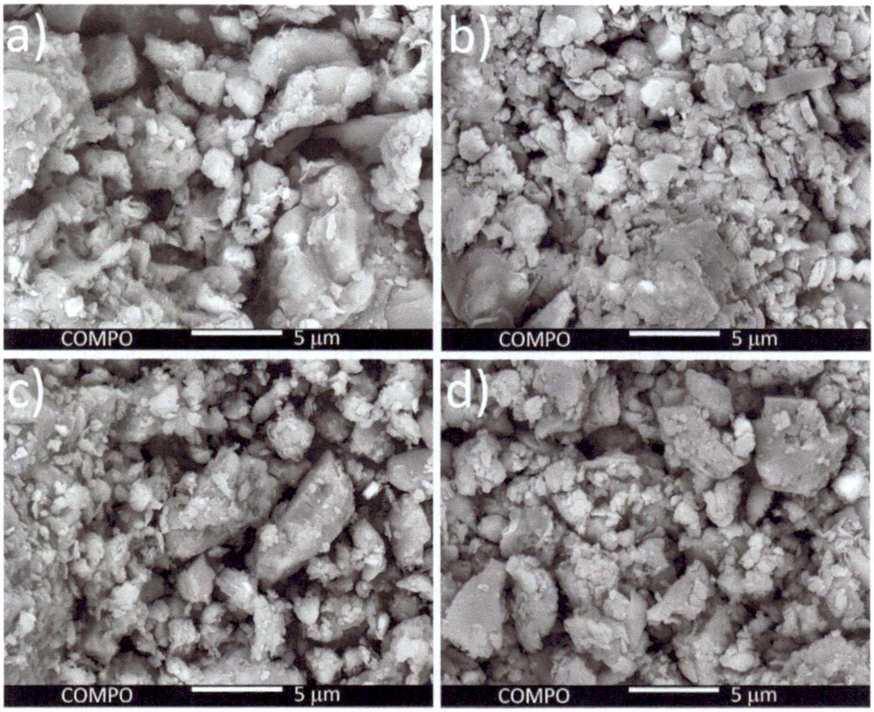

Fig. 4 Micrographs applying the backscattered electron technique to the mineral samples after 500 h of leaching in marine broth. **a** and **b** TH; **c** and **d** HTwas

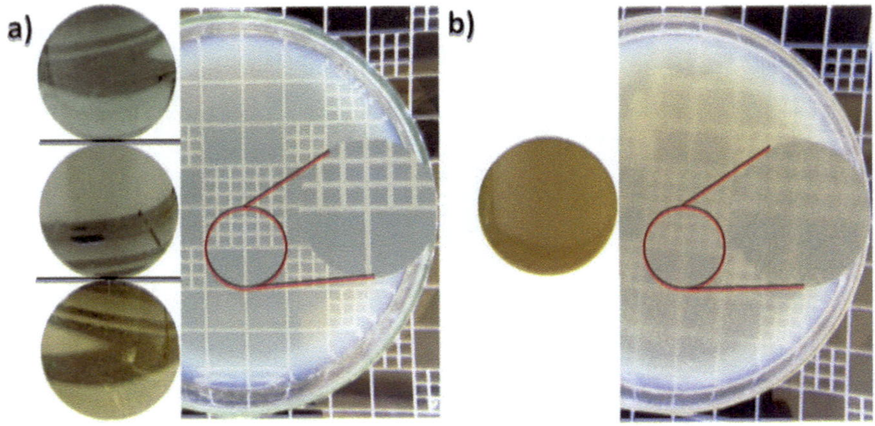

Fig. 5 **a** Effect of dissolved copper from tailings on the growth of *Shewanella putrefaciens* ATCC® 8071 and sterility control; **b** Growth of the positive control in the marine agar.

sterility control indicated that aseptic conditions were maintained during the experiment. Regarding the section b, the marine broth inoculated as a positive control is shown, which presented abundant growth, showing that this medium is favorable for the development of *Shewanella putrefaciens* ATCC® 8071.

3.4 Determination of the Maximum Tolerable Concentration of Copper

Figure 6 section a shows the massive growth of the bacterium in the marine agar without the presence of copper; however, as the concentration of this element increases, the development of colonies decreases, Such is the case of the 19.06 ppm Cu concentration (section b), where 782 $\pm$ 15 CFUs were counted, the Maximum Tolerance Concentration (MTC) being 34.94 ppm Cu, in which 425 $\pm$ 9 CFUs were counted (section c); from the concentration 38.13 to 228.77 ppm of Cu, no bacterial growth was observed, as observed in section d. Therefore, when the *Shewanella putrefaciens* ATCC® 8071 bacterium grows below 50.84 ppm Cu, it is considered to be sensitive to this metal, confirming the inhibitory effect of copper solubilized by the marine broth, where the lowest concentration obtained in the HTwas of 59.73 ppm Cu is much higher than the value obtained in the MTC.

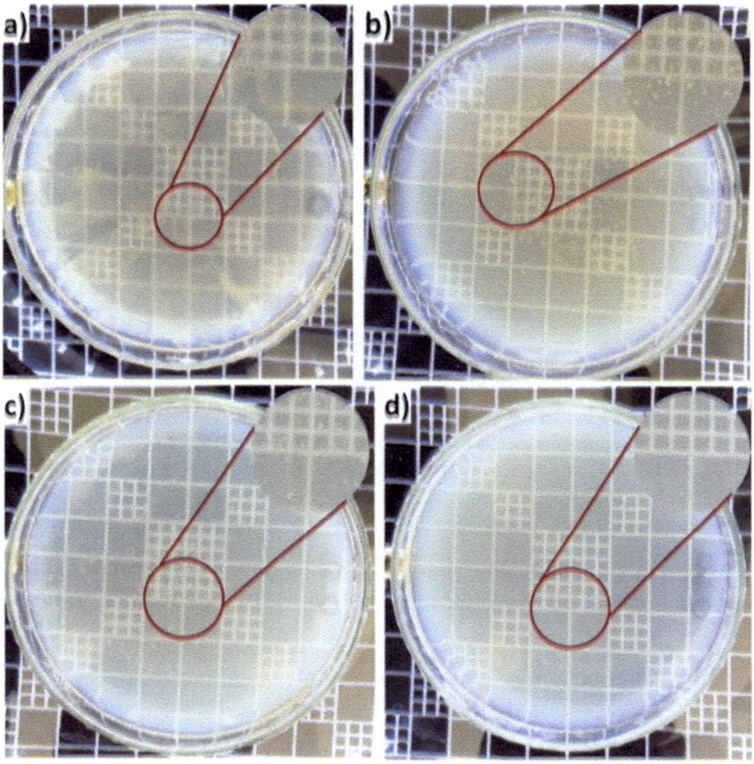

Fig. 6 Marine agar plates to determine the MTC to copper. **a** Control plate without addition of cupric sulfate; **b** Plate with 19.06 ppm Cu and 782 CFU; **c** Plate with 34.95 ppm Cu and 425 ± 9 CFU; **d** Example of plates without growth for concentrations between 38.13 to 228.77 ppm of Cu

4 Conclusions

Derived from this research focused on a biotechnological application in the area of biomining, the following is concluded:

The *Shewanella putrefaciens* ATCC® 8071 bacterium presented a Maximum Tolerance Concentration (MTC) for copper of 34.95 ppm. When carrying out the bioreduction process of the tailings in marine broth with copper contents greater than the mentioned MTC, the leaching of copper will inhibit the growth of the bacterial colony. It is then necessary to exhaustively reduce the presence of copper in the tailings so that it does not generate a biocidal effect on the bacteria or transform the microorganism by introducing a plasmid that generates greater resistance to higher concentrations of copper, since *Shewanella putrefaciens* ATCC® 8071 showed good adaptation to the anaerobic environment and, therefore, to iron consumption.

Acknowledgements The financial support for this work from the Consejo Nacional de Ciencia y Tecnología (CONACyT) of Mexico. Hugo Jhonathan Marín-García deeply thanks to CONACyT for the scholarship during his Ph.D. studies through the grant (444401).

Conflict of Interest The authors declare that there is no conflict of interest.

References

1. Alcalde, J., Kelm, U., Vergara, D.: Historical assessment of metal recovery potential from old mine tailings: A study case for porphyry copper tailings, Chile, Miner. Eng. **127**, 334–338 (2018) https://doi.org/10.1016/j.mineng.2018.04.022
2. Sethurajan, M., van Hullebusch, E.D., Nancharaiah, Y.V.: Biotechnology in the management and resource recovery from metal bearing solid wastes: Recent advances, J. Environ. Manage. **211**, 138–153 (2018) https://doi.org/10.1016/j.jenvman.2018.01.035
3. Moreno, R., Monroy, M.G., Castañeda, E.P.: Evaluación geoquímica de reiduos mineros (jales o colas) de mineralización tipo epitermal, Hidalgo, México. Rev. Geológica América Cent. **41**, 79–98 (2009)
4. de Andrade Lima, L.R.P., Bernardez, L.A., Barbosa, L.A.D.: Characterization and treatment of artisanal gold mine tailings. J. Hazard. Mater. **150**, 747–753 (2008) https://doi.org/10.1016/j.jhazmat.2007.05.028
5. Bingöl, D., Canbazoğlu, M.: Dissolution kinetics of malachite in sulphuric acid, Hydrometallurgy. **72**, 159–165 (2004) https://doi.org/10.1016/j.hydromet.2003.10.002
6. Nicol, M.J.: The kinetics of the dissolution of malachite in acid solutions. Hydrometallurgy. **177**, 214–217 (2018) https://doi.org/10.1016/j.hydromet.2018.03.017
7. Marín-García, H.J., Escudero-García, R., Cortés-Penagos, C., Cholico-González, D.F.: Chemical characterization and quantification of metallic gold from copper tailings in the western region of Mexico, MRS Adv. (2022) https://doi.org/10.1557/s43580-022-00437-6
8. Mandigan, M.T., Martinko, J.M., Brock, J.P.: Biología de los microorganismos. 12va ed., Pearson-Prentice Hall, Madrid España, (2009)
9. Bird, L.J., Bonnefoy, V., Newman, D.K.: Bioenergetic challenges of microbial iron metabolims. Trends Microbiol.Microbiol. **19**, 330–340 (2011)
10. Bonneville, S., Behrends, T., Cappellan, P.V., Hyacinthe, C., Röling, W.F.M.: Reduction of Fe(III) colloids by Shewanella putrefaciens: A kinetic model. Geochim. Cosmochim. Acta. Cosmochim. Acta **70**, 5842–5854 (2006)
11. Castro, L., Blázquez, M.L., González, F., Muñoz, J.A., Ballester, A.: Anaerobic bioleaching of jarosites by Shewanella putrefaciens, influence of chelators and biofilm formation. Hydrometallurgy. **168**, 56–63 (2017) https://doi.org/10.1016/j.hydromet.2016.08.002
12. Zhang, S., Wang, Y., Song, H., Lu, J., Yuan, Z., Guo, J.: Copper nanoparticles and copper ions promote horizontal transfer of plasmid-mediated multi-antibiotic resistance genes across bacterial genera. Environ. Int. **129**, 478–487 (2019) https://doi.org/10.1016/j.envint.2019.05.054
13. Pal, C., Asiani, K., Arya, S., Rensing, C., Stekel, D.J., Larsson, D.G.J., Hobman, J.L.: Chapter seven - metal resistance and its association with antibiotic resistance, in: R.K.B.T.-A. In: Poole, M.P. (ed.) Microbiol. Met. Ions, pp. 261–313. Academic Press, (2017) https://doi.org/10.1016/bs.ampbs.2017.02.001.
14. Kaksonen, A.H., Boxall, N.J., Gumulya, Y., Khaleque, H.N., Morris, C., Bohu, T., Cheng, K.Y., Usher, K.M., Lakaniemi, A.-M.: Recent progress in biohydrometallurgy and microbial characterisation. Hydrometallurgy. **180**, 7–25 (2018) https://doi.org/10.1016/j.hydromet.2018.06.018

15. Nies, D.H.: Microbial heavy-metal resistance. Appl. Microbiol. Biotechnol.Microbiol. Biotechnol. **51**, 730–750 (1999). https://doi.org/10.1007/s002530051457
16. Martínez-Bussenius, C., Navarro, C.A., Jerez, C.A.: Microbial copper resistance: importance in biohydrometallurgy. Microb. Biotechnol. **10**, 279–295 (2017) https://doi.org/10.1111/1751-7915.12450
17. da Costa, W.F., Giambiagi-deMarval, M., Laport, M.S.: Shewanella harboring antimicrobial and copper resistance genes in sea urchins (Paracentrotus lividus) from the Crozon peninsula (Brittany, France), Infect. Genet. Evol. **85**, 104437 (2020) https://doi.org/10.1016/j.meegid.2020.104437
18. Bondarczuk, K., Piotrowska-Seget, Z.: Molecular basis of active copper resistance mechanisms in Gram-negative bacteria. Cell Biol. Toxicol.Toxicol. **29**, 397–405 (2013)
19. Puig, S., Rees, E.M., Thiele, D.J.: The ABCDs of periplasmic copper trafficking. Structure. **10**, 1292–1295 (2002)
20. Kang, C.-H., So, J.-S.: Antibiotic and heavy metal resistance in Shewanella putrefaciens strains isolated from shellfishes collected from West Sea, Korea, Mar. Pollut. Bull. **112**, 111–116 (2016) https://doi.org/10.1016/j.marpolbul.2016.08.033
21. ATCC Medium: 2 Marine Agar/Broth 2216, (n.d.) https://www.atcc.org/~/media/B5025AB990724A91B4491C391E054448.ashx (accessed September 2, 2020).
22. Sutton, S.: Measurement of Microbial Cells by Optical Density. J. Valid. Technol. **17**, 46–49 (2011)
23. Lozano-Guzmán, E., Santos-Ramirez, B., Cervantes-Flores, M., Nieto-Pescador, M.G., Moreno-Cruz, F.J.: Low Accuracy of the McFarland Method for Estimation of Bacterial Populations, African. J Microbiol. Res. **12**, 736–740 (2018)
24. Beltran, M.F., Sagredo, B., Millas, P.: Resistencia-susceptibilidad al cobre de aislamientos bacterianos asociados al cancer bacterial, Bol. INIA-Instituto Investig. Agropecu. (2020)
25. Koneman, E.W.: Diagnóstico microbiológico: texto y atlas a color. 6ta (ed.) Médica Panamericana, Buenos Aires, Argentina, (2008)
26. Forbes, B.S., Daniel, F., Weissfeld, A.S.: Bailey & Scott; diagnóstico microbiológico, 12va (ed.) Médica Panamericana, Argentina, (2009)
27. Janda, J.M., Abbott, S.L.: The genus Shewanella: from the briny depths below to human pathogen. Crit. Rev. Microbiol.Microbiol. **40**, 293–312 (2014). https://doi.org/10.3109/1040841X.2012.726209

Hugo J. Marin García He is a full-time doctoral student in the Extractive Metallurgy Department of the Metallurgy and Materials Research Institute of the Universidad Michoacana de San Nicolás de Hidalgo. He has experience in food safety, sanitary surveillance program, clinical analysis, implementation and updating of the Quality and Occupational Health and Safety Management Systems, as well as Waste Management, audit attention, human capital training, implementation of biotechnological and thermochemical techniques for the extraction of valuable metals, effluent treatment, and biodegradation of vegetable oils. He has directed some bachelor thesis.

Ramiro Escudero García He is Chief Professor of the Extractive Metallurgy Department at Universidad Michoacana de San Nicolás de Hidalgo in México. Is professor and researcher at the Institute of Research in Metallurgy and Materials in the same University. He received his Ph.D. in Mineral Engineering from McGill University in Canada. He worked during five years in the blast furnace department of a steelmaking plant located in Mexico. He has about 150 publications, including two books, and book chapters in the areas of Mineral Processing, Phytomining, and Biomining, Wastewater Treatment, and Gas Disperser Design using Computer Simulators. He is a member of the National Research System in México. He is a member of editorial committees of several international journals. He has directed many master's and doctoral thesis.

Carlos Cortés Penagos He is Professor of the Faculty of Chemistry and Pharmaceutic Biology at Universidad Michoacana de San Nicolás de Hidalgo, México. He is a researcher at the Graduate Divisi6n in the School of Medicine at Universidad Michoacana de San Nicolás de Hidalgo, México. He received his Ph.D. in Cell Biology from CINVESTAV-IPN, México. He was a postdoctoral researcher at Texas A&M University and has been working as a Quality Manager at Mendel Laboratory. He has about 25 publications, including two chapter books, in the areas of Molecular Biology and Clinical Research. He was a member of the National Research System in México. He has been a thesis advisor of several graduate and undergraduate students.

Ricardo Morales Estrella In 1999, he worked at the R&D department of HYL headquarters in Mexico. In 2003, Professor Ricardo received his Ph.D. in Metallurgy Processes from The Royal Institute of Technology in Sweden. His research is in the following areas: chemical kinetics of gas-solid reactions; thermochemistry of inorganic materials; additive manufacturing; ironmaking and steelmaking. In 2016, he was awarded with the Fulbright Scholarship to do a sabbatical leave in the Department of Metallurgy at the University of Utah. His impact metrics are: H = 13 and i10 = 15.

Examining the Problems and Possibility of Immunological Control for Engineered AAV as a CRISPR Vector and Other Genetic Transfers

Asra Hamidi

Abstract The adeno-associated virus (AAV) is a small, non-pathogenic virus commonly used as a carrier for in vivo and ex vivo gene transfer and genome editing projects. However, the healthy immune system is capable of identifying foreign antigens (non-self) and mounting an immune response to eliminate them. Therefore, it is crucial to evaluate and study the potential mechanisms of preventing rejection of synthetic delivery modules, especially during in vivo transfer. For gene therapy utilizing viral vectors to be successful, it is crucial to ensure that the method of gene delivery is safe. This chapter will first examine the problem, applications, and benefits of using AAV as well as its immunological challenges in genetic engineering projects conducted in living organisms and/or in isolated cells/tissues. Then, it will discuss proposed solutions, emphasizing immunological solutions to increase the efficiency of using AAV as a vector. Additionally, the development process in this field will be explored. The development of strategic solutions that address the challenges and limitations associated with the use of this widely used vector can provide a foundation for progress in the fields of genome editing and programming, vaccine design, and targeted gene transfer, particularly in the context of in vivo techniques. These advancements are expected to create a significant revolution in the treatment and prevention of diseases, bioengineering, and the production of transgenic and artificial biomaterials.

Keywords AAV · Vector · CRISPR-Cas9 · In vivo · Epigenome editing · DNA editing · Genome editing · Gene transferring · Gene therapy

A. Hamidi (✉)
Free Biotechnology Researcher, Erbil, Kurdistan, Iraq
e-mail: Asra.Mahmood.Reza@gmail.com

R. Malviya and S. Sundram (eds.), *Engineered Biomaterials*, Engineering Materials,
https://doi.org/10.1007/978-981-99-6698-1_21

1 Introduction

"The adeno-associated virus (AAV) is a tiny virus that primarily affects mammals, but there is currently no evidence to suggest that it causes any illnesses. The engineered type of adeno-associated virus (AAV) that cannot replicate has emerged as a frequently utilized carrier for genome editing and gene transfer projects, particularly those utilizing in vivo methods" [1, 2]. "Treatments based on genome transfer, editing and programming are basic and new treatments that generally try to repair congenital or acquired genetic mutations. Newer studies show the direction and progress of treatments based on genetic material, from the laboratory phase to the creation of practical and real solutions. Meanwhile, AAV vectors, with high delivery efficiency, specific target tissues, replication defects, and significant immunological success rates, have been used repeatedly and have gained relatively good validity" [3]. "The results of research have indicated that AAV is associated with fewer immunological issues compared to most other viruses in the context of in vivo genetic engineering when used as a vector. This has contributed to AAV becoming the most popular viral vector for genetic engineering. However, there is still a possibility that the immune system may eliminate the virus or that it may mutate, which raises concerns about the safety of in vivo applications. Another potential issue is the size limit on genetic material that can be packaged into AAV, which may limit its usefulness in certain applications That is why some researchers prefer to use separate AAVs to deliver sgRNA and Cas9 and co-inject them, while others have investigated the use of smaller Cas proteins or target fragments for gene editing" [4].

"A suitable viral vector should not only have mass production capability but also exhibit precise delivery efficiency, reasonable price, acceptable packaging constraints, and low immunological challenges" [3].

"It is crucial to assess the levels of anti-adeno antibodies in individuals to ensure the safety and effectiveness of genetic treatments utilizing AAV. This is particularly important as a considerable fraction of humans already have pre-existing immunity to adenoviruses. Screening tests have been developed to identify patients with significant antibody levels and prevent adverse outcomes during the treatment process. These tests not only help prevent transduction inhibition but also allow for the measurement of total antibody levels, which can be a valuable predictor of AAV treatment safety" [5]. "In addition, the matters of large-scale production and the expensive nature of using AAV vectors for private and commercial purposes are significant factors to take into account when considering the practical utilization of this carrier in genetic therapies" [6].

This chapter will begin by discussing the significance and applications of AAV vectors in genome editing, programming, and gene transfer. While providing an overview of the potential hurdles in utilizing this vector, we will delve into the specific immunological obstacles that may arise during in vivo projects. We will then explore various solutions that have been proposed to overcome these challenges. Finally, we will discuss the future possibilities and progress in the utilization of AAV vectors, specifically in gene transfers, DNA editing, and programming.

2 Research Methodology

This study examines the immunological challenges associated with using AAV vectors in genetic engineering and proposes solutions to overcome them. We conducted a systematic review of 58 scientific articles published between 2018 and 2023, searching for information on AAV vectors, gene therapy, genome editing, epigenome editing, and immunology. We synthesized the information and analyzed it by categorizing the challenges and proposed solutions into themes. The study's limitations include potential bias and the need for further validation of proposed solutions through experiments. However, the study provides a comprehensive and systematic approach to understanding the immunological challenges of using AAV vectors in genetic engineering.

3 AAV Vectors for Gene Transfer, DNA Editing, and Programming Applications

"Adeno-associated virus carriers have exhibited encouraging outcomes in gene transfer and genome therapy, as evidenced by the approval of FDA-sanctioned treatments like Luxturna, Zolgensma, and HEMGENIX, which are employed for the treatment of Leber's congenital amaurosis, SMA, and hemophilia B. Moreover, the EMA has authorized an AAV5-Factor 8 vector for the treatment of hemophilia A. However, AAV immunogenicity remains a challenge for clinical success in these trials" [7].

"Prokaryotes have a defense mechanism known as CRISPR, which is an acquired immune system that aids in warding off viral infections. The CRISPR loci include spacers that are obtained from extrachromosomal components and are interspersed with short, repetitive sequences that code for non-coding mRNA molecules. The spacer sequences are capable of adapting and changing the bacteriophage resistance in prokaryotic cells. The CRISPR/Cas mechanism may be utilized as a method of modifying the genome across different organisms. By altering the gRNA sequence, the CRISPR/Cas9 genome editing system's target can be modified. The type II CRISPR/Cas system is the most widely used genetic materials editing tool, as it depends on an individual Cas to accurately bind to a particular DNA sequence" [3]. "Furthermore, CRISPR holds the potential to modify specific genes' transcription or translation by altering the non-coding genome, which may help in mitigating diseases. Apart from genome manipulation, CRISPR technology has the potential to be utilized for the treatment and prevention of communicable illnesses" [8].

"However, depending on Double Stranded Breaking (DSB) for genome editing can potentially jeopardize its safety. Therefore, an alternative approach called base editors has been devised, which may enable precise single nucleotide mutations into DNA without DSBs" [9]. "Moreover, scientists have developed an innovative method of genome editing known as prime editing, capable of introducing various changes in DNA sequences such as insertion, deletion, and base substitution. Prime editing does not rely on double-stranded DNA cleavage or the use of donor DNA templates. Instead, it uses an engineered sgRNA connected to reverse transcriptase and nCas9 to synthesize new DNA strands and introduce them to the targeted site. Since prime editing can address a wide range of genetic mutations associated with human genetic problems, it holds great potential as a clinical therapy tool" [10].

"The potential of large DNA fragment insertion for treating genetic diseases is significant, but current gene editing technologies can only effectively insert short fragments. However, a new method for gene editing, known as GRAND editing, has been developed to enable the precise insertion of big DNA pieces devoid of the need for DNA donors. Unlike prime editors, which require templates that can hybridize with the target sequence, GRAND editing uses a pair of guide-RNAs with reverse transcriptase templates that are not homologous to the intended region but instead match each other. This method has achieved a success rate of up to 63.0% for a 150-bp insertion with minimal by-products, and 28.4% for a 250-bp insertion. Although it can accommodate insertions of up to ~1 kb, the potency decreases for fragments larger than 400 bp" [11].

"Safe and effective delivery methods for gene editing agents are essential for successful in vivo gene editing therapies aimed at treating genetic diseases. Various delivery technologies have been studied, including viral vectors, lipid nanoparticles, and virus-like particles, to achieve this goal. These methods play a crucial role in enabling therapeutic in vivo gene editing" [4].

"Eukaryotic cells have a defense mechanism against foreign DNA different from the CRISPR system used by prokaryotes. They use DNA sensors that add chemical markers to exogenous DNA, inhibiting its transcription and reducing its impact on the cell. However, this poses a challenge in gene therapy. The DNA-sensing machinery of the host can detect certain elements of Recombinant AAV vectors, which are frequently utilized in gene therapy, resulting in decreased expression of the transgene. In addition, the HUSH complex can deposit repressive histone marks on viral DNA, and the NP220 DNA sensor can bind to double-stranded DNA from either host or foreign viral genomes. Following this interaction, the HUSH complex is recruited, which in turn facilitates the recruitment of histone-modifying enzymes, such as HDAC and SETDB1. These enzymes remove acetyl marks and add repressive methyl marks, respectively. In summary, NP220-HUSH complex plays a role in epigenetic silencing of DNA by modifying histones which should be taken into consideration when planning genome editing and gene therapy projects based on AAV" [12].

"Epigenome editing is a method used to manipulate the epigenome of specific DNA to determine the role of the epigenome in a particular gene. This technique involves using a catalytically inactive mutant of Cas9 (dCas9) to recruit epigenetic modifying enzymes to specific target sequences. While forward genetics is a phenotype-to-gene approach, reverse genetics analyzes the phenotype caused by the mutation of specific genes. Epigenetic modifications are crucial for the activation of genes regulation in the course of embryogenesis and are linked to different maladies. DNA methylation and histone modifications are responsible for transmitting information for controlling gene expression in a manner unrelated to alterations in DNA sequence. Epigenome editing technology has the potential to expand the use of epigenetic therapy in solid tumors" [13]. In the meantime, "AAV vectors are considered one of the most interesting options for epigenome editing" [14].

Furthermore, "Epic Bio, a company established by Stanley Qi, a bioengineer from Stanford, has created a technology known as Gene Expression Modulation System (GEMS), which enables the modification of gene expression without making permanent alterations to DNA. GEMS comprises DNA-binding proteins, guide RNAs, and modulators, which can activate or repress gene expression and modify the gene locus. They use the largest library of novel modulators, which can be customized for specific effects upon a target gene. CasMINI is a modified Cas DNA-binding protein that has been engineered to operate strongly within mammalian cells. Additionally, it is smaller in size than both Cas9 and Cas12a, measuring less than half of their dimensions. Epic has licensed CasMINI for human use from Stanford and can deliver it to target organs via adeno-associated virus (AAV), lipid nanoparticles (LNPs), or lentiviruses. They want to de-risk its initial programs by using AAV as it has more clinical experience, but it is not tied to AAV and aims to validate the efficacy of the constructs with other delivery methods" [15].

4 Exploring the Advantages and Challenges of Using AAV Vectors in Genetic Engineering

"AAVs are tiny viruses that do not have an envelope and contain a genome consisting of single-stranded DNA. They have a capsid structure that is composed of three proteins arranged in a 1:1:10 ratio, which creates a symmetrical $T = 1$ icosahedron structure. The AAV genome is compact and contains two open-reading frames that are bordered by palindromic inverted terminal repeats arranged in a T-shape. These repeats enable the genome to be replicated and packaged. The cap gene codes for the three capsid proteins, while the rep gene encodes four replication proteins" [6] (Fig. 1).

"AAV vectors are commonly used in genetic engineering as they are highly effective at transporting to specific cells. Also, they can sustain persistent transgene expression" [17]. " AAV vectors can transduce cells that are either actively dividing or non-dividing, and they can persist as an extrachromosomal element without integrating into the host cell's genome" [18]. However, "they still pose some immunological challenges and carry a slight possibility of causing mutagenesis by insertion" [19]. "AAV vectors have demonstrated high delivery efficiency, replication defects, specificity toward target tissues, and notable immunological success rates, and have therefore gained considerable credibility in their repeated use. Compared to other viral vectors, AAV poses fewer immunological challenges when utilized for in vivo gene transfer and DNA edit, and has thus become the preferred vector in this field" [3]. "Despite

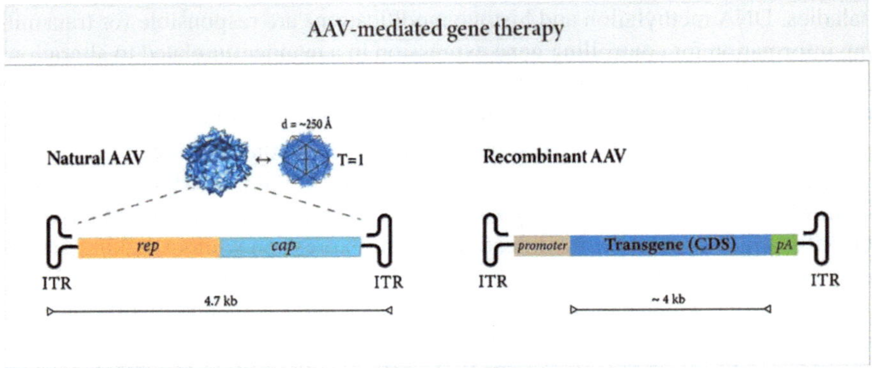

Fig. 1 The diagram shows the use of AAVs in gene replacement therapy, where the therapeutic sequence takes the place of the rep and cap genes. The AAV capsid structure, which is highly symmetrical, consists of three proteins and has been crafted to optimize non-covalent interactions to accommodate the genetic material carried by the virus [6, 16]

this, there remain potential concerns regarding the safety of in vivo applications, such as the possibility of the immune system eliminating or mutating the virus. AAV's effectiveness for specific applications is limited by the size constraints of the genetic material it can accommodate, which requires to be addressed. Researchers have proposed various solutions, including the use of separate AAVs to attach sgRNA and Cas9, the exploration of smaller Cas proteins, and the targeting of smaller genetic fragments for transfer" [4].

In addition to the advantages and challenges of using AAV vectors, there are some other important aspects to consider.

4.1 AAV Serotypes

"AAV vectors offer versatility in targeting different tissues due to their diverse capsids. Researchers can modify the tropism of these vectors by combining natural serotype capsids" [11]. "More than 13 naturally occurring AAV serotypes have been identified, in addition to over 108 capsid variants that have been classified" [7]. "To construct an AAV vector, the therapeutic gene, and genetic material must be placed between two sets of inverted terminal repeats (ITRs) obtained from AAV serotype 2. The size of the entire construct must not exceed 4.6 kb" [18]. For example, "AAV9 has been observed to be effective in systemically delivering therapeutic agents to the nervous system. This includes the capacity to penetrate natural obstacles, such as the cerebral vascular barrier, to reach its target" [6].

4.2 Generation and Refinement of Vectors

"The production and purification of AAV vectors require careful handling and elimination of any probable impurities. A commonly used method for AAV production is transient transfection of mammalian cell lines, but it has limitations in terms of scalability and downstream processing. An alternative method uses Sf9 insect cells and baculovirus but this has limitations, so TIPS cells are utilized to overcome these issues. AAV vectors are manufactured through a process that involves the use of serum-free suspension culture, where TIPS cells and new Sf9 cells are co-cultured. The resulting AAV particles are obtained by lysing the producer cells" [18] (Fig. 2).

"Presently, the preferred way of purifying AAV for clinical and research purposes is through ultracentrifugation on a cesium chloride or iodixanol density. Nevertheless, this process has certain disadvantages, such as being time-consuming and difficult to expand, and it may also lead to impurities being co-purified. On the other hand, liquid chromatography, particularly high-performance liquid chromatography (HPLC), is

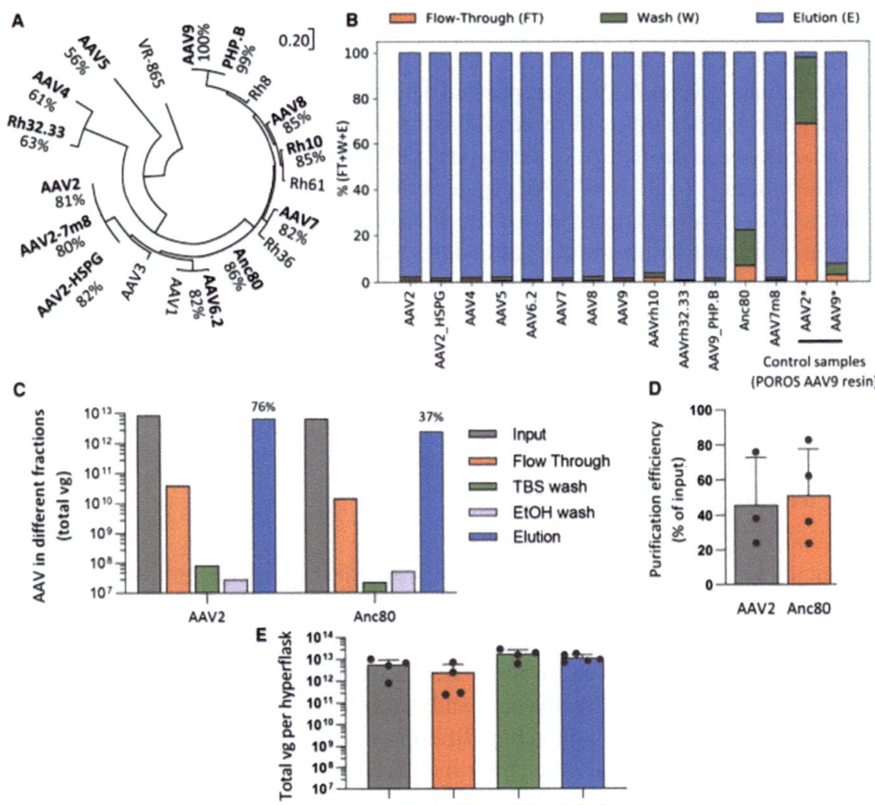

Fig. 2 The process of purifying AAV through AAVX affinity chromatography [19]

a more effective and scalable method for purifying AAV. However, this technique requires optimization for different AAV serotypes to work efficiently. Recently, commercial AAV binding resins have been introduced that provide a more flexible and efficient method for purifying AAV for research purposes than ultracentrifugation or traditional chromatographic methods. These resins require serotype-specific optimization and may have lower binding efficiency for certain AAV serotypes, but overall they offer promising results" [19].

4.3 Previously Established Immunity

"AAV-based genome editing may not work effectively in individuals with existing immune responses to certain AAV variations" [5]. "The body's exposure to natural AAV serotypes can produce specific antibodies, such as neutralizing antibodies (NAb), which can prevent AAV vectors from entering the objective cells, reducing Delivery effectiveness and lowering the synthesis of therapeutic agents. Patient exclusion from clinical trials is prevalent due to high titers of neutralizing antibodies against AAV vectors, which can vary depending on demographic factors. It is essential to evaluate several AAV serotypes and determine the collective occurrence of the primary serotypes employed in AAV-mediated treatment because cross-reactivity can arise between distinct AAV serotypes" [20] (Table 1).

"The binding of anti-AAV AAV capsid antibodies can impact the delivery of these vectors. This effect can occur in two ways: Neutralizing antibodies can block vector transfer, whereas non-neutralizing antibodies have no effect. Both types of antibodies can also cause the redirection of the vector to secondary lymphoid organs, as illustrated in Fig. 3. However, administering rAAV vectors directly into protected sites might be less susceptible to the impact of non-systemic AAV-targeting antibodies" [21].

4.4 Vector Standardization

"Maintaining a consistent level of rAAV properties is crucial for comprehensive medical trials. The initial phase in producing recombinant adeno-associated virus involves reinforcement, which is commonly done using a triple transfection method in clinical practice. However, this approach is limited in scalability, leading to low productivity. To enhance effectiveness and durability, different manufacturing platforms have been developed, including different viral vectors, such as baculovirus or herpes simplex virus, as well as cell lines such as HEK293 and adenovirus-transfected HEK293 cells" [1].

Table 1 Comparing tests for transduction inhibition and total antibody detection [5]

Evaluation	TI evaluation	TAb evaluation
Fundamental	A reporter gene encoding rAAV vector is used in a cell-based test	A capture-based method that uses immunochemical analysis (For example, ELISA or ECLA)
	Cell transduction inhibition tests can be performed using blood-derived samples containing neutralizing antibodies and Non-immunoglobulin components	Detects the existence of antibodies against AAV (with no regard to their neutralizing capacity)
Benefits	The method determines the extent of cell transduction inhibition by identifying neutralizing antibodies and additional neutralizing agents that may impact transduction	This unique approach is less complex and more amenable to automation than TI tests. Therefore, it holds promise aimed at the progress of diagnostic tests that could be utilized in diverse medical laboratory settings
	A frequent design involves selecting the same rAAV capsid as gene therapy with options for cell lines and reporter genes	The detection of specific immunoglobulin isotypes is possible using an antigen-capture format
Possible drawbacks	Necessary specialized experts	It is difficult to standardize due to the various available platforms and format options
	Can be difficult to develop into readily available commercial test kits	It is possible that low-affinity non-specific antibodies could be detected, but their presence potentially does not impact the clinical outcome of gene therapy

4.5 *Safety Concerns*

"AAV vectors are generally considered safe but require thorough pre-clinical testing and patient monitoring to ensure their safety and effectiveness" [5]. "AAV vectors have elements that trigger the activation of toll-like receptor (TLR)2 and TLR9, leading to the discharge of pro-inflammatory signaling molecules and immune system proteins that promote inflammation and antiviral responses. AAV vector suspensions may contain impurities from the production process, which can also affect their immunogenicity. The detection of AAV vectors is greatly influenced by the presence of plasmacytoid dendritic cells (pDCs) and can secrete and produce substantial quantities of type I interferon and irritating cytokines in response to viral challenges. Non-packaged viral DNA in AAV vector production preparations can influence their immunogenic properties in human pDCs" [23]. "The prevalence of anti-AAV antibodies raises safety concerns for gene therapy. Excluding positive patients from clinical trials could hinder progress without sufficient evidence of safety risks" [24]. "AAV vectors are a powerful tool for genetic engineering with great potential in various applications, but their use requires careful consideration of serotype selection, vector production, pre-existing immunity, and safety concerns" [1].

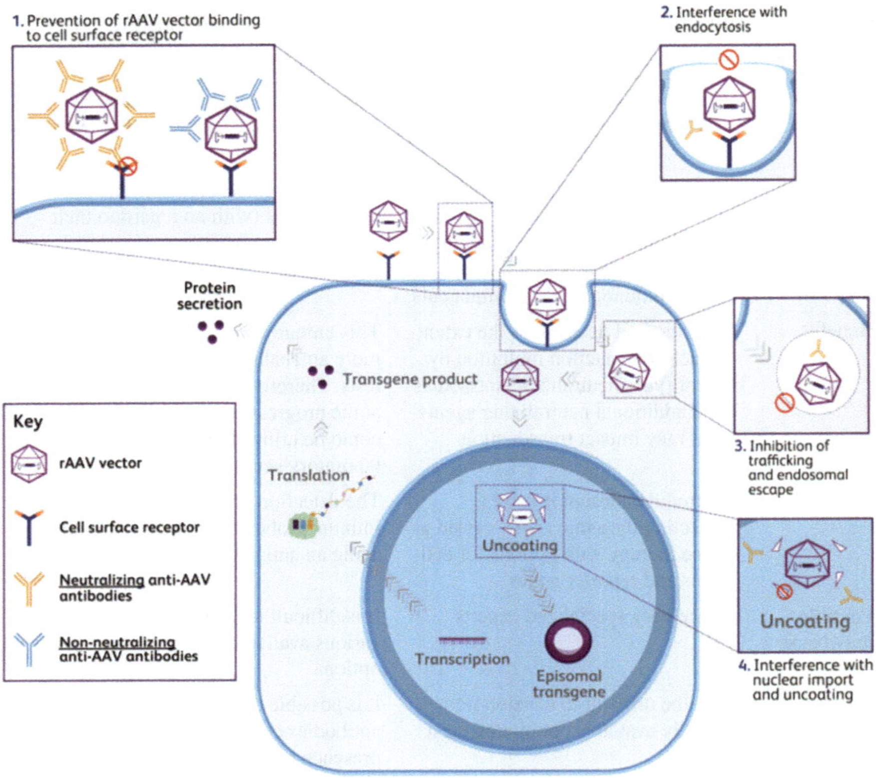

Fig. 3 Non-inhibitory antibodies against AAV may Interact with the protein shell of the recombinant adeno-associated virus, but they do not obstruct attachment of the vector or transduction. The mechanism of neutralizing antibodies against AAV is depicted [22]

"AAV has been successful in delivering therapeutic genes for illnesses like hemophilia, spinal muscular atrophy, and inherited retinal disorders during testing on patients" [25]. "Genome editing tools like CRISPR-Cas9 can be delivered by adeno-associated viruses (AAVs) to target cells, enabling accurate genome modifications" [2]. Additionally, " AAV vectors are frequently utilized in fundamental research to investigate genetic mechanisms, control, create preclinical models of human illnesses using animals" [26, 27], and "apply newer genome editing techniques such as prime editing and base editing" [28].

"AAV vectors have been used in agriculture to deliver genetic modifications to crops, improving resistance to pests and environmental stressors" [29].

"AAV vectors can target specific cells by incorporating cell-specific promoters and can mediate extended gene function due to their potential to incorporate into the recipient cell's DNA" [30].

"AAV vectors are promising for treating neurological disorders in the central nervous system like Alzheimer's, Huntington's, and Parkinson's illness" [1]. "AAV vectors are capable of being used to regulate the body's defense response" [31] and "have potential in cancer therapy by delivering genes that regulate immune function, express tumor suppressor genes, or induce apoptosis in cancer cells" [32]."AAV vectors have limited cargo capacity" [13], They can serve as in vivo carriers, particularly for CRISPR-based systems such as Cas9, Cas12a, and Cas13a, which find prominent application in the field of cancer therapy [33] and "pre-existing immunity can reduce their effectiveness" [5], but "efforts to improve AAV vector technology show promise for overcoming these challenges" [13]. "Despite these challenges, AAV vectors offer a valuable tool for genetic engineering with various uses in genome therapy, immune modulation, and cancer therapy, among others" [1].

5 Immunogenicity and AAV Vectors: Challenges and Solutions

"Adeno-associated viruses have demonstrated significant potential in genome engineering applications, but one of the challenges associated with their use is the potential for immune responses" [1]. Here are some key points to consider.

5.1. **Prior immunity**: "A considerable fraction of the population carries a Prior immune response to the viral vectors, which may hinder their efficacy in gene therapy or other applications. This is caused by the widespread prevalence of natural AAV infections among individuals, leading to the generation of neutralizing antibodies against the vectors" [5]. "Addressing the issue of prior immune responses to viral vectors is crucial for the advancement of genome editings. Ongoing research and innovative strategies hold promise in overcoming this hurdle and unlocking the full potential of viral vector-based treatments for a wide range of genetic mutations" [34] (Fig. 4; Table 2).

"Perocheau et al. investigated the prevalence of inhibitory antibodies to AAV-LK03, 3B, and AAV8 between populations and observed that it is influenced by various factors like region, environmental factors, crowdedness, immune system, and genetic characteristics. The occurrence of antibodies that neutralize was found to be lower during the later stages of childhood and teenage years, suggesting that AAV applications may be more suitable for this age group. The study suggests that AAV-LK03, AAV5, AAV8, and AAV3B may not be suitable for subsequent administration following the primary therapy. The prevalence of IgG and neutralizing antibodies to AAV-LK03, AAV-3B, and AAV-8 was presented in different age groups and sexes. The study also found antibody response memory is intricate and might be independently regulated by circulating memory B cells and plasma cells that produce antibodies" [36] (Fig. 5).

"The serological prevalence rates of AAV-specific antibodies can be affected by factors like donor health status and the use of immunosuppressive drugs. There is a significant ability of antibodies to react with multiple various

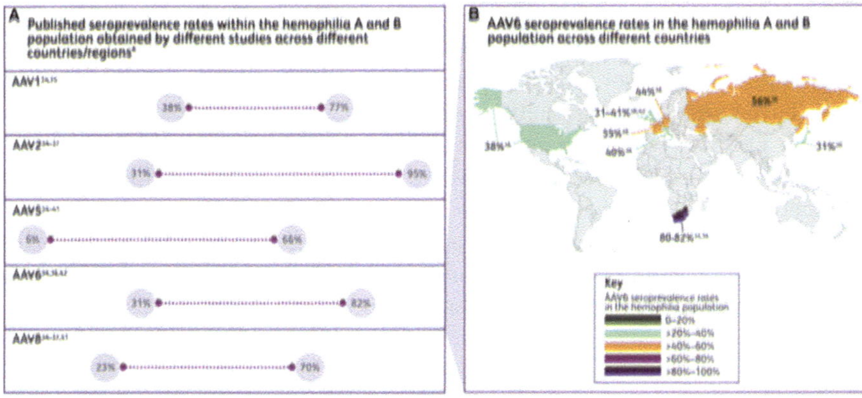

Fig. 4 Describing the frequency of existing immunity to viral variants and their distribution across different geographic regions, using the population of patients with hemophilia A and B as an illustrative instance [34]

Table 2 Neutralizing antibodies' cross-reactivity with AAV-LK03, 3B, and 8 in populations [35]

	AAV-LK03	AAV3B	AAV8
AAV-LK03	–	97%	61%
AAV3B	67%	–	47%
AAV8	79%	87%	–

AAV strains, likely due to the high degree of sequence similarity in AAV capsid proteins. Reporter vectors with genes like GFP or luciferase are used to measure neutralizing activity against recombinant AAV (rAAV), but the type of reporter gene and the cell type used can affect the results. TAb tests detect all capsid-bound antibodies but not non-antibody neutralizing factors. Choosing recombinant adeno-associated virus capsid-based tools is crucial for developing antibody tests" [5].

"Factors like donor health and immunosuppressive drugs affect AAV-specific antibody prevalence. Antibodies can interact with multiple AAV strains due to a high degree of sequence homology among capsid proteins. Reporter vectors with genes like GFP or luciferase measure neutralizing activity, but the choice of gene and cell type impacts results. TAb tests detect capsid-bound antibodies but not non-antibody neutralizing factors. Selecting recombinant vector capsid-based tools is crucial for accurate antibody test development" [38].

5.2. **Immunological reactions to the vector**: "The viral vectors used in genome therapy can still induce immune responses in patients without existing immunity, leading to clearance of AAV and decreased gene expression" [39]. "The immune response involves both natural and acquired immunity, with the detection of foreign nucleic acids triggering the secretion of inflammatory signaling

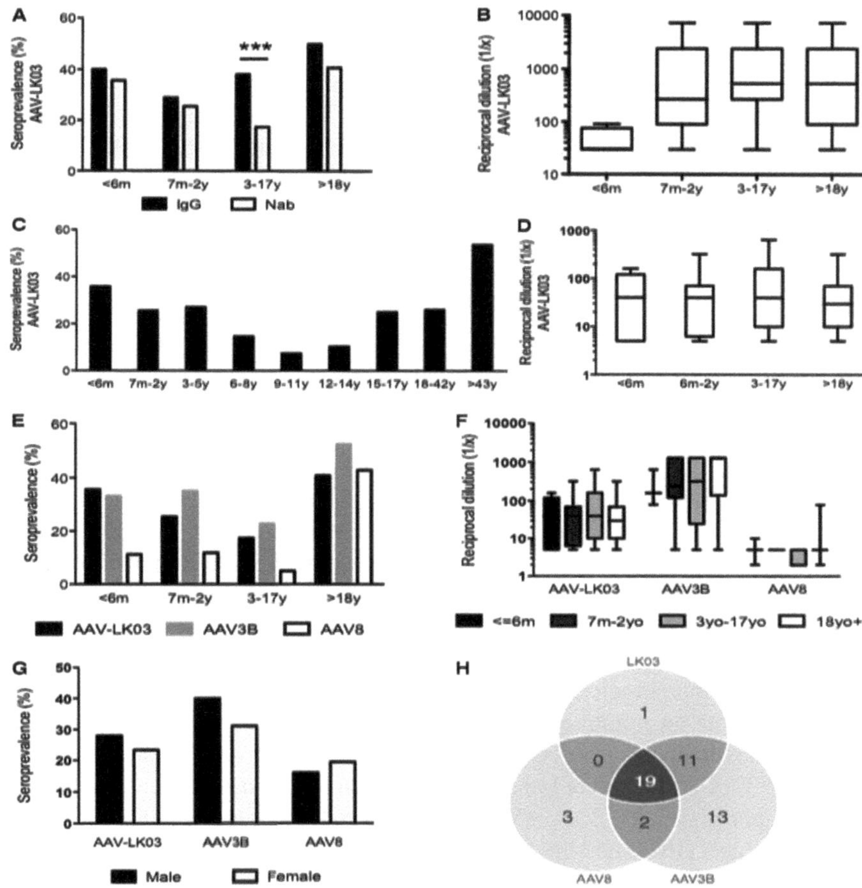

Fig. 5 The Perocheau et al. study examined the prevalence and titers of neutralizing antibodies and IgG to AAV serotypes AAV-LK03, AAV-3B, and AAV-8 in populations. The results were presented in eight sections, including the prevalence and titers of antibodies based on age, sex, and cross-reactivity between serotypes. The results were shown using vertical bars and box-and-whisker plots [37]

molecules and the stimulation of T and B cells" [40]. "Monitoring immunogenicity throughout drug development and providing sufficient information to support label claims is necessary, and the cytotoxic cell-mediated immune response, involving Cytotoxic T lymphocytes, helper T cells, and NK cells, is particularly relevant to the sustained response to genome therapies delivered by AAV vectors" [41]. "Various factors can impact immune reactions against AAV vectors, and it is essential for closely monitoring them to guarantee the well-being and efficacy of genome therapy" [42].

5.3. **Epigenetics challenges**: "The terminal inverted sequences in recombinant AAV (rAAV) genomes can activate the host cell's foreign DNA-sensing machinery, including the DNA damage response (DDR) proteins, and induce host-mediated silencing of the transgene. This silencing can occur through various mechanisms such as DNA methylation and histone modifications. The selection of AAV shell may also impact host-mediated silencing. The human silencing hub (HUSH) complex, made up of MPP8, TASOR, and PPHLN1, is involved in silencing retroelements and can also silence rAAV genomes. The NP220 protein can recruit the human silencing hub complex and the enzyme SETDB1 that adds suppressive histone modifications on retroviral genomes, but without requiring the HUSH complex, it can repress the gene performance from the extrachromosomal DNA of retroviruses that are closely relevant, such as HIV-1 and Mason-Pfizer monkey virus. The selection of the AAV capsid (serotype) has a significant impact on these mechanisms and can ascertain the gene repression of integrated viral DNA mediated by the host" [12].

5.4. **Adjuvants and immune modulators**: "The use of adjuvants or immune modulators is a potential approach to boost the immunological reaction to AAV vectors in gene therapy" [43]. "Adjuvants can stabilize pathogens and raise the immunological function, especially in older adults and immunocompromised individuals. The most widely used adjuvant is alum, but research is ongoing to develop combined adjuvant formulations that elicit strong humoral and cellular immune responses" [44]. "Recombinant adenoviruses are gaining popularity as COVID-19 vaccines due to their ability to stimulate B and T cell-initiated immunological function. However, endeavors to enhance their ability to stimulate an immune response by administering them with conventional immunostimulants have been mostly ineffective. The use of STING ligands, like 2′3′-cGAMP, before transgene expression can reduce immunity, while the parallel activity of 2′3′-cGAMP with the introduced gene can enhance induction of the STING signaling pathway in advanced stages of viral contamination. The timing of adjuvant effect is crucial and may have broader implications for the application of other non-biological immunostimulants" [45].

5.5. **Alternative vector systems**: "Alternative vector systems have been explored to address the immune challenges associated with AAV vectors. Lentiviral vectors and non-viral gene delivery methods are potential alternatives to AAV vectors" [46]. "There are various drawbacks associated with using AAV vectors as a means of delivering genes, including immunogenicity, cytotoxicity, and the risk of gene disruption through viral integration. When viral genetic material integrates into the wrong location on a chromosome, it can lead to abnormal changes in cells leading to cancer development. Research in populations has also shown that this process is often associated with an Immunological reaction. Techniques that utilize physical means to facilitate the delivery of nucleic acids are restricted to DNA applications. Nonviral vectors are more attractive than viral vectors since they may address issues related to potential adverse effects on the immune system and cellular health. Nonviral vectors are also a less expensive option and are easy to generate at a massive level" [47].

5.6. **Immune evasion strategies**: "There are ongoing efforts to develop immune evasion strategies for AAV vectors. For example, engineering AAV vectors

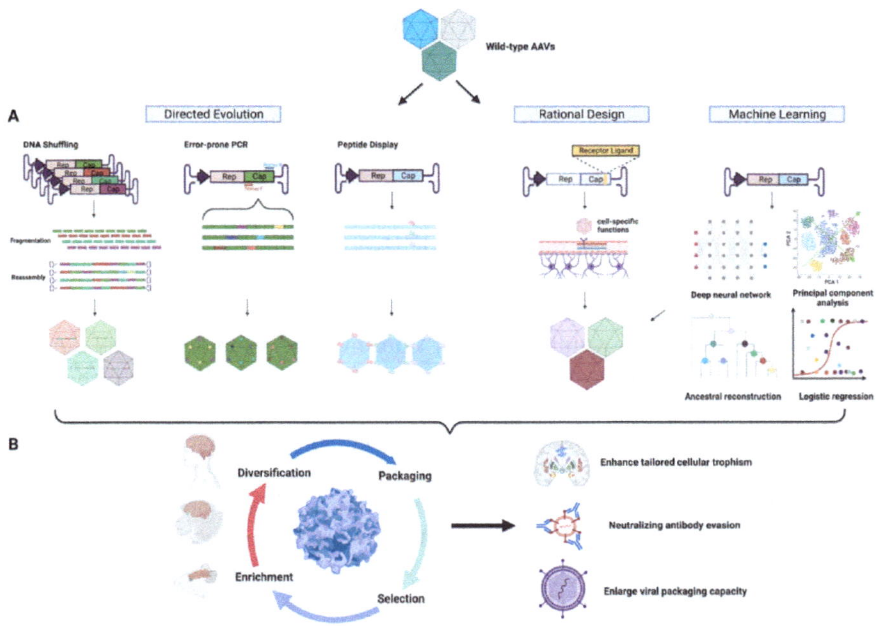

Fig. 6 Illustration of manipulating capsids to enhance their capability to target the brain and spinal cord. **a** Techniques for generating diverse AAV capsids. **b** Procedure for modifying capsids, which involves selecting from multiple species and customizing their capability to selectively target particular cells evade the immune system, and package the virus [48]

with modified capsid proteins or using novel AAV serotypes can help to evade pre-existing immunity and enhance gene expression" [48] (Fig. 6).

"The structure of AAV capsids can be altered to improve vector delivery efficiency and avoid triggering immunological defenses. The development of engineered AAV capsids that target special tissue involves three main techniques: natural discovery, deliberate engineering, and evolutionary optimization. Natural discovery involves obtaining AAV from various hosts and identifying a distinct genetic code for the capsids. The code is cloned and analyzed to confirm the modified capsids' ability to target the special tissue. Rational design involves utilizing current understanding of capsid biology to insert binding proteins into specific regions of the AAV capsid protein, thereby enhancing the therapeutic properties of the resulting capsids. Directed evolution uses mutagenesis techniques to create a large number of capsid libraries, which are then exposed to specific conditions that favor the development of the desired characteristics.

While these methods offer great potential, they also present several challenges. Natural discovery is limited by low delivery efficiency and the time-consuming process of isolating and characterizing novel capsids. Intelligent design requires an in-depth knowledge of capsid biology, encompassing attachment, entry, delivery, disassembly, genome manifestation, and structural alteration. Directed evolution relies on animal models that closely replicate the properties of the target cells, thus imposing constraints.

Although capsids obtained through natural discovery are the most commonly employed and safest in present medical treatments, rational design and mutagenesis-based optimization offer opportunities to enhance capsid function. Nonetheless, these methods also have limitations that need to be considered" [1].

Moreover, "Significant advances have been made in cassette engineering to improve the safety features, cellular tropism, expression levels, and duration of action of transgenes in clinical applications. Altering different elements of the adeno-associated virus cassette is a technique utilized in this approach, including poly(A), ITRs, and promoters, among additional constituents. One example of cassette manipulation that improves gene delivery effectiveness is double-stranded AAV, which bypasses the need for second-strand synthesis. Furthermore, promoter selection can impact gene expression rates and the cells targeted to decrease the adaptive immunological reactions. Moreover, the use of poly(A) and opposite poly(A) sequences can promote gene transfer by enhancing cytoplasmic transportation or bypassing ITR-mediated transcription. In clinical settings, cassettes are commonly used to transfer complementary DNAs for treating autosomal recessive diseases. Studies conducted in animal models are investigating various approaches that utilize peptide fragments, endogenous small non-coding RNA, and short nucleic acid sequences to modulate transgene expression and enhance suppression of the immune response and gene transfer efficiency" [1].

6 Specialized Solutions for Overcoming Immunological Challenges of AAV Vectors

As discussed in the preceding section, potential solutions to address the immunological hurdles associated with AAV vectors can be summarized as the research and development of the following strategies.

6.1. **Engineering AAV vectors**: "One approach to addressing immunological challenges is to engineer AAV vectors to have reduced immunogenicity. This can be achieved through modifications such as altering the capsid protein or incorporating immunomodulatory agents into the vector" [1].

6.2. **Immunosuppression**: "Immunosuppressive drugs can be used to reduce immune responses to AAV vectors. However, this approach is not ideal as it can compromise the immune system's ability to respond to other threats" [43].

6.3. **Co-administration of immunomodulatory agents**: "Co-administration of immunomodulatory agents, such as cytokines or co-stimulatory molecules, can help to regulate the immunological responses to the viral vectors and improve their effectiveness" [45].

6.4. **Use of immune-privileged sites**: "Administering Adeno-associated-virus vectors to immunologically protected areas, like CNS or eye, can help to avoid immune responses. However, this approach is not always feasible, and the effectiveness of the therapy may be limited by the site of administration" [49].

6.5. **Selection of AAV serotypes**: "Selecting AAV serotypes that are less prone to inducing immune responses can also help to mitigate immunological challenges" [19].

6.6. **Addressing pre-existing immunity**: "Strategies to address pre-existing immunity, such as selecting AAV serotypes that are less commonly encountered or employing methods to deplete or improve neutralizing antibodies, can also enhance the impact of gene therapy using AAV vectors" [20].

"rAAV has surfaced as a potential gene transfer mechanism for addressing monogenic diseases in various tissues. When compared to adenoviral vectors, rAAV exhibits lower immunogenicity due to their weaker activation of natural defense mechanisms and specialized immune cells. However, recombinant adeno-associated virus vectors carrying diverse transgenes can provoke strong immune reactions in different conditions, resulting in the removal of genetically modified cells and decreased activity of the engineered gene. Delivery of rAAV in humans and primates may activate considerable cytotoxic CD8+T cell reactions to both vector and transgene-encoded protein, ultimately resulting in the removal of transduced cells and complete loss of transgene expression. Moreover, the development of antibodies that bind to and inactivate the capsid proteins can hinder subsequent vector administration and speed up the degradation of the therapeutic agent. Consequently, crafting approaches to overcome immunological reactions and ensuring the sustained production of therapeutic proteins is the primary challenge for translating rAAV vectors to clinical settings" [50].

"Although AAV vectors have size limitations for delivering transgenes, these limitations can be addressed by modifying the genes to remove unnecessary protein structures and shorten the cDNA. For instance, AAV vectors can be used for Duchenne muscular dystrophy by eliminating protein segments from the dystrophin gene sequence, which is over 11 kilobases in size. A similar approach has been used in gene therapies for hemophilia A, where the normal FVIII protein-gene sequence is approximately 7 kb. By using B-domain-deleted FVIII protein isoforms, the gene sequence for the FVIII protein has been decreased to just over the maximum size for encapsidation. This allows successful packaging of vector genomes in the viral capsids, modifying the size of vector genomes can lead to changes in gene expression when relative to the conventional-sized ones. Research conducted on mice utilizing AAV-8 found that a larger vector, containing FVIII and exceeding the standard size at 5.1 kb, resulted in 2–3 times less expression of FVIII protein in plasma and decreased vector genome levels in hepatocytes after 3 months, as compared to a typical-sized vector at 4.6 kb" [51].

"Scientists have developed a more efficient and effective delivery system for prime editing, a gene editing technique that corrects mutations without inducing ds-DNA cleavage. The team made improvements to the reverse transcriptase (RT) moiety and truncated the prime editing components to overcome the viral vector size restrictions, which at present, there is limited support for effective transfer of the bulky prime editing tools. Consequently, a smaller prime editing system with a gene sequence optimized for expression and size advantages was developed. The team also optimized the split intein prime editing system and identified Cas9 breaking points that coupled with the shortened prime editing components transferred by double-stranded AAVs triggered improved AAV levels and prime editing productivity. Furthermore, they demonstrated the split prime editing components transferred by double-stranded-AAV1, particularly altered prime editing system 1024, where able to induce changes in nucleotides and insert genetic material at four natural locations within human tissue cultures and they also demonstrated successful editing of the Dnmt1 gene in the retina of live adult mice" [52]. "Another group of researchers constructed four split prime editing systems using Rma intein and identified two efficient split prime editing systems that were able to facilitate point substitution and addition of endogenous areas in human cells when delivered by dual-AAV1. They then successfully edited Dnmt1 in adult mouse retina in vivo using one of the identified efficient split prime editing systems" [53].

"Epigenome editing techniques have proven effective in manipulating specific gene targets, but their delivery through adeno-associated virus (AAV) vectors has been constrained by the small assembly restrictions of these vectors. Researchers have addressed the issue by developing a compact deactivated Cas9 repression system that can be packaged into a single optimized adeno-associated virus. The system uses a smaller variant of Cas9, derived from *Staphylococcus aureus* (Sa), and a novel repressor composed of the condensed transcription repression domain (TRD) from MeCP2 fused with the KRAB inhibitory domain. This construct, along with its associated gRNA, may efficiently fit into adeno-associated viruses. The platform was successfully used to silence APOE, a gene associated with Alzheimer's disease, in cell culture as well as in living organisms. The method offers a more efficient and effective delivery system for epigenome editing" [54].

"The modification of histones by the Human Silencing Hub (HUSH) complex and NP220 is essential for regulating AAV gene expression by silencing them. Deletion of NP220 had the most significant effect on the expression of AAV transgenes, suggesting a crucial role in controlling the epigenetic suppression of AAV genomes. However, NP220 may recruit other epigenetic regulators besides the HUSH complex. NP220 is present and highly conserved in diverse vertebrate species, but the molecules responsible for inhibiting viral gene activity may vary based on the virus and host organism. For example, Rous Sarcoma virus is unresponsive to NP220-induced suppression, perhaps due to the relatively limited amount of cytidine present in the long terminal repeat regions. An alternative approach to enhance AAV transgene expression is to use low biomolecular weight inhibitors of HDACs, which can promote histone deacetylation and inhibit host-silencing mechanisms. The impact of

HUSH complex-mediated chromatin suppression on gene expression under in vivo-like conditions is still under investigation and can potentially be modeled in mice with gene knockouts or through pharmacological modulation in preclinical genome editing experiments. Investigating the relationship between NP220 and the HUSH complex and how they operate differently in various cell populations within living organisms and affect genome editing treatments using different AAV variants is crucial for further research. In summary, selecting and engineering the appropriate AAV serotype is essential to overcome epigenetic obstacles when using AAV vectors" [12].

7 Looking Ahead: The Future of AAV Vectors

"The future of AAV vectors in gene editing and transfer is promising with continued development likely resulting in safer and more efficient delivery of therapeutic genes or proteins. Researchers are exploring ways to optimize AAV vectors for specific applications, enhance their specificity, reduce immunogenicity, and increase their cargo capacity" [1].

"Researchers are focusing on designing innovative AAV serotypes to more effectively target particular cell populations, which may lead to new treatments for diseases that are currently difficult to target" [48].

"Scientists are studying methods to combine AAV vectors with other genome editing tools like CRISPR/Cas9 for better precision and efficacy in gene editing with fewer off-target effects" [3]. Moreover, "they are constantly improving the technology of AAV or its substitutes such as Base editing, Prime editing, Grand editing, and epigenome editing, to achieve more effective gene editing in "in vivo projects". The refinement and enhancement of these viral vectors play a crucial role in advancing these associated tools" [4, 9–12, 14, 15].

"Improving AAV vector design to penetrate the blood–brain barrier and deliver therapeutic genes or proteins for the treatment of neurodegenerative diseases like Parkinson's and Huntington's is a current research focus. This involves engineering AAV vectors that can cross the blood–brain barrier and transport the curative agents directly to the brain" [6].

"Researchers are working to improve the safety and efficiency of AAV vectors by exploring strategies to reduce the chances of immune responses. One approach is using small molecules to inhibit the immunological response" [43–45]. "They are investigating methods to optimize the packaging AAV productivity to increase the delivery of therapeutic genes or proteins to target cells" [6, 51, 54].

"New techniques are being developed to enhance the manufacturing and scalability of AAV vectors. This is crucial for making AAV-based gene therapies more widely available and affordable" [18].

"Scientists are developing viral vectors that can target stem cells, which exhibit the capacity to differentiate into multiple cell types within the organism and are desirable for gene therapy applications. These vectors can selectively deliver therapeutic genes to stem cells, creating new opportunities for regenerative healthcare" [55].

"Current research is centered on the development of adeno-associated viruses that can treat genetic disorders resulting from mutations in non-coding areas. These areas make up most of the human genome, and their role in gene expression and regulation is not fully understood. Scientists are investigating the use of AAV vectors to deliver therapeutic genes or RNA molecules to non-coding areas, aiming to correct gene expression and treat conditions such as diabetes, obesity, and heart disease" [8].

Furthermore, "AAV vectors are being investigated as a potential therapy for viral infections such as hepatitis B and HIV. These vectors can be customized to deliver therapeutic RNA molecules or genes that disrupt viral replication, potentially leading to a functional cure for the illness" [23].

The ongoing advancement of AAV vector technology is expected to produce even more robust tools for gene editing and transfer, which hold the promise of transforming the management of diverse illnesses.

References

1. Kang, L., Jin, S., Wang, J., Lv, Z., Xin, C., Tan, C., Zhao, M., Wang, L., Liu, J.: AAV vectors applied to the treatment of CNS disorders: clinical status and challenges. J. Control. Release **355**, 458–473 (2023). https://doi.org/10.1016/j.jconrel.2023.01.067. PMID: 36736907
2. Huang, J., Zhou, Y., Li, J., Lu, A., Liang, C.: CRISPR/Cas systems: delivery and application in gene therapy. Front. Bioeng. Biotechnol. **22**(10), 942325 (2022). https://doi.org/10.3389/fbioe.2022.942325. PMID:N/A;PMCID:N/A. [Review]
3. Wang, S.W., Gao, C., Zheng, Y.M., et al.: Current applications and future perspective of CRISPR/Cas9 gene editing in cancer. Mol. Cancer **21**(1), 57 (2022). https://doi.org/10.1186/s12943-022-01518-8
4. Raguram, A., Banskota, S., Liu, D.R.: Therapeutic in vivo delivery of gene editing agents. Cell **185**(15), 2806–2827 (2022). https://doi.org/10.1016/j.cell.2022.03.045
5. Schulz, M., Levy, D.I., Petropoulos, C.J., Bashirians, G., Winburn, I., Mahn, M., et al.: Binding and neutralizing anti-AAV antibodies: detection and implications for rAAV-mediated gene therapy. Mol. Ther. **31**(3), 616–630 (2023). Table 1. https://doi.org/10.1016/j.ymthe.2023.01.010
6. Marrone, L., Marchi, P.M., Azzouz, M.: Circumventing the packaging limit of AAV-mediated gene replacement therapy for neurological disorders. Expert Opin. Drug Deliv. **19**(10):1163–1176 (2022). Figure 1. https://doi.org/10.1080/14712598.2022.2012148. PMID: 35144839
7. Arjomandnejad, M., Dasgupta, I., Flotte, T.R., et al.: Immunogenicity of Recombinant Adeno-Associated Virus (AAV) vectors for gene transfer. BioDrugs (2023). https://doi.org/10.1007/s40259-023-00585-7. Published March 2 2023. Accessed April 30 2023
8. Chavez, M., Chen, X., Finn, P.B., et al.: Advances in CRISPR therapeutics. Nat. Rev. Nephrol. **19**, 9–22 (2023). https://doi.org/10.1038/s41581-022-00636-2. Epub 2022 Oct 24 PMID: 34812125
9. Porto, E.M., Komor, A.C., Slaymaker, I.M., et al.: Base editing: advances and therapeutic opportunities. Nat. Rev. Drug Discov. **19**, 839–859 (2020). https://doi.org/10.1038/s41573-020-0084-6
10. Kweon, J., Hwang, H.Y., Ryu, H., Jang, A.H., Kim, D., Kim, Y.: Targeted genomic translocations and inversions generated using a paired prime editing strategy. Mol. Ther. **31**(1), 249–259 (2023). https://doi.org/10.1016/j.ymthe.2022.09.008. PMID: 34598185

11. Wang, J., He, Z., Wang, G., Zhang, R., Duan, J., Gao, P., Lei, X., Qiu, H., Zhang, C., Zhang, Y., Yin, H.: Efficient targeted insertion of large DNA fragments without DNA donors. Nat. Methods **19**(10), 1093–1098 (2022). https://doi.org/10.1038/s41592-022-01399-1. PMID: 34616111
12. Das, A., Vijayan, M., Walton, E.M., Stafford, V.G., Fiflis, D.N., Asokan, A.: Epigenetic silencing of recombinant adeno-associated virus genomes by NP220 and the HUSH complex. J. Virol. **96**(4), e02039–e2121 (2022). https://doi.org/10.1128/JVI.02039-21. PMID: 34918051
13. Hatada, I., Morita, S., Horii, T.: Epigenome editing in mice: the dawn of the reverse epigenetics era. Gene Genome Editing 3–4, 100012 (2022). ISSN 2666-3880. https://doi.org/10.1016/j.ggedit.2022.100012
14. Lau, C.H., Suh, Y.: In vivo epigenome editing and transcriptional modulation using CRISPR technology. Transgenic Res. **27**(6), 489–509 (2018). https://doi.org/10.1007/s11248-018-0096-8. Epub 2018 Oct 4 PMID: 30288626
15. Philippidis, A.: This is epic: epigenomic CRISPR startup goes big by thinking small. GEN Edge **4**(1), 539–546 (2022). https://doi.org/10.1089/genedge.4.1.89
16. Ren, D., Fisson, S., Dalkara, D., Ail, D.: Immune responses to gene editing by viral and non-viral delivery vectors used in retinal gene therapy. Pharmaceutics **14**(9), 1973 (2022). https://doi.org/10.3390/pharmaceutics14091973
17. Nasimuzzaman, M., Villaveces, S., van der Loo, J.C.M., Alla, S.: Process development for the production and purification of adeno-associated virus (AAV)2 vector using baculovirus-insect cell culture system. J. Vis. Exp. **179**, e62829 (2022). https://doi.org/10.3791/62829
18. Kenjo, E., Hozumi, H., Makita, Y., et al.: Low immunogenicity of LNP allows repeated administrations of CRISPR-Cas9 mRNA into skeletal muscle in mice. Nat. Commun. **12**, 7101 (2021). https://doi.org/10.1038/s41467-021-26714-w
19. Florea, M., Nicolaou, F., Pacouret, S., Zinn, E.M., Sanmiguel, J., Andres-Mateos, E., Unzu, C., Wagers, A.J., Vandenberghe, L.H.: High-efficiency purification of divergent AAV serotypes using AAVX affinity chromatography. Mol. Ther. Methods Clin. Dev. **28**, 146–159 (2023). https://doi.org/10.1016/j.omtm.2022.12.009. Figure 2. PMID:35694980;PMCID:PMC8742189.fsdfsd
20. Sierra-Delgado, J.A., Likhite, S., Bautista, P.K., Gómez-Ochoa, S.A., Echeverría, L.E., Guío, E., Vargas, C., Serrano, N.C., Meyer, K.C., Rincon, M.Y.: Prevalence of neutralizing antibodies against adeno-associated virus serotypes 1, 2, and 9 in non-injected latin american patients with heart failure—ANVIAS study. Int. J. Mol. Sci. **24**(6), 5579 (2023). https://doi.org/10.3390/ijms24065579
21. Ronzitti, G., Gross, D.A., Mingozzi, F.: Human immune responses to Adeno-Associated Virus (AAV) vectors. Front. Immunol. **17**(11), 670 (2020). https://doi.org/10.3389/fimmu.2020.00670. PMID:32362844;PMCID:PMC7190319
22. Schulz, M., Levy, D.I., Petropoulos, C.J., Somanathan, S., Cheng, S.H., Byrne, B.J.: Mechanisms for the inhibition of vector transduction and transgene expression by neutralizing anti-AAV antibodies. Mol. Ther. **31**(3), 616–630 (2023). https://doi.org/10.1016/j.ymthe.2023.01.010. PMID:34610483. Figure 3. Adapted from Schulz et al. Mol Ther. 2023 Mar;31(3):616–630
23. Bucher, K., Rodríguez-Bocanegra, E., Wissinger, B., et al.: Extra-viral DNA in adeno-associated viral vector preparations induces TLR9-dependent innate immune responses in human plasmacytoid dendritic cells. Sci. Rep. **13**(1), 1890 (2023). https://doi.org/10.1038/s41598-023-28830-7
24. McNally, J.: Pre-existing immunogenicity: how will cell and gene therapies change our testing approach? Int. Pharm. Ind. **14**(4), 34 (2022)
25. Earley, J., Piletska, E., Ronzitti, G., Piletsky, S.: Evading and overcoming AAV neutralization in gene therapy. Trends Biotechnol. **S0167–7799**(22), 00303–00311 (2022). https://doi.org/10.1016/j.tibtech.2022.11.006
26. Nyberg, W.A., Ark, J., To, A., Clouden, S., Reeder, G., Muldoon, J.J., Chung, J.Y., Xie, W.H., Allain, V., Steinhart, Z., Chang, C., Talbot, A., Kim, S., Rosales, A., Havlik, L.P., Pimentel, H., Asokan, A., Eyquem, J.: An evolved AAV variant enables efficient genetic engineering of murine T cells. Cell **186**(2), 446–460.e19 (2023). https://doi.org/10.1016/j.cell.2022.12.022. PMID: 34987467

27. Seo, Y.E., Baine, S.H., Kempton, A.N., Rogers, O.C., Lewis, S., Adegboye, K., Haile, A., Griffin, D.A., Peterson, E.L., Pozsgai, E.R., Potter, R.A., Rodino-Klapac, L.R.: Systemic γ-sarcoglycan AAV gene transfer results in dose-dependent correction of muscle deficits in the LGMD 2C/R5 mouse model. Mol. Ther. Methods Clin. Dev. **28**, 284–299 (2023). https://doi.org/10.1016/j.omtm.2023.01.004. PMID:33738450;PMCID:PMC7975753
28. Sichani, A., Saber, R., Ranjbar, M., Baneshi, M., Sanati-Nezhad, A.: A Review on advanced CRISPR-based genome-editing tools: base editing and prime editing. Mol. Biotechnol. (2022). https://doi.org/10.1007/s12033-022-00639-1
29. Gao, X., Yang, J., Tsang, S.H., Garg, S.: Adeno-associated virus-mediated gene delivery for the treatment of retinal diseases. Expert Opin. Drug Deliv. **16**(8), 833–845 (2019). https://doi.org/10.1080/17425247.2019.1628199. PMID: 31189496
30. Wang, D., Tai, P.W., Gao, G.: Adeno-associated virus vector as a platform for gene therapy delivery. Nat. Rev. Drug Discov. **18**(5), 358–378 (2019)
31. Gottwick, C., Carambia, A., Herkel, J.: Harnessing the liver to induce antigen-specific immune tolerance. Semin Immunopathol. **44**(4), 475–484 (2022). https://doi.org/10.1007/s00281-022-00942-8
32. Perocheau, D.P., Cunningham, S.C., Lee, J., Antinao Diaz, J., Waddington, S.N., Gilmour, K., Eaglestone, S., Lisowski, L., Thrasher, A.J., Alexander, I.E., Gissen, P., Baruteau, J.: Age-related seroprevalence of antibodies against AAV-LK03 in a UK population cohort. Human Gene Ther. **30**(1), 79–87 (2019). https://doi.org/10.1089/hum.2018.098. [Epub ahead of print: October 19, 2018; Epub ahead of editing: July 20, 2018]
33. Ahmadi, S.E., Soleymani, M., Shahriyary, F. et al.: Viral vectors and extracellular vesicles: innate delivery systems utilized in CRISPR/Cas-mediated cancer therapy. Cancer Gene Ther. (2023). https://doi.org/10.1038/s41417-023-00597-z
34. Schulz et al.: Binding and neutralizing anti-AAV antibodies: detection and implications for rAAV-mediated gene therapy. Mol Ther. **31**(3), 616–630 (2023). Figure 4
35. Perocheau DP, Cunningham SC, Lee J, Diaz JA, Waddington SN, Gilmour K, Eaglestone S, Lisowski L, Thrasher AJ, Alexander IE, Gissen P, Baruteau J. Age-related seroprevalence of antibodies against AAV-LK03 in a UK population cohort. Human Gene Ther. **30**(1), 79–87 (2019). Table 2. Seroprevalence of antibodies against AAV-LK03 in a UK population cohort by age group. https://doi.org/10.1089/hum.2018.098
36. Perocheau, D.P., Cunningham, S.C., Lee, J., Diaz JA, Waddington SN, Gilmour K, et al. Age-related seroprevalence of antibodies against AAV-LK03 in a UK population cohort [Figure 4]. Human Gene Ther. **30**(1), 79-87 (2019). https://doi.org/10.1089/hum.2018.098. PMID: 30346857
37. Perocheau, D.P., Cunningham, S.C., Lee, J., Diaz, J.A., Waddington, S.N., Gilmour, K., et al.: Age-related seroprevalence of antibodies against AAV-LK03 in a UK population cohort [Figure 5]. Human Gene Ther. **30**(1), 79-87 (2019). https://doi.org/10.1089/hum.2018.098. PMID: 30346857
38. Rapti, K., Grimm, D.: Adeno-Associated Viruses (AAV) and host immunity—A race between the hare and the hedgehog. Front. Immunol. **29**(12), 753467 (2021). https://doi.org/10.3389/fimmu.2021.753467
39. Mingozzi, F., High, K.A.: Immune responses to AAV vectors: overcoming barriers to successful gene therapy. Blood **122**(1), 23–36 (2013). https://doi.org/10.1182/blood-2013-01-306647
40. Kang, Q., Sun, W., Zhang, J., et al.: Adeno-associated virus vectors: current status and future perspectives in gene therapy. Int. J. Mol. Sci. **22**(10), 5171 (2021). https://doi.org/10.3390/ijms22105171
41. Gorovits, B., Azadeh, M., Buchlis, G., et al.: Evaluation of cellular immune response to adeno-associated virus-based gene therapy. AAPS J. **25**(3), 47 (2023). https://doi.org/10.1208/s12248-023-00814-5. PMID: 32346741
42. Whitehead, M., Osborne, A., Yu-Wai-Man, P., Martin, K.: Humoral immune responses to AAV gene therapy in the ocular compartment. Biol. Rev. (2021). https://doi.org/10.1111/brv.12718
43. Nidetz, N.F., McGee, M.C., Tse, L.V., Li, C., Cong, L., Li, Y., Huang, W.: Adeno-associated viral vector-mediated immune responses: understanding barriers to gene delivery. Pharmacol. Ther. **207**, 107453 (2020). https://doi.org/10.1016/j.pharmthera.2019.107453

44. Retnakumar, S.V., Bonam, S.R., Hu, H., Bayry, J.: Vaccines and vaccine adjuvants/immunomodulators for infectious diseases. Vaccines **11**(2), 383 (2023). https://doi.org/10.3390/vaccines11020383
45. Padron-Regalado, E., Ulaszewska, M., Douglas, A.D., et al.: STING-pathway modulation to enhance the immunogenicity of adenoviral-vectored vaccines. Sci. Rep. **12**(1), 14464 (2022). https://doi.org/10.1038/s41598-022-18750-3
46. Farzanehpour, M., Miri, A., Alvanegh, A.G., Esmaeili, G.H.: Viral vectors, exosomes, and vexosomes: potential armamentarium for delivering CRISPR/Cas to cancer cells. Biochem. Pharmacol. **212**, 115555 (2023). https://doi.org/10.1016/j.bcp.2023.115555. PMID: 34119860
47. Df Kanvinde, S., Kulkarni, T., Deodhar, S., Bhattacharya, D., Dasgupta, A.: Non-viral vectors for delivery of nucleic acid therapies for cancer. Biotech **11**(1), 6 (2022). https://doi.org/10.3390/biotech11010006
48. Ghauri, M.S., Ou, L.: AAV engineering for improving tropism to the central nervous system. Biology **12**(2), 186 (2023). Figure 6. https://doi.org/10.3390/biology12020186
49. Bucher, K., Rodríguez-Bocanegra, E., Dauletbekov, D., Fischer, M.D.: Immune responses to retinal gene therapy using adeno-associated viral vectors—Implications for treatment success and safety. Prog Retin Eye Res. **83**, 100915 (2021). ISSN 1350-9462. https://doi.org/10.1016/j.preteyeres.2020.100915
50. Muhuri, M., Levy, D.I., Schulz, M., McCarty, D., Gao, G.: Durability of transgene expression after rAAV gene therapy. Mol. Ther. **30**(4), 1364–1380 (2022). https://doi.org/10.1016/j.ymthe.2022.03.004. PMID: 35261636
51. Handyside, B., Ismail, A.M., Zhang, L., Yates, B., Xie, L., Sihn, C.R., Murphy, R., Bouwman, T., Kim, C.K., De Angelis, R., Karim, O.A., McIntosh, N.L., Doss, M.X., Shroff, S., Pungor, E., Bhat, V.S., Bullens, S., Bunting, S., Fong, S.: Vector genome loss and epigenetic modifications mediate decline in transgene expression of AAV5 vectors produced in mammalian and insect cells. Mol. Ther. **30**(12), 3570–3586 (2022). https://doi.org/10.1016/j.ymthe.2022.11.001. PMID: 34857541
52. Gao, Z., Ravendran, S., Mikkelsen, N.S., Haldrup, J., Cai, H., Ding, X., Paludan, S.R., Thomsen, M.K., Mikkelsen, J.G., Bak, R.O.: A truncated reverse transcriptase enhances prime editing by split AAV vectors. Mol. Ther. **30**(9), 2942–2951 (2022). https://doi.org/10.1016/j.ymthe.2022.07.001. PMID: 34304054
53. Zhi, S., Chen, Y., Wu, G., Wen, J., Wu, J., Liu, Q., Li, Y., Kang, R., Hu, S., Wang, J., Liang, P., Huang, J.: Dual-AAV delivering split prime editor system for in vivo genome editing. Mol. Ther. **30**(1), 283–294 (2022). https://doi.org/10.1016/j.ymthe.2021.07.011
54. Kantor, B., Odonovan, B., Rittiner, J., Lindner, N., Dong, W., Zhang, A., Nicholls, P., Chiba-Falek, O.: All-in-one AAV-delivered epigenome-editing platform: proof-of-concept and therapeutic implications for neurodegenerative disorders. bioRxiv (2023). https://doi.org/10.1101/2023.04.14.536951
55. Ferrari, S., Jacob, A., Cesana, D., Laugel, M., Beretta, S., Varesi, A., Unali, G., Conti, A., Canarutto, D., Albano, L., Calabria, A., Vavassori, V., Cipriani, C., Castiello, M.C., Esposito, S., Brombin, C., Cugnata, F., Adjali, O., Ayuso, E., Merelli, I., Villa, A., Di Micco, R., Kajaste-Rudnitski, A., Montini, E., Penaud-Budloo, M., Naldini, L.: Choice of template delivery mitigates the genotoxic risk and adverse impact of editing in human hematopoietic stem cells. Cell Stem Cell **29**(10), 1428-1444.e9 (2022). https://doi.org/10.1016/j.stem.2022.09.001

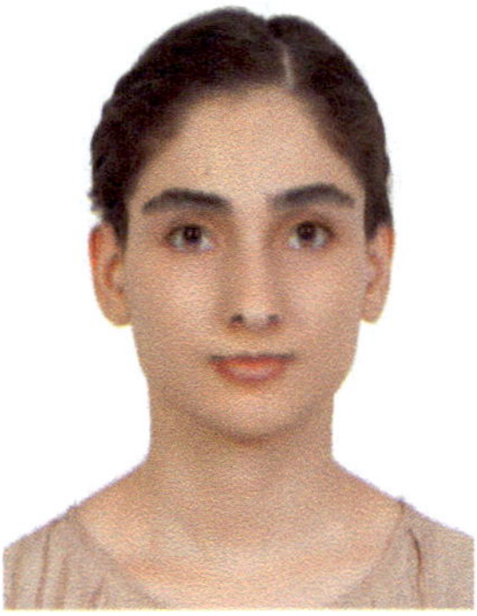

Asra Hamidi also Known as Jivam or Ataran, is a talented and passionate young researcher and innovator in biotechnology. Her journey began at the age of 12 when she conducted groundbreaking research on a breast cancer DNA vaccine, successfully defending her research proposal at Sistan and Baluchistan University in Iran. Asra Hamidi (Jivam) consistently demonstrates exceptional creativity and resourcefulness in her work. Her research interests focus on genome editing, innovative genetic vaccines, and quantum biotechnology. She is dedicated to advancing biotechnology and contributing to the development of novel healthcare solutions.

Subject Index

R. Malviya and S. Sundram (eds.), *Engineered Biomaterials*, Engineering Materials,
https://doi.org/10.1007/978-981-99-6698-1

MIX
Papier aus verantwortungsvollen Quellen
Paper from responsible sources
FSC® C105338

If you have any concerns about our products,
you can contact us on
ProductSafety@springernature.com

In case Publisher is established outside the EU,
the EU authorized representative is:
Springer Nature Customer Service Center GmbH
Europaplatz 3, 69115 Heidelberg, Germany

Printed by Libri Plureos GmbH
in Hamburg, Germany